TABLE OF INVERSE LAPLACE TRANSFORMS

$F(s)$	$f(t) = \mathcal{L}\{F(s)\}$
1. $\dfrac{1}{(s-a)(s-b)}$	$\dfrac{1}{a-b}(e^{at} - e^{bt})$
2. $\dfrac{s}{(s-a)(s-b)}$	$\dfrac{1}{a-b}(ae^{at} - be^{bt})$
3. $\dfrac{1}{s(s^2 + a^2)}$	$\dfrac{1}{a^2}(1 - \cos at)$
4. $\dfrac{1}{s^2(s^2 + a^2)}$	$\dfrac{1}{a^3}(at - \sin at)$
5. $\dfrac{1}{(s^2 + a^2)^2}$	$\dfrac{1}{2a^3}(\sin at - at \cos at)$
6. $\dfrac{s}{(s^2 + a^2)^2}$	$\dfrac{t}{2a}\sin at$
7. $\dfrac{s^2}{(s^2 + a^2)^2}$	$\dfrac{1}{2a}(\sin at + at \cos at)$
8. $\dfrac{s^2 - a^2}{(s^2 + a^2)^2}$	$t \cos at$
9. $\dfrac{1}{(s-a)^2 + b^2}$	$\dfrac{1}{b}e^{at}\sin bt$
10. $\dfrac{s-a}{(s-a)^2 + b^2}$	$e^{at}\cos bt$

PROPERTIES OF THE LAPLACE TRANSFORM: $F(s) = \mathcal{L}\{f\} = \displaystyle\int_0^\infty e^{-st} f(t)\, dt$

1. $\mathcal{L}\{f + g\} = \mathcal{L}\{f\} + \mathcal{L}\{g\}$

2. $\mathcal{L}\{cf\} = c\mathcal{L}\{f\}$

3. $\mathcal{L}\{f'\} = s\mathcal{L}\{f\} - f(0)$

4. $\mathcal{L}\{f''\} = s^2\mathcal{L}\{f\} - sf(0) - f'(0)$

5. $\mathcal{L}\{f^{(n)}\} = s^n\mathcal{L}\{f\} - s^{n-1}f(0) - s^{n-2}f'(0) - \cdots - f^{(n-1)}(0)$

6. $\mathcal{L}\{e^{at}f(t)\} = F(s - a)$

7. $\mathcal{L}\{t^n f(t)\} = (-1)^n \dfrac{d^n}{ds^n} F(s)$

8. $\mathcal{L}\{f(at)\} = \dfrac{1}{a}F\left(\dfrac{s}{a}\right)$

9. $\mathcal{L}\{f * g\} = \mathcal{L}\{f\}\mathcal{L}\{g\}$

10. $\mathcal{L}\left\{\displaystyle\int_0^t f(\tau)\, d\tau\right\} = \dfrac{1}{s}F(s)$

11. $\mathcal{L}\left\{\dfrac{f(t)}{t}\right\} = \displaystyle\int_s^\infty F(\xi)\, d\xi$

12. $\displaystyle\lim_{s\to\infty} sF(s) = f(0)$

13. $\displaystyle\lim_{s\to 0} sF(s) = f(\infty)$

AN INTRODUCTION TO
DIFFERENTIAL
EQUATIONS
AND THEIR APPLICATIONS

AN INTRODUCTION TO
DIFFERENTIAL
EQUATIONS
AND THEIR APPLICATIONS

STANLEY J. FARLOW
University of Maine

DOVER PUBLICATIONS, INC.
Mineola, New York

Bibliographical Note

This Dover edition, first published in 2006, is an unabridged republication of
the work originally published by McGraw-Hill, Inc., New York, in 1994. All color
materials have been reprinted in black and white.

International Standard Book Number

ISBN-13: 978-0-486-44595-3
ISBN-10: 0-486-44595-X

Manufactured in the United States by Courier Corporation
44595X04   2014
www.doverpublications.com

ABOUT THE AUTHOR

Stanley J. Farlow has academic degrees from Iowa State University, the University of Iowa, and Oregon State University. He is a former Lieutenant Commander and Public Health Service Fellow who served for several years as a computer analyst at the National Institutes of Health in Bethesda, Maryland. In 1968 he joined the faculty of the University of Maine, where he is currently Professor of Mathematics. He is also the author of *Partial Differential Equations for Scientists and Engineers.* (currently being published by Dover Publications, Inc.), *Finite Mathematics* (McGraw-Hill, 1988, 1994), *Applied Mathematics* (McGraw-Hill, 1988), *Introduction to Calculus* (McGraw-Hill, 1990), and *Calculus and Its Applications* (McGraw-Hill, 1990). He has also edited *The GMDH Method: Self-Organizing Methods in Modeling* (Marcel Dekker, 1984).

To Dorothy and Susan

C O N T E N T S

Preface   xv

CHAPTER 1 **INTRODUCTION TO DIFFERENTIAL EQUATIONS**   1

Prologue   2
1.1   Basic Definitions and Concepts   9
1.2   Some Basic Theory   14

CHAPTER 2 **FIRST-ORDER DIFFERENTIAL EQUATIONS**   29

2.1   First-Order Linear Equations   30
2.2   Separable Equations   38
2.3   Growth and Decay Phenomena   47
2.4   Mixing Phenomena   60
2.5   Cooling and Heating Phenomena   67
2.6   More Applications   76
2.7   The Direction Field and Euler's Method   91
2.8   Higher-Order Numerical Methods   100

CHAPTER 3 **SECOND-ORDER LINEAR EQUATIONS**   109

3.1   Introduction to Second-Order Linear Equations   110
3.2   Fundamental Solutions of the Homogeneous Equation   118
3.3   Reduction of Order   124
3.4   Homogeneous Equations with Constant Coefficients:
Real Roots   130
3.5   Homogeneous Equations with Constant Coefficients:
Complex Roots   135
3.6   Nonhomogeneous Equations   141
3.7   Solving Nonhomogeneous Equations: Method of Undetermined
Coefficients   148
3.8   Solving Nonhomogeneous Equations: Method of Variation
of Parameters   158
3.9   Mechanical Systems and Simple Harmonic Motion   163

3.10 Unforced Damped Vibrations   173
3.11 Forced Vibrations   181
3.12 Introduction to Higher-Order Equations (Optional)   189

CHAPTER 4   **SERIES SOLUTIONS**   199

4.1 Introduction: A Review of Power Series   200
4.2 Power Series Expansions about Ordinary Points: Part I   207
4.3 Power Series Expansions about Ordinary Points: Part II   216
4.4 Series Solutions about Singular Points: The Method of Frobenius   223
4.5 Bessel Functions   232

CHAPTER 5   **THE LAPLACE TRANSFORM**   243

5.1 Definition of the Laplace Transform   244
5.2 Properties of the Laplace Transform   254
5.3 The Inverse Laplace Transform   260
5.4 Initial-Value Problems   268
5.5 Step Functions and Delayed Functions   276
5.6 Differential Equations with Discontinuous Forcing Functions   284
5.7 Impulse Forcing Functions   292
5.8 The Convolution Integral   303

CHAPTER 6   **SYSTEMS OF DIFFERENTIAL EQUATIONS**   313

6.1 Introduction to Linear Systems: The Method of Elimination   314
6.2 Review of Matrices   323
6.3 Basic Theory of First-Order Linear Systems   334
6.4 Homogeneous Linear Systems with Real Eigenvalues   341
6.5 Homogeneous Linear Systems with Complex Eigenvalues   354
6.6 Nonhomogeneous Linear Systems   359
6.7 Nonhomogeneous Linear Systems: Laplace Transform (Optional)   367

6.8 Applications of Linear Systems 374
6.9 Numerical Solution of Systems of Differential Equations 385

C H A P T E R 7 **DIFFERENCE EQUATIONS** 395

7.1 Introduction to Difference Equations 396
7.2 Homogeneous Equations 403
7.3 Nonhomogeneous Equations 410
7.4 Applications of Difference Equations 420
7.5 The Logistic Equation and the Path to Chaos 429
7.6 Iterative Systems: Julia Sets and the Mandelbrot Set
(Optional) 440

C H A P T E R 8 **NONLINEAR DIFFERENTIAL EQUATIONS AND CHAOS** 451

8.1 Phase Plane Analysis of Autonomous Systems 452
8.2 Equilibrium Points and Stability for Linear Systems 469
8.3 Stability: Almost Linear Systems 480
8.4 Chaos, Poincaré Sections and Strange Attractors 489

C H A P T E R 9 **PARTIAL DIFFERENTIAL EQUATIONS** 503

9.1 Fourier Series 504
9.2 Fourier Sine and Cosine Series 515
9.3 Introduction to Partial Differential Equations 523
9.4 The Vibrating String: Separation of Variables 529
9.5 Superposition Interpretation of the Vibrating String 539
9.6 The Heat Equation and Separation of Variables 545
9.7 Laplace's Equation Inside a Circle 555

Appendix: Complex Numbers and Complex-Valued Functions 567
Answers to Problems 575
Index 605

PREFACE

An Introduction to Differential Equations and Their Applications is intended for use in a beginning one-semester course in differential equations. It is designed for students in pure and applied mathematics who have a working knowledge of algebra, trigonometry, and elementary calculus. The main feature of this book lies in its exposition. The explanations of ideas and concepts are given fully and simply in a language that is direct and almost conversational in tone. I hope I have written a text in differential equations that is more easily read than most, and that both your task and that of your students will be helped.

Perhaps in no other college mathematics course is the interaction between mathematics and the physical sciences more evident than in differential equations, and for that reason I have tried to exploit the reader's physical and geometric intuition. At one extreme, it is possible to approach the subject on a highly rigorous "lemma-theorem-corollary" level, which, for a course like differential equations, squeezes out the life-blood of the subject, leaving the student with very little understanding of how differential equations interact with the real world. At the other extreme, it is possible to wave away all the mathematical subtleties until neither the student nor the instructor knows what's going on. The goal of this book is to balance mathematical rigor with intuitive thinking.

FEATURES OF THE BOOK
Chaotic Dynamical Systems

This book covers the standard material taught in beginning differential equations courses, with the exception of Chapters 7 and 8, where I have included optional sections relating to chaotic dynamical systems. The period-doubling phenomenon of the logistic equation is introduced in Section 7.5 and Julia sets and the Mandelbrot set are introduced in Section 7.6. Then, in Section 8.4, the chaotic behavior of certain nonlinear differential equations is summarized, and the Poincaré section and strange attractors are defined and discussed.

Problem Sets

One of the most important aspects of any mathematics text is the problem sets. The problems in this book have been accumulated over 25 years of teaching differential equations and have been written in a style that, I hope, will pique the student's interest.

Because not all material can or should be included in a beginning textbook, some problems are placed within the problem sets that serve to introduce additional new topics. Often a brief paragraph is added to define relevant terms. These problems can be used to provide extra material for special students or to introduce new material the instructor may wish to discuss. Throughout the book, I have included numerous computational problems that will allow the students to use computer software, such as DERIVE, MATHEMATICA, MATHCAD, MAPLE, MACSYMA, PHASER, and CONVERGE.

Writing and Mathematics

In recent years I have joined the "Writing Across the Curriculum" crusade that is sweeping U.S. colleges and universities and, for my own part, have required my students to keep a scholarly journal. Each student spends five minutes at the end of each lecture writing and outlining what he or she does or doesn't understand. The idea, which is the foundation of the "Writing Across the Curriculum" program, is to learn through writing. At the end of the problem set in Section 1.1, the details for keeping a journal are outlined. Thereafter, the last problem in each problem set suggests a journal entry.

Historical Notes

An attempt has been made to give the reader some appreciation of the richness and history of differential equations through the use of historical notes. These notes, are intended to allow the reader to set the topic of differential equations in its proper perspective in the history of our culture. They can also be used by the instructor as an introduction to further discussions of mathematics.

DEPENDENCE OF CHAPTERS AND COURSE SUGGESTIONS

Since one cannot effectively cover all nine chatpers of this book during a one-semester or quarter course, the following dependence of chapters might be useful in organizing a course of study. Normally, one should think of this text as a one-semester book, although by covering all the material and working through a sufficient number of problems, it could be used for a two-semester course.

I often teach an introductory differential equations course for students of engineering and science. In that course I cover the first three chapters on first- and second-order equations, followed by Chapter 5 (the Laplace transform), Chapter 6 (systems), Chapter 8 (nonlinear equations), and part of Chapter 9 (partial differential equations). I generally spend a couple of days giving a rough overview of the omitted chapters: series solutions (Chapter 4) and difference equations (Chapter 7). For classes that contain mostly physics students who intended to take a follow-up course in partial differential equations, I cover Chapter 4 (series solutions) at the expense of some material on the Laplace transform.

I have on occasion used this book for a problems course in which I cover only Chapters 1, 2, and 3. Chapter 2 (first-order equations) contains a wide variety of problems that will keep any good student busy for an entire semester (some students have told me a lifetime).

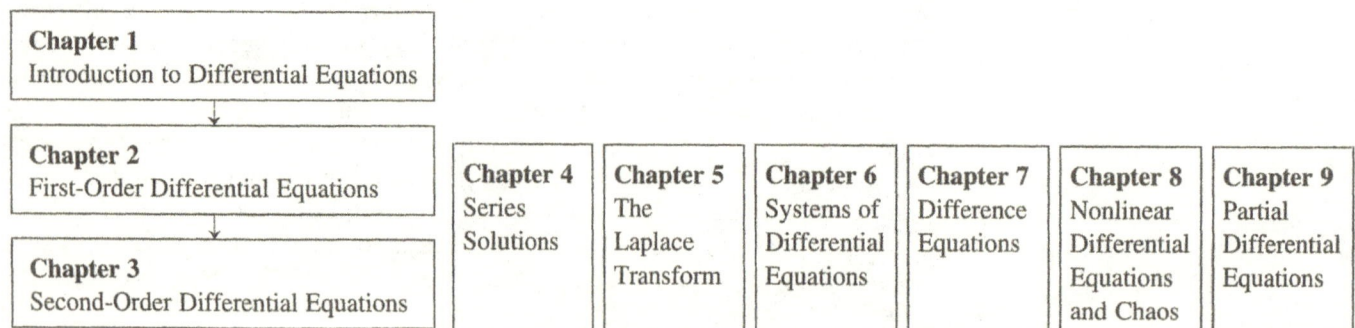

ACKNOWLEDGMENTS   No textbook author can avoid thanking the authors of the many textbooks that have come before. A few of the textbooks to which I am indebted are: *Differential Equations* by Ralph Palmer Agnew, McGraw-Hill, New York, (1942), *Differential Equations* by Lester R. Ford, McGraw-Hill, New York, (1955), and *Differential Equations* by Lyman Kells, McGraw-Hill, New York, (1968).

I would also like to thank my advisor of more than 25 years ago, Ronald Guenther, who in addition to teaching me mathematics, taught me the value of rewriting.

I am grateful to the many people who contributed to this book at various stages of the project. The following people offered excellent advice, suggestions, and ideas as they reviewed the manuscript: Kevin T. Andrews, Oakland University; William B. Bickford, Arizona State University; Juan A. Gatica, University of Iowa; Peter A. Griffin, California State University, Sacramento; Terry L. Herdman, Virginia Polytechnic Institute and State University; Hidefumi Katsuura, San Jose State University; Monty J. Strauss, Texas Tech University; Peter J. Tonellato, Marquette University.

Finally, I am deeply grateful to the McGraw-Hill editors Jack Shira and Maggie Lanzillo, for their leadership and encouragement, and to Margery Luhrs and Richard Ausburn who have contributed to the project and worked so hard throughout the production process.

All errors are the responsibility of the author and I would appreciate having these brought to my attention. I would also appreciate any comments or suggestions from students and instructors.

Stanley J. Farlow

AN INTRODUCTION TO
DIFFERENTIAL EQUATIONS
AND THEIR APPLICATIONS

1

Introduction to Differential Equations

PROLOGUE
1.1 BASIC DEFINITIONS AND CONCEPTS
1.2 SOME BASIC THEORY

1.0 PROLOGUE

GALLOPING GERTIE

There was a lot of excitement in the air when on July 1, 1940, local dignitaries cut the ribbon that opened the Tacoma Narrows Bridge over Puget Sound in the state of Washington, but the excitement didn't stop there. Because it tended to experience undulating vibrations in the slightest breeze, the bridge gained a great deal of attention and was nicknamed "Galloping Gertie." Although one might have thought that people would have been afraid to cross the bridge, this was not so. People came from hundreds of miles just for the thrill of crossing "Gertie." Although a few engineers expressed concern, authorities told the public that there was "absolutely nothing to worry about." They were so sure of this that they even planned to drop the insurance on the bridge.

When Galloping Gertie collapsed into Puget Sound on November 7, 1940, bridge designers gained a new respect for nonlinear differential equations. (AP/Wide World Photos)

However, at about 7:00 A.M. on November 7, 1940, Gertie's undulations became more violent, and entire portions of the bridge began to heave wildly. At one time, one side of the roadway was almost 30 feet higher than the other. Then, at 10:30 A.M. the bridge began to crack up. Shortly thereafter it made a final lurching and twisting motion and then crashed into Puget Sound. The only casualty was a pet dog owned by a reporter who was crossing the bridge in his car. Although the reporter managed to reach safety by crawling on his hands and knees, clinging to the edge of the roadway, the dog lost its life.

Later, when local authorities tried to collect the insurance on the bridge, they discovered that the agent who had sold them the policy hadn't told the insurance company and had pocketed the $800,000 premium. The agent, referring to the fact that authorities had planned on canceling all policies within a week, wryly observed that if the ''damn thing had held out just a little longer no one would have been the wiser.'' The man was sent to prison for embezzlement. The collapse also caused embarrassment to a local bank, whose slogan was ''As safe as the Tacoma Bridge.'' After the bridge collapsed into Puget Sound, bank executives quickly sent out workers to remove the billboard.

Of course, after the collapse the government appointed all sorts of commissions of inquiry. The governor of the State of Washington made an emotional speech to the people of Washington proclaiming that ''we are going to build the exact same bridge, exactly as before.'' Upon hearing this, the famous engineer Theodor von Karman rushed off a telegram stating, ''If you build the exact same bridge, exactly as before, it will fall into the same river, exactly as before.''

After the politicians finished their analysis of the bridge's failure, several teams of engineers from major universities began a technical analysis of the failure. It was the general consensus that the collapse was due to resonance caused by an aerodynamical phenomenon known as ''stall flutter.''

Roughly, this phenomenon has to do with frequencies of wind currents agreeing with natural frequencies of vibration of the bridge. The phenomenon can be analyzed by comparing the driving frequencies of a differential equation with the natural frequencies of the equation.

FISHES, FOXES, AND THE NORWAY RAT

Although at one time Charlie Elton suspected that sunspot activity might be the cause of the periodic fluctuation in the rodent population in Norway, he later realized that this fluctuation probably had more to do with the ecological balance between the rats and their biological competitors.

The populations of many species of plants, fish, mammals, insects, bacteria, and so on, vary periodically due to boom and bust cycles in which they alternately die out and recover in their constant struggle for existance against their ecological adversaries. (Leonard Lee Rue III/Photo Researchers)

At about the same time, in the 1920s an Italian marine biologist, Umberto D'Ancona, observed that certain populations of fish in the northern Adriatic varied periodically over time. More specifically, he noted that when the population of certain *predator* fish (such as sharks, skates, and rays) was up, the population of their *prey* (herbivorous fish) was down, and vice versa. To better understand these ''boom and bust'' cycles, D'Ancona turned to the famous Italian mathematician and differential equations expert

Vito Volterra. What Volterra did was to repeat for biology what had been done in the physical sciences by Newton 300 years earlier. In general, he developed a mathematical theory for a certain area of biology; in particular, he developed a mathematical framework for the cohabitation of organisms. One might say that he developed the mathematical theory for the "struggle for existence" and that current research in ecological systems had its beginnings in the differential equations of Volterra.

WHERE WERE YOU WHEN THE LIGHTS WENT OUT?

Most readers of this book were probably pretty young during the New York City power failure of 1977 that plunged the entire northeastern section of the United States and a large portion of Canada into total darkness. Although the lessons learned from that disaster have led to more reliable power grids across the country, there is always the (remote) possibility that another failure will occur at some future time.

The problem is incredibly complicated. How to match the energy needs of the millions of customers with the energy output from the hundreds of generating stations? And this must be done so that the entire network remains synchronized at 60 cycles per second and the customer's voltage levels stay at acceptable levels! Everything would not be quite so difficult if demand remained constant and if there were never any breakdowns. As one system engineer stated, "It's easy to operate a power grid if nothing breaks down. The trick is to keep it working when you have failures." However, there will always be the possibility of a generator breaking down or lightning hitting a transformer. And when this happens, there is always the possibility that the entire network may go down with it.

In any large scale system there is always the possibility that a failure in one part of the system can be propagated throughout the system. Systems of differential equations can be used to help understand the total dynamics of the system and prevent disasters. (Bill Gallery/Stock, Boston)

To help design large-scale power grids to be more reliable (stable), engineers have constructed mathematical models based on systems of differential equations that describe the dynamics of the system (voltages and currents through power lines). By simulating random failures the engineers are able to determine how to design reliable systems. They also use mathematical models to determine after the fact how a given failure can be prevented in the future. For example, after a 1985 blackout in Colombia, South America, mathematical models showed that the system would have remained stable if switching equipment had been installed to trip the transmission lines more quickly.

DIFFERENTIAL EQUATIONS IN WEATHER PATTERNS

Meteorologist Edward Lorenz was not interested in the cloudy weather outside his M.I.T. office. He was more interested in the weather patterns he was generating on his new Royal McBee computer. It was the winter of 1961, and Lorenz had just constructed a mathematical model of convection patterns in the upper atmosphere based on a system of three nonlinear differential equations. In the early 1960s there was a lot of optimism in the scientific world about weather forecasting, and the general consensus was that it might even be possible in a few years to modify and control the weather. Not only was weather forecasting generating a great deal of excitement, but the techniques used in meteorology were also being used by physical and social scientists hoping to make predictions about everything from fluid flow to the flow of the economy.

Anyway, on that winter day in 1961 when Edward Lorenz came to his office, he decided to make a mathematical shortcut, and instead of running his program from the beginning, he simply typed into the computer the numbers computed from the previous day's run. He then turned on the computer and left the room to get a cup of coffee. When he returned an hour later, he saw something unexpected—something that would change the course of science.

The new run, which should have been the same as the previous day's run, was completely different. The weather patterns generated on this day were completely different from the patterns generated on the previous day, although their initial conditions were the same.

Initially, Lorenz thought he had made a mistake and keyed in the wrong numbers, or maybe his new computer had a malfunction. How else could he explain how two weather patterns had diverged if they had the same initial conditions? Then it came to him. He realized that the computer was using *six-place* accuracy, such as 0.209254, but only *three places* were displayed on the screen, such as 0.209. So when he typed in the new numbers, he had entered only *three decimal places,* assuming that one part in a thousand was not important. As it turned out insofar as the differential equations were concerned, it was *very* important.

The ''chaotic'' or ''randomlike behavior'' of those differential equations was so sensitive to their initial conditions that *no* amount of error was tolerable. Little did Lorenz know it at the time, but these were the differential equations that opened up the new subject of *chaos.* From this point on scientists realized that the prediction of such complicated physical phenomena as the weather was impossible using the classical methods of differential equations and that newer theories and ideas would be required. Paradoxically, chaos theory provides a way to see the *order* in a chaotic system.

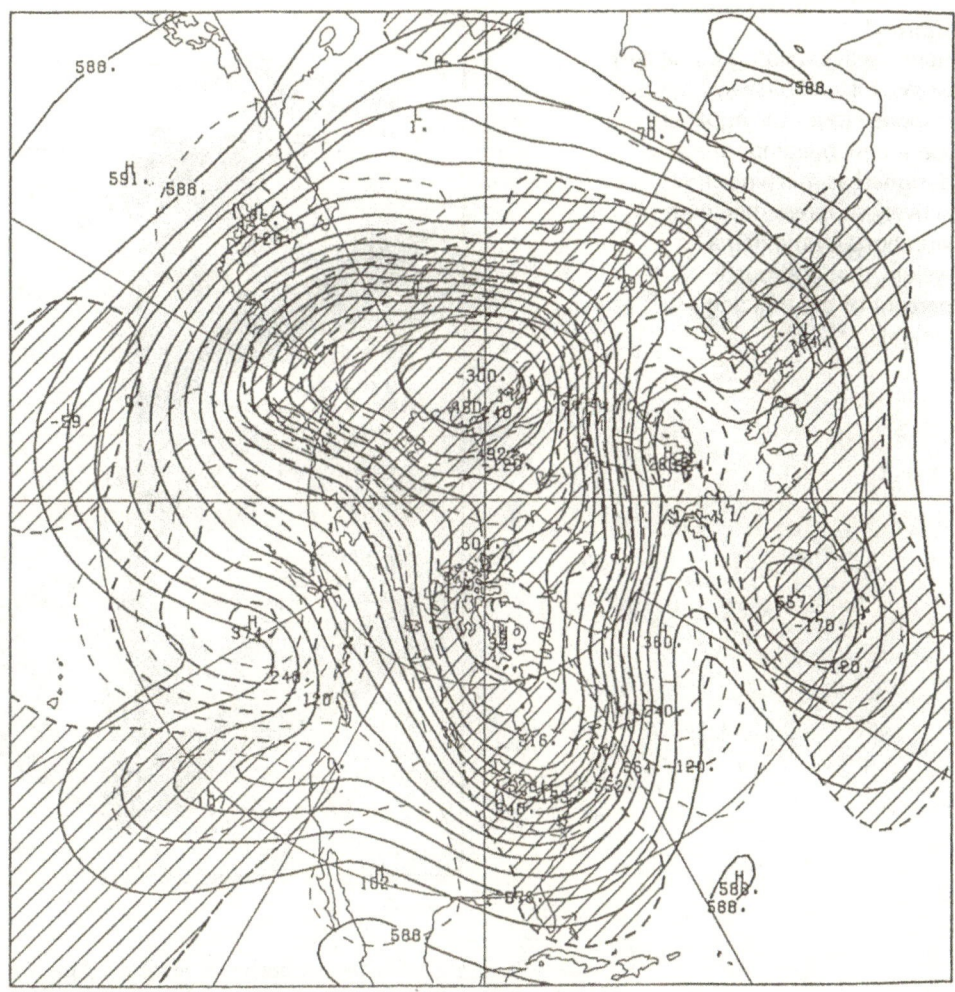

The future success of accurate long-run weather predictions is not completely clear. The accurate determination of long-term weather patterns could well depend on new research in dynamical systems and differential equations. (Courtesy National Meteorological Center)

ENGINEERS TEACH SMART BUILDING TO FOIL QUAKES*

Engineers and applied mathematicians are now designing self-stabilizing buildings that, instead of swaying in response to an earthquake, actively suppress their own vibrations with computer-controlled weights. (See Figure 1.1.) In one experimental building, the sway was said to be reduced by 80 percent.

During an earthquake, many buildings collapse when they oscillate naturally with the same frequency as seismic waves traveling through the earth, thus amplifying their effect, said Dr. Thomas Heaton, a seismologist at the U.S. Geological Survey in Pasadena, California. Active control systems might prevent that from happening, he added.

* Based on an article taken from the *New York Times,* July 2, 1991.

Figure 1.1
How a self-stabilizing building works. Instead of swaying in response to an earthquake, some new buildings are designed as machines that actively suppress their own vibrations by moving a weight that is about 1 percent of the building's weight.

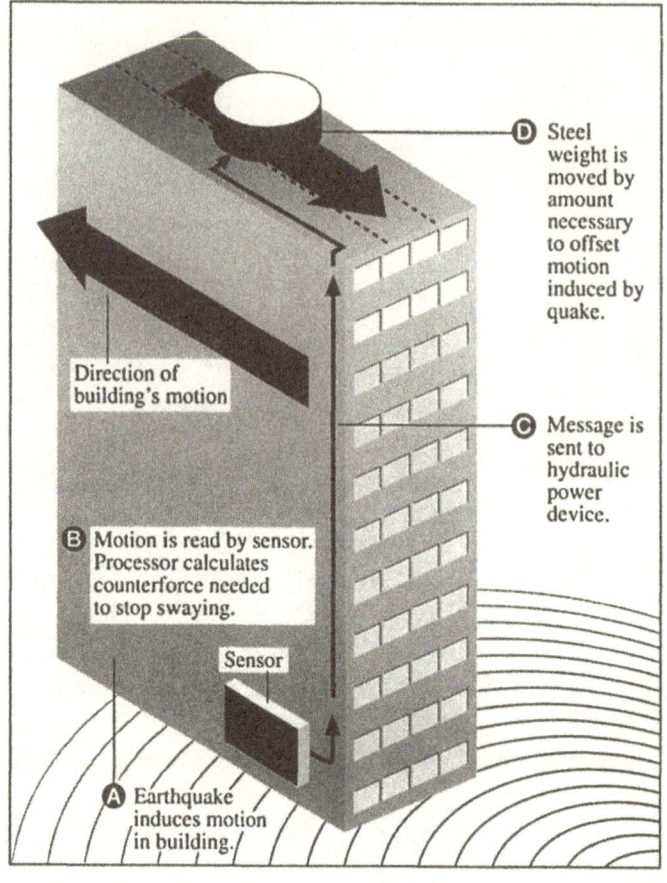

D Steel weight is moved by amount necessary to offset motion induced by quake.

C Message is sent to hydraulic power device.

Direction of building's motion

B Motion is read by sensor. Processor calculates counterforce needed to stop swaying.

Sensor

A Earthquake induces motion in building.

One new idea for an active control system is being developed by the University of Southern California by Dr. Sami Masri and his colleagues in the civil engineering department. When wind or an earthquake imparts energy to the building, Dr. Masri said, it takes several seconds for the oscillation to build up to potentially damaging levels. Chaotic theory of differential equations, he said, suggests that a random source of energy should be injected into this rhythmic flow to disrupt the system.

At the present time, two new active stabilizing systems are to be added to existing buildings in the United States that sway excessively. Because the owners do not want their buildings identified, the names of the buildings are kept confidential.

Bridges and elevated highways are also vulnerable to earthquakes. During the 1989 San Francisco earthquake (the ''World Series'' earthquake) the double-decker Interstate 880 collapsed, killing several people, and the reader might remember the dramatic pictures of a car hanging precariously above San Francisco Bay where a section of the San Francisco–Oakland Bay Bridge had fallen away. Less reported was the fact that the Golden Gate Bridge might also have been close to going down. Witnesses who were on the bridge during the quake said that the roadbed underwent wavelike motions in which the stays connecting the roadbed to the overhead cables alternately loosened and tightened ''like spaghetti.'' The bridge oscillated for about a minute, about four times as long as the actual earthquake. Inasmuch as an earthquake of up to *ten times* this magnitude (the ''big one'') is predicted for California sometime in the future, this experience reinforces our need for a deeper understanding of nonlinear oscillations in particular and nonlinear differential equations in general.

■ 1.1   BASIC DEFINITIONS AND CONCEPTS

PURPOSE

To introduce some of the basic terminology and ideas that are necessary for the study of differential equations. We introduce the concepts of

- ordinary and partial differential equations,
- order of a differential equation,
- linear and nonlinear differential equations.

THE ROLE OF DIFFERENTIAL EQUATIONS IN SCIENCE

Before saying what a differential equation *is,* let us first say what a differential equation *does* and how it is used. Differential equations can be used to describe the amount of money in a savings bank, the orbit of a spaceship, the amount of deformation of elastic structures, the description of radio waves, the size of a biological population, the current or voltage in an electrical circuit, and on and on. *In fact, differential equations can be used to explain and predict new facts for about everything that changes continuously.* In more complex systems we don't use a single differential equation, but a *system* of differential equations, as in the case of an electrical network of several circuits or in a chemical reaction with several interacting chemicals.

The *process* by which scientists and engineers use differential equations to understand physical phenomena can be broken down into three steps. First, a scientist or engineer defines a real problem. A typical example might be the study of shock waves along fault lines caused by an earthquake. To understand such a phenomenon, the scientist or engineer first *collects data,* maybe soil conditions, fault data, and so on. This first step is called **data collection.**

The second step, called the **modeling process,** generally requires the most skill and experience on the part of the scientist. In this step the scientist or engineer sets up an *idealized problem,* often involving a differential equation, which describes the real phenomenon as precisely as possible while at the same time being stated in such a way that mathematical methods can be applied. This idealized problem is called a **mathematical model** for the real phenomenon. In this book, mathematical models refer mainly to differential equations with initial and boundary conditions. There is generally a dilemma in constructing a good mathematical model. On one hand, a mathematical model may describe accurately the phenomenon being studied, but the model may be so complex that a mathematical analysis is extremely difficult. On the other hand, the model may be easy to analyze mathematically but may not reflect accurately the phenomenon being studied. The goal is to obtain a model that is sufficiently accurate to explain all the facts under consideration and to enable us to predict new facts but at the same time is mathematically tractable.

The third and last step is to solve mathematically the ideal problem (i.e., the differential equation) and compare the solution with the measurements of the real phenomenon. If the mathematical solution agrees with the observations, then the scientist or engineer is entitled to claim with some confidence that the physical problem has been ''solved mathematically,'' or that the theory has been verified. If the solution does not agree with the observations, either the observations are in error or the model is inaccurate and should be changed. This entire process of how mathematics (differential equations in this book) is used in science is described in Figure 1.2.

Figure 1.2
Schematic diagram of a mathematical analysis of physical phenomena

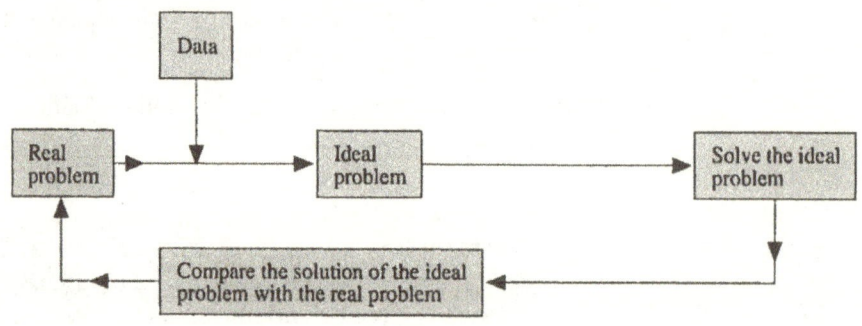

WHAT IS A DIFFERENTIAL EQUATION?

Quite simply, a **differential equation** is an equation that relates the derivatives of an unknown function, the function itself, the variables by which the function is defined, and constants. If the unknown function depends on a single real variable, the differential equation is called an **ordinary differential equation.** The following equations illustrate four well-known ordinary differential equations.

$$\frac{dy}{dx} + y = y^2 \qquad \text{(Bernoulli's equation)} \qquad (1a)$$

$$\frac{d^2y}{dx^2} = xy \qquad \text{(Airy's equation)} \qquad (1b)$$

$$x^2\frac{d^2y}{dx^2} + x\frac{dy}{dx} + (x^2 - 4)y = 0 \qquad \text{(Bessel's equation)} \qquad (1c)$$

$$\frac{d^2y}{dx^2} - (1 - y^2)\frac{dy}{dx} + y = 0 \qquad \text{(Van der Pol's equation)} \qquad (1d)$$

In these differential equations the unknown quantity $y = y(x)$ is called the **dependent variable,** and the real variable, x, is called the **independent variable.**

In this book, derivatives will be often represented by primes and higher derivatives sometimes by superscripts in parentheses. For example,

$$\frac{dy}{dx} = y', \qquad \frac{d^2y}{dx^2} = y'', \qquad \frac{d^3y}{dx^3} = y''' = y^{(3)}, \qquad \ldots \tag{2}$$

Differential equations are as varied as the phenomena that they describe.* For this reason it is convenient to classify them according to certain mathematical properties. In so doing, we can better organize the subject into a coherent body of knowledge.

In addition to ordinary differential equations,[†] which contain ordinary derivatives with respect to a single independent variable, a **partial differential equation** is one that contains partial derivatives with respect to more than one independent variable. For example, Eqs. (1a)–(1d) above are ordinary differential equations, whereas Eqs. (3a)–(3d) below are partial differential equations.

$$\frac{\partial u}{\partial x} + \frac{\partial u}{\partial y} = 0 \qquad \text{(flux equation)} \tag{3a}$$

$$\frac{\partial u}{\partial t} = \frac{\partial^2 u}{\partial x^2} \qquad \text{(heat equation)} \tag{3b}$$

$$\frac{\partial^2 u}{\partial t^2} = \frac{\partial^2 u}{\partial x^2} \qquad \text{(wave equation)} \tag{3c}$$

$$\frac{\partial^2 u}{\partial x^2} + \frac{\partial^2 u}{\partial y^2} = 0 \qquad \text{(Laplace's equation)} \tag{3d}$$

HOW DIFFERENTIAL EQUATIONS ORIGINATE

Inasmuch as derivatives represent rates of change, acceleration, and so on, it is not surprising to learn that differential equations describe many phenomena that involve motion. The most common models used in the study of planetary motion, the vibrations of a drumhead, or evolution of a chemical reaction are based on differential equations. In summary, differential equations originate whenever some universal law of nature is expressed in terms of a mathematical variable and at least one of its derivatives.

ORDER OF A DIFFERENTIAL EQUATION

Differential equations are also classified according to their order. The **order** of a differential equation is simply the *order* of the highest derivative that occurs in the equation.

---

* A differential equation is often named after the person who first studied it. For example, Van der Pol's equation listed here was first investigated by the Dutch radio engineer Balthasar Van der Pol (1889–1959), in studying oscillatory currents in electric circuits.

[†] Although differential equations should probably be called *derivative equations,* inasmuch as they contain derivatives, the term ''differential equations'' (*aequatio differentialis*) was coined by Gotfried Leibniz in 1676 and is used universally today.

For example,
$$\frac{dy}{dx} - 3y = 2 \qquad \text{(first-order)} \tag{4a}$$

$$\frac{d^2y}{dx^2} + x\frac{dy}{dx} - 3y = 0 \qquad \text{(second-order)} \tag{4b}$$

$$y\left(1 + \left(\frac{dy}{dx}\right)^2\right) = 0 \qquad \text{(first-order)} \tag{4c}$$

$$\frac{d^4y}{dx^4} - y = 0 \qquad \text{(fourth-order)} \tag{4d}$$

$$\frac{\partial u}{\partial t} = \frac{\partial^2 u}{\partial x^2} \qquad \text{(second-order)} \tag{4e}$$

DEFINITION: Ordinary Differential Equation

An nth-order ordinary differential equation is an equation that has the general form

$$F(x, y, y', y'', ..., y^{(n)}) = 0 \tag{5}$$

where the primes denote differentiation with respect to x, that is, $y' = dy/dx$, $y'' = d^2y/dx^2$, and so on.

LINEAR AND NONLINEAR DIFFERENTIAL EQUATIONS

Some of the most important and useful differential equations that arise in applications are those that belong to the class of **linear differential equations.** Roughly, this means that they do not contain products, powers, or quotients of the unknown function and its derivatives. More precisely, it means the following.

DEFINITION: Linear Differential Equation

An nth-order ordinary differential equation is **linear** when it can be written in the form

$$a_0(x)\frac{d^ny}{dx^n} + a_1(x)\frac{d^{n-1}y}{dx^{n-1}} + \cdots + a_{n-1}(x)\frac{dy}{dx} + a_n(x)y = f(x) \quad (a_0(x) \neq 0)$$

The functions $a_0(x), a_1(x), ..., a_n(x)$ are called the **coefficients** of the differential equation, and $f(x)$ is called the **nonhomogeneous term.** When the coefficients are constant functions, the differential equation is said to have **constant coefficients.** Unless it is otherwise stated, we shall always assume that the coefficients are continuous functions and that $a_0(x) \neq 0$ in any interval in which the equation is defined. Furthermore, the differential equation is said to be **homogeneous** if $f(x) \equiv 0$ and **nonhomogeneous** if $f(x)$ is *not* identically zero.

Finally, an ordinary differential equation that cannot be written in the above general form is called a **nonlinear ordinary differential equation.**

Some examples of linear and nonlinear differential equations are the following:

$$\frac{dy}{dx} = xy + 1 \qquad \text{(linear)} \tag{6a}$$

$$\frac{d^2y}{dx^2} + x\frac{dy}{dx} + y^2 = 0 \qquad \text{(nonlinear)} \tag{6b}$$

$$a_0(x)\frac{dy}{dx} + a_1(x)\,y = g(x) \qquad \text{(linear)} \tag{6c}$$

$$a_0(x)\frac{d^2y}{dx^2} + a_1(x)\frac{dy}{dx} + a_2(x)\,y = g(x) \qquad \text{(linear)} \tag{6d}$$

$$\frac{dy}{dx} = -\frac{x}{y} \qquad \text{(nonlinear)} \tag{6e}$$

$$yy'' + y' + y = 1 \qquad \text{(nonlinear)} \tag{6f}$$

Table 1.1 summarizes the above ideas. Note that in Table 1.1 the concepts of being homogeneous or having constant coefficients have no relevance for nonlinear differential equations.

Table 1.1
Classification of Differential Equations

Differential equation	Linear or nonlinear	Order	Homogeneous or nonhomogeneous	Constant or variable coefficients
$\dfrac{dy}{dx} + xy = 1$	Linear	1	Nonhomogeneous	Variable
$\dfrac{d^2y}{dx^2} + y\dfrac{dy}{dx} + y = x$	Nonlinear	2	*	*
$\dfrac{d^2y}{dx^2} + x\dfrac{dy}{dx} + y^2 = 0$	Nonlinear	2	*	*
$\dfrac{d^2y}{dx^2} + 3\dfrac{dy}{dx} + 2y = 0$	Linear	2	Homogeneous	Constant
$\dfrac{d^2y}{dx^2} + y = \sin y$	Nonlinear	2	*	*
$\dfrac{d^4y}{dx^4} + 3y = \sin x$	Linear	4	Nonhomogeneous	Constant

PROBLEMS: Section 1.1

For Problems 1–10, classify each differential equation according to the following categories: order; linear or nonlinear; constant or variable coefficients; homogeneous or nonhomogeneous.

1. $\dfrac{dy}{dx} + xy^2 = 1$

2. $x\dfrac{dy}{dx} + y = \sin x$

3. $e^x\dfrac{d^2y}{dx^2} + 2\dfrac{dy}{dx} + y = 0$

4. $y\dfrac{d^2y}{dx^2} + \dfrac{dy}{dx} + xy = 1$

5. $\dfrac{d^3y}{dx^3} + \dfrac{d^2y}{dx^2} + \dfrac{dy}{dx} + y = 0$

6. $\dfrac{d^3y}{dx^3} + y = \sin x$

7. $\dfrac{d^2u}{dt^2} + t\dfrac{du}{dt} + 3u = 1$

8. $\dfrac{d^2w}{dt^2} - w^2\dfrac{dw}{dt} + w = 0$

9. $\dfrac{d^2v}{dt^2} = t^2v$

10. $\dfrac{d^2y}{dt^2} + y^2 = 0$

Problems 11–17 list differential equations that arise in the mathematical formulation of pure and applied science. Classify each equation according to order, number of independent variables, whether it is ordinary or partial, and whether it is linear or nonlinear. For linear equations, tell whether the equation has constant coefficients and whether it is homogeneous or nonhomogeneous.

11. $y' = ky$ (unrestricted growth or decay equation)

12. $y'' + \omega^2 y = 0$ (simple harmonic motion equation)

13. $EI\, y^{(4)} = w(x)$ (deflection of beam equation)

14. $u_{rr} + \dfrac{1}{r} u_r + \dfrac{1}{r^2} u_{\theta\theta} = 0$ (Laplace's equation, u_r, u_{rr}, and $u_{\theta\theta}$ denote partial derivatives)

15. $u_t + u_{xxx} - 6uu_x + \dfrac{1}{2t} u = 0$ (KdV equation from fluid dynamics)

16. $y'' + y + \epsilon y^3 = 0$ (Hill's equation for vibrating systems)

17. $(1 - x^2)y'' - 2xy' + n(n + 1)y = 0$ (Legendre's wave equation)

18. Gotcha Beginning students of differential equations are often confused as to whether a linear differential equation is homogeneous or nonhomogeneous. Problems (a)–(d) often confuse the beginner. Can you say for sure which of the linear equations in (a)–(d) are homogeneous?

(a) $y' - xy - 1 = 0$ (c) $y'' + x + y = 0$

(b) $xy' + y + \sin x = 0$ (d) $xy'' + xe^x + y' = 0$

Keeping a Scholarly Journal (Read This)

One cannot help but be impressed with the large number of important English naturalists who lived during the nineteenth century. Of course, there was Darwin, but there were also Wallis, Eddington, Thompson, Haldane, Fisher, Jevons, Fechner, Galton, and many more. If one studies the works of these eminent scholars, one cannot help but be impressed with the manner in which they paid attention to scientific details. Part of that attention to scientific details was the keeping of detailed journals in which they recorded their observations and impressions. These journals provided not only a means for storing data, but a means for exploring their thoughts and ideas. *They in fact learned through writing.*

Over the past 100 years, journal keeping has declined in popularity, but in recent years there has been a renaissance in the "learning through writing" movement. A few people are beginning to realize that writing is an important *learning* tool as well as a means of communication.

In this book we give the reader the opportunity to explore thoughts and ideas through writing. We only require that the reader *possess* a bound journal* in which daily entries are made. Each entry should be *dated* and, if useful, given a short title. There are no rules telling you what to include in your journal or how to write. The style of writing is strictly free form—don't worry about punctuation, spelling, or form. You will find that if you make a conscientious effort to make a daily entry, your writing style will take care of itself.

The best time to make your entry is immediately after class. Some professors allow their students the last five minutes of class time for journal entries. You might spend five minutes writing about the day's lecture. You might focus on a difficult concept. Ask yourself what you don't understand. Realize that you are writing for yourself. No one cares about your journal except you.

19. Your First Journal Entry Spend ten minutes exploring your goals for this course. Do you think differential equations will be useful to you? How does the material relate to your career goals as you see them? Maybe summarize in your own language the material covered in this first section. Be sure to date your entry. There will be a journal entry suggestion at the end of each problem set. Good luck.

* A good leather-bound journal can be purchased in any office supply store for about $10. Regular notebook paper can be inserted into these journals.

1.2 SOME BASIC THEORY

PURPOSE

To introduce more concepts that are necessary to the study of differential equations. In particular, we will study

- explicit and implicit solutions,
- initial-value problems,
- the existence and uniqueness of solutions.

SOLUTIONS OF DIFFERENTIAL EQUATIONS

The general form of an nth-order ordinary differential equation can be written as

$$F\left(x,\ y,\ \frac{dy}{dx},\ \frac{d^2y}{dx^2},\ ...,\ \frac{d^ny}{dx^n}\right) = 0 \tag{1}$$

where F is a function of the independent variable x, the dependent variable y, and the derivatives of y up to order n. We assume that x lies in an interval I that can be any of the usual types: (a, b), $[a, b]$, (a, ∞), $(-\infty, b)$, $(-\infty, \infty)$, and so on. Often, it is possible to solve algebraically for the highest-order derivative in the differential equation and write it as

$$\frac{d^ny}{dx^n} = f\left(x,\ y,\ \frac{dy}{dx},\ \frac{d^2y}{dx^2},\ ...,\ \frac{d^{n-1}y}{dx^{n-1}}\right) \tag{2}$$

where f is a function of x, y, y', ..., $y^{(n)}$.

One of the main reasons for studying differential equations is to learn how to ''solve'' a differential equation.

> **HISTORICAL NOTE**
>
> The origins of differential equations go back to the beginning of the calculus to the work of Isaac Newton (1642–1727) and Gottfried Wilhelm von Leibniz (1646–1716). Newton classified first-order differential equations according to the forms $dy/dx = f(x)$, $dy/dx = f(y)$, or $dy/dx = f(x, y)$. He actually developed a method for solving $dy/dx = f(x, y)$ using infinite series when $f(x, y)$ is a polynomial in x and y. A simple example would be $dy/dx = 1 + xy$. Can you find an infinite series $y = y(x)$ that satisfies this equation?

This brings us to the concept of the solution of a differential equation.

> **DEFINITION: Solution of a Differential Equation***
>
> A **solution** of an nth-order differential equation is an n times differentiable function $y = y(x)$, which, when substituted into the equation, satisfies it identically over some interval $a < x < b$. We would say that the function y is a solution of the differential equation **over the interval** $a < x < b$.

Example 1 **Verifying a Solution** Verify that the function

$$y(x) = \sin x - \cos x + 1 \tag{3}$$

is a solution of the equation

$$\frac{d^2y}{dx^2} + y = 1 \tag{4}$$

for all values of x.

---

* Solutions of differential equations are sometimes called *integrals* of the differential equations, since they are more or less an extension of the process of integration in calculus.

Solution   Clearly, $y(x)$ is defined on $(-\infty, \infty)$, and $y'' = -\sin x + \cos x$. Substituting y'' and y into the differential equation (4) yields the identity

$$\underbrace{(-\sin x + \cos x)}_{y''} + \underbrace{(\sin x - \cos x + 1)}_{y} = 1$$

for all $-\infty < x < \infty$.   ■

Example 2   Verifying a Solution   Verify that the function

$$y(x) = 3e^{2x}$$

is a solution of the differential equation

$$\frac{dy}{dx} - 2y = 0 \tag{5}$$

for all x.

Solution   Clearly, the function $y(x)$ is defined for all real x, and substitution of $y(x) = 3e^{2x}$ and $y'(x) = 6e^{2x}$ into the differential equation yields the identity

$$\underset{y'}{(6e^{2x})} - 2\underset{y}{(3e^{2x})} = 0 \tag{6}$$

In fact, note that *any* function of the form $y(x) = Ce^{2x}$, where C is a constant, is a solution of this differential equation.   ■

Example 3   Verifying a Solution   Verify that both functions

$$y_1(x) = e^{5x} \qquad \text{and} \qquad y_2(x) = e^{-3x} \tag{7}$$

are solutions of the second-order equation

$$\frac{d^2y}{dx^2} - 2\frac{dy}{dx} - 15y = 0 \tag{8}$$

for all real x.

Solution   Substituting $y_1(x) = e^{5x}$ into the equation gives

$$y'' - 2y' - 15y = \underset{y''}{(25e^{5x})} - 2\underset{y'}{(5e^{5x})} - 15\underset{y}{(e^{5x})} = 0$$

Substituting $y_2(x) = e^{-3x}$ into the equation, we get

$$y'' - 2y' - 15y = (9e^{-3x}) - 2(-3e^{-3x}) - 15(e^{-3x}) = 0$$

$$\uparrow \qquad\qquad \uparrow \qquad\qquad \uparrow$$

$$y'' \qquad\qquad y' \qquad\qquad y$$

Hence both functions satisfy the equation for all x. ■

IMPLICIT SOLUTIONS

We have just studied solutions of the form $y = y(x)$ that determine y directly from a formula in x. Such solutions are called **explicit solutions**, since they give y directly, or *explicitly*, in terms of x. On some occasions, especially for nonlinear differential equations, we must settle for the less convenient form of solution, $G(x, y) = 0$, from which it is impossible to deduce an *explicit representation* for y in terms of x. Such solutions are called **implicit solutions.**

> **DEFINITION: Implicit Solution**
>
> A relation $G(x, y) = 0$ is said to be an **implicit solution** of a differential equation involving x, y, and derivatives of y with respect to x if $G(x, y) = 0$ *defines* one or more explicit solutions of the differential equation.*

Example 4   Implicit Solution   Show that the relation

$$x + y + e^{xy} = 0 \tag{9}$$

is an implicit solution of

$$(1 + xe^{xy})\frac{dy}{dx} + 1 + ye^{xy} = 0 \tag{10}$$

Solution   First, note that we cannot solve Eq. (9) for y in terms of x. However, a change in x in Eq. (10) results in a change in y, so we would expect that on some interval, Eq. (9) would define at least one function[†] $y = y(x)$. This is true in this case, and such a function $y = y(x)$ is also differentiable. See Figure 1.3.

---

* Realize that even though you may not be able to actually solve the equation $G(x, y) = 0$ for y, thus obtaining a formula in x, nevertheless, any change in x still results in a corresponding change in y. Thus the expression $G(x, y) = 0$ gives rise to at least one function $y = y(x)$, even though you cannot find a formula for it. The exact conditions under which $G(x, y) = 0$ gives rise to a function $y = y(x)$ are known as the **implicit function theorem.** Details of this theorem can be found in most textbooks of advanced calculus.

† The implicit function theorem provides exact conditions under which an implicit relation $G(x, y) = 0$ defines y as a function of x. See Problem 35 at the end of this section.

Figure 1.3
Note that $x + y + e^{xy} = 0$
defines a function $y = y(x)$
on certain intervals and that
this function is an explicit
solution of the differential
equation. Also, it can be
shown that the slope dy/dx of
the tangent line at each point
of the curve satisfies Eq. (9).

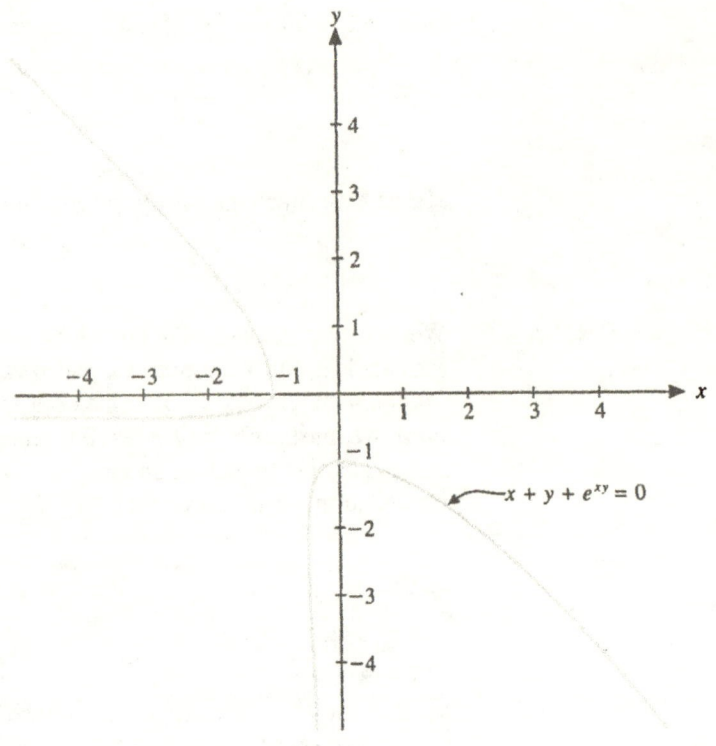

Once we know that the implicit relationship in Eq. (9) defines a differentiable function of x, we differentiate implicitly with respect to x. Doing this, we get

$$1 + \frac{dy}{dx} + e^{xy}\left(y + x\frac{dy}{dx}\right) = 0 \tag{11}$$

which is equivalent to the differential equation

$$(1 + xe^{xy})\frac{dy}{dx} + 1 + ye^{xy} = 0 \tag{12}$$

Hence Eq. (9) is an implicit solution of Eq. (10).   ■

Example 5   Implicit Solution   Show that the relation

$$x^2 + y^2 - c = 0 \tag{13}$$

where c is a positive constant and is an implicit solution of the differential equation

$$\frac{dy}{dx} = -\frac{x}{y} \tag{14}$$

on the open interval $(-c, c)$.

Solution   By differentiating Eq. (13) with respect to x we get

$$2x + 2y\frac{dy}{dx} = 0 \tag{15}$$

or $y' = -x/y$, which shows that Eq. (13) is an implicit solution of Eq. (14).

The geometric interpretation of this implicit solution is that the tangent line to the circle $x^2 + y^2 = c$ at the point (x, y) has slope $dy/dx = -x/y$. Also, note that there are many functions $y = y(x)$ that satisfy the implicit relation $x^2 + y^2 - c = 0$ on $(-c, c)$, and some of them are shown in Figure 1.4. However, the only ones that are continuous (and hence possibly differentiable) are

$$y(x) = +\sqrt{c - x^2} \quad \text{and} \quad y(x) = -\sqrt{c - x^2}$$

Note that $y'(-c)$ and $y'(c)$ do not exist, and so $y' = -x/y$ is an implicit relation only on the open interval $(-c, c)$. Hence from the implicit solution we are able to find two *explicit* solutions on the interval $(-c, c)$.   ■

Figure 1.4
By taking portions of either the upper or lower semicircle of the circle $x^2 + y^2 = c^2$, one obtains a function $y = y(x)$ that satisfies the relationship $x^2 + y^2(x) = c^2$ on the interval $(-c, c)$. A few of them are shown here.

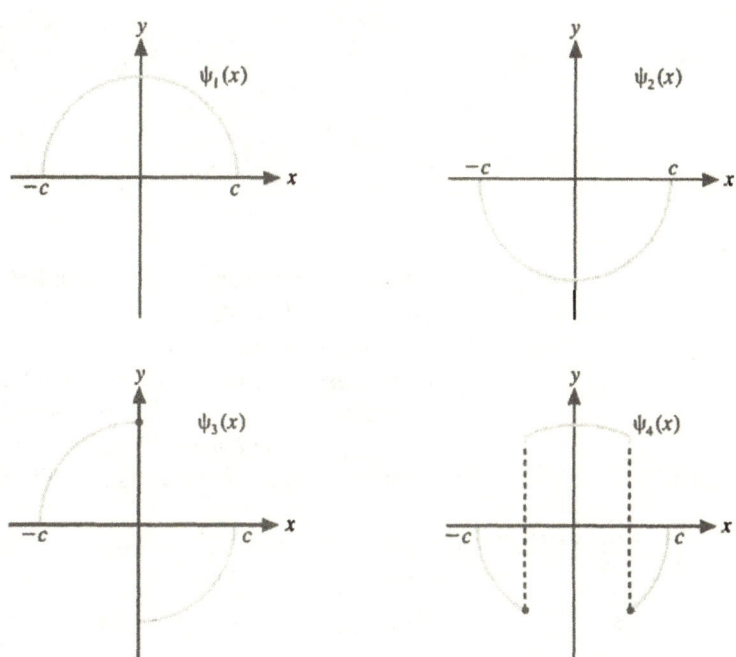

COMMENT ON EXPLICIT VERSUS IMPLICIT SOLUTIONS

To better appreciate the use of the words "explicit" and "implicit," we would say that $y = x + 1$ states *explicitly* that y *is* $x + 1$ and thus is called an explicit solution. On the other hand, the equation $y - x = 1$ does not state explicitly that y *is* $x + 1$ but only *implies* or states *implicitly* that y is $x + 1$. When a differential equation is solved, it is generally an explicit solution* that is desired.

* The importance of implicit solutions lies in the fact that some methods of solution do not lead to explicit solutions, but only to implicit solutions. Thus implicit solutions might be thought of as "better than nothing." In some areas of nonlinear differential equations, implicit solutions are the *only* solutions obtainable and in this context are referred to simply as solutions.

**THE INITIAL-VALUE
PROBLEM**

When solving differential equations in science and engineering, one generally seeks a solution that also satisfies one or more supplementary conditions such as initial or boundary conditions. The general idea is to first find *all* the solutions of the differential equation and then pick out the particular one that satisfies the supplementary condition(s).

DEFINITION: Initial-Value Problem

An **initial-value problem** for an *n*th-order equation

$$F\left(x, y, \frac{dy}{dx}, \frac{d^2y}{dx^2}, ..., \frac{d^ny}{dx^n}\right) = 0$$

consists in finding the solution to the differential equation on an interval *I* that also satisfies the *n* **initial conditions**

$$y(x_0) = y_0$$
$$y'(x_0) = y_1$$
$$y''(x_0) = y_2$$
$$\vdots$$
$$y^{(n-1)}(x_0) = y_{n-1}$$

where $x_0 \in I$ and $y_0, y_1, ..., y_{n-1}$ are given constants.

Note that in the special case of a first-order equation the only initial condition is $y(x_0) = y_0$, and in the case of a second-order equation the initial conditions are $y(x_0) = y_0$ and $y'(x_0) = y_1$.

The reason it is natural to specify *n* "side" conditions to accompany the *n*th-order linear differential equation lies in the fact that the general solution of the *n*th-order linear equation contains *n* arbitrary constants. Hence the *n* initial conditions will determine the constants, giving a unique solution to the initial-value problem.

Example 6   **First-Order Initial-Value Problem**   Verify that $y(x) = e^{-x} + 1$ is a solution of the initial-value problem

$$y' + y = 1 \qquad y(0) = 2 \tag{16}$$

Solution   Computing $y'(x) = -e^{-x}$ and substituting $y(x)$ and $y'(x)$ into the differential equation, we get

$$y' + y = -e^{-x} + (e^{-x} + 1) = 1$$

Hence $y(x)$ satisfies the differential equation. To verify that $y(x)$ also satisfies the initial condition, we observe that

$$y(0) = e^0 + 1 = 2$$

This solution is shown in Figure 1.5.   ∎

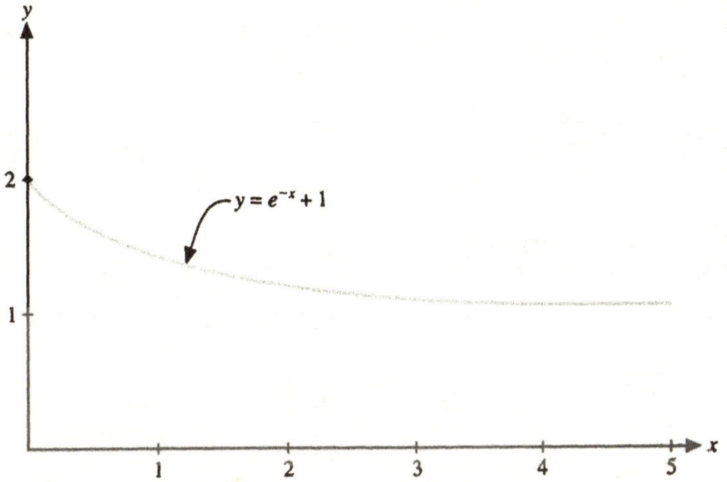

Figure 1.5
In Chapter 2 we will learn that $y(x) = ce^{-x} + 1$, where c is any constant, constitutes *all* the solutions of the equation $y' + y = 1$. However, the only one of these solutions that satisfies $y(0) = 2$ is $y(x) = e^{-x} + 1$.

$y = e^{-x} + 1$

Example 7   **Initial-Value Problem**   Verify that $y(x) = \sin x + \cos x$ is a solution of the initial-value problem

$$y'' + y = 0 \qquad y(0) = 1 \qquad y'(0) = 1 \tag{17}$$

Solution   Computing $y'(x) = \cos x - \sin x$ and $y''(x) = -\sin x - \cos x$ and substituting these values into the differential equation, we get

$$y'' + y = (-\sin x - \cos x) + (\sin x + \cos x) = 0$$

Hence $y(x)$ satisfies the differential equation. To verify that $y(x)$ also satisfies the initial conditions, we observe that

$$y(0) = \sin 0 + \cos 0 = 1$$
$$y'(0) = \cos 0 - \sin 0 = 1$$

This solution is drawn in Figure 1.6.   ∎

EXISTENCE AND UNIQUENESS OF SOLUTIONS

Although differential equations studied in applied work normally have solutions, it is clear that the equation

$$\left(\frac{dy}{dx}\right)^2 + 1 = 0$$

has none.* Inasmuch as some differential equations have solutions and some do not, it is important to know the conditions under which we know a solution exists. We state a fundamental *existence* and *uniqueness* result for the first-order initial-value problem.

* Although the equation has no real-valued solution, it does have a complex-valued solution. A study of complex-valued solutions of differential equations is beyond the scope of this book.

Figure 1.6
Although it is clear that any function of the general form $y(x) = c_1 \sin x + c_2 \cos x$, where c_1 and c_2 are any constants, is a solution of $y'' + y = 0$, only $y(x) = \sin x + \cos x$ satisfies the conditions $y(0) = 1$ and $y'(0) = 1$.

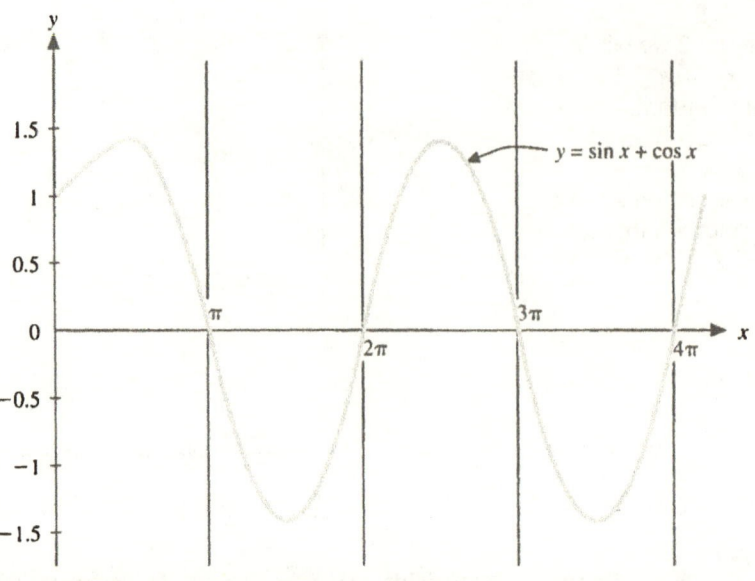

$y = \sin x + \cos x$

THEOREM 1.1 (PICARD'S THEOREM): Existence and Uniqueness for First-Order Equations

Assume that for the initial-value problem

$$\frac{dy}{dx} = f(x, y) \qquad y(x_0) = y_0$$

the functions f and $\partial f/\partial y$ are continuous on some rectangle

$$R = \{(x, y): a < x < b, c < y < d\}$$

that contains the initial point (x_0, y_0). Under these conditions the initial-value problem has a unique solution $y = \phi(x)$ on some interval $(x_0 - h, x_0 + h)$, where h is some positive number.

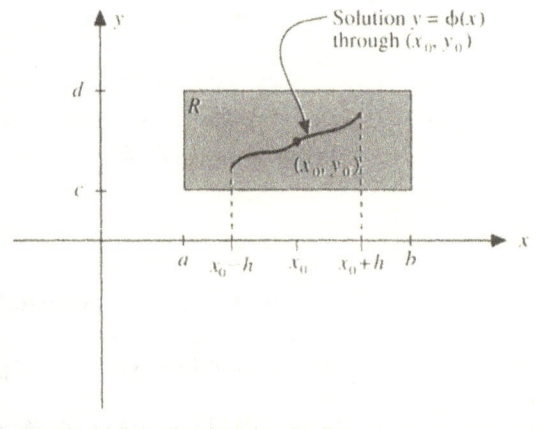

Note: Picard's theorem is one of the more popular existence and uniqueness theorems, since one only has to check the continuity of f and $\partial f/\partial y$, which is generally easy to do. A weakness of the theorem lies in the fact that it doesn't specify the size of the interval on which a solution exists without actually solving the differential equation.

Example 8   **Picard's Theorem**   What does Picard's theorem guarantee about a solution to the initial-value problem

$$y' = y + e^{2x} \quad y(0) = 1 \tag{18}$$

Solution   Since $f(x, y) = y + e^{2x}$ and $\partial f/\partial y = 1$ are continuous in any rectangle containing the point $(0, 1)$, the hypothesis of Picard's theorem is satisfied. Hence Picard's theorem guarantees that a unique solution of the initial-value problem exists in some interval $(-h, h)$, where h is a positive constant. We will learn how to solve this equation in Chapter 2 and see that the solution is $y(x) = e^{2x}$ for all $-\infty < x < \infty$. Picard's theorem tells us that this is the only solution of this initial-value problem.   ∎

HISTORICAL NOTE

Theorem 1.1 and other closely related existence theorems are generally associated with the name of Charles Émile Picard (1856–1941), one of the greatest French mathematicians of the nineteenth century. He is best known for his contributions to complex variables and differential equations. It is interesting to note that in 1899, Picard lectured at Clark University in Worcester, Massachusetts.

Example 9   **Picard's Theorem**   What does Picard's theorem imply about a unique solution of the initial-value problem

$$y' = y^{1/3} \quad y(0) = 0$$

Solution   Here we have $f(x, y) = y^{1/3}$ and $\partial f/\partial y = \frac{1}{3}\bar{y}^{2/3}$. Although $f(x, y)$ is continuous in the entire xy-plane, the partial derivative $\partial f/\partial y$ is not continuous on any rectangle that intersects the line $y = 0$. Hence the hypothesis of Picard's theorem fails, and so we cannot be *guaranteed** that there exists a unique solution to any problem whose initial value of y is zero. In Problem 45 at the end of this section we will see that there are in fact solutions to this problem—in fact, an infinite number of solutions.   ∎

---

* Remember, the conditions stated in Picard's theorem are *sufficient* but not *necessary*. If the conditions stated in Picard's theorem hold, then there will exist a unique solution. However, if the conditions stated in the hypothesis of Picard's theorem do *not* hold, then nothing is known: The initial-value problem may have either (a) no solution, (b) more than one solution, or (c) a unique solution.

GENERAL SOLUTION OF A DIFFERENTIAL EQUATION

The nature of the solutions of a differential equation is somewhat reminiscent of finding antiderivatives in calculus. Remember that when finding an antiderivative, one obtains a function containing an arbitrary constant.* In solving a first-order differential equation $F(x, y, y') = 0$, the standard strategy is to obtain a *family* of curves $G(x, y, c) = 0$ containing one arbitrary constant c (called a parameter) such that each member of the family satisfies the differential equation. In the general case when solving an nth-order equation $F(x, y, y', ..., y^{(n)}) = 0$, we generally obtain an **n-parameter family of solutions** $G(x, y, c_1, c_2, ..., c_n) = 0$. A solution of a differential equation that is free of arbitrary parameters is called a specific or **particular solution.**

The theory of differential equations would be simplified if one could say that each differential equation of order n has an n-parameter family of solutions and that those are the only solutions. Normally, this is true as we will see for linear differential equations. However, there are some nasty nonlinear equations of order n that have an n-parameter family of solutions but still have a few more **singular solutions** hanging around the fringes, so to speak. For example, the nonlinear equation $(y')^2 + xy' = y$ has a one-parameter family of solutions $y(x) = cx + c^2$, but it still has one more *singular solution*, $y(x) = -x^2/4$. (See Problem 46 at the end of this section.) It is called a singular solution, since it cannot be obtained by assigning a specific value of c to $y(x) = cx + c^2$.

The fact that for some nonlinear differential equations, not all solutions of an nth-order equation are members of an n-parameter family has given rise to two schools of thought concerning the definition of a ''general solution'' of a differential equation. Some people say that the general solution of an nth-order differential equation is a family of solutions consisting of n essential† parameters. In this book we use a slightly broader definition of the general solution. We define the general solution of a differential equation to be the collection of *all* solutions of a differential equation. Period. If the only solutions of a differential equation consist of an n-parameter family of solutions, then the two definitions are the same. For those nasty nonlinear equations that have an n-parameter family of solutions *plus* a few more singular solutions, we will call the general solution the collection of both these types of solutions.

---

* For example, the antiderivative of $f(x) = 2x$ is $F(x) = x^2 + c$, where c is an arbitrary constant.

† By *essential* parameters or constants we mean just that. For instance, we could write the solution of the equation $y' = 1$ as either $y = x + c$ or $y = x + c_1 + c_2$. The two constants in second form are ''fraudulent,'' since we could let $c = c_1 + c_2$. It is more difficult to determine the number of essential constants in other equations, such as $y = c_1 + \ln c_2 x$. How many do you think there are? Be careful, there is really only one! See Problem 47.

PROBLEMS: Section 1.2

For Problems 1–11, show that each function is a solution of the given differential equation. Assume that a and c are con- *stants. For what values of the independent variable(s) is your solution defined?*

Differential equation	Function
1. $\dfrac{dy}{dx} = ay$	$y = e^{ax}$
2. $\dfrac{dy}{dx} = y + e^x$	$y = xe^x$
3. $\dfrac{dy}{dx} = \dfrac{x}{\sqrt{x^2 + a^2}}$ $(a \neq 0)$	$y = \sqrt{x^2 + a^2}$
4. $\dfrac{d^2y}{dx^2} + a^2 y = 0$	$y = c \sin ax$
5. $\dfrac{1}{4}\left(\dfrac{d^2y}{dx^2}\right)^2 - x\dfrac{dy}{dx} + y = 1 - x^2$	$y = x^2$
6. $\dfrac{1}{4}\left(\dfrac{d^2y}{dx^2}\right) - x\dfrac{dy}{dx} + y = -2$	$y = 2(x - 1)$
7. $(1 + x^2)\dfrac{d^2y}{dx^2} - 2x\dfrac{dy}{dx} + 2y = 2$	$y = x^2$
8. $x(x - 2)\dfrac{d^2y}{dx^2} - 2(x - 1)\dfrac{dy}{dx} + 2y = 12x^5 - 30x^4$	$y = x^5$
9. $\dfrac{\partial^2 u}{\partial x^2} + \dfrac{\partial^2 u}{\partial y^2} = 0$	$u(x, y) = \tan^{-1}\left(\dfrac{y}{x}\right)$
10. $\dfrac{\partial^2 u}{\partial x^2} + \dfrac{\partial^2 u}{\partial y^2} + \dfrac{\partial^2 u}{\partial z^2} = 0$	$u(x, y, z) = \dfrac{1}{\sqrt{x^2 + y^2 + z^2}}$
11. $\dfrac{\partial^2 u}{\partial t^2} = a^2 \dfrac{\partial^2 u}{\partial x^2}$	$u(x, t) = f(x - at) + g(x + at)$ (where f and g are differentiable functions)

For Problems 12–16, show that the following relation defines an implicit solution of the given differential equation.

Diffential equation	Relation
12. $yy' = e^{2x}$	$y^2 = e^{2x}$
13. $y' = \dfrac{xy}{x^2 + y^2}$	$2y^2 \ln y - x^2 = 0$
14. $2xyy' = x^2 + y^2$	$y^2 = x^2 - cx$
15. $y' = \dfrac{y^2}{xy - x^2}$	$y = ce^{y/x}$
16. $y' = \dfrac{xy}{x^2 - 1}$	$x^2 + cy^2 = 1$

Test Your Intuition

For Problems 17–25, see whether you can make an educated guess to find a solution of the given equation. After you have selected your candidate, check to see whether it satisfies the equation. Have you found all the solutions of the equation?

17. $\dfrac{dy}{dx} = y$		(When is the derivative equal to the function itself?)
18. $\dfrac{dy}{dx} = y^2$		(When is the derivative equal to the function squared?)
19. $\dfrac{dy}{dx} + y = 1$		(There is a solution staring you in the face.)
20. $\dfrac{dy}{dx} + y = e^x$	(Almost staring you in the face)	
21. $\dfrac{dy}{dx} + y = ae^x$	(More interesting)	
22. $\dfrac{dy}{dx} + ay = e^x$	(Kind of like Problem 21)	

23. $\dfrac{dy}{dx} + \dfrac{1}{x}y = 0$ (A little tougher)

24. $\dfrac{d^2y}{dx^2} + y = 0$ (Do you have all of them?)

25. $\dfrac{d^2y}{dx^2} - y = 0$ (Compare with Problem 24.)

26. Simplest Differential Equations The simplest of all differential equations are equations of the form

$$\frac{dy}{dx} = f(x), \quad \frac{d^2y}{dx^2} = f(x), \quad \dots, \quad \frac{d^ny}{dx^n} = f(x), \quad \dots \quad (19)$$

that are studied in calculus. Recall that the solution of the first-order equation $y' = f(x)$ is simply the collection of antiderivatives of $f(x)$, or

$$y(x) = \int f(x)\, dx + c \quad (20)$$

For Problems (a)–(e), solve the given first- or second-order differential equation for all $-\infty < x < \infty$. Where initial conditions are specified, solve the initial-value problem.

(a) $y' = 3$
(b) $y' = x^2$
(c) $y' = \sin x$ $y(0) = 0$
(d) $y'' = 1$ $y(0) = 1$ $y'(0) = 0$
(e) $y'' = \sin x$ $y(0) = 0$ $y'(0) = 1$

Initial-Value Problems

For Problems 27–33, verify that the specified function is a solution of the given initial-value problem.

Differential equation	Initial condition(s)	Function
27. $y' + y = 0$	$y(0) = 2$	$y(x) = 2e^{-x}$
28. $y' = y^2$	$y(0) = 0$	$y(x) = 0$
29. $y' = y^2$	$y(1) = -1$	$y(x) = -1/x,\ x \in (0, \infty)$
30. $y' = -x/y$	$y(0) = 2$	$y(x) = \sqrt{4 - x^2},\ x \in (-2, 2)$
31. $y'' + 4y = 0$	$y(0) = 1$ $y'(0) = 0$	$y(x) = \cos 2x$
32. $y'' - y = 0$	$y(0) = 1$ $y'(0) = 0$	$y(x) = \cosh x$
33. $y'' + 3y' + 2y = 0$	$y(0) = 0$ $y'(0) = 1$	$y(x) = e^{-x} - e^{-2x}$

34. No Solutions Why don't the following differential equations have real-valued solutions on *any* interval?

(a) $\left|\dfrac{dy}{dx}\right| + |x| + |y| + 1 = 0$

(b) $\left(\dfrac{dy}{dx}\right)^2 + 1 = -e^x$

(c) $|y'| + y^2 = -1$

(d) $\sin y' = 2$

35. Implicit Function Theorem The implicit function theorem states that if $G(x, y)$ has continuous first partial derivatives in a rectangle $R = \{(x, y): a < x < b, c < y < d\}$ containing a point (x_0, y_0), and if $G(x_0, y_0) = 0$ and $\partial G(x_0, y_0)/\partial y$ is not zero, then there exists a differentiable function $y = \phi(x)$, defined on an interval $I = (x_0 - h, x_0 + h)$, that satisfies $G(x, \phi(x)) = 0$ for all $x \in I$. The implicit function theorem provides the conditions under which $G(x, y) = 0$ defines y implicitly as a function of x. Use the implicit function theorem to verify that the relationship $x + y + e^{xy} = 0$ defines y implicitly as a function of x near the point $(0, -1)$.

Existence of Solutions

For Problems 36–43, determine whether Picard's theorem implies that the given initial-value problem has a unique solution on some interval containing the initial value of x.

36. $y' = x^2$ $y(0) = 0$
37. $y' - y = 1$ $y(0) = 3$
38. $y' = x^2 + y^2$ $y(0) = 1$
39. $y' = x^3 - y^3$ $y(0) = 0$
40. $y' = y^2$ $y(1) = 1$
41. $yy' = x^2 + y^2$ $y(0) = 1$
42. $y' = -x/y$ $y(1) = 0$
43. $y' = y/x$ $y(0) = 1$

44. A Strange Differential Equation The initial-value problem

$$y' = 2\sqrt{y} \qquad y(0) = 0 \qquad (21)$$

has an infinite number of solutions on the interval $[0, \infty)$.

(a) Show that $y(x) = x^2$ is a solution.

Figure 1.7
An infinite number of
solutions of the initial-value
problem $y' = 2\sqrt{y}$, $y(0)$
$= 0$. This problem
has long been a popular
example for illustrating when
Picard's theorem fails.

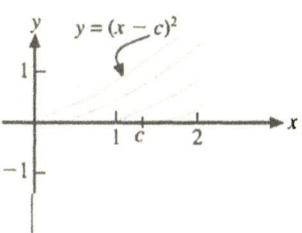

(b) Show that *any* function of the form

$$y(x) = \begin{cases} 0 & 0 \le x < c \\ (x - c)^2 & c \le x \end{cases}$$

where $c \ge 0$ is a solution of Eq. (21). See Figure 1.7.

(c) How do these results relate to Picard's existence and uniqueness theorem?

45. Hubbard's Empty Bucket* If you are given an empty bucket with a hole in it, can you determine when the bucket was full of water? Of course, the answer is no, but did you know that the reason is that a certain differential equation does not have a unique solution? It's true—The differential equation that describes the height h of the water satisfies **Torricelli's law,**

$$\frac{dh}{dt} = -k\sqrt{h} \tag{22}$$

where $k > 0$ is a constant depending on the cross section of the hole and the bucket.

(a) Show that Eq. (22) does not satisfy Picard's theorem.

(b) Find an infinite number of solutions of Eq. (22).

(c) Sketch the graphs of several solutions of Eq. (22) and discuss why you can't determine when the bucket was full if it is currently empty.

46. Singular Solution Given the first-order nonlinear equation

$$(y')^2 + xy' = y \tag{23}$$

verify the following.

(a) Each member of the one-parameter family of functions $y(x) = cx + c^2$, where c is a real constant, is a solution of Eq. (23).

(b) The function $y(x) = -\frac{1}{4}x^2$ cannot be obtained from $y = cx + c^2$ by any choice of c, yet it satisfies Eq. (23). It is a singular solution of Eq. (23).

47. Oniy One Parameter Show that it is possible to rewrite

$$y = c_1 + \ln(c_2 x)$$

in terms of only one parameter.

48. Delay Differential Equation A delay differential equation differs from the usual differential equation by the presence of a shift $x - x_0$ in the argument of the unknown function $y(x)$. These equations are much more difficult to solve than the usual differential equation, although much is known about them. Show that the simple delay differential equation

$$\frac{dy}{dx} = ay(x - b) \tag{24}$$

where $a \ne 0$ and b are given constants, has a solution of the form $y = Ce^{kx}$ for any constant C, provided that the constant k satisfies the transcendental equation $k = ae^{-bk}$. If you have access to a computer with a program to approximate solutions of transcendental equations, find an approximate solution to the equation $y' = y(x - 1)$ and sketch the graph of this solution.

49. Journal Entry—How's Your Intuition? Using your intuition, spend a few minutes and try to answer one of the following questions. How many solutions should there be to the equation $y' = ky$? What are all the solutions of the differential equation $y'' + y = 0$? Are there more solutions to second-order differential equations than to first-order equations? Can you find a solution of a differential equation by just finding the antiderivative of each term in the equation?

---

* This interesting problem is based on an example taken from *Differential Equations: A Dynamical Systems Approach* by J. H. Hubbard and B. H. West (Springer-Verlag, New York, 1989).

2

First-Order
Differential Equations

2.1 FIRST-ORDER LINEAR EQUATIONS
2.2 SEPARABLE EQUATIONS
2.3 GROWTH AND DECAY PHENOMENA
2.4 MIXING PHENOMENA
2.5 COOLING AND HEATING PHENOMENA
2.6 MORE APPLICATIONS
2.7 THE DIRECTION FIELD AND EULER'S METHOD
2.8 HIGHER-ORDER NUMERICAL METHODS

2.1   FIRST-ORDER LINEAR EQUATIONS

PURPOSE

To solve the general first-order linear differential equation

$$y' + p(x)y = f(x)$$

using the integrating factor method. We will construct a function $\mu(x)$ that satisfies $\mu(y' + py) = (\mu y)'$, thus allowing the equation to be integrated.

The general first-order linear differential equation can be written as

$$a_0(x)\frac{dy}{dx} + a_1(x)y = F(x) \tag{1}$$

where $a_0(x)$, $a_1(x)$, and $F(x)$ are given functions of x defined on some given interval* I. Inasmuch as we always assume that $a_0(x) \neq 0$ for all $x \in I$, it is convenient to divide by $a_0(x)$ and rewrite the equation as

$$y' + p(x)y = f(x) \tag{2}$$

where $p(x) = a_1(x)/a_0(x)$ and $f(x) = F(x)/a_0(x)$. If we assume that the functions $p(x)$ and $f(x)$ are continuous for x belonging to the interval I, then Picard's theorem guarantees the existence of a unique solution to Eq. (2) in some subinterval satisfying arbitrary initial conditions $y(x_0) = y_0$, where x_0 belongs to I. (See Problem 37 at the end of this section.) The goal of this section is to find the general solution (all solutions) of Eq. (2). In the case of first-order equations the general solution will contain one arbitrary constant.

INTEGRATING FACTOR METHOD (Constant Coefficients)

Before solving Eq. (2), however, we will solve the simpler equation

$$y' + ay = f(x) \tag{3}$$

where we have replaced $p(x)$ by the constant a. The idea behind the integrating factor method is the simple observation that

$$e^{ax}(y' + ay) = \frac{d}{dx}(e^{ax}y) \tag{4}$$

which turns the differential equation (3) into a "calculus problem." To see how this method works, multiple each side of Eq. (3) by e^{ax}, getting

$$e^{ax}(y' + ay) = f(x)e^{ax}$$

which, using the fundamental property (4), reduces to

$$\frac{d}{dx}(e^{ax}y) = f(x)e^{ax} \tag{5}$$

---

* Often the interval I over which the differential equation is defined is not specified if it is not relevant to the discussion. Just keep in mind that it is some interval, such as $(-\infty, \infty)$, $[0, \infty)$, $(0, 1)$, and so on.

This equation can now be integrated directly, and we get

$$e^{ax}y = \int f(x)e^{ax}dx + c$$

where c is an arbitrary constant and the integral sign refers to any antiderivative of $f(x)e^{ax}$. Solving for y gives

$$y(x) = e^{-ax}\int f(x)e^{ax}\,dx + ce^{-ax} \tag{6}$$

We now solve the general first-order equation

$$y' + p(x)y = f(x) \tag{7}$$

where $p(x)$ is assumed to be a continuous function. The general idea is motivated by the constant coefficient equation; we seek a function $\mu(x)$, called an **integrating factor,** that satisfies

$$\mu(x)\,[y' + p(x)y] = \frac{d}{dx}\{\mu(x)y(x)\} \tag{8}$$

To find $\mu(x)$, we carry out the differentiation on the right-hand side and simplify, getting

$$\mu(x)y' + \mu(x)p(x)y = \mu'(x)y + \mu(x)y'$$

If we now assume that $y(x) \neq 0$, we arrive at

$$\mu'(x) = p(x)\mu(x)$$

But we can find a solution $\mu(x) > 0$ by separating variables, getting

$$\frac{\mu'(x)}{\mu(x)} = p(x)$$

$$\ln \mu(x) = \int p(x)\,dx$$

$$\mu(x) = e^{\int p(x)\,dx} \tag{9}$$

Note: Since $\int p(x)\,dx$ denotes the *collection* of *all* antiderivatives of $p(x)$, it contains an arbitrary additive constant. Hence $\mu(x)$ contains an arbitrary *multiplicative* constant. However, since we are interested in finding only one integrating factor, we will pick the multiplicative constant to be 1.

Now that we *know* the integrating factor, we simply multiply each side of Eq. (7) by the integrating factor (9), getting

$$\mu(x)[y' + p(x)y] = \mu(x)\,f(x)$$

But from the property $\mu(x)\,[y' + p(x)\,y] = [\mu(x)y]'$ we have

$$[\mu(x)\,y]' = \mu(x)\,f(x) \tag{10}$$

---

* The integrating factor method was first introduced into mathematics by the Swiss mathematician, Leonhard Euler (1707–1783).

We can now integrate Eq. (10), getting

$$\mu(x)\, y(x) = \int \mu(x)\, f(x)\, dx + c \tag{11}$$

and since $\mu(x) \neq 0$, we can solve for $y(x)$ algebraically, getting

$$y(x) = \frac{1}{\mu(x)} \int \mu(x)\, f(x)\, dx + \frac{c}{\mu(x)} \tag{12}$$

We summarize these ideas, which give rise to the **integrating factor method** for solving the general first-order linear equation.

Integrating Factor Method

To solve the first-order linear differential equation

$$y' + p(x)y = f(x)$$

on a given interval I, perform the following steps.

Step 1   (Find the Integrating Factor). Find the integrating factor

$$\mu(x) = e^{\int p(x)\, dx}$$

where $\int p(x)\, dx$ represents *any* antiderivative of $p(x)$. Normally, pick the arbitrary constant in the antiderivative to be zero. Note that $\mu(x) \neq 0$ for $x \in I$.

Step 2   (Multiply by the Integrating Factor). Multiply each side of the differential equation by the integrating factor to get

$$e^{\int p(x)\, dx}\,(y' + p(x)\, y) = f(x)\, e^{\int p(x)\, dx}$$

which will always reduce to

$$\frac{d}{dx}\left(e^{\int p(x)\, dx}\, y(x)\right) = f(x)\, e^{\int p(x)\, dx}$$

Step 3   (Find the Antiderivative). Integrate the equation from Step 2 to get

$$e^{\int p(x)\, dx}\, y(x) = \int f(x)\, e^{\int p(x)\, dx}\, dx + c$$

Step 4   (Solve for y). Solve the equation found in Step 3 for $y(x)$ to get the general solution

$$y = e^{-\int p(x)\, dx} \int f(x)\, e^{\int p(x)\, dx}\, dx + ce^{-\int p(x)\, dx} \tag{13}$$

Notes:

1. We have shown that if $y' + p(x)\, y = f(x)$ has a solution, it must be of the form in Eq. (13). Conversely, it is a straightforward matter to verify that Eq. (13) also constitutes a one-parameter family of solutions of the differential equation.

2. One could memorize the formula for the general solution of the general first-order linear equation. However, it is easier to simply remember that multiplication of the differential equation by $\mu(x)$ turns the differential equation into an "ordinary" antiderivative problem of the type one studies in calculus. The following examples illustrate the integrating factor method.

3. For first-order *linear* differential equations a one-parameter family of solutions constitutes *all* the solutions of the equation and is called the general solution of the equation. Hence the one-parameter family in Eq. (13) constitutes all the solutions of $y' + p(x)\,y = f(x)$ and is called the general solution.

Example 1   Integrating Factor Method   Find the general solution of

$$y' - 2xy = e^{x^2} \qquad (-\infty < x < \infty) \qquad (14)$$

Figure 2.1
The one-parameter family of curves $y = (x + c)e^{x^2}$, where c is an arbitrary constant, represents the entire collection of solutions of $y' - 2xy = e^{x^2}$

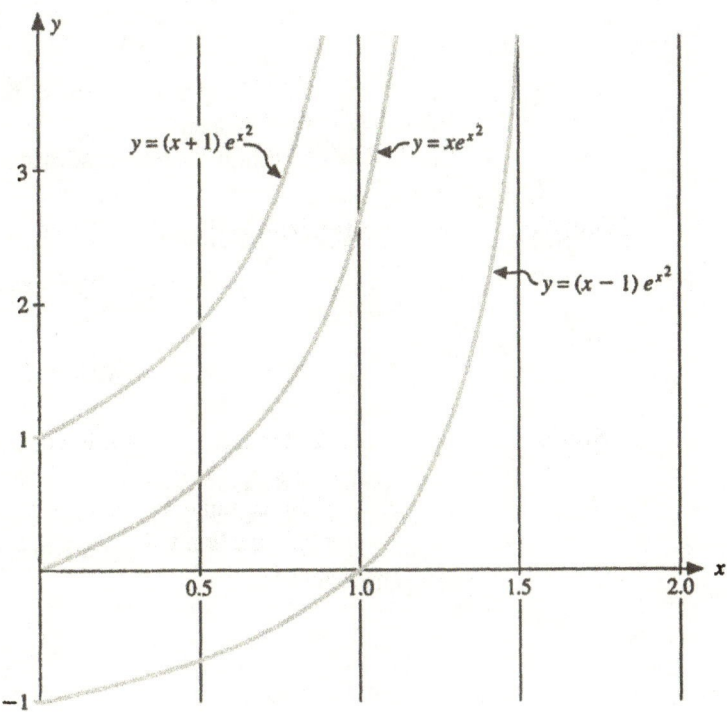

Solution   Here $p(x) = -2x$, and so the integrating factor is

$$\mu(x) = e^{-\int 2x\,dx} = e^{-x^2} \qquad (15)$$

Multiplying each side of the differential equation by the integrating factor, we get

$$e^{-x^2}\,(y' - 2xy) = 1$$

which can be rewritten as

$$\frac{d}{dx}(e^{-x^2}y(x)) = 1$$

Integrating, we get

$$e^{-x^2}y(x) = x + c$$

Solving for y, we find the solutions

$$y(x) = (x + c)\,e^{x^2} \tag{16}$$

where c is an arbitrary constant. A few of these solutions are drawn in Figure 2.1.   ∎

INITIAL-VALUE PROBLEM FOR FIRST-ORDER EQUATIONS

We are often interested in finding the single solution of a first-order equation that passes through a given point (x_0, y_0). This is the initial-value problem for first-order equations

$$y' = f(x, y) \tag{17}$$
$$y(x_0) = y_0$$

The strategy for solving this problem is first to find all the solutions of the differential equation and then to determine which solution satisfies the initial condition $y(x_0) = y_0$. The following example illustrates this idea.

Example 2   Initial-Value Problem   Solve the initial value problem

$$\frac{dy}{dx} + \frac{3}{x}y = \frac{\sin x}{x^3} \qquad (x > 0) \tag{18}$$
$$y(\pi/2) = 1$$

Solution   Note that the differential equation is not defined when $x = 0$, and so we restrict the equation to the interval $(0, \infty)$, over which the coefficient $3/x$ and the nonhomogeneous term $(\sin x)/x^3$ are continuous. To solve this probem, we first find the general solution of the differential equation using the integrating factor method. Since $p(x) = 3/x$, the integrating factor is

$$\begin{aligned}
\mu(x) &= e^{\int (3/x)\,dx} \\
&= e^{3\ln x} \\
&= e^{\ln x^3} \\
&= x^3
\end{aligned} \tag{19}$$

Multiplying by $\mu(x)$ gives

$$\frac{d}{dx}(x^3 y) = \sin x$$

and by direct integration we find

$$x^3\,y(x) = -\cos x + c$$

Finally, dividing by x^3 gives the general solution

$$y(x) = \frac{c}{x^3} - \frac{\cos x}{x^3} \qquad (x > 0) \tag{20}$$

To determine which curve passes through the initial point $(\pi/2, 1)$, we simply solve the equation $y(\pi/2) = 1$ for c. Doing this gives

$$1 = \frac{c}{(\pi/2)^3} - \frac{\cos (\pi/2)}{(\pi/2)^3}$$

or

$$c = \frac{\pi^3}{8}$$

Hence the solution of the initial-value problem is

$$y(x) = \frac{\pi^3 - 8 \cos x}{8x^3} \qquad (x > 0) \tag{21}$$

The graph of this solution is shown in Figure 2.2.

Figure 2.2
Solution of the initial-value
problem

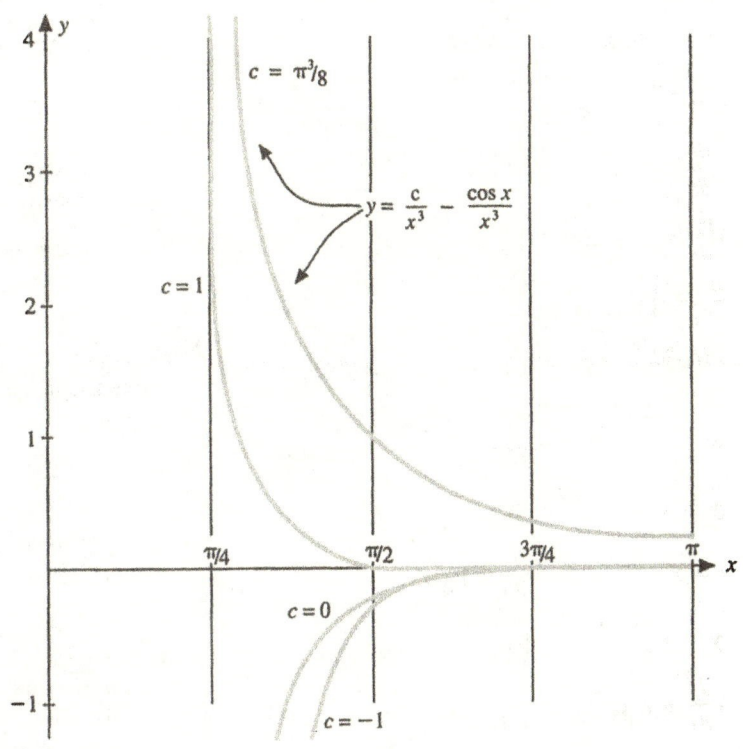

PROBLEMS: Section 2.1

For Problems 1–15, find the general solution to the indicated equation.

1. $\dfrac{dy}{dx} + 2y = 0$

2. $\dfrac{dy}{dx} + 2y = 3e^x$

3. $\dfrac{dy}{dx} - y = e^{3x}$

4. $\dfrac{dy}{dx} + y = \sin x$

5. $\dfrac{dy}{dx} + y = \dfrac{1}{1 + e^{2x}}$

6. $\dfrac{dy}{dx} + 2xy = x$

7. $\dfrac{dy}{dx} + 3x^2y = x^2$

8. $\dfrac{dy}{dx} + \dfrac{1}{x}y = \dfrac{1}{x^2}$

9. $x\dfrac{dy}{dx} + y = 2x$

10. $\cos x \dfrac{dy}{dx} + y \sin x = 1$

11. $\dfrac{dy}{dx} - \dfrac{2y}{x} = x^2 \cos x$

12. $\dfrac{dy}{dx} + \dfrac{3}{x}y = \dfrac{\sin x}{x^3} \ (x \neq 0)$

13. $(1 + e^x)\dfrac{dy}{dx} + e^x y = 0$

14. $(x^2 + 9)\dfrac{dy}{dx} + xy = 0$

15. $\dfrac{dy}{dx} + \left(\dfrac{2x + 1}{x}\right)y = e^{-2x}$

For Probems 16–20, find the solution of the given initial-value problem.

16. $\dfrac{dy}{dx} - y = 1 \qquad\qquad y(0) = 1$

17. $\dfrac{dy}{dx} + 2xy = x^3 \qquad\quad y(1) = 1$

18. $\dfrac{dy}{dx} - \dfrac{3}{x}y = x^3 \qquad\quad y(1) = 4$

19. $\dfrac{dy}{dx} + 2xy = x \qquad\quad y(0) = 1$

20. $(1 + e^x)\dfrac{dy}{dx} + e^x y = 0 \quad y(0) = 1$

21. **The Integrating Factor Identity** Verify the fundamental integrating factor identity

$$\frac{d}{dx}\left(e^{\int p(x)\,dx}\,y(x)\right) = e^{\int p(x)\,dx}\left(\frac{dy}{dx} + p(x)y\right) \qquad (22)$$

22. **Interchanging x and y to Get Linearity** Solve the non-linear differential equation

$$\frac{dy}{dx} = \frac{1}{x + y} \qquad y(-1) = 0$$

by considering the inverse function and writing x as a function of y. *Hint*: Using the basic identity from calculus that $dy/dx = \dfrac{1}{dx/dy}$, rewrite the given equation $y' = f(x, y)$ as $dx/dy = 1/f(x, y)$.

23. **A Tough Problem Made Easy** The differential equation

$$\frac{dy}{dx} = \frac{y^2}{e^y - 2xy}$$

would seem to be impossible to solve. However, if one treats y as the independent variable and x as the dependent variable and uses the relationship $dy/dx = \dfrac{1}{dx/dy}$, one can find an implicit solution. Find this implicit solution.

24. **Use of Transformations** Often a difficult problem is quite easy if it is viewed in the proper perspective.

 (a) Solve the nonlinear equation

$$\frac{dy}{dx} + ay = by \ln y$$

 where a and b are constants and can be solved by transforming the dependent variable* to $z = \ln y$.

 (b) Use the result from part (a) to solve

$$\frac{dy}{dx} + y = y \ln y$$

25. **Bernoulli Equation** An equation that is not linear but can be transformed into a linear equation is the *Bernoulli equation*

$$\frac{dy}{dx} + p(x)y = q(x)y^n \qquad (n = 2, 3, 4, \ldots) \quad (23)$$

 (a) Show that the transformation $v = y^{1-n}$ reduces the Bernoulli equation to a linear equation in v.

---

* By changing dependent variables from an old variable y to a new variable z we are essentially using a different "scale" to measure the dependent variable. Often the differential equation that relates the new dependent variable to the independent variable is much simpler than the original differential equation.

(b) Use the transformation in part (a) to solve the Bernoulli equation

$$y' - y = y^3$$

26. The Riccati Equation The equation

$$\frac{dy}{dx} = a(x) + b(x)y + c(x)y^2 \qquad (24)$$

is known as the *Riccati equation*.

(a) Show that if one solution $y_1(x)$ of the Riccati equation is known, then a more general solution containing an arbitrary constant can be found by making the substitution

$$y = y_1(x) + \frac{1}{v(x)} \qquad (25)$$

and showing that $v(x)$ satisfies the linear equation

$$\frac{dv}{dx} = -(b + 2cy_1)\, v - c$$

(b) Verify that $y_1(x) = 1$ satisfies the Riccati equation

$$y' = -1 + 2y - y^2$$

and use this fact to find the general solution.

27. General Theory of First-Order Equations Show that if y_1 and y_2 are two different solutions of

$$\frac{dy}{dx} + p(x)y = f(x) \qquad (26)$$

then $y_1 - y_2$ is a solution of the homogeneous equation

$$\frac{dy}{dx} + p(x)y = 0$$

28. Discontinuous Coefficients There are phenomena in which the coefficient $p(x)$ in the linear first-order equation is not continuous but has jump discontinuities. However, it is often possible to solve these types of problems with a little effort. Consider the initial-value problem

$$\frac{dy}{dx} + p(x)y = 1 \qquad y(0) = 0$$

where

$$p(x) = \begin{cases} 1 & (0 \le x \le 1) \\ 2 & (1 < x) \end{cases}$$

(a) Find the solution of the initial value problem in the interval $0 \le x \le 1$.

(b) Find the solution of the problem for $1 < x$.

(c) Sketch the graph of the solution for $0 \le x \le 4$.

29. Discontinuous Right-Hand Side Often the right-hand side $f(x)$ of the first-order linear equation is not continuous but has jump discontinuities. It is still possible to solve problems of this type with a little effort. For example, consider the problem

$$\frac{dy}{dx} + y = f(x) \qquad y(0) = 0$$

where

$$f(x) = \begin{cases} 1 & (0 \le x \le 1) \\ 0 & (1 < x) \end{cases}$$

(a) Find the solution of the initial value problem in the interval $0 \le x \le 1$.

(b) Find the solution of the problem for $1 < x$.

(c) Sketch the graph of the solution for $0 \le x \le 4$.

30. Comparing a Linear and Nonlinear Equation

(a) Verify the solutions:

$$\begin{array}{lll} y' = y & \Diamond\ y = e^x & \text{(linear equation)} \\ y' = -y^2 & \Diamond\ y = x^{-1} \quad (x \ne 0) & \text{(nonlinear equation)} \end{array}$$

(b) Verify

$$\begin{array}{lll} y' = y & \Diamond\ \ y = ce^x & \text{for all } c \\ y' = y^2 & \Diamond\ \ y = cx^{-1} & \text{for only } c = 0, -1 \end{array}$$

31. Error Function Express the solution of $y' = 1 + 2xy$ in terms of the **error function**

$$\operatorname{erf}(x) = \frac{2}{\sqrt{\pi}} \int_0^x e^{-t^2}\, dt \qquad (27)$$

32. Computer Problem—Sketching Solutions Use a graphing calculator or a graphing package for your computer* to plot some of the solutions of the one-parameter family of solutions drawn in Figure 2.1. You might try sketching some solutions for different values of the parameter c than have been drawn in the text. The author used the computer package *MICRO CALC* to draw the curves shown in Figure 2.1.

33. Computer Problem—Sketching a Solution Redraw the solution of the initial-value problem drawn in Figure 2.2 using either a graphing calculator or a computer.

34. Today's Journal Entry Spend ten minutes exploring your thoughts about some aspect of the integrating factor. Is it possible to solve *algebraic equations* using an integrating factor? Were you disappointed in the complicated-looking solution to the first-order equation? What did you expect? Will the integrating factor method always work? Can you think of another way to solve the first-order equation? Do you think this course is going to be worthwhile? Date your entry.

---

* There are many good computer packages that would help you in your study of differential equations. A few of them are *DERIVE, MAPLE, MICRO CALC, MATHCAD, CONVERGE, PHASER,* and *MATHEMATICA.* Ask your professor whether your college or university has a site license for any such package.

2.2   SEPARABLE EQUATIONS

PURPOSE

To solve the class of first-order differential equations of the form

$$\frac{dy}{dx} = \frac{f(x)}{g(y)}$$

known as separable equations by a method known as separation of variables. The importance of this class of equations lies in the fact that many important nonlinear equations are separable and hence solvable.

SOLVING SEPARABLE EQUATIONS

The very simplest differential equation is the one studied in calculus,

$$\frac{dy}{dx} = f(x) \tag{1}$$

where $f(x)$ is a given continuous function. In calculus we learned that we can solve this equation by essentially ''integrating both sides'' of the equation, getting

$$y(x) = \int f(x)\, dx + c \tag{2}$$

where c is an arbitrary constant and the integral sign denotes any single antiderivative of $f(x)$. We now see that this procedure can be applied to a broader class of differential equations, known as **separable equations,** having the form

$$\frac{dy}{dx} = \frac{f(x)}{g(y)} \tag{3}$$

Clearly, any separable equation* reduces to the simpler form in Eq. (1) when we have $g(y) = 1$. To solve a separable differential equation, we rewrite it as

$$g(y)\frac{dy}{dx} = f(x) \tag{4}$$

At this point, we can observe that the left-hand side of the equation involves only the variable y and its derivative, while the right-hand side involves only x, hence the name ''separated equation.'' We can now solve the separated equation by simply integrating each side of the equation with respect to x, getting

$$\int g(y(x))\frac{dy}{dx}\, dx = \int f(x)\, dx \tag{5}$$

---

* Of course, equations of the form $y' = F(x)G(y)$ are also separable, since we can identify $F(x) = f(x)$ and $G(y) = 1/g(y)$.

The rule for integration by substitution in calculus then permits the left-hand side of this equation to be rewritten as an integral with respect to y, giving

$$\int g(y)\, dy = \int f(x)\, dx \tag{6}$$

which we choose to write as*

$$\int g(y)\, dy = \int f(x)\, dx + c \tag{7}$$

in order to emphasize the constant of integration c. Since all antiderivatives differ by a constant, the two forms (6) and (7) are equivalent. But Eq. (7) is an algebraic equation that equates an antiderivative of $g(y)$ with respect to y to an antiderivative of $f(x)$ with respect to x. (Each antiderivative contains an arbitrary constant, and we have lumped the constants as a single constant c and placed it on the right-hand side.) In other words, we have found an implicit solution of the separable equation (3). If we can solve for y in terms of x, we will also have found an explicit solution.

The above discussion motivates the procedure for solving separable differential equations.

Separation of Variables Method[†]

Step 1. Rewrite the separable equation

$$\frac{dy}{dx} = \frac{f(x)}{g(y)}$$

in separated (or differential) form

$$g(y)\, dy = f(x)\, dx$$

Step 2. Integrate each side of this equation with respect to its respective variable, obtaining the implicit solution

$$\int g(y)\, dy = \int f(x)\, dx + c$$

Step 3. If possible, solve for y in the implicit solution, getting the explicit solution.

---

* More precisely, the integral signs in Eq. (6) denote the family of antiderivatives of their respective integrands, whereas the integral signs in Eq. (7) denote *specific* antiderivatives. Since all antiderivatives of a given function differ by at most a constant, the two equations (6) and (7) are equivalent.

[†] As one might suspect, the method of separation of variables in ordinary differential equations goes all the way back to almost the beginnings of the calculus. Leibniz used the method of separation of variables implicitly in 1691 when he solved the *inverse problem of tangents*. However, it was John Bernoulli who developed the explicit process and the name *"separation of variables"* in 1694 in a letter to Leibniz. After developing the method, John Bernoulli discovered that the method broke down when he tried to solve the equation $ax\, dy - b\, dx = 0$. This was because the differential dx/x had not yet been integrated.

It is generally an easy matter to determine whether a first-order differential equation is separable or not separable. The following example illustrates this idea.

Example 1   Separable Equations   Equations (a), (b), and (c) are separable differential equations and their separated form. Equation (d) is not separable.

(a) $\dfrac{dy}{dx} = -\dfrac{x}{y}$ ⟹ $y\,dy = -x\,dx$ (separable)

(b) $\dfrac{dy}{dx} = x^2 y$ ⟹ $\dfrac{1}{y}\,dy = x^2\,dx$ (separable)

(c) $\dfrac{dy}{dx} = y + 1$ ⟹ $\dfrac{1}{y+1}\,dy = dx$ (separable)

(d) $\dfrac{dy}{dx} = x + y$ (not separable)   ∎

Example 2   Separation of Variables   Solve

$$\frac{dy}{dx} = 1 + y^2 \tag{8}$$

Solution   Although the differential equation is nonlinear, it is separable. Separating variables, we write the equation in separated form

$$\frac{dy}{1 + y^2} = dx \tag{9}$$

Integrating each side of this equation with respect to its respective variable, we obtain the implicit solution

$$\tan^{-1} y = x + c \tag{10}$$

Solving for y, we obtain the explicit solution

$$y = \tan(x + c) \tag{11}$$

∎

Example 3   Separation of Variables   Solve the initial-value problem

$$\frac{dy}{dx} = -\frac{x}{y} \tag{12}$$

$$y(0) = 1$$

Solution   We first solve the separable (nonlinear) differential equation. Separating variables, we get

$$y\,dy = -x\,dx$$

Integrating gives the implicit solution

$$\frac{1}{2}y^2 = -\frac{1}{2}x^2 + c$$

or

$$\frac{1}{2}x^2 + \frac{1}{2}y^2 = c \tag{13}$$

Figure 2.3

Solution of the initial-value problem $y' = -x/y$, $y(0) = 1$

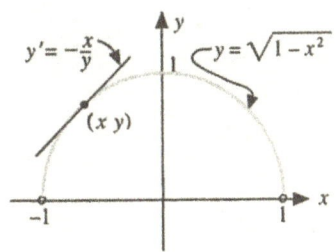

Since it is convenient to find c at this point, we do so. Substituting the initial conditions $x = 0$, $y = 1$ into this equation gives $c = 1/2$. Hence the implicit solution is the relationship $x^2 + y^2 = 1$. To obtain the explicit solution, we solve Eq. (13) for y, getting

$$y = \pm \sqrt{1 - x^2} \tag{14}$$

Although we have *two* solutions, only the positive square root function in Eq. (14) satisfies the initial condition $y(0) = 1$. See Figure 2.3. Also note that the interval over which the solution is valid is $(-1, 1)$, which is the interval where the radical is nonnegative and the differential equation is defined.   ■

Example 4   Separation of Variables   Solve the initial-value problem

$$\frac{dy}{dx} = xy \tag{15}$$

$$y(0) = 1$$

Solution   Separating variables, we find

$$\frac{dy}{y} = x\, dx \tag{16}$$

Integrating gives the implicit solution

$$\ln |y| = \frac{x^2}{2} + c$$

$$|y| = e^{x^2/2 + c} \tag{17}$$

$$= e^c\, e^{x^2/2}$$

Solving for y, we get the explicit solution

$$y = Ce^{x^2/2} \tag{18}$$

where $C = \pm e^c$ represents an arbitrary positive or negative constant. Substituting this general solution into the initial condition $y(0) = 1$, we find $C = 1$, and so the solution to the initial-value problem is

$$y = e^{x^2/2} \tag{19}$$

See Figure 2.4.   ■

Figure 2.4
Solution of the initial-value
problem $y' = xy$, $y(0) = 1$

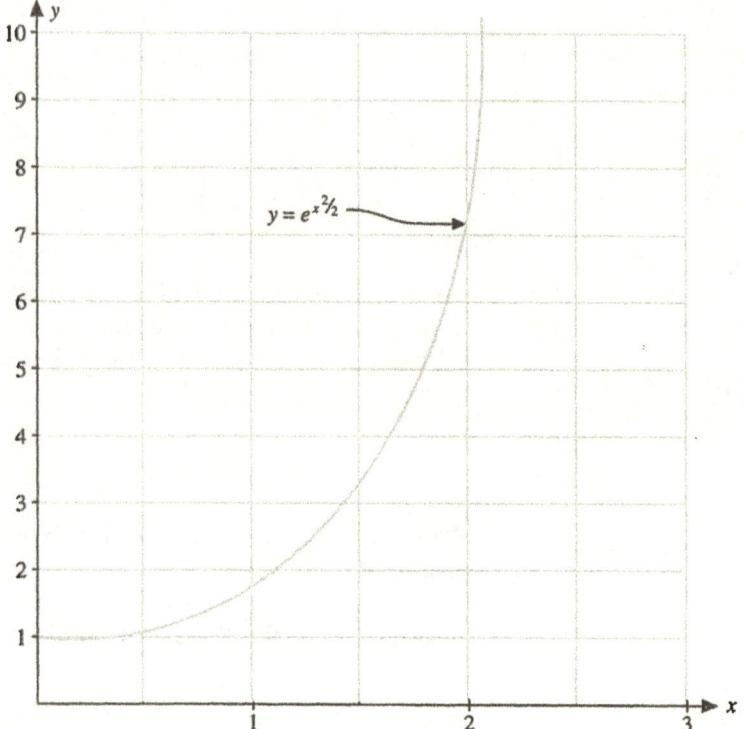

$$y = e^{x^2/2}$$

PROBLEMS: Section 2.2

For Problems 1–10, determine which differential equations are separable.

1. $\dfrac{dy}{dx} = 1 + y$

2. $\dfrac{dy}{dx} = y - y^3$

3. $\dfrac{dy}{dx} = \sin(x + y)$

4. $\dfrac{dy}{dx} = \ln(xy)$

5. $\dfrac{dy}{dx} = e^{x+y}$

6. $\dfrac{dy}{dx} = \dfrac{y + 1}{xy} + y$

7. $\dfrac{dy}{dx} = \dfrac{e^{x+y}}{x + 1}$

8. $\dfrac{dy}{dx} = x \ln(y^{2x}) + x^2$

9. $\dfrac{dy}{dx} = \dfrac{y}{x} + \dfrac{x}{y}$

10. $x\dfrac{dy}{dx} = 1 + y^2$

For Problems 11–25, solve the given differential equation.

11. $\dfrac{dy}{dx} = x - x^2$

12. $\dfrac{dy}{dx} = \dfrac{2y}{x}$

13. $\dfrac{dy}{dx} = x^2 y$

14. $\dfrac{dy}{dx} = x^2 y^3$

15. $\dfrac{dy}{dx} = e^{x+y}$

16. $\dfrac{dy}{dx} = -\dfrac{1}{\sqrt{x}}$

17. $\dfrac{dy}{dx} = \sqrt{\dfrac{y}{x}}$

18. $\dfrac{dy}{dx} = \dfrac{2x(y + 1)}{y}$

19. $\dfrac{dy}{dx} = \dfrac{1}{x - x^3}$

20. $\dfrac{dy}{dx} = \dfrac{y^2 - 1}{x^3}$

21. $\dfrac{dr}{d\theta} = \dfrac{r \sin \theta}{1 - \cos \theta}$

22. $\dfrac{dy}{dx} = \dfrac{x}{y^2 \sqrt{1 + x^2}}$

23. $\dfrac{dy}{dx} = y - y^2$ *Hint:* $\dfrac{1}{y(1 - y)} = \dfrac{1}{y} + \dfrac{1}{1 - y}$

24. $\dfrac{dy}{dx} = x\sqrt{\dfrac{1 - y}{1 - x^2}}$

25. $\dfrac{dy}{dx} = -\dfrac{1 + \ln x}{1 + \ln y}$

For Problems 26–30, solve the given initial-value problem.

26. $\dfrac{dy}{dx} = \dfrac{y}{1 + x}$ $y(0) = 1$

27. $\dfrac{dy}{dx} = -\dfrac{x}{y}$ $\qquad\qquad y(0) = 1$

28. $\dfrac{dy}{dx} = y^2 - 4$ $\qquad\qquad y(0) = -6$

29. $x\dfrac{dy}{dx} - y = 2x^2 y$ $\qquad y(1) = e$

30. $\dfrac{dy}{dx} = \dfrac{x + xy^2}{4y}$ $\qquad\quad y(1) = 0$

Solutions Not Expressible as Elementary Functions

Certain antiderivatives in calculus, such as $\int e^{-x^2}\,dx$, cannot be expressed as a finite sum of elementary functions. However, it is possible to solve differential equations involving such expressions, such as

$$\frac{dy}{dx} = \frac{e^{x^2}}{y^2} \qquad y(0) = 1$$

by first multiplying each side of the equation by y^2, getting

$$y^2 \frac{dy}{dx} = e^{x^2}$$

In this equation, however, instead of writing the equation in differential form (which would get us nowhere), we integrate each side of the equation from $x = 0$ (the initial value of x) to an arbitrary $x = x_1$, getting

$$\int_0^{x_1} y^2(x)\frac{dy}{dx}\,dx = \frac{1}{3}\Big(y^3(x_1) - y^3(0)\Big) = \int_0^{x_1} e^{x^2}\,dx$$

We now substitute the initial condition $y(0) = 1$ into this equation, use the variable t for the dummy variable of integration, and replace the upper limit x_1 by x. All this gives

$$y(x) = \sqrt[3]{1 + 3\int_0^x e^{t^2}\,dt}$$

For Problems 31–35, use the described method to find the explicit solution to the given initial-value problem.

31. $\dfrac{dy}{dx} = \dfrac{2}{\sqrt{\pi}}\,e^{-x^2}$ $\qquad\qquad y(0) = 0$

32. $\dfrac{dy}{dx} = y^2\,e^{x^2}$ $\qquad\qquad\quad y(1) = 1$

33. $\dfrac{dy}{dx} = \sqrt{1 + \sin x\,(1 + y^2)}$ $\quad y(0) = 1$

34. $\dfrac{dy}{dx} = e^{x^2}\,y^{-2}$ $\qquad\qquad\quad y(0) = 1$

35. $\dfrac{dy}{dx} = \dfrac{\sin x}{xy^2}$ $\qquad\qquad\quad y(0) = 1$

36. Linear Versus Separable Equations Students often think that all separable equations are nonlinear. This is not always true.
 (a) Find a differential equation that is both linear and separable.
 (b) Find a differential equation that is linear but not separable.
 (c) Find a differential equation that is separable but not linear.
 (d) Find a differential equation that is neither linear nor separable.

37. Double Whammy The first-order linear equation of the form

$$\frac{dy}{dx} + ay = b \tag{20}$$

where a and b are constants, can be solved both by the integrating factor method and by separation of variables. Solve this equation using both methods to see that you get the same solution.

Transforming Into Separable Equations

There are many nonlinear differential equations that are not separable, but by making a proper substitution of either the independent or dependent variable it is possible to transform the equation into a separable equation. The following problems illustrate a few such equations.

38. Homogeneous Equation Differential equations that can be written in the form

$$\frac{dy}{dx} = F\!\left(\frac{y}{x}\right) \tag{21}$$

are called **homogeneous equations** (no relationship to the homogeneous linear equations).
 (a) Show that the transformation $v = y/x$ from the old dependent variable y to a new variable v gives rise to a separable differential equation.
 (b) Use the transformation described in part (a) to solve the homogeneous equation

$$\frac{dy}{dx} = \frac{x^2 + 2xy}{x^2}$$

39. Another Nonlinear Class It is possible to transform any nonlinear equation of the form

$$\frac{dy}{dx} = F(ax + by + c) \tag{22}$$

to a new differential equation with dependent variable $v(x)$ by the transformation

$$v = ax + by + c$$

(a) Find the transformed separable equation in v.

(b) Use the above transformation to solve

$$\frac{dy}{dx} = e^{x+y-1} - 1$$

(c) Use the above transformation to solve

$$\frac{dy}{dx} = \cos(x + y)$$

Hint: $\int\left(\frac{1}{1 + \sin x}\right)dx = -\tan\left(\frac{\pi}{4} - \frac{x}{2}\right) + c$

40. More Equations Transformed to Separable Equations
It is possible to transform any differential equation of the form

$$\frac{x}{y}\frac{dy}{dx} = F(xy) \tag{23}$$

to a separable equation in v and x by the transformation

$$v = xy$$

(a) Find the transformed separable equation in v and x.

(b) Use the above transformation to solve

$$\frac{x}{y}\frac{dy}{dx} = xy$$

Orthogonal Trajectories

41. Orthogonal Trajectories The problem of finding a family of curves, called **orthogonal trajectories,** that intersects a given family of curves perpendicularly (or **orthogonally**) occurs in many areas of applied mathematics. Consider the family of curves $f(x, y) = c$ where c is a parameter.

(a) Differentiating $f(x, y) = c$ implicitly with respect to x, show that the slope of each curve in the family is given by

$$\frac{dy}{dx} = -\frac{\partial f/\partial x}{\partial f/\partial y}$$

(b) Use the fact that the slopes of perpendicular lines are *negative reciprocals* of each other to show that the curves orthogonal to the given family $f(x, y) = c$ satisfy the differential equation

$$\frac{dy}{dx} = +\frac{\partial f/\partial y}{\partial f/\partial x}$$

This differential equation can then be solved to find the orthogonal trajectories of the given family $f(x, y) = c$.

(c) Use the differential equation found in part (b) to show that the orthogonal trajectories of the family of circles

$$x^2 + y^2 = c^2$$

is simply the family of straight lines $y = kx$ shown in Figure 2.5.

Figure 2.5
The orthogonal trajectories of the family of circles defined by $x^2 + y^2 = c^2$ is the family of straight lines $y = kx$

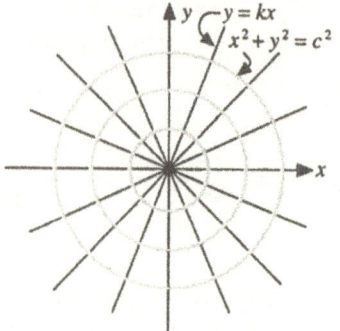

42. Orthogonal Trajectories of Well-Known Families Use the strategy shown in Problem 41 to find the orthogonal trajectories of the following well-known families of curves. Graph both the given family and its orthogonal trajectories.

(a) $y = c$ (family of horizontal lines)

(b) $xy = c$ (family of hyperbolas)

(c) $4x^2 + y^2 = c$ (family of ellipses)

(d) $y = cx^2$ (family of parabolas)

(e) $x^2 = 4cy^3$ (family of cubics)

43. Self Orthogonality Show that the family of parabolas

$$y^2 = 4cx + 4cx^2 + c$$

is "self-orthogonal."

44. The Simplest Differential Equation There is a famous differential equation, known as **Clairaut's equation:**

$$y = xy' + f(y') \tag{24}$$

where $f(y')$ simply means any function involving only y' that has straight line solutions. In fact, they are the straight lines given by

$$y = cx + f(c) \tag{25}$$

where c is an arbitrary constant.

(a) Verify that Eq. (25) is a solution of the general Clairaut equation (24).

(b) Find the family of straight lines that satisfy the Clairaut equation

$$y = xy' + \tfrac{1}{2}(y')^2 \qquad (26)$$

(c) Plot the straight line solutions of Eq. (26) for different values of c to get the one-parameter family drawn in Figure 2.6.

(d) Show that the **envelope curve** $y = -\tfrac{1}{2}x^2$ also satisfies the Clairaut equation (26). Do you see why this curve is called the envelope curve?

Applications Related to Separable Equations

45. Hmmmmm The sine function, $\sin x$, has the property that the square of itself plus the square of its derivative is identically equal to 1. Find the most general function that has this property.

46. The Disappearing Mothball The rate at which a mothball evaporates from a solid to a gaseous state is proportional to the surface area of the mothball. Suppose a mothball has been observed to have a radius of 0.5 inch and after 6 months a radius of 0.25 inch.

(a) Express the radius of the mothball as a function of time.

(b) When will the mothball disappear completely?

47. The Famous Snowplow Problem One day it started snowing at a steady rate. A snowplow started out at noon; it went 2 miles during the first hour and 1 mile during the second hour. At what time did it start snowing? *Hint*: Let t measure time in hours from the time the snowplow started plowing. Assume that the rate the snowplow travels is inversely proportional to the depth of the snow and that the depth of the snow is directly proportional to the time it has been snowing, which is best written as $d = k(t + t_s)$, where t_s is the time it started snowing. After solving the differential equation for $y(t)$, the distance the snowplow travels after t hours, use the two conditions $y(0) = 0$ and $y(1) = 2$ to find the two arbitrary constants k and t_s.

48. The Tractrix A child initially stands at the point $(b, 0)$ and grasps a rope of length 1 that is attached to a wagon located at the origin $(0, 0)$. The child then starts walking at a constant speed in the positive y direction along the vertical line $x = b$. See Figure 2.7. The path that the wagon (assumed to be a point) traces out as it follows the child is

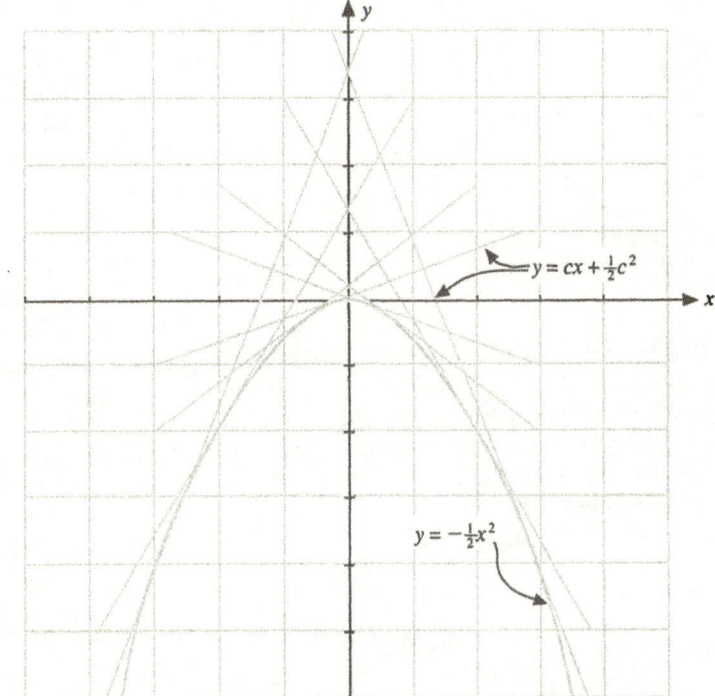

Figure 2.6
All members of the family
$y = cx + \tfrac{1}{2}c^2$, where c is an
arbitrary constant, are
solutions of $y = xy' + \tfrac{1}{2}(y')^2$.

Also, $y = -\tfrac{1}{2}x^2$ is not a
member of the family of
straight lines, but it is a
solution. It is a singular
solution.

$y = cx + \tfrac{1}{2}c^2$

$y = -\tfrac{1}{2}x^2$

Figure 2.7
If a child walks in a straight line and pulls a wagon attached at the end of a rope, the wagon traces out a curve called the tractrix. It is the same curve that is traced out by an airplane as it chases another airplane moving in a straight line. Tractrices also occur in many other areas of applied mathematics.

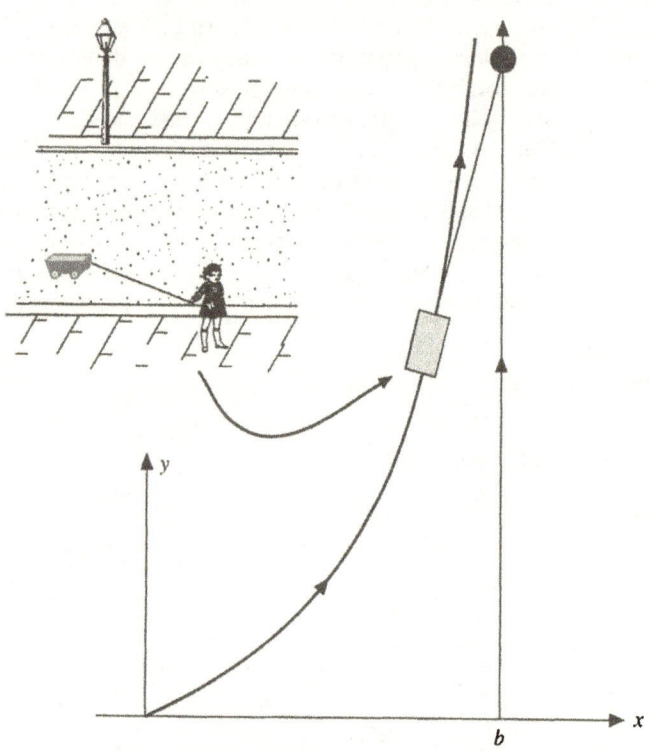

called a **tractrix.*** You can find the equation of this famous curve by solving the following problems.

(a) Since both the child and the wagon are traveling at speed 1 and the wagon is always headed straight toward the child, the slope dy/dx at an arbitrary point (x, y) on the tractrix will be given by

$$\frac{dy}{dx} = \frac{y - t}{x - b} \qquad (27a)$$

* The tractrix or equitangential curve was first studied by the Dutch scientist Christian Huygens in 1692. The word "tractrix" is derived from the Latin "*tractum*," meaning "drag."

(b) Since the wagon travels t feet in t seconds, this distance is also the *length* of the tractrix $y = y(x)$ from $(0, 0)$ to an arbitrary point (x, y). Use the arc length formula from calculus to show that

$$t = \int_0^x \sqrt{1 + [y'(s)]^2}\, ds \qquad (27b)$$

(c) Setting the values of t in Eqs. (27a) and (27b) equal to each other, we conclude that

$$y - (x - b)\frac{dy}{dx} = \int_0^x \sqrt{1 + [y'(s)]^2}\, ds \qquad (27c)$$

(d) Differentiate each side of Eq. (27c) with respect to x to show that

$$(x - b)\frac{dw}{dx} = -\sqrt{1 + w^2} \qquad (27d)$$

where $w = dy/dx$.

(e) Solving for $w(x)$ in Eq. (27d) by separation of variables and using the initial conditions $w(0) = w'(0) = 0$, we get

$$w(x) = \frac{dy}{dx} = \frac{1}{2}\left(\left(1 - \frac{x}{b}\right)^{-1} - \left(1 - \frac{x}{b}\right)\right) \qquad (27e)$$

(f) Solve for y in Eq. (27e), using separation of variables and the initial condition $y(0) = 0$ to obtain the equation for the tractrix:

$$y(x) = \frac{b}{2}\left\{\frac{1}{2}\left(\left(1 - \frac{x}{b}\right)^2 - 1\right) - \ln\left(1 - \frac{x}{b}\right)\right\} \qquad (27f)$$

If the reader has access to a computer with graphing capabilities, it would be useful to graph this tractrix. (Use the value $b = 1$.)

49. Computer Problem Use a graphing calculator or a computer to sketch the graph of the solution of the initial-value problem $y' = xy$, $y(0) = 1$ shown in Figure 2.4.

50. Journal Entry Do you think the method of separation of variables can be extended to higher-order differential equations? Would you call the second-order equation $y'' = xy$ separable, and can you solve it by some type of separation of variables technique? Can the second-order derivative y'' be interpreted as the quotient of two "second-order" differentials? If you can find an interpretation of this, you will be the first person to do so. Normally, separable equations are associated with first-order equations, but who knows?

2.3 GROWTH AND DECAY PHENOMENA

PURPOSE
To show how several growth and decay phenomena that occur in the physical and natural sciences can be described by two simple first-order differential equations known as the growth and decay equations. We will illustrate their use by examples.

Two differential equations that are basic to the study of many growth and decay phenomena are the **growth and decay equations:**

$$\frac{dy}{dt} = ky \qquad \text{(growth equation)} \tag{1a}$$

$$\frac{dy}{dt} = -ky \qquad \text{(decay equation)} \tag{1b}$$

where the **constant of proportionality** k is always positive. We could, of course, write these two equations collectively as a single equation by simply letting k take on both positive and negative values, but it is notationally more convenient to denote a special equation for each phenomenon. Note too that we have denoted the independent variable as t to represent time.

SOLVING THE GROWTH AND DECAY EQUATIONS

We know two methods that can be used to solve the growth and decay equations: the integrating factor method and the method of separation of variables. To solve the growth equation $y' = ky$ by the integrating factor method,* we carry out the following steps:

$$y' - ky = 0 \qquad \text{(growth equation)}$$

$$e^{-kt}(y' - ky) = 0 \qquad \text{(multiply by the integrating factor)}$$

$$\frac{d}{dt}(e^{-kt}y) = 0 \qquad \text{(rewrite the left-hand side of the equation)}$$

$$e^{-kt}y(t) = c \qquad \text{(find the antiderivative)}$$

$$y(t) = ce^{kt} \qquad \text{(solve for } y\text{)}$$

The general solution $y(t) = ce^{kt}$ of the growth equation is called the **exponential growth curve.** One can obtain the solution of the decay equation (1b) using the same steps, getting the **exponential decay curve** $y(t) = ce^{-kt}$.

---

* The student may think it is easier to solve this equation by separation of variables rather than by the integrating factor method. However, when one considers the fact that separation of variables finds only nonzero solutions (one must assume that the function y is different from zero in order to separate variables), the integrating factor has some appeal. Also, when one uses the separation of variables method, one must carry out manipulations involving the absolute value of y that occur in $\ln |y|$. Although not difficult, this does require some care.

INITIAL-VALUE PROBLEMS FOR GROWTH AND DECAY

One normally specifies initial conditions to accompany the growth and decay equations. These ideas are summarized below.

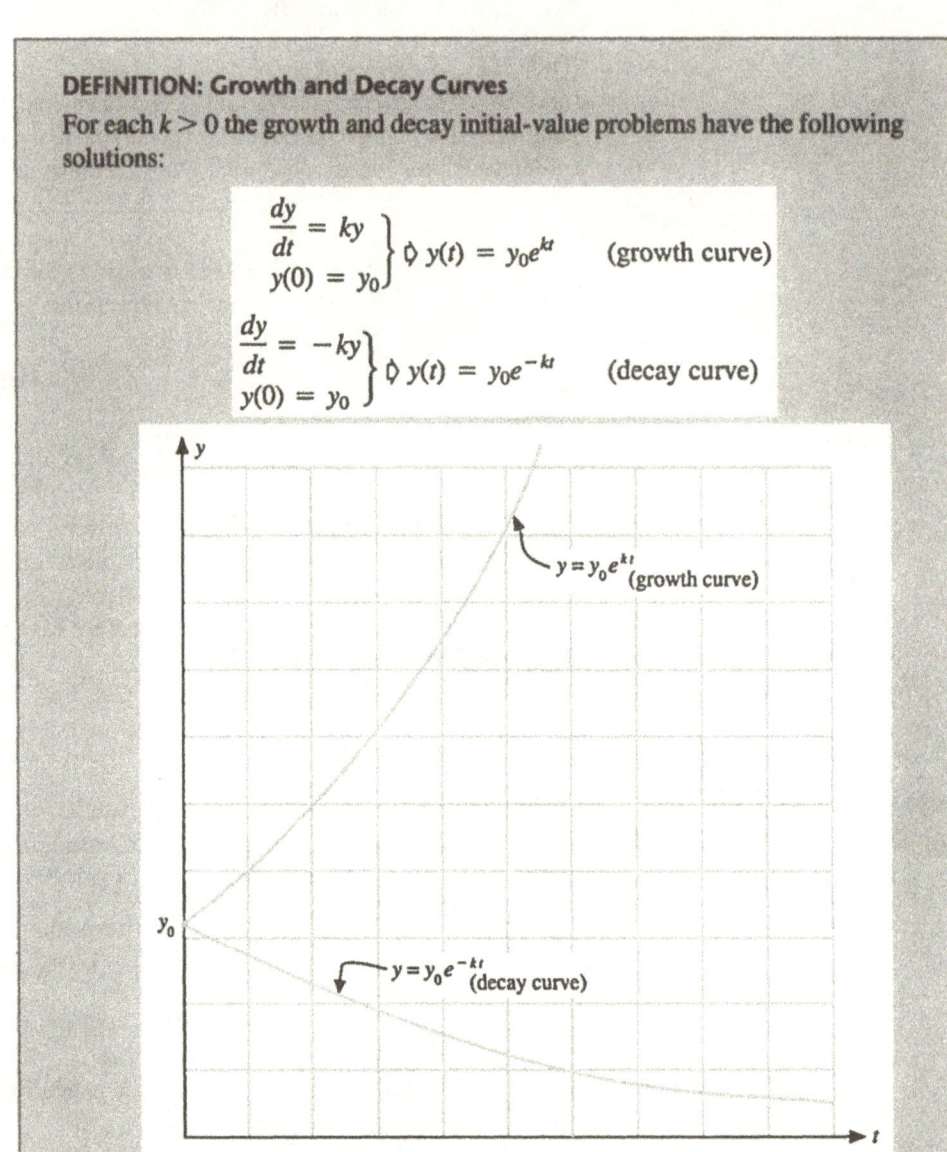

DEFINITION: Growth and Decay Curves

For each $k > 0$ the growth and decay initial-value problems have the following solutions:

$$\left.\begin{array}{l} \dfrac{dy}{dt} = ky \\ y(0) = y_0 \end{array}\right\} \Diamond \; y(t) = y_0 e^{kt} \quad \text{(growth curve)}$$

$$\left.\begin{array}{l} \dfrac{dy}{dt} = -ky \\ y(0) = y_0 \end{array}\right\} \Diamond \; y(t) = y_0 e^{-kt} \quad \text{(decay curve)}$$

$y = y_0 e^{kt}$ (growth curve)

$y = y_0 e^{-kt}$ (decay curve)

An important application of the decay equation lies in the prediction of radioactive decay.

RADIOACTIVE DECAY

In 1940 a group of boys were walking in the woods near the village of Lascaux, France, when they suddenly realized that their dog was missing. Later, the dog was found in a deep hole from which he was unable to climb out. One of the boys was lowered into the hole and, in the process, stumbled upon one of the greatest archeological discoveries of all time. What they discovered was the remnants of a cave whose walls were covered with drawings of wild horses, cattle, and even a fierce-looking beast that resembles the present-day bull. One of the findings that made the discovery exciting to scientists was the charcoal remains of a small fire. It was from these remains that scientists were able to determine that cave dwellers inhabited this cave 15,000 years ago.

To understand why the charcoal remains of a small fire were so important to archeologists, it is important to realize that charcoal is simply burnt wood and that over time a fundamental change takes place in all dead organic matter. Before death, all living matter contains a tiny but fixed amount of the radioactive isotope carbon-14 (C-14). After death, however, the amount of C-14 decays at a rate proportional to the amount present. On the basis of these physical principles the American chemist Willard Libby (1908–1980) developed the technique of **radiocarbon dating,** which resulted in his winning the Nobel prize for chemistry in 1960.

The following example illustrates the important concepts involved in radiocarbon dating.

Example 1   **Radiocarbon Dating**   By means of a chemical analysis it was determined that the residual amount of C-14 present in samples of charcoal taken from the Lascaux cave was 15% of the original amount at the time the tree died. It is well-known that the *half-life* of C-14, or time required for a given amount of C-14 to be reduced to one-half its original amount, is approximately 5600 years. It is also known that the quantity Q of C-14 in a sample of charcoal satisfies the decay equation

$$\frac{dQ}{dt} = -kQ \tag{2}$$

(a) Find the decay constant k.

(b) Find $Q(t)$ at any time t if the initial amount is $Q(0) = Q_0$.

(c) Find the age of the charcoal remains and hence the approximate age of the Lascaux cave paintings.

Solution   (a) If $Q(0) = Q_0$ denotes the initial amount of C-14 in a sample of charcoal, then the amount $Q(t)$ of C-14 at any time $t \geq 0$ is given by the decay curve $Q(t) = Q_0 e^{-kt}$. See Figure 2.8. After the half-life of 5600 years the amount of C-14 will decrease to one-half the original amount Q_0. Hence to find k, we use the equation

$$Q(5600) = Q_0 e^{-5600k} = \frac{Q_0}{2} \tag{3}$$

Solving for k gives

$$e^{-5600k} = \frac{1}{2}$$

$$-5600k = -\ln 2$$

$$k = \frac{\ln 2}{5600} \doteq 0.00012378 \tag{4}$$

Figure 2.8
The decay of C-14 decreases exponentially according to the law $Q(t) = Q_0 e^{-kt}$

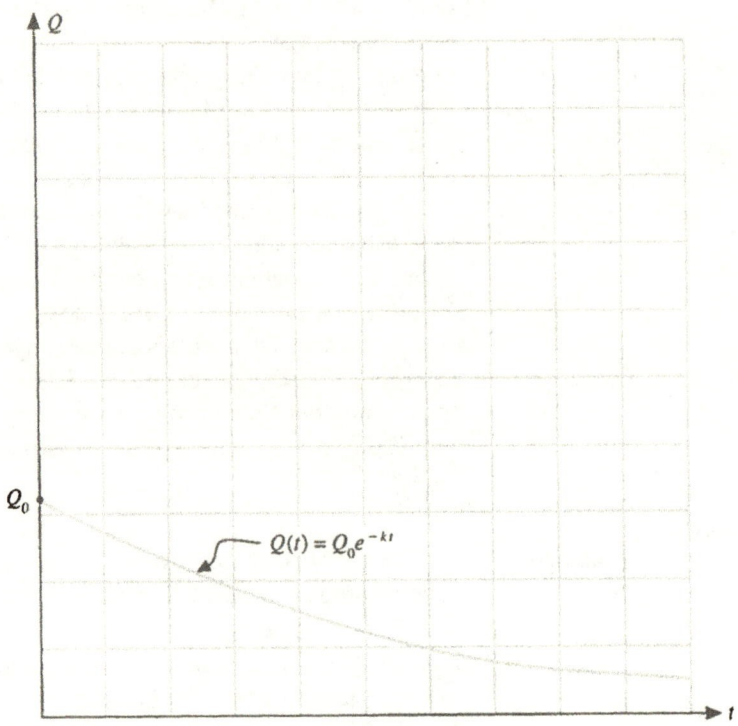

(b) The amount of C-14 at any time t is given by the decay curve

$$Q(t) = Q_0 e^{-kt} = Q_0 e^{0.00012378t} \tag{5}$$

(c) Since 15% of the original C-14 remains in the sample at the time of discovery, the age of the charcoal can be found by solving for t in the equation

$$Q_0 e^{-0.00012378t} = 0.15 Q_0 \tag{6}$$

We get

$$e^{-0.00012378t} = 0.15$$

$$-0.00012378t = \ln(0.15)$$

$$t = \frac{\ln(0.15)}{-0.00012378} \doteq 15{,}336 \text{ years} \tag{7}$$

The decay of C-14 in a sample of charcoal is illustrated in Figure 2.9. ∎

Note: In general, the relationship between the rate constant k of the differential equation $dy/dt = ky$ and the half-life k of the solution is $k = \frac{\ln 2}{\text{half-life}}$.

Compound Interest

A useful application of the growth equation $y' = ky$ lies in the prediction of the future value of a savings account. When one deposits a sum of money, S_0, in a bank account, the bank pays the depositor *interest* depending on the amount of the deposit and the length of time the money is deposited. When interest is **compounded,** the interest paid by the bank is put back into the account, along with the current balance. After another time period has elapsed, interest is paid again on this new balance. This process is then continued again and again.

Figure 2.9
The decay curve for C-14 with a half-life of 5600 years

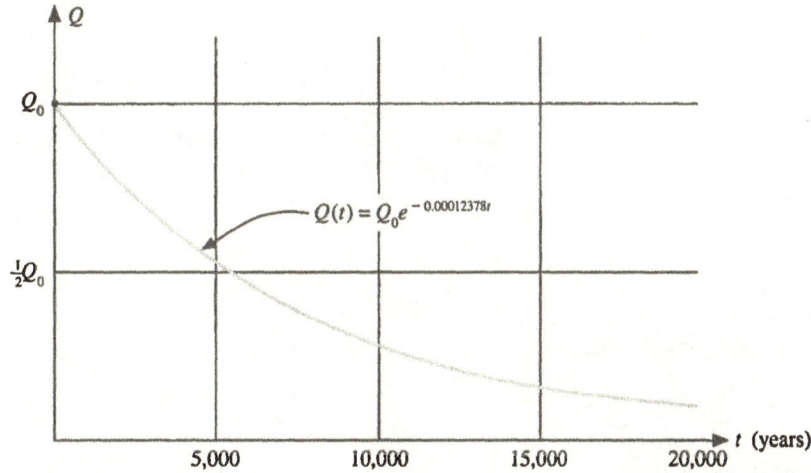

The **future value,** $S(t)$, of a deposit will depend on the **rate of interest** r that a bank pays as well as the **frequency** at which the bank makes the interest payments. If the bank pays an annual interest rate of r and makes payments *once* a year (compounded annually), then after 1, 2, 3, ..., t years the value of the account will be

$$
\begin{aligned}
\text{Intial deposit:} \quad & S_0 \\
\text{After one year:} \quad & S_0 + rS_0 & = S_0(1 + r) \\
\text{After two years:} \quad & S_0(1 + r) + r \cdot S_0(1 + r) & = S_0(1 + r)^2 \\
\text{After three years:} \quad & S_0(1 + r)^2 + r \cdot S_0(1 + r)^2 & = S_0(1 + r)^3 \\
& \quad \vdots \\
\text{After } t \text{ years:} \quad & & = S_0(1 + r)^t
\end{aligned}
$$

If the same interest rate r is compounded *twice* a year, then at the end of six months (one compounding) the value of the account will be

$$
\text{After 6 months:} \quad S_0\!\left(1 + \frac{r}{2}\right)
$$

and after one year (two compoundings) it will be

$$
\text{After one year:} \quad S_0\!\left(1 + \frac{r}{2}\right)^2
$$

Thus after t years (and $2t$ compoundings) the value of the account will have increased to

$$
S(t) = S_0\!\left(1 + \frac{r}{2}\right)^{2t}
$$

In general, if interest is compounded n times a year, then after t years the value of the account will be

$$S(t) = S_0\left(1 + \frac{r}{n}\right)^{nt} \tag{8}$$

When banks compound interest on a daily basis, the value of n is 365. Figure 2.10 compares the growth of an initial deposit of $100 when interest is compounded once (annually), twice (semiannually), four times (quarterly), and twelve times (monthly) per year. In each case we have assumed that the annual interest rate is $r = 0.10$ (10%).

Figure 2.10

A comparison of the growth of $100 in á bank that pays interest at an annual rate of 10%. The different curves compare the growth when interest is compounded annually, semiannually, quarterly, and monthly.

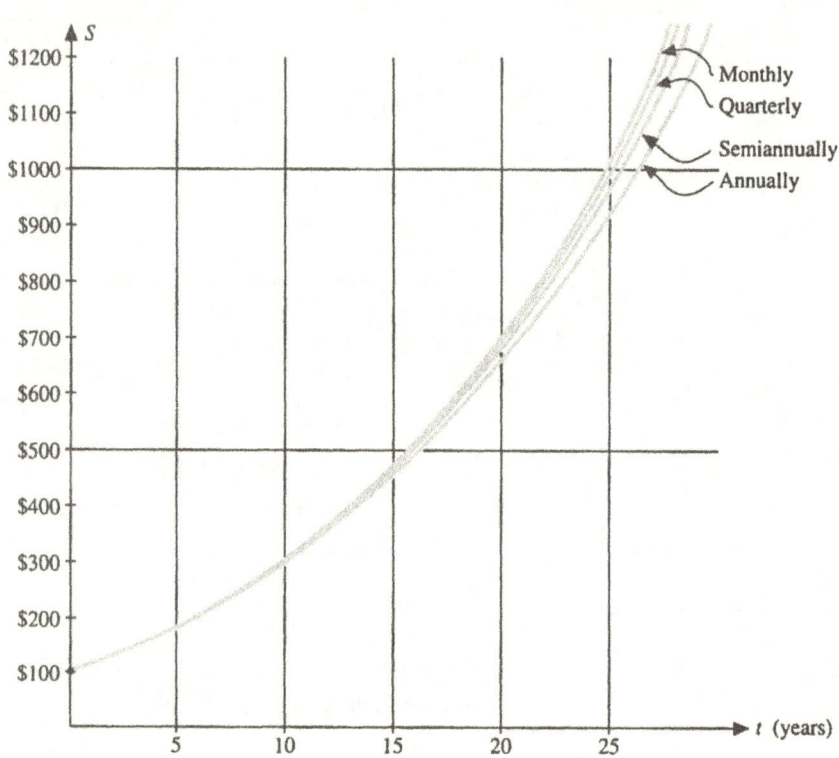

When the future value $S(t)$ of an account is computed from the limit

$$S(t) = \lim_{n \to \infty} S_0\left(1 + \frac{r}{n}\right)^{nt} = S_0\, e^{rt} \tag{9}$$

we say that interest is **compounded continuously.** Of course, most banks do not compound interest continuously, but continuous compounding is an accurate approximation of daily compounding ($n = 365$). These ideas are summarized on the facing page.

Continuous Compounding of Interest

If an initial amount of S_0 dollars is deposited in a bank that pays an annual interest of r, compounded continuously, then the future value, $S(t)$, of the account satisfies the initial-value problem

$$\frac{dS}{dt} = rS$$

$$S(0) = S_0 \tag{10}$$

The solution of this initial-value problem, and hence the future value $S(t)$ of the account, is

$$S(t) = S_0\, e^{rt} \tag{11}$$

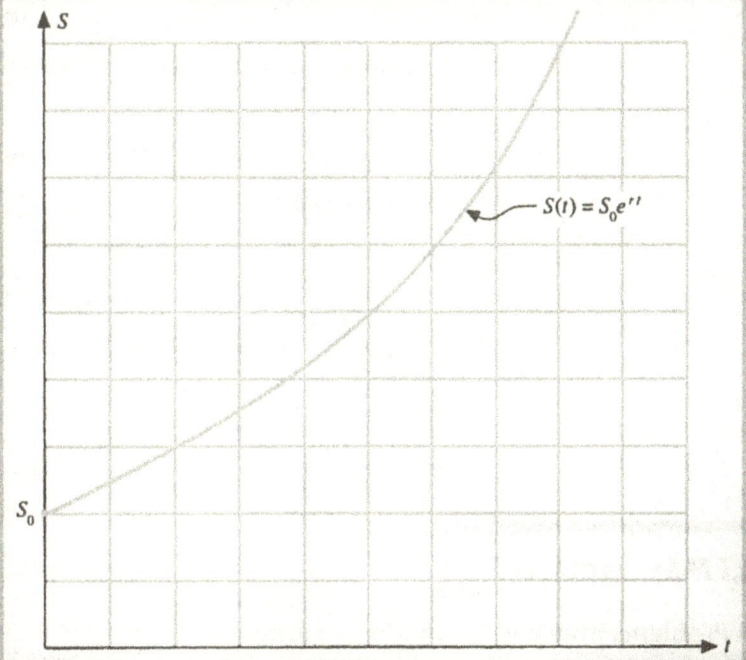

Moreover, if deposits amounting to d dollars per year (made uniformly throughout the year) are made into the above account, then the future value $S(t)$ of this new account, called an **annuity,** will satisfy

$$\frac{dS}{dt} = rS + d$$

$$S(0) = S_0 \tag{12}$$

The solution of this initial-value problem, which describes the **future value** of an annuity, is easily found to be

$$S(t) = S_0\, e^{rt} + \frac{d}{r}\left(e^{rt} - 1\right) \tag{13}$$

Example 2 **Saving Your Cigarette Money** Jenna has just entered college and has decided to quit smoking. She wants to regain her health and save some money in the process. She smokes two packs a day, which translates into roughly $30 per week. Suppose she makes weekly deposits of this money into a bank account that pays annual interest of 10%, compounded continuously. What will be the balance of Jenna's account when she retires in 47 years?

Solution Jenna's weekly deposit of $30 into her account translates into an annual deposit of $(30)(52) = \$1560$. Hence the initial-value problem that describes the amount $S(t)$ of money (dollars) in Jenna's account after t years is

$$\frac{dS}{dt} - 0.10S = 1560$$
$$S(0) = 0 \qquad (14)$$

The solution of Eq. (14) can easily be found by using the integrating factor method, giving

$$S(t) = 15{,}600\,[e^{0.1t} - 1] \qquad (15)$$

Hence the amount of money in Jenna's account after 47 years will be

$$S(47) = 15{,}600\,(e^{4.7} - 1)$$
$$= \$1{,}699{,}575.90$$

In other words, Jenna will have saved over 1.5 million dollars by the time she retires. ∎

PROBLEMS: Section 2.3

General Problems Involving Growth and Decay

1. **Alternative Solution for the Growth Equation** In the text we solved the growth equation $y' = ky$ using the integrating factor method. Solve this equation using the separation of variables method.
2. **Decay Equation** Solve the decay equation $y' = -ky$ by both separation of variables and the integrating factor method.
3. **Half-Life Problem** The time required for the solution y of the decay equation.

$$y' = -ky \qquad (16)$$
$$y(0) = y_0$$

to be reduced from the initial value y_0 to one-half of y_0 is called the **half-life** t_h.

(a) Find the half-life t_h in terms of the constant k.
(b) Show that if the solution of the decay equation is some constant C at some t_1, then the solution one half-life later will be $C/2$.
4. **Doubling Time** The time required for the initial value y_0 of the growth equation

$$y' = ky \qquad (17)$$
$$y(0) = y_0$$

to double in value is called the **doubling time** t_d. Find the doubling value of this equation in terms of the constant k.
5. **The Annuity Equation** Solve the annuity problem

$$\frac{dS}{dt} = rS + d$$
$$S(0) = S_0$$

Radioactive Decay

6. **Interpretation of 1/k in the Decay Equation** The reciprocal $1/k$ (which has units of time) of k that appears in the decay equation $y' = -ky$ has a rough interpretation that it is the time required for the solution $y(t)$ to fall *two-thirds* of the distance from the initial value y_0 to 0. See Figure 2.11. Why does this claim have merit? *Hint*: Evaluate the solution of the initial-value problem $y' = -ky$, $y(0) = y_0$ at $t = 1/k$ and approximate $1/e \doteq \frac{1}{3}$.

7. **Radioactive Decay** A certain radioactive material is known to decay at a rate proportional to the amount present. Initially, 100 grams of the substance are present, but after 50 years the mass decays to 75 grams. Find an expression for the mass of the material at any time. What is the half-life of the material?

8. **Determining Decay from Half-Life** A certain radioactive substance has a half-life of 5 hours. Find the time required for the material to decay to one-tenth of the original mass.

9. **Radioactive Thorium-234** Thorium-234 is a radioactive substance that decays at a rate proportional to the amount present. Suppose 1 gram of this material is reduced to 0.80 gram in one week.
   (a) Find an expression for the amount of T-234 present at any time.
   (b) Find the half-life of T-234.
   (c) Find the amount of T-234 present after 10 weeks.

10. **Radioactive Dating in Sneferu's Tomb** A cypress beam found in the tomb of Sneferu in Egypt contained 55% of the radioactive carbon-14 that is found in living cypress wood. Estimate the age of the tomb.

11. **Newspaper Announcement** A 1960 New York Times article announced, *"Archeologists Claim Sumerian Civilization Occupied the Tigris Valley 5,000 Years Ago."* Assuming that the archeologists used carbon-14 to date the site, determine the percentage of carbon-14 found in the samples.

12. **Radium Decay** Radium decays at a rate proportional to the amount present and has a half-life of 1600 years. What percentage of an original amount will be present after 6400 years? *Hint*: This problem is *very* easy.

13. **General Equation for Half-Life** If Q_1 and Q_2 are the amounts of a radioactive substance present at times t_1 and t_2, respectively ($t_1 < t_2$), show that the half-life of the substance is given by

$$t_h = \frac{(t_2 - t_1) \ln 2}{\ln\left(\dfrac{Q_1}{Q_2}\right)} \tag{18}$$

Figure 2.11
A rough interpretation of $1/k$ is that it is the time required for the decay curve $y_0 e^{-kt}$ to fall two-thirds of the way from the initial value of y_0 to the limiting value of 0

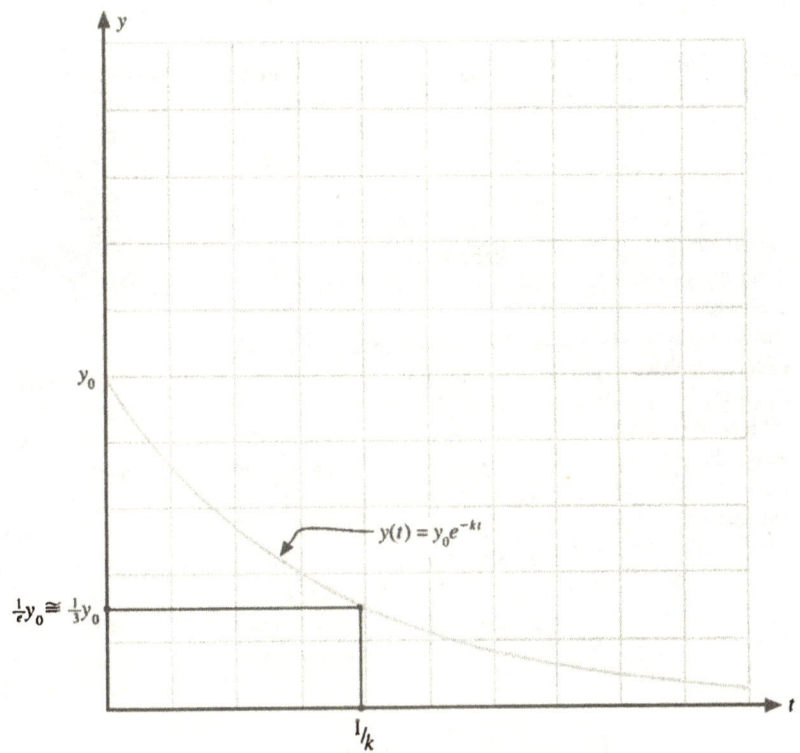

$$y(t) = y_0 e^{-kt}$$

$$\frac{1}{e} y_0 \cong \frac{1}{3} y_0$$

$$\frac{1}{k}$$

14. **Nuclear Waste** The U.S. government has dumped roughly 100,000 barrels of radioactive waste into the Atlantic and Pacific oceans. The waste is mixed with concrete and encased in steel drums. The drums will eventually rust, and seawater will gradually leach the radioactive material from the concrete and diffuse it throughout the ocean. It is assumed that the leached radioactive material will be so diluted that no environmental damage will result. However, scientists have discovered that one of the pollutants, americium 241, is sticking to the ocean floor near the drums. Given that americium 241 has a half-life of 258 years, how long will it take for the americium 241 to be reduced to 5% of its present amount?

15. **Bombarding Plutonium** In 1964, Soviet scientists made a new element with atomic number 104, called E104, by bombarding plutonium with neon ions. The half-life of this new element is 0.15 second, and it was produced at a rate of 2×10^{-5} micrograms per second. Assuming that none was present initially, how much E104 is present after t seconds?

Decay in Biology

The growth and decay of many phenomena in biology satisfy the growth and decay equations discussed in this section. The following problems illustrate typical growth and decay phenomena.

16. **Blood Alcohol Levels** In many states it is illegal to drive with a blood alcohol level greater than 0.10% (one part of alcohol per 1000 parts of blood). Suppose someone who was involved in an automobile accident had blood alcohol tested at 0.20% at the time of the accident. Assume that the percentage of alcohol in the bloodstream decreases exponentially at the rate of 10% per hour.
    (a) Find the percentage of alcohol in the bloodstream at any time.
    (b) How long will it be until this person can legally drive?

17. **The Exxon Valdez Problem*** In the tragic 1989 accident in which the Exxon Valdez dumped 240,000 barrels of oil into Prince William Sound, the National Safety Board determined that blood tests of Captain Joseph Hazelwood showed a blood alcohol content of 0.06%. It was also revealed that blood testing did not take place until at least nine hours after the accident. Blood alcohol is eliminated from the system at a rate of about 0.015 percentage points per hour. If the permissible level of alcohol is 0.10%, can the National Safety Board determine that the captain was liable?

* Material taken from the *New York Times*, March 31, 1989.

18. **Elimination of Sodium Pentobarbital** Ed is undergoing surgery for an old football injury and must be anesthetized. The anesthesiologist knows that Ed will be anesthetized when the concentration of sodium pentobarbital in Ed's blood is at least 50 milligrams per kilogram of body weight. Suppose Ed weighs 100 kilograms (220 lb) and that sodium pentobarbital is eliminated from the bloodstream at a rate proportional to the amount present. If the half-life of the sodium pentobarbital decay is 10 hours, what single dose should be given in order to anesthetize Ed for 3 hours?

19. **Moonlight at High Noon** The fact that sunlight is absorbed by water is well known to any diver who has dived to a depth of 100 feet. If 25 feet of water absorb 15% of the light that strikes the surface, at what depth will the light at noon be as bright as a full moon, which is 1/300,000th as bright as the noonday sun? See Figure 2.12.

Figure 2.12
The intensity of light in a lake or ocean falls exponentially with depth. The decay constant k depends on the clarity of the water

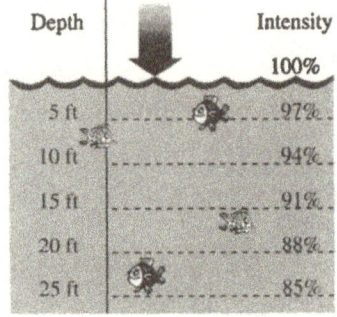

20. **Flattened by a Differential Equation** Campus police are investigating a hit-and-run accident in which the driver of a late-model sports car has run down the statue of a popular mathematics professor. The accident took place at 8:00 P.M., and using a tip from a bystander, police go to a dormitory and at 10:00 P.M. make an arrest. At that time they measure the percent of alcohol in the suspect's blood and find it to be 0.07%. Two hours later, at 12:00 P.M., while the suspect is being dried out at police headquarters, the police again measure the percentage of alcohol in the suspect's blood and determine it to be 0.05%. See Figure 2.13. If the legal limit for alcohol content in the blood is 0.10%, was the suspect legally drunk at the time of the accident?

Figure 2.13
The chronicle of events in the hit-and-run case

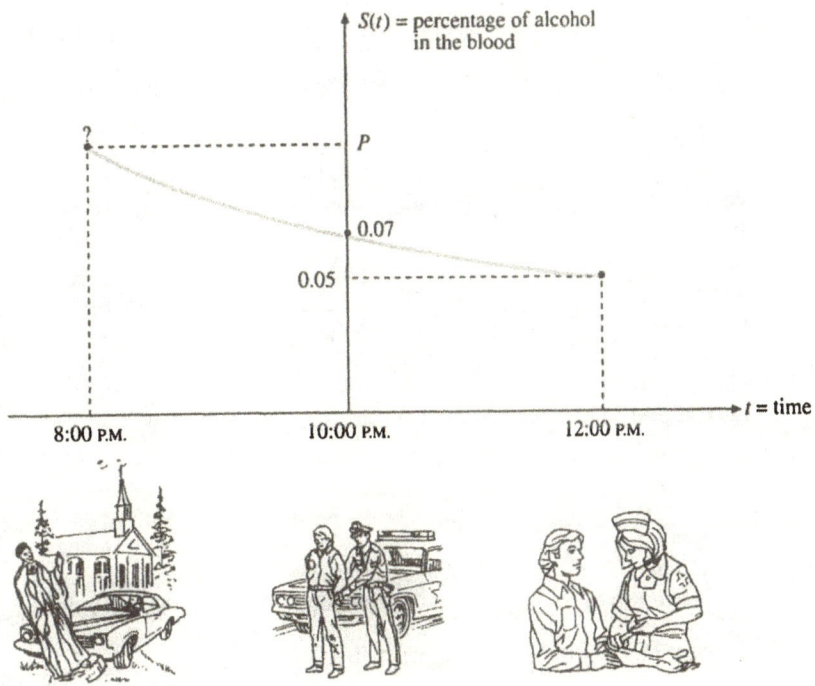

Growth In Biology

21. Converting Doubling Time to Tripling Time The number of bacteria in a colony increases at a rate proportional to the number present. If the number of bacteria doubles every 10 hours, how long will it take for the colony to triple in size?

22. Extrapolating the Past If the number of bacteria in a culture is 5 million at the end of 6 hours and 8 million at the end of 9 hours, how many bacteria were present initially?

23. Helene's Big Decision Helene has just accepted a position as a microbiologist for the Board of Health. It is known that when a certain food product is packaged, the number of organisms present in the food is N, and after 30 days the number of organisms has grown to $100N$. How many days will it take for the number of organisms to reach $200N$?

24. Unrestricted Yeast Growth The number of bacteria in a yeast culture grows at a rate proportional to the number present. If the population of a colony of yeast bacteria doubles in 1 hour, and if 5 million are present initially, find the number of bacteria in the colony after 4 hours.

25. Unrestricted Bacterial Growth A certain colony of bacteria grows at a rate proportional to the amount of bacteria present. Suppose the number of bacteria doubles every 12 hours. How long will it take this colony of bacteria to grow to five times the original amount?

26. Growth of *Escherichia Coli* A culture of *Escherichia coli* grows at a rate proportional to its size. A researcher has determined that every hour the culture is 2% larger than the hour before and that initially there are 100 organisms present. How many organisms are present at any time t?

27. Growth of the Earth's Population It is estimated that in the year 1500 A.D. ($t = 0$) the size of the earth's population was 100 million. In 1990 ($t = 490$) the population was estimated to be 2.6 billion. Assume that the rate of increase is proportional to the current population size.
(a) Find the expression for the size of the earth's population at any time.
(b) Assuming that the earth can support at most 50 billion people, when will this occur? Do you trust this mathematical model? Why or why not?

28. The Logistic Growth Model* In 1837 the Belgium biomathematician P. F. Verhulst used the **logistic growth equation**

$$\frac{dN}{dt} = rN\left(1 - \frac{N}{k}\right) \qquad (19)$$

to predict the population of the United States, where r and k are constants. The constant k is called the **carrying capacity** of the environment and represents the steady state

---

* The geologist M. King Hubbert became somewhat famous because of his use of the logistic equation in his accurate prediction of the decline of U.S. domestic oil production in the 1970's.

population size. The constant r is a measure of the rate at which it is attained. Verhulst estimated the parameters to be $r = 0.03$ and $k = 200$ million. Using 4 million people as the 1790 population of the United States:

(a) Solve Eq. (19) with the initial condition $N_0 = 4$ (million) to find $N(t)$. *Hint:* Use separation of variables followed by a partial fraction decomposition. Then use properties of logarithms.

(b) If you have access to a computer, plot the solution of $N(t)$ for $0 \leq t \leq 200$ (from 1790 to 1990).

(c) Evaluate $N(200)$, which will give Verhulst's estimate of the U.S. population in 1990. *Note:* The actual population in 1990 was 250 million.

29. The Famous Cat and Mouse Problem On an island that had no cats, the mouse population doubled during the years from 1980 to 1990, reaching 50,000. At that time the islanders imported several cats, who thereafter have killed 6000 mice per year.

(a) What is the number of mice on the island at year t?

(b) How many mice will be on the island in the year 2000?

Growth of Money (Continuous Compounding)

30. Banker's Interpretation of e A banker once gave the interpretation of the constant e as the future value of a bank account after 10 years if $1 is initially deposited and the bank pays an annual interest of 10% compounded continuously. Show why the banker made this claim.

31. Rule of 70 In banking circles there is a "Rule of 70" that says that the time (in years) required for the value of a bank account to double in value can be approximated by dividing 70 by the annual interest rate paid (in percent). What is the reasoning behind this rule?

32. Power of Continuous Compounding In 1820 a William Record of London, England, deposited $0.50 (or the equivalent in English pounds sterling) for his granddaughter in the Bank of London. Unfortunately, he died before he could tell his granddaughter about the account. One hundred and sixty years later, in 1980, the granddaughter's heirs discovered the account. What was the value of the account if the bank paid an annual interest of 6% compounded continuously?

33. Compound Interest Thwarts Hollywood Stunt In 1944 Hollywood publicists were going to dramatize the opening of the motion picture *Knickerbocker Holiday* by arranging a publicity stunt in which three bottles of whiskey, the amount originally thought to have been given to the Canarsie Indians for the island of Manhattan, were to have been returned to the mayor of New York City, plus 8% interest compounded annually. To their horror, just before the gala event the agents discovered that the compound interest on the whiskey over a period of 320 years would be more than 100 million bottles. As one agent put it, "The stunt just ain't worth it." Exactly how many bottles of whiskey should have been given to the mayor of New York City in 1944?

34. Canarsie Indian Problem In 1626 (so the story goes), Dutch explorer Peter Minuit paid the Canarsie Indians $24 for what is now New York City. If the Canarsies could have deposited this money in a savings account that paid an annual interest of 8% compounded continuously, how much would the value of this account have been worth in 1992? See Figure 2.14.

35. It Ain't Like It Used to Be John's grandfather tells John that 50 years ago the average cost of a new car was only $1000, whereas today the average cost is $18,000. What is the continuous rate of increase over the past 50 years to produce this change?

36. How to Become a Millionaire Upon graduating from college, Donna has no initial capital. However, during each year she makes deposits amounting to $d = $1000 in a bank that pays interest at an annual rate of $r = 8\%$, compounded continuously.

(a) Find the future value, $S(t)$, of Donna's account at any time.

(b) What should be the value of the annual deposit d in order that the balance of Donna's account will be 1 million dollars when she retires in 40 years?

(c) If $d = $2500, what should be the value of r in order that Donna's account will have a balance of 1 million dollars in 40 years?

37. Marlboros for Millions Ann has just entered college and has decided to quit smoking. She wants to regain her health and save money in the process. Ann smokes one pack a day, which translates into roughly $15 per week. Suppose she makes weekly deposits in a bank that pays an annual interest of 8% compounded continuously.

(a) What will be the balance in Ann's account in week t?

(b) What will be the balance of Ann's account when she graduates in 4 years?

38. How Sweet It Is John has won the Mega-Bucks lottery, which pays $1,000,000. Suppose he deposits the money in a savings account that pays an annual interest of 8% compounded continuously. How long will this money last if he makes annual withdrawals of $100,000?

39. Living Off Your Money Suppose a rich uncle has left you S_0 dollars, which you deposit in a bank that pays annual interest at a rate of r, compounded continuously. Show that if you make withdrawals amounting to d dollars per year $(d > rS_0)$, then the time required to deplete the money in the bank is

$$t = \frac{1}{r} \ln \left(\frac{d}{d - rS_0} \right) \tag{20}$$

Figure 2.14
If $24 had been deposited in 1624 in a bank account that paid 8% annual interest, compounded continuously, then the future value of this account in the year 1992 would have been worth more than the 1992 real estate value of New York City

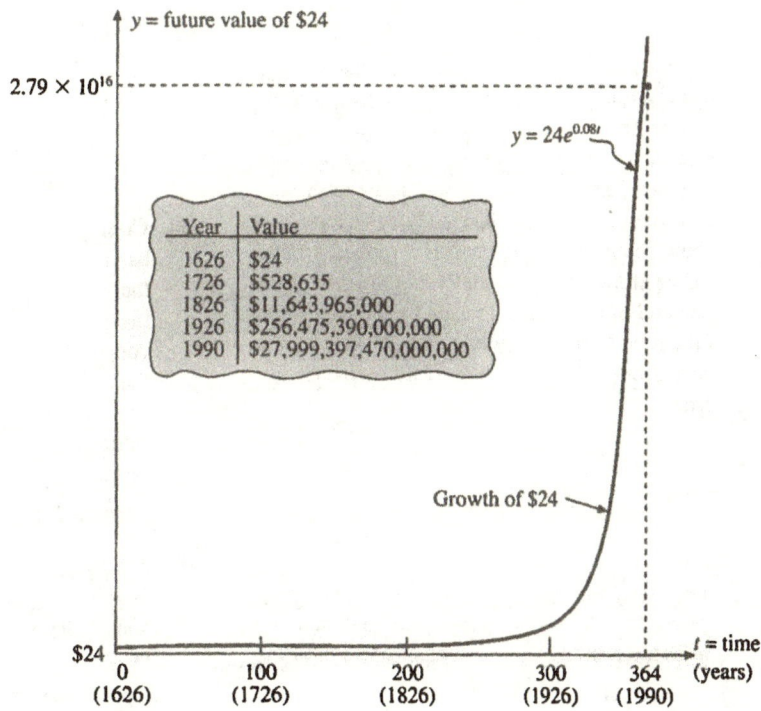

y = future value of $24

2.79×10^{16}

$y = 24e^{0.08t}$

Year	Value
1626	$24
1726	$528,635
1826	$11,643,965,000
1926	$256,475,390,000,000
1990	$27,999,397,470,000,000

Growth of $24

t = time

$24

| 0 | 100 | 200 | 300 | 364 | (years) |
| (1626) | (1726) | (1826) | (1926) | (1990) | |

What happens when the annual amount withdrawn d is less than or equal to rS_0?

40. How Much Is the Lottery Really Worth? You have to be careful about money. Lottery winners sometimes think they are millionaires when in fact they are not as rich as they think. Suppose a state runs a lottery in which the Grand Prize is $1 million dollars. What that means is that the winner is paid $1 million dollars continuously over the next 20 years. Assuming that banks pay interest of 10% on borrowed money over the next 20 years, how much is the lottery worth in today's dollars? *Hint*: Letting t denote time in years, first solve $S' = 0.10S - 50,000$, $S(0) = S_0$, and then set $S(20) = 0$ and solve for S_0.

41. General Compound Interest Problem The general first-order linear differential equation

$$\frac{dS}{dt} = p(t)\,S + f(t) \qquad S(0) = S_0 \qquad (21)$$

can be interpreted as describing the future value $S(t)$ of a bank account at time t in which S_0 represents the initial amount of money in the account. The function $p(t)$ represents the variable interest rate, and $f(t)$ is the rate of continuous deposits (positive values) or withdrawals (negative values). Show that the solution of this initial-value problem

can be written as

$$S(t) = S_0 w(t_0, t) + w(t_0, t) \int_{t_0}^{t} w(s, t_0)\,f(s)\,ds \qquad (22)$$

where

$$w(s, t) = e^{\int_s^t p(\tau)\,d\tau}$$

represents the amount that a dollar will grow by time t if deposited at time $s < t$. The first term on the right-hand side of Eq. (22) depends only on S_0 and p, and it represents the value of the account due to the initial amount S_0, whereas the second term depends only on p and f, and it represents the value of the account due to the continuous deposits or withdrawals.

42. Continuous Compounding Most savings banks advertise that they compound interest continuously, meaning that the amount of money $S(t)$ in an account satisfies the differential equation $dS/dt = rS$, where r is the annual interest rate and t is time measured in years.

(a) Show that an annual interest rate of 8% compounded continuously is the same as an annual interest rate of 8.33% compounded annually.

(b) Show that an annual interest rate of r compounded continuously is the same as an annual interest rate of $e^r - 1$ compounded annually.

An Extra

43. The Tug-of-War Problem Everyone knows that when a rope is wrapped around a post or tree, a single person holding onto one end can resist a much larger force or *tension* exerted at the other end. The tug-of-war problem consists in finding just how much force is necessary to resist a given force of F_0 at the other end of the rope as a function of how many times the rope is wrapped around the post. It can be shown experimentally that the force $F(\theta)$ required at one end of the rope to resist a force of F_0 at the other end of the rope, where θ is the number of degrees the rope is wrapped around the tree, is described by the initial-value problem

$$\frac{dF}{d\theta} = -kF$$

$$F(0) = F_0$$

where k is the *coefficient of friction* between a given rope and post. Suppose $k = 0.25$ (when θ is measured in radians) and the Notre Dame football team pulling at one end of the rope exerts a force of $F_0 = 5000$ lb.

Figure 2.15
Can Mary beat the Notre Dame football team in a tug of war?

(a) Find the force required to resist these players as a function of θ.

(b) If Mary can exert a 125-lb force, how many times would she have to wrap the rope around the post to resist the force of the football players? See Figure 2.15.

44. Computer Problem Upon graduating from college, Ed has no money. However, he immediately obtains a job and thereafter makes annual deposits of $d = \$750$ in a bank that pays interest at an annual rate of $r = 8\%$, compounded continuously. Find the future value of Ed's account over the next 40 years, and use a graphing calculator or computer to sketch the graph of this future value function. Will this future value reach 1 million dollars before Ed retires in 40 years? When will Ed's account reach a million dollars if the bank pays 10% annual interest?

45. Journal Entry Do you think that the population of the United States grows exponentially according to the law $Q_0 e^{kt}$? Did the population of the United States ever grow according to this law? What kinds of populations might be predicted by this law?

2.4	**MIXING PHENOMENA**

PURPOSE
To show how a certain first-order linear differential equation, known as the continuity equation, can be used to predict the amount of a substance in a container in which a mixture of the substance is constantly being added and extracted. The basic mixing problem discussed in this section is a prototype for more general phenomena found in chemistry and biology.

In this section we consider mixing problems involving the mixing of substances that have significance in chemistry and biology. More specifically, we are interested in

Figure 2.16
Single tank configuration

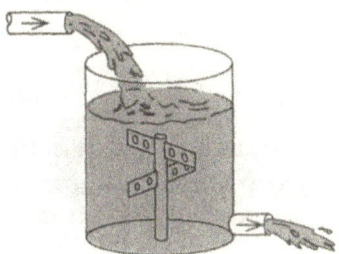

predicting the amount of a substance (salt, drugs, hormones, etc.) in a container in which a mixture of the substance flows into the container, is mixed with the ingredients in the container, and then flows out of the container at some given rate. See Figure 2.16.

The goal is to find the quantity of the substance in the container at any time. To determine this, we let $Q(t)$ denote the amount of the substance in the tank (say in pounds) at time t. We then have that dQ/dt is the *rate of change* (say in pounds per minute) in $Q(t)$ and is given by the difference between the *rate of inflow* (RATE IN) into the container and the *rate of outflow* (RATE OUT) out of the container. This basic law, called the **continuity equation,** can be written as

$$\frac{dQ}{dt} = \text{RATE IN} - \text{RATE OUT} \tag{1}$$

We can now find expressions for RATE IN and RATE OUT from

$$
\begin{array}{ccc}
\text{RATE IN} & = \text{(CONCENTRATION IN)} & \text{(FLOW RATE IN)} \\
\uparrow & \uparrow & \uparrow \\
\text{(lb/min)} & \text{(lb/gal)} & \text{(gal/min)}
\end{array} \tag{2}
$$

$$
\begin{array}{ccc}
\text{RATE OUT} & = \text{(CONCENTRATION OUT)} & \text{(FLOW RATE OUT)} \\
\uparrow & \uparrow & \uparrow \\
\text{(lb/min)} & \text{(lb/gal)} & \text{(gal/min)}
\end{array}
$$

where CONCENTRATION IN and CONCENTRATION OUT are the concentrations (in pounds per gallon) of the substance entering and leaving the container, respectively. The expressions FLOW RATE IN and FLOW RATE OUT are the flow rates (in gallons per minute) of the liquid entering and leaving the container, respectively. The following example illustrates these ideas.

Example 1   **Standard Mixing Problem**   A tank initially contains 50 gal of pure water. A salt solution containing 2 pounds of salt per gallon of water is poured into the tank at a rate of 3 gal/min. The mixture is stirred and is drained out of the tank at the same rate.

(a) Find the initial-value problem that describes the amount Q of salt in the tank at any time.
(b) Find the amount of salt in the tank at any time.
(c) Find the amount of salt in the tank after 20 minutes.
(d) Find the amount of salt in the tank after a "long time."

Solution   (a) **Initial-Value Problem.** If we let $Q(t)$ denote the amount of salt in the tank at time t, the continuity equation becomes

$$\frac{dQ}{dt} = \text{RATE IN (lb/min)} - \text{RATE OUT (lb/min)}$$

$$= \left(2 \text{ lb/gal}\right)\left(3 \text{ gal/min}\right) - \left(\frac{Q}{50} \text{ lb/gal}\right)\left(3 \text{ gal/min}\right)$$

$$= 6 \text{ lb/min} - \frac{3}{50}Q \text{ lb/min} \tag{3}$$

Since initially the tank contains no salt, we have $Q(0) = 0$. Hence the initial-value problem is

$$\frac{dQ}{dt} + \frac{3}{50}Q = 6$$

$$Q(0) = 0 \tag{4}$$

(b) **Amount of Salt in the Tank.** The differential equation (4) can be solved either by separating variables or by the integrating factor method. We use the integrating factor method and multiply each side of the differential equation (4) by its integrating factor, $\mu(x) = e^{3t/50}$. This gives

$$e^{3t/50}\left(\frac{dQ}{dt} + \frac{3}{50}Q\right) = 6e^{3t/50}$$

or

$$\frac{d}{dt}\left(e^{3t/50}Q(t)\right) = 6e^{3t/50} \tag{5}$$

Taking the antiderivative of each side of this equation gives

$$e^{3t/50}Q(t) = 100e^{3t/50} + c$$

and solving for $Q(t)$, we get the general solution

$$Q(t) = ce^{-3t/50} + 100 \tag{6}$$

The initial condition $Q(0) = 0$ gives

$$0 = c + 100 \not{c} \; c = -100$$

Hence the amount of salt $Q(t)$ in the tank at time t is given by

$$Q(t) = 100\left(1 - e^{-3t/50}\right) \tag{7}$$

See Figure 2.17.

Figure 2.17
The amount of salt in the tank starts at zero and grows like a "concave downward exponential curve" approaching the limiting value of 100 pounds

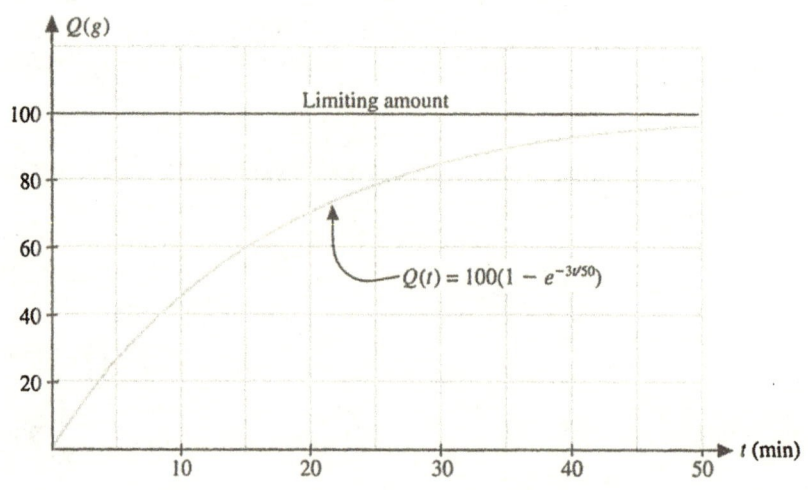

(c) **Value of $Q(20)$.** The amount of salt in the tank after 20 minutes is

$$Q(20) = 100\,(1 - e^{-60/50}) \doteq 69.9 \text{ lb} \tag{8}$$

(d) **Value of $Q(t)$ After a Long Time.** After a long period of time has elapsed, it is clear that the exponential term in the solution (7) approaches 0, and so the solution will approach a steady state solution of $Q(\infty) = 100$. For example, after two hours ($t = 120$ min) the solution will be $Q(120) = 100(1 - e^{-360/50}) \doteq 99.92$ lb. ∎

UNEQUAL INPUT AND OUTPUT RATE OF FLOW

Often, the flow rate of the liquid entering the tank is not the same as the flow rate leaving the tank, and so the tank either fills or empties. We illustrate the situation in which the input flow rate is *less* than the output flow rate.

Example 2 **RATE IN < RATE OUT** A tank initially contains 100 gallons of water in which 10 pounds of salt are dissolved. A salt solution containing 0.5 pound of salt per gallon is poured into the tank at a rate of 1 gal/min. The mixture in the tank is stirred and drained off at the rate of 2 gal/min.

(a) Find the initial-value problem that describes the amount of salt $Q(t)$ in the tank until the tank is empty.

(b) Find the amount of salt $Q(t)$ in the tank until the tank is empty.

(c) Find the concentration $c(t)$ of salt in the tank until the tank is empty.

(d) Find the concentration of salt in the tank at the exact time the tank becomes empty.

Solution (a) **Initial-Value Problem.**

Finding expressions for RATE IN and RATE OUT, we get

$$\text{RATE IN (lb/min)} = \left(\frac{1}{2}\text{ lb/gal}\right)\!\left(1\text{ gal/min}\right) = \frac{1}{2}\text{ lb/min}$$

$$\text{RATE OUT (lb/min)} = \left(\frac{Q}{100 - t}\text{ lb/gal}\right)\!\left(2\text{ gal/min}\right) = \frac{2Q}{100 - t}\text{ lb/min}$$

Note that the expression $100 - t$ in the denominator of RATE OUT is due to the fact that the *volume* of solution in the tank is not fixed at 100 gal but is decreasing by 1 gal/min. Substituting these values for RATE IN and RATE OUT into the continuity equation and using the initial condition $Q(0) = 10$, we have

$$\frac{dQ}{dt} + \frac{2Q}{100 - t} = \frac{1}{2} \qquad (0 \le t < 100)$$
$$Q(0) = 10 \tag{9}$$

Note that the tank will run dry after 100 minutes.

(b) **Finding the General Solution.** Finding the integrating factor, we have

$$\mu(t) = e^{\int \left(\frac{2}{100 - t}\right) dt}$$
$$= e^{-2 \ln (100 - t)}$$
$$= e^{\ln (100 - t)^{-2}}$$
$$= (100 - t)^{-2}$$
$$= \frac{1}{(100 - t)^2} \tag{10}$$

Multiplying by $\mu(t)$, we get

$$\frac{1}{(100 - t^2)} \left(\frac{dQ}{dt} + \frac{2Q}{100 - t}\right) = \frac{1}{2(100 - t)^2}$$

or

$$\frac{d}{dt} \left(\frac{Q}{(100 - t)^2}\right) = \frac{1}{2(100 - t)^2}$$

Integrating, we get

$$\frac{Q}{(100 - t)^2} = \frac{1}{2(100 - t)} + c$$

or

$$Q(t) = c (100 - t)^2 + \frac{1}{2} (100 - t) \tag{11}$$

Using the initial condition $Q(0) = 10$, we see that

$$10 = 10^4 c + 50 \Rightarrow c = -\frac{1}{250}$$

Hence

$$Q(t) = -\frac{1}{250} (100 - t)^2 + \frac{1}{2}(100 - t)$$
$$= -\frac{1}{250} (t - 100)(t + 25) \qquad (0 \le t < 100) \tag{12}$$

(c) **Find the Concentration.** The concentration $c(t)$ of the salt in the tank is simply $Q(t)$ divided by the amount of solution in the tank. Hence we have

$$c(t) = \frac{Q(t)}{100 - t}$$
$$= \frac{t + 25}{250} \qquad (0 \le t < 100) \tag{13}$$

The graphs of the quantity $Q(t)$ and concentration $c(t)$ are shown in Figures 2.18(a) and 2.18(b), respectively.

(d) **Limiting Concentration.** Since the tank will drain dry after 100 minutes, we can see from Eq. (13) that the concentration of salt in the tank will be 0.5 lb/gal just at the time the tank empties. ∎

Figure 2.18
Graphs of the amount of salt
$Q(t)$ and concentration $c(t)$ of
salt in a tank when the level
of solution is falling

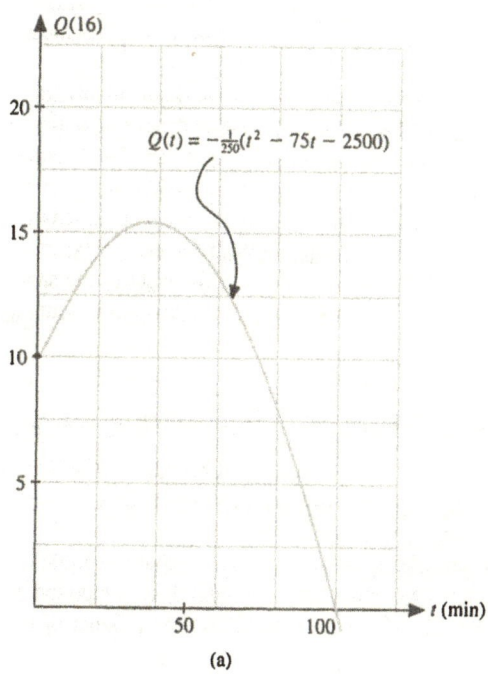

$$Q(t) = -\tfrac{1}{250}(t^2 - 75t - 2500)$$

(a)

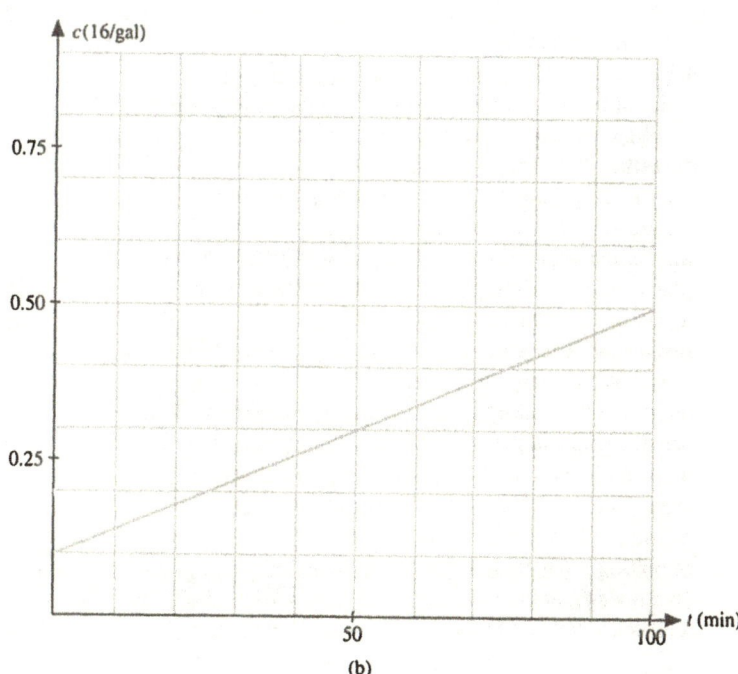

(b)

PROBLEMS: Section 2.4

Mixing Problems

1. **Standard Tank Problem** Initially, 50 pounds of salt are dissolved in a 300-gallon tank, and then a salt solution with a concentration of 2 pounds of salt per gallon of solution flows into a tank at a rate of 3 gal/min. The solution inside the tank is kept well-stirred and flows out of the tank at the same rate as that at which it flows in.
   (a) Find the amount of salt in the tank at any time.
   (b) Find the concentration of salt in the tank at any time.
   (c) Find the limiting amount of salt in the tank.
   (d) Find the limiting concentration of the salt in the tank.

2. **Standard Tank Problem** Initially, a 100-liter tank contains a salt solution that has a concentration of 0.5 kg/liter. A less concentrated salt solution with a concentration of 0.1 kg/liter flows into a tank at a rate of 4 liters/min. The solution inside the tank is kept well-stirred and flows out of the tank at the same rate as that at which it flows into the tank.
   (a) Find the amount of salt in the tank at any time.
   (b) Find the concentration of salt in the tank at any time.
   (c) Find the limiting amount of salt in the tank.
   (d) Find the limiting concentration of salt in the tank.

3. **Rate In > Rate Out** Initially, a large tank with a capacity of 100 gallons contains 50 gallons of pure water. A salt solution with a concentration of 0.1 lb/gal flows into the tank at a rate of 4 gal/min. The mixture is kept well-stirred and flows out of the tank at the rate of 2 gal/min.
   (a) Find the initial-value problem that describes the amount of salt in the tank.
   (b) Find the amount of salt in the tank until the tank overflows.
   (c) Find the concentration of salt in the tank until the tank overflows.
   (d) What is the initial-value problem that describes the amount of salt in the tank after the tank overflows?

4. **Rate In < Rate Out** A tank initially contains 100 gallons of brine whose salt concentration is 3 lb/gal. Fresh water is poured into the tank at a rate of 3 gal/min, and the well-stirred mixture flows out of the tank at the rate of 5 gal/min.

   (a) Find the initial-value problem that describes the amount of salt in the tank.

   (b) Find the amount of salt in the tank after time t.

   (c) If you have access to a computer with graphing capabilities, sketch the graph of the solution found in (b).

5. **Interesting Problem** Initially, 5 pounds of salt are dissolved in 20 gallons of water. A salt solution with a concentration of 2 lb/gal is added to the tank at a rate of 3 gal/min, and the well-stirred mixture is drained from the tank at the same rate. How long should this mixture be added in order to raise the amount of salt in the tank to 25 lb?

6. **Hmmmmm** A tank initially contains 200 gallons of fresh water, but thereafter an unknown salt concentration is poured into the tank at the rate of 2 gal/min. The mixture enters the tank and the well-stirred mixture flows out of the tank at the same rate. After 120 min the concentration of salt in the tank is 1.4 lb/gal. What is the concentration of the entering brine?

7. **Correcting a Mistake** Into a tank containing 100 gallons of fresh water, Julie was supposed to add 10 lb of salt but accidentally added 20 lb. To correct her mistake she started adding fresh water at a rate of 3 gal/min while drawing off salt solution from the tank at the same rate. How long will it take until the tank contains the correct amount of salt?

8. **Cleaning Up Lake Erie** Lake Erie has a volume of roughly 100 cubic miles whose inflow and outflow rates are the same at 40 cubic miles per year. Suppose at the year $t = 0$ a certain pollutant has a volume concentration of 0.05% but thereafter the concentration of pollutant flowing into the lake drops to 0.01%. Assuming that the pollutant leaving the lake is well mixed with lake water:

   (a) What is the initial-value problem that describes the volume V (cubic miles) of pollutant in the lake?

   (b) What is the volume V of pollutant in the lake at time t?

   (c) How long will it take to reduce the pollution concentration to 0.02% in volume?

9. **Cascading Tanks** Fresh water is poured into tank A whose volume is 100 gallons at a rate of 2 gal/min, which initially contains a salt solution with a salt concentration of 0.5 lb/gal. The stirred mixture flows out of tank A at the same rate and into tank B, which initially contains 100 gallons of fresh water. The mixture in tank B is also stirred and flows from tank B at the same rate.

   (a) Find the initial-value problem that describes the amount of salt in the first tank.

   (b) Find the amount of salt in the first tank at any time.

   (c) Find the initial-value problem that describes the amount of salt in the second tank.

   (d) Find the amount of salt in the second tank at any time.

10. **Many Cascading Tanks** A cascade of several tanks is shown in Figure 2.19. Initially, Tank 0 contains 1 gallon of alcohol and 1 gallon of water, while the other tanks contain 2 gallons of pure water. Fresh water is pumped into Tank 0 at the rate of 1 gal/min, and the varying mixture in each tank is pumped into the tank to its right at the same rate. Letting $x_n(t)$ denote the amount of alcohol in Tank n at time t:

    (a) Show that $x_0(t) = e^{-t/2}$.

    (b) Show by induction $x_n(t) = \dfrac{t^n\, e^{-t/2}}{n!\, 2^n}$ for $n = 1, 2, \ldots$.

    (c) Show the maximum value of $x_n(t)$ is $M_n = n^n\, e^{-n}/n!$

    (d) Use **Stirling's approximation n!** $\doteq \sqrt{2\pi n}\; n^n e^{-n}$ to show that $M_n \doteq (2\pi n)^{-1/2}$.

11. **Law of Mass Action** Chemists observe that in certain instances when two chemicals are mixed, a compound is formed at a rate that is proportional to the product of the weights of the unmixed parts of the two substances. In such cases the **law of mass action** applies, giving rise to what are called **second-order chemical reactions.** For instance, let a grams of substance A and b grams of substance B be mixed, forming a compound C, and suppose the newly formed compound C is formed from m parts by weight of A and n parts by weight of B. If the law of mass action applies, find the differential equation that describes the amount x of compound C produced. *Hint*: When x grams of C have been formed, show that $a - [m/(m + n)]x$ grams of A remain and $b - [n/(m + n)]x$ grams of B remain. See Figure 2.20.

12. **Drug Metabolism** The rate at which a drug is absorbed into the bloodstream is governed by the first-order differential equation

$$\frac{dC}{dt} = a - bC(t)$$

Figure 2.19
Cascading tanks

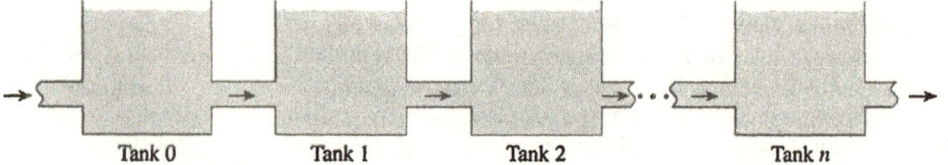

Tank 0   Tank 1   Tank 2   Tank n

Figure 2.20
Law of mass action giving rise to a second-order chemical reaction

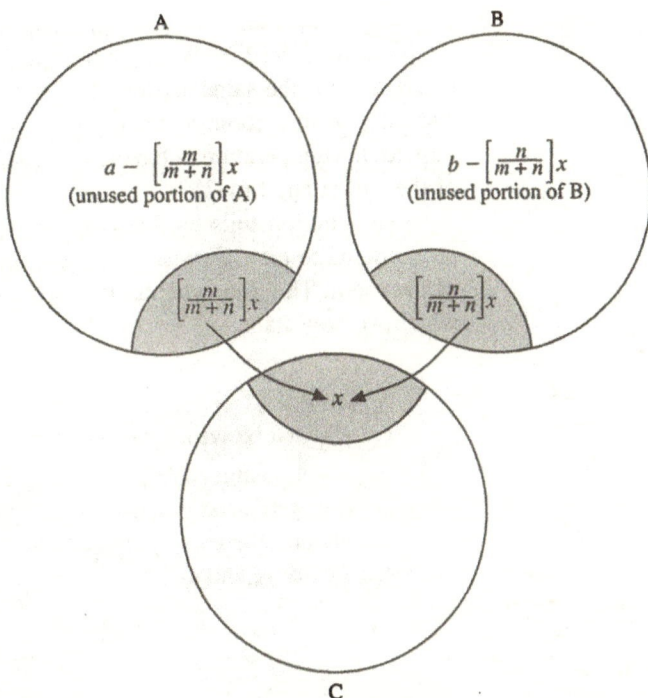

where a and b are constants and $C(t)$ denotes the concentration of drug in the bloodstream at time t. Assuming that no drug is initially present in the bloodstream, find the limiting concentration of drug in the bloodstream when $t \to \infty$, and determine how long it takes for the concentration to reach one-half this limiting value.

13. **Computer Problem** Initially, 50 pounds of salt are dissolved in a 300-gallon tank, and then a salt solution with a concentration of 2 pounds of salt per gallon of solution flows into a tank at a rate of 3 gal/min. The solution inside the tank is kept well-stirred and flows out of the tank at the same rate as that at which it flows in. Use a graphing calculator or a computer to sketch the graphs of the future amount of salt in the tank and the concentration of salt in the tank.

14. **Computer Problem** Use a graphing calculator or a computer to sketch the graphs of the functions $x_0(t)$, $x_1(t)$, $x_2(t)$, and $x_3(t)$ from Problem 10. Do the shapes of the curves agree with your intuition?

15. **Journal Entry** Since differential equations have so many applications in the physical sciences, could the topic be taught better by a physics, engineering, or chemistry professor? Are there any advantages in having a course in differential equations taught by a mathematics professor?

2.5   COOLING AND HEATING PHENOMENA

PURPOSE

To show how a certain first-order linear differential equation, called the **heating and cooling equation,** can be used as a predictor of temperature. The equation is a consequence of the physical principle known as **Newton's law of cooling.** We illustrate the basic ideas with examples.

NEWTON'S LAW OF COOLING

Suppose a steel ball is placed in a pan of boiling water so that the ball is heated to a temperature of 212°F. We assume that heat flows so fast through the ball that the temperature is the same at all points in the ball. We now take the ball out of the water and place it in a room whose temperature is a constant 70°F. Suppose that after 10 minutes the temperature of the ball has fallen to 150°F. How can we find the temperature of the ball at any time?

Our intuition tells us that the rate at which the temperature changes is directly proportional to the *difference* between the temperature of the ball and the temperature of the room. This fundamental law of physics is known as **Newton's law of cooling**, which we now state.

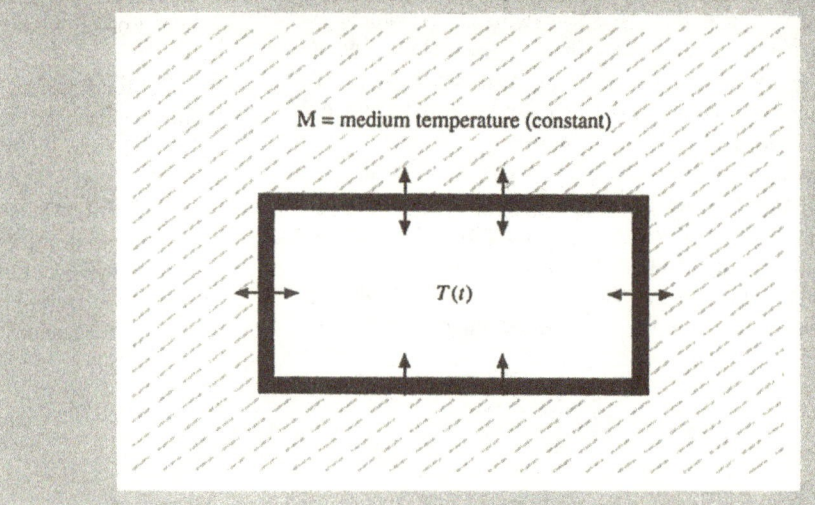

DEFINITION: Newton's Law of Cooling

The rate of change, dT/dt, in the temperature T of a body placed in a medium of temperature M is proportional to the difference between the temperature of the body and the temperature of the medium. This law is known as **Newton's law of cooling** and can be written mathematically as

$$\frac{dT}{dt} = -k(T - M)$$

where $k > 0$ is a constant of proportionality.

M = medium temperature (constant)

$T(t)$

Note that Newton's law of cooling says that the temperature of the object will *fall* (rise) if the temperature of an object is *greater than* (less than) the temperature of the surrounding medium.

CONSTANT MEDIUM TEMPERATURE

Consider an object that has an initial temperature of $T(0) = T_0$ that is placed in a medium of constant temperature M. The temperature $T(t)$ of the object satisfies the initial-value problem

$$\frac{dT}{dt} = -k(T - M)$$

$$T(0) = T_0 \tag{1}$$

This problem is easily solved by using the integrating factor method, getting

$$T(t) = T_0 e^{-kt} + M(1 - e^{-kt}) \tag{2}$$

Note that the temperature of the object $T(t)$ changes exponentially from the initial temperature T_0 of the object to the limiting temperature M. See Figure 2.21.

Figure 2.21
Temperature of an object with an initial temperature of T_0 placed in a room of temperature M

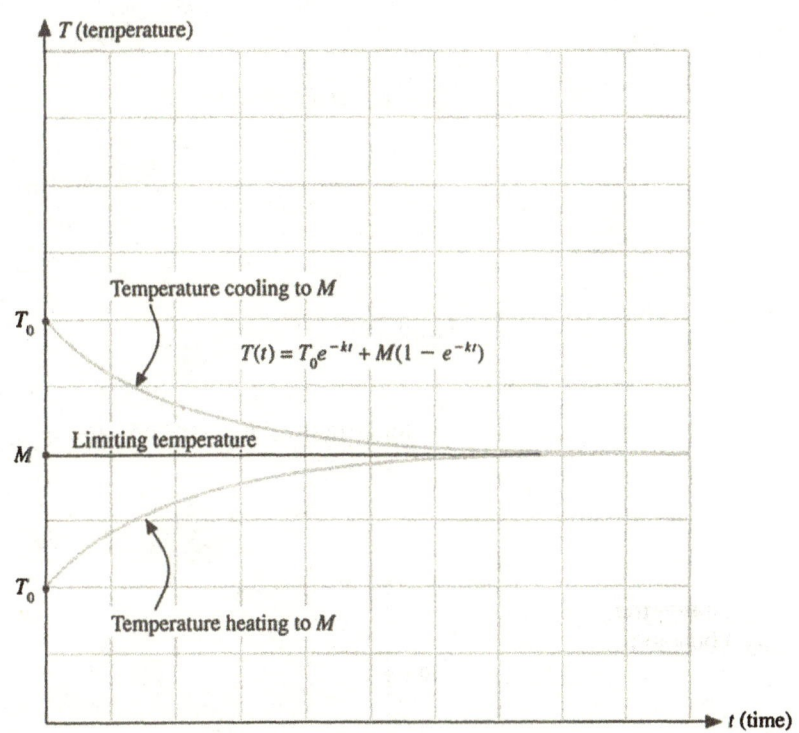

Example 1 **Constant Medium Temperature** At 12:00 midnight, with the temperature inside the mathematics building at 70°F and the outside temperature at 20°F, the furnace breaks down. Two hours later the temperature in the building has fallen to 50°F.

(a) Find the initial-value problem that describes the temperature inside the building for the remainder of the night. We assume that the outside temperature remains constant at 20°F.

(b) Determine the temperature in the building for the remainder of the night.

(c) Determine when the temperature in the building will fall to 40°F.

Solution   (a) The initial-value problem that describes T is

$$\frac{dT}{dt} = -k(T - 20)$$

$$T(0) = 70°F \tag{3}$$

(b) The solution of the initial-value problem is

$$T(t) = T_0 e^{-kt} + M(1 - e^{-kt})$$
$$= 70e^{-kt} + 20(1 - e^{-kt})$$
$$= 20 + 50e^{-kt} \tag{4}$$

To find k, substitute Eq. (4) into the given condition $T(2) = 50$, getting

$$50 = 20 + 50e^{-2k}$$

Solving for k gives

$$50e^{-2k} = 30$$
$$e^{-2k} = 0.60$$
$$-2k = \ln(0.60)$$
$$k = -\frac{\ln(0.60)}{2} \quad (\doteq 0.255) \tag{5}$$

Hence

$$T(t) = 20 + 50e^{-0.255t} \tag{6}$$

This temperature curve is shown in Figure 2.22.

Figure 2.22
Temperature inside the
mathematics building

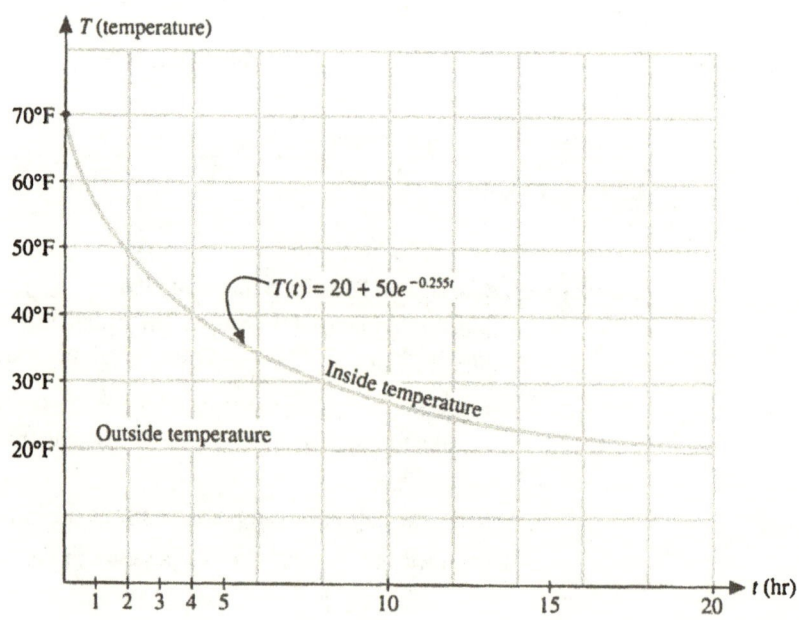

(c) To determine when the temperature falls to 40°F, we solve for t in the equation

$$20 + 50e^{-0.255t} = 40$$

This gives

$$e^{-0.255t} = 0.40$$

$$-0.255t = \ln(0.40) \doteq -0.916$$

$$t \doteq 3.592 \text{ hours} \quad (3 \text{ hr}, 36 \text{ min}) \tag{7}$$

Hence the temperature inside the building will fall to 40°F by about 3:36 A.M.   ∎

INTERPRETATION OF THE TIME CONSTANT 1/k

The reciprocal $1/k$ of the constant k that occurs in the Newton's law of cooling formula

$$\frac{dT}{dt} = -k(T - M) \tag{8}$$

has units of time (hours, minutes, seconds, etc.) and is called the *time constant* of the equation. Its value reflects how fast heat is transferred between the object in question and the surrounding medium. The larger (smaller) the time constant $1/k$, the faster (slower) the rate of heat transfer. To give meaning to $1/k$, the temperature $T(k)$ of an object is described by Eq. (8) with an initial temperature of T_0. The solution of this initial-value problem is

$$T(t) = T_0 e^{-kt} + M(1 - e^{-kt}) \tag{9}$$

If we evaluate Eq. (9) at $t = 1/k$, we find

$$T(1/k) = T_0 e^{-1} + M(1 - e^{-1})$$

$$= M + \frac{T_0 - M}{e}$$

$$\doteq M + \frac{1}{3}(T_0 - M) \tag{10}$$

Equation (10) now gives us an interpretation of the time constant $1/k$: The time constant $1/k$ is the time required for the temperature T to fall (roughly) two-thirds* of the way from the initial temperature T_0 to the final (limiting) temperature M. (See Figure 2.23.)

Example 2   **Use of the Time Constant 1/k**   John and Mary have just built a new house, and the builder informs them that the insulation in the house provides for a *time constant* of 5 hours. Suppose that at midnight the furnace fails with the outside temperature at a constant 10°F and the inside temperature at 70°F.

(a) Determine the initial-value problem that describes the future temperature inside the house.

---

* Note that in Eq. (10) we used the crude approximation that $1/e \doteq 1/3$.

(b) Determine the temperature inside the house for the remainder of the night.

(c) Determine the temperature inside the house at 5:00 A.M. Does your outcome from part (b) agree with the rule of thumb interpretation of the time constant?

Figure 2.23
The interpretation of the time constant $1/k$ as the length of time required for the temperature to fall approximately two-thirds of the way from the initial temperature to the limiting room temperature

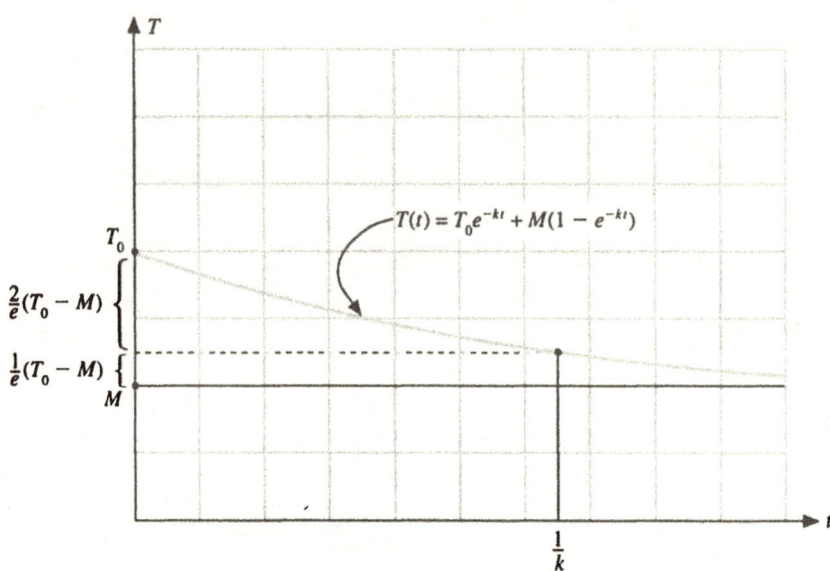

Solution   (a) Since the time constant is $1/k = 5$ hours, we have $k = 0.20$, and so the initial-value problem is

$$\frac{dT}{dt} = -0.20(T - 10)$$

$$T(0) = 70 \tag{11}$$

(b) The solution of the initial-value problem can easily be found by using the integrating factor method, which gives

$$T(t) = 70e^{-0.20t} + 10(1 - e^{-0.20t}) \tag{12}$$

(c) The temperature after one time period of $1/k = 5$ hours is

$$T(5) = 70e^{-1} + 10(1 - e^{-1})$$

$$\doteq 32°F \tag{13}$$

In other words, after 5 hours the temperature has fallen 38 degrees from a high of 70°F toward the low of 10°F. This drop of 38 degrees is pretty close to two-thirds of the total temperature drop of 60 degrees. See Figure 2.24.

Figure 2.24
Typical falling temperature curve after you shut off the heat

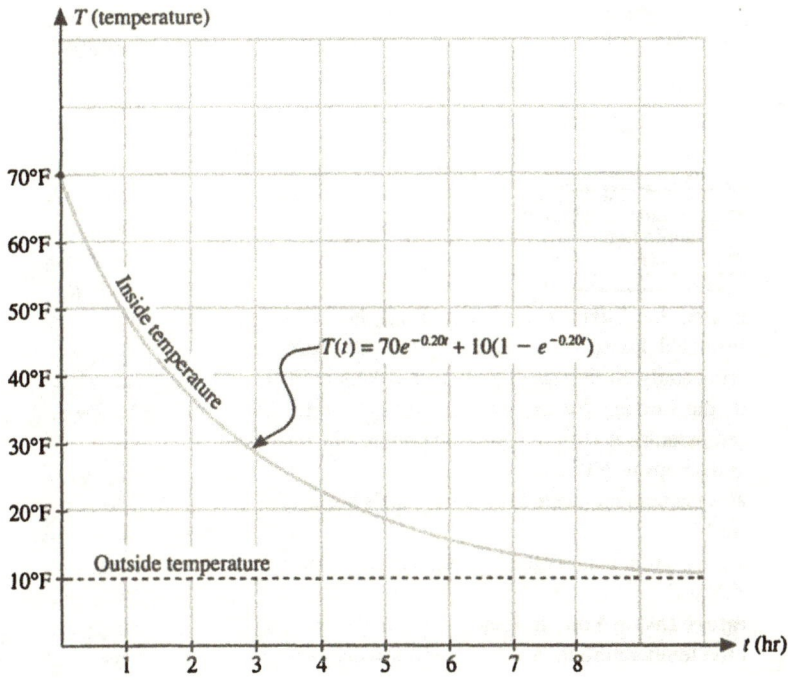

$$T(t) = 70e^{-0.20t} + 10(1 - e^{-0.20t})$$

Inside temperature

Outside temperature

PROBLEMS: Section 2.5

General Temperature Problems

1. **Solution of the Temperature Equation** Solve the heating and cooling problem

$$\frac{dT}{dt} = -k(T - M)$$

$$T(0) = T_0$$

when the temperature M of the medium is a constant.
   (a) Use the separation of variables method.
   (b) Use the integrating factor method.

2. **Interesting Observation** Introduce the new unknown variable $y(t)$ defined by $y(t) = T(t) - M$ in place of $T(t)$ in the equation

$$\frac{dT}{dt} = -k(T - M)$$

and find the differential equation for $y(t)$ in terms of t.

3. **Stephan's Law** A law that holds over greater temperature ranges than Newton's Law is **Stephan's Law,** which states that

$$\frac{dT}{dt} = -k(T^4 - M^4) \tag{14}$$

where T, M, and k are the same as defined in Newton's Law. Show that the general solution of this equation is

$$\ln\left(\frac{T + M}{T - M}\right) + 2\tan^{-1}\left(\frac{T}{M}\right) = 4M^3kt + c \tag{15}$$

where c is an arbitrary constant. *Hint*: Separate variables and then factor the fourth-order expression that occurs in the denominator into a product of two quadratic factors. Then find a partial fraction decomposition and integrate.

Constant Medium Temperature

4. **Using the Time Constant** At noon, with the temperature in Professor Snarf's office at 75°F and the outside temperature at 95°F, the air-conditioner breaks down. Suppose the time constant for Professor Snarf's office is 4 hours. See Figure 2.25.
   (a) Determine the temperature in Professor Snarf's office at 2:00 P.M.
   (b) Determine when the temperature in Professor Snarf's office will reach 80°F.

Figure 2.25

The heating of Professor Snarf's office

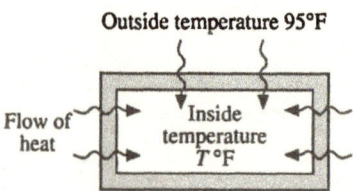

5. **Turning Off the Furnace** Professor Snarf has a very poorly insulated house and in winter keeps the furnace running constantly so that the temperature is kept at 70°F. However, the furnace breaks down at midnight with the outside temperature at 10°F, and after 30 minutes the inside temperature drops to 50°F.

   (a) What is the temperature in Professor Snarf's house after 1 hour?

   (b) How long will it take for the inside temperature to drop to 15°F?

6. **Temperature Inside Your Refrigerator** Here's how you can find the temperature in your refrigerator *without* actually putting a thermometer in the refrigerator. Take a can of soda from your refrigerator and let it warm for 0.5 hour and then record its temperature. Then let it warm for another 0.5 hour and take a second reading. Assume that you

know the room temperature and it is 70°F. What is the temperature in the refrigerator?

7. **Warm or Cold Beer?** A cold beer with an initial temperature of 35°F warms up to 40°F in 10 minutes while sitting in a room of temperature 70°F. What will be the temperature of the beer after *t* minutes? After 20 minutes?

8. **Case of the Cooling Corpse** In a murder investigation a corpse was found by Inspector Tousteau at exactly 8:00 P.M. Being alert, he measures the temperature of the body and finds it to be 70°F. Two hours later, Inspector Tousteau again measures the temperature of the corpse and finds it to be 60°F. If the room temperature is 50°F, when did the murder occur? See Figure 2.26.

9. **Professor Snarf's Coffee** Professor Snarf always has a cup of coffee before his 8:00 A.M. class. Suppose the temperature of the coffee is 200°F when it is freshly poured at 7:30 A.M. and 15 minutes later it cools to 120°F in a room whose temperature is 70°F. However, Professor Snarf never drinks his coffee until it cools to 90°F. When will Professor Snarf be able to drink his coffee?

10. **The Famous Coffee and Cream Problem** John and Mary are having dinner, and each orders a cup of coffee. John cools his coffee with some cream. They wait 10 minutes, and then Mary cools her coffee with the same amount of cream. The two then begin to drink. Who drinks the hotter coffee?

Figure 2.26

The case of the cooling corpse

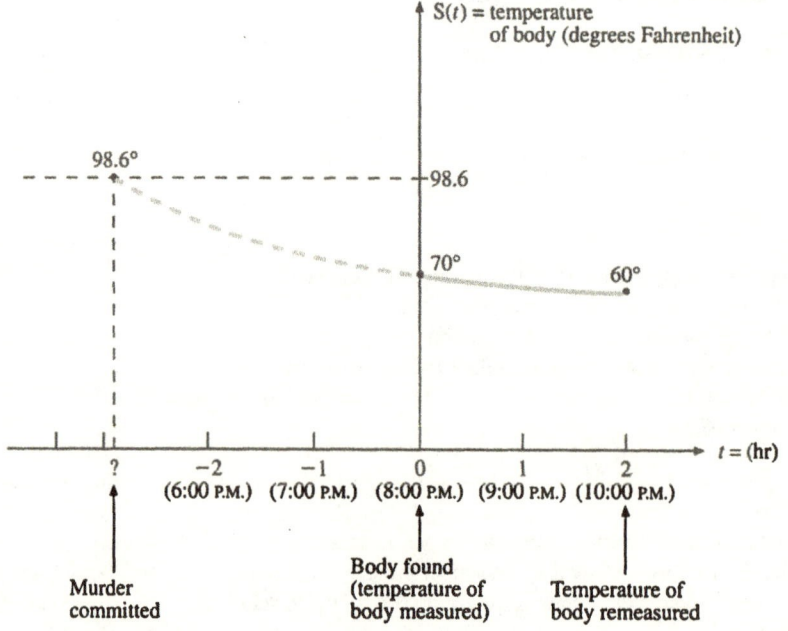

11. A Real Mystery At 1:00 P.M., Sally puts into a refrigerator a can of soda that has been sitting in a room of temperature 70°F. The temperature in the refrigerator is 40°F. Fifteen minutes later, at 1:15 P.M., the temperature of the soda has fallen to 60°F. At *some* later time, Sally removes the soda from the refrigerator to the room, where at 2:00 P.M. the temperature of the soda is 60°F. At what time did Mary remove the soda from the refrigerator?

Variable Medium Temperature

12. Cooling Corpse Revisited In Problem 8, Inspector Tousteau concluded that the time of the murder was 2.6 hours before he took the first temperature reading, or at 5:24 P.M. However, someone points out that Inspector Tousteau's analysis is faulty because the room temperature in which the corpse was found was not constant but decreased exponentially according to the law $50e^{-0.05t}$, where t is time in hours starting at 8:00 P.M. See Figure 2.27. Assume that the temperature falls according to this law.

(a) What is the initial-value problem that Inspector Tousteau must solve?
(b) What is the temperature of the body at any time t?
(c) When was the time of the murder?

13. Varying Weather Conditions A large building with a time constant $1/k = 1$ day has neither heating nor cooling. The outside temperature varies as a sine function, reaching a minimum of 40°F at 2:00 A.M. and a maximum of 90°F at 2:00 P.M.

(a) Find the initial-value problem for the temperature inside the building. *Hint*: Let t denote time in days with $t = 0$ starting at 8:00 A.M.
(b) Find the steady state solution of the differential equation found in part (a).
(c) How hot will it get inside the building?
(d) How cold will it get inside the building?

Figure 2.27
How does the temperature in the room affect the temperature of the corpse?

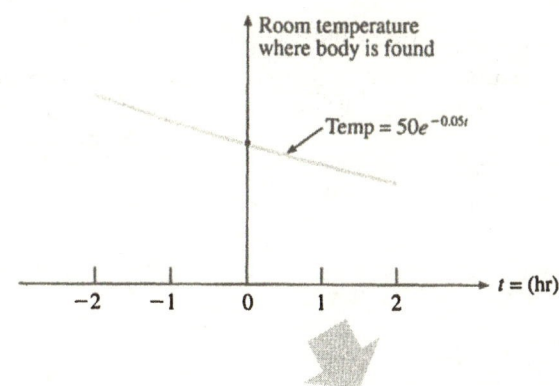

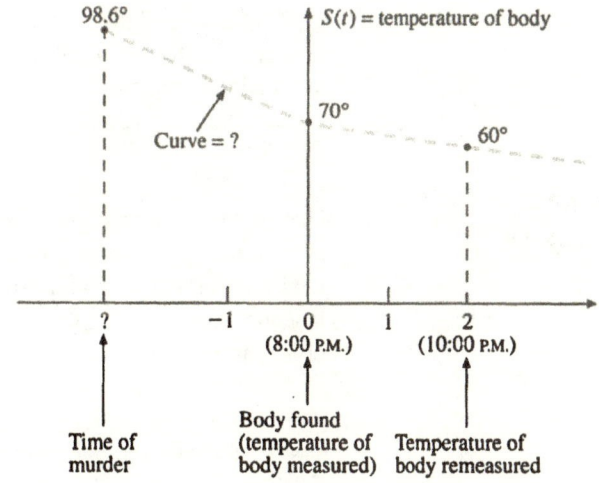

Temperature Problems with Heating and Cooling

We now solve temperature problems that have heating and cooling sources. Consider the equation

$$\frac{dT}{dt} = -k(T - M) + H(t) \tag{16}$$

where T, M, and k are defined as before and H(t) denotes either a heating or a cooling source, depending on whether it is positive (heating) or negative (cooling). The following problems illustrate this important equation.

14. **Solar Heat** A building has a solar heating system that consists of a solar panel and a hot water tank. The tank is well insulated and has a time constant of $1/k = 50$ hr. Under sunlight the energy generated by the solar panel will increase the water temperature in the tank at the *rate* of 2°F per hour, provided that there is no heat loss from the tank. Suppose at 9:00 A.M. the water temperature is 100°F and the room temperature where the tank is stored is a constant 70°F.
    (a) Find the initial-value problem that describes the temperature of the water in the tank.
    (b) Find the temperature of water in the tank at any time *t*.

(c) Find the temperature of the water in the tank after 8 hours of sunlight.

15. **A Good Air-Conditioner?** John and Mary install an air-conditioner in their house that has a given time constant of $1/k = 10$ hr. The air-conditioner will lower the temperature in the house at a rate of 4 deg/hr provided that no heat enters the house from the outside. Suppose the air-conditioner is turned on at 10:00 A.M. when the temperature in the house is 75°F and the outside temperature is a hot 100°F.
    (a) Determine the initial-value problem that describes the temperature inside the house.
    (b) Determine the future temperature inside the house.

16. **Computer Problem** Use a graphing calculator or a computer to sketch the graph of the future temperature of Professor Snarf's office discussed in Problem 4. Does your curve agree with your intuition? Try changing the parameters of the problem (initial temperature, time constant, outside temperature) and see how the temperature curves change.

17. **Journal Entry** Is Newton's law of cooling a mathematical theorem that can be proven, or is it just a law that everyone thinks is true because it tends to agree with physical observations? Does it make any difference? How would you go about proving such a law?

2.6   MORE APPLICATIONS

PURPOSE

This section (and the problems at the end of the section) presents more applications of first-order differential equations. The variety of examples and problems should convince the reader of the usefulness of differential equations in many areas of science.

ELEMENTARY MECHANICS

One of the most important uses of first-order differential equations lies in the area of mechanics. As a typical example, consider an object of mass m dropped from rest in a medium that offers a resistance proportional to the absolute magnitude of the instantaneous velocity $|v|$ of the object.* The goal is to find the position and velocity of the object at any time t.

---

* The absolute value of v is used because resistance is a function of the *speed* of the object. Normally, the assumption of a linear term in $|v|$ is reasonable only for small velocities. For larger velocities the resistance is often proportional to higher powers of $|v|$ or some function of $|v|$.

Figure 2.28
A mass falling in a resistive medium

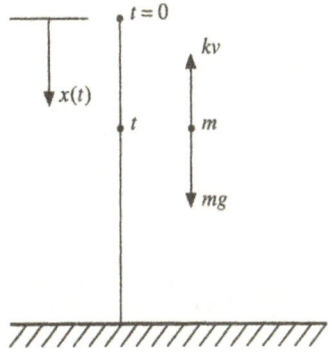

To solve this problem, it is convenient to draw an *x*-axis pointing downward with the origin taken as the point where the object is released. See Figure 2.28.

To determine the differential equation that describes the motion of the object, we use **Newton's second law of motion,** which states that the mass ($m > 0$) times the acceleration (dv/dt) of an object is equal to the sum of the forces acting on the object. Here there are two forces acting on the falling object: the downward force of gravity, *mg,* and the upward resistive force of the medium, $-kv$, where *k* is a positive constant, called the **coefficient of resistance,*** and $v(t)$ is the velocity of the object at time *t*. Using this notation, Newton's second law takes the form:

$$m \frac{dv}{dt} = mg - kv$$

or

$$\frac{dv}{dt} + \frac{k}{m}v = g \qquad (1)$$

Equation (1) is a first-order linear equation that has an integrating factor $e^{kt/m}$. The solution is found to be

$$v(t) = \frac{mg}{k} + ce^{-kt/m} \qquad (2)$$

By using the initial condition $v(0) = 0$, the constant of integration, *c*, is found to have the value $c = -mg/k$. Hence the velocity of the object is

$$v(t) = \frac{mg}{k}(1 - e^{-kt/m}) \qquad (3)$$

To determine the position $x(t)$ of the object, we replace *v* by dx/dt in Eq. (3) and integrate this equation, using the initial condition $x(0) = 0$. Doing this gives[†]

$$x(t) = \frac{mg}{k}t - \frac{m^2 g}{k}(1 - e^{-kt/m}) \qquad (4)$$

See Figure 2.29.

Note: Note that as $t \to \infty$, the velocity $v(t)$ approaches the value $v_L = mg/k$. This limiting velocity is directly proportional to the mass of the object and inversely proportional to the coefficient of resistance. Also, it is interesting to note that it does not depend on the initial conditions of the object.

---

* The value of *k* depends on the shape of the object and the nature of the medium the object travels through. For a person falling out of an airplane without a parachute, the coefficient of friction might be $k = 0.5$, whereas for a person with a parachute the coefficient of friction might be $k = 10$.

[†] In this problem, $x(t)$ represents the distance the object has fallen at time *t*.

Figure 2.29
Position of a falling object that weighs $mg = 100$ lb ($m = 100/32$ slugs) dropped from rest in different mediums

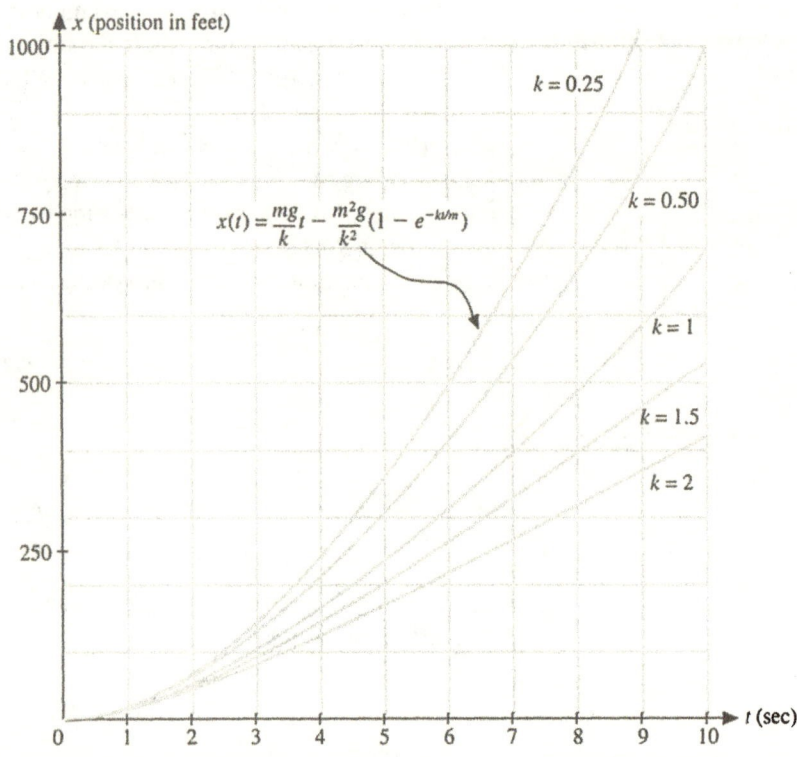

$$x(t) = \frac{mg}{k}t - \frac{m^2g}{k^2}(1 - e^{-kt/m})$$

THE SUBMARINE SEARCH PROBLEM

A destroyer spots an enemy submarine on the surface of the water d miles away. To avoid capture, the submarine dives and proceeds at full speed along a straight line. Although the destroyer knows that the submarine is traveling along some straight line, it doesn't know which line. The submarine search problem consists of finding a path the destroyer should follow that will *ensure* that it will pass *directly over* the submarine, assuming that the velocity v_d of the destroyer is greater than the velocity v_s of the submarine. The idea is that once the destroyer is directly over the submarine, it can detect it with regular sonar equipment.

To solve this problem, it is convenient to choose the origin at the point where the submarine was first spotted and to measure time from that instant. See Figure 2.30(a).

We first observe that since the submarine has set off along a straight line with constant velocity v_s, then after time t the submarine will lie *somewhere* on the circle of radius $v_s t$. See Figure 2.30(b). Hence if the destroyer follows a *spiral path* around the point where the submarine was first spotted so that its distance from that point is always $v_s t$, then at some time before one complete spiral is completed, it will lie directly over the submarine. Of course, before the destroyer can begin on this spiral path, it must first reach a point where its distance from the origin is the *same* as the submarine's distance from the origin. But this happens when the distance $v_d t$ the destroyer travels is equal to $d - v_s t$ or

$$v_d t = d - v_s t \tag{5}$$

Figure 2.30
What is the path of the destroyer that will ensure that it eventually lies directly over the submarine?

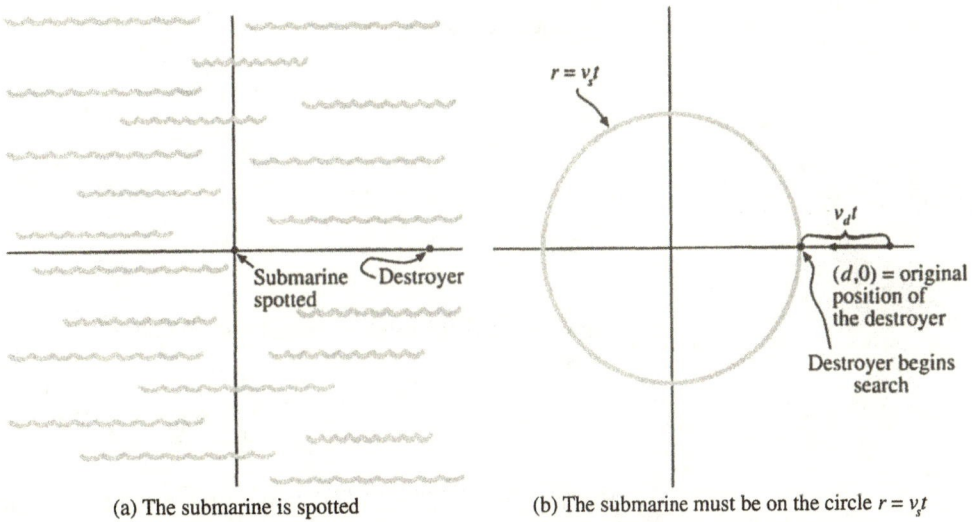

(a) The submarine is spotted   (b) The submarine must be on the circle $r = v_s t$

Solving Eq. (5) for t and calling the solution t_1, we have

$$t_1 = \frac{d}{v_d + v_s} \tag{6}$$

In other words, until time t_1 the destroyer continues cruising straight ahead in the direction in which the submarine was spotted. After time t_1 the destroyer begins its spiral search. The goal then is to find the spiral path, which has the general form in polar coordinates

$$x = r(t) \cos \theta(t)$$
$$y = r(t) \sin \theta(t) \tag{7}$$

To find $r(t)$ and $\theta(t)$, we use the fact that the velocity of the destroyer v_d can be written as

$$\sqrt{\left(\frac{dx}{dt}\right)^2 + \left(\frac{dy}{dt}\right)^2} = v_d \tag{8}$$

Substituting dx/dt and dy/dt determined from Eq. (7) into Eq. (8), we find after a little simplification

$$\sqrt{r^2 \left(\frac{d\theta}{dt}\right)^2 + \left(\frac{dr}{dt}\right)^2} = v_d \tag{9}$$

Since the destroyer must always lie on the same circle as the submarine, we can write

$$\frac{dr}{dt} = v_s \tag{10}$$

or

$$r = v_s t \tag{11}$$

Substituting these values into Eq. (9) and solving for $d\theta/dt$, we have

$$\frac{d\theta}{dt} = \left(\frac{v_d^2 - v_s^2}{v_s^2}\right)\frac{1}{t} \tag{12}$$

Solving this differential equation with the initial condition $\theta(t_1) = 0$, we find

$$\theta(t) = \sqrt{\frac{v_d^2 - v_s^2}{v_s^2}} \ln\left(\frac{t}{t_1}\right) \qquad (t \geq t_1) \tag{13}$$

Finally, substituting the values of $r(t)$ from Eq. (11) and $\theta(t)$ from Eq. (13) into the general spiral equation (1), we find the spiral path

$$x(t) = v_s t \cos\left(\sqrt{\frac{v_d^2 - v_s^2}{v_s^2}} \ln\left(\frac{t}{t_1}\right)\right)$$

$$\qquad (t \geq t_1) \tag{14}$$

$$y(t) = v_s t \sin\left(\sqrt{\frac{v_d^2 - v_s^2}{v_s^2}} \ln\left(\frac{t}{t_1}\right)\right)$$

where $t_1 = d/(v_d + v_s)$. See Figure 2.31.

It is also useful to determine the polar form $r = r(\theta)$ of the spiral path that will not depend on t. This can be done by making the change in variables

$$\frac{d\theta}{dt} = \frac{d\theta}{dr}\frac{dr}{dt} = v_s \frac{d\theta}{dr} \tag{15}$$

and rewrite Eq. (13) in terms of r and θ. Substituting $r = v_s t$ and $d\theta/dt$ found from Eq. (14) into Eq. (13), we arrive at

$$v_s \frac{d\theta}{dr} = \sqrt{\frac{v_d^2 - v_s^2}{v_s^2}}\left(\frac{v_s}{r}\right) \tag{16}$$

Figure 2.31
The path that will ensure that
the destroyer will eventually
lie directly over the submarine

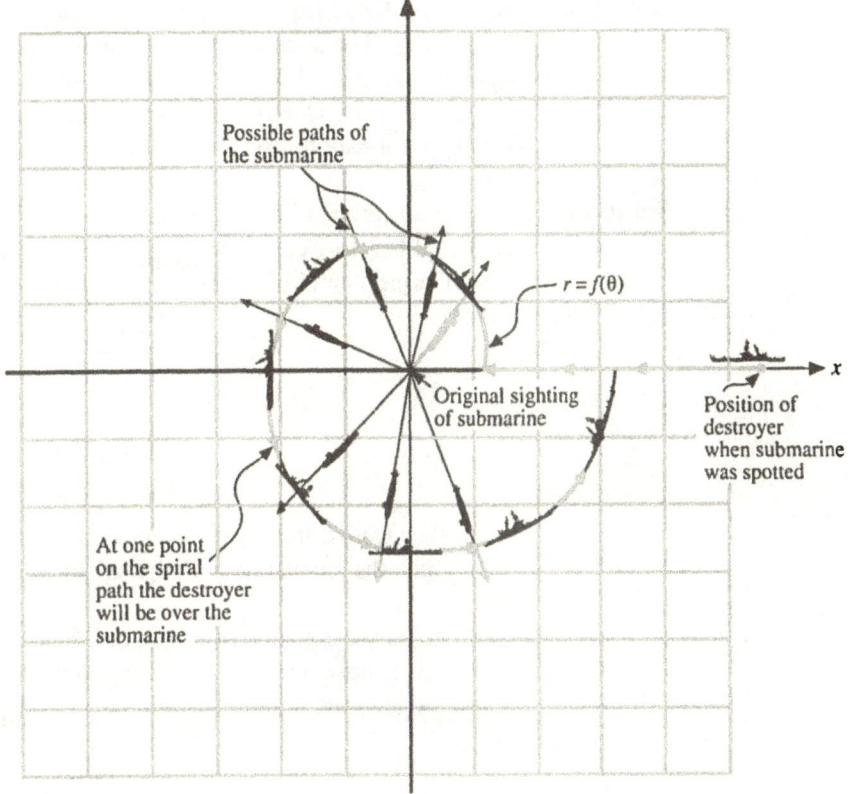

Solving this differential equation for $\theta(r)$ with initial condition $\theta(v_s t_1) = 0$ gives

$$\theta(r) = \sqrt{\frac{v_d^2 - v_s^2}{v_s^2}} \ln\left(\frac{r}{v_s t_1}\right) \tag{17}$$

Finally, solving this equation for r in terms of θ, we get the *exponential* spiral

$$r = v_s t_1\, e^{k\theta} \tag{18}$$

where

$$k = \sqrt{\frac{v_s^2}{v_d^2 - v_s^2}}$$

Note: We have determined the *path* that the destroyer should follow; we have not determined *when* the destroyer will lie above the submarine. Our only claim is that sometime before one complete revolution of the spiral, the destroyer will lie directly over the submarine, nothing more. Hopefully, the destroyer's sonar equipment will be able to detect when it lies directly over the submarine.

Example 1 **Submarine Chasing** A destroyer's radar system spots a submarine on the surface 6 miles away. The submarine immediately dives and heads off at full speed in some unknown direction. The captain of the destroyer knows that the top speed of the submarine is $v_s = 15$ mph and that the destroyer has a top velocity of $v_d = 45$ mph. With this information, what is the captain's strategy for catching the submarine?

Solution We solve this problem in two steps.

Step 1 (When to Begin the Spiral Search). We first compute the time t_1 when the submarine should begin its spiral search. We have

$$t_1 = \frac{d}{v_s + v_d}$$
$$= \frac{6}{15 + 45}$$
$$= 0.10 \text{ hour} \quad (6 \text{ minutes}) \tag{19}$$

In other words, the captain continues steaming under full power in the direction in which the submarine was sighted for 6 minutes. At that time the spiral search will begin.

Step 2 (Find the Equation of the Spiral). Since we solved the general problem earlier, we now simply substitute v_s, v_d, and t_1 into the general spiral equation (11). Doing this, we find the desired position (x, y) of the destroyer after t hours to be

$$x(t) = v_s t \cos\left(\sqrt{\frac{v_d^2 - v_s^2}{v_s^2}} \ln\left(\frac{t}{t_1}\right)\right)$$
$$= 15t \cos\left(\sqrt{8} \ln(10t)\right) \qquad (t \geq 0.10 \text{ hr}) \tag{20a}$$

$$y(t) = v_s t \sin\left(\sqrt{\frac{v_d^2 - v_s^2}{v_s^2}} \ln\left(\frac{t}{t_1}\right)\right)$$
$$= 15t \sin\left(\sqrt{8} \ln(10t)\right) \qquad (t \geq 0.10 \text{ hr}) \tag{20b}$$

The graph of this spiral is shown in Figure 2.32.

We can also find r in terms of θ by using Eq. (18). We get

$$r = v_s t_1 e^{k\theta}$$
$$= 1.5 e^{\theta/\sqrt{8}} \qquad (0 \leq \theta) \tag{21}$$

Note: Of course, if the spiral "opens out" very fast, it will take the destroyer longer to complete one complete revolution, and the submarine may make it to a safe harbor before being caught. This happens when the speed of the destroyer is only slightly greater than that of the submarine. Realize, too, that the captain of the submarine does not have to follow a straight line course or travel at full speed and may use a more sophisticated avoidance strategy. ∎

Figure 2.32
By following this spiral path
the destroyer can be sure that
it will lie over the submarine
once during the first
revolution of the spiral

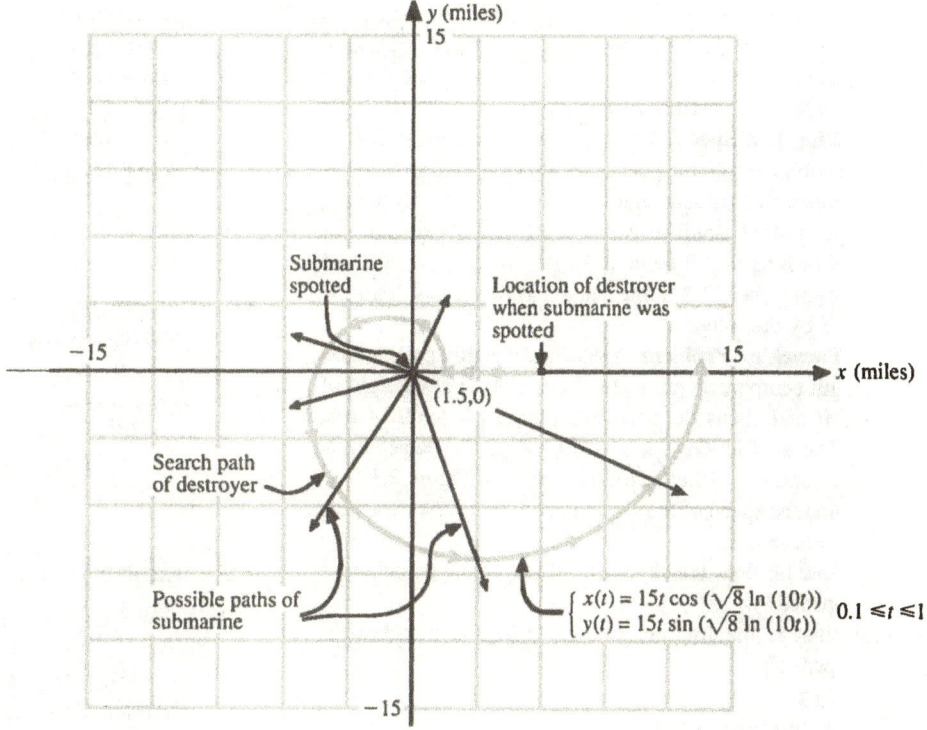

PROBLEMS: Section 2.6

Exact Differential Equations

If a derivative dy/dx is interepreted as the ratio of two differentials, dy divided by dx, then any first-order differential equation can be written in differential form

$$M(x, y) \, dx + N(x, y) \, dy = 0 \qquad (22)$$

If ∂M/∂y = ∂N/∂x is satisfied, then Eq. (22) is called an exact differential equation. Exact differential equations can often be solved by inspection by rewriting the equation as d(g) = 0, where d is a differential of some algebraic expression g. If g can be identified, then the solution of the differential equation is simply g = c. For problems 1–10, verify that the given equation is exact, identify the function g, and find the solution.

1. $y \, dx + x \, dy = 0$
2. $2xy \, dx + x^2 \, dy = 0$
3. $y^2 \, dx + 2xy \, dy = 0$
4. $2xy^2 \, dx + 2x^2y \, dy = 0$
5. $\sin y \, dx + x \cos y \, dy = 0$
6. $e^x \sin y \, dx + e^x \cos y \, dy = 0$
7. $ye^{xy} \, dx + xe^{xy} \, dy = 0$
8. $\dfrac{y \, dx - x \, dy}{y^2} = 0$
9. $\dfrac{x \, dx + y \, dy}{x^2 + y^2} = 0$
10. $\dfrac{x \, dy - y \, dx}{x^2 + y^2} = 0$

Motion in a Resistive Medium

11. **Boston's Grape-Catching Man*** According to an article in the *Boston Globe*, a man named Paul Tavilla stood at the base of Boston's 60-story John Hancock Tower, his neck bent back, mouth open, as grapes were dropped from the observation tower, 600 ft above. On his 70th try, he caught one in his mouth as a crowd of 150 spectators burst into applause. "Catching grapes is no picnic," Tavilla insists. "If you don't know what you're doing, you could

---

* From the *Boston Globe*, 1990.

put an eye out.'' If a grape weighs $mg = 0.02$ lb and the air friction is proportional to the speed of the grape with a coefficient of resistance 0.00002, what is the speed of the grape when Mr. Tavilla catches it?

12. **Reaching Top Speed** A skydiver weighing $mg = 200$ lb (with equipment) jumps out of an airplane. Before the skydiver opens the parachute, the air friction is proportional to the speed of the skydiver with a coefficient of resistance of 0.5. How long will it be until the skydiver reaches 90% of the limiting speed? Assume that the skydiver hasn't hit the ground by that time.

13. **The Parachute Problem** A skydiver weighing $mg = 200$ lb (with equipment) jumps from an airplane at a height of 2500 ft and opens the parachute after 10 seconds of free fall. The air friction is $k = 0.50$ before the parachute is opened and $k = 10$ after it is opened. See Figure 2.33.
    (a) Find the speed of the skydiver at the time the parachute is opened.
    (b) How far does the skydiver fall before the parachute is opened?
    (c) What is the limiting velocity after the parachute is opened?

Figure 2.33
The parachute problem

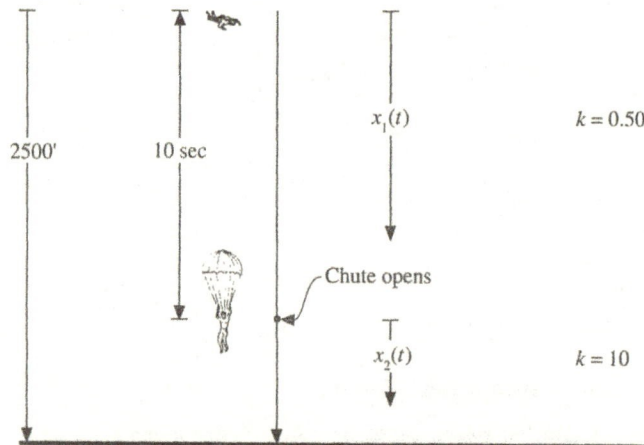

14. **The Flight of the Parachutist** A parachutist jumps from an airplane with the parachute closed. With k near zero the parachute reaches a certain limiting velocity. When the parachute opens, the resistance factor k jumps to a much larger value. Sketch a rough graph that illustrates the general nature of the velocity $v(t)$ of the falling parachutist.

15. **Hmmmmm** Two parachutists jump from an airplane flying at a great height. One parachutist carries extra equipment resulting in a total weight that is twice that of the other parachutist. Assuming that the wind resistance is the same for both parachutists, compare the descents of the two parachutists.

16. **Maximum Height** An object is thrown upward with initial velocity v_0. If air resistance is proportional to its instantaneous velocity, the constant of proportionality being k, show that the maximum height of the object is
$$M = \frac{mv_0}{k} - \frac{m^2 g}{k^2} \ln\left(1 + \frac{kv_0}{mg}\right)$$

17. **Maximum Speed of a Ship** A 42,000-ton (1 ton = 2000 lb) freighter is powered by propellers exerting a force of 120,000 lb. Assume that the ship starts from rest and that resistance through the water is $7000v$, where v is the velocity of the ship.
    (a) What is the velocity of the ship at time t?
    (b) Find the limiting velocity of the ship.

Submarine Search Problems

18. **Submarine Search** A destroyer spots a submarine on the surface of the water 4 miles away. The submarine dives and heads off in a straight line. Suppose the destroyer's maximum speed is 30 mph and the submarine has a top speed of 10 mph.
    (a) Find the time when the destroyer begins its spiral path.
    (b) Find the equation of the spiral path in terms of time.
    (c) Find the equation of the spiral path in terms of θ.

19. **Submarine Search Project** If you have access to a computer with graphing capabilities, draw the spiral search path for the given values of v_s, v_d, and d. One must first compute t_1, the time when the spiral search begins, and then draw the spiral starting from $t \geq t_1$.
    (a) $v_s = 10$ mph, $v_d = 50$ mph, $d = 4$ miles
    (b) $v_s = 10$ mph, $v_d = 30$ mph, $d = 4$ miles
    (c) $v_s = 10$ mph, $v_d = 15$ mph, $d = 4$ miles
    (d) $v_s = 10$ mph, $v_d = 12$ mph, $d = 4$ miles

Torricelli's Equation

*Torricelli's equation describes the rate at which the level of a fluid drops from a leaking tank. More specifically, if a tank has a hole with the area a at its bottom and if $A(y)$ denotes the horizontal cross-sectional area $A(y)$ of the tank at depth y, then the rate dy/dt at which the liquid drops is given by **Torricelli's equation:***
$$A(y)\frac{dy}{dt} = -a\sqrt{2gy} \tag{23}$$
where g is acceleration due to gravity. See Figure 2.34. Problems 20–21 deal with Torricelli's equation.

Figure 2.34
Illustration of Torricelli's
equation

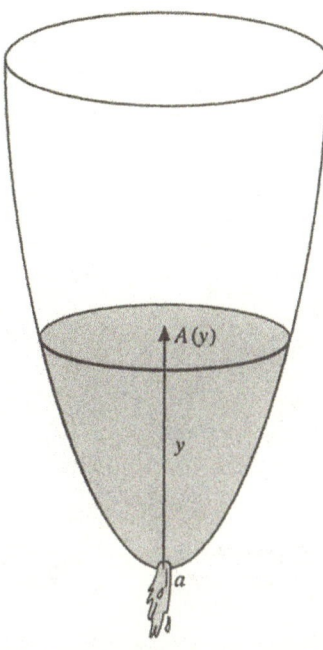

20. **Draining Water from a Hemispherical Tank** A hemi-spherical tank as shown in Figure 2.35 with a radius of R is initially full of water. A leak is formed when a circular hole of radius r_0 is punctured at the bottom of the tank.

Figure 2.35
Draining water from a
hemispherical tank

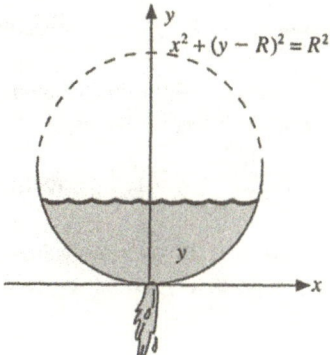

(a) Show that the differential equation that describes the height y of the water in the tank is

$$\frac{dy}{dt} = -\frac{\pi r_0^2 \sqrt{2gy}}{\pi x^2}$$

(b) From the relationship $x^2 = R^2 - (y - R)^2 = 2yR - y^2$, show that the implicit solution of this equation can be found by separation of variables to be

$$\frac{4}{3}Ry^{3/2} - \frac{2}{5}y^{5/2} = -r_0^2\sqrt{2gt} + c$$

(c) Use the condition $y = R$ when $t = 0$ to find the constant of integration c.

(d) Set $y = 0$ and solve for t in the equation found in part (b) to determine the time it takes to drain all the water from the tank to be

$$t = \frac{14}{15}\frac{R^{5/2}}{r_0^2\sqrt{2g}}$$

(e) How long will it take to drain a hemispherical tank of radius 1 ft if a 1-inch hole is punctured in the bottom?

21. **Making Your Own Clepsydra** Jane is making a water clock or "clepsydra," as it was known in ancient times. The goal is to build a tank so that the water drops at a constant rate. Show that the cross-sectional area $A(h)$ of the desired tank at height h above the bottom of the tank should be proportional to $\sqrt{h}$. Use this fact to determine the exact shape $y = f(x)$ of Jane's water clock with overall dimensions shown in Figure 2.36.

Figure 2.36
The clepsydra

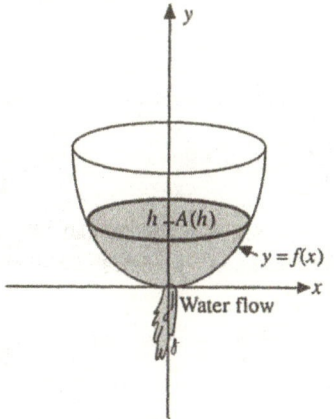

Pursuit Problems

22. **Dog and Rabbit Problem** A rabbit starting at the origin is chased up the y-axis by a dog that starts at $(1, 0)$. See Figure 2.37. The velocity of the dog, v_d, is assumed to be greater than the velocity of the rabbit, v_r. To find the *pursuit path* of the dog, solve the following problems.

Figure 2.37
The dog and rabbit pursuit
problem

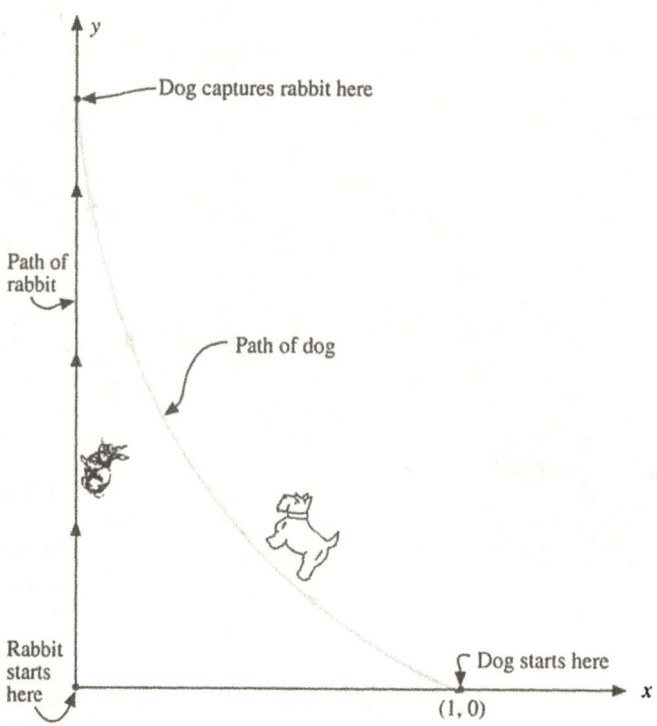

(a) Since the dog is always running directly toward the rabbit, show that the tangent to the dog's pursuit path $y = y(x)$ will be

$$\frac{dy}{dx} = \frac{y - v_r t}{x} \quad \text{or} \quad xy' = y - v_r t \quad (24a)$$

(b) Differentiate Eq. (24a) with respect to x to get

$$x\frac{d^2y}{dx^2} = -v_r\frac{dt}{dx} \quad (24b)$$

(c) Calling s the length along the dog's path, we have the relation $ds/dt = v_d$. Hence making a change of variables, we get

$$\frac{dt}{dx} = \frac{dt}{ds}\frac{ds}{dx} = -\frac{1}{v_d}\sqrt{1 + [y'(x)]^2} \quad (24c)$$

Show that the differential equation for the pursuit curve $y(x)$ in terms of x is

$$xy'' = \frac{v_r}{v_d}\sqrt{1 + [y'(x)]^2} \quad (24d)$$

(d) Letting $v = dy/dx$, show that Eq. (24d) becomes

$$x\frac{dv}{dx} = \frac{v_r}{v_d}\sqrt{1 + v^2} \quad (24e)$$

(e) Solve the differential equation (24e) for $v(x)$ with the initial condition $v(a) = 0$, getting

$$\ln(v + \sqrt{1 + v^2}) = \ln x^k \quad (24f)$$

where $k = v_r/v_d$.

(f) Solve Eq. (24f) for $v = dy/dx$ in terms of x, getting

$$\frac{dy}{dx} = \frac{1}{2}(x^k - x^{-k}) \quad (24g)$$

(g) Realizing that $k < 1$, integrate Eq. (24g) with the initial condition $y(1) = 0$, getting the pursuit curve

$$y(x) = \frac{x}{2}\left(\frac{x^k}{k + 1} - \frac{x^{-k}}{k - 1}\right) \quad (24h)$$

(h) Plot this pursuit curve for $0 \le x \le 1$ for various values of (v_r, v_d), and a, using a computer.

Interesting Problems

23. Digging to the South Indian Ocean Contrary to popular belief, a person in the United States who digs a hole through the center of the earth wouldn't arrive in China but thousands of miles away in the South Indian Ocean. In any case, if an object were dropped down a hole through the center of the earth, it would be attracted toward the center with a force directly proportional to the distance from the center. We assume that there is no air friction and that the radius of the earth is 4000 miles.

(a) Find the velocity of the ball when it passes through the center of the earth.

(b) Find the time it takes the ball to reach the other end of the hole.

Hint: Newton's universal law of attraction gives the differential equation for the velocity v of the ball as

$$\frac{dv}{dt} = -\frac{-gr}{R}$$

where

R = radius of the earth (4000 miles)

g = acceleration due to gravity (32.2 ft/sec^2)

r = distance of the ball from the center of the earth

Since we wish to find v as a function of r, we make the change in variable

$$dv/dt = (dv/dr)(dr/dt) = v\, dv/dr$$

See Figure 2.38.

Figure 2.38
Dropping a ball through the center of the earth

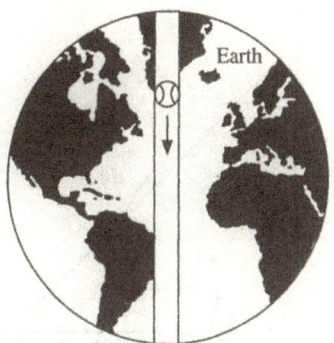

24. **Interesting Property** Can you find a curve passing through the point (0, 1) that has the property that the length of any segment of its graph $y = y(x)$ is equal to the area above the x-axis and under that section of graph? See Figure 2.39. *Hint*: The length of the graph of a function $y = y(x)$ for x between a fixed abscissa a and a variable x is

$$L = \int_a^x \sqrt{1 + y'^2\,(s)}\, ds \qquad (25)$$

Figure 2.39
A curve whose length is equal to the area under the curve

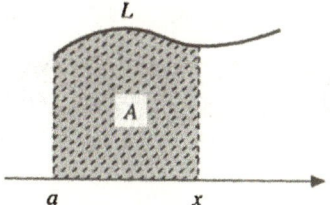

25. **Curves of Constant Curvature** In calculus, one learns that the radius of curvature of a curve is given by

$$R = \frac{[1 + (y')^2]^{3/2}}{|y''|} \qquad (26)$$

Solve this differential equation when R is a constant to show that the family of curves that have constant curvature is the set of all circles in the xy-plane.

26. **The Path Across the River** A boat crosses a river from a point located at the point (1, 0), always heading toward the origin. The boat has a velocity v_b relative to the water, and the current of the river has a uniform velocity v_r in the negative y-direction. See Figure 2.40.

Figure 2.40
Finding the path of a boat crossing a river or an airplane flying in a wind

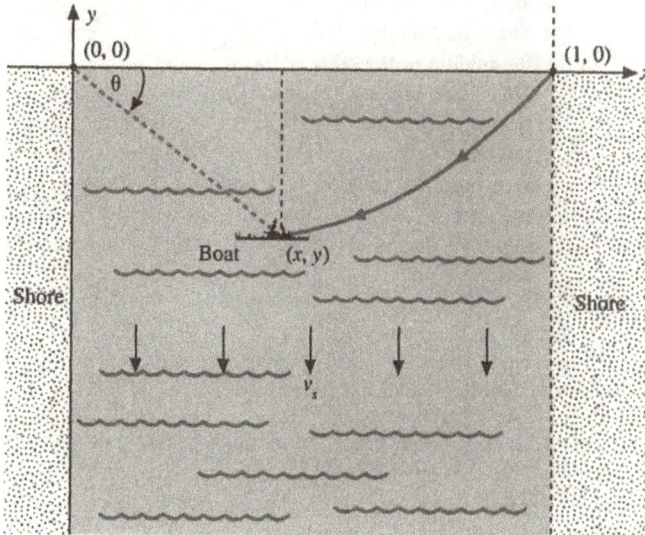

(a) Show that the equations that describe the components of the boat's velocity are

$$\frac{dx}{dt} = -v_b \cos \theta \qquad \frac{dy}{dt} = -v_r + v_b \sin \theta \quad (27)$$

where θ is the angle between the boat and the x-axis.

(b) Show that equations that describe the path $y = y(x)$ of the boat are

$$\frac{dy}{dx} = \frac{v_r \sqrt{x^2 + y^2} + v_b y}{v_b x} \qquad y(1) = 0 \quad (28)$$

(c) Solve the differential equation in part (b) by transforming to the new dependent variable $v = y/x$ ($y = xv$, $y' = v + xv'$) and finding the implicit solution of the resulting separable equation for $v(x)$. Then use the relationship $y = xv$ to get

$$y + \sqrt{x^2 + y^2} = x^{k+1}$$

where $k = v_r/v_b$.

(d) Solve the equation in part (c) for y, getting

$$y = \frac{x}{2}(x^k - x^{-k})$$

(e) Using the equation for the path $y = y(x)$ in part (d), under what conditions on v_r and v_b will the boat actually reach the origin? If you have access to a computer with graphing facilities, sketch the graph of the path for different values of k.

27. **Four-Bug Problem** Four bugs sit at the corners of a square table, each side having length $L = 100$ inches. Simultaneously, each starts walking at the same rate of 1 inch per second toward the bug on its right. See Figure 2.41(a).

(a) Show that the time elapsed before the bugs collide at the middle of the table is 100 seconds.

Hint: Each bug always moves in a perpendicular direction from the line of sight of the bug behind it. Hence the rate of change of the distance between two successive bugs will always be decreasing at the rate of 1 inch per second. Note that the bugs will always lie on a square that is shrinking and rotating counterclockwise.*

* In general, if n bugs are initially located at the vertices of a polygon with n corners, then the time it takes the bugs to collide is given by $L/[r(1 + \cos \alpha)]$, where L is the length of each side of the polygon, r is the speed the bugs travel, and α is the interior angle of the polygon ($\alpha = \pi/2$ for a square, $\alpha = \pi/3$ for a triangle, and so on). The path of the bug that starts at $(r, \theta) = (1, 0)$ is the logarithmic spiral defined by $r = \exp[-(1 + \cos \alpha)\theta]$.

Figure 2.41
The bug problem (or the famous colliding airplanes problem)

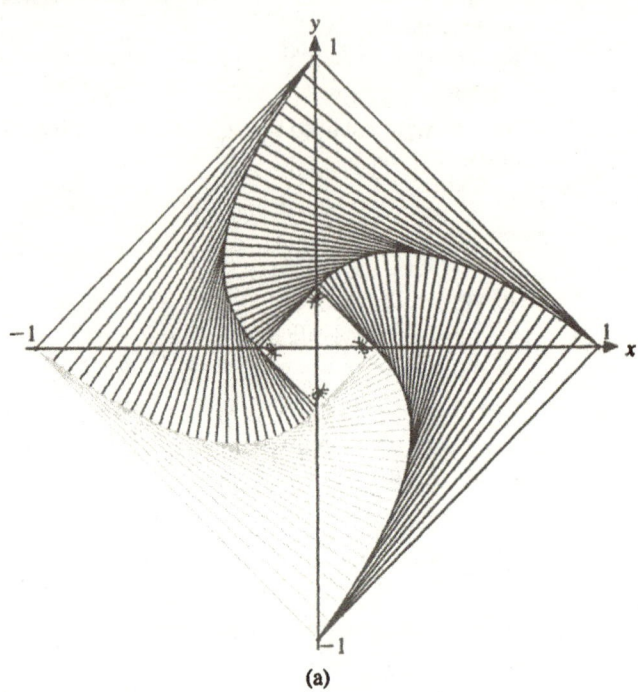

(a)

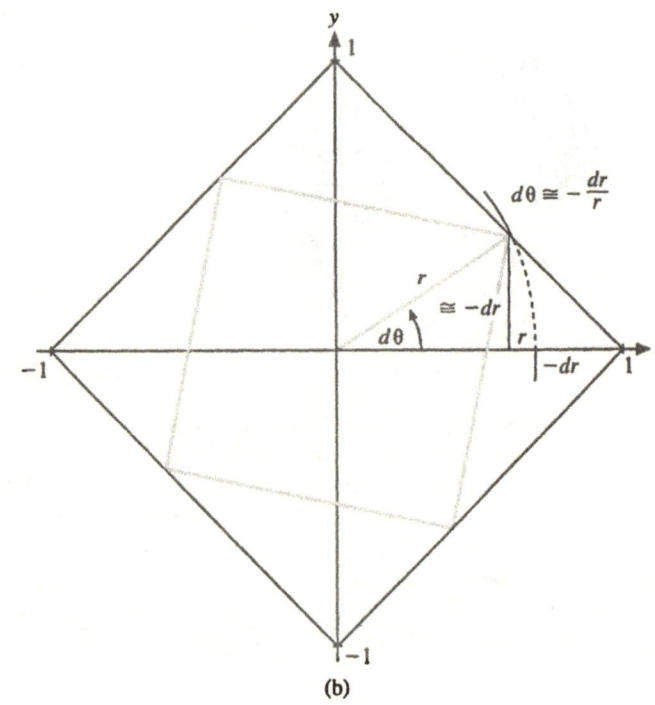

(b)

(b) Using the result from part (a), how *far* will each bug travel? *Hint*: This is really easy—no calculus please.

(c) Here's the only part that uses differential equations. Find the paths of the bugs. *Hint*: Let the bugs start at the four points $(\pm 1, 0)$, $(0, \pm 1)$ and find the *path* of the bug that starts at $(1, 0)$. Also use Figure 2.41(b) to show the relationship $dr = -r\, d\theta$.

28. **Can You Fill This Tank?** A conical tank 15 ft deep with an open top has a radius of 15 ft. See Figure 2.42. Initially the tank is empty, but water is added at a rate of π ft^3/hr. Water evaporates from the tank at a rate proportional to the surface area, the constant of proportionality being 0.01.

(a) Show that the differential equation that describes the volume V of water in the tank is

$$\frac{dV}{dt} = \pi - 0.01\pi h^2$$

(b) Use the relation $V = \frac{1}{3}\pi h^3$, where h is the depth of the water, to drive the differential equation

$$\frac{dh}{dt} = \frac{1 - 0.01h^3}{h^2}$$

(c) Use the fact that the equation in part (b) is separable and $h(0) = 0$ to show that the implicit solution is

$$500 \ln\left(\frac{1 + 0.1h}{1 - 0.1h}\right) - 100h = t$$

(d) Using the equation in part (*c*), will the tank ever fill?

Figure 2.42
There are many situations, such as reservoirs behind dams, that are never completely filled owing to a large surface area resulting in a high evaporation rate.

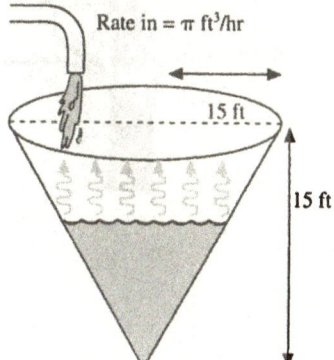

Rate in $= \pi$ ft^3/hr

15 ft

15 ft

29. **The Art of Substitution** It is sometimes possible to solve difficult differential equations by changing variables, either the independent or dependent variable or both. Generally, there is no guarantee in advance that a given substitution will work, but if the equation is important, then a trial-and-error approach may be worthwhile. Consider the second-order equation

$$2xy\frac{dy}{dx} = 1 + y^2 - x^2 \qquad (x > 0)$$

(a) Let $v = y^2$ ($y = v^{1/2}$) and transform the equation to

$$x\frac{dv}{dx} = 1 + v - x^2$$

(b) Let $u = x^2$ ($x = u^{1/2}$) and transform the equation in part (a) to

$$2u\frac{dv}{du} = 1 + v - u$$

(c) Let $w = 1 + v$ ($v = w - 1$) and transform the equation in part (b) to

$$2u\frac{dw}{du} = w - u$$

(d) Solve the equation in part (c) for $w = w(u)$ and then use the previous substitutions to find in successive order: $v = v(u)$, $v = v(x)$, and finally an implicit relationship $f(x, y) = 0$. Do you recognize the solution?

30. **Optional Control Theory** Harry is growing a plant in his apartment where the height, $x(t)$, of the plant is described by the differential equation

$$\frac{dx}{dt} = 1 + u(t) \qquad x(0) = 0$$

where t is time measured in years. The function $u(t)$, called a **control function,** measures the amount of light Harry shines on his plant (the larger $u(t)$ is, the faster the plant grows). Suppose the plant has an initial height of 0 at the beginning of the school year (hence $x(0) = 0$), and Harry wants the height to be 2 feet at the end of the school year (final condition $x(1) = 2$). To minimize the cost of electricity, he wants to find the control function $u(t)$ that minimizes the **objective function**

$$J(u) = \int_0^1 u^2(t)\, dt$$

(a) Show that the control function

$$u(t) = \begin{cases} 0 & (0 \le t < 0.5) \\ 2 & (0.5 \le t \le 1) \end{cases}$$

will "drive" the solution $x(t)$ from $x(0) = 0$ to the final condition $x(1) = 2$ with cost $J = 2$.

(b) Show that if the final condition $x(1) = 2$ is satisfied, then u must satisfy

$$\int_0^1 u(t)\, dt = 1$$

(c) Use the result from part (b) to show that the objective function J can be written as

$$J(u) = \int_0^1 (u - 1)^2\, dt + 1$$

(d) Use the result of part (c) to show that the **optimal control,** that is, the control function $u(t)$ that minimizes $J(u)$, is $u(t) \equiv 1$ and that the minimum value of $J(u)$ is 1.

(e) Substitute the optimal control $u^*(t) \equiv 1$ in the state equation

$$\dot{x} = 1 + u \qquad x(0) = 0$$

to find the **optimal solution** $x^*(t)$, which describes the growth curve of the plant.

31. The Monkey Problem Karin is an ecologist and is catching monkeys to be tagged.* Suppose Karin fires the dart at an angle θ with the horizontal. See Figure 2.43.

(a) Assuming only the force of gravity acting on the dart, show that the height of the dart as a function of time is given by

$$y_d(t) = (v_0 \sin \theta)\, t - \frac{1}{2} g t^2$$

where v_0 is the initial velocity of the dart and g is acceleration due to gravity.

(b) Assuming that the monkey jumps (straight down) at the moment when Karin fires the dart, show that the height of the monkey will be

$$y_m(t) = h - \frac{1}{2} g t^2$$

where h is the height of the monkey in the tree.

(c) If Karin is to hit the monkey, then $y_d(t) = y_m(t)$ at some value of time, say t_h. Show that this hit will take place at time

$$t_h = \frac{d}{v_0}$$

where d is the initial distance Karin was from the monkey.

(d) Show that to hit the monkey, Karin should aim directly at the monkey. In other words, show that the relationship $\sin \theta = h/d$ holds when $t_h = d/v_0$. See Figure 2.43.

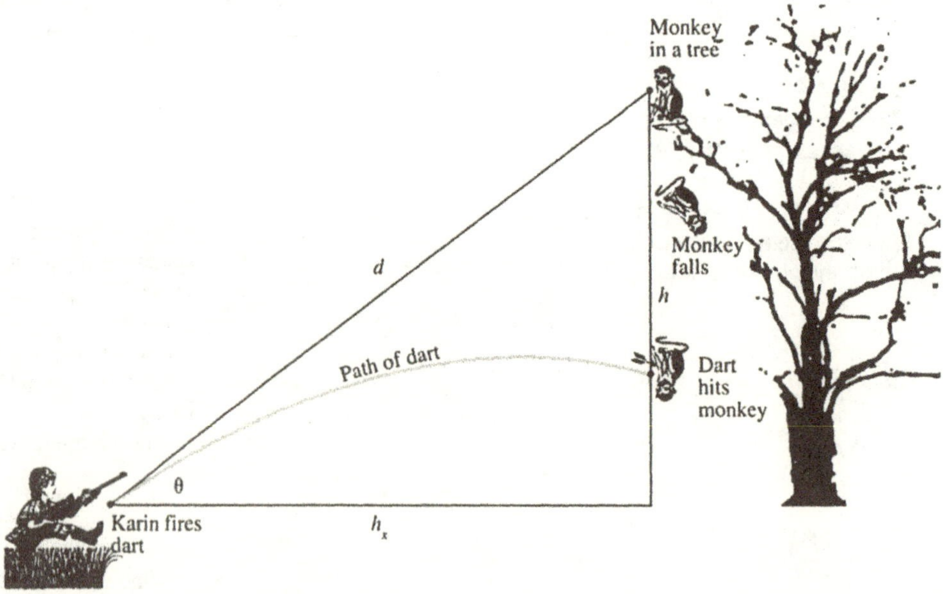

Figure 2.43
How should Karin aim the dart gun in order to hit a monkey that jumps from the tree at the exact moment when Karin fires the dart?

* A demonstration of this interesting problem can be seen on television on the education channel.

32. **Computer Problem** Use a graphing calculator or computer to sketch the graphs of the velocity and distance an object falls in the presence of air friction given by Eqs. (3) and (4) in the text. Sketch the graphs for different values of the parameters m and k.

33. **Computer Problem** Use a graphing calculator or computer to sketch the graph of a boat across a river as found in Problem 26. Sketch the graph for different values of the parameter $k = v_r/v_b$.

34. **Journal Entry** We have ended our discussion of applications of first-order differential equations. There are, of course, many more applications, but you have seen a representative sample of these applications. Spend a few minutes summarizing to yourself how first-order differential equations can be used to describe real-world phenomena. Think of explaining these applications to someone who has never studied differential equations.

2.7   THE DIRECTION FIELD AND EULER'S METHOD

PURPOSE

To introduce the geometric concept of the direction field of a first-order differential equation and show how it can aid in the general understanding of solutions of differential equations. Second, to show how **numerical solutions** can approximate the true solution when the solution cannot be found. This section introduces a numerical method known as **Euler's method.**

THE DIRECTION FIELD

Consider the first-order equation

$$\frac{dy}{dx} = f(x, y) \tag{1}$$

Geometrically, this equation says that at any point (x, y) the slope dy/dx of the solution has the value $f(x, y)$. This fact can be illustrated graphically by drawing a small line segment, called a **direction element,** through (x, y) with slope $f(x, y)$. A collection of these direction elements drawn at various points is called a **direction field** of the differential equation. A solution $y = y(x)$ of the differential equation (1) has the property that at every point (x, y) its graph is tangent to the direction element at that point. For example, the direction field of the differential equation

$$\frac{dy}{dx} = 1 + xy \tag{2}$$

is shown in Figure 2.44. Although a direction field will not yield an *analytic* solution or a formula for a differential equation, it does provide useful *qualitative* information about the shapes and behavior of the solutions. This is often true for nonlinear equations when it may be difficult or impossible to obtain an analytical solution.

Figure 2.44
Direction field of $y' = 1 + xy$

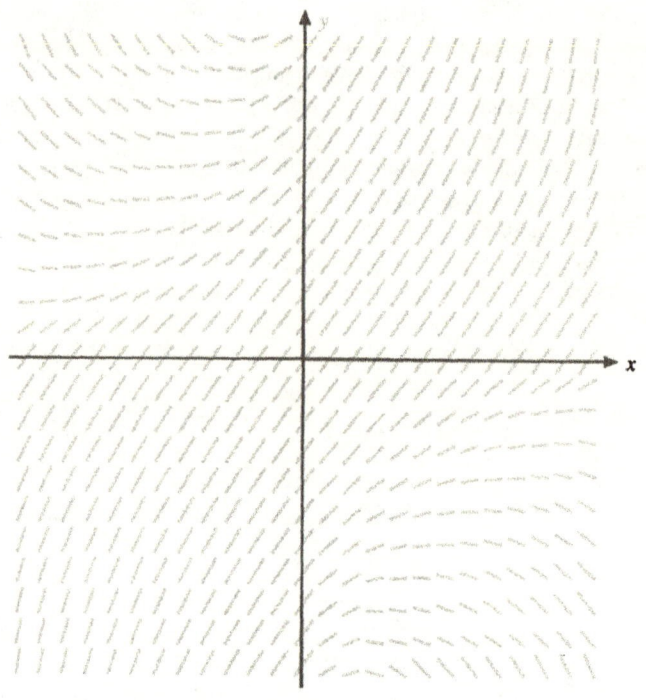

GENERAL INTRODUCTION TO NUMERICAL SOLUTIONS

By a **numerical solution** of a first-order differential equation we mean a table of values, given in either tabular or graphical form, that approximate the solution of a differential equation at a collection of points. By a numerical solution of an initial-value problem $y' = f(x, y)$, $y(x_0) = y_0$, we would mean values $y_0, y_1, y_2, \cdots, y_n, \cdots$ that would approximate the analytic solution at given points $x_0 < x_1 < x_2 < \cdots < x_n < \cdots$. See Figure 2.45.

Figure 2.45
Numerical solution of an initial-value problem

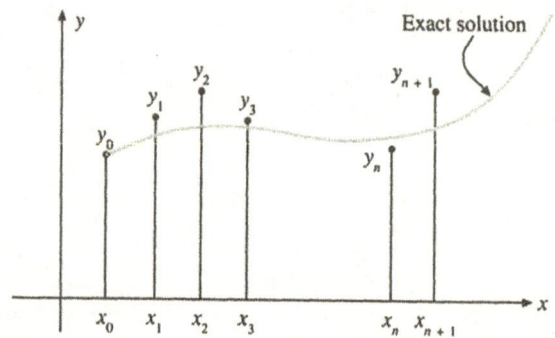

For instance, a numerical solution of the initial-value problem

$$y' = y \qquad y(0) = 1 \tag{3}$$

is graphed in Figure 2.46 and compared to the analytic solution $y(x) = e^x$.

This section considers the general initial-value problem

$$y' = f(x, y)$$
$$y(x_0) = y_0 \tag{4}$$

where we assume that there exists a unique solution to Eq. (4) on some interval about x_0. For some cases, $f(x, y)$ may be so simple that Eq. (4) can be integrated directly. For many problems in the applied sciences, however, this is not the case, and it is natural to consider numerical methods. Then, too, even if Eq. (4) can be integrated directly, the solution may be in the form of a complicated implicit relation $F(x, y) = 0$. For those situations a numerical solution is inviting.

Figure 2.46
The analytic and a numerical solution of the initial-value problem $y' = y$, $y(0) = 1$

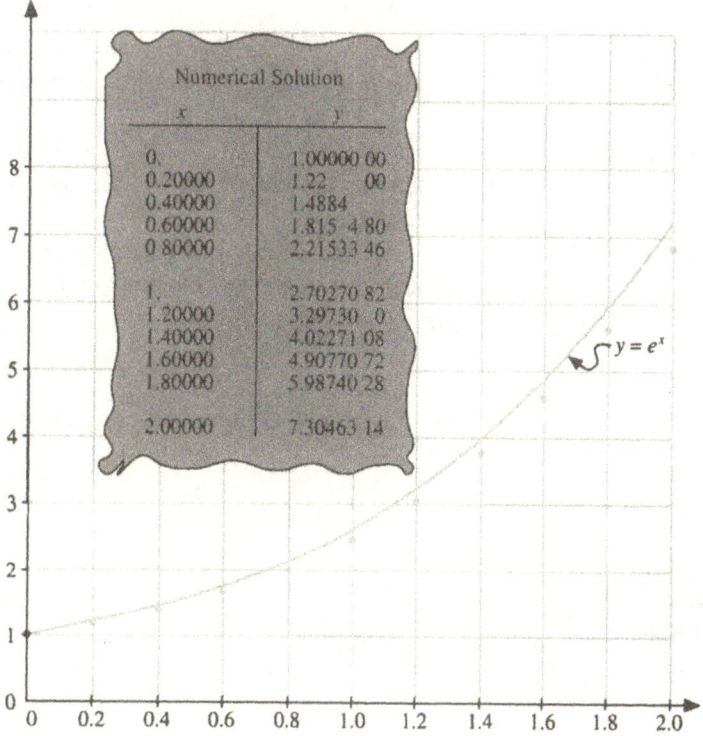

However, one should not abandon the search for solutions and start using numerical methods and the computer exclusively. In fact, an analytic solution is generally the solution of preference. It provides important information about the problem under study, such as the dependence of the solution on physical parameters and initial conditions. For instance, in Section 2.6 we found the analytic solution

$$v(t) = \frac{mg}{k}\left(1 - e^{-kt/m}\right) \tag{5}$$

of the initial-value problem

$$v' = mg - kv \qquad v(0) = 0 \tag{6}$$

which describes the velocity v of an object falling in a resistive medium. From this analytic solution, one can easily see how the velocity depends on the mass m and the coefficient of resistance k. We can also see that the limiting velocity is mg/k.

One of the simplest methods for approximating the solution of an initial-value problem is Euler's method.

EULER'S METHOD
(Tangent Line Method)

Consider the initial-value problem

$$y' = f(x, y) \qquad y(x_0) = y_0 \tag{7}$$

in which we assume that there exists a unique solution $y(x)$ on some interval around x_0. The goal is to approximate the solution $y = y(x)$ at a *finite set* of points

$$x_n = x_0 + nh \qquad (n = 1, 2, 3, \ldots, K) \tag{8}$$

where the **step size,** $h = x_n - x_{n-1}$, is a given positive number and K is predetermined. See Figure 2.47.

Starting at the initial point (x_0, y_0), recall that the slope of the solution passing through that point is simply $y' = f(x_0, y_0)$. Hence the tangent line to the solution passing through (x_0, y_0) is

$$y = y_0 + (x - x_0)f(x_0, y_0) \tag{9}$$

By evaluating this linear function at $x = x_1$ we obtain the approximate solution

$$y_1 = y_0 + hf(x_0, y_0) \tag{10}$$

Figure 2.47
Euler's method

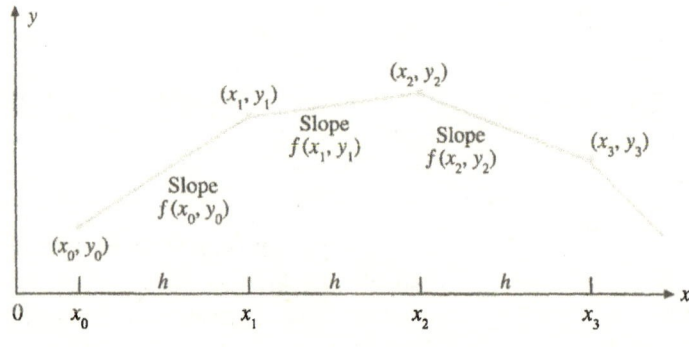

If we repeat this process again, starting from (x_1, y_1) and constructing a new line having slope $f(x_1, y_1)$, we get $y = y_1 + (x - x_1) f(x_1, y_1)$. We then evaluate this line at the next point $x = x_2$, which gives the next approximation,

$$y_2 = y_1 + hf(x_1, y_1) \tag{11}$$

Continuing this process we obtain the sequence of points

$$\begin{aligned} x_{n+1} &= x_n + h \\ y_{n+1} &= y_n + hf(x_n, y_n) \end{aligned} \qquad (n = 0, 1, 2, \ldots) \tag{12}$$

This procedure is known as **Euler's method** or the **method of tangents.** We summarize these ideas here.

DEFINITION: Euler's Method

Given the initial-value problem

$$y' = f(x, y) \qquad y(x_0) = y_0$$

Euler's method consists of using the iterative formula

$$y_{n+1} = y_n + hf(x_n, y_n) \qquad (n = 0, 1, 2, \ldots)$$

to compute $y_1, y_2, y_3, \ldots$, which approximates the true solution at the respective points $x_1 = x_0 + h$, $x_2 = x_0 + 2h$, $x_3 = x_0 + 3h$, ...

Example 1   **Euler's Method**   Use Euler's method with step size $h = 0.1$ to approximate the solution of the initial-value problem

$$y' = x + y \qquad y(0) = 1 \tag{13}$$

Solution   Using the given information $f(x, y) = x + y$, $x_0 = 0$, $y_0 = 1$, and $h = 0.1$, we find

$$\begin{cases} x_1 = x_0 + h = 0 + 0.1 = 0.1 \\ y_1 = y_0 + hf(x_0, y_0) = y_0 + h(x_0 + y_0) = 1 + 0.1(0 + 1) = 1.1 \end{cases}$$

$$\begin{cases} x_2 = x_1 + h = 0.1 + 0.1 = 0.2 \\ y_2 = y_1 + hf(x_1, y_1) = y_1 + h(x_1 + y_1) = 1.1 + 0.1(0.1 + 1.1) = 1.22 \end{cases}$$

$$\begin{cases} x_3 = x_2 + h = 0.2 + 0.1 = 0.3 \\ y_3 = y_2 + hf(x_2, y_2) = y_2 + h(x_2 + y_2) = 1.22 + 0.1(0.2 + 1.22) = 1.362 \end{cases}$$

$$\begin{cases} x_4 = x_3 + h = 0.3 + 0.1 = 0.4 \\ y_4 = y_3 + hf(x_3, y_3) = y_3 + h(x_3 + y_3) = 1.362 + 0.1(0.3 + 1.362) = 1.5282 \end{cases}$$

The initial-value problem (13) can be solved exactly by using the integrating factor method and has the solution $y = 2e^x - x - 1$. Table 2.1 compares the values obtained using Euler's method with the actual solution.

Table 2.1

Euler's Method for
$y' = x + y$, $y(0) = 1$

	Euler's method	Solution (8 places)	
x_n	y_n	$y(x)$	Error
0	1	1	
0.1	1.1	1.1103418	0.0103418
0.2	1.22	1.2428055	0.0228055
0.3	1.362	1.3997176	0.0377176
0.4	1.5282	1.5836494	0.0554493

ERROR IN EULER'S METHOD

So far we have avoided the important question concerning the accuracy of the computed values y_1, y_2, ... in approximating the true solution. We now provide a basis for the understanding of this important facet of numerical analysis.

There are two kinds of errors that arise when Euler's method is used. The first kind is the *roundoff error*, which is simply the error introduced as a result of rounding off numbers at each stage of a computation. In practice, all computers have limitations in computational accuracy. Even when calculations are made using ten decimal places of accuracy, roundoff errors may cause serious problems after many calculations. A common strategy for monitoring roundoff error is to perform all computations keeping a given number of decimal places, then repeat the computations keeping twice as many decimal places. If the results of the two experiments are close together, one can be reasonably sure that roundoff error is insignificant.

The second kind of error, known as the *discretization error* or *truncation error*, is the error introduced as a result of using the *linear approximation*

$$y_{n+1} = y_n + hy_n' \tag{14}$$

to compute y_1, y_2, ... when in fact the exact formula that relates y_{n+1} to y_n is the Taylor series expansion of the solution

$$y_{n+1} = y_n + hy_n' + \frac{h^2}{2!} y_n'' + \frac{h^3}{3!} y_n''' + \cdots \tag{15}$$

When the step size h is small, the error or difference between the true solution $y(x_{n+1})$ and the approximate solution y_{n+1} using Eq. (14), *assuming that y_n is known exactly* is approximately the first neglected term in Taylor's series (15), or $(h^2/2) y_n''$. In other words, it is proportional to h^2. This error in computing y_{n+1} over a *single step* from y_n, when we assume that y_n is correct, is called the *local truncation error*. Of course, if we use Euler's method to advance all the way from x_0 to x_n, the number of repeated applications of Euler's method is n, which is *inversely* proportional to the step size, h. Hence the *cumulative truncation error* that arises over n steps is proportional to h^2/h, or simply h. We now summarize these ideas.

> **Error in Euler's Method**
>
> Assume that the solution $y = y(x)$ of the initial-value problem
>
> $$y' = f(x, y) \qquad y(x_0) = y_0 \qquad (16)$$
>
> has a continuous second derivative in the interval of interest. (This can be assured by assuming that f, f_x, and f_y are continuous in the region of interest.) If Newton's method with step size h is used to compute N approximations
>
> $$y_{n+1} = y_n + hy_n' \qquad (n = 0, 1, ..., N - 1) \qquad (17)$$
>
> to the exact solution $y = y(x)$, then there exists a constant K such that the **cumulative discretization error,** or difference between the exact solution $y(x_{n+1})$ and the approximate solution y_n, satisfies
>
> $$|y_{n+1} - y(x_{n+1})| \le Kh \qquad (18)$$
>
> for each $n = 0, 1, ..., N - 1$.

COMPARISON OF ROUNDOFF AND DISCRETIZATION ERRORS

We have just seen that the cumulative discretization error in using Euler's method is proportional to the step size, h. This means that if the step size h is halved, then the error is halved. However, on the other hand, the smaller the step size, the more steps are required to cover a given interval and hence there is a larger cumulative *roundoff error*. In other words, we are faced with the dilemma that a large value of h results in a large discretization error and a small value of h results in a large roundoff error. Hence it is clear that there is an optimal value of h, neither too big nor too small, that results in the **total error,** or sum of the discretization and roundoff errors, being a minimum. See Figure 2.48. To find the optimal step size, one often resorts to experimentation and experience.

Figure 2.48
The total error is a minimum when the step size h is neither too big nor too small

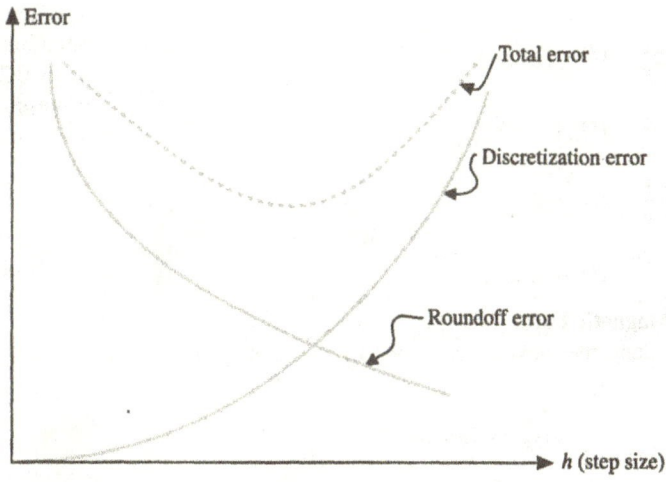

PROBLEMS: Section 2.7

For Problems 1–10, sketch the direction field of the given differential equation. Use this direction field to sketch a few individual solutions. Can you conclude that there is a simple solution passing through any given initial point? Are there any initial points where there might be more than one solution passing through the point?

1. $y' = -xy$
2. $y' = -x/y$
3. $y' = -\sqrt{y}$ $(y \geq 0)$
4. $y' = xy$
5. $y' = y^2 - x$
6. $y' = xy^2$
7. $y' = 2y^{2/3}$
8. $y' = y^2 - x^2$
9. $y' = \sin x - y$
10. $y' = e^{-y}$

Linear Fractional Equations

A differential equation of the form

$$y' = f\left(\frac{ax + by}{cx + dy}\right) \tag{19}$$

*is called a **linear fractional equation**. The solutions of this equation are very diverse, depending on the values of a, b, c, and d. The solutions change qualitatively depending on the value of D = ad − bc. For Problems 11–15, use a computer to draw a direction field of the following fractional equations.*

11. $y' = \dfrac{x}{y}$ $\qquad (D = 0)$

12. $y' = \dfrac{x}{x + y}$ $\qquad (D = 1)$

13. $y' = \dfrac{x + y}{x}$ $\qquad (D = -1)$

14. $y' = \dfrac{x - y}{x + y}$ $\qquad (D = 2)$

15. $y' = \dfrac{x + 2y}{x}$ $\qquad (D = -2)$

16. **Lines of Magnetic Force** The lines of force due to a magnet lying along the *y*-axis is described by the first-order equation

$$3xyy' = 2y^2 - x^2$$

Use a computer to sketch the directional field of this equa-

tion. The solutions of this equation are the curves on which iron filings would align themselves.

17. **The Empty Pail and the Direction Field*** The height *h* of the water in a pail with a hole in the bottom can be shown to satisfy the differential equation

$$\frac{dh}{dt} = -k\sqrt{h}$$

where $k > 0$ is a constant that depends on the size of the hole. Sketch the direction field of this equation $(k = 1)$, and from it say why it is possible to determine *when* the pail is empty given that you know that the pail currently contains a given amount but not possible to determine when the pail *was* full given that you know that the pail is currently empty?

18. **What Can You Say About This Equation?** Researchers have become interested in the equation

$$\frac{dy}{dx} = e^{-xy}$$

which arose in an area of physics. Obtain some information about the equation by sketching the direction field of the equation over some domain.

Euler's Method

For the following problems, either write your own computer program or use existing software.

19. **Comparison of Numerical and Analytic Solutions** Consider the initial-value problem

$$y' = y - 2 \qquad y(0) = 1$$

(a) Find the approximate solution at $x = 0.1, 0.2,$ and 0.3 using Euler's method with $h = 0.1$.
(b) Repeat the calculations in part (a) using $h = 0.05$.
(c) Find the analytic solution and compare the exact value $y(0.2)$ with your results from parts (a) and (b).

20. **Comparison of Numerical and Analytic Solutions** Consider the initial-value problem

$$\frac{dy}{dx} = xy \qquad y(0) = 1$$

(a) Find the approximate solution at $t = 1$ using Euler's method with the different step sizes of $h = 1, 1/2, 1/4, 1/8$.
(b) Find the analytic solution and compare the exact value of $y(1)$ with your results from part (a).

* This problem is based on an example taken from the text *Differential Equations: A Dynamical Approach* by J. H. Hubbard and B. H. West (Springer-Verlag, 1990).

For Problems 21–25, use existing software to solve the given differential equation using Euler's method with $h = 0.1$.

Differential equation	Initial condition	Interval
21. $y' = x - y$	$y(0) = 1$	$[0, 5]$
22. $y' = 3x^2 - y$	$y(0) = 1$	$[0, 1]$
23. $y' = x^2 + e^{-y}$	$y(0) = 0$	$[0, 2]$
24. $y' = \sqrt{x + y}$	$y(1) = 1$	$[1, 5]$
25. $y = x^2 - y^2$	$y(0) = 1$	$[0, 5]$

26. Stephan's Law of Cooling Stephan's law of cooling states that the rate of change in temperature T of a body in a medium of temperature M is proportional to $T^4 - M^4$. That is,

$$\frac{dT}{dt} = -k(T^4 - M^4) \qquad (20)$$

Assume that the time constant is $1/k = 20$ hours, the medium temperature is $M = 50°F$, and the initial temperature is $T(0) = 100°F$. Use Euler's method with $h = 0.25$ to approximate $T(1)$.

27. A Nasty Solution to Approximate Using Euler's method, approximate the solution of

$$y' = y^2 \qquad y(0) = 1$$

at $x = 0.25, 0.50, 0.75, 1.0$ with step size $h = 0.25$. Use the exact solution $y(x) = 1/(1 - x)$ to find the error at each point.

28. Approximating e Estimate the value of e using Euler's method to approximate the solution of

$$y' = y \qquad y(0) = 1$$

at $x = 1$. Use a step size of $h = 0.1$. If you have access to a computer, experiment by approximating e using smaller and smaller values of h. Note that as h gets smaller, the approximation of e becomes better, but after a while the approximation becomes worse.

29. Trouble Ahead The initial-value problem

$$\frac{dy}{dx} = y^{1/3} \qquad y(0) = 0$$

has an infinite number of solutions, two of which are the functions $y(x) = 0$ and $y(x) = (2x/3)^{3/2}$ for $x \geq 0$. See Figure 2.49.

(a) What happens if Euler's method is applied to this problem?

(b) What happens if the initial condition is changed to the value $y(0) = 0.01$?

(c) If you have access to a computer with Euler's method already programmed, find an approximate solution to this problem for x in the interval $[0, 6]$ using $h = 0.1$.

Figure 2.49
The problem $y' = y^{1/3}$, $y(0) = 0$, cannot be solved numerically, since the functions $y(x) = 0$ and $y(x) = (2x/3)^{3/2}$ are both solutions passing through $(0, 0)$

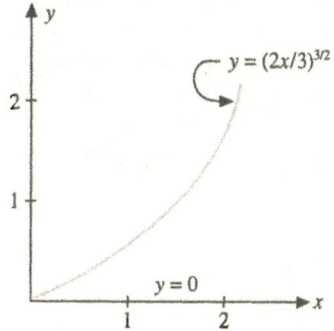

30. Roundoff Error Problems The solution of

$$y' = y \qquad y(0) = A$$

is $y(x) = Ae^x$. If a roundoff error of ϵ occurs when the value of A is read into a computer, how will this affect the solution at $x = 1$? At $x = 10$? At $x = 20$?

Richardson's Extrapolation

We have seen that when one uses Euler's method starting from x_0, the accumulated discretization error in the approximation to the true solution at $\bar{x} = x_0 + nh$ is bounded by a constant times the step size h. In other words, the true solution $y(\bar{x})$ at $x_0 + nh$ can be written as

$$y(\bar{x}) = y_n(h) + Kh + O(h^2) \qquad (21)$$

where the "big oh" notation $O(h^2)$ simply refers to terms involving h that are of order h^2 and higher (the remainder of the Taylor series). Note that we have written the approximation to the solution as $y_n(h)$ to denote explicitly that it depends on the step-size h. Now suppose the computations in Euler's method are repeated with a halved step size $h/2$. Remembering that $2n$ steps are now required, we can rewrite Eq. (21) as

$$y(\bar{x}) = y_{2n}(h/2) + K(h/2) + O(h^2) \qquad (22)$$

We have left the "big oh" unchanged, since it still refers to terms involving h^2 and larger. Now if Eq. (22) is multiplied by

2 and then subtracted from Eq. (21), the term involving K will be eliminated, and we get

$$y(\bar{x}) = [2y_{2n}(h/2) - y_n(h)] + O(h^2) \qquad (23)$$

*The first two terms on the right-hand side of Eq. (23) give a new approximation to $y(\bar{x})$, and it should be noted that the first neglected term is proportional to h^2, in contrast to Euler's approximation, whose first neglected term was proportional to h. This technique of approximating the solution $y(\bar{x})$ in terms of both $y_n(h)$ and the "halved" approximation $y_{2n}(h/2)$ is known as **Richardson's extrapolation** or Richardson's deferred approach to the limit, named after the mathematician L. F. Richardson, who first studied the technique in 1927. Solve Problems 31–34 using Richardson's extrapolation to compute new estimates for the true solution at x = 0.2 by making use of results with h = 0.1 and h = 0.2. If possible, compare the results with the exact solution.*

31. $y' = y$ $y(0) = 1$
32. $y' = xy$ $y(0) = 1$
33. $y' = -y + 1$ $y(0) = 1$
34. $y' = y^2$ $y(0) = 1$

35. Other Richardson Extrapolation A numerical method is said to have *cumulative accuracy h^p* (Euler's method has cumulative accuracy h^1) if the error at $\bar{x} = x_0 + nh$ is proportional to h^p. This can be written as

$$y(\bar{x}) = y_n(h) + Kh^p + O(h^{p+1}) \qquad (24)$$

Show that Richardson's extrapolation leads to the improved approximation

$$y(\bar{x}) = y_{2n}(h/2) + \frac{y_{2n}(h/2) - y_n(h)}{2^p - 1} \qquad (25)$$

with cumulative accuracy h^{p+1}.

36. Useful Sign Analysis The graphical solution of a first-order differential equation of the form $dy/dx = f(y)$ can be analyzed by the following procedure.
(a) Find all roots r of the equation $f(r) = 0$, and graph the functions $y = r$. These are the constant solutions of the differential equation.
(b) Determine the sign of the function $f(y)$ between the horizontal solutions. The solution of $y' = f(y)$ is increasing when $f(y) > 0$ and decreasing when $f(y) < 0$. Use this procedure to analyze the solution of

$$y' = -y^3 + 3y^2 - 2y$$

37. Journal Entry Do you think that analytic solutions are overrated and that we should be spending more time studying numerical solutions? Which types of solutions are more important in your own area of interest?

2.8   HIGHER-ORDER NUMERICAL METHODS

PURPOSE

To show how Euler's method is only the first of an entire family of methods based on approximating solutions by a finite number of terms of a Taylor series. This section introduces the three-term Taylor series method and the four-term Runge-Kutta method.

THE THREE-TERM TAYLOR SERIES METHOD

In studying Euler's method we saw how repeated use of the formula

$$y_{n+1} = y_n + hf(x_n, y_n) \qquad (1)$$

can be used to find an approximate solution of the initial-value problem

$$y' = f(x, y) \qquad y(x_0) = y_0 \qquad (2)$$

Since the basic Euler formula (1) consists in using the first two terms of the Taylor series expansion

$$y(x_n + h) = y(x_n) + hy'(x_n) + \frac{1}{2!} h^2 y''(x_n) + \frac{1}{3!} h^3 y'''(x_n) + \cdots \qquad (3)$$

of the solution about $x = x_n$, we see that Euler's method is simply the first in an entire *family* of numerical methods, each method using a given number of terms. For example, if we retain three terms in the series (3), we have the **three-term Taylor series**

$$y(x_n + h) = y(x_n) + hy'(x_n) + \frac{1}{2!} h^2 y''(x_n) \tag{4}$$

However, using more terms in the Taylor series requires knowledge of higher-order derivatives and not just the *given* first derivative $y' = f(x, y)$. In theory, this poses no problem, since one can always differentiate $y' = f(x, y)$ repeatedly with respect to x. For example, we can find y'' by differentiating $y' = f(x, y)$ with respect to x, getting

$$\begin{aligned}
y'' &= \frac{d}{dx} f(x\ y) \\
&= f_x + f_y y' \\
&= f_x + f_y f
\end{aligned} \tag{5}$$

Substituting this value into the second derivative in Eq. (4), we obtain the **three-term Taylor series approximation**

$$\begin{aligned}
y_{n+1} &= y_n + hy'(x_n, y_n) + \frac{h^2}{2!} y''(x_n, y_n) \\
&= y_n + hf(x_n, y_n) + \frac{h^2}{2!} \left(f_x (x_n, y_n) + f_y (x_n, y_n) f(x_n, y_n) \right)
\end{aligned} \tag{6}$$

This equation can then be used to find approximate values y_n at $x_n = x_0 + nh$.

Example 1   **Three-Term Taylor Series Approximation**   Use the three-term Taylor series approximation (6) to approximate the solution of the initial-value problem

$$y' = 1 + x + 2y \qquad y(0) = 1 \tag{7}$$

in the interval $[0, 1]$ using $h = 0.1$.

Solution   We have

$$\begin{aligned}
f(x, y) &= 1 + x + 2y \\
f_x(x, y) &= 1 \\
f_y(x, y) &= 2
\end{aligned}$$

Hence

$$\begin{aligned}
y_n' &= f(x_n, y_n) \\
&= 1 + x_n + 2y_n
\end{aligned}$$

$$\begin{aligned}
y_n'' &= f_x(x_n, y_n) + f_y(x_n, y_n) f(x_n, y_n) \\
&= 1 + 2(1 + x_n + 2y_n) \\
&= 3 + 2x_n + 4y_n
\end{aligned}$$

The three-term Taylor series approximation is

$$y_{n+1} = y_n + hy_n' + \frac{h^2}{2} y_n''$$

$$= y_n + h(1 + x_n + 2y_n) + \frac{h^2}{2}(3 + 2x_n + 4y_n) \qquad (8)$$

By using this formula, y_1 is

$$y_1 = 1 + 0.1(1 + 0 + 2 \cdot 1) + \frac{0.01}{2}\left(3 + 2(0) + 4(1)\right) = 1.3350$$

Table 2.2 compares the three-term Taylor series approximation with the solution

$$y(x) = \frac{1}{4}(7e^{2x} - 2x - 3)$$

Table 2.2

Three-Term Taylor Series
Approximation

n	x_n	y_n (Taylor series)	Solution (6 places)	Absolute error $|y_n - \text{solution}|$
0	0	1.0	1.0	0.0
1	0.1	1.3350	1.337455	0.002455
2	0.2	1.7547	1.760693	0.005993
3	0.3	2.2777	2.288708	0.011008
4	0.4	2.9268	2.944697	0.017897
5	0.5	3.7297	3.756993	0.027293
6	0.6	4.7293	4.760205	0.030905
7	0.7	5.9397	5.996601	0.056901
8	0.8	7.4385	7.517807	0.079307
9	0.9	9.2780	9.386886	0.108886
10	1.0	11.5331	11.680850	0.147750

The graph of the solution and the plotted points of the three-term Taylor series approximation are shown in Figure 2.50. ∎

The three-term Taylor series method can be summarized as follows.

DEFINITION: Three-Term Taylor Series Method

The **three-term Taylor series** method for approximating the solution of

$$y' = f(x, y) \qquad y(x_0) = y_0$$

at the points $x_n = x_0 + nh$ $(n = 1, 2, ..., N)$ consists in computing a table of approximate values $(x_1, y_1), (x_2, y_2), ..., (x_N, y_N)$ using the formula

$$y_{n+1} = y_n + hf(x_n, y_n) + \frac{h^2}{2}\left(f_x(x_n, y_n) + f_y(x_n, y_n)f(x_n, y_n)\right)$$

Figure 2.50
Comparing the three-term
Taylor series method with the
solution

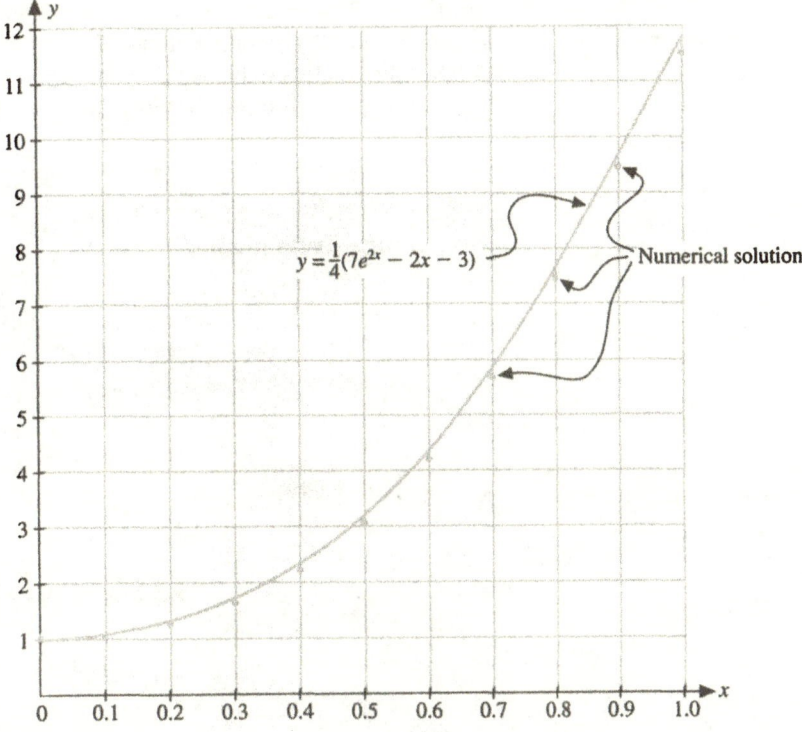

THE RUNGE-KUTTA METHOD

The difficulty in using the Taylor series to obtain higher-order approximations lies in the fact that higher derivatives often become complicated and unwieldy. Of course, with the widespread availability of computer programs to carry out symbolic and analytic calculations, this is becoming less of a problem. However, it is possible to develop formulas that have the *accuracy* of higher-order approximations without actually having to compute higher derivatives. Such a formula is the **Runge-Kutta formula:**

$$y_{n+1} = \frac{h}{6}\left(k_{n1} + 2k_{n2} + 2k_{n3} + k_{n4}\right) \tag{9}$$

where k_{n1}, k_{n2}, k_{n3}, and k_{n4} are given as

$$k_{n1} = f(x_n, y_n) \tag{10a}$$

$$k_{n2} = f(x_n + \tfrac{1}{2}h, y_n + hk_{n1}) \tag{10b}$$

$$k_{n3} = f\left(x_n + \tfrac{1}{2}h, y_n + \tfrac{1}{2}hk_{n2}\right) \tag{10c}$$

$$k_{n4} = f(x_n + h, y_n + hk_{n3}) \tag{10d}$$

The constants k_{n1}, k_{n2}, k_{n3}, and k_{n4} are simply first derivatives, or slopes of the direction field, taken at four different points in the interval $x_n \leq x \leq x_{n+1}$. Also, the expression

$(1/6) (k_{n1} + 2k_{n2} + 2k_{n3} + k_{n4})$ is simply a *weighted average* of derivatives (slopes) evaluated at four different locations. Hence while Euler's formula computes y_{n+1} using the slope at the single point (x_n, y_n), the Runge-Kutta formula finds y_{n+1} from a weighted average of slopes at four different points. See Figure 2.51.

DEFINITION: Runge-Kutta Method

The **Runge-Kutta method** for approximating the solution of

$$y' = f(x, y) \qquad y(x_0) = y_0$$

at the points $x_n = x_0 + nh$ $(n = 1, 2, ..., N)$ consists in computing a table of approximate values $(x_1, y_1), (x_2, y_2), ..., (x_N, y_N)$ using the formula

where

$$y_{n+1} = y_n + \frac{h}{6}\left(k_{n1} + 2k_{n2} + 2k_{n3} + k_{n4}\right)$$

$$k_{n1} = f(x_n, y_n)$$

$$k_{n2} = f\left(x_n + \frac{1}{2}h, y_n + \frac{1}{2}hk_{n1}\right)$$

$$k_{n3} = f\left(x_n + \frac{1}{2}h, y_n + \frac{1}{2}hk_{n2}\right)$$

$$k_{n4} = f\left(x_n + h, y_n + hk_{n3}\right)$$

Figure 2.51
Geometric interpretation of the Runge-Kutta formula

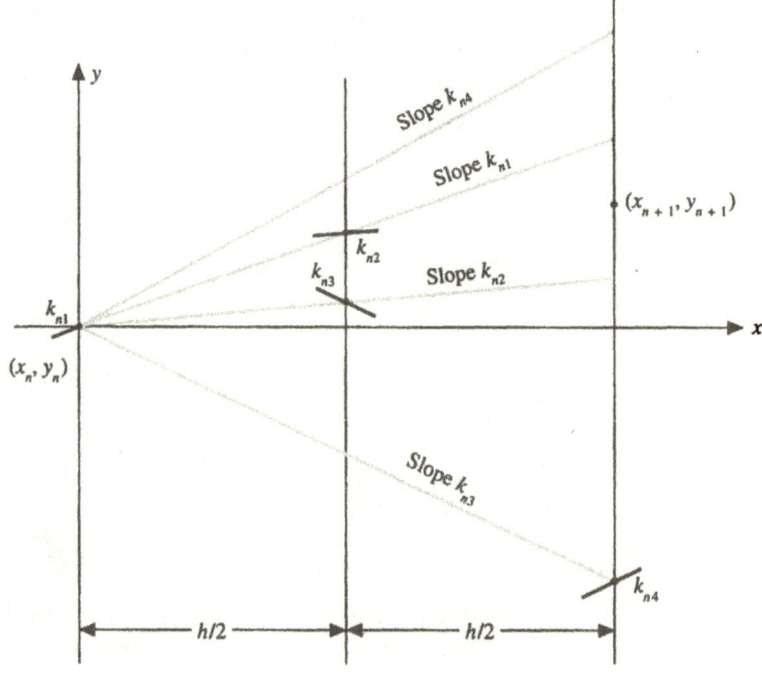

Note: In Euler's method the derivative $y' = f(x, y)$ is treated as if it were a constant across a single step from x_n to x_{n+1} and equal to the value at the beginning of that step. The Runge-Kutta method averages four derivatives at each step to estimate the solution.

The following BASIC program performs the steps of the Runge-Kutta method.

```
Runge-Kutta Method
REM This program finds the numerical solution of the initial
REM value problem y' = f(x, y), y(x0) = y0 using the Runge-
REM Kutta method. Input values are x0, y0, the step size
REM and the maximum value of x.
REM
REM Enter the function f(x, y) in the function statement
REM
      DEF FN f(x, y) = (enter the function f(x, y) here)
REM
      INPUT "enter initial x and y"; x, y
      INPUT "enter step size and the maximum value of x,"; h, xmax
      n = xmax/h
      FOR i = 1 TO n STEP h
         k1 = FN f(x, y)
         xx = x + h/2
         yy = y + h * k1/2
         k2 = FN f(xx, yy)
         yy = y + h * k2/2
         k3 = FN f(xx, yy)
         xx = x + h
         yy = y + h * k3/2
         k4 = FN f(xx, yy)
         x = x + h
         y = y + h * (k1 + 2 * k2 + 2 * k3 + k4)/6
         PRINT x, y
      NEXT i
      END
```

Example 2 **Runge-Kutta Method**

Use the Runge-Kutta method to approximate the solution of the initial-value problem

$$y' = 1 + x + y \qquad y(0) = 1 \tag{11}$$

in the interval [0, 1] using a step size of $h = 0.1$.

Solution We are given $f(x, y) = 1 + x + y$, $x_0 = 0$, $y_0 = 1$, and $h = 0.1$. Hence to find y_1 at $x = 0.1$, we have

$$\begin{aligned}
k_{01} &= f(x_0, y_0) \\
&= f(0, 1) \\
&= 2
\end{aligned}$$

$$\begin{aligned}
k_{02} &= f\left(x_0 + \frac{1}{2}h, y_0 + \frac{1}{2}hk_{01}\right) \\
&= f(0.05, 1.1) = 1 + 0.05 + 1.1 \\
&= 2.15
\end{aligned}$$

$$\begin{aligned}
k_{03} &= f\left(x_0 + \frac{1}{2}h, y_0 + \frac{1}{2}hk_{02}\right) \\
&= f(0.05, 1.1075) \\
&= 1 + 0.05 + 1.1075 \\
&= 2.1575
\end{aligned}$$

$$\begin{aligned}
k_{04} &= f\left(x_0 + h, y_0 + hk_{03}\right) \\
&= f(0.1, 1.21575) \\
&= 1 + 0.1 + 1.21575 \\
&= 2.31575
\end{aligned}$$

Hence

$$\begin{aligned}
y_1 &= y_0 + \frac{h}{6}\left(k_{01} + 2k_{02} + 2k_{03} + k_{04}\right) \\
&= 1 + \frac{0.1}{6}\left(2 + 2(2.15) + 2(2.1575) + 2.31575\right) \\
&= 1.215513
\end{aligned}$$

The remainder of the values were found by using a computer and are shown in Table 2.3. The table compares the approximate values obtained by using the Runge-Kutta method with the analytic solution.

Table 2.3
Runge-Kutta Solution of
$y' = 1 + x + y$, $y(0) = 1$

n	x_n	y_n (Runge-Kutta)	Analytic solution (7 places)	Absolute error $\|y_n - \text{solution}\|$
0	0	1.0		
1	0.1	1.215513	1.215513	2.38×10^{-7}
2	0.2	1.464208	1.464208	5.96×10^{-7}
3	0.3	1.749576	1.749576	8.34×10^{-7}
4	0.4	2.075473	2.075474	9.54×10^{-7}
5	0.5	2.446162	2.446164	16.69×10^{-7}
6	0.6	2.866354	2.866356	23.82×10^{-7}
7	0.7	3.341255	3.341259	40.53×10^{-7}
8	0.8	3.876619	3.876623	42.92×10^{-7}
9	0.9	4.478805	4.478810	52.45×10^{-7}
10	1.0	5.154840	5.154846	66.76×10^{-7}

PROBLEMS: Section 2.8

For Problems 1–10, use a step size of $h = 0.1$ and compare the approximate solutions at $x = 0.1$, 0.2, and 0.3 of the given initial-value problem using (a) Euler's methed, (b) the three-term Taylor series method, and (c) the Runge-Kutta method. Compare (when possible) the approximate solutions with the exact solution.

1. $y' = y + 1$ $y(0) = 1$
2. $y' = y + x$ $y(0) = 1$
3. $y' = 1 + x + y$ $y(0) = -1$
4. $y' = x^2 + y$ $y(0) = 2$
5. $y' = x^2 + y^2$ $y(0) = 2$
6. $y' = \sqrt{y}$ $y(0) = 1$
7. $y' = 2x + e^{-xy}$ $y(0) = 1$
8. $y' = \sqrt{x + y}$ $y(0) = 1$
9. $y' = x/y$ $y(0) = 1$
10. $y' = e^y$ $y(0) = 1$

For Problems 11–15, find the four-term Taylor approximation

$$y_{n+1} = y_n + hy_n' + \frac{h}{2!}y_n'' + \frac{h^2}{3!}y_n''' \qquad (12)$$

for the given differential equation. Hint: Compute y'' and y''' from the differential equation.

11. $y' = 1 + x$
12. $y' = 1 + x + y$
13. $y' = x^2 + y^2$
14. $y' = e^{xy}$
15. $y' = \sin y$

Using Your Own Computer

For Problems 16–18, use your own computer and existing software to solve the given problem.

16. **Write Your Own Program** Write a computer program to carry out the steps of the Runge-Kutta method.
17. **Calculating π** Use the Runge-Kutta method with step size $h = 0.01$ to approximate the solution of the initial-value problem

$$y' = \frac{4}{1 + x^2} \qquad y(0) = 0$$

at $x = 1$. Try experimenting with other step sizes to see how close you can get to the exact value of $y(1) = \pi$.

18. Calculating ln 2 Use the Runge-Kutta method with step size $h = 0.01$ (and 100 steps) to approximate the solution of the initial-value problem

$$y' = \frac{1}{y} \qquad y(1) = 0$$

at $x = 2$. Try experimenting with different step sizes to see how close you get to the exact value of $y(2) = \ln 2$.

Picard's Approximation Method

It is possible to approximate the solution of the initial-value problem

$$y' = f(x, y) \qquad y(x_0) = y_0 \qquad (13)$$

by integrating each side of the equation with respect to x from x_0 to x, getting

$$\int_{x_0}^{x} y'(x) \, dx = y(x) - y(x_0) = \int_{x_0}^{x} f(x, y(x)) \, dx \qquad (14)$$

If we now solve for $y(x)$, we get

$$y(x) = y_0 + \int_{x_0}^{x} f(t, y(t)) \, dt \qquad (15)$$

Now, selecting an initial guess of $\phi_0(x)$ for $y(t)$, we can use Eq. (15) to find a new approximation $\phi_1(x)$ using

$$\phi_1(x) = y_0 + \int_{x_0}^{x} f(t, \phi_0(t)) \, dt$$

In general, we can find a sequence of approximations $\phi_2(x)$, $\phi_3(x)$, ... using **Picard's formula***

$$\phi_{n+1}(x) = y_0 + \int_{x_0}^{x} f(t, \phi_n(t)) \, dt \qquad (16)$$

for $n = 0, 1, 2, \dots$. See Figure 2.52.

For Problems 19–21, use Picard's method to find the first three successive approximations $\phi_1(x)$, $\phi_2(x)$, and $\phi_3(x)$ to the given initial-value problem using the given guess $\phi_0(x)$.

Initial-value problem		Initial guess
19. $y' = x - y$	$y(0) = 1$	$\phi_0(x) = y(0)$
20. $y' = x + y$	$y(0) = 0$	$\phi_0(x) = y(0)$
21. $y' = 1 + y^2$	$y(0) = 0$	$\phi_0(x) = y(0)$

22. Journal Entry What is your opinion concerning the analytic method versus numerical methods? In the future, will all scientific problems be solved by using numerical methods and computers, leaving analytic methods to wither and die?

Figure 2.52
Picard's method for approximate solutions

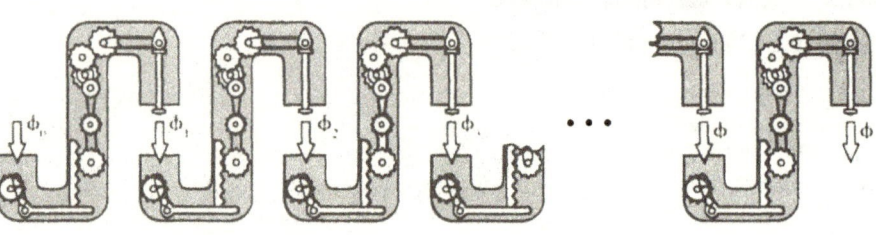

---

* Although we study Picard's method here as a means for finding approximate solutions to differential equations, it can sometimes be used to prove the *existence* of a solution to a differential equation by showing that the limiting function $\phi(x)$ of the sequence $\{\phi_n(x)\}$ exists as $n \to \infty$, and this function is a solution of the equation.

3
Second-Order Linear Equations

3.1   Introduction to Second-Order Linear Equations
3.2   Fundamental Solutions of the Homogeneous Equation
3.3   Reduction of Order
3.4   Homogeneous Equations with Constant Coefficients: Real Roots
3.5   Homogeneous Equations with Constant Coefficients: Complex Roots
3.6   Nonhomogeneous Equations
3.7   Solving Nonhomogeneous Equations: Method of Undetermined Coefficients
3.8   Solving Nonhomogeneous Equations: Method of Variation of Parameters
3.9   Mechanical Systems and Simple Harmonic Motion
3.10 Unforced Damped Vibrations
3.11 Forced Vibrations
3.12 Introduction to Higher-Order Equations (Optional)

3.1   INTRODUCTION TO SECOND-ORDER LINEAR EQUATIONS

PURPOSE

To introduce some of the basic concepts and ideas related to the second-order linear differential equation

$$y'' + p(x)y' + q(x)y = f(x)$$

We introduce

- the corresponding homogeneous equation,
- the principle of superposition,
- the initial-value problem,
- the existence and uniqueness of solutions.

In Chapter 2 we saw how first-order differential equations were useful in the description of physical systems relating to heat flow, mixing problems, and a wide variety of other phenomena often relating to chemistry, physics, and biology. We now focus our attention on second-order differential equations. It could be argued that second-order differential equations are even more important than first-order equations.* Certainly, one can say that without an appreciation of second-order differential equations, one cannot obtain a deep understanding of many of the laws of physics. In this chapter we will solve various kinds of second-order equations and see how they are used to model various kinds of motion. Our main focus will be on the **second-order linear differential equation**

$$a_0(x)\frac{d^2y}{dx^2} + a_1(x)\frac{dy}{dx} + a_2(x)y = g(x) \tag{1}$$

where $a_0(x) \neq 0$, $a_1(x)$, $a_2(x)$, and $g(x)$ are continuous functions on some interval of interest I. When the functions $a_0(x)$, $a_1(x)$, and $a_2(x)$ are constants, the differential equation is said to have **constant coefficients;** otherwise, it is said to have **variable coefficients.** Inasmuch as we are interested in those differential equations for which $a_0(x) \neq 0$ for all $x \in I$, we rewrite Eq. (1) in **standard form** as

$$\frac{d^2y}{dx^2} + p(x)\frac{dy}{dx} + q(x)y = f(x) \tag{2}$$

where $p(x) = a_1(x)/a_0(x)$, $q(x) = a_2(x)/a_0(x)$, and $f(x) = g(x)/a_0(x)$. Often Eq. (2) is written in the compact form

$$y'' + p(x)y' + q(x)y = f(x) \tag{3}$$

---

* We probably should be careful, since later we will see that any second-order differential equation can be rewritten as a system of two first-order equations.

Associated with the nonhomogeneous Eq. (3) is the **homogeneous equation**

$$y'' + p(x)y' + q(x)y = 0 \tag{4}$$

which is obtained from Eq. (3) by simply replacing $f(x)$ by zero. We call Eq. (4) the corresponding homogeneous equation to Eq. (3).

INTEGRATING FACTOR METHOD FAILS FOR SECOND-ORDER EQUATIONS

The reader may wonder whether we are going to solve the second-order equation (3) in a manner similar to the way we solved the first-order equation by multiplying by an integrating factor, thereby turning the equation into a simple integration. If this idea occurred to you, then you are in good company, since the idea also occurred to the eminent mathematician Joseph Louis Lagrange (1736–1813), who proved the *existence* of such an integrating factor over 200 years ago. Although Lagrange showed that under certain conditions such a factor existed, he couldn't actually find it; he proved that it satisfied *another* differential equation that was just as difficult to solve as the original equation. Basically, Lagrange tried to find a function $\mu(x)$ so that multiplication of the equation

$$y'' + p(x)y' + q(x)y = f(x) \tag{5}$$

by $\mu(x)$ would yield an equation of the form

$$\frac{d}{dx}[\mu(x)y' + g(x)y] = \mu(x)f(x) \tag{6}$$

where the function $g(x)$ could be determined* from the coefficients $p(x)$ and $q(x)$. If such a function $\mu(x)$ could be found, then it would be possible to integrate Eq. (6) directly, arriving at the first-order equation.

$$\mu(x)y' + g(x)y = \int \mu(s)f(s)\,ds \tag{7}$$

This first-order equation (7) could then be solved by using the ''first-order'' integrating factor method studied in Chapter 2. The only flaw in this method is that to find $\mu(x)$, one must solve

$$\mu'' - p(x)\mu' + [q(x) - p'(x)]\mu = 0 \tag{8}$$

which, called the **adjoint equation,** is every bit as hard to solve as the original equation. In other words, Lagrange essentially replaced solving one differential equation by another one of equal difficulty. If you think about it, the only reason Lagrange's integrating factor method is successful in solving the first-order equation

$$y' + p(x)y = f(x) \tag{9}$$

is because the adjoint equation is

$$\mu' - p(x)\mu = 0 \tag{10}$$

which can be solved by separation of variables, which gives

$$\mu(x) = e^{\int p(x)\,dx} \tag{11}$$

---

* We won't bother you with the formula for $g(x)$. The important point is that it can be determined from $p(x)$ and $q(x)$.

SEARCH FOR THE GENERAL SOLUTION OF THE SECOND-ORDER EQUATION

The failure of the integrating factor method in solving the second-order equation means that we must start, so to speak, from square one. We begin our search for the general solution of the second-order equation by introducing some important properties of solutions of second-order equations. We start with one of the most important properties of linear homogeneous differential equations.

PRINCIPLE OF SUPERPOSITION FOR HOMOGENEOUS LINEAR EQUATIONS

Two of the most important properties of the homogeneous linear equation are that the *sum* of any two solutions is again a solution and that any *constant multiple* of a solution is a solution. This principle is called the **principle of superposition** for homogeneous equations.

THEOREM 3.1: Principle of Superposition for Homogeneous Equations

If $y_1(x)$ and $y_2(x)$ are any two solutions of the linear homogeneous equation

$$y'' + p(x)y' + q(x)y = 0 \tag{12}$$

on some interval I, then any **linear combination**

$$y(x) = c_1 y_1(x) + c_2 y_2(x) \tag{13}$$

where c_1 and c_2 are arbitrary constants, is also a solution on the interval. I.

PROOF: The conclusion follows from elementary properties of the derivative. That is, if

$$y = c_1 y_1 + c_2 y_2$$

then

$$y' = c_1 y_1' + c_2 y_2'$$
$$y'' = c_1 y_1'' + c_2 y_2''$$

Substituting Eq. (13) into the differential equation (12), we get

$$
\begin{aligned}
y'' + p(x)y' &+ q(x)y \\
&= [c_1 y_1 + c_2 y_2]'' + p[c_1 y_1 + c_2 y_2]' + q[c_1 y_1 + c_2 y_2] \\
&= [c_1 y_1'' + c_2 y_2''] + p[c_1 y_1' + c_2 y_2'] + q[c_1 y_1 + c_2 y_2] \\
&= c_1 [y_1'' + p y_1' + q y_1] + c_2 [y_2'' + p y_2' + q y_2] \\
&= c_1(0) + c_2(0) \\
&= 0
\end{aligned} \tag{14}
$$

This verifies that Eq. (13) is a solution of Eq. (12).

Example 1   **Superposition Example for Homogeneous Equations**   It is a simple matter to verify that both $y_1(x) = \sin x$ and $y_2(x) = \cos x$ are solutions of $y'' + y = 0$. Find five other solutions.

Solution   The principle of superposition for linear homogeneous equations says that any linear combination of $y_1(x)$ and $y_2(x)$ is a solution. Hence we arbitrarily select the five of the linear combinations $y = c_1 \cos x + c_2 \sin x$:

(a) $y(x) = 3 \sin x$ $c_1 = 0$ $c_2 = 3$

(b) $y(x) = -\cos x$ $c_1 = -1$ $c_2 = 0$

(c) $y(x) = -2 \cos x + \sin x$ $c_1 = -2$ $c_2 = 1$

(d) $y(x) = \pi \cos x - \sin x$ $c_1 = \pi$ $c_2 = -1$

(e) $y(x) = 4 \cos x - 3.5 \sin x$ $c_1 = 4$ $c_2 = -3.5$ ∎

THE INITIAL-VALUE PROBLEM FOR SECOND-ORDER EQUATIONS

We have seen that the general solution of the first-order differential equation contains *one* arbitrary constant. Although we have not yet solved the general second-order equation, we know from calculus that the general solution of the trivial second-order equation $y'' = 1$ is the *two-parameter family* of solutions, $y(x) = c_1 + c_2 x$. This leads us to suspect that two side conditions are required to uniquely solve a second-order equation. We summarize these ideas by stating the initial-value problem for second-order differential equations.

DEFINITION: Initial-Value Problem for Second-Order Equations

By an **initial-value problem for a second-order differential equation**

$$F(x, y, y', y'') = 0$$

we mean the problem of finding the solution of the differential equation on an interval that also satisfies the initial conditions

$$y(x_0) = y_0$$
$$y'(x_0) = y_1$$

where y_0 and y_1 are given constants.

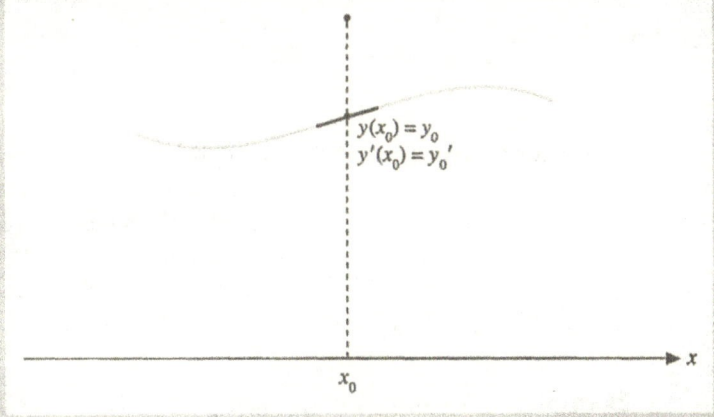

EXISTENCE AND UNIQUENESS FOR AN INITIAL-VALUE PROBLEM

Before we solve the second-order linear equation, it is useful to know if and when a solution actually exists and, if a solution does exist, on what interval it will exist. Also, how many solutions are there: zero, one, two, three, ...? One of the interesting properties of a linear equation, whether it is a linear differential equation, a linear difference equation, a linear algebraic equation, or even a system of linear equations, is that the number of solutions is always *zero, one,* or *infinity.* There will never be two, three, four, ..., solutions. The following **existence and uniqueness theorem** gives conditions *sufficient* for the second-order initial-value problem to have exactly one solution.

THEOREM 3.2: Existence and Uniqueness Theorem

Let $p(x)$, $q(x)$, and $f(x)$ be continuous functions on an open interval (a, b) containing a point x_0. For any two numbers y_0 and y_1 there is *one and only one function* satisfying

$$y'' + p(x)y' + q(x)y = f(x)$$

on the entire interval (a, b) that also satisfies the initial conditions

$$y(x_0) = y_0$$
$$y'(x_0) = y_1$$

Example 2   **Verifying the Existence Theorem**   Determine the largest interval that contains $x = 0$ on which a unique solution of

$$y'' + \left(\frac{1}{x-1}\right)y = e^{-x}$$
$$y(0) = 1 \qquad\qquad\qquad (15)$$
$$y'(0) = 0$$

is certain to exist.

Solution   We identify the functions $p(x) = 0$, $q(x) = 1/(x - 1)$, and $f(x) = e^{-x}$. The only value of x for which these functions are not all continuous is $x = 1$, where $q(x)$ is undefined. Since the initial conditions are defined at the initial point $x_0 = 0$, Theorem 3.2 guarantees that a unique solution exists on the interval $(-\infty, 1)$. We may not know the solution, but we know that one exists. We also know that a solution does not exist on any interval that contains $x = 1$.   ▩

Example 3   **Finding the Arbitrary Constant**   Find the solution of the initial-value problem

$$y'' + y = 0$$
$$y(0) = 1 \tag{16}$$
$$y'(0) = 1$$

Solution It is easy to verify that both $\sin x$ and $\cos x$ satisfy the differential equation. Using the principle of superposition, we know that any linear combination

$$y(x) = c_1 \cos x + c_2 \sin x \tag{17}$$

also satisfies the equation. To find c_1 and c_2, we substitute $y(x)$ into the initial conditions, getting

$$y(0) = c_1 \cos (0) + c_2 \sin (0) = c_1 = 1$$
$$y'(0) = -c_1 \sin (0) + c_2 \cos (0) = c_2 = 1$$

Hence we have $c_1 = c_2 = 1$, and so the unique solution of the initial-value theorem is

$$y(x) = \cos x + \sin x \tag{18}$$

See Figure 3.1 ∎

Figure 3.1
The unique solution of the initial-value problem
$y'' + y = 0$, $y(0) = 1$,
$y'(0) = 1$

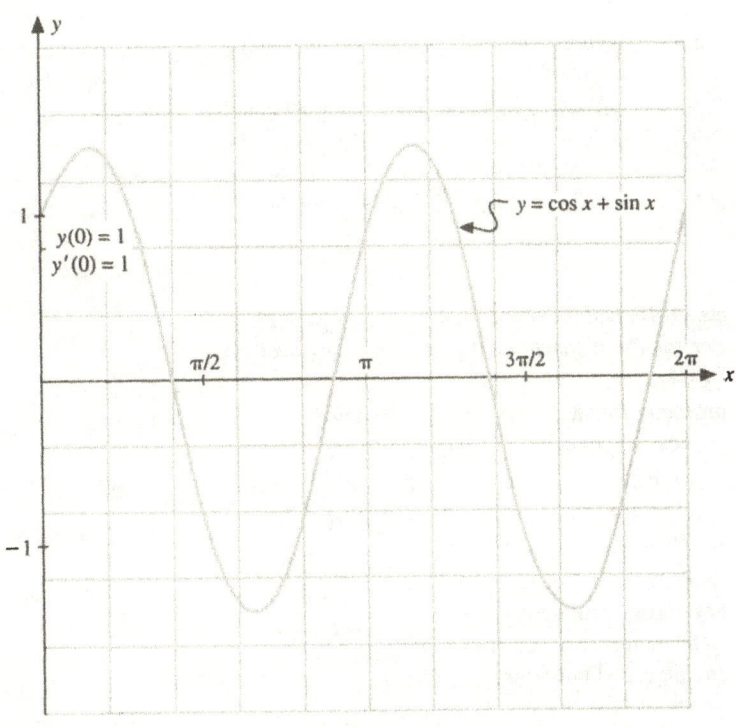

PROBLEMS: Section 3.1

For Problems 1–10, determine whether the given equation is linear or nonlinear. If it is linear, classify it as homogeneous or nonhomogeneous and with constant or variable coefficients.

1. $y'' + xy' - y = \sin x$
2. $xy'' + y' - xy = 1$
3. $x^2 y'' + y' + y^2 = 0$
4. $y'' + (y')^2 + 2y = 1$
5. $y'' - (x + 1) + xy = \sin y$
6. $y'' + 3y' + \sqrt{y} = 0$
7. $y'' + y = \tan x$
8. $yy'' + (y')^2 = 0$
9. $y'' + yy' = \sin x$
10. $y'' + (\sin x)y' + xy = e^x$
11. **Solutions of Second-Order Equations** Verify that e^{-x} and e^{2x} and any linear combination $c_1 e^{-x} + c_2 e^{2x}$ are all solutions of the differential equation

$$y'' - y' - 2y = 0$$

12. **The Principle of Superposition** Verify that e^x and e^{-x} and any linear combination $c_1 e^x + c_2 e^{-x}$ are solutions of the differential equation

$$y'' - y = 0$$

Show that the hyperbolic sine and cosine functions, sinh x and cosh x, are also solutions.
13. **Interesting Comparison** Show that the circular sine and cosine functions, sin x and cos x, are solutions of the equation $y'' + y = 0$, whereas the hyperbolic sine and cosine functions, sinh x and cosh x, are solutions of $y'' - y = 0$.

For Problems 14–18, find the specific function among the given two-parameter family of functions that satisfies the specified initial conditions.

Two-parameter family	Initial conditions
14. $y(x) = c_1 \cos x + c_2 \sin x$	$y(0) = 0 \; y'(0) = 1$
15. $y(x) = c_1 + c_2 x$	$y(0) = 1 \; y'(0) = -1$
16. $y(x) = c_1 e^x + c_2 e^{-x}$	$y(0) = 1 \; y'(0) = 0$
17. $y(x) = c_1 e^{2x} + c_2 x e^{2x}$	$y(0) = 0 \; y'(0) = 2$
18. $y(x) = c_1 e^x \cos x + c_2 e^x \sin x$	$y(0) = 1 \; y'(0) = 1$

19. **Boundary-Value Problems** When the solution of a differential equation is specified at two *different points*, these conditions are called **boundary conditions.** This is in contrast to conditions that are specified at a single point, which are called initial conditions. Given that every solution of

$$y'' + y = 0 \qquad (19)$$

has the form

$$y(x) = c_1 \cos x + c_2 \sin x$$

where c_1 and c_2 are arbitrary constants, verify the following.
(a) Show that there is a *unique solution* to Eq. (19) that satisfies the boundary conditions

$$y(0) = 1$$
$$y(\pi/2) = 0$$

(b) Show that there is *no solution* to Eq. (19) that satisfies the boundary conditions

$$y(0) = 1$$
$$y(\pi) = 0$$

(c) Show that there are an *infinite number* of solutions to Eq. (19) that satisfy the boundary conditions

$$y(0) = 1$$
$$y(\pi) = -1$$

Finding Differential Equations from Solutions

The solution of a second-order equation $y'' = F(x, y, y')$ generally involves two arbitrary constants. Conversely, if a two-parameter family of functions $y(x) = G(x, c_1, c_2)$ with parameters c_1 and c_2 is differentiated two times and the parameters are eliminated from the three resulting equations, one will obtain a second-order differential equation that will be satisfied by all the functions in the family. For Problems 20–24, find the differential equation satisfied by each of the given families.

20. $y = c_1 + c_2 x$
21. $y = c_1 \sin 2x + c_2 \cos 2x$
22. $y = c_1 e^x + c_2 e^{-x}$
23. $y = c_1 e^{2x} + c_2 e^{-x} + x$
24. $y = c_1 \sin 3x + c_2 \cos 3x + \sin x$

Second-Order Equations with Missing y

It is possible to solve a second-order differential equation of the form $y'' = F(x, y')$ by letting $v = y'$, $v' = y''$ and arriving at a first-order equation of the form $v' = F(x, v)$. If this new equation in v can be solved, it is then possible to find y by integrating $dy/dx = v(x)$. Note that a constant will be obtained in solving the first-order equation in v, and another constant will be obtained in integrating dy/dx. For Problems 25–30, *solve the given differential equations with missing y.*

25. $y' y'' = 1$

26. $xy'' + y' = x$

27. $y'' = 1 + (y')^2$

28. $xy'' + y' = 1$

29. $y'' + y' = e^{-x}$

30. $2yy'' = 1 + (y')^2$

31. The Hanging Cable A cable hanging with its ends fastened can be shown to satisfy the equation

$$y'' = k\sqrt{1 + [y'(x)]^2} \qquad (20)$$

where $y = y(x)$ describes the shape of the cable, called a *catenary*. The constant $k > 0$ is related to the size of the cable. If the ends of the cable are fastened at the two points $(a, f(a))$ and $(b, f(b))$, find the shape of the cable.

Second-Order Equations with Missing *x*

If a second-order equation has the form $y'' = F(y, y')$, where the independent variable x does not appear explicitly in the equation, then it is possible to make the substitution $y = v'$ and obtain a first-order equation $dv/dx = F(y, v)$. Since the new right-hand side depends on v and y, the trick is to think of y as the new independent variable and use the chain rule to rewrite the derivative as

$$\frac{dv}{dx} = \frac{dv}{dy}\frac{dy}{dx} = v\frac{dv}{dy}$$

For Problems 32–37, solve the given differential equations with missing x. When initial conditions are given, find the solution satisfying the stated conditions.

32. $y' y'' = 1$

33. $y'' + y = 0$

34. $y'' - y = 0$

35. $y'' = (y')^3 y \quad y(0) = 1 \quad y'(0) = -2$

36. $y'' = (y')^2$

37. $y'' = e^{-y'}$

38. Constants of Motion Given the second-order equation

$$y'' = f(x, y, y') \qquad (21)$$

a function $F(x, y, y')$ is called a **constant of motion*** (or **integral**) of Eq. (21) if

$$\frac{\partial F}{\partial x} + \frac{\partial F}{\partial y}\frac{dy}{dx} + \frac{\partial F}{\partial y'}f \equiv 0$$

on a given domain.

---

* In practice, constants of motion are difficult to find and have limited use. One use lies in the reduction of order of a differential equation. In many physical systems, constants of motion represent the total energy of the system.

(a) Show that if $y = \phi(x)$ is a solution of Eq. (21) and F is a constant of motion of Eq. (21), then the expression $F(x, \phi(x), \phi'(x))$ is a constant.

(b) Show that

$$F(y, y') = \frac{1}{2}ky^2 + \frac{1}{2}my'^2$$

is a constant of motion of $my'' + ky = 0$.

(c) Constants of motion are generally difficult to find and are relatively rare. One use of constants of motion is that they can sometimes be used to reduce the order of the equation by one. To illustrate this idea, solve the equation $F(y, y') = c$ for y in part (b), and substitute this into the differential equation $my'' + ky = 0$, getting a first-order equation.

Intervals of Existence and Uniqueness

For Problems 39–45, use the existence and uniqueness theorem (Theorem 3.2) to determine intervals for which a unique solution to the differential equation with initial conditions $y(x_0) = y_0, y'(x_0) = y'$ is certain to exist.

39. $y'' + y = 0$

40. $y'' + 4y' + 4y = 1$

41. $y'' + xy' + 2y = \sin x$

42. $xy'' + y = 1$

43. $x(x - 1)y'' + 2xy' + y = 0$

44. $y'' + x(1 - x)y' + y = 0$

45. $y'' + (\sin x)y' + y = 1$

46. Lagrange's Adjoint Equation Prove that if the function $\mu(x)$ satisfies

$$\mu(x)[y'' + p(x)y' + q(x)y] = \frac{d}{dx}[y' + g(x)y] \qquad (22)$$

for some function $g(x)$, then $\mu(x)$ must satisfy the *adjoint equation*

$$\mu'' - p(x)\mu' + [q(x) - p'(x)]\mu = 0 \qquad (23)$$

Hint: By equating coefficients in Eq. (22) and eliminating $g(x)$, one arrives at Eq. (23).

47. Computer Problem Use a graphing calculator or computer to sketch the graph of the solution of part (a) of Problem 19. Is the graph consistent with the differential equation and boundary conditions?

48. Computer Problem Use a graphing calculator or computer to sketch the shape of a cable as found in Problem 31. Use different values of the parameter k. Is the shape of the cable consistent with your intuition?

49. Journal Entry Summarize the results of this section.

3.2   FUNDAMENTAL SOLUTIONS OF THE HOMOGENEOUS EQUATION

PURPOSE
To show that all solutions of the second-order linear homogeneous equation

$$y'' + p(x)y' + q(x)y = 0$$

can be written as a linear combination

$$y = c_1 y_1 + c_2 y_2$$

of two linearly independent solutions y_1 and y_2.

Linear Independence of Two Functions

This section provides much of the theory for the remainder of the chapter. Before actually solving a second-order equation, it is important to know what one is looking for—the nature of the beast, so to speak. We begin with the important concept of linear independence (and linear dependence). This concept is important not only in the study of linear differential equations, but in the study of other areas of applied mathematics as well.*

DEFINITION: Linear Independence and Dependence

Two functions f and g are said to be **linearly dependent** on an interval I if there exist two constants k_1 and k_2, not both zero, that satisfy

$$k_1 f(x) + k_2 g(x) = 0 \tag{1}$$

for all x in the interval I. Two functions f and g are said to be **linearly independent** on an interval I if they are not linearly dependent on I. That is, Eq. (1) holds for all x in I only for $k_1 = k_2 = 0$.

Note: The above definitions of linear independence and dependence can be extended to more than two functions. In the special case of two functions, linear *dependence* on an interval I simply means that one function is a constant multiple of the other function, and linear *independence* means that neither function is a constant multiple of the other. (See Problem 21 at the end of this section.)

Example 1   **Linearly Dependent Functions**   Determine whether the following pairs of functions are linearly independent or dependent on $(-\infty, \infty)$.

(a) $f(x) = e^x$ $g(x) = x + 1$
(b) $f(x) = \sin 2x$ $g(x) = \sin x \cos x$

---

* Linear independence is intimately related to the concept of the dimension of a vector space and hence is essential to the study of many areas of mathematics.

Solution (a) By observing the graphs $y = f(x)$ and $y = g(x)$ we can see that neither function is a constant multiple of the other. Indeed, if a constant k did exist that satisfied

$$e^x = k(x + 1)$$

for all $-\infty < x < \infty$, then simply by letting $x = 0$ and $x = 1$ we would get the contradiction

$$e^0 = k(0 + 1) \, \Diamond \, k = 1$$
$$e^1 = k(1 + 1) \, \Diamond \, k = e/2$$

Hence e^x and $x + 1$ are linearly independent on $(-\infty, \infty)$. In fact, they are linearly independent on any interval.

(b) Since $f(x) = \sin 2x = 2 \sin x \cos x$, we have $f(x) = 2g(x)$ for all x. Hence they are linearly dependent on $(-\infty, \infty)$. In fact, they are linearly dependent on any interval. ∎

Example 2 **Linearly Independent Functions** The following functions are linearly independent on any interval:

(a) $\sin x$ and $\cos x$

(b) e^x and xe^x

(c) $x \sin x$ and $\sin x$

(d) x^2 and $x + 1$

(e) e^{-x} and e^x

(f) $e^x \sin x$ and $e^{-x} \sin x$

(g) $e^{2x} \sin x$ and $e^{2x} \cos x$ ∎

Example 3 **Linearly Dependent Functions** The following functions are linearly dependent on any interval:

(a) $2x$ and $-x$

(b) 0 and $f(x)$ (any function) (since $0 = 0 \cdot f(x)$)

(c) x^2 and $4x^2$ ∎

The Wronskian

Although it is generally a simple matter to determine whether two functions are linearly dependent, there is a useful tool for determining linear independence or dependence of two functions.

> **DEFINITION: The Wronskian**
> Given two differentiable functions y_1 and y_2, the function
>
> $$W(y_1, y_2) = \begin{vmatrix} y_1 & y_2 \\ y_1' & y_2' \end{vmatrix} = y_1 y_2' - y_1' y_2 \qquad (2)$$
>
> is called the **Wronskian** of y_1 and y_2.

Since the Wronskian is a function of x, we often denote it by $W[y_1, y_2](x)$, or simply by $W(x)$ when no confusion exists.

Example 4 **Sample Wronskian** (a) The Wronskian of $y_1(x) = \sin x$ and $y_2(x) = \cos x$ is

$$W(\sin x, \cos x) = \begin{vmatrix} \sin x & \cos x \\ \cos x & -\sin x \end{vmatrix} = -\sin^2 x - \cos^2 x = -1 \tag{3}$$

(b) The Wronskian of $y_1(x) = e^x$ and $y_2(x) = e^{-x}$ is

$$W(e^x, e^{-x}) = \begin{vmatrix} e^x & e^{-x} \\ e^x & -e^{-x} \end{vmatrix} = -1 - 1 = -2 \tag{4}$$

■

The importance of the Wronskian in the study of differential equations lies in the following theorem, which we state without proof.

THEOREM 3.3: Wronskian Test for Linear Independence

Assume that the coefficients $p(x)$ and $q(x)$ in the homogeneous equation

$$y'' + p(x)y' + q(x)y = 0 \tag{5}$$

are continuous on (a, b) and that y_1 and y_2 are two given solutions. If *any one* of the following statements is true, then all of the others are also true.

- y_1 and y_2 are linearly independent solutions on (a, b).
- $W[y_1, y_2](x) \neq 0$ for all x in (a, b).
- $W[y_1, y_2](x_0) \neq 0$ for at least one x_0 in (a, b).

Note: Although we will not prove this theorem, note the interesting fact that if the Wronskian of two solutions is different from zero at a *single point* on an interval, then it will be different from zero at each point of the interval. (The reader can prove this in Problem 25 at the end of this section.) In other words, two solutions y_1 and y_2 of Eq. (5) are linearly independent on an interval I if and only if the Wronskian of y_1 and y_2 is not identically zero on I.

HISTORICAL NOTE

The Wronskian is named after the Polish mathematician Józef Maria Hoene-Wroński (1778–1853). The Wronskian determinant was his sole contribution to mathematics. More a philosopher than a mathematician, he insisted that absolute truth could be attained only through mathematical reasoning. A gifted but troubled soul, his life was marked by heated disputes with individuals and institutions. He eventually went insane.

Example 5   **Wronskian Test for Linear Independence**   Show that the two functions $y_1(x) = \sin x$ and $y_2(x) = \cos x$ are linearly independent solutions of

$$y'' + y = 0 \tag{6}$$

on $(-\infty, \infty)$

Solution   By direct verification, y_1 and y_2 both satisfy Eq. (6). The Wronskian of y_1 and y_2 is

$$W(\sin x, \cos x) = \begin{vmatrix} \sin x & \cos x \\ \cos x & -\sin x \end{vmatrix} = -\sin^2 x - \cos^2 x = -1 \neq 0 \tag{7}$$

Since the Wronskian is never zero on $(-\infty, \infty)$, Theorem 3.3 guarantees that y_1 and y_2 are linearly independent on $(-\infty, \infty)$.

We have seen that if two functions y_1 and y_2 are solutions of the second-order linear homogeneous equation, then *every* linear combination $c_1 y_1 + c_1 y_2$ of these functions is a solution. We now ask whether this collection of linear combinations includes *all* solutions of the equation. The answer to this important question is given in the following theorem.

THEOREM 3.4:   Finding *All* Solutions of the Homogeneous Equation

If y_1 and y_2 are two **linearly independent solutions** of the linear homogeneous equation

$$y'' + p(x)y' + q(x)y = 0 \tag{8}$$

on an interval (a, b), then *any* solution Y of the differential equation on the interval (a, b) can be expressed uniquely as a linear combination

$$Y(x) = c_1 y_1(x) + c_2 y_2(x)$$

for some constants c_1 and c_2.

PROOF: We must show that if y_1 and y_2 are linearly independent solutions of Eq. (8), then for *any* solution $Y(x)$ of Eq. (8) it is possible to find two constants c_1 and c_2 such that $Y(x) = c_1 y_1(x) + c_2 y_2(x)$. The idea behind the proof is to show there exist constants c_1 and c_2 so that both $Y(x)$ and $c_1 y_1(x) + c_2 y_2(x)$ satisfy the *same* initial conditions, and hence by uniqueness they must be the same function. To find such constants c_1 and c_2, we select an arbitrary point x_0 in (a, b) and find c_1 and c_2 that satisfy

$$c_1 y_1(x_0) + c_2 y_2(x_0) = Y(x_0) \tag{9}$$
$$c_1 y_1'(x_0) + c_2 y_2'(x_0) = Y'(x_0)$$

These two equations uniquely determine the constants c_1 and c_2, since by hypothesis, y_1 and y_2 are linearly independent on (a, b) and thus

$$W[y_1, y_2](x_0) = \begin{vmatrix} y_1(x_0) & y_2(x_0) \\ y_1{}'(x_0) & y_2{}'(x_0) \end{vmatrix} = y_1(x_0)\,y_2{}'(x_0) - y_1{}'(x_0)\,y_2(x_0) \neq 0 \quad (10)$$

Now, since both $Y(x)$ and $c_1 y_1(x) + c_2 y_2(x)$ satisfy the same differential equation (8) and have the same *initial conditions* (9), Theorem 3.2 from Section 3.1 says that we must have $Y(x) = c_1 y_1(x) + c_2 y_2(x)$ for all x in (a, b). This proves the theorem.

Theorem 3.4 leads us to the following definition.

DEFINITION: Fundamental Set of Solutions

Any two linearly independent solutions y_1 and y_2 of

$$y'' + p(x)y' + q(x)y = 0 \quad (11)$$

is called a **fundamental set of solutions,** and the collection of all linear combinations $c_1 y_1 + c_2 y_2$ that constitute the collection of all solutions of Eq. (11) is called the **general solution.**

Example 6   **Fundamental Set of Solutions**   Show that the two functions $y_1(x) = e^x$ and $y_2(x) = e^{-2x}$ constitute a fundamental set of solutions of

$$y'' + y' - 2y = 0 \quad (12)$$

and find the general solution.

Solution   First, we verify that both functions satisfy Eq. (12). Evaluating the Wronskian, we find

$$W(e^x, e^{-2x}) = \begin{vmatrix} e^x & e^{-2x} \\ e^x & -2e^{-2x} \end{vmatrix} = -2e^{-x} - e^{-x} = -3e^{-x} \neq 0 \quad (13)$$

Since the Wronskian is nonzero at some point (in fact it's never zero), we conclude that e^x and e^{-2x} are linearly independent on $(-\infty, \infty)$. Hence e^x and e^{-2x} constitute a fundamental set of solutions of Eq. (12). Hence the general solution (all solutions) of Eq. (11) is

$$y(x) = c_1 e^x + c_2 e^{-2x} \quad (14)$$

where c_1 and c_2 are arbitrary constants.   ▣

PROBLEMS: Section 3.2

For Problems 1–10, determine whether the functions f and g are linearly independent on $(-1, 1)$. Compute the Wronskian $W(f, g)$.

1. $f(x) = 1$ $g(x) = -1$
2. $f(x) = x$ $g(x) = 0$
3. $f(x) = 1 - x$ $g(x) = x - 1$
4. $f(x) = e^{mx}$ $g(x) = e^{nx} \quad (m \neq n)$
5. $f(x) = e^x \cos x$ $g(x) = e^x \sin x$
6. $f(x) = \sin^2 x + \cos^2 x$ $g(x) = 1$
7. $f(x) = e^x \cos x$ $g(x) = e^{2x} \cos x$
8. $f(x) = x^2$ $g(x) = x^3$
9. $f(x) = x$ $g(x) = |x|$
10. $f(x) = \ln x$ $g(x) = \ln x^2$

For Problems 11–20, carry out the following steps.
(a) Verify that the functions y_1 and y_2 are solutions of the specified differential equation.
(b) Verify that y_1 and y_2 are linearly independent.
(c) Find the general solution of the differential equation.
(d) Find the solution of the indicated initial-value problem.

11 $y'' = 0$ $y_1(x) = 1$ $y_2(x) = x$ $y(0) = 0$ $y'(0) = 1$

12. $y'' - 5y' + 6y = 0$ $y_1(x) = e^{2x}$ $y_2(x) = e^{3x}$ $y(0) = 0$ $y'(0) = 0$

13. $y'' + 5y' + 6y = 0$ $y_1(x) = e^{-2x}$ $y_2(x) = e^{-3x}$ $y(0) = 0$ $y'(0) = 1$

14. $y'' - 4y = 0$ $y_1(x) = e^{2x}$ $y_2(x) = e^{-2x}$ $y(0) = 1$ $y'(0) = 1$

15. $y'' - 5y = 0$ $y_1(x) = e^{\sqrt{5}\,x}$ $y_2(x) = e^{-\sqrt{5}\,x}$ $y(0) = 1$ $y'(0) = 0$

16. $y'' + 4y = 0$ $y_1(x) = \sin 2x$ $y_2(x) = \cos 2x$ $y(0) = 1$ $y'(0) = 1$

17. $y'' - y = 0$ $y_1(x) = \sinh x$ $y_2(x) = \cosh x$ $y(0) = 0$ $y'(0) = 1$

18. $y'' - 2y' + y = 0$ $y_1(x) = e^x$ $y_2(x) = xe^x$ $y(0) = 1$ $y'(0) = 1$

19. $x^2 y'' + xy' - y = 0$ $y_1(x) = x$ $y_2(x) = x^{-1}$ $y(1) = 0$ $y'(1) = 0$

20. $x^2 y'' - 2y = 0$ $y_1(x) = x^2$ $y_2(x) = x^{-1}$ $y(1) = 1$ $y'(1) = 1$

21. Linear Dependence and Independence of Two Functions

(a) Show that if two functions f and g are linearly dependent on an interval, then one is a constant multiple of the other.

(b) Show that if neither of two functions is a multiple of the other on an interval, then the two functions are linearly independent on the interval.

22. More Than One Fundamental Set Consider the differential equation

$$y'' - y = 0 \tag{15}$$

(a) Show that both collections of functions $\{e^x, e^{-x}\}$ and $\{\sinh x, \cosh x\}$ are fundamental sets of solutions.

(b) Are there any other fundamental sets of solutions?

23. Complex Fundamental Set of Solutions Consider the differential equation

$$y'' + y = 0 \tag{16}$$

Show that both collections of functions $\{e^{ix}, e^{-ix}\}$ and $\{\sin x, \cos x\}$ are fundamental sets of solutions, where i is the complex constant $i = \sqrt{-1}$. Use *Euler's formulas:*

$$e^{ix} = \cos x + i \sin x$$
$$e^{-ix} = \cos x - i \sin x \tag{17}$$

24. Finding New Solutions from Old Ones Show that if $y_1(x)$ and $y_2(x)$ are linearly independent solutions of

$$y'' + p(x)y = 0$$

then $y_3(x) = y_1(x) + y_2(x)$ and $y_4(x) = y_1(x) - y_2(x)$ are also linearly independent solutions.

25. Wronskian Identically Zero or Never Zero* A useful property of the Wronskian of two solutions $y_1(x)$ and $y_2(x)$ of

$$y'' + p(x)y' + q(x)y = 0 \tag{18}$$

is that it is either *identically* zero for all x or *never* zero. Solve the following problems to verify this important property.

(a) First note that since $y_1(x)$ and $y_2(x)$ are solutions of Eq. (18), we have

$$\begin{aligned} y_1'' + p(x)y_1' + q(x)y_1 &= 0 \\ y_2'' + p(x)y_2' + q(x)y_2 &= 0 \end{aligned} \tag{19}$$

Multiply the first equation in Eq. (19) by $-y_2$ and the second by y_1, and add. Now, noting that the derivative of the Wronskian can be written

$$W'(x) = y_1 y_2'' - y_1'' y_2$$

show that the Wronskian $W(x)$ satisfies the first-order equation

$$W' + p(x)W = 0 \tag{20}$$

(b) Solve Eq. (20), getting the formula[†]

$$W(x) = ce^{\int p(t)\,dt} \tag{21}$$

---

* This problem was originally stated and proved by the Norwegian mathematician Niels Henrik Abel (1802–1829) and is often called *Abel's theorem.*

[†] This formula is called Abel's formula.

Hence, depending on whether the constant c is zero or nonzero, the Wronskian is either always zero or never zero.

26. Converting to Normal Form Show that the substitution $y(x) = u(x)P(x)$, where

$$P(x) = \exp\left(-\frac{1}{2}\int p(x)\,dx\right) \qquad (22)$$

will transform the standard second-order linear homogeneous equation

$$y'' + p(x)y' + q(x)y = 0 \qquad (23)$$

into the second-order linear homogeneous equation in so-called **normal form:**

$$u'' + g(x)\,u = 0 \qquad (24)$$

27. Exact Equations The differential equation

$$y'' + p(x)\,y' + q(x)\,y = 0$$

is called an **exact equation** if it can be written in the form

$$y'' + [g(x)\,y]' = 0$$

where $g(x)$ is determined from $p(x)$ and $q(x)$. An exact equation can be integrated directly, getting a first-order linear equation. The first-order equation can then be integrated by using the integrating factor method from Chapter 2. For Problems 28–30, solve the following exact equations.

28. $y'' + \dfrac{1}{x}y' - \dfrac{1}{x^2}y = 0$

29. $y'' + \dfrac{2}{x}y' - \dfrac{2}{x^2}y = 0$

30. $(x^2 - 2x)\,y'' + 4(x - 1)\,y' + 2y = e^{2x}$

Linear Independence of Several Functions

The concept of linear independence can be extended to more than two functions. We say that n functions y_1, y_2, ..., y_n are **linearly dependent** *on an interval if there exist constants k_1, k_2, ..., k_n, not all zero, such that*

$$k_1y_1(x) + k_2y_2(x) + \cdots + k_ny_n(x) = 0 \qquad (25)$$

for every x in the interval. Otherwise, they are called **linearly independent.**

For Problems 31–37, determine whether the following sets of functions are linearly independent or dependent.

31. $\{1, x, x^2\}$

32. $\{x, x + 1, 2x - 3\}$

33. $\{1, x, 2x + 3\}$

34. $\{x, x^2, \sin x\}$

35. $\{1, e^x, e^{-x}\}$

36. $\{e^x, e^{-x}, \cosh x\}$

37. $\{\sin^2 x, \cos^2 x, 1\}$

38. Sturm's Claim In the 1800's the French mathematician Jacques Charles Sturm (1803–1855) studied the number of zeros (roots) of solutions of differential equations. Using nothing more than your intuition, can you say why Sturm claimed that *any* solution of the differential equation

$$y'' + x^2\,y = 0$$

would have an infinite number of roots?

39. Computer Problem Use a graphing calculator or computer to give you some idea whether the functions $\sin 2x$ and $3\sin x \cos x$ are linearly independent or linearly dependent on the interval $[0, 2\pi]$.

40. Journal Entry Summarize to yourself the concepts of linear independence and dependence. Can you relate these ideas to the study of vectors in the plane that you studied in calculus?

3.3 REDUCTION OF ORDER

PURPOSE

To show how knowledge of one solution of the homogeneous linear equation

$$y'' + p(x)y' + q(x)y = 0$$

allows us to obtain a first-order equation in which one can (sometimes) find a second linearly independent solution.

In elementary algebra, knowledge that $x = 1$ is one root of the equation

$$x^3 - 7x + 6 = 0$$

allows one to find the other two roots by factoring out $x - 1$, getting

$$x^3 - 7x + 6 = (x - 1)(x^2 + x - 6) = 0$$

One can then solve the resulting quadratic equation, obtaining the second and third roots, $x = -3, 2$. Often in mathematics, knowledge of one solution of an equation provides means for finding other solutions.

THE REDUCTION OF ORDER METHOD

In differential equations, **reduction of order** is a method whereby knowledge of one solution of a differential equation can be used to transform the equation to an equation of lower order from which the remaining solution(s) can (hopefully) be found. For instance, if we know one (nonzero) solution to a second-order linear homogeneous equation, the reduction of order method provides a systematic way to transform the second-order equation to a *first-order* equation. From this first-order equation it is possible to find a second linearly independent solution to the original equation.

We know that if y_1 is a solution of

$$y'' + p(x)y' + q(x)y = 0 \tag{1}$$

then so is cy_1, where c is any constant. However, if we wish to find a second solution that is linearly independent of y_1, we must pick a solution that is not of the form cy_1. This suggests that we might replace the constant c by a *function* and seek solutions of the form

$$y = v(x)\, y_1(x)$$

where v is a differentiable function to be determined. We begin by computing y' and y'', which gives

$$y' = v(x)\, y_1'(x) + v'(x)y_1(x)$$
$$y'' = v(x)y_1''(x) + 2v'(x)y_1'(x) + v''(x)y_1(x)$$

HISTORICAL NOTE

The reduction of order method is due to the French mathematician, Jean d'Alembert (1717–1783), who was a contemporary of Euler and Daniel Bernoulli. In 1743 he published his *Traité de dynamique*, based on the now-famous principle of kinetics that bears his name. In 1747 he studied vibrating strings, which led to the solution of the wave equation in partial differential equations. Any student who studies partial differential equations knows of the contributions of d'Alembert.

Substituting y, y', and y'' into Eq. (1) and collecting terms, we arrive at

$$v(y_1'' + py_1' + qy_1) + v'(2y_1' + py_1) + v''y_1 = 0 \tag{2}$$

Now since $y_1(x)$ is a solution of Eq. (1), the expression in the first set of parentheses in Eq. (2) is zero. Hence on any interval over which $y_1(x)$ does not vanish we can divide by $y_1(x)$, getting

$$v'' + \left(p + 2\frac{y_1{}'}{y_1}\right)v' = 0 \tag{3}$$

Although Eq. (3) is a second-order equation in v, it is in fact a *first-order* equation in the derivative v'. Letting $w = v'$, we rewrite Eq. (3) as the first-order equation

$$w' + \left(p + 2\frac{y_1{}'}{y_1}\right)w = 0 \tag{4}$$

But Eq. (4) is a first-order separable equation that can be solved by using either the integrating factor method or separation of variables. Using separation of variables, we arrive at

$$\int \frac{dw}{w} = -2\int \frac{y_1{}'(s)}{y_1(s)}\,ds - \int p(s)\,ds \tag{5}$$

or

$$w(x) = \frac{1}{y_1{}^2(x)}\,e^{\int p(s)\,ds} \tag{6}$$

Finally, replacing w by v' in Eq. (6) and integrating, we arrive at

$$v(x) = \int \frac{e^{\int p(s)\,ds}}{y_1{}^2(r)}\,dr \tag{7}$$

This completes the method. We summarize the above discussion in the following theorem.

THEOEM 3.5: D'Alembert's Reduction of Order Method

If a nonzero solution $y_1(x)$ of

$$y'' + p(x)y' + q(x)y = 0 \tag{8}$$

is known, then a second linearly independent solution $y_2(x)$ can be found by substituting $y_2 = v(x)y_1(x)$ into Eq. (8) and solving the resulting first-order separable equation in v'. Integrating v' then gives $v(x)$, from which the solution $y_2 = v(x)\,y_1(x)$ can be found. Carrying out these steps, one arrives at

$$y_2(x) = y_1(x)\int \frac{e^{\int p(s)\,ds}}{y_1{}^2(r)}\,dr \tag{9}$$

Normally, it is easier to carry out the steps described in Theorem 3.5 to find the second linearly independent solution $y_2(x)$ than to use the formula (9).

Example 1   **Reduction of Order**   Given that $y_1(x) = e^{3x}$ is a solution of

$$y'' - 6y' + 9y = 0 \quad (-\infty < x < \infty) \tag{10}$$

find a second linearly independent solution. From the two linearly independent solutions, find the general solution.

Solution   One can easily verify that $y_1(x) = e^{3x}$ is a solution of Eq. (10). To find a second linearly independent solution, let $y = v(x)e^{3x}$. Differentiating, we get

$$y' = 3ve^{3x} + v'e^{3x}$$
$$y'' = 9ve^{3x} + 6v'e^{3x} + v''e^{3x}$$

Substituting y, y', and y'' into Eq. (10) gives

$$(9ve^{3x} + 6v'e^{3x} + v''e^{3x}) - 6(3ve^{3x} + v'e^{3x}) + 9(ve^{3x}) = 0$$

which simplifies to $v'' = 0$. Solving this trivial differential equation for $v(x)$ gives the expression $v(x) = c_1x + c_2$. And since we are interested in finding only *one* linearly independent solution, we simply pick $c_1 = 1$, $c_2 = 0$, which gives $v(x) = x$. Hence a second linearly independent solution is

$$y_2(x) = v(x)e^{3x} = xe^{3x}$$

which gives the general solution of Eq. (10) as

$$y(x) = c_1e^{3x} + c_2xe^{3x} \tag{11}$$

■

Example 2   **Reduction of Order**   Given that $y_1 = e^x$ is a solution of

$$xy'' - (2x + 1)y' + (x + 1)y = 0 \tag{12}$$

on any interval not containing the origin, find a second solution. What is the general solution of this equation?

Solution   Letting

$$y = ve^x$$

we compute

$$y' = e^xv + e^x v'$$
$$y'' = e^xv + 2e^xv' + e^xv''$$

Substituting y, y', and y'' into Eq. (12), we find

$$xe^xv'' - e^xv' = 0$$

or

$$v'' - \frac{1}{x}v' = 0 \tag{13}$$

By letting $w = v'$, Eq. (13) can be written

$$\frac{dw}{w} = \frac{dx}{x}$$

Integrating, we obtain $w = cx$. Thus $v' = cx$, and hence

$$v(x) = \frac{c}{2}x^2 + d$$

Again, since we are interested only in finding one solution, we pick $c = 2$ and $d = 0$ for convenience, getting

$$y_2(x) = v(x)e^x = x^2 e^x \tag{14}$$

Hence the general solution of Eq. (12) is

$$y(x) = c_1 e^x + c_2 x^2 e^x \tag{15}$$

PROBLEMS: Section 3.3

For Problems 1–10, a differential equation and one solution are given. Use d'Alembert's reduction of order method to find a second linearly independent solution. What is the general solution of the differential equation?

Differential equation	Solution
1 $y'' - y = 0$	$y_1(x) = e^x$
2. $y'' + y = 0$	$y_1(x) = \sin x$
3. $y'' - 4y' + 4y = 0$	$y_1(x) = e^{2x}$
4. $y'' + y' = 0$	$y_1(x) = 1$
5. $xy'' + y' = 0$	$y_1(x) = 1$
6. $xy'' - 2(x + 1)y' + 4y = 0$	$y_1(x) = e^{2x}$
7. $x^2 y'' - 6y = 0$	$y_1(x) = x^3$
8. $x^2 y'' - xy' + y = 0$	$y_1(x) = x$
9. $(x^2 + 1)y'' - 2xy' + 2y = 0$	$y_1(x) = x$
10. $y'' + \dfrac{1}{x}y' + \left(1 - \dfrac{1}{4x^2}\right)y = 0$	$y_1(x) = \dfrac{1}{\sqrt{x}}\sin x$

11. Reducing the General Equation One solution of the equation

$$y'' - 2by' + b^2 y = 0$$

is $y_1(x) = e^{bx}$. Find a second linearly independent solution. What is the general solution?

Solving Nonhomogeneous from Homogeneous Ones

It is possible to use the reduction of order method to find a single solution of the nonhomogeneous equation

$$y'' + p(x)y' + q(x)\,y = f(x) \tag{16}$$

knowing a nonzero solution $y_1(x)$ of the corresponding homogeneous equation

$$y'' + p(x)y' + q(x)y = 0 \tag{17}$$

We use the same technique and substitute $y(x) = v(x)y_1(x)$ into Eq. (16), finding the unknown function $v(x)$. For Problems 12–17, use this technique to find a solution of the given nonhomogeneous equation given the single solution $y_1(x)$ of the corresponding homogeneous equation.

Nonhomogeneous equation	Homogeneous solution
12. $y'' = 1$	$y_1(x) = 1$
13. $y'' - y = e^x$	$y_1(x) = e^{-x}$
14. $y'' + y' = e^x$	$y_1(x) = 1$
15. $y'' + y = \csc x$	$y_1(x) = \sin x$
16. $x^2 y'' - xy' + y = x$ $(x \neq 0)$	$y_1(x) = x$
17. $x^2 y'' + xy' - y = x$ $(x \neq 0)$	$y_1(x) = x$

The equations in Problems 18–21 are some of the most famous differential equations in physics. Use the given solution $y_1(x)$ to find a second linearly independent solution of these equations.

18. Hermite's Equation of Order 2

$$y'' - 2xy' + 4y = 0 \quad y_1(x) = 1 - 2x^2$$

19. Chebychev's Equation of Order 1

$$(1 - x^2)y'' - xy' + y = 0 \quad y_1(x) = x$$

20. Laguerre's Equation of Order 1

$$xy'' + (1 - x)y' + y = 0 \quad y_1(x) = x - 1$$

21. Bessel's Equation of Order 1/2

$$x^2y'' + xy' + \left(x^2 - \frac{1}{4}\right)y = 0 \quad y_1(x) = x^{-1/2}\sin x$$

Reduction of Higher-Order Equations

For Problems 22–23, use the method of reduction to reduce the given third-order differential equation to a homogeneous linear equation of degree one less than the given degree. If possible, solve the lower-degree differential equation to find as many linearly independent solutions as the degree of the differential equation.

22. $y''' - 3y'' + 2y' = 0 \qquad y_1(x) = 1$

23. $y''' - 2y'' - y' + 2y = 0 \quad y_1(x) = e^{2x}$

24. Interesting Equation A fascinating differential equation is

$$xy'' - (x + n)y' + ny = 0 \qquad (x > 0)$$

where n is a nonnegative integer.

(a) Verify that $y_1(x) = e^x$ is a solution.

(b) Use the method of reduction to find a second solution of the form

$$y_2(x) = ce^x \int x^n e^{-x}\, dx$$

(c) Calculate $y_2(x)$ for $n = 1$ and 2. Convince yourself (when $c = -1/n!$) that $y_2(x)$ is the polynomial

$$y_2(x) = 1 + \frac{x}{1!} + \frac{x^2}{2!} + \cdots + \frac{x^n}{n!}$$

This equation is interesting in that one solution is e^x while the second solution is the first $n + 1$ terms of the Maclaurin series for e^x. (One wonders whether there are other differential equations that have one transcendental solution ($\ln x$, $\sin x$, ...) and a second solution that is a Taylor series polynomial approximation to the first solution.)

25. Another Way to Find a Second Solution Suppose $y_1(x)$ is a nonzero solution of

$$y'' + p(x)y' + q(x)y = 0$$

and $y_2(x)$ is a second linearly independent solution of the differential equation.

(a) Verify the identity

$$\frac{d}{dx}\left(\frac{y_2}{y_1}\right) = \frac{W(y_1, y_2)}{y_1^2}$$

where $W(y_1, y_2)$ is the Wronskian of $y_1(x)$ and $y_2(x)$.

(b) Use the formula in part (a) and Abel's formula

$$W(y_1, y_2) = ce^{-\int p(t)\, dt} \qquad (18)$$

to find a second solution $y_2(x)$.

(c) Use the method discussed in parts (a) and (b) to find a second solution of

$$y'' - y = 0$$

given that one solution is $y_1(x) = e^x$.

26. Solving Higher-Order Equations with Reduction of Order Let $y_1(x)$ be a solution of the nth-order equation

$$a_0(x)\frac{d^ny}{dx^n} + a_1(x)\frac{d^{n-1}y}{dx^{n-1}} + \cdots$$
$$+ a_{n-1}(x)\frac{dy}{dx} + a_n(x)y = 0 \quad (19)$$

Substituting $y(x) = y_1(x)v(x)$ into Eq. (19) reduces the equation to a linear homogeneous equation of order $n - 1$ in $v'(x)$. One can then solve this lower-order equation for $v'(x)$ and then integrate to find $v(x)$.

(a) Verify that $y_1(x) = x$ is a solution of the third-order equation.

$$x^3y''' - 3x^2y'' + x(6 - x^2)y' - (6 - x^2)y = 0 \quad (20)$$

and show that by substituting $y = y_1v$ into Eq. (20), one arrives at the second-order equation in v':

$$(v')'' - v' = 0 \qquad (21)$$

(b) Solve Eq. (21) for $v'(x)$ and integrate to find two linearly independent solutions.

(c) Use the results from parts (a) and (b) to find the general solution to Eq. (20).

27. Computer Problem Use a graphing calculator or computer to sketch the graphs of the solutions $y_1(x)$ and $y_2(x)$ on $0 < x \le 1$ of the differential equation in Problem 24 for $n = 1, 2, 3$, and 4.

28. Journal Entry Discuss how differential equations are used in your major area of interest. Are differential equations essential for an understanding of your major area of interest? Remember, your journal entry is for your own benefit.

3.4 HOMOGENEOUS EQUATIONS WITH CONSTANT COEFFICIENTS: REAL ROOTS

PURPOSE

To find the general solution of the second-order linear homogeneous equation with constant coefficients

$$ay'' + by' + cy = 0$$

when the roots of the characteristic polynomial

$$am^2 + bm + c = 0$$

are real and unequal and when they are real and equal.

We have seen that the general solution of the second-order linear homogeneous equation is a linear combination of two linearly independent solutions. In this and the next section we will actually find the general solution for second-order linear equations with *constant coefficients*. The corresponding problem with variable coefficients is more difficult, and in Chapter 4 we will study the special case in which the coefficient functions are analytic.

Consider solving the second-order linear homogeneous equation

$$ay'' + by' + cy = 0 \tag{1}$$

where *a*, *b*, and *c* are constants with $a \neq 0$. That is, we seek a function $y(x)$ whose second derivative times *a*, plus its first derivative times *b*, plus itself times *c* is identically zero. Stated in this way, it seems natural to seek a function whose derivatives are constant multiples of itself. That is, we look for solutions* of the form

$$y(x) = e^{mx} \tag{2}$$

We have that $y(x) = e^{mx}$ is a solution of Eq. (1) if and only if

$$a(e^{mx})'' + b(e^{mx})' + c(e^{mx}) = 0$$

or, equivalently,

$$e^{mx}(am^2 + bm + c) = 0$$

But since e^{mx} is never zero, we conclude that $y(x) = e^{mx}$ is a solution of Eq. (1) if and only if *m* satisfies the **characteristic equation**

$$am^2 + bm + c = 0 \tag{3}$$

---

* Looking for exponential solutions of this equation was the original idea of Leonhard Euler (1707–1783). Euler and Joseph Louis Lagrange (1736–1813) are considered to be the two greatest mathematicians of the eighteenth century.

Solving the characteristic equation, we get

$$r_1 = \frac{-b + \sqrt{b^2 - 4ac}}{2a}, \qquad r_2 = \frac{-b - \sqrt{b^2 - 4ac}}{2a} \tag{4}$$

We must consider three cases: when the roots are *real and distinct,* when the roots are *real and equal,* and when the roots are *complex.* The reader will recall that the nature of the roots depends on whether the **discriminant,** $D \equiv b^2 - 4ac$, is positive, zero, or negative. In this section we consider the cases in which the roots are real and unequal ($D > 0$) and real and equal ($D = 0$).

REAL AND UNEQUAL ROOTS

We first consider the case in which the roots of the characteristic polynomial are real and unequal.

> **CASE I: Real and Unequal Roots ($b^2 - 4ac > 0$)**
> If the characteristic equation of
>
> $$ay'' + by' + cy = 0 \tag{5}$$
>
> has real and distinct roots r_1 and r_2, then $e^{r_1 x}$ and $e^{r_2 x}$ are linearly independent solutions of Eq. (5), and the general solution is
>
> $$y(x) = c_1 e^{r_1 x} + c_2 e^{r_2 x}$$
>
> where c_1 and c_2 are arbitrary constants.

Example 1   Real and Unequal Roots   Find the general solution of

$$y'' + 5y' - 6y = 0 \tag{6}$$

Solution   The characteristic equation of Eq. (6) is

$$m^2 + 5m - 6 = (m - 1)(m + 6) = 0$$

which has two real roots $r_1 = -6$, $r_2 = 1$. Hence two linearly independent solutions are

$$y_1(x) = e^{-6x} \qquad \text{and} \qquad y_2(x) = e^x$$

and the general solution is

$$y(x) = c_1 e^{-6x} + c_2 e^x \tag{7}$$

Example 2   Initial-Value Problem   Solve the initial-value problem

$$y'' - y = 0$$
$$y(0) = 1 \tag{8}$$
$$y'(0) =$$

Solution   The characteristic equation is $m^2 - 1 = 0$, which has roots $r_1 = -1$ and $r_2 = 1$. Hence the general solution is

$$y(x) = c_1 e^{-x} + c_2 e^x \tag{9}$$

To obtain c_1 and c_2, we first differentiate $y(x)$, getting

$$y'(x) = -c_1 e^{-x} + c_2 e^x$$

Substituting $y(x)$ and $y'(x)$ into the initial conditions, we obtain

$$y(0) = c_1 + c_2 = 1$$
$$y'(0) = -c_1 + c_2 = 0$$

Solving this linear system gives $c_1 = c_2 = 1/2$. Hence the solution is

$$y(x) = \frac{e^{-x} + e^x}{2} = \cosh x \tag{10}$$

Repeated Roots   We now consider the equation

$$ay'' + by' + cy = 0 \tag{11}$$

in which the roots of the characteristic equation are real and equal. This happens when $D = b^2 - 4ac = 0$, which results in the single root $r = -b/2a$, which we call a repeated root. From this root we obtain *one* nonzero solution, $y = e^{-(b/2a)x}$. To find a second linearly independent solution, it is possible to use the *reduction of order* method and seek a second solution of the form

$$y = v(x)e^{-(b/2a)x} \tag{12}$$

Differentiating Eq. (12), we obtain

$$y' = v'(x)\, e^{-(b/2a)x} - \frac{b}{2a}\, v(x)e^{-(b/2a)x}$$

$$y'' = \left(v''(x) - \frac{b}{a}v'(x) + \frac{b^2}{4a^2}v(x)\right)e^{-(b/2a)x}$$

Substituting y, y', and y'' into

$$ay'' + by' + cy = 0 \tag{13}$$

gives

$$a\left(v'' - \frac{b}{a}v' + \frac{b^2}{4a^2}v\right) + b\left(v' - \frac{b}{2a}v\right) + cv = 0$$

Collecting terms, we find

$$av'' - \left(\frac{b^2}{4a} - c\right)v = 0$$

However, since $b^2 - 4ac = 0$, this equation reduces to

$$v'' = 0$$

Hence

$$v(x) = c_1x + c_2 \tag{14}$$

where c_1 and c_2 are arbitrary constants. Since we are interested only in finding one solution, we pick $c_1 = 1$ and $c_2 = 0$, which gives $v(x) = x$. Hence we obtain a second solution

$$y_2(x) = v(x)e^{-(b/2a)x} = xe^{-(b/2a)x} \tag{15}$$

Thus the general solution of Eq. (13) is

$$y(x) = c_1e^{-(b/2a)x} + c_2xe^{-(b/2a)x} \tag{16}$$

These ideas are summarized below.

CASE II: Repeated Roots ($b^2 - 4ac = 0$)

If the characteristic equation of

$$ay'' + by' + cy = 0 \tag{17}$$

has **repeated roots** $r = -b/2a$, then Eq. (17) has two linearly independent solutions:

$$y_1(x) = e^{-(b/2a)x} \quad \text{and} \quad y_2(x) = xe^{-(b/2a)x}$$

The general solution is

$$y(x) = c_1e^{-(b/2a)x} + c_2xe^{-(b/2a)x}$$

where c_1 and c_2 are arbitrary constants.

Example 3   Repeated Roots   Find the general solution of

$$y'' + 4y' + 4y = 0 \tag{18}$$

Solution   The characteristic equation

$$m^2 + 4m + 4 = (m + 2)^2 = 0$$

has repeated roots $r_1 = r_2 = -2$. Hence two linearly independent solutions are

$$y_1(x) = e^{-2x} \quad \text{and} \quad y_2(x) = xe^{-2x}$$

and the general solution is

$$y(x) = c_1 e^{-2x} + c_2 x e^{-2x} \tag{19}$$

PROBLEMS: Section 3.4

For Problems 1–19, find the general solution of the given differential equation. When initial conditions are given, find the solution that satisfies the stated conditions.

1. $y'' = 0$
2. $y'' - y' = 0$
3. $y'' - 9y = 0$
4. $4y'' - y = 0$
5. $y'' - 3y' + 2y = 0$
6. $y'' - y' - 2y = 0$
7. $y'' + 2y' + y = 0$
8. $4y'' - 4y' + y = 0$
9. $2y'' - 3y' + y = 0$
10. $y'' - 6y' + 9y = 0$
11. $y'' - 8y' + 16y = 0$
12. $y'' - 25y = 0$ $y(0) = 0$ $y'(0) = 0$
13. $y'' + y' - 2y = 0$ $y(0) = 1$ $y'(0) = 0$
14. $y'' + 2y' + y = 0$ $y(0) = 0$ $y'(0) = 1$
15. $y'' - 9y = 0$ $y(0) = 1$ $y'(0) = 0$
16. $y'' - 6y' + 9y = 0$ $y(0) = 1$ $y'(0) = 0$
17. $y'' - 4y' + 4y = 0$ $y(0) = 1$ $y'(0) = 1$
18. $4y'' - 4y' - 3y = 0$ $y(0) = 0$ $y'(0) = 1$
19. $y'' - 2y' + y = 0$ $y(1) = 0$ $y'(1) = 1$
20. **Alternative Way of Writing the General Solution** Show that the general solution of

$$y'' - y = 0$$

can be written as

$$y(x) = c_1 \sinh x + c_2 \cosh x$$

Euler's Equation

One of the most famous differential equations is Euler's equation:

$$ax^2 y'' + bxy' + cy = 0 \tag{20}$$

where $a \neq 0$, b, and c are constants. Show that Euler's equation has solutions of the form x^r, which can be found by substituting x^r into Eq. (20) and obtaining the characteristic equation

$$ar(r - 1) + br + c = 0 \tag{21}$$

Hence if Eq. (21) has two distinct roots r_1 and r_2, then the general solution has the form

$$y(x) = c_1 x^{r_1} + c_2 x^{r_2} \tag{22}$$

where c_1 and c_2 are arbitrary constants. Problems 21–23 involve Euler's equation.

21. **Finding Euler's Characteristic Equation** Substitute x^r into Euler's equation and show that r satisfies the characteristic equation (21).
22. **Solving Euler's Equation** Solve the following Euler equations.
    (a) $x^2 y'' + xy' = 0$
    (b) $x^2 y'' + 2xy' - 12y = 0$
    (c) $4x^2 y'' + 8xy' - 3y = 0$
23. **Transforming Euler's Equation**
    (a) Show that by making a change in the independent variable, Euler's equation (20) can be transformed into the constant coefficient equation

    $$\frac{d^2 y}{dt^2} + (b - 1)\frac{dy}{dt} + cy = 0 \tag{23}$$

    by the substitution $x = e^t$. *Hint*: Show $x\, dy/dx = dy/dt$ and $x^2\, d^2y/dt^2 = d^2y/dt^2 - dy/dt$.
    (b) Use the result from part (a) to solve the Euler equation

    $$x^2 \frac{d^2 y}{dx^2} + 2x \frac{dy}{dx} - 6y = 0$$

24. **Just a Final Test of Your Intuition** We have two curves. The first curve starts at $y(0) = 1$, and its *rate of increase* is equal to its height, that is, it satisfies the first-order equation $y' = y$. The second curve starts at the same point $y(0) = 1$ with $y'(0) = 0$, and its *upward curvature* is equal to its height, that is, it satisfies the second-order equation $y'' = y$. Which curve lies above the other? Make a guess before resolving the question mathematically.
25. **Journal Entry** Without looking at the text, summarize this section to yourself in your own language.

3.5   HOMOGENEOUS EQUATIONS WITH CONSTANT COEFFICIENTS: COMPLEX ROOTS

PURPOSE
To find the general solution of the equation

$$ay'' + by' + cy = 0$$

when the roots of the characteristic equation are complex. If the roots are given by $r_1 = p + iq$ and $r_2 = p - iq$, then we shall show that the general solution is

$$y(x) = c_1 e^{px} \cos qx + c_2 e^{px} \sin qx$$

where c_1 and c_2 are arbitrary constants.

Until now, solutions of the differential equation

$$ay'' + by' + cy = 0 \tag{1}$$

have consisted mostly of growth and decay type curves, such as e^{2x}, e^{-x}, or xe^{-3x}. Now things are going to change, and we will obtain solutions that one often associates with second-order phenomena such as oscillating circuits, vibrating springs, and periodic motion of a pendulum.

CHARACTERISTIC EQUATION WITH COMPLEX ROOTS

Consider the case in which the discriminant, $D = b^2 - 4ac$, of the characteristic equation

$$am^2 + bm + c = 0 \tag{2}$$

is negative. In this case the two roots are the *complex conjugate* numbers*

$$r_1 = p + iq = -\frac{b}{2a} + i\frac{\sqrt{4ac - b^2}}{2a}$$

$$r_2 = p - iq = -\frac{b}{2a} - i\frac{\sqrt{4ac - b^2}}{2a}$$

where i is the complex constant $i = \sqrt{-1}$. Hence the general solution of Eq. (1) is

$$y(x) = c_1 e^{(p + iq)x} + c_2 e^{(p - iq)x} \tag{3}$$

where c_1 and c_2 are arbitrary constants. We should point out that this function is a **complex-valued function** of the real variable x, whereas up to this point, all functions

---

* Readers who are uncomfortable with complex numbers and complex-valued functions can read the Appendix to obtain a brief overview.

considered have been real-valued. Again, we advise readers who feel uncomfortable with complex numbers and functions to spend a brief time reading the Appendix at the back of the book.

Assuming now more familiarity with complex functions, we will show how it is possible to rewrite the complex-valued function in Eq. (3) in a more convenient form using (arguably) the most famous equation in mathematics, **Euler's formula:**

$$e^{ix} = \cos x + i \sin x \tag{4}$$

Using Euler's formula, we rewrite Eq. (3), getting

$$
\begin{aligned}
y(x) &= c_1 e^{(p + iq)x} + c_2 e^{(p - iq)x} \\
&= c_1 e^{px} e^{iqx} + c_2 e^{px} e^{-iqx} \\
&= c_1 e^{px} [\cos qx + i \sin qx] + c_2 e^{px} [\cos qx - i \sin qx] \\
&= (c_1 + c_2) e^{px} \cos qx + (ic_1 - ic_2) e^{px} \sin qx \\
&= C_1 e^{px} \cos qx + C_2 e^{px} \sin qx
\end{aligned}
$$

where $C_1 = c_1 + c_2$ and $C_2 = ic_1 - ic_2$. Since c_1 and c_2 are arbitrary constants, the new constants C_1 and C_2 are also arbitrary. We summarize these ideas.

CASE III: Characteristic Equation with Complex Roots

If the characteristic equation of the differential equation

$$ay'' + by' + cy = 0 \tag{5}$$

has **complex roots** $p \pm iq$, then Eq. (5) has two linearly independent solutions

$$y_1(x) = e^{px} \cos qx \qquad \text{and} \qquad y_2(x) = e^{px} \sin qx$$

The general solution of Eq. (5) is

$$y(x) = c_1 e^{px} \cos qx + c_2 e^{px} \sin qx$$

where c_1 and c_2 are arbitrary constants.*

Because of the importance of Euler's formula in mathematics, we present its proof.

THEOREM 3.6: Euler's Formula

If x is any real number, then

$$e^{ix} = \cos x + i \sin x$$

PROOF: It is possible to expand the complex function e^{ix} by using the Taylor series expansion from calculus to get

---

* In fact the constants c_1 and c_2 can be complex constants. However, if we are interested in real solutions, then we should pick c_1 and c_2 to be real constants.

$$e^{ix} = 1 + (ix) + \frac{(ix)^2}{2!} + \frac{(ix)^3}{3!} + \cdots + \frac{(ix)^n}{n!} + \cdots$$

$$= 1 + ix - \frac{x^2}{2!} + \frac{ix^3}{3!} + \frac{x^4}{4!} + \frac{ix^5}{5!} + \cdots$$

$$= \left(1 - \frac{x^2}{2!} + \frac{x^4}{4!} - \cdots\right) + i\left(x - \frac{x^3}{3!} + \frac{x^5}{5!} - \cdots\right)$$

$$= \cos x + i \sin x \qquad\blacksquare$$

This completes the proof. We can also replace x by $-x$ to obtain a second formula*

$$e^{-ix} = \cos(-x) + i \sin(-x) = \cos x - i \sin x$$

Example 1 **Complex Roots**   Find a general solution to

$$y'' + 2y' + 4y = 0 \tag{6}$$

Solution   The roots of the characteristic equation

$$m^2 + 2m + 4 = 0 \tag{7}$$

are

$$r = \frac{-2 \pm \sqrt{4 - 16}}{2} = \frac{-2 \pm \sqrt{-12}}{2} = -1 \pm i\sqrt{3} \tag{8}$$

Since $p = -1$ and $q = \sqrt{3}$, the general solution is

$$y(x) = c_1 e^{-x} \cos\left(\sqrt{3}x\right) + c_2 e^{-x} \sin\left(\sqrt{3}x\right) \tag{9}$$

$$\blacksquare$$

HISTORICAL NOTE

Leonhard Euler (1707–1783) was born in Basel, Switzerland, and studied under Johann Bernoulli. Euler, one of the greatest mathematicians of all time, was responsible for relating the three most famous constants of mathematics in a single equation: $e^{2\pi i} = 1$. Today, students of mathematics often wear T-shirts bearing the slogan "Mathematicians—We're Number $e^{2\pi i}$."

* In previous discussions in this section we glossed over many details concerning complex numbers and complex-valued functions. For example, in the proof of Euler's formula (Theorem 3.6) we used the fact that it is possible to rearrange the terms of a power series with complex coefficients. For a quick overview of some of the ideas of complex numbers and functions, see the Appendix at the back of the book. For a more thorough discussion of complex numbers, see *Complex Variables for Scientists and Engineers*, Second Edition, by John D. Paliouras and Douglas S. Meadows (Macmillan, 1990).

Example 2   Complex Roots   Find the solution to the initial-value problem

$$y'' + 4y = 0$$
$$y(0) = 1$$
$$y'(0) = -1 \tag{10}$$

Solution   The roots of the characteristic equation $m^2 + 1 = 0$ are $r = \pm 2i$. Hence we identify $p = 0$ and $q = 2$, getting the general solution

$$y(x) = c_1 \cos 2x + c_2 \sin 2x \tag{11}$$

where c_1 and c_2 are arbitrary constants. Substituting Eq. (11) into the initial conditions, we get

$$y(0) = c_1 \cos (0) + c_2 \sin (0) = 1$$
$$y'(0) = -2c_1 \sin (0) + 2c_2 \cos (0) = -1$$

or

$$c_1 = 1$$
$$2c_2 = -1 \Rightarrow c_2 = -1/2$$

Hence the solution to the initial-value problem is

$$y(x) = \cos 2x - \frac{1}{2} \sin 2x$$

The graph of this function is shown in Figure 3.2.

Figure 3.2
Solution of the initial-value problem

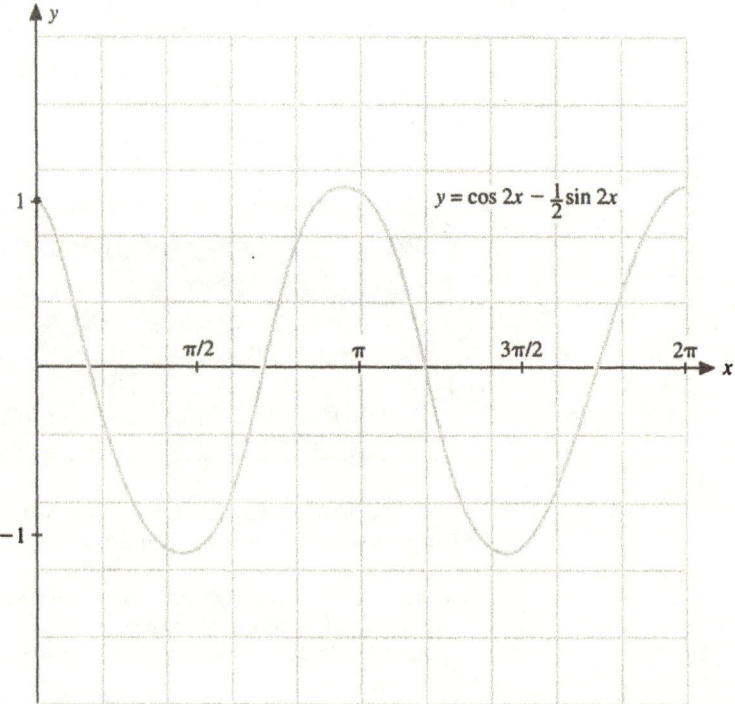

Finally, Table 3.1 summarizes the general solution of the second-order linear differential equation with constant coefficients.

Table 3.1
General Solution
of the Equation
$ay'' + by' + cy = 0$

Nature of roots of the characteristic equation	Conditions on the coefficients	General solution
Real and unequal $(r_1 \neq r_2)$	$b^2 - 4ac > 0$	$y(x) = c_1 e^{r_1 x} + c_2 e^{r_2 x}$
Real and equal $(r_1 = r_2)$	$b^2 - 4ac = 0$	$y(x) = c_1 e^{rx} + c_2 x e^{rx}$
Complex roots $(p + iq, p - iq)$	$b^2 - 4ac < 0$	$y(x) = e^{px}(c_1 \cos qx + c_2 \sin qx)$

PROBLEMS: Section 3.5

For Problems 1–15, determine the general solution of the given differential equation. If initial conditions are given, find the solution satisfying the stated conditions.

1. $y'' + 9y = 0$
2. $y'' + y' + y = 0$
3. $y'' - 4y' + 5y = 0$
4. $y'' + 2y' + 8y = 0$
5. $y'' + 2y' + 4y = 0$
6. $y'' - 4y' + 7y = 0$
7. $y'' - 10y' + 26y = 0$
8. $3y'' + 4y' + 9y = 0$
9. $y'' - y' + y = 0$
10. $y'' + y' + 2y = 0$
11. $y'' + 4y = 0$ $\quad y(0) = 1 \quad y'(0) = -1$
12. $y'' - 4y' + 13y = 0$ $\quad y(0) = 1 \quad y'(0) = 0$
13. $y'' + 2y' + 2y = 0$ $\quad y(0) = 1 \quad y'(0) = 0$
14. $y'' - y' + y = 0$ $\quad y(0) = 1 \quad y'(0) = 0$
15. $y'' - 4y' + 7y = 0$ $\quad y(0) = -1 \quad y'(0) = 0$
16. **Change of Variables Method** Solve

$$\frac{d^2 y}{dx^2} + \omega^2 y = 0 \tag{12}$$

by letting $v = dy/dx$ and rewriting

$$\frac{d^2 y}{dx^2} = \frac{dv}{dx} = \frac{dv}{dy}\frac{dy}{dx} = v\frac{dv}{dy} \tag{13}$$

Show that the general solution can be written in the form

$$y(x) = A \sin(\omega x + \alpha) \tag{14}$$

where A and α are arbitrary constants.

17. **An Interesting Equation*** Solve

$$y''(x) = y(-x) \tag{15}$$

Hint: Differentiate Eq. (15) with respect to x and then substitute Eq. (15) into the differentiated equation to find an equation you can solve.

18. **Nature of Solutions** Figure 3.3 shows the qualitative nature of the solutions of

$$ay'' + by' + cy = 0$$

as a function of the roots r_1 and r_2 of the characteristic equation.

The solutions can be categorized as belonging to one of eight different cases:

(i) $r_1 < 0, r_2 < 0$
(ii) $r_1 < 0, r_2 = 0$
(iii) $r_1 = 0, r_2 = 0$
(iv) $r_1 = 0, r_2 > 0$
(v) $r_1 > 0, r_2 > 0$
(vi) $r_1, r_2 = p \pm iq \quad (p < 0)$
(vii) $r_1, r_2 = p \pm iq \quad (p = 0)$
(viii) $r_1, r_2 = p \pm iq \quad (p > 0)$

(a) Which cases approach zero as $t \to \infty$?
(b) Which cases approach ∞ at $t \to \infty$?
(c) Which cases have damped oscillations?
(d) Which cases have oscillations that get larger?

* This problem was taken from *Math. Magazine*, vol. 47, no. 5 (1974).

Figure 3.3
Qualitative nature of the
solution as a function of the
roots of the characteristic
equation

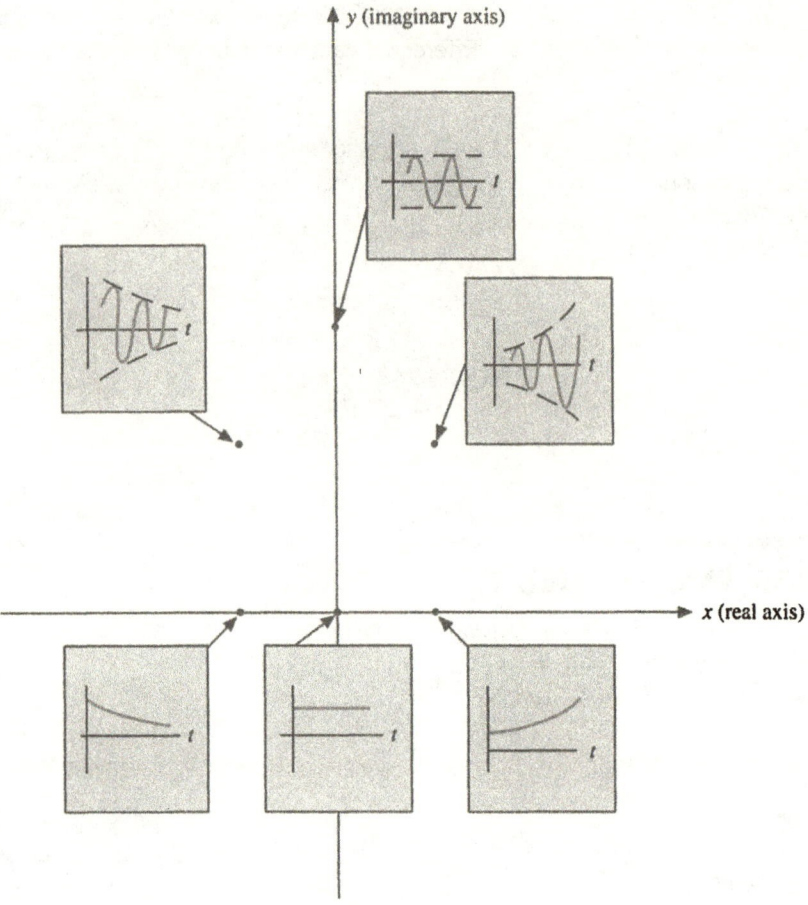

19. Riccati Equation

(a) **Show that the substitution**

$$y = e^{\int v(x)\, dx} \tag{16}$$

transforms the equation

$$y'' + p(x)y' + q(x)y = 0 \tag{17}$$

into the **Riccati equation**

$$v' = -q(x) - p(x)v - v^2 \tag{18}$$

Hence if one can solve the Riccati equation for v, then one solution of Eq. (17) can be found by simply evaluating y in Eq. (16).

(b) Transform

$$xy'' + y' - x^3 y = 0 \tag{19}$$

to a Riccati equation and solve the Riccati equation by inspection.

(c) Find one solution of Eq. (19).

(d) Find a second solution of Eq. (19).

20. Lesson in Complex Numbers

Use Euler's formula to show that any complex number $x + iy$ can be written in exponential form $re^{i\theta}$, where $r \geq 0$ and $-\pi < \theta \leq \pi$. Use this result to write the following complex numbers in exponential form.

(a) i　　　　(d) $1 + i$

(b) $-i$　　　(e) $1 - i$

(c) -3　　　(f) $-1 + i$

Boundary-Value Problems

*When the solution of a differential equation is specified at more than one point, these conditions are called **boundary conditions**, and the problem of finding the solution of the differential equation that also satisfies the boundary conditions is called a **boundary-value problem**.*

21. The Easiest Boundary-Value Problem

Anyone can solve this boundary-value problem. Plot the two points $(0, 0)$ and $(1, 1)$ in the xy-plane and connect them with a straight line. Congratulations! You have just solved the boundary-value problem

$$y'' = 0 \quad \text{(differential equation)}$$

$$\begin{aligned} y(0) &= 0 \\ y(1) &= 1 \end{aligned} \quad \text{(boundary conditons)}$$

Can you find the formula for your solution?

22. **Typical Boundary-Value Problem** Given that the general solution of differential equation

$$y'' + y = 0 \qquad (20)$$

is

$$y = c_1 \cos x + c_2 \sin x$$

where c_1 and c_2 are constants, verify the following.

(a) There is *exactly one* solution of Eq. (20) that satisfies the boundary conditions $y(0) = 1$ and $y(\pi/2) = 0$.

(b) There are *infinitely many* solutions of Eq. (20) that satisfy the two boundary conditions $y(0) = 1$ and $y(\pi) = -1$.

(c) There are *no* solutions of Eq. (20) that satisfy the boundary conditions $y(0) = 1$ and $y(\pi) = 0$.

23. **Journal Entry** Summarize what you know about the solution of $ay'' + by' + cy = 0$.

3.6   NONHOMOGENEOUS EQUATIONS

PURPOSE

To show that the general solution of the second-order linear nonhomogeneous equation

$$y'' + p(x)y' + q(x)y = f(x)$$

has the form

$$y(x) = c_1 y_1(x) + c_2 y_2(x) + y_p(x)$$

where y_1 and y_2 are linearly independent solutions of the corresponding homogeneous equation

$$y'' + p(x)y' + q(x)y = 0$$

and $y_p(x)$ is *any* single solution of the nonhomogeneous equation.

In applied mathematics the differential equation

$$y'' + p(x)y' + q(x)y = f(x) \qquad (1)$$

is often thought of as an *input-output system* in which the nonhomogeneous term $f(x)$ represents the *input* to the system and the solution $y(x)$ represents the *output* of the system resulting from $f(x)$. See Figure 3.4.

Figure 3.4
A nonhomogeneous linear differential equation can be interpreted as a linear input-output system in which the nonhomogeneous term is the input and the solution of the equation is the output of the system

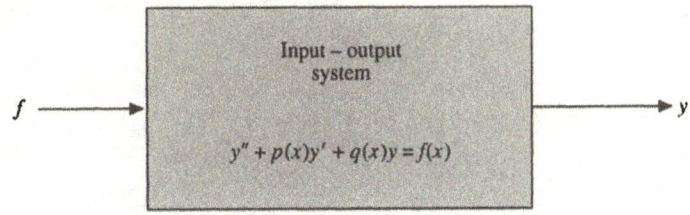

PRINCIPLE OF SUPERPOSITION FOR NONHOMOGENEOUS EQUATIONS

We have seen that linear homogeneous differential equations obey the **principle of superposition.** We now introduce a version of this principle that is satisfied for nonhomogeneous linear equations. In terms of linear input-output systems the principle of superposition says the following.

DEFINITION: Superposition Principle for Input-Output Systems

1. **Scalar Property of Superposition.** If an input f gives rise to an output y, then the input kf gives rise to the output ky, where k is any constant.

2. **Additive Property of Superposition.** If the input f_1 gives rise to an output y_1 and f_2 gives rise to the output y_2, then $f_1 + f_2$ will give rise to the output of $y_1 + y_2$.

Roughly stated, the scalar property says that if you double (triple, etc.) the input, you will double (triple, etc.) the output, and the additive property says that the sum of two inputs gives rise to the sum of the outputs. Stated in terms of differential equations, we have the following.

THEOREM 3.7: Superposition Principle for Nonhomogeneous Linear Equations

Scalar Property: If y is a solution of

$$y'' + p(x)\,y' + q(x)\,y = f(x)$$

then ky is a solution of

$$y'' + p(x)\,y' + q(x)\,y = kf(x)$$

for any constant k.

Additive Property: If $y_1(x)$ is a solution of

$$y'' + p(x)y' + q(x)y = f_1(x)$$

and $y_2(x)$ is a solution of

$$y'' + p(x)y' + q(x)y = f_2(x)$$

then $y_1(x) + y_2(x)$ is a solution of

$$y'' + p(x)y' + q(x)y = f_1(x) + f_2(x)$$

PROOF: Both properties can be verified by using properties of the derivative. See Problem 11 at the end of this section.

Note: The scalar and additive properties of the principle of superposition can be combined by stating that if y_1 and y_2 are both solutions of the equation $y'' + p(x)y' + q(x)y = f_1(x)$ and $y'' + p(x)y' + q(x)y = f_2(x)$, respectively, then $c_1 y_1 + c_2 y_2$ is a solution of $y'' + p(x)y' + q(x)y = c_1 f_1(x) + c_2 f_2(x)$, where c_1 and c_2 are arbitrary constants.

Example 1   Principle of Superposition   If

$$y'' + y' + 2y = x \tag{2}$$

has a solution $y_1(x) = \dfrac{1}{2}x - 1$ and

$$y'' + y' + 2y = e^x \tag{3}$$

has a solution $y_2(x) = \dfrac{1}{4}e^x$, find a solution of

$$y'' + y' + 2y = 5x + 3e^x \tag{4}$$

Solution   Note that the right-hand side, $5x + 3e^x$, of Eq. (4) can be written as $5f_1(x) + 3f_2(x)$, where $f_1(x)$ and $f_2(x)$ are the right-hand sides of Eqs. (2) and (3), respectively. Hence by the principle of superposition we have

$$
\begin{aligned}
5y_1(x) + 3y_2(x) &= 5\left(\frac{1}{2}x - 1\right) + 3\left(\frac{1}{4}e^x\right) \\
&= \frac{5}{2}x - 5 + \frac{3}{4}e^x
\end{aligned} \tag{5}
$$

is a solution of Eq. (4). ∎

SOLUTION OF THE NONHOMOGENEOUS EQUATION

Thus far we have studied mainly the homogeneous linear equations. We now pursue a comprehensive understanding of the **nonhomogeneous linear equation**

$$y'' + p(x)y' + q(x)y = f(x) \tag{6}$$

Since the form of the general solution of Eq. (6) often causes some bewilderment to beginning students, we start with a brief discussion of the general solution of a linear system of two algebraic equations.

General Solution of Nonhomogeneous Linear Systems

Consider for a moment the two simultaneous equations

$$x_1 + x_2 = 1 \tag{7}$$
$$2x_1 + 2x_2 = 2$$

At first glance you may think that something is wrong, since the equations are in fact the same. This is intentional, however, and in fact the system is mathematically equivalent to the *single* equation $x_1 + x_2 = 1$. Hence the general solution (all solutions) of this trivial system (7) consists of the points on the line $x_1 + x_2 = 1$. See Figure 3.5(a).

Figure 3.5
The general solution of a linear system is the sum of all the homogeneous solutions plus any nonhomogeneous solution

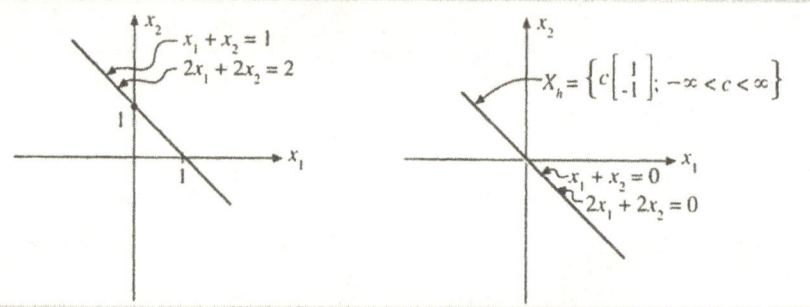

(a)   To find all the solutions of a nonhomogeneous linear system of equations ...

(b)   First find all the solutions of the corresponding homogeneous system. Call these solutions X_h.

Now suppose we solve the corresponding homogeneous system to Eqs. (7). In other words,

$$x_1 + x_2 = 0 \tag{8}$$
$$2x_1 + 2x_2 = 0$$

The general solution (all solutions) of this homogeneous system (8) consists of all points on the line $x_1 + x_2 = 0$, which we can write in vector form as

$$X_h = c \begin{bmatrix} 1 \\ -1 \end{bmatrix} \tag{9}$$

where c is an arbitrary constant. See Figure 3.5(b). Now, let us find just *one solution* of the original nonhomogeneous system (7). We don't care which one—select any one. Picking one particular solution from an infinite number of solutions, we have

$$X_p = \begin{bmatrix} 0 \\ 1 \end{bmatrix} \tag{10}$$

(continued)

See Figure 3.5(c). If we now add *all* the solutions X_h of the homogeneous linear system (8) to the single *particular solution* X_p of the original system

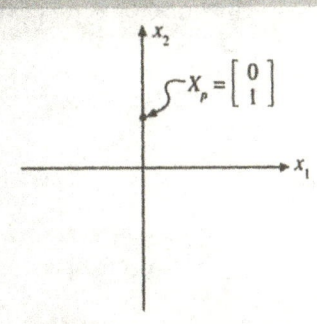

(c)   Then find *any* solution of the nonhomogeneous linear system of equations (it makes no difference which one). Call this solution, X_p.

(7), we get the general solution (all solutions) $X = X_h + X_p$ of the original system (7), which is

$$ X = \begin{bmatrix} x_1 \\ x_2 \end{bmatrix} = c \begin{bmatrix} 1 \\ -1 \end{bmatrix} + \begin{bmatrix} 0 \\ 1 \end{bmatrix} \tag{11} $$

Of course, this equation is simply the straight line $x_1 + x_2 = 1$, which is drawn in Figure 3.5(d). *What we have seen is that the general solution of the nonhomogeneous linear system (7) can be written as the general solution of the corresponding homogeneous system, plus any particular solution of the original nonhomogeneous system.* The important observation in this example is that solutions of *all* linear systems, whether they be linear algebraic equations or linear differential equations, have the same general form.

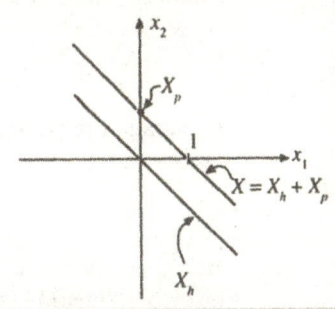

(d)   Add to *each* solution of the homogeneous linear system the particular solution X_p. This collection of sums gives the entire set of solutions of the nonhomogeneous linear system.

Figure 3.5
(continued)

Figure 3.5
(continued)

The above discussion now leads to the general solution of the nonhomogeneous differential equation.

THEOREM 3.8: Solution of the Nonhomogeneous Equation

If $y_p(x)$ is a *particular* solution (any single solution) of the nonhomogeneous equation

$$y'' + p(x)y' + q(x)y = f(x) \qquad (12)$$

on an interval (a, b) and if $y_1(x)$ and $y_2(x)$ are linearly independent solutions of the corresponding homogeneous equation

$$y'' + p(x)y' + q(x)y = 0 \qquad (13)$$

then *any solution* $Y(x)$ of the nonhomogeneous equation (12) can be written in the form

$$Y(x) = c_1 y_1(x) + c_2 y_2(x) + y_p(x) \qquad (14)$$

where c_1 and c_2 are arbitrary constants. The collection of all solutions of the nonhomogeneous equation is called the **general solution** of the equation. Since the general solution of the corresponding homogeneous equation can be written as

$$y_h(x) = c_1 y_1(x) + c_2 y_2(x) \qquad (15)$$

one can also write the general solution of the nonhomogeneous equation in the form

$$Y(x) = y_h(x) + y_p(x) \qquad (16)$$

PROOF: Since both $Y(x)$ and $y_p(x)$ satisfy Eq. (12), we have

$$Y'' + p(x)Y' + q(x)Y = f(x) \qquad (17a)$$
$$y_p'' + p(x)y_p' + q(x)y_p = f(x) \qquad (17b)$$

Subtracting (17b) from (17a), and after using a few simple properties of the derivative, we get

$$[Y - y_p]'' + p(x)[Y - y_p]' + q(x)[Y - y_p] = 0 \qquad (18)$$

Hence we have proven that the *difference* between any two solutions of the nonhomogeneous equation (12) is a solution of the corresponding homogeneous equation (13). This is *very* important, since we know from Theorem 3.4 in Section 3.2 that *any* solution of the homogeneous equation (13) can be expressed uniquely as a linear combination of $y_1(x)$ and $y_2(x)$. Hence we can write

$$Y(x) - y_p(x) = c_1 y_1(x) + c_2 y_2(x)$$

or

$$Y(x) = c_1 y_1(x) + c_2 y_2(x) + y_p(x) \qquad (19)$$

This proves the theorem. ∎

Using Theorem 3.8 as a guide, we now list the steps required to find the general solution of the nonhomogeneous equation.

Method for Solving Nonhomogeneous Equations

To solve the nonhomogeneous linear equation

$$y'' + p(x)y' + q(x)y = f(x) \tag{20}$$

carry out the following steps.

Step 1. Find the general solution

$$y_h(x) = c_1 y_1(x) + c_2 y_2(x)$$

of the corresponding homogeneous equation

$$y'' + p(x)y' + q(x)y = 0$$

Step 2. Find any solution $y_p(x)$ of Eq. (20).

Step 3. Add the results from Steps 1 and 2, getting

$$y(x) = y_h(x) + y_p(x)$$

Example 2 **General Solution of a Nonhomogeneous Equation** From inspection we see that $y_p(x) = 1$ is a (particular) solution of

$$y'' + y = 1 \tag{21}$$

Find the general solution.

Solution The corresponding homogeneous linear equation.

$$y'' + y = 0$$

has the general solution

$$y_h(x) = c_1 \cos x + c_2 \sin x \tag{22}$$

Hence the general solution of the nonhomogeneous solution is

$$y(x) = c_1 \cos x + c_2 \sin x + 1 \tag{23}$$

■

PROBLEMS: Section 3.6

For Problems 1–10, find the general solution (all solutions) of the given nonhomogeneous equation. A single solution is given for each equation.

1. $y'' = 1$ $\qquad$ $y_p(x) = \dfrac{1}{2}x^2$

2. $y'' = 1$ $\qquad$ $y_p(x) = \dfrac{1}{2}x^2 + 1$

3. $y'' + y' = 1$ $\qquad$ $y_p(x) = x$

4. $y'' + y = 1$ $\qquad$ $y_p(x) = 1$

5. $y'' + y = x$ $\qquad$ $y_p(x) = x$

6. $y'' + 3y' + 2y = \sin x$ $\quad$ $y_p(x) = \dfrac{1}{10}\left(\sin x - 3\cos x\right)$

7. $y'' + y' + y = x^2$ $\qquad$ $y_p(x) = x^2 - 2x$

8. $y'' + y' = x^2 + 2x$ $\qquad$ $y_p(x) = \dfrac{1}{3}x^3$

9. $y'' - 2y' - 8y = 9xe^x$ $\quad$ $y_p(x) = -xe^x$

10. $y'' + 2y' + y = x^2 e^{-x}$ $\quad$ $y_p(x) = \dfrac{1}{12}x^4 e^{-x}$

11. Proving the Superposition Principle Prove that if $y_1(x)$ is a solution of

$$y'' + p(x)y' + q(x)y = f_1(x)$$

and $y_2(x)$ is a solution of

$$y'' + p(x)y' + q(x)y = f_2(x)$$

then $y(x) = c_1 y_1(x) + c_2 y_2(x)$ is a solution of

$$y'' + p(x)y' + q(x)y = c_1 y_1(x) + c_2 y_2(x)$$

12. Applying the Superposition Principle Given that

$$y'' + y = \cos 2x \text{ has a solution } y_1(x) = -\frac{1}{3}\cos 2x$$

$$y'' + y = x^2 + 2 \text{ has a solution } y_2(x) = x^2$$

find solutions of the following equations:

(a) $y'' + y = 3\cos 2x$

(b) $y'' + y = -2x^2 - 4$

(c) $y'' + y = \cos 2x + 3x^2 + 6$

13. Reduction of Order for Nonhomogeneous Equations This problem shows how the reduction of order method can also be applied to the nonhomogeneous equation

$$y'' + p(x)y' + q(x)y = f(x) \qquad (24)$$

(a) Make the substitution $y = vy_1$, where y_1 is a solution of the corresponding homogeneous equation, and reduce Eq. (24) to a first-order linear homogeneous equation in v.

(b) Given that $y_1(x) = x$ is a solution of

$$x^2 y'' + xy' - y = 0$$

use the method of substitution described in part (a) to find the general solution of

$$x^2 y'' + xy' - y = 1$$

General Solution to Linear Nonhomogeneous Equations

We have seen that the general solution of a nonhomogeneous linear equation can be written as the sum of the solution of the corresponding homogeneous equation, plus any particular solution of the nonhomogeneous equation. For Problems 14–19, find the general solution of the given nonhomogeneous equation by adding the general solution of the corresponding homogeneous equation and any particular solution of the nonhomogeneous equation.

14. $y' = 1$

15. $y'' = 1$

16. $y' + y = 1$

17. $\quad x + 2y = 1$
$\qquad 2x + 4y = 2$

18. $x + y = 1$
$\qquad x - y = 1$

19. $\qquad x + y + z = 1$
$\qquad 2x + 2y + 2z = 2$
$\qquad 3x + 3y + 3z = 3$

20. Journal Entry Summarize in your own words the theory of nonhomogeneous linear systems.

3.7 SOLVING NONHOMOGENEOUS EQUATIONS: METHOD OF UNDETERMINED COEFFICIENTS

PURPOSE

To present a simple procedure for finding a particular solution $y_p(x)$ of the linear nonhomogeneous equation

$$ay'' + by' + cy = f(x)$$

when the nonhomogeneous term $f(x)$ is either x^n, e^{ax}, $\sin \beta x$, $\cos \beta x$, or products of these functions.

In the previous section we saw that the general solution of the nonhomogeneous equation is the sum of the solution of the corresponding homogeneous equation plus any particular solution of the nonhomogeneous equation. In this section we will learn how to find a particular solution to a nonhomogeneous equation with *constant coefficients* when the nonhomogeneous term $f(x)$ is itself a solution of some linear differential equation with constant coefficients; that is, it is either a *polynomial*, an *exponential, a sine* or *cosine*, or some product of these functions. The method studied is called the **method of undetermined coefficients.** *

Example 1   Finding a Solution by Inspection   Find a particular solution of the differential equation

$$y'' + 3y = 2 \tag{1}$$

Solution   It is sometimes (but not often) possible to find a single solution by simple inspection. For this equation it is obvious that the constant function $y_p(x) = 2/3$ is a solution. If we suspect that the equation has a constant solution but do not know *what* constant, we can try substituting a *trial* solution, $y_p(x) = A$, into Eq. (1). If the constant A, called an *undetermined coefficient*, is to satisfy the equation, it must have the value $0 + 3A = 2$, or $A = 2/3$.   ∎

Example 2   An Exponential Nonhomogeneous Term   Find a single solution of

$$y'' + 3y' + 2y = 2e^{3x} \tag{2}$$

Solution   Since the derivatives of the nonhomogeneous term $2e^{3x}$ are multiples of the term (i.e., $6e^{3x}$, $18e^{3x}$,...), we suspect the equation should have a solution of the form

$$y_p(x) = Ae^{3x} \tag{3}$$

To find the value of A that makes Ae^{3x} a solution, we substitute it into Eq. (2) and find

$$9Ae^{3x} + 3(3Ae^{3x}) + 2(Ae^{3x}) = 2e^{3x}$$

Eliminating the common factor e^{3x}, we find $A = 1/10$. Hence one solution to Eq. (2) is

$$y_p(x) = \frac{1}{10}e^{3x} \tag{4}$$

∎

Example 3   A Polynomial Nonhomogeneous Term   Find a single solution of

$$y'' - 4y' + 2y = 2x^2 \tag{5}$$

Solution   Since the first and second derivatives of the nonhomogeneous term $2x^2$ are polynomials of degree one and zero, we seek a solution of the form

$$y_p = Ax^2 + Bx + C \tag{6}$$

* We could also call the method of undetermined coefficients ''the method of judicious guessing.''

Differentiating, we obtain

$$y_p' = 2Ax + B$$
$$y_p'' = 2A$$

Substituting y_p, y_p', and y_p'' into Eq. (5) gives

$$(2A) - 4(2Ax + B) + 2(Ax^2 + Bx + C) = 2x^2$$

Rewriting this equation in terms of 1, x, and x^2 gives

$$(2A - 4B + 2C) \cdot 1 + (-8A + 2B)x + (2A)x^2 = 2x^2 \tag{7}$$

Setting the coefficients of 1, x, and x^2 equal to one another,* we have

$$\begin{aligned} 2A - 4B + 2C &= 0 \\ -8A + 2B &= 0 \\ 2A &= 2 \end{aligned} \tag{8}$$

Solving for A, B, and C from Eq. (8) gives $A = 1$, $B = 4$, and $C = 7$. Hence one solution is

$$y_p(x) = x^2 + 4x + 7 \tag{9}$$

∎

The method of undetermined coefficients can also be used when the nonhomogeneous term involves a sine or cosine function.

Example 4 **Sine or Cosine Nonhomogeneous Term**   Find a particular solution of

$$y'' - 3y' + 2y = 10 \sin 2x \tag{10}$$

Solution   Since the derivatives of $\sin 2x$ are constant multiples of $\sin 2x$ and $\cos 2x$, we seek a solution of the form

$$y_p = A \sin 2x + B \cos 2x \tag{11}$$

Differentiating gives

$$y_p' = 2A \cos 2x - 2B \sin 2x$$
$$y_p'' = -4A \sin 2x - 4B \cos 2x$$

Substituting y_p, y'_p, and y_p'' into the differential equation gives

$$(A \sin 2x + B \cos 2x) - 3(2A \cos 2x - 2B \sin 2x) \\ + 2(-4A \sin 2x - 4B \cos 2x) = 10 \sin 2x$$

Rewriting this, we obtain

$$(A + 6B - 8A) \sin 2x + (B - 6A - 8B) \cos 2x = 10 \sin 2x \tag{12}$$

---

* We can set the coefficients of 1, x, and x^2 equal to each other, since the functions are linearly independent. Although we have discussed linear independence only for two functions, linear independence has meaning for any number of functions. Problem 43 at the end of this section will allow the reader to gain more experience with this important concept.

Setting the coefficients of the linearly independent functions $\sin 2x$ and $\cos 2x$ equal to each other gives

$$A + 6B - 8A = 10$$
$$B - 6A - 8B = 0$$

Solving for A and B, we get $A = -14/17$, $B = 12/17$. Hence we have a solution

$$y_p(x) = \frac{1}{17}\left(-14 \sin 2x + 12 \cos 2x\right) \qquad (13)$$

■

Example 5 **Combination Nonhomogeneous Term** Find a particular solution of

$$y'' - 3y' + 4y = xe^{2x} \qquad (14)$$

Solution Using our experience from the previous examples, we seek a solution of the form

$$y_p(x) = e^{2x}(Ax + B)$$

where we keep not only the term Axe^{2x} but the term Be^{2x} as well. Differentiating y_p, we find

$$y_p' = e^{2x}(A) + 2e^{2x}(Ax + B) = e^{2x}\left(2Ax + (A + 2B)\right)$$
$$y_p'' = e^{2x}(2A) + 2e^{2x}\left(2Ax + (A + 2B)\right) = 4e^{2x}\left(Ax + (A + B)\right)$$

Substituting y_p, y_p', and y_p'' into the differential equation (14), we find

$$4e^{2x}\left(Ax + (A + B)\right) - 3\left(e^{2x}\left(2Ax + (A + 2B)\right)\right) + 4\left(e^{2x}(Ax + B)\right) = xe^{2x}$$

Rewriting in terms of e^{2x} and xe^{2x}, we find

$$\left(4A - 6A + 4A\right)xe^{2x} + \left(4A + 4B - 3A - 6B + 4B\right)e^{2x} = xe^{2x}$$

Setting the coefficients of the linearly independent functions e^{2x} and xe^{2x} equal to each other gives

$$4A - 6A + 4A = 1$$
$$4A + 4B - 3A - 6B + 4B = 0$$

Solving for A and B gives $A = 1/2$ and $B = -1/4$. Hence we have found a solution

$$y_p(x) = \frac{1}{4}e^{2x}\left(2x - 1\right) \qquad (15)$$

■

The following example is not quite so easy.

Example 6 **Nonhomogeneous Term Agrees with a Homogeneous Solution** Find a solution of the differential equation

$$y'' - 3y' - 4y = e^{-x} \tag{16}$$

Solution We try

$$y_p = Ae^{-x}$$

Differentiating, we get

$$y_p' = -Ae^{-x}$$
$$y_p'' = Ae^{-x}$$

Substituting y_p, y_p', and y_p'' into Eq. (16) gives

$$\left(Ae^{-x}\right) - 3\left(-Ae^{-x}\right) - 4\left(Ae^{-x}\right) = e^{-x}$$

or

$$\left(1 + 3 - 4\right)Ae^{-x} = e^{-x}$$

or simply

$$0 = 1$$

Hence the equation has no solution of the form $y_p = Ae^{-x}$. The difficulty comes about from the fact that the nonhomogeneous term e^{-x} is *also* a solution of the corresponding homogeneous equation $y'' - 3y' - 4y = 0$, which has the solution $y_h = c_1e^{-x} + c_2e^{4x}$. One might suspect that a trial solution of the form Ae^{-x} would fail in this case, since it does not contribute anything ''new'' to the homogeneous solution y_h. After trial and error, however, we discover that the equation does have a solution of the general form

$$y_p = Axe^{-x} \tag{17}$$

Differentiating, we get

$$y_p' = e^{-x}\left(-Ax + A\right)$$
$$y_p'' = e^{-x}\left(Ax - 2A\right)$$

Substituting into the differential equation (16), we find

$$\left(e^{-x}\left(Ax - 2A\right)\right) - 3\left(e^{-x}\left(-Ax + A\right)\right) - 4\left(Axe^{-x}\right) = e^{-x}$$

or

$$\left(A + 3A - 4A\right)xe^{-x} + \left(-2A - 3A\right)e^{-x} = e^{-x}$$

Setting coefficients of e^{-x} and xe^{-x} equal, we find

$$0 = 0$$
$$-5A = 1$$

from which we conclude that $A = -1/5$. Hence Eq. (16) has a solution

$$y_p(x) = -\frac{1}{5}xe^{-x} \tag{18}$$

∎

Table 3.2 summarizes the general rules we have alluded to in this section.

Table 3.2
Form of a Particular Solution y_p of the Differential Equation $ay'' + by' + cy = f(x)$

Type	$f(x)$	$y_p(x)$
1	a (constant)	$x^s \cdot A$ (constant)
2	$P_n(x) = a_0 x^n + \cdots + a_{n-1}x + a_n$	$x^s \cdot (A_0 x^n + \cdots + A_{n-1}x + A_n)$
3	$ae^{\alpha x}$	$x^s \cdot Ae^{\alpha x}$
4	$a \cos kx + b \sin kx$	$x^s \cdot (A \cos kx + B \sin kx)$
5	$P_n(x)e^{\alpha x}$	$x^s (A_0 x^n + A_1 x^{n-1} + \cdots + A_{n-1}x + A_n)e^{\alpha x}$
6	$P_n(x) \cos kx$	$x^s \cdot \Big((A_0 x^n + A_1 x^{n-1} + \cdots + A_{n-1}x + A_n) \cos kx$
7	$P_n(x) \sin kx$	$\quad + (B_0 x^n + B_1 x^{n-1} + \cdots + B_{n-1}x + B_n) \sin kx\Big)$
8	$P_n(x)e^{\alpha x} \cos kx$	$x^s \cdot \Big((A_0 x^n + A_1 x^{n-1} + \cdots + A_{n-1}x + A_n)e^{\alpha x} \cos kx$
9	$P_n(x)e^{\alpha x} \sin kx$	$\quad + (B_0 x^n + B_1 x^{n-1} + \cdots + B_{n-1}x + B_n)e^{\alpha x} \sin kx\Big)$

Note: The exponent s in x^s is the *smallest* of the integers 0, 1, or 2 (in other words, x^s will be either 1, x, or x^2) that ensures that no term in the particular solution $y_p(x)$ is also a solution of the corresponding homogeneous equation.

Example 7 **Using the Table** Find the general solution of

$$y'' - 2y' + y = e^x \tag{19}$$

Solution Finding the solution of the corresponding homogeneous equation, we get

$$y_h(x) = c_1 e^x + c_2 x e^x \tag{20}$$

To find a single solution of the nonhomogeneous equation (19), it is clear that neither function of the form Ae^x or Axe^x is a solution, since both e^x and xe^x are solutions of the *homogeneous* equation. However, when we look at Type 3 in Table 3.2, we are instructed to try a solution of the form

$$y_p(x) = Ax^2 e^x \tag{21}$$

Substituting $y_p(x)$ into Eq. (19) results (after some computation) in $A = 1/2$. Hence a particular solution is

$$y_p(x) = \frac{1}{2}x^2 e^x \tag{22}$$

∎

Example 8 **Correct Form for a Particular Solution** Determine the form of a solution $y_p(x)$ of the equation

$$y'' + 2y' - 3y = f(x) \qquad (23)$$

where $f(x)$ is given as

(a) $x^2 + x - 3$

(b) e^{-x}

(c) xe^x

(d) $2x \sin 3x + \cos 3x$

(e) $xe^{-2x} \sin x$

Solution First, it is necessary to find the solution of the homogeneous equation

$$y'' + 2y' - 3y = 0 \qquad (24)$$

which is

$$y_h(x) = c_1 e^x + c_2 e^{-3x} \qquad (25)$$

(a) $[f(x) = x^2 + x - 3] \, \lozenge \, [y_p = A_0 x^2 + A_1 x + A_2]$

Referring to Table 3.2, we recognize that $f(x)$ is of Type 2. Since no terms in $f(x)$ agree with any terms in $y_h(x)$, it is not necessary to multiply by x or x^2. In other words, $s = 0$. Hence $y_p(x)$ has the form $y_p(x) = A_0 x^2 + A_1 x + A_2$.

(b) $[f(x) = e^{-x}] \, \lozenge \, [y_p = A e^{-x}]$

The function $f(x)$ is of Type 3 with $a = 1$ and $\alpha = -1$. And since e^{-x} does not occur in the homogeneous solution (25), it is not necessary to multiply by x or x^2. In other words, $s = 0$. Hence a solution can be found having the form $y_p(x) = A e^{-x}$.

(c) $[f(x) = xe^x] \, \lozenge \, [y_p = (A_0 x^2 + A_1 x) e^x]$

The function $f(x)$ is of Type 5. However, since the second term of $(A_0 x + A_1)e^x$ also occurs in the homogeneous solution (25), we must multiply by x ($s = 1$), getting $y_p(x) = (A_0 x^2 + A_1 x) e^x$.

(d) $[f(x) = 2x \cos 3x + \sin 3x \, \lozenge \, y_p = (A_0 x + A_1) \cos 3x + (B_0 x + B_1) \sin 3x]$

The terms of $f(x)$ are of Types 6 and 7; both of these terms give rise to the same particular solution. Hence we choose ($s = 0$)

$$y_p(x) = (A_0 x + A_1) \cos 3x + (B_0 x + B_1) \sin 3x$$

(e) $[f(x) = xe^{-2x} \cos x \, \lozenge \, y_p = (A_0 x + A_1) e^{-2x} \cos 3x + (B_0 x + B_1) e^{-2x} \sin 3x]$

The terms of $f(x)$ are of Type 8, and so we choose ($s = 0$)

$$y_p(x) = (A_0 x + A_1) e^{-2x} \cos 3x + (B_0 x + B_1) e^{-2x} \sin 3x \qquad \blacksquare$$

We summarize the method of undetermined coefficients.

Method of Undetermined Coefficients

The following are steps for finding a particular solution of the second-order linear equation with constant coefficients

$$ay'' + by' + cy = f(x)$$

where the nonhomogeneous term $f(x)$ is a polynomial, exponential, sine, or cosine function or some product of these functions.

Step 1. Solve the corresponding homogeneous equation.

Step 2. Pick the form for a particular solution $y_p(x)$ depending on the non-homogeneous term $f(x)$. Remember that no term in the particular solution may also be a solution of the homogeneous equation. If agreement occurs, multiply $y_p(x)$ by x.

Step 3. Substitute $y_p(x)$ and its derivatives into the differential equation and set the coefficients of the algebraic expressions on each side of the equation equal to each other, getting a system of equations for which one can solve for the undetermined coefficients.

Step 4. Solve the system of equations for the undetermined coefficients.

It is possible, using the principle of superposition, to use the method of undetermined coefficients when the nonhomogeneous term contains *sums* of the forms described in the previous rules. The following example illustrates this idea.

Example 9   Using the Principle of Superposition   Find a particular solution to the equation

$$y'' + 4y = x^2 + xe^x \tag{26}$$

Solution   Observe that the solution of the corresponding homogeneous equation is

$$y_h(x) = c_1 \cos 2x + c_2 \sin 2x \tag{27}$$

To construct a particular solution of the nonhomogeneous equation (26), we use the principle of superposition and find particular solutions $y_{p_1}(x)$ and $y_{p_2}(x)$ of the two equations

$$y'' + 4y = x^2 \tag{28}$$

$$y'' + 4y = xe^x \tag{29}$$

Particular solutions of Eqs. (28) and (29) are

$$y_{p_1}(x) = Ax^2 + Bx + C$$

$$y_{p_2}(x) = (Dx + E)e^x$$

respectively. Hence a particular solution $y_p(x)$ of Eq. (26) is

$$y_p(x) = y_{p_1} + y_{p_2} = Ax^2 + Bx + C + (Dx + E)e^x \tag{30}$$

PROBLEMS: Section 3.7

For each differential equation in Problems 1–21, find the general solution by finding the homogeneous solution and a particular solution.

1. $y' = 1$
2. $y' + y = 1$
3. $y' + y = x$
4. $y'' = 1$
5. $y'' + 4y' = 1$
6. $y'' + 4y = 1$
7. $y'' + 4y' = x$
8. $y'' + y' - 2y = 3 - 6x$
9. $y'' + y = 6e^x + 3$
10. $y'' - y' - 2y = 6e^x$
11. $y'' + y' = 6 \sin 2x$
12. $y'' + 4y' + 5y = 2e^x$
13. $y'' + 3y' = \sin x + 2 \cos x$
14. $y'' + 4y' + 4y = xe^{-x}$
15. $y'' - y = x \sin x$
16. $y'' - 3y' + 2y = e^x \sin x$
17. $y'' - 4y' + 4y = xe^{2x}$
18. $y'' + y = 12 \cos^2 x$
19. $y'' - 4y' + 3y = 20 \cos x$
20. $y'' - y = 8xe^x$
21. $y'' - 5y' + 6y = \cosh x$

For Problems 22–28, find the solution to the initial-value problem.

22. $y' - y = 1$ $\qquad$ $y(0) = 0$
23. $y'' + y = 2x$ $\qquad$ $y(0) = 1$ $\quad$ $y'(0) = 2$
24. $y'' + y' - 2y = 2x$ $\qquad$ $y(0) = 0$ $\quad$ $y'(0) = 1$
25. $y'' - 5y' + 6y = e^x(2x - 3)$ $\quad$ $y(0) = 1$ $\quad$ $y'(0) = 3$
26. $y'' - 4y' + 4y = e^{2x}$ $\qquad$ $y(0) = 0$ $\quad$ $y'(0) = 0$
27. $y'' + 16y = 5 \sin x$ $\qquad$ $y(0) = 0$ $\quad$ $y'(0) = 0$
28. $y'' + 3y' + 2y = 20 \cos 2x$ $\quad$ $y(0) = -1$ $\quad$ $y'(0) = 6$

For Problems 29–38, determine the form of the particular solution $y_p(x)$. Do not actually determine the values of the coefficients.

29. $y'' - y' = x^2 e^x$
30. $y'' + 4y = x^2 e^x \sin x$
31. $y'' + 4y = (x^2 + 2x - 1)e^x \sin 2x$
32. $y'' - 9y = x^2 e^{3x} + \sin x$
33. $y'' - 4y' + 5y = \sin x$
34. $y'' - 3y' + 2y = e^x + e^{-x} + x + 1$
35. $y'' + 2y' + y = x^2 e^{-x} \sin 2x$
36. $y'' - 3y' + 2y = x^2 + \sin x + xe^{-x}$
37. $y'' - 2y' + y = x^2 e^x + \sin x$
38. $y'' - 2y' + y = x \sin x + x \cos x$

Solutions Using Complex Variables

There is a nice way to solve linear nonhomogeneous equations with constant coefficients whose right-hand sides consist of sine and cosine functions.

$$ay'' + by' + cy = \begin{cases} R \cos qx \\ R \sin qx \end{cases} \tag{31}$$

The idea is to solve a modified equation in which we replace the sine or cosine term on the right-hand side of the original equation by the complex exponential function Re^{iqx}:

$$ay'' + by' + cy = Re^{iqx} \tag{32}$$

Equation (32) is then solved in the usual manner: A particular solution to the nonhomogeneous equation is sought of the form $y_p(x) = Ae^{iqx}$, where the complex constant $i = \sqrt{-1}$ is treated the same as any real constant. The important fact here is that the real part of the solution to Eq. (32) is the solution to Eq. (31) with the cosine term on the right-hand side and the complex part of the solution to Eq. (32) is the solution to Eq. (31) with the sine term on the right-hand side. For Problems 39–41, use this idea to find the solutions to the given equation.

39. $y'' - 2y' + y = 2 \sin x$

40. $y'' + 25y = 6 \sin x$

41. $y'' + 25y = 20 \sin 5x$

42. Importance of Complex Exponents Solve the differential equation

$$y'' - 3y' + 2y = 3e^{i2x} \tag{33}$$

and verify the real and complex parts of the solution are the solutions obtained when the right-hand side is replaced by $3 \cos 2x$ and $3 \sin 2x$, respectively.

43. In Honor of Charles Proteus Steinmetz* In electrical engineering it is known that if a periodic electromotive force $E = E_0 \cos \omega t$, acts on a simple circuit containing a resistor with resistance R, an inductor with inductance L, and a capacitor with capacitance C, then the charge q on the capacitor is governed by the second-order equation

$$L\frac{d^2q}{dt^2} + R\frac{dq}{dt} + \frac{1}{C}q = F_0 \cos \omega t \tag{34}$$

Show that a particular solution of this equation is

$$q(t) = \frac{E_0}{\sqrt{(1/C - \omega^2 L)^2 + \omega^2 R^2}} \cos (\omega t - \phi)$$

where

$$\tan \phi = \omega R/(1/C - \omega^2 L)$$

by finding a particular solution of the related equation

$$L\frac{d^2q}{dt^2} + R\frac{dq}{dt} + \frac{1}{C}q = F_0 e^{i\omega t} \tag{35}$$

and then taking the real part.

44. Useful Property of Linear Independence Show that if the functions $\{f_1(x), f_2(x), f_3(x), \ldots, f_n(x)\}$ are linearly independent on an interval I, then

* The person who pioneered the use of complex numbers in differential equations was the brilliant electrical engineer Charles Steinmetz (1865–1923). Steinmetz came to the United States from his home in Germany in 1889 and became the brains of the newly founded General Electric Company. It was Steinmetz who, with power of his own mind, determined ways to mass-produce electric motors and generators. Physically, he was a dwarf who was crippled by a congenital deformity, but mentally, he had no peer as an electrical engineer and scientist.

$$a_1 f_1(x) + \cdots + a_n f_n(x) = b_1 f_1(x) + \cdots + b_n f_n(x) \tag{36}$$

implies $a_i = b_i$ for $i = 1, 2, \ldots, n$. This is why it is valid to set the coefficients of terms equal to each other in finding the undetermined coefficients.

45. The Surefire Strategy Neglecting air friction, the planar motion $(x(t), y(t))$ of an object in a gravitational field is governed by the two second-order differential equations

$$\ddot{x} = 0 \qquad \ddot{y} = -g \tag{36}$$

where the "double dot" notation denotes the second derivative with respect to time and g is acceleration due to gravity. See Figure 3.6. Suppose you fire a dart gun, located at the origin, directly at a target located at (x_0, y_0). Suppose the dart has a velocity v and at the exact instant the dart is fired, the target starts falling.

(a) Find the heights of the dart and the target.

(b) Show that the dart will always hit the target (assuming that negative values of y are valid).

(c) At what height will the projectile hit the target?

46. Journal Entry Summarize the method of undetermined coefficients. Do you think it would be possible to write a computer program to evaluate the undetermined coefficients so that one could simply enter into a computer a certain differential equation and the computer would print out the solution?

Figure 3.6

The correct way to fire a projectile at an object that begins to fall at the instant of firing is to aim *directly* at the object

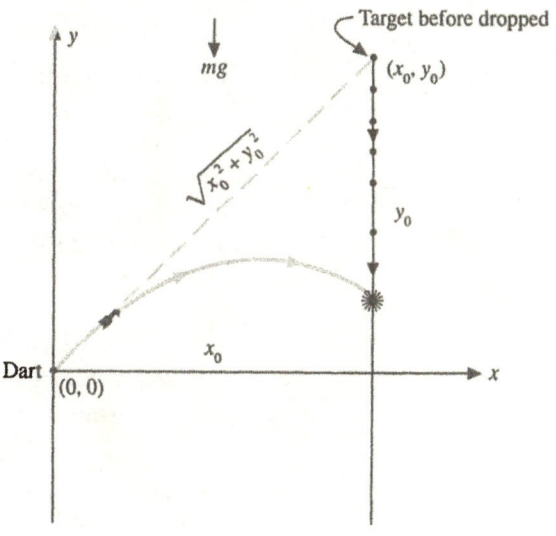

3.8 SOLVING NONHOMOGENEOUS EQUATIONS: METHOD OF VARIATION OF PARAMETERS

PURPOSE

To present a powerful technique, known as the method of variation of parameters, for finding a particular solution of the form

$$y_p(x) = v_1(x)y_1(x) + v_2(x) y_2(x)$$

to the nonhomogeneous differential equation

$$y'' + p(x)y' + q(x)y = f(x)$$

where $y_1(x)$ and $y_2(x)$ are linearly independent solutions of the corresponding homogeneous equation and $v_1(x)$ and $v_2(x)$ are functions to be determined.

VARIATION OF PARAMETERS

In the previous section we discussed a simple method for finding a particular solution to a linear nonhomogeneous equation with constant coefficients, provided that the nonhomogeneous term had a special form. In this section we consider a method for finding a particular solution to the more general equation.

$$y'' + p(x)y' + q(x)y = f(x) \tag{1}$$

where $p(x)$, $q(x)$, and $f(x)$ are continous functions. In addition to allowing variable coefficients $p(x)$ and $q(x)$, the nonhomogeneous term $f(x)$ is not restricted as it is in the method of undetermined coefficients.

As did the method of undetermined coefficients, the **method of variation of parameters** requires knowledge of two linearly independent solutions $y_1(x)$ and $y_2(x)$ of the corresponding homogeneous equation

$$y'' + p(x)y' + q(x)y = 0 \tag{2}$$

from which one can find the general solution*

$$y_h(x) = c_1y_1(x) + c_2y_2(x) \tag{3}$$

where c_1 and c_2 are arbitrary constants. The method of variation of parameters seeks to find a solution of the nonhomogeneous equation (1) by replacing the constants c_1 and c_2 in Eq. (3) by unknown *functions* $v_1(x)$ and $v_2(x)$, thus obtaining an expression of the form

$$y_p(x) = v_1(x)y_1(x) + v_2(x)y_2(x) \tag{4}$$

The unknown functions $v_1(x)$ and $v_2(x)$ are often called *parameters,* since they were initially intended by Lagrange to be "constants that vary." Lagrange's strategy was to

---

* Be aware that there is no general method for finding the general solution for equations with *variable coefficients*. However, it is often possible to find solutions by inspection, transformations, reduction of order methods, series methods, and on and on. We will study series methods in Chapter 4.

substitute Eq. (4) into the nonhomogeneous equation (1) and solve for $v_1(x)$ and $v_2(x)$. It turns out that $v_1(x)$ and $v_2(x)$ can be found quite easily.

HISTORICAL NOTE

The method of variation of parameters (or variation of constants) was formulated by the great Italian-French mathematician, Joseph Louis Lagrange (1736–1813). The two great mathematicians of the seventeenth century were Lagrange and the Swiss mathematician Leonhard Euler (1707–1783). Although Euler discovered the general solution to the homogeneous differential equation $ay'' + by' + cy = 0$, it was Lagrange who found a particular solution to the nonhomogeneous equation.

Carrying out Lagrange's strategy we differentiate Eq. (4), getting

$$y_p' = (v_1' y_1 + v_2' y_2) + (v_1 y_1' + v_2 y_2') \tag{5}$$

We now *choose* v_1 and v_2 so that they satisfy

$$v_1' y_1 + v_2' y_2 = 0 \tag{6}$$

thus making the second derivative y_p'' more manageable.* With this condition imposed on v_1 and v_2, the second derivative becomes

$$y_p'' = v_1' y_1' + v_2' y_2' + v_1 y_1'' + v_2 y_2'' \tag{7}$$

As a result of condition (6), only the *first derivatives* or v_1 and v_2 appear in the expression for y_p''. Substituting y_p' and y_p'' into Eq. (1) leads to the second condition on v_1 and v_2:

$$v_1' y_1' + v_2' y_2' = f(x) \tag{8}$$

Rewriting Eqs. (6) and (8), we arrive at the system of equations

$$\begin{aligned} y_1 v_1' + y_2 v_2' &= 0 \\ y_1' v_1' + y_2' v_2' &= f(x) \end{aligned} \tag{9}$$

in which we have interchanged the order of the factors to denote that the two derivatives v_1' and v_2' are the unknown functions. Note that all the other quantities are known. Solving this system of equations gives

$$v_1' = \frac{-y_2 f}{W(y_1, y_2)} \quad \text{and} \quad v_2' = \frac{y_1 f}{W(y_1, y_2)} \tag{10}$$

where $W(y_1, y_2) = y_1 y_2' - y_1' y_2$ is the Wronskian of y_1 and y_2. Note that since y_1 and y_2 are linearly independent solutions of the homogeneous equation, we know that

---

* Inasmuch as we seek two functions $v_1(x)$ and $v_2(x)$, we can specify *two* conditions. The main condition, of course, is that we require that $y_p(x)$ satisfies the differential equation. The second condition can be imposed arbitrarily, and Lagrange imposed the condition $v_1' y_1 + v_2' y_2 = 0$ so that the second derivative y_p'' will be simplified.

$W(y_1, y_2)$ will not vanish on the interval of interest. Finally, integrating Eq. (10) and substituting v_1 and v_2 into Eq. (4), we find a particular solution to Eq. (1). We summarize this discussion in the following theorem.

THEOREM 3.9: Method of Variation of Parameters

If $p(x)$, $q(x)$, and $f(x)$ are continuous functions and if $y_1(x)$ and $y_2(x)$ are linearly independent solutions of the homogeneous equation corresponding to the nonhomogeneous equation

$$y'' + p(x)y' + q(x)y = f(x) \tag{11}$$

then a particular solution of Eq. (11) is

$$y_p(x) = -y_1(x) \int \frac{y_2(t) f(t)}{W[y_1, y_2](t)} \, dt + y_2(x) \int \frac{y_1(t) f(t)}{W[y_1, y_2](t)} \, dt \tag{12}$$

The following examples illustrate the method of variation of parameters.

Example 1   Variation of Parameters   Find the general solution of

$$y'' + y = \sec x \qquad (0 < x < \pi/2) \tag{13}$$

Solution   The two linearly independent solutions y_1 and y_2 of the corresponding homogeneous equation $y'' + y = 0$ are $y_1(x) = \cos x$ and $y_2(x) = \sin x$. Note that we cannot use the method of undetermined coefficients, since the right-hand side is not of the proper form. Using the method of variation of parameters, we write

$$y_p(x) = v_1(x) \cos x + v_2(x) \sin x \tag{14}$$

For Eq. (14) to satisfy Eq. (13) we must have that v_1 and v_2 satisfy

$$\begin{aligned} \cos x \, v_1' + \sin x \, v_2' &= 0 \\ -\sin x \, v_1' + \cos x \, v_2' &= \sec x \end{aligned} \tag{15}$$

Solving for v_1' and v_2', we obtain

$$v_1'(x) = -\tan x \qquad \text{and} \qquad v_2'(x) = 1 \tag{16}$$

Integrating this gives

$$v_1(x) = \ln \cos x \qquad \text{and} \qquad v_2(x) = x$$

Finally, substituting these values into y_p, we have the particular solution

$$y_p(x) = x \sin x + (\cos x) \ln \cos x \tag{17}$$

Hence the general solution is

$$y(x) = c_1 \cos x + c_2 \sin x + x \sin x + (\cos x) \ln \cos x \tag{18}$$

∎

Example 2   **Variation of Parameters**   Find the general solution of the nonhomogeneous Euler equation

$$x^2 y'' - 2xy' + 2y = x \sin x \tag{19}$$

given that $y_1(x) = x$ and $y_2(x) = x^2$ are linearly independent solutions of the corresponding homogeneous Euler equation.

Solution   Substituting

$$y_p(x) = x v_1(x) + x^2 v_2(x) \tag{20}$$

into Eq. (19) gives the two equations

$$\begin{aligned} x v_1' + x^2 v_2' &= 0 \\ v_1' + 2x v_2' &= x \sin x \end{aligned} \tag{21}$$

Solving these equations for v_1' and v_2' gives

$$\begin{aligned} v_1' &= -x \sin x \\ v_2' &= \sin x \end{aligned} \tag{22}$$

or

$$\begin{aligned} v_1(x) &= x \cos x - \sin x \\ v_2(x) &= -\cos x \end{aligned} \tag{23}$$

Hence we find the particular solution

$$y_p(x) = x(x \cos x - \sin x) - x^2(\cos x) = -x \sin x \tag{24}$$

Adding this to the homogeneous solution, we obtain the general solution

$$y(x) = c_1 x + c_2 x^2 - x \sin x \tag{25}$$

∎

SUMMARY OF THE SECOND-ORDER LINEAR EQUATION WITH CONSTANT COEFFICIENTS

We have now come to the end of the line in our search for the general solution to the differential equation

$$ay'' + by' + cy = f(x) \tag{26}$$

To use the method of variation of parameters, we assume without loss of generality* that $a = 1$. Table 3.3 summarizes the steps we have used in this chapter.

* In using the method of variation of parameters we first divide the differential equation by the coefficient of the highest derivative. Hence for second-order differential equations we assume that the equation has the form $y'' + by' + cy = f(x)$.

Table 3.3
Steps for Finding the General
Solution to the Differential
Equation $y'' + by' + cy = f(x)$

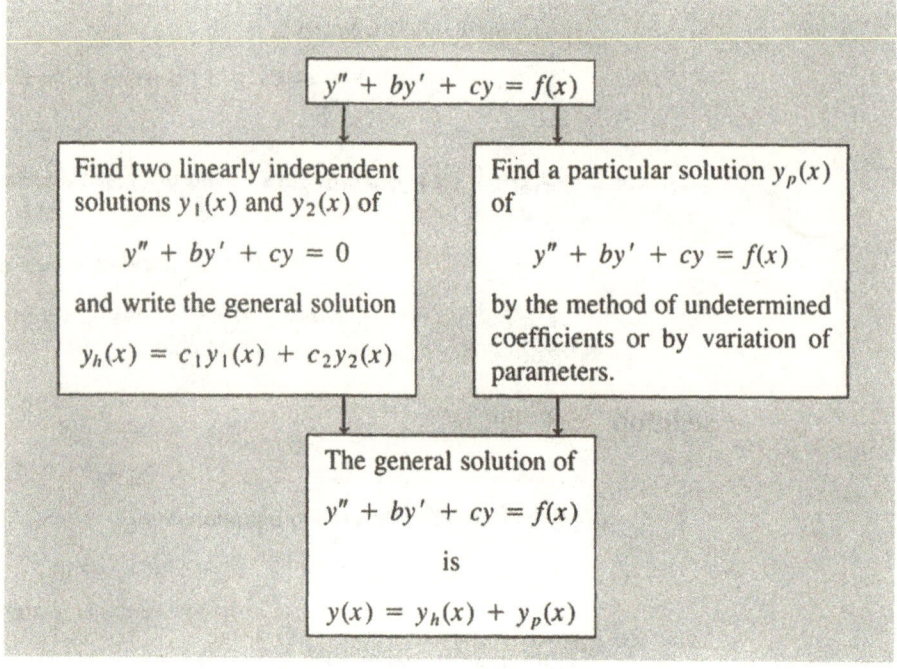

PROBLEMS: Section 3.8

For Problems 1–12, use the method of variation of parameters to find a particular solution of the given nonhomogeneous equation. Then find the general solution of the equation.

1. $y'' = 1$
2. $y'' + y = 1$
3. $y'' - y = x + 1$
4. $y'' - y = 5e^x$
5. $y'' + y' = 4x$
6. $y'' + y' = e^{-x}$
7. $y'' + y = \sin x$
8. $y'' - 2y' + 2y = e^x \sin x$
9. $y'' + y = \sec x \tan x$
10. $y'' + y = \csc x$
11. $y'' - 3y' + 2y = \dfrac{1}{1 + e^{-x}}$
12. $y'' + 2y' + y = e^{-x} \ln x$

For Problems 13–17, find a particular solution of the nonhomogeneous equation, given that the functions $y_1(x)$ and $y_2(x)$ are linearly independent solutions of the corresponding homogeneous equation. Note: The coefficient of y'' must always be 1, and hence a preliminary division may be required.

13. $x^2 y'' - 2xy' + 2y = x^3 \sin x$
    $y_1(x) = x$ $\qquad$ $y_2(x) = x^2$

14. $x^2 y'' + xy' - 4y = x(x + x^3)$
    $y_1(x) = x^2$ $\qquad$ $y_2(x) = x^{-2}$
15. $(1 - x)y'' + xy' - y = 2(x - 1)^2 e^{-x}$
    $y_1(x) = x$ $\qquad$ $y_2(x) = e^x$
16. $y'' + \dfrac{1}{x}y' + \left(1 - \dfrac{1}{4x^2}\right)y = x^{-1/2}$
    $y_1(x) = x^{-1/2} \sin x$ $\quad$ $y_2(x) = x^{-1/2} \cos x$
17. $y'' - y = e^{-x^2}$
    $y_1(x) = e^x$ $\qquad$ $y_2(x) = e^{-x}$

Why Didn't We Use This Method Before?

The perceptive reader might have asked why we didn't add the solution of the homogeneous equation to a particular solution when we solved the first-order equation $y' + p(x)y = f(x)$ back in Chapter 2. We could have; however, the integrating factor method found both of these components at the same time. Nevertheless, for Problems 18–23, solve the given first-order linear equation both by the integrating factor method and by adding the solution to the corresponding homogeneous equation to a particular solution of the nonhomogeneous equation.

18. $y' = 1$ $\qquad\qquad$ $y = c + x$

19. $y' + 2y = 1$ $\qquad$ $y = ce^{-2x} + \dfrac{1}{2}$

20. $y' + 4y = e^x$ $\qquad$ $y = ce^{-4x} + \dfrac{1}{5}e^x$

21. $y' + \dfrac{2}{x}y = 4x$ $\qquad$ $y = \dfrac{c}{x^2} + x^2$

22. $y' + \dfrac{1}{x}y = 4x^2 + 2$ $\quad$ $y = \dfrac{c}{x} + x^3 + x$

23. $y' + \dfrac{1}{x}y = \dfrac{\sin x}{x}$ $\qquad$ $y = \dfrac{c}{x} - \dfrac{\cos x}{x}$

24. Green's Function Representation of the Solution Show that a particular solution of

$$y'' + y = f(x)$$

can be written in the *integral form*

$$y(x) = \int_0^x \sin(x - s) f(s) \, ds$$

Hint: Combine the two integrals that appear in the general variation of parameters formula given in the text. Denoting

the dummy variable of integration by s and using a trigonometric identity, one obtains the desired integral form. The function $\sin(x - s)$ that appears inside the integral is called the **Green's function** for the differential equation. All linear equations have a Green's function, and once Green's function can be found, the solution can be written in integral form.

25. Higher-Order Equation Solve the third-order equation

$$y''' - 2y'' - y' + 2y = e^x$$

using the method of variation of parameters.

26. Journal Entry We have come to the end of the line in solving the second-order linear equation with constant coefficients. (The remainder of this chapter deals with applications.) Give your own impressions of the subject of differential equations thus far.

3.9   MECHANICAL SYSTEMS AND SIMPLE HARMONIC MOTION

PURPOSE
To show how second-order differential equations are useful in the description of dynamical systems. We study one of the simplest dynamical systems, the unforced undamped vibrating spring, described by

$$m\ddot{u} + ku = 0$$

and study its motion, known as simple harmonic motion. We will study various properties of simple harmonic motion, such as frequency, period, and amplitude.

Many objects, when moved from their place of rest, exhibit a vibratory motion, oscillating back and forth about a fixed point. We could mention the vibrations of a tuning fork, the swinging of a pendulum, the motion of electric current moving back and forth through a circuit, or the vibrations of an electron in an electric field. An idealization of all these systems is the small vibrations of a mass attached to a spring.

THE EQUATION OF MOTION OF A MASS ON A SPRING

We start with an unstretched spring of length L hanging from the ceiling. See Figure 3.7(a). To this spring we attach a mass m, which stretches the spring an additional amount ΔL to an equilibrium position. See Figure 3.7(b). We now let u measure the vertical displacement of the mass from this equilibrium position with the positive direction taken downward. See Figure 3.7(c).

To study the motion of the mass, we isolate the forces acting on the mass. When this is done, we can then resort to **Newton's Second Law** of motion, $F = ma$, to find an equation that describes the motion of the mass. There are basically four forces acting on the mass that one must consider.

Figure 3.7
Simple mass-spring system

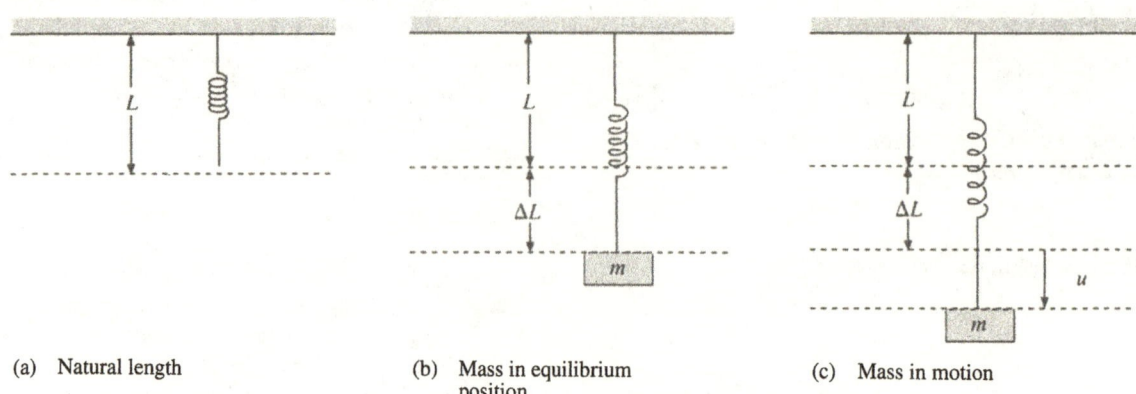

(a)   Natural length

(b)   Mass in equilibrium
       position

(c)   Mass in motion

FORCES ACTING UPON THE MASS

Figure 3.8
Mass-spring system with damping

1. $F_g = mg$ *(force due to gravity).* The force due to gravity F_g always acts downward with magnitude mg. This force is called the *weight* of the object. See Figure 3.8.

2. $F_s = -k(\Delta L + u)$ *(restoring force of the spring).* We assume that the restoring force F_s caused by the spring is directly proportional to the elongation $\Delta L + u$ and always acts to restore the mass to its equilibrium position.* Calling $k > 0$ the constant of proportionality of this force, we have the relationship $F_s = -k(\Delta L + u)$.

3. $F_d = -d(du/dt)$ *(damping force).* We assume that the friction or resistance that the medium exerts on the mass, called the **damping force,** is proportional to the velocity of the mass and always acts in a direction opposite to the velocity of the mass.[†] Hence we have $F_d = -d(du/dt)$, where $d > 0$ is called the constant of proportionality. Note that this equation says that the damping force F_d always acts in a direction opposite to the velocity, du/dt, of the mass. We have followed the common convention and have denoted the damping force by the usual *dashpot mechanism.*

4. $F_e = F(t)$ *(external forces).* Often an external force, $F_e = F(t)$, directed either upward or downward, is applied. The force might be applied directly to the mass or to the mount to which the spring is attached.

---

* A spring is called *linear* if the restoring force is proportional to the elongation. In other words, it takes twice the force to stretch a spring twice the amount. Most springs, however, are *hard springs,* which means that it takes more than twice the force to stretch a spring twice the distance.

[†] For small velocities the damping force is often nearly proportional to the velocity. For larger velocities it may be closer to the square of the velocity or some other function of the velocity.

We now apply Newton's Second Law, which states that *the product of the mass times acceleration is equal to the sum of the external forces*. If we use the common convention of denoting the first and second time derivatives of u by $\dot{u}$ and $\ddot{u}$, respectively, in honor of Sir Isaac Newton, who originated the notation, Newton's Second Law can be written as

$$m\ddot{u} = F_g + F_s + F_d + F_e \tag{1a}$$

or

$$m\ddot{u} = mg - k(\Delta L + u) - d\dot{u} + F(t) \tag{1b}$$

But when the mass is in equilibrium position, the restoring force $(k\Delta L)$ of the spring is equal to the force of gravity (mg). Hence we have the relationship $k\Delta L = mg$. These two terms in Eq. (1b) cancel, giving

$$m\ddot{u} + d\dot{u} + ku = F(t) \tag{2}$$

This second-order differential equation with constant coefficients describes the motion of the vibrating spring. When $F(t) = 0$ for all t, the equation is called **unforced;** otherwise, it is called **forced.** When $d = 0$, the equation is called **undamped;** when the constant $d > 0$, it is called **damped.**

UNITS OF MEASURE

We will express the units of length (L), weight (w), and time (t) in this chapter mostly in the English or engineering system, which measures length in *feet,* weight or force in *pounds,* and time in *seconds*. The unit mass (m) is the *slug,* and it is defined to be that mass for which a force of 1 lb will give an acceleration of 1 ft/sec^2. The conversion factor between the mass m of an object and its weight w (in pounds) is $w = mg$, where g is the acceleration due to gravity, which we take to be 32 ft/sec^2. Other systems of units used in this book are shown in Table 3.4.

Table 3.4
Units of Measure in Different Systems

Quantity	Engineering	mks	cgs
Force	pound (lb)	newton (N)	dyne
Mass	slug	kilogram (kg)	gram (g)
Length	foot (ft)	meter (m)	centimeter (cm)
Value of g	32 ft/sec^2	9.8 m/sec^2	980 cm/sec^2

Example 1   **Finding the Differential Equation**   A 16-lb weight is attached to a spring, which in turn is attached to the ceiling. The weight stretches the spring 4 inches and comes to rest in its equilibrium position. It is also known that friction in the spring gives rise to a damping force of 0.05 lb for a velocity of 3 in./sec. Find the differential equation that describes the motion of this vibrating mass.

Solution We compute the three basic parameters *m, d,* and *k* that define the mass-spring-dashpot system. It is important that all units are commensurate.

$$m = \frac{w}{g} = \frac{16 \text{ lb}}{32 \text{ ft/sec}^2} = \frac{1}{2} \frac{\text{lb-sec}^2}{\text{ft}} = \frac{1}{2} \text{ slug}$$

$$d = \frac{0.05 \text{ lb}}{3 \text{ in./sec}} = \frac{0.05 \text{ lb}}{(1/4) \text{ ft/sec}} = 0.20 \frac{\text{lb-sec}}{\text{ft}}$$

$$k = \frac{16 \text{ lb}}{4 \text{ in.}} = \frac{16 \text{ lb}}{(1/3) \text{ ft}} = 48 \frac{\text{lb}}{\text{ft}}$$

Hence we have

$$\frac{1}{2}\ddot{u} + 0.02\dot{u} + 48u = 0 \tag{3}$$

If the mass was initially pulled down 6 inches from the equilibrium position and released without imparting an initial velocity, the initial conditions would be given by $u(0) = 1/2$ and $\dot{u}(0) = 0$. ∎

SIMPLE HARMONIC MOTION

Many objects have a natural vibratory motion in which they oscillate back and forth about a fixed point of equilibrium. If the particle is attracted toward the equilibrium position with a force proportional to the displacement, the particle is said to undergo **simple harmonic motion.** Possibly, the best known occurrence of simple harmonic motion is the undamped movement of a mass attached to a spring that satisfies the differential equation

$$m\ddot{u} + ku = 0 \tag{4}$$

To find this motion, we first divide by *m*, arriving at

$$\ddot{u} + \omega_0^2 u = 0 \tag{5}$$

where $\omega_0 = \sqrt{k/m}$. The characteristic equation for this equation* is

$$r^2 + \omega_0^2 = 0 \tag{6}$$

which has the complex conjugate roots $\pm i\omega_0$. Hence the general solution of Eq. (4), and the simple harmonic motion that describes the mass-spring system, is

$$u(t) = c_1 \cos \omega_0 t + c_2 \sin \omega_0 t \tag{7}$$

If we now multiply and divide the right-hand side of Eq. (7) by $\sqrt{c_1^2 + c_2^2}$, we can rewrite this simple harmonic motion as

$$u(t) = \sqrt{c_1^2 + c_2^2} \left(\frac{c_1}{\sqrt{c_1^2 + c_2^2}} \cos \omega_0 t + \frac{c_2}{\sqrt{c_1^2 + c_2^2}} \sin \omega_0 t \right) \tag{8}$$

---

* We have called the variable in the characteristic equation *r* since we have used *m* for the mass.

If we now define δ, called the **phase angle,** by the two equations

$$\sin \delta = \frac{c_2}{\sqrt{c_1^2 + c_2^2}} \quad \text{and} \quad \cos \delta = \frac{c_1}{\sqrt{c_1^2 + c_2^2}} \qquad (9)$$

Equation (8) becomes

$$u(t) = R(\cos \delta \cos \omega_0 t + \sin \delta \sin \omega_0 t)$$
$$= R \cos (\omega_0 t - \delta) \qquad (10)$$

The quantity R is called the **amplitude** of the simple harmonic motion and can be determined from the formula $R = \sqrt{c_1^2 + c_2^2}$. The quantity $\omega_0 = \sqrt{k/m}$ is called the **(circular) frequency** of the simple harmonic motion, and it represents the number of oscillations the spring makes every 2π units of time.* More useful than the circular frequency is the **natural frequency** (or just **frequency**), defined as $f = \omega_0/2\pi$, which represents the number of oscillations the spring makes every unit of time. Finally, the reciprocal of the frequency, $T = 1/f = 2\pi/\omega_0 = 2\pi \sqrt{m/k}$, is the **period** of the motion and is the time required for the motion to undergo one complete oscillation. These ideas are illustrated in Figure 3.9.

In other words, the vibratory motion of an undamped spring is a simple sinusoidal curve.[†]

Figure 3.9
Simple harmonic motion

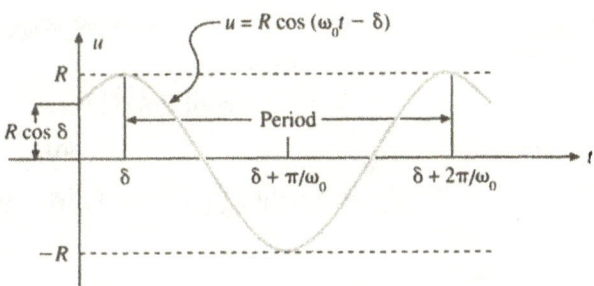

Example 2   Simple Harmonic Motion   An 8-lb weight is attached to a spring, which in turn is suspended from the ceiling. The weight comes to rest in equilibrium position, stretching the spring 6 inches in the process. The weight is then pulled down an additional 1 ft

---

* Although mathematically, $\omega_0 = \sqrt{k/m}$ represents the frequency of a vibrating spring, no person would be gullible enough to think a mass of 10^{50} lb attached to a spring constant with constant 1 lb/in. would vibrate with a frequency of 10^{-25} oscillations every 2π seconds or that a mass of 10^{-50} lb would vibrate with a frequency of 10^{25} oscillations every 2π seconds. One must be sure of the assumptions made in setting up a differential equation to know whether or not to have confidence in the equation and its solution.

[†] If we had interchanged the roles of c_1 and c_2, we would have written Eq. (9) as $R \sin (\omega_0 t + \phi)$, where $\tan \phi = c_2/c_1$. In other words, the only difference between a sine curve and a cosine curve is a phase angle.

and released with an upward velocity of 8 ft/sec. Assume that the resistance of the medium is negligible.

(a) Find the initial-value problem that describes the motion of the weight.

(b) Find the motion of the weight.

(c) Find the amplitude, phase angle, frequency, and period of the weight.

Solution   (a) We have

$$m = \frac{w}{g} = \frac{8 \text{ lb}}{32 \text{ ft/sec}^2} = \frac{1}{4} \text{ slug}$$

$$d = 0 \quad \text{(no friction)}$$

$$k = \frac{w}{\Delta L} = \frac{8 \text{ lb}}{(1/2) \text{ ft}} = 16 \frac{\text{lb}}{\text{ft}}$$

Hence the equation of motion is

$$\frac{1}{4}\ddot{u} + 16\,u = 0$$

or

$$\ddot{u} + 64\,u = 0 \tag{11}$$

where u is measured in feet and t in seconds. The initial conditions are $u(0) = 1$, $\dot{u}(0) = -8$.

(b) The solution of Eq. (11) is

$$u(t) = c_1 \cos 8t + c_2 \sin 8t \tag{12}$$

Substituting Eq. (12) into the initial conditions gives

$$u(0) = c_1 = 1$$
$$\dot{u}(0) = 8c_2 = -8$$

or $c_1 = 1$, $c_2 = -1$. Hence the simple harmonic motion is described by

$$u(t) = \cos 8t - \sin 8t \tag{13}$$

However, it is possible and is more illustrative to rewrite Eq. (13) in terms of a single sine or cosine curve. Using Figure 3.10 as a guide, we set

$$R = \sqrt{c_1^2 + c_2^2} = \sqrt{(1)^2 + (-1)^2} = \sqrt{2}$$

$$\tan \delta = \frac{c_1}{c_2} = -1$$

Since c_1 is positive and c_2 is negative, we see from Figure 3.10 that the phase angle δ lies in the second quadrant. Hence the phase angle is $\delta = 3\pi/4$. Hence the polar form of the simple harmonic motion is

$$u(t) = \sqrt{2} \cos\left(8t - \frac{3\pi}{4}\right) \tag{14}$$

Figure 3.10
Geometric illustration of the transformation from Cartesian to polar coordinates

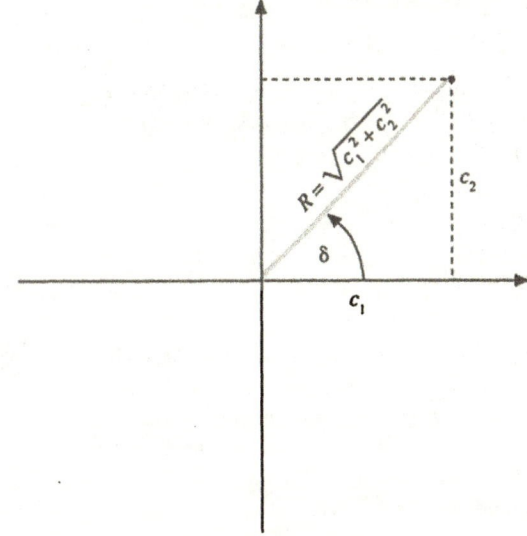

The graph of Eq. (14) is shown in Figure 3.11.

(c) From Eq. (14) we see that the amplitude is $\sqrt{2}$ feet, the phase angle is $\delta \doteq 3\pi/4$ radians, the natural frequency is $f = \omega_0/2\pi = 8/2\pi = 4/\pi$ oscillations per second, and the period is $T = 2\pi/\omega_0 = 2\pi/8 = \pi/4$ seconds.

Figure 3.11
Simple harmonic motion

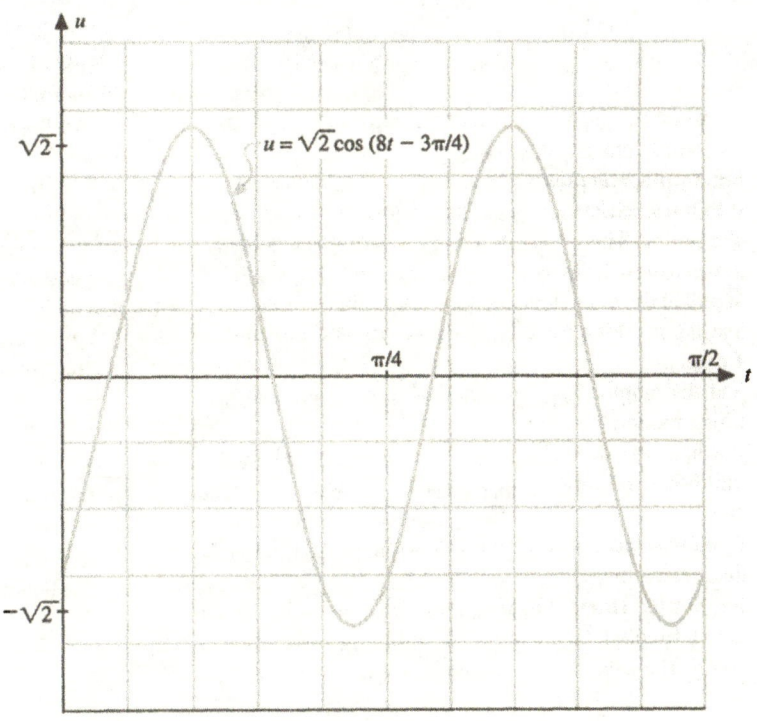

PROBLEMS: Section 3.9

For Problems 1–5, find the simple harmonic motion described by the given initial-value problem. Determine the amplitude, phase angle, frequency, and period of the motion.

1. $\ddot{u} + u = 0$ $u(0) = 1$ $\dot{u}(0) = 0$
2. $\ddot{u} + u = 0$ $u(0) = 0$ $\dot{u}(0) = 1$
3. $\ddot{u} + u = 0$ $u(0) = 1$ $\dot{u}(0) = 1$
4. $\ddot{u} + 9u = 0$ $u(0) = 1$ $\dot{u}(0) = 1$
5. $\ddot{u} + 4u = 0$ $u(0) = 1$ $\dot{u}(0) = -2$

6. **Finding the Differential Equation** An 8-lb weight is attached to a frictionless spring, which in turn is suspended from the ceiling. The weight stretches the spring 6 inches and comes to rest in its equilibrium position. The weight is then pulled down an additional 3 inches and released with a downward velocity of 1 ft/sec.
   (a) Find the initial-value problem that describes the motion of the weight.
   (b) Find the motion of the weight.
   (c) Find the amplitude, phase angle, frequency, and period of the motion.

7. **Finding the Initial-Value Problem** A 16-lb weight is attached to a frictionless spring, which in turn is suspended from the ceiling. The weight stretches the spring 6 inches and comes to rest in its equilibrium position. Find the initial-value problem that describes the motion of the weight under the following conditions.
   (a) The weight is pulled down 4 inches below its equilibrium position and released with an upward initial velocity of 4 ft/sec.
   (b) The weight is pushed up 2 inches and released with a downward velocity of 1 ft/sec.

8. **Finding Simple Harmonic Motion** A 12-lb weight is attached to a frictionless spring, which in turn is suspended from the ceiling. The weight stretches the spring 1.5 inches and comes to rest in its equilibrium position. The weight is then pulled down an additional 2 inches and released.
   (a) Find the resulting motion of the weight as a function of time.
   (b) Find the amplitude, period, and frequency of the resulting motion.
   (c) At what time does the weight first pass through the equilibrium position, and what is its velocity at that time?
   (d) If you have access to a computer with graphing facilities, sketch the graph of the motion of the weight.

9. **Finding Simple Harmonic Motion** A 12-lb weight is attached to a frictionless spring, which in turn is attached to the ceiling. The weight stretches the spring 6 inches before coming to equilibrium. Find the equation of motion of the weight if it is initially pushed upward to a position 4 inches above equilibrium and given an upward velocity of 2 ft per sec.

10. **Simple Harmonic Motion in Polar Form** Rewrite the following simple harmonic motions in the new polar form $R \cos(\omega_0 t - \delta)$.
    (a) $\cos t + \sin t$
    (b) $\cos t - \sin t$
    (c) $-\cos t + \sin t$
    (d) $-\cos t - \sin t$

11. **Simple Harmonic Motion as a Sine Function** It was shown in the text that simple harmonic motion could be written as a pure cosine function by means of the formula

    $$c_1 \cos \omega_0 t + c_2 \sin \omega_0 t = R \cos(\omega_0 t - \delta)$$

    where R and δ were determined in terms of c_1 and c_2. Show that simple harmonic motion can also be written as a pure sine function

    $$c_1 \cos \omega_0 t + c_2 \sin \omega_0 t = R \sin(\omega_0 t - \phi)$$

    Find R and ϕ in this formula in terms of c_1 and c_2.

12. **An Alternative Equation for the Period** Show that the period of motion for an undamped vibrating mass attached to a spring is $T = 2\pi\sqrt{\Delta L/g}$, where ΔL is the elongation of the spring due to the weight.

13. **Simple Harmonic Motion from Conservation of Energy** The conservation of energy says that if no energy is lost through friction, then in any mechanical system the sum of the kinetic and potential energies remains constant over time. Use the fact that the kinetic energy of a vibrating spring is $KE = 1/2\, m\dot{u}^2$ and the potential energy is given by $PE = 1/2\, ku^2$ to derive the differential equation

    $$m\ddot{u} + ku = 0 \qquad (15)$$

    Hint: Differentiate the equation $KE + PE = $ constant with respect to time.

14. **Factoring Out the Friction** We have seen that a mass-spring system in which friction is present is described by the differential equation

    $$m\ddot{u} + d\dot{u} + ku = 0 \qquad (16)$$

    An old trick is to "factor out" the friction component of the solution by introducing the new variable $U(t)$, defined by $u(t) = e^{-(d/2m)t} U(t)$.
    (a) Show that $U(t)$ satisfies the "frictionless" differential equation

    $$m\ddot{U} + \left(k - \frac{d^2}{4m}\right) U = 0 \qquad (17)$$

(b) Assuming that $k - \dfrac{d^2}{4m} > 0$, solve Eq. (17) for $U(t)$ and then show that the solution of the damped Eq. (16) is

$$u(t) = e^{-(d/2m)t}(c_1 \cos \omega_0 t + c_2 \sin \omega_0 t) \quad (18)$$

where $\omega_0 = \dfrac{1}{2m}\sqrt{4mk - d^2}$.

15. **An Interesting Property of Simple Harmonic Motion** A mass attached to a spring is pulled down a given distance and released. The same mass is then pulled down twice as far and released. What is the relationship between the periods of the two simple harmonic motions? Will the period of the second simple harmonic be twice as large as that of the first?

16. **A Pendulum Experiment** A rod of length L is suspended from the ceiling so that it can swing freely. A weight (the bob) is attached to the lower end of the rod, where the weight of the rod is assumed to be negligible compared to the weight of the bob. Let θ (radians) be the angular displacement from the vertical (plumb line), as shown in Figure 3.12. It can be shown that the differential equation that describes the motion of the pendulum is the **pendulum equation**

$$\ddot{\theta} + \frac{g}{L}\sin\theta = 0 \quad (19)$$

Determine the natural period for a pendulum of length L for small oscillations by approximating $\sin\theta \doteq \theta$.

Figure 3.12
A simple pendulum

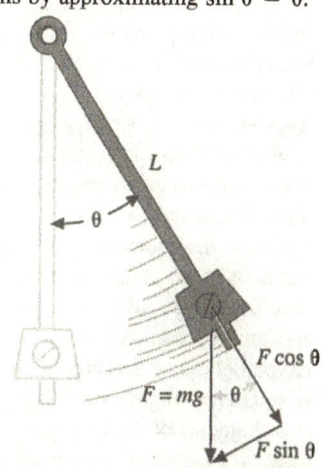

17. **Simple Harmonic Motion** A particle moves around the circle $x^2 + y^2 = r^2$ with a constant angular velocity of ω_0 radians per unit time. Show that the projection of the particle on the x-axis satisfies the equation $\ddot{x} + \omega_0 x = 0$.

18. **Other Mass-Spring Systems** It can be shown that the differential equation that describes the vibrations of the mass-spring-pulley system shown in Figure 3.13 is given by

$$\ddot{u} + \left(\frac{kR^2}{mR^2 + I}\right)u = 0 \quad (20)$$

Figure 3.13
Mass-spring-pulley system

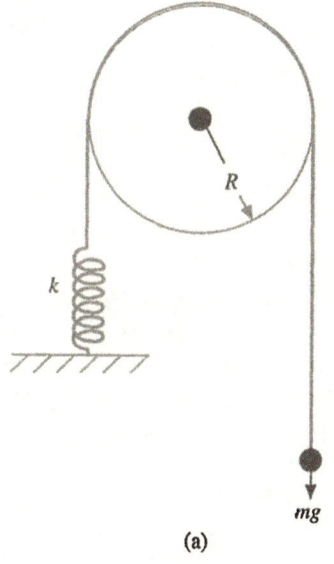

(a)

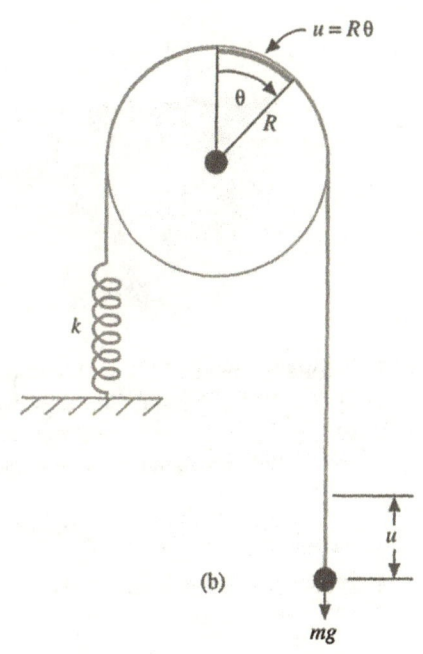

(b)

where u is the displacement from equilibrium of a weight of mass m; the values R and I are the radius and moment of inertia, respectively, of the pulley; and k is the spring constant.* Use the values of these quantities given in Eq. (20) to find the frequency of the mass.

19. Attractive Versus Repulsive Forces Simple harmonic motion of an object is the result of an *attractive* force that is proportional to the displacement of the object from equilibrium, that is, $m\ddot{u} = -ku$. On the other hand, if the force is *repulsive*, the equation of motion is $m\ddot{u} = kx$. Show that in the case of an attractive force, the displacement is described by the (circular) sine or cosine functions, whereas if the force is repulsive, the displacement is described by the hyperbolic sine and cosine functions.

20. Simple Harmonic Motion of a Buoy[†] A cylindrical buoy with a diameter of 18 inches floats in water with its axis vertical. See Figure 3.14. When it is depressed slightly and released, its period of vibration is found to be 2.7 sec. Find the weight of the cylinder. Archimedes' Principle says that an object submerged in water is buoyed (forced) up by a force equal to the weight = volume × density of the water displaced. The density of water is taken to be 62.5 lb/ft³.

Figure 3.14
Simple harmonic motion of a buoy

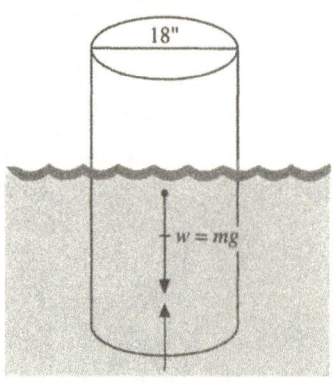

21. Simple Harmonic Motion from Los Angeles to Tokyo It can be shown that the force on a particle inside a homogeneous spherical mass is directed toward the center of the sphere with a magnitude proportional to the first power of

the distance from the center of the sphere. By using this principle, a train starting at rest and traveling through a straight line tunnel from Los Angeles to Tokyo can be shown to satisfy the initial-value problem

$$m\ddot{x} + kx = 0$$
$$x(0) = \sqrt{R^2 - d^2}$$
$$\dot{x}(0) = 0 \qquad (21)$$

where x is the distance of the train from the center of the tunnel, R is the radius of the earth (4000 miles), and k is the constant of proportionality in the equation giving the force moving the train, $F = kr$. See Figure 3.15.

(a) Find the distance of the train from the center of the tunnel at any time t.

(b) If a train starts from *any* city on earth, find the time it takes to go to *any other city* on the earth through a tunnel of this type.

Figure 3.15
Tunnel from Los Angeles to Tokyo

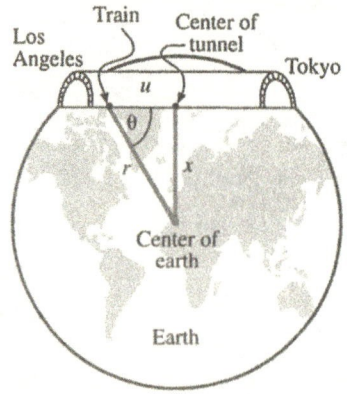

22. Computer Problem Use a graphing calculator or computer to sketch the graphs of simple harmonic motion as given by Eq. (7) for different values of c_1, c_2, and ω_0.

23. Journal Entry Make a list of some of the concepts that we have studied thus far that you do not understand.

---

* The derivation of this differential equation can be found in *Differential Equations* by C. Ray Wylie (McGraw-Hill, 1979), p. 137.

[†] This interesting problem is taken from the classic text *Differential Equations* by Lyman M. Kells. (McGraw-Hill, 1937).

3.10   UNFORCED DAMPED VIBRATIONS

PURPOSE

To study phenomena described by the unforced damped equation

$$m\ddot{u} + d\dot{u} + ku = 0$$

and to illustrate these phenomena by studying the damped vibrating spring. We study three cases: underdamped, critically damped, and overdamped, which depend on the numerical value of the *discriminant* $d^2 - 4mk$.

In the previous section we studied undamped vibrating systems. Since in reality practically all vibrations have retarding forces acting on the moving mass, we now consider the vibration of a particle subject to a resistance or damping force. For simplicity we consider only damping forces that are proportional to the first power of the velocity. Although frictional forces are often proportional to higher powers of the velocity, these types of friction lead to nonlinear differential equations and require more advanced analyses.

We have seen that the motion of a mass attached to a damped vibrating spring is

$$m\ddot{u} + d\dot{u} + ku = 0 \tag{1}$$

and that the roots of the corresponding characteristic equation

$$mr^2 + dr + k = 0 \tag{2}$$

are

$$r_1 = -\frac{d}{2m} + \frac{\sqrt{d^2 - 4mk}}{2m} \quad \text{and} \quad r_2 = -\frac{d}{2m} - \frac{\sqrt{d^2 - 4mk}}{2m} \tag{3}$$

Hence the motion of an unforced damped system depends on the nature of these roots. We consider separately the cases in which the discriminant, $d^2 - 4mk$, or quantity under the square root, is negative, zero, and positive.

UNDERDAMPED MOTION ($d^2 < 4mk$)

Here we are considering the case in which the resistance d is relatively weak in comparison with the spring constant k, and so we expect an oscillatory motion. If the quantity $d^2 - 4mk < 0$, then the roots of the characteristic equation are complex conjugate numbers, and the general solution of Eq. (1) gives rise to **damped oscillatory motion:**

$$u(t) = e^{-(d/2m)t}\left(c_1 \cos \mu t + c_2 \sin \mu t\right) \tag{4}$$

where $\mu = \dfrac{1}{2m} \sqrt{4mk - d^2}$.

We can also express Eq. (4) in the more useful polar form

$$u(t) = Re^{-(d/2m)t} \cos\left(\mu t - \delta\right) \tag{5}$$

where $R = \sqrt{c_1^2 + c_2^2}$ and $\tan \delta = c_2/c_1$. The graph of Eq. (5) is shown in Figure 3.16.

Figure 3.16
Damped oscillatory motion

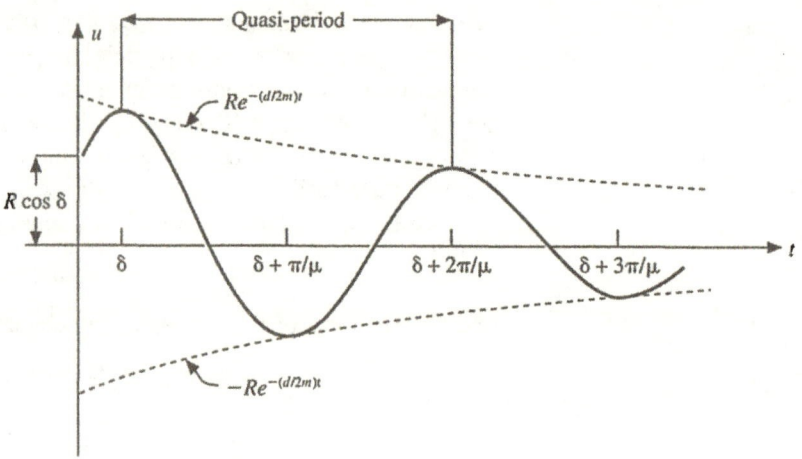

Since the cosine factor in Eq. (5) oscillates between $+1$ and -1 with the period $2\pi/\mu$, the function $u(t)$ oscillates between $-Re^{-(d/2m)t}$ and $Re^{-(d/2m)t}$. While the motion of $u(t)$ is not truly periodic, we define the *quasi-frequency* or **damped (circular) frequency** as $\omega_d = \mu$ and the **damped (natural) frequency** as $f = \mu/2\pi$. Related to the frequency is the *quasi-period* or **damped period,** defined by

$$T_d = \frac{2\pi}{\mu} = \frac{4\pi m}{\sqrt{4mk - d^2}} \tag{6}$$

This damped period is the time elapsed between successive maxima or minima in the position of the mass. The factor $Re^{-(d/2m)t}$ is called the **damped amplitude** of the oscillation. Finally, the **time constant** τ of the oscillation is defined as $\tau = 2m/d$ and is the time required for the damped amplitude, $Re^{-(d/2m)t}$, to decay from its starting value of R to R/e—in other words, about two-thirds of the way from its starting value of R to its limiting value of zero.

CRITICALLY DAMPED MOTION ($d^2 = 4mk$)

When $d^2 - 4mk = 0$, the characteristic equation (2) has the repeated root $-d/2m$. Hence the general solution to Eq. (1) gives rise to what is called **critically damped motion:**

$$u(t) = (c_1 + c_2 t)e^{-(d/2m)t} \tag{7}$$

To understand this type of motion, we use L'Hôpital's rule to find its behavior as $t \to \infty$. We find

$$\lim_{t\to\infty} u(t) = \lim_{t\to\infty}\left(\frac{c_1 + c_2 t}{e^{(d/2m)t}}\right) = \lim_{t\to\infty}\left(\frac{c_2}{(d/2m)e^{(d/2m)t}}\right) = 0 \tag{8}$$

It can easily be shown that $u(t)$ has at most one local maximum or minimum value for $t > 0$. Hence $u(t)$ does not oscillate, and this leaves, qualitatively, only two possibilities for the motion of $u(t)$, which are shown in Figure 3.17.

Figure 3.17
Critically damped or overdamped motion crosses the equilibrium point either once or never, depending on the initial conditions

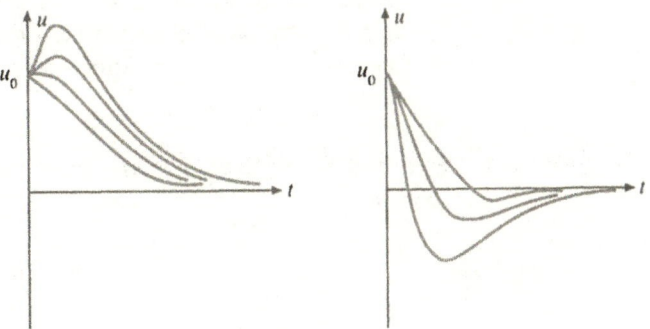

(a) If the initial slope is positive or not too negative, the curve will *not* cross the $u = 0$ line but only approach it.

(b) If the initial slope is sufficiently negative, the curve will cross the $u = 0$ line once.

OVERDAMPED MOTION ($d^2 > 4mk$)

Here we are considering the case in which the friction d is relatively large in comparison with the spring constant k, and so we expect no oscillations in the motion of the spring. Mathematically, when $d^2 - 4mk > 0$, the characteristic equation (2) has two distinct real roots:

$$r_1 = -\frac{d}{2m} + \frac{1}{2m}\sqrt{d^2 - 4mk} \quad r_2 = -\frac{d}{2m} - \frac{1}{2m}\sqrt{d^2 - 4mk} \tag{9}$$

Here the general solution to Eq. (1) gives rise to **overdamped motion:**

$$u(t) = c_1 e^{r_1 t} + c_2 e^{r_2 t} \tag{10}$$

To understand better the behavior of overdamped motion, first realize the obvious fact that $r_2 < 0$. With a little more thought, it is also not difficult to understand that r_1 is also negative. Hence as $t \to \infty$, we have $u(t) \to 0$. Also, since

$$u'(t) = c_1 r_1 e^{r_1 t} + c_2 r_2 e^{r_2 t} = e^{r_1 t}\left(c_1 r_1 + c_2 r_2 e^{(r_2 - r_1)t}\right) \tag{11}$$

it follows that $u'(t) = 0$ only when $c_1 r_1 + c_2 r_2 e^{(r_2 - r_1)t} = 0$, and hence a nontrivial solution $u(t)$ can have *at most one* maximum and minimum value for $t > 0$. Thus we conclude that, qualitatively, the possible motions of the spring are the same as for the critically damped case, as are drawn in Figure 3.17.

> *Note:* For critically damped and overdamped motions the displacement $u(t)$ "creeps" back to the equilibrium position. Depending on the initial conditions, this movement will cross the equilibrium point either *exactly once* or not at all. In other words, it will never cross the equilibrium point more than once. (See Problem 16 at the end of this section.) One should realize too that from a practical or engineering point of view the critically damped case has little meaning, serving only to separate the two major cases of underdamping and overdamping systems.
>
> The reader has no doubt observed overdamped motion in the suspension of a car. Above each wheel are a large spring and a damper (shock absorber). When the suspension system is working properly, any disturbance will quickly die out. In a car with old shock absorbers, the damping may be underdamped, and the car may rock for some time after hitting a bump.

Example 1   **Effect of Damping**   A mass-spring system with damping satisfies the initial-value problem

$$\ddot{u} + d\dot{u} + u = 0$$
$$u(0) = 1$$
$$\dot{u}(0) = 0 \tag{12}$$

Compare the solution for the damping constants of $d = 1, 2,$ and 4.

Solution   The characteristic equation for Eq. (12) is

$$r^2 + dr + 1 = 0 \tag{13}$$

whose roots are

$$r = -\frac{d}{2} \pm \frac{1}{2}\sqrt{d^2 - 4} \tag{14}$$

■

UNDERDAMPED CASE
($d = 1$)

When $d = 1$, the roots of Eq. (14) are $-\frac{1}{2} \pm i\frac{\sqrt{3}}{2}$. Hence the underdamped motion is given by

$$u(t) = e^{-t/2}\left(c_1 \cos\left(\frac{\sqrt{3}}{2}t\right) + c_2 \sin\left(\frac{\sqrt{3}}{2}t\right)\right) \tag{15}$$

Setting $u(0) = 1$ and $\dot{u}(0) = 0$ gives

$$u(t) = e^{-t/2}\left(\cos\left(\frac{\sqrt{3}}{2}t\right) + \frac{\sqrt{3}}{3}\sin\left(\frac{\sqrt{3}}{2}t\right)\right)$$

$$= \frac{2}{\sqrt{3}}e^{-t/2}\cos\left(\frac{\sqrt{3}}{2}t - \frac{\pi}{6}\right) \tag{16}$$

See Figure 3.18.

Figure 3.18
Comparison of underdamped, critically damped, and overdamped motion

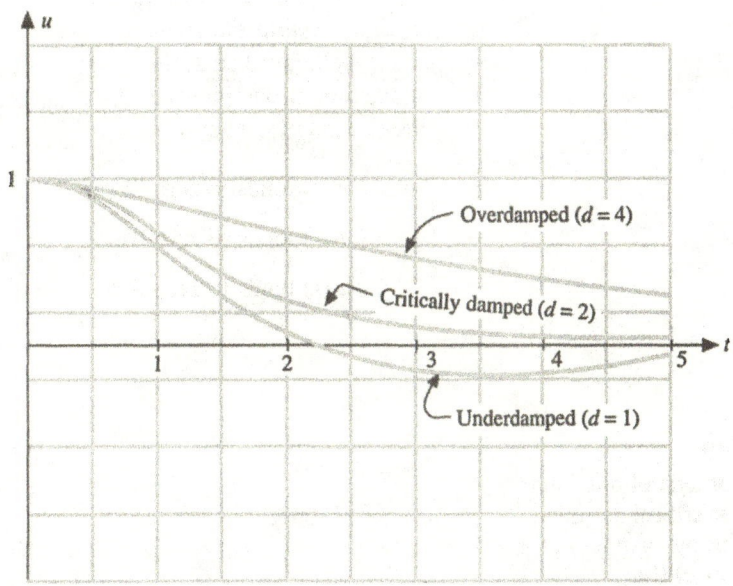

Overdamped ($d = 4$)

Critically damped ($d = 2$)

Underdamped ($d = 1$)

CRITICALLY DAMPED CASE ($d = 2$)

When $d = 2$, the characteristic equation (14) has a double root of -1. Hence the damped motion is described by

$$u(t) = (c_1 + c_2 t)e^{-t} \tag{17}$$

Using the initial conditions $u(0) = 1$ and $\dot{u}(0) = 0$, we find $c_1 = c_2 = 1$. Hence

$$u(t) = (1 + t)e^{-t} \tag{18}$$

See Figure 3.18.

OVERDAMPED CASE ($d = 4$)

When $d = 4$, the roots of the characteristic equation (14) are $-2 \pm \sqrt{3}$. Hence the overdamped motion is described by

$$u(t) = c_1 e^{(-2+\sqrt{3})t} + c_2 e^{(-2-\sqrt{3})t} \tag{19}$$

Using the initial conditions $u(0) = 1$ and $\dot{u}(0) = 0$ to find c_1 and c_2, we find

$$u(t) = \frac{1}{2\sqrt{3}}\left((2 + \sqrt{3})\,e^{(-2+\sqrt{3})t} + (-2 + \sqrt{3})\,e^{(-2-\sqrt{3})t}\right) \tag{20}$$

See Figure 3.18.

Note: Figure 3.19 shows how the roots of the characteristic equation change as the damping constant d increases from 0 to ∞. Note that for $d = 0$ there is no damping. However, with increasing d, the oscillations of $u(t)$ damp to zero faster and faster until d reaches 2, when motion becomes critically damped. When d is larger than 2, the overdamped motion is described essentially by the exponential term containing the *larger* real root (the one that is less negative). Hence as d becomes larger than 2, the solution damps to zero *more slowly*. This phenomenon of maximum damping when the system is critically damped can be seen in the above example, when

$d = 1$ ▷ $u(t)$ tends to zero like $e^{-0.5t}$

$d = 2$ ▷ $u(t)$ tends to zero like e^{-t} (decays fastest)

$d = 4$ ▷ $u(t)$ tends to zero like $e^{(-2 + \sqrt{3})t} \doteq e^{-0.27t}$ (decays the slowest)

Figure 3.19
The phenomena of maximum damping at critical damping can be seen by examining the dependence of the characteristic roots on the damping constant. When damping is $d = 0$, the roots are $\pm i$ and there is no damping. When $0 < d < 2$, the roots are complex conjugate moving along the circle in the complex plane. When $d = 2$, there is a double root at -1. When $2 < d < \infty$, both roots are real and negative, the more negative root approaching $-\infty$ and the larger root approaching 0.

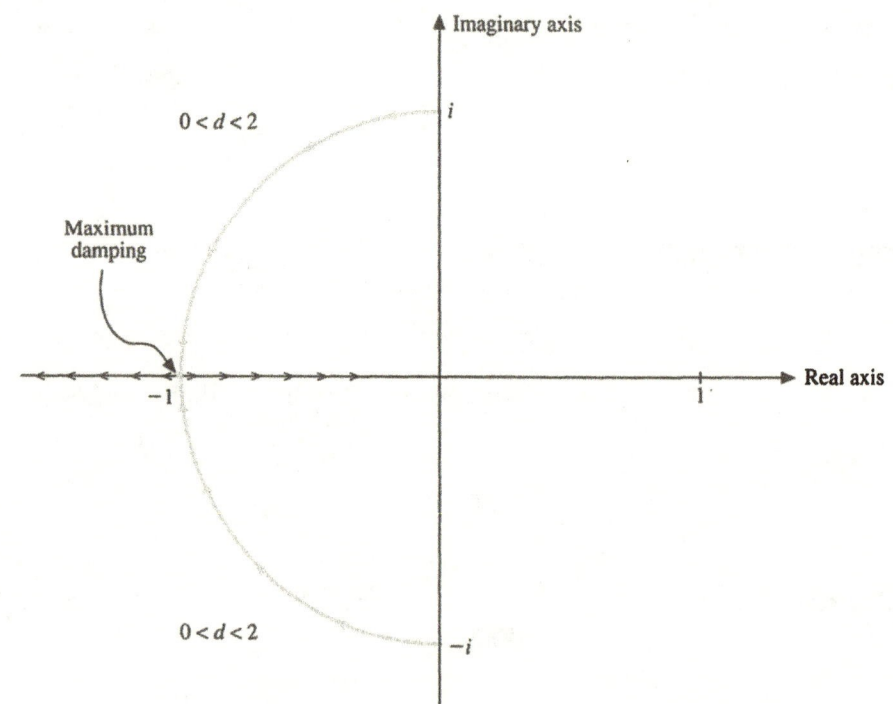

PROBLEMS: Section 3.10

Properties of Damped Oscillations

For Problems 1–4, determine the damped amplitude, the damped natural frequency, the damped period, and the time constant. Sketch the graphs of the functions.

1. $u(t) = 2e^{-t} \cos\left(2t - \pi\right)$

2. $u(t) = 3e^{-2t} \cos\left(\sqrt{3}t - \dfrac{\pi}{3}\right)$

3. $u(t) = 5e^{-0.25t} \cos\left(t + \pi\right)$

4. $u(t) = e^{-2t} \cos\left(4t - \pi\right)$

Underdamped, Critically Damped, or Overdamped?

For Problems 5–11, determine whether the motion described by the given differential equation is underdamped, critically damped, or overdamped.

5. $\ddot{u} + 4u = 0$
6. $\ddot{u} - 2\dot{u} + u = 0$
7. $\ddot{u} + 4\dot{u} + 4u = 0$
8. $\ddot{u} + 2d\dot{u} + d^2 u = 0 \quad (d > 0)$
9. $\ddot{u} + 2d\dot{u} + k^2 u = 0 \quad (d > 0 \text{ and } d^2 = k^2)$
10. $\ddot{u} + 2d\dot{u} + ku = 0 \quad (d^2 > k \text{ and } k < 0)$
11. $\ddot{u} + d\dot{u} + u = 0 \quad (0 < d < 2)$

12. **Unforced Damped Vibrations** A 32-lb weight is attached to the lower end of a coil spring, which in turn is suspended from the ceiling. The weight stretches the spring 2 ft in the process. The weight is then pulled down 6 inches below its equilibrium position and released. The resistance of the medium is given as 4 lb/(ft/sec).
    (a) Determine the motion of the weight.
    (b) Determine the damped amplitude, damped frequency, damped period, and time constant of the subsequent motion.
    (c) Sketch the graph of the motion.

13. **Unforced Damped Vibrations** A 10-lb weight is attached to a spring suspended from the ceiling. When the weight comes to rest, the spring is stretched by 1.5 inches. The damping constant for the system is $5\sqrt{2}$ lbs/(ft/sec). The weight is pulled down 3 inches from the equilibrium position and released.

    (a) Determine the motion of the weight.
    (b) Determine the damped amplitude, damped natural frequency, and damped period of the subsequent motion.
    (c) Sketch the graph of the motion.

14. **Maximum Displacement** An 8-lb weight stretches a spring 6 ft, thereby reaching its equilibrium position. Assuming a damping constant for the system of 4 lb/(ft/sec), the weight is pulled down 3 inches below its equilibrium position and given a downward velocity of 2 ft/sec. When will the mass attain its maximum displacement below equilibrium? What is the maximum displacement? Sketch the graph of the motion.

15. **Finding the Differential Equation for Observations** John starts in motion a vibrating mass attached to a spring. He observes that after 10 seconds the damping amplitude has decreased by 75% and that the damped period is 2 seconds. How does John determine the differential equation of motion from these two observations?

16. **Interesting Phenomena** Show that if the motion of a mass that is described by

$$m\ddot{u} + d\dot{u} + ku = 0 \qquad (21)$$

    is critically damped or overdamped, then the mass can pass through the equilibrium position *at most once*, regardless of the initial conditions.

17. **Decrease in the Velocity** For an underdamped mass-spring system, show that if v is the velocity of the mass at *any time*, then the velocity one (damping) period later is $ve^{-(d/2m)T_d}$.

18. **System Identification of the Damping Constant d** It is generally almost impossible to measure the damping constant d directly. One of the important uses of differential equations is to find system parameters such as this *indirectly*. Starting with the equation for underdamped motion

$$u(t) = Re^{-(d/2m)t} \cos(\mu t - \delta) \qquad (22)$$

    verify the following mathematical steps that will allow one to estimate d for underdamped motion.
    (a) Show that the ratio of two successive maximum displacements at time t and $t + T_d$ is $e^{(d/2m)T_d}$. Thus successive maximum displacements form a geometric sequence with ratio $e^{-(d/2m)T_d}$.
    (b) Show that the natural logarithm of this ratio, called the **logarithmic decrement,** is given by

$$\Delta = \frac{d}{2m} T_d = \frac{\pi d}{m\mu} \qquad (23)$$

    (c) How would you use the logarithmic decrement to find the damping constant d?

19. **Need New Shocks?** The vibrations of an automobile weighing 1600 lb (50 slugs) are controlled by a spring-shock system in which the shock absorbers provide the damping. Suppose the spring constant is $k = 4800$ lb/ft and the dashpot mechanism provides a damping constant of $c = 200$ lb-slug/ft. Find the speed (in miles per hour) at which resonance vibrations will occur if the car is driven on a washboard road surface with an amplitude of 2 inches and a wavelength of $L = 30$ ft. See Figure 3.20.

Figure 3.20
Shock-spring system

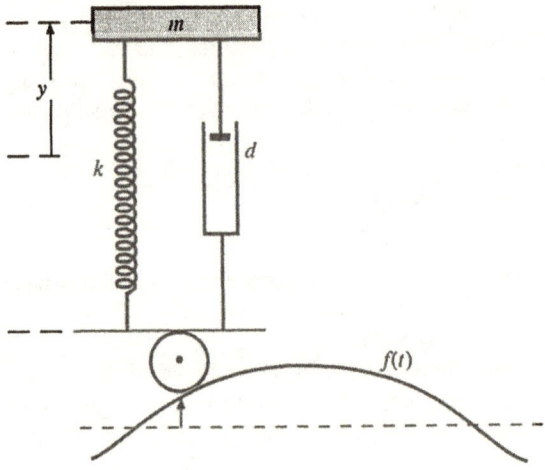

Electric Circuits

20. **RLC Circuit Problem** The charge q on the capacitor in an RLC series electrical circuit is described by the differential equation

$$L\frac{d^2q}{dt^2} + R\frac{dq}{dt} + \frac{1}{C}q = E(t) \tag{24}$$

$$q(0) = q_0 \qquad \frac{dq(0)}{dt} = i_0$$

where L is the inductance of the inductor (henrys), R is the resistance of the resistor (ohms), C is the capacitance of the capacitor (farads), and $E(t)$ is the electromotive force (volts). The initial conditions q_0 and i_0 are the initial charge across the capacitor and the initial current flowing through the circuit (amperes). See Figure 3.21. The current $i(t)$ in the circuit is given by the derivative dq/dt. Find the charge $q(t)$ on the capacitor for the LRC circuit described by the system when $L = 0.25$ henrys, $R = 10$ ohms, and $C = 0.001$ farads, $E(t) = 0$, $q(0) = q_0$ coulombs, $i(0) = \dot{q}(0) = 0$.

Figure 3.21
RLC series electrical circuit

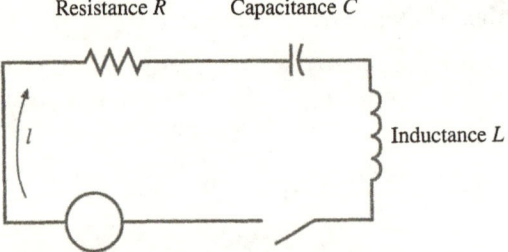

Impressed voltage $E(t)$

Calculus of Variations and Euler's Equation

Assuming that there exists a differentiable function $y = y(x)$ defined on an interval $[x_1, x_2]$ that minimizes the integral

$$J(y) = \int_{x_1}^{x_2} F(x, y, y')\, dx \tag{25}$$

where F is a given function, Leonhard Euler showed in 1744 that y must satisfy the ordinary differential equation

$$\frac{\partial F}{\partial y} - \frac{d}{dx}\left(\frac{\partial F}{\partial y'}\right) = 0 \tag{26}$$

*called **Euler's equation**. The class of problems associated with finding the minimum of a function of a function such as Eq. (25) is called the **calculus of variations**.* Often a function y also satisfies an additional set of boundary conditions such as $y(x_1) = y_1$ and $y(x_2) = y_2$. For Problems 21–22, find the Euler equation associated with the given functional, and verify that the given function $y = y(x)$ satisfies the Euler equation. Can you interpret the physical meaning of each minimizing curve $y = y(x)$?*

21. **Length of Curve Functional**

$$J(y) = \int_0^1 \sqrt{1 + (y')^2}\, dx \quad y(0) = 0 \quad y(1) = 1$$

Minimizing curve: $y(x) = x$

22. **Surface of Revolution Functional**

$$J(y) = \int_0^1 2\pi y \sqrt{1 + (y')^2}\, dx$$

$$y(0) = 1 \quad y(1) = \frac{1}{2}(e + e^{-1})$$

Minimizing curve: $y = \cosh x$

---

* The function of a function $J(y)$ is called a **functional** and is a mapping from a class of functions or vectors to a field of scalars. Although functions y that minimize $J(y)$ were studied by Newton and the Bernoullis in the late seventeenth century, the subject of calculus of variations was launched in 1744 by Leonhard Euler with his discovery of the differential equation that now bears his name.

23. Hamilton's Principle One of the basic principles of mechanics, **Hamilton's principle,** states that if a system of particles is governed by their mutual gravitational attractions, then their paths of motion will minimize the integral over time of the difference between the kinetic and potential energies of the system. For the undamped vibrating spring problem this integral is

$$J(y) = \int_{t_1}^{t_2} (T - V)\, dt \qquad (27)$$

where the kinetic and potential energies are given by $T = \frac{1}{2}m\dot{x}^2$ and $V = \frac{1}{2}kx^2$, respectively. Find Euler's equation for the functional $J(y)$, and find its general solution.

Computer Experiments

24. Varying Damping Constant Solve the initial-value problem

$$\ddot{u} + d\dot{u} + u = 0$$
$$u(0) = 1$$
$$\dot{u}(0) = 0$$

in terms of d, and sketch the graphs of $u(t)$ for $d = 0, 2, 4$, and 6.

25. Varying Mass Solve the initial-value problem

$$m\ddot{u} + d\dot{u} + u = 0$$
$$u(0) = 1$$
$$\dot{u}(0) = 0$$

in terms of the mass m, and sketch the graphs of $u(t)$ for $m = 0.1, 0.25$, and 1.

26. Varying Spring Constant Solve the initial-value problem

$$\ddot{u} + \dot{u} + ku = 0$$
$$u(0) = 1$$
$$\dot{u}(0) = 0$$

in terms of k, and sketch the graphs of $u(t)$ for $k = 0.1$, 0.25, and 0.50.

27. Running Time Backward In addition to using differential equations to determine the future of physical systems, we can use differential equations to determine the *past* of physical systems. If we introduce a *backward* time τ that runs in the *opposite* direction from the usual time t, in other words $\tau = -t$, then for any "forward running" differential equation with independent variable t, we can find the corresponding "backward running" differential equation by simply letting $\tau = -t$. For Problems (a)–(c), find the "backward running" differential equation in τ corresponding to the given differential equation in t and observe that the solution of the "backward running" equation is the "backward" solution of the "forward running" equation.

(a) $\dfrac{dy}{dt} + ay = 0 \qquad\qquad y(0) = 1$

(b) $\dfrac{d^2y}{dt^2} + y = 0 \qquad\qquad y(0) = 1 \quad y'(0) = 0$

(c) $\dfrac{d^2y}{dt^2} + 3\dfrac{dy}{dt} + 2y = 0 \quad y(0) = 0 \quad y'(0) = 1$

28. Computer Problem Use a graphing calculator or computer to sketch the graphs of the damped vibrations of a linear spring given by Eq. (5) for different values of the parameters d, m, and k.

29. Journal Entry Do you think that critically damped motions can occur in the real world, since they occur only when the discriminant is exactly zero?

▐3.11▌ FORCED VIBRATIONS

PURPOSE

To show how forced vibrations of the undamped equation

$$m\ddot{u} + ku = F_0 \cos \omega t$$

give rise to beats when $\omega \neq \omega_0 = \sqrt{k/m}$ and to resonance when $\omega = \omega_0$. We then show how forced vibrations to the damped equation

$$m\ddot{u} + d\dot{u} + ku = F_0 \cos \omega t$$

give rise to a steady state solution that has the same frequency, but is out of phase, and has a different amplitude as the driving force.

Consider the vibrations of a damped mass-spring system that acts under the influence of an external periodic force. Mathematically speaking, we will study vibrating systems governed by the differential equation

$$m\ddot{u} + d\dot{u} + ku = F_0 \cos \omega t \tag{1}$$

where m, d, and k have already been defined and F_0 and ω are nonnegative constants describing the amplitude and frequency of the periodic input. First, however, we will study the important subcase in which there is no damping, or when Eq. (1) reduces to

$$m\ddot{u} + ku = F_0 \cos \omega t \tag{2}$$

In the case of no damping, we consider two further subcases: the subcase in which the frequency of the forcing function is *different* from the natural frequency of the system ($\omega \neq \omega_0$), which gives rise to the phenomenon of *beats*, and the case in which the frequency of the forcing function is *equal to* the natural frequency ($\omega = \omega_0$), which gives rise to the phenomenon of *resonance*.

BEATS

When $\omega \neq \omega_0 = \sqrt{k/m}$, the general solution of Eq. (2) is

$$u(t) = \underbrace{c_1 \cos \omega_0 t + c_2 \sin \omega_0 t}_{\text{Transient solution}} + \underbrace{\frac{F_0}{m(\omega_0{}^2 - \omega^2)} \cos \omega t}_{\text{Steady state solution}} \tag{3}$$

where the constants c_1 and c_2 are determined from the initial conditions. In other words, the resultant motion described by Eq. (3) is the sum of *two periodic motions* with different frequencies. The first two terms in Eq. (3), which is the solution of the homogeneous equation corresponding to Eq. (2), is called the **transient solution** by engineers and applied scientists. Although for undamped systems the transient solution will not approach zero, it is that part of the solution that depends on the initial conditions. The last term in Eq. (3), which is a particular solution of Eq. (2), is called the **steady state solution** and is that part of the solution that depends on the external forcing term.

If we now assume the mass to be initially at rest, that is, $u(0) = \dot{u}(0) = 0$, the solution $u(t)$ becomes

$$u(t) = \frac{F_0}{m(\omega_0{}^2 - \omega^2)} \left(\cos \omega t - \cos \omega_0 t \right) \tag{4}$$

Note that this expression is the sum of two different oscillations with the same amplitude but different frequencies. Using the trigonometric identities

$$\cos (A \pm B) = \cos A \cos B \mp \sin A \sin B \tag{5}$$

we can rewrite Eq. (4) as

$$u(t) = \left(\frac{2F_0}{m(\omega_0{}^2 - \omega^2)} \sin \frac{(\omega_0 - \omega)t}{2} \right) \sin \frac{(\omega_0 + \omega)t}{2} \tag{6}$$

Note in this equation that when $\omega \doteq \omega_0$, the factor $\sin[(\omega_0 + \omega)t/2]$ oscillates at a *much faster rate* than does $\sin[(\omega_0 - \omega)t/2]$. Consequently, Eq. (6) describes a rapidly oscillating function with (circular) frequency $(\omega_0 + \omega)/2$, oscillating *inside* a slower sinusoidal function

$$\text{Sinusoidal amplitude} = \frac{2F_0}{m(\omega_0{}^2 - \omega^2)} \sin\left(\frac{(\omega_0 - \omega)t}{2}\right) \qquad (7)$$

This equation can be interpreted as a *sinusoidal amplitude* of the more rapidly oscillating function. This type of motion, which characterizes systems in which the driving frequency approximately equals the natural frequency of the system, exhibits the phenomenon of **beats.** See Figure 3.22.

The phenomenon of beats occurs in acoustics when two tuning forks vibrate with approximately the same frequency. One can often hear a definite rising and falling in the noise level. Also, each note of a musical instrument has a definite frequency associated with it. When a standard frequency and its corresponding musical note are played at the same time, beats will result if their frequencies are not in tune. The idea in tuning the instrument is to make the beats disappear. The musical note of the instrument then has the same frequency as the standard note.

Figure 3.22
Phenomenon of beats. We say that the faster oscillating wave is *amplitude modulated* by the slower varying wave.

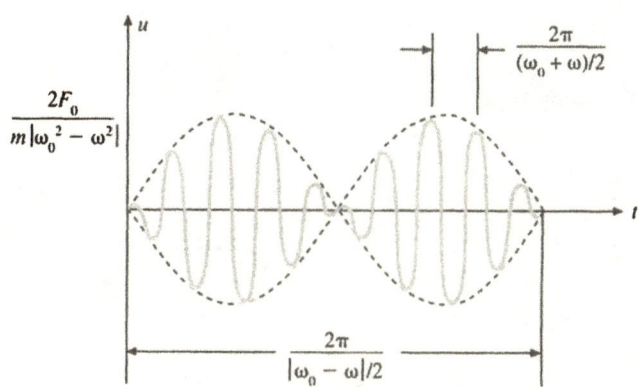

RESONANCE

When the external frequency ω of the forcing function is the *same* as the natural vibrating frequency ω_0 of a vibrating spring, the system gives rise to the phenomenon of **resonance.** To understand this phenomenon, we study the equation

$$\ddot{u} + ku = F_0 \cos \omega t \qquad (8)$$

with $\omega = \omega_0$. The important fact here is that the forcing term also appears in the solution of the corresponding homogeneous equation. We solved this equation earlier, using the method of undetermined coefficients. In this case the general solution of Eq. (8) is

$$u(t) = c_1 \cos \omega_0 t + c_2 \sin \omega_0 t + \frac{F_0}{2m\omega_0} t \sin \omega_0 t \qquad (9)$$

Note that in Eq. (9), regardless of the values of c_1 and c_2, the particular solution (last term) contains the factor of t and hence oscillates with increasing amplitude as $t \to \infty$. This type of motion, illustrated in Figure 3.23, represents the phenomenon of **resonance.**

The phenomenon of resonance is common in everyday life. Soldiers crossing a bridge are asked to break step so as not to create resonance for fear that their cadence is the same as, or near, the natural vibrating frequency of the bridge. The walls of Jericho, some assert, came tumbling down because the sound of the trumpets caused a wave motion equal to the natural frequency of the walls of the city. A diver jumping on a diving board can make the board vibrate with larger and larger amplitude by jumping with the same frequency as the natural frequency of the board. A ship will roll and pitch more wildly in a storm if the natural rolling frequency of the ship matches the frequency of the waves. And, annoyingly, the reader no doubt has heard rattles in his or her automobile caused by resonance.

Resonance is generally not a desired phenomenon for engineers. The engineers who designed the Tacoma Narrows Bridge in the state of Washington learned this the hard way when, on November 7, 1940, external periodic forces were generated by the periodic shearing of von Karman vortices. The bridge started shaking, and the roadway began to pitch and roll. A few minutes later the bridge was resting on the bottom of Puget Sound.

Figure 3.23
Phenomenon of resonance

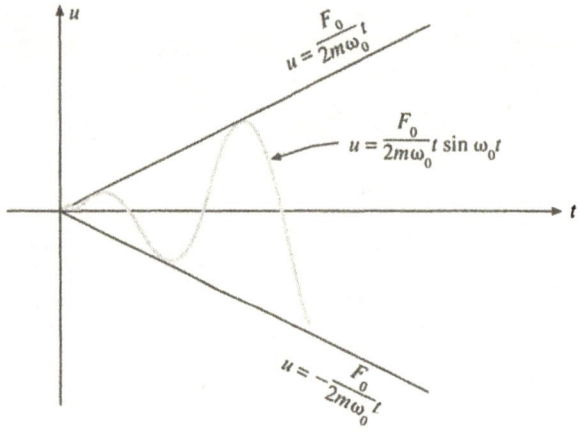

Example 1   **Ten Miles of Bad Road**   Sally is rushing to her differential equations exam and must travel on a bumpy road. Her car weighs $w = 2400$ lb (75 slugs), and the springs on her car have a collective constant of $k = 7200$ lb/ft. Unfortunately, her shock absorbers are worn out, and so her springs have no damping. Suppose the road has a sinusoidal washboard surface with an amplitude of $A = 1$ ft and a given wavelength of $L = 25$ ft. See Figure 3.24. At what speed will resonance occur?

Solution   We think of the car as a mass-spring system as drawn in Figure 3.24, in which u measures the displacement of the mass from its equilibrium position. If Sally drives at v ft/sec, then the car bouncing over the road surface will give rise to the forcing term

$$f(t) = A \cos\left(\frac{2\pi vt}{L}\right) \tag{10}$$

Hence the differential equation that describes the vertical motion of the car is

$$m\ddot{u} + ku = \cos A \cos\left(\frac{2\pi vt}{L}\right)$$

or

$$75\ddot{u} + 7200u = \cos\left(\frac{2\pi vt}{25}\right) \tag{11}$$

For this equation to have resonance, it is necessary that $\omega = \omega_0$ or

$$\frac{2\pi v}{25} = \sqrt{\frac{7200}{75}} \doteq 9.8$$

Solving for v gives $v \doteq 39$ ft/sec (27 mph). Hence if Sally drives faster than this speed, she will avoid resonance and even arrive for her differential equations exam on time. ■

FORCED DAMPED VIBRATIONS

We now consider vibrations of a damped mass-spring system driven by a periodic force. That is, we study the system

$$m\ddot{u} + d\dot{u} + ku = F_0 \cos \omega t \tag{12}$$

Figure 3.24
Sinusoidal washboard surface

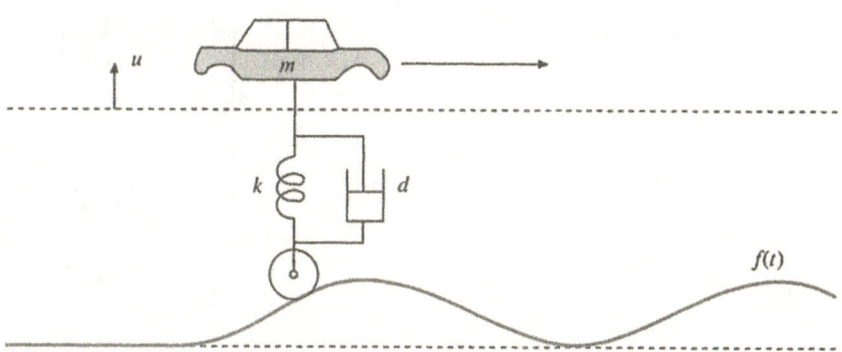

where $d^2 - 4mk < 0$ (underdamped case). The general solution of Eq. (12) can be found by using the method of undetermined coefficients to obtain a particular solution. After several computations we obtain

$$u(t) = \underbrace{c_1 e^{r_1 t} + c_2 e^{r_2 t}}_{\text{Transient solution, } u_T} + \underbrace{\frac{F_0}{\sqrt{m^2(\omega_0^2 - \omega^2)^2 + d^2\omega^2}} \cos(\omega t - \delta)}_{\text{Steady state solution, } u_{ss}} \tag{13}$$

where r_1 and r_2 are the roots of the characteristic equation and δ is determined by $\cos \delta = m(\omega_0^2 - \omega^2)/\Delta$ and $\sin \delta = d\omega/\Delta$ with $\Delta = \sqrt{m^2(\omega_0^2 - \omega^2)^2 + d^2\omega^2}$. Again, as in the undamped case, the solution depends on two terms. The first term, u_T (homogeneous solution), is called the **transient solution.** It depends on the initial condition, and it is important to note that it approaches zero as $t \to \infty$. Hence after a period of time (sometimes almost instantaneously) the oscillations (13) will approach the steady state solution

$$u_{ss}(t) = \frac{F_0}{\sqrt{m^2(\omega_0^2 - \omega^2)^2 + d^2\omega^2}} \cos(\omega t - \delta) \tag{14}$$

This second term, u_{ss} (a particular solution), is called the **steady state solution.** It is the consequence of the forcing term, $F_0 \cos \omega t$, and like the forcing term has the same frequency ω. However, it is out of phase with the forcing term, and its amplitude is changed by the **amplification factor***

$$M(\omega) = \frac{1}{\sqrt{m^2(\omega_0^2 - \omega^2)^2 + d^2\omega^2}} \tag{15}$$

When $M(\omega)$ is graphed as a function of input frequency ω, it is called the **frequency response curve** for the system, and it represents the *gain* in amplitude of the steady state response as a function of the frequency of the forcing term. For natural frequency $\omega_0 = \sqrt{k/m} = 1$, various frequency response curves are drawn in Figure 3.25.

Figure 3.25
Frequency response curves

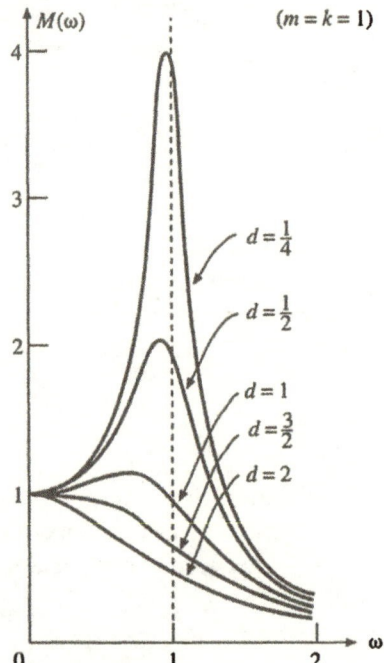

---

* Athough we mention $M(\omega)$ only for underdamped vibrations, the function $M(\omega)$ is also well defined for overdamped and critically damped vibrations.

PROBLEMS: Section 3.11

1. Express $\cos 3t - \cos t$ in the form $A \sin \alpha t \sin \beta t$.

2. Express $\sin 3t - \sin t$ in the form $A \sin \alpha t \cos \beta t$.

For Problems 3–5, find the steady state solution having the form $u_{ss} = A \cos (\omega t - \delta)$.

3. $\ddot{u} + 4\dot{u} + 4u = \cos t$

4. $\ddot{u} + 2\dot{u} + 2u = 2 \cos t$

5. $\ddot{u} + \dot{u} + u = 4 \cos 3t$

6. Resonance A mass of one slug is hanging at rest on a spring whose constant is 12 lb/ft. At time $t = 0$ an external force of $f(t) = 16 \cos \omega t$ lb is applied to the system.

(a) What is the frequency of the forcing function that is in resonance with the system?

(b) Find the equation of motion of the mass with resonance.

7. Good and Bad Resonance Sometimes resonance in a physical system is desirable, and sometimes it is undesirable. Discuss whether the resulting resonance would be helpful or destructive in the following systems.

(a) Soldiers marching on a bridge with the same frequency as the natural frequency of the bridge.

(b) A person rocking a car stuck in the snow with the same frequency as the natural frequency of the stuck car.

(c) A child pumping a swing.

(d) Vibrations caused by air passing over an airplane wing having the same frequency as the natural flutter of the wing.

(e) Acoustic vibrations having the same frequency as the natural vibrations of a wine glass (the "Memorex experiment").

8. Ed's Buoy* Ed is sitting on the dock and observes a cylindrical buoy bobbing vertically in calm water. He observes that the period of oscillation is 5 sec and that 4 ft of the buoy are above water when it reaches its maximum height and 2 ft above water when it is at its minimum height. An old seaman tells Ed that the buoy weighs 2000 lb.

(a) How will this buoy behave in rough waters with the waves 6 ft from crest to trough and with a period of 7 sec?

(b) Will the buoy ever be submerged?

* This problem is based on a problem taken from Engineering Mathematics *by Robert E. Gaskell (Dryden Press, 1958).*

9. Forced Damped Motion Find the steady state motion of a mass that vibrates according to the law

$$\ddot{u} + 8\dot{u} + 36u = 72 \cos 6t$$

10. Forced Damped Motion A 32-lb weight is attached to a spring suspended from the ceiling, stretching the spring by 1.6 ft before coming to rest. At time $t = 0$ an external force of $f(t) = 20 \cos 2t$ is applied to the system. Assume that the mass is acted upon by a damping force of $4 \, du/dt$, where du/dt is the instantaneous velocity in feet per second. Find the displacement of the weight.

11. General Forced Damped Oscillations Show that the general solution of the forced damped system

$$m\ddot{u} + d\dot{u} + ku = F_0 \cos \omega t \qquad (16)$$

is

$$u(t) = c_1 e^{r_1 t} + c_2 e^{r_2 t}$$
$$+ \frac{F_0}{\sqrt{m^2(\omega_0{}^2 - \omega^2)^2 + d^2\omega^2}} \cos (\omega t - \delta) \qquad (17)$$

where r_1 and r_2 are the roots of the characteristic equation, c_1 and c_2 are arbitrary constants, and

$$\omega_0 = \sqrt{k/m}, \qquad \delta = \tan^{-1}\left(\frac{d\omega}{m(\omega_0{}^2 - \omega^2)}\right)$$

12. How to Pump Up a Swing A child is sitting in a swing trying to pump it up by kicking his legs. Could you explain to this child, in terms of the amplitude amplification factor $M(\omega)$, how often to kick?

Computer Experiments in Graphing

If you have access to a computer with graphing capabilities, draw the graphs of the vibratory motions given in Problems 13–14.

13. Beats The following functions are typical vibratory motions for an undamped system when the forcing frequency is close to the natural frequency of the system.

(a) $u(t) = \sin t \sin 3t$

(b) $u(t) = 5 \sin t \sin 10t$

(c) $u(t) = 2 \sin (0.05t) \sin 10t$

(d) $u(t) = 10 \sin (0.025t) \sin 20t$

(e) $u(t) = 10 \sin (0.0005t) \sin 20t$

14. Resonance The following functions are typical vibratory motions for an undamped system when the forcing frequency is equal to the natural frequency of the system.

(a) $u(t) = t \sin 2\pi t$

(b) $u(t) = 2t \sin t$

(c) $u(t) = 0.25t \sin 4t$

(d) $u(t) = 0.50t \sin 10t$

(e) $u(t) = 0.25t \sin 20t$

Figure 3.26
The game of bob

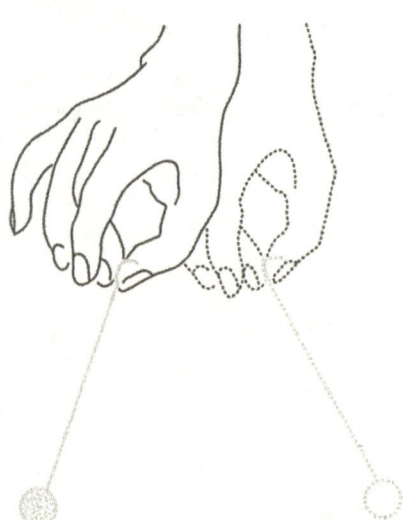
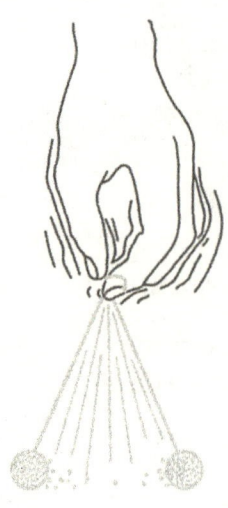
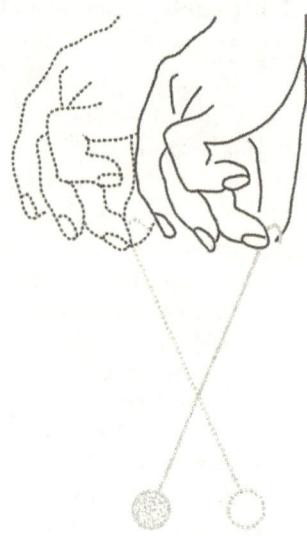

(a)   Find the natural frequency of the bob and then move your hand back and forth at a lower frequency than the natural frequency of the bob. Note that as you increase the frequency of your hand, the amplitude of the bob increases.

(b)   Move your hand back and forth at roughly the same frequency as the natural frequency of the bob. Note that the amplitude of the bob is large.

(c)   Move your hand at a higher frequency than the natural frequency of the bob. Note that the amplitude of the bob decreases with increasing frequency of your hand.

15. Graphing the Amplification Factor The function

$$M(\omega) = \frac{1}{\sqrt{m^2(\omega_0^2 - \omega^2)^2 + d^2\omega^2}}$$

represents the ratio of the output to input amplitude in the damped forced system

$$m\ddot{u} + d\dot{u} + ku = F_0 \cos \omega t$$

where $\omega_0^2 = k/m$. Sketch the graph of the amplification factor for $m = \omega_0 = 1$ and $d = 0, 1, 2, 3, 4$, and 5. In each, $M(\omega)$ should reach its maximum value at the value $\omega = \omega_0$.

16. Amplitude Modulation When the function $\sin \omega t$ is multiplied by an "amplitude" function $A(t)$, producing the function $A(t) \sin \omega t$, the wave $\sin \omega t$ is said to be **amplitude modulated.** In acoustics the fluctuations in amplitude are called beats, with the sound being loud for large amplitude. Sketch the graphs of the following amplitude-modulated functions.

(a) $f(t) = e^{-t} \sin t$ (c) $f(t) = \cos t \sin t$
(b) $f(t) = \sin 2t \sin t$

17. The Game of Bob* The game of bob requires only a small weight tied to a string. The game illustrates how the frequency and amplitude of the output of a damped linear system depend on ("track") the frequency and amplitude of the input to the system. (Although we have studied only the vibrating spring, the pendulum obeys the same second-order linear differential equation for small amplitudes.)

Part 1 (finding ω_0): Begin your experiment by giving your hand a small jerk and then holding it still. See Figure 3.26(a). The bob will swing with a frequency of the unforced (damped) pendulum. Measure this frequency.

Part 2 ($\omega < \omega_0$): Now that you know the natural frequency of the bob, move your hand back and forth with a frequency (ω) that is *slower* than the natural frequency (ω_0). Note that the pendulum "follows" your hand, which means that the oscillations of the bob and your hand are close to being "in phase." Note that as you increase the frequency of your hand movement, the amplitude of the bob also increases. See Figure 3.26(a). This is in agreement with the graph of the amplitude factor $M(\omega)$ shown in Figure 3.25.

---

* This game was extracted from the delightful little book *Dynamics: The Geometry of Behavior, Part One: Periodic Behavior* by Ralph Abraham and Christopher Shaw (The Visual Mathematics Library, Aerial Press, 1982).

Part 3 ($\omega \doteq \omega_0$): When you increase the frequency of your hand movement so that it is close to the natural frequency of the bob, the amplitude of the bob becomes large. See Figure 3.26(b).

Part 4 ($\omega > \omega_0$): Finally, move your hand back and forth *faster* than the natural frequency of the bob. Note that the bob swings *opposite* to the motion of your hand. This means that the bob is way out of phase from the movement of your hand. Also, note that the amplitude of the bob decreases with increasing frequency of your hand movement. See Figure 3.26(c). This, of course, is a physical verification of the fact that the amplitude factor $M(\omega)$ approaches zero as $\omega \to \infty$.

18. **Journal Entry** Describe the phenomenon of resonance that results when you drive your car over a bumpy road. Do you ever try to hit the bumps faster than the resonance speed?

3.12 INTRODUCTION TO HIGHER-ORDER EQUATIONS (Optional)

PURPOSE
To present a brief introduction of higher-order differential equations and show how the deflection of a beam can be described by a certain fourth-order equation.

This final section of Chapter 3 is a continuation from second-order differential equations to higher-order differential equations. Although we could have included an entire chapter on higher-order equations, the theory and methodology for solving higher-order linear equations closely parallel those of second-order equations, and so we have included only a brief introduction to the subject.

The following theorem, which gives the higher-order version of the existence and uniqueness result, should come as no surprise.

THEOREM 3.10: Existence and Uniqueness Theorem

If $a_0(x), a_1(x), \cdots, a_n(x)$ and $f(x)$ are continuous on an interval I, and if x_0 is a point in I, then the initial-value problem

$$a_0(x)\,y^{(n)} + a_1(x)\,y^{(n-1)} + \cdots + a_{n-1}(x)\,y' + a_n(x)\,y = f(x)$$

$$y(x_0) = y_0$$
$$y'(x_0) = y_0'$$
$$y''(x_0) = y_0''$$
.
.
.
$$y^{(n-1)}(x_0) = y_0^{(n-1)}$$

has exactly one solution on I.

THE HOMOGENEOUS EQUATION

The following theorem is a straightforward extension of a theorem that we saw in Chapter 2.

THEOREM 3.11: General Solution of the Homogeneous Equation

If $y_1(x)$, $y_2(x)$, ..., $y_n(x)$ are solutions of the nth-order linear homogeneous equation

$$a_0(x)\, y^{(n)} + a_1(x)\, y^{(n-1)} + \cdots + a_{n-1}(x)\, y' + a_n(x)\, y = 0$$

on an interval I, then so is

$$y = c_1\, y_1(x) + c_2 y_2(x) + \cdots + c_n\, y_n(x)$$

where c_1, c_2, ..., c_n are arbitrary constants. What's more, the functions $y_1(x)$, $y_2(x)$, ..., $y_n(x)$ are linearly independent on I if and only if there is some x_0 in I where $W(y_1, y_2, ..., y_n) \neq 0$, where W is the n-dimensional **Wronskian** defined by

$$W(y_1, y_2, ..., y_n) = \begin{vmatrix} y_1 & y_2 & \cdots & y_n \\ y_1' & y_2' & \cdots & y_n' \\ \cdot & \cdot & \cdot & \cdot \\ \cdot & \cdot & \cdot & \cdot \\ \cdot & \cdot & \cdot & \cdot \\ y_1^{(n-1)} & y_2^{(n-1)} & \cdots & y_n^{(n-1)} \end{vmatrix}$$

Such a set of solutions is called a **fundamental set of solutions.**

Example 1 **A Fundamental Set of Solutions** Show that $\cos x$, $\sin x$, and e^x constitute a fundamental set of solutions of

$$y''' - y'' + y' - y = 0$$

on $(-\infty, \infty)$.

Solution By inspection it is easy to see that the three functions satisfy the differential equation. Computing the Wronskian, we find

$$W = \begin{vmatrix} \cos x & \sin x & e^x \\ -\sin x & \cos x & e^x \\ -\cos x & -\sin x & e^x \end{vmatrix} = 2e^x \neq 0$$

Hence the functions are linearly independent solutions of the equation, and the general solution is

$$y = c_1 \cos x + c_2 \sin x + c_2 e^x \qquad \blacksquare$$

SOLVING HOMOGENEOUS EQUATIONS

We now solve the nth-order homogeneous equation with constant coefficients

$$a_0 y^{(n)} + a_1 y^{(n-1)} + \cdots + a_{n-1} y' + a_n y = 0 \tag{1}$$

by trying a function of the form e^{mx}. Doing this, we arrive at the characteristic equation

$$a_0 m^n + a_1 m^{n-1} + \cdots + a_{n-1} m + a_n = 0 \tag{2}$$

as we did for second-order equations. As it was in the case of second-order equations, the solution depends on the nature of the roots of the characteristic equation. Although we know from the theory of equations that (2) has exactly n roots (possibly repeated), the problem of actually finding the roots may require the use of approximate methods and a computer. The following examples illustrate these ideas.

Example 2   Real Unequal Roots   Find the general solution of

$$y''' - 2y'' - y' + 2y = 0$$

Solution   The characteristic equation

$$m^3 - 2m^2 - m + 2 = 0$$

has the roots $r_1 = -1$, $r_2 = 1$, $r_3 = 2$, and hence e^{-x}, e^x, and e^{2x} are solutions. It is an easy matter to show that these functions are linearly independent, either from the definition of linear independence or by showing that the Wronskian is nonzero (at any point). Hence the general solution is

$$y = c_1 e^{-x} + c_2 e^x + c_3 e^{2x} \qquad \blacksquare$$

Example 3   Real Repeated Roots   Solve

$$y''' + 3y'' + 3y' + y = 0$$

Solution   The roots of the characteristic equation

$$m^3 + 3m^2 + 3m + 1 = 0$$

are -1, -1, and -1. Hence the general solution is

$$y = c_1 e^{-x} + c_2 x e^{-x} + c_3 x^2 e^{-x} \qquad \blacksquare$$

Example 4   Complex Roots   Solve

$$y^{(4)} - y = 0$$

Solution   The roots of the characteristic equation

$$m^4 - 1 = (m^2 - 1)(m^2 + 1) = 0$$

are -1, $+1$, $\pm i$. Hence the general solution is

$$y = c_1 e^{-x} + c_2 e^x + c_3 \cos x + c_4 \sin x \qquad \blacksquare$$

THE NONHOMOGENEOUS EQUATION

The following theorem summarizes the theory for the higher-order nonhomogeneous linear equation.

> **THEOREM 3.12:** Nonhomogeneous Solution
>
> If $y_p(x)$ is any solution of the nonhomogeneous equation
> $$a_0(x)y^{(n)} + a_1(x)y^{(n-1)} + \cdots + a_{n-1}(x)y' + a_n(x)y = f(x) \qquad (3)$$
> and if $y_1, y_2, \ldots, y_n$ are linearly independent solutions of the corresponding homogeneous equation
> $$a_0(x)y^{(n)} + a_1(x)y^{(n-1)} + \cdots + a_{n-1}(x)y' + a_n(x)y = 0 \qquad (4)$$
> then the general solution of Eq. (3) is
> $$y = c_1 y_1(x) + c_2 y_2(x) + \cdots + c_n y_n(x) + y_p(x) \qquad (5)$$

FINDING NONHOMOGENEOUS SOLUTIONS

As we did for second-order equations, we use the method of undetermined coefficients and variation of parameters to find a particular solution of the nonhomogeneous equation.

Example 5 Undetermined Coefficients Find the general solution of
$$y''' - 2y'' - y' + 2y = 2x - 1$$

Solution We saw in Example 2 that the homogeneous solution is
$$y_h(x) = c_1 e^{-x} + c_2 e^x + c_3 e^{2x}$$

To find a particular solution, the rules are essentially the same as for second-order equations, so we try $y_p(x) = Ax + B$, from which we find $A = 1, B = 0$. Hence
$$y = c_1 e^{-x} + c_2 e^x + c_3 e^{2x} + x \qquad \blacksquare$$

Example 6 Undetermined Coefficients Find the general solution of
$$y^{(4)} - 5y''' + 3y'' + 19y' - 30y = e^{-2x}$$

Solution The characteristic equation
$$m^4 - 5m^3 + 3m^2 + 19m - 30 = 0$$

has roots $r = -2, 3, 2 \pm i$. Hence the homogeneous equation has the four solutions
$$e^{-2x}, e^{3x}, e^{2x} \cos x, e^{2x} \sin x$$

One can show that the Wronskian of this function is nonzero (at any x) and so the homogeneous solution is
$$y_h(x) = c_1 e^{-2x} + c_2 e^{3x} + c_3 e^{2x} \cos x + c_4 e^{2x} \sin x$$

To find a particular solution, we would normally try Ae^{-2x} except that this function appears in the homogeneous solution, and so we try $y_p(x) = Axe^{-2x}$. Substituting this function into the nonhomogeneous equation, we eventually find $A = -1/95$. Hence the general solution is

$$y = c_1 e^{-2x} + c_2 e^{3x} + c_3 e^{2x} \cos x + c_4 e^{2x} \sin x - \tfrac{1}{95} xe^{-2x}$$ ∎

VARIATION OF PARAMETERS

The method of variation of parameters can also be extended to higher-order equations. To find a particular solution of the nth-order equation

$$a_0(x)y^{(n)} + a_1(x)y^{(n-1)} + \cdots + a_{n-1}(x)y' + a_n(x)y = f(x) \tag{6}$$

where $y_1, y_2, ..., y_n$ are n linearly independent solutions of the homogeneous equation are known, one seeks a solution of the form

$$y_p = v_1(x)\,y_1(x) + v_2(x)y_2(x) + \cdots + v_n(x)y_n(x)$$

Substituting this expression into Eq. (6), one arrives at the system of equations

$$y_1 v_1' + y_2 v_2' + \cdots + y_n v_n' = 0$$
$$y_1' v_1' + y_2' v_2' + \cdots + y_n' v_n' = 0$$
$$y_1'' v_1' + y_2'' v_2' + \cdots + y_n'' v_n' = 0$$
$$\vdots$$
$$y_1^{(n-2)} v_1' + y_2^{(n-2)} v_2' + \cdots + y_n^{(n-2)} v_n' = 0$$
$$y_1^{(n-1)} v_1' + y_2^{(n-1)} v_2' + \cdots + y_n^{(n-1)} v_n' = f(x) \tag{7}$$

from which one can solve for $v_1', v_2', ..., v_n'$. After these values have been found, then by integration one can (hopefully) find $v_1, v_2, ..., v_n$ and hence y_p.

Example 7   **Variation of Parameters**   Find a particular solution of

$$y''' + y' = \tan x \tag{8}$$

Solution   The homogeneous solution has three linearly independent solutions 1, $\cos x$, and $\sin x$. Hence we try a particular solution of the form

$$y_p = v_1(x) + v_2(x) \cos x + v_3(x) \sin x$$

Substituting this expression into Eq. (8) yields the three equations

$$v_1' + \cos x\, v_2' + \sin x\, v_3' = 0$$
$$0 - \sin x\, v_2' + \cos x\, v_3' = 0$$
$$0 - \cos x\, v_2' - \sin x\, v_3' = \tan x \tag{9}$$

Solving these equations gives

$$v_1' = \tan x$$
$$v_2' = -\sin x$$
$$v_3' = -\sin x \tan x$$

Integration of these expressions gives

$$v_1 = \ = \ -\ln|\cos x|$$
$$v_2 = \cos x$$
$$v_3 = \sin x \ - \ \ln|\tan x + \sec x|$$

Hence we have a particular solution

$$y_p = v_1 + v_2 \cos x + v_3 \sin x$$
$$= -\ln|\cos x| + 1 - \sin x \ln|\tan x + \sec x| \qquad (10)$$

We summarize the method for solving a higher-order equation in Figure 3.27.

Figure 3.27
Summary of methods for
solving a higher-order
equation

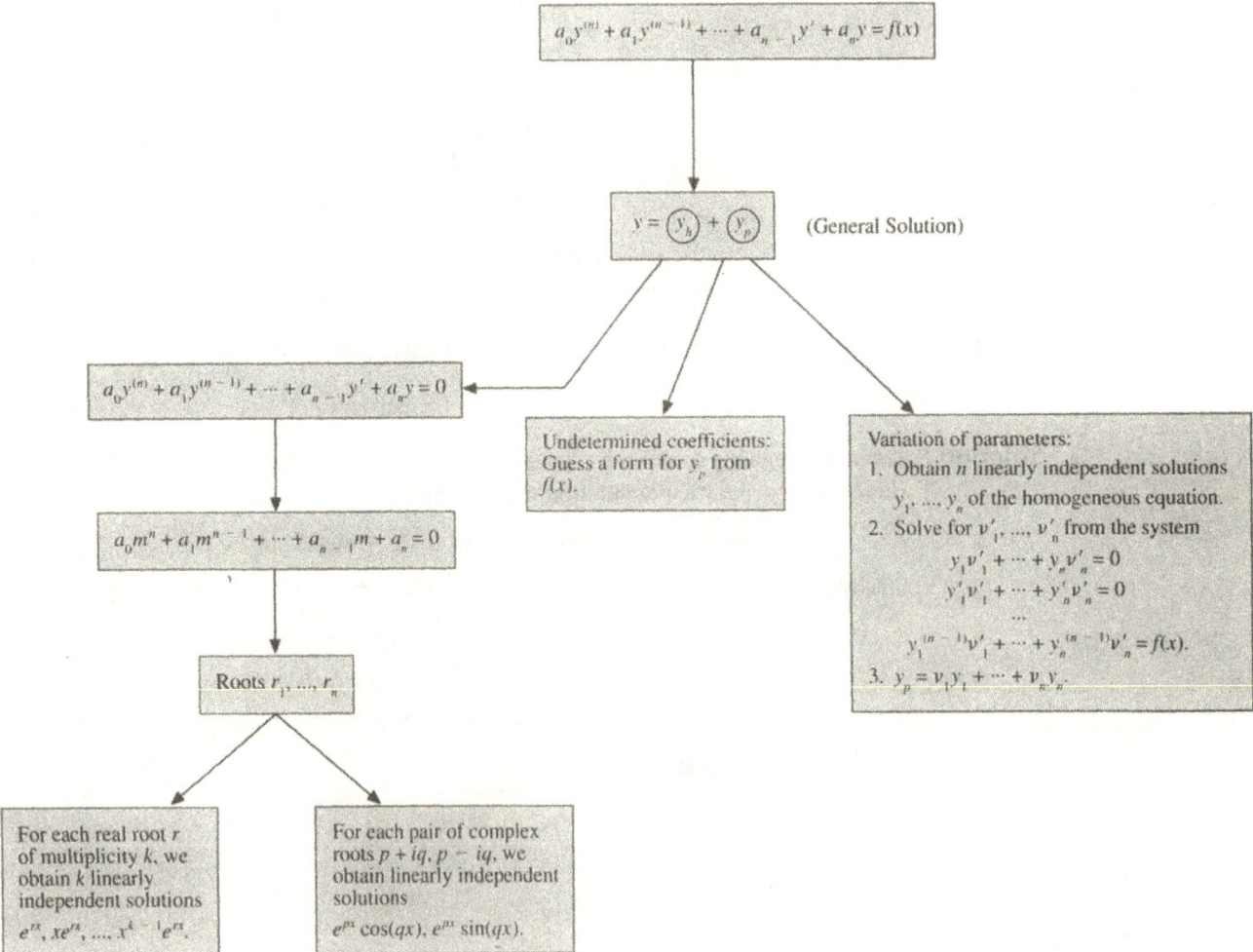

DEFLECTION OF A UNIFORMLY LOADED BEAM

One of the useful applications of higher-order differential equations lies in the study of beams that have a uniform cross section (such as an *I* beam). Without going into the derivation,* one can show that the *static deflection* $y(x)$ of a beam of length L carrying a load $w(x)$ per unit length is governed by the fourth-order equation

$$EI \frac{d^4 y}{dx^4} = w(x) \qquad (0 \le x \le L) \tag{11}$$

where E is **Young's modulus of elasticity,** I is a moment of inertia of a cross section of the beam, and y is the *downward* deflection of the beam. See Figure 3.28.

Figure 3.28
One of the applications of fourth-order differential equations lies in the study of static and vibrating beams

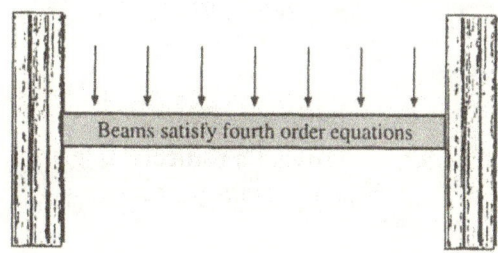

Beams satisfy fourth order equations

Although all beams with uniform cross section satisfy the same differential equation (11), the boundary conditions that the beam satisfies at the two ends $x = 0$ and $x = L$ depend on how the beam is attached at the ends. A few common types of attachments are the following.

Three Common Beams

Clamped Beam

If both ends of the beam are clamped, the boundary conditions are
$$y(0) = y'(0) = y(L) = y'(L) = 0$$

Cantilever Beam

If the left end is clamped and the right end is free, then the boundary conditions are
$$y(0) = y'(0) = y''(L) = y'''(L) = 0$$

Hinged Beam

If both ends are hinged, then the boundary conditions are
$$y(0) = y''(0) = y(L) = y''(L) = 0$$

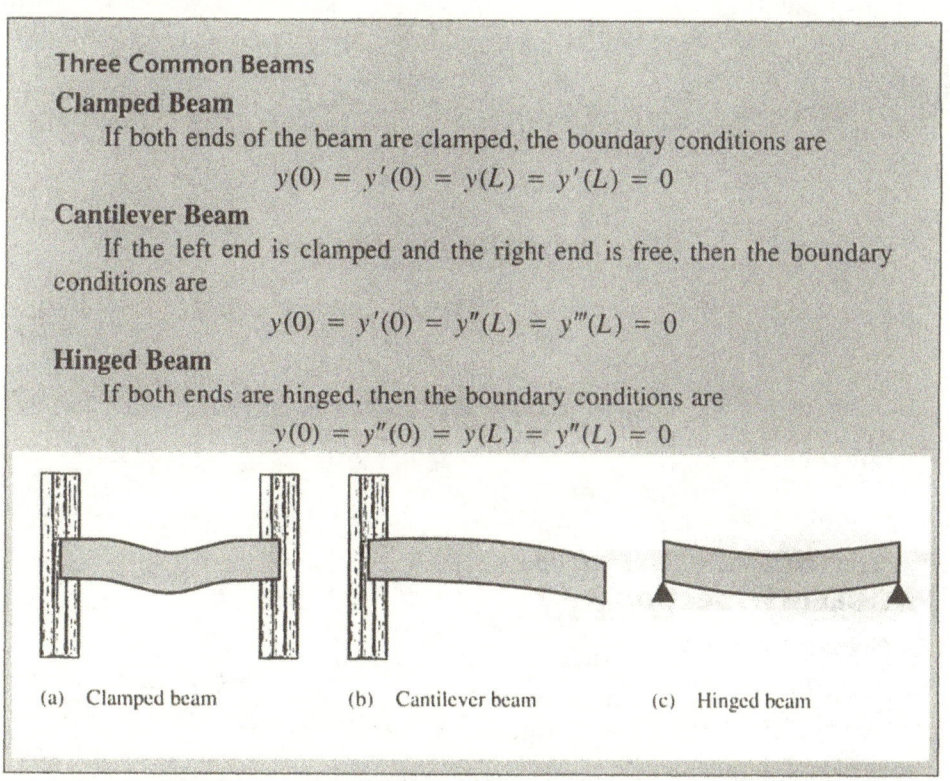

(a)   Clamped beam   (b)   Cantilever beam   (c)   Hinged beam

---

* A derivation of the beam equation can be found in the excellent book *Advanced Engineering Mathematics* by Peter O'Neil (Wadsworth, New York, 1983).

Example 8 **Clamped Beam** Find the static deflection of a beam that is clamped at both ends when a constant load w_0 is distributed uniformly along its length $0 \le x \le 1$.

Solution The boundary-value problem that governs this problem is

$$EI\frac{d^4y}{dx^4} = w_0$$
$$y(0) = y'(0) = y(1) = y'(1) = 0$$

We first find the homogeneous solution by observing that the characteristic equation is $m^4 = 0$, which has a root at 0 with multiplicity four. Hence the homogeneous solution is

$$y_h(x) = c_1 + c_2 x + c_3 x^2 + c_4 x^3$$

To find a particular solution, we try $y_p(x) = Ax^4$, getting $A = \dfrac{w_0}{24EI}$. Hence the general solution of the nonhomogeneous equation is

$$y = c_1 + c_2 x + c_3 x^2 + c_4 x^3 + \frac{w_0}{24EI}x^4 \tag{12}$$

We find the constants c_1, c_2, c_3, and c_4 by substituting y into the four boundary conditions, getting

$$y(0) = c_1 = 0$$
$$y'(0) = c_2 = 0$$
$$y(1) = c_1 + c_2 + c_3 + c_4 + \frac{w_0}{24EI} = 0$$
$$y'(1) = c_2 + 2c_3 + 3c_4 + \frac{w_0}{6EI} = 0$$

Solving these equations gives

$$c_1 = 0, \quad c_2 = 0, \quad c_3 = \frac{1}{24}\left(\frac{w_0}{EI}\right), \quad c_4 = -\frac{1}{12}\left(\frac{w_0}{EI}\right)$$

Substituting these values into Eq. (12) and simplifying give

$$y(x) = \frac{w_0 x^2}{24EI}(x - 1)^2 \tag{13}$$

∎

PROBLEMS: Section 3.12

For Problems 1–4, find the Wronskian of the given functions.

1. $\{1, x, x^2\}$

2. $\{1, x, x^2, x^3, ..., x^n\}$

3. $\{e^x, xe^x, e^{-x}\}$

4. $\{1, \cos x, \sin x\}$

For Problems 5–21, find the general solution of the given equation. Compute the Wronskian to determine that the solutions are linearly independent.

5. $y''' = 0$

6. $y^{(4)} = 0$

7. $y''' + y' = 0$

8. $y^{(iv)} + y'' = 0$

9. $y''' - y'' = 0$

10. $y''' + y = 0$

11. $y''' - 3y'' + 2y' = 0$

12. $y''' - 5y'' + 2y' + 8y = 0$

13. $y''' - 2y'' - 7y' - 4y = 0$

14. $y''' + y'' - 7y' - 15y = 0$

15. $y''' - y'' - y' + y = 0$
16. $y''' - 3y'' + 3y' - y = 0$
17. $y''' - 6y'' + 11y' - 6y = 0$
18. $y^{(iv)} + y = 0$
19. $y^{(4)} - y''' - 9y'' - 11y' - 4y = 0$
20. $y^{(iv)} - 3y'' - 4y = 0$
21. $y^{(iv)} - 2y'' + y = 0$

For Problems 22–27, find a particular solution of the given equation using the method of undetermined coefficients.

22. $y''' + y = 2x + 1$
23. $y''' + y'' + y' + y = x^2 + 2x + 2$
24. $y''' + y' + y = \sin x$
25. $y''' + y'' + y = \sin x$
26. $y''' + y'' + y' + y = 4e^x$
27. $y^{(4)} + y = 2 \sin x$

For Problems 28–29, find a particular solution of the given equation using the method of variation of parameters.

28. $y''' - y' = x$
29. $y''' - 2y'' + y' = x$
30. **Cauchy-Euler Equation** Find a particular solution of the nonhomogeneous Cauchy-Euler equation

$$x^3 y''' - 3x^2 y'' + 6xy' - 6y = x^{-1} \qquad (x > 0)$$

given that $\{x, x^2, x^3\}$ is a fundamental set for the corresponding homogeneous equation.

Method of Annihilators

*A procedure for finding the proper form of a particular solution to be used in the method of undetermined coefficients is based on the fact that the polynomial, exponential, and sinusoidal functions (and sums of products of such terms) are solutions of linear homogeneous differential equations with constant coefficients. If we let D denote d/dx, then for example the equation $(D - 1) y = 0$ has the solution e^x, and we say that the differential operator $D - 1$ annihilates or is an **annihilator** of e^x. Also, we would say that $D^2 + 1$ annihilates $\sin x$ and $\cos x$, that the operator $(D + 1)^2 = D^2 + 2D + 1$ annihilates e^{-x} and xe^{-x}, and so on. Problems 31–33 show how annihilators can be used to find particular solutions.*

31. **Using Annihilators** Suppose we wish to find a particular solution y_p of

$$(D - 3)^2 (D + 1)y_p = 5xe^x + 3e^{2x} \qquad (14)$$

(a) Show that $(D - 1)^2$ and $D - 2$ annihilate the first and second terms, respectively, on the right-hand side and that the combined operator $(D - 1)^2 (D - 2)$ annihilates the entire right-hand side.

(b) Apply $(D - 1)^2 (D - 2)$ to Eq. (14) to obtain

$$(D - 1)^2 (D - 2)(D - 3)^2 (D + 1)y_p = 0$$

Hence y_p has the form

$$y_p = c_1 e^x + c_2 xe^x + c_3 e^{2x} + c_4 e^{3x} + c_5 xe^{3x} + c_6 e^{-x}$$

Since the homogeneous solution is

$$y_h = c_1 e^x + c_2 xe^x + c_3 e^{2x}$$

this means that y_p should be chosen to be

$$y_p = c_4 e^{3x} + c_5 xe^{3x} + c_6 e^{-x}$$

32. **Annihilator Method** Use the annihilator method to find the form of a particular solution for

$$y''' - 2y'' + y' = x^3 + 2e^x$$

33. **Annihilator Method** Use the annihilator method to find the form of a particular solution for

$$y^{(4)} - y''' - y'' + y' = x \sin x + x^2 + 1$$

34. **Integral Formulas** Find a formula involving integrals for the solution of

$$y^{(4)} - y = f(x)$$

Hint: The functions $\cos x, \sin x, \cosh x,$ and $\sinh x$ constitute a fundamental set of solutions of the homogeneous equation.

35. **Cantilever Beam** A cantilever beam (of length $L = 1$) is clamped at its left end ($x = 0$) and is free at its right end ($x = 1$). Along with the fourth-order differential equation $EIy^{(4)} = w(x)$, it satisfies the given boundary conditions $y(0) = y'(0) = y''(1) = y'''(1) = 0$. If the load $w(x) = w_0$ is distributed uniformly, find the displacement $y(x)$. How much does the free end sag below the clamped end?

36. **Hinged Beam** A beam of length $L = 1$ is hinged at both ends $x = 0$ and $x = 1$. Along with the differential equation $EIy^{(4)} = w(x)$, it satisfies the given boundary conditions $y(0) = y''(0) = y(1) = y''(1) = 0$. If the load is given by $w(x) = w_0$ is distributed uniformly, find the displacement $y(x)$. How much does the center sag below the ends?

37. **Hmmmmm** Solve the differential equations $y' = y$, $y'' = y, y''' = y, y^{(iv)} = y,$ and so on.

38. **Computer Problem** Use a graphing calculator or computer to sketch the graph of the clamped beam as found in Example 8 for different values of the parameters $E, I,$ and w_0.

39. **Journal Entry** Summarize the similarities and differences between first-, second-, and higher-order linear differential equations.

Series Solutions

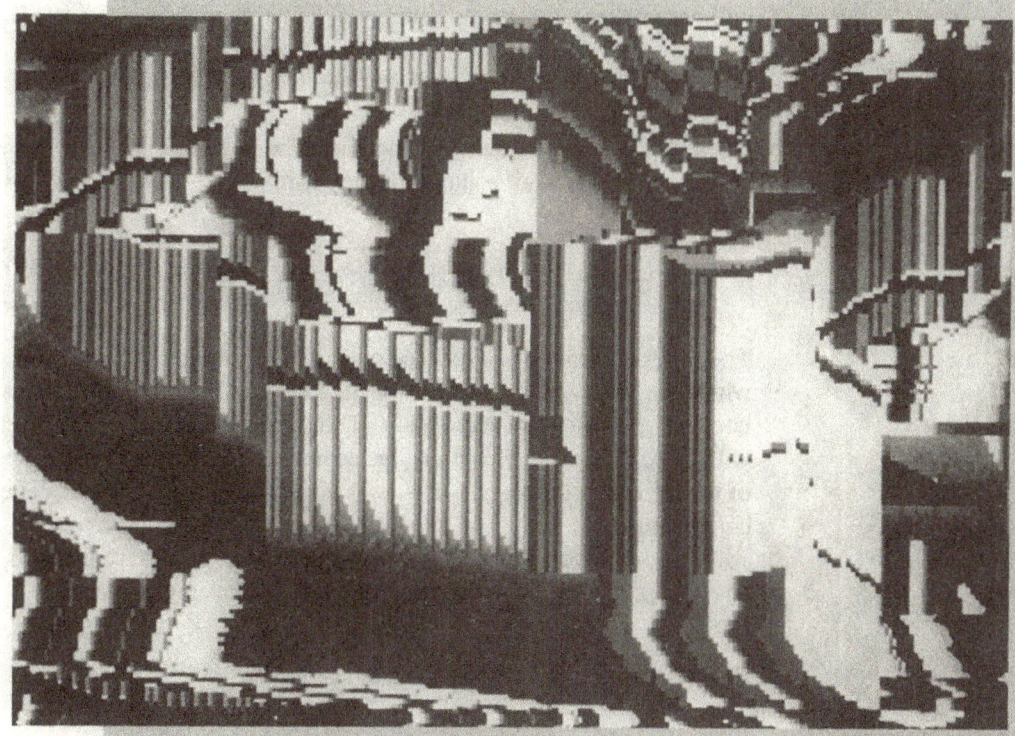

4.1 INTRODUCTION: A REVIEW OF POWER SERIES

4.2 POWER SERIES EXPANSIONS ABOUT ORDINARY
POINTS: PART I

4.3 POWER SERIES EXPANSIONS ABOUT ORDINARY
POINTS: PART II

4.4 SERIES SOLUTIONS ABOUT SINGULAR POINTS:
THE METHOD OF FROBENIUS

4.5 BESSEL FUNCTIONS

4.1 INTRODUCTION: A REVIEW OF POWER SERIES

PURPOSE

To present the most important properties of power series that will be needed in later sections of this chapter to solve linear variable-coefficient differential equations.

In Chapter 3 we saw that linear differential equations with *constant coefficients* often have solutions of the form e^{mx}. Unfortunately, such simple exponential functions will not in general satisfy a differential equation with *variable coefficients,* such as

$$y'' + p(x)y' + q(x)y = 0 \tag{1}$$

where p and q are continuous functions. The new strategy for solving Eq. (1) is to try a more general type of solution of the form*

$$y(x) = a_0 + a_1 (x - x_0) + a_2 (x - x_0)^2 + \cdots = \sum_{n=0}^{\infty} a_n(x - x_0)^n \tag{2}$$

The series (2), which is of prime concern to us in this chapter, is called a **power series,** where the constants a_0, a_1, $\cdots$ are called the **coefficients** of the series. The constant x_0 is called the **center of expansion** of the series. Normally, we select $x_0 = 0$, but we will allow for $x_0 \neq 0$. Generally, the center of expansion is chosen to be the x-coordinate of the initial conditions $y(x_0) = y_0$, $y'(x_0) = y_0'$ generally associated with the differential equation.

Although we assume that most readers have been introduced to power series in a course in calculus, we will review the basic ideas in this section.

CONVERGENCE OF A POWER SERIES

A power series of the form of Eq. (2) is said to **converge** at a point $x = r$ if the limit of the partial sums

$$\lim_{N \to \infty} \sum_{n=0}^{N} a_n(x - x_0)^n \tag{3}$$

exists. The limiting value is called the **sum** of the power series. It is clear that the power series converges when $x = x_0$, since

$$a_0 + a_1(x_0 - x_0) + a_2(x_0 - x_0)^2 + \cdots = a_0 \tag{4}$$

---

* Sir Isaac Newton used a similar strategy when he found an infinite series solution of the differential equation $y' = 1 + xy$.

An interesting and somewhat surprising property of a power series is the fact that a power series converges in an interval about x_0 and diverges outside the interval.* Moreover, in the interior of the interval of convergence, the power series converges *absolutely* in the sense that

$$a_0 + a_1|x - x_0| + a_2|x - x_0|^2 + \cdots$$

converges. These ideas are summarized in the following theorem.

THEOREM 4.1: Radius of Convergence of a Power Series

The region of convergence for a power series

$$\sum_{n=0}^{\infty} a_n (x - x_0)^n$$

is always an interval of one of the following types:

- A single point $x = x_0$
- An interval $(x_0 - R, x_0 + R)$ plus possibly one or both endpoints
- The whole real line

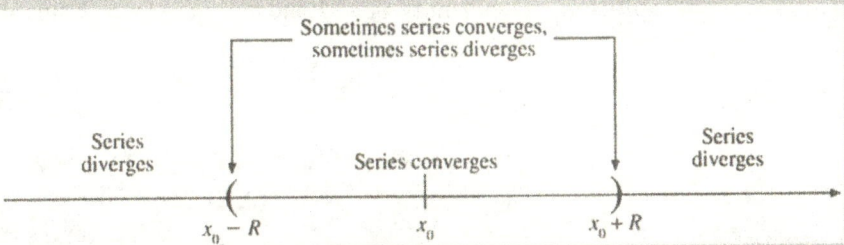

The constant R is called the **radius of convergence** of the interval. The region of convergence is called the **interval of convergence.**

TEST OF CONVERGENCE   When the power series converges only at $x = x_0$, we say that the radius of convergence R is zero, while if the series converges for all x, we say that the radius of convergence is infinity. Often, the radius of convergence R can be determined by the **ratio test.**

* This type of ''simple'' region of convergence associated with power series contrasts with the horribly complex (but beautiful) regions of convergence that are associated with other infinite sequences, such as the regions of convergence associated with Newton's method and the repeated iteration of complex numbers. In those cases the regions of ''convergence'' (for sequences of points in the plane) are often fractal sets, such as the Mandelbrot set. We will study fractal sets in Section 7.6.

> **THEOREM 4.2:** Ratio Test
>
> Suppose the power series
>
> $$a_0 + a_1(x - x_0) + a_2(x - x_0)^2 + \cdots + a_n(x - x_0)^n + \cdots$$
>
> has all nonzero coefficients a_n. If
>
> $$L = \lim_{n \to \infty} \left| \frac{a_{n+1}}{a_n} \right|$$
>
> where $0 \leq L \leq \infty$, then the radius of convergence R of the series is given by $R = 1/L$. We interpret $L = 0$ to mean $R = \infty$ and $L = \infty$ to mean $R = 0$.

Example 1   **Ratio Test**   Find the radius of convergence R and the corresponding interval of convergence $(x_0 - R, x_0 + R)$ of the power series

$$\sum_{n=0}^{\infty} (-1)^n n(x - 1)^n$$

Solution   We identify $x_0 = 1$ and $a_n = (-1)^n n$. Using the ratio test, we find

$$L = \lim_{n \to \infty} \left| \frac{a_{n+1}}{a_n} \right| = \lim_{n \to \infty} \left| \frac{(-1)^{n+1} (n + 1)}{(-1)^n n} \right| = \lim_{n \to \infty} \left(\frac{n + 1}{n} \right) = 1 \qquad (5)$$

Hence the radius of convergence is $R = 1/L = 1$, and the interval of convergence is $(x_0 - R, x_0 + R) = (0, 2)$. Hence the series converges for $0 < x < 2$ and diverges for $x < 0$ and $x > 2$. It is also clear that the power series diverges at the endpoints 0 and 2 of the interval of convergence, since in each case the terms of the power series do not approach zero.

POWER SERIES AS A MEANS OF DEFINING A FUNCTION

For each value of x in the interval of convergence of a power series, the series converges to a number. Hence the sum defines a function f given by

$$f(x) = \sum_{n=0}^{\infty} a_n(x - x_0)^n \qquad (6)$$

For example, the *geometric series* $\sum_{n=0}^{\infty} x^n$ converges on $(-1, 1)$ to the function defined by $f(x) = 1/(1 - x)$. Hence we write

$$\frac{1}{1 - x} = 1 + x + x^2 + \cdots + x^n + \cdots \qquad (-1 < x < 1) \qquad (7)$$

Note that this infinite series does not converge at the endpoints $x = \pm 1$.

IMPORTANT PROPERTIES OF POWER SERIES

One reason power series are so useful in mathematics is because they "behave" like polynomials and in fact possess many properties of polynomials.* We list several important properties possessed by power series. For simplicity we state these properties for $x_0 = 0$, although they hold for arbitrary x_0.

1. **Addition and Subtraction.** Power series can be added and subtracted term by term in the interval of convergence. That is,

$$\sum_{n=0}^{\infty} a_n x^n \pm \sum_{n=0}^{\infty} b_n x^n = \sum_{n=0}^{\infty} (a_n \pm b_n) x^n \qquad (8)$$

2. **Multiplication of Power Series.** Two power series can be multiplied, giving the third power series

$$\left(\sum_{n=0}^{\infty} a_n x^n \right) \left(\sum_{n=0}^{\infty} b_n x^n \right) = \sum_{n=0}^{\infty} c_n x^n \qquad (9)$$

where

$$c_n = a_0 b_n + a_1 b_{n-1} + a_2 b_{n-2} + \cdots + a_n b_0$$

This product power series is called the **Cauchy product,** and it will converge for all x in the common open interval of convergence for the two multiplied power series.[†]

3. **Term-by-Term Differentiation.** A power series defines a continuous function in its interval of convergence and can be differentiated any number of times by term-by-term differentiation, giving

$$\frac{d}{dx} \left(a_0 + a_1 x + a_2 x^2 + \cdots \right) = a_1 + 2 a_2 x + 3 a_3 x^2 + \cdots \qquad (10a)$$

$$\frac{d^2}{dx^2} \left(a_0 + a_1 x + a_2 x^2 + \cdots \right) = 2 a_2 + 6 a_3 x + \cdots \qquad (10b)$$

and so on.

4. **Term-by-Term Integration.** A power series has an antiderivative that can be found by using term-by-term integration:

$$\int \left(a_0 + a_1 x + a_2 x^2 + \cdots \right) dx = a_0 x + \frac{a_1}{2} x^2 + \frac{a_2}{3} x^3 + \cdots + C \qquad (11)$$

---

* Although power series contain an infinite number of terms, they possess many properties that are usually reserved for finite sums. For example, within their intervals of convergence the terms of a power series can be rearranged without changing the sum of the series, they can be integrated and differentiated term by term, and so on. A power series representation of a function also provides deeper information about the function.

† In fact, the Cauchy product may have a radius of convergence that is *larger* than those of the two multiplied power series.

ANALYTIC FUNCTIONS

If a function f can be written as a power series centered at x_0, then there is only one way to do it, namely,

$$f(x) = \sum_{n=0}^{\infty} \frac{f^{(n)}(x_0)}{n!}(x - x_0)^n \tag{12}$$

If a function f can be written as a power series centered at x_0, then f is said to be **analytic** at x_0 or f is said to be *representable* by a power series expansion about x_0.

DEFINITION: Analytic Functions

A function f is called **analytic** at x_0 if there exists a power series of the form $\sum_{n=0}^{\infty} a_n(x - x_0)^n$ with positive radius of convergence R such that for each x in the interval of convergence the series converges to $f(x)$.

A polynomial $a_0 + a_1 x + a_2 x^2 + \cdots + a_n x^n$ is an analytic function at every x_0, since it can be rewritten in the form $b_0 + b_1(x - x_0) + b_2(x - x_0)^2 + \cdots$. Also, rational functions $P(x)/Q(x)$, where $P(x)$ and $Q(x)$ are polynomials without a common factor, are analytic functions for all x except when $Q(x) = 0$. The elementary functions of calculus—$\sin x$, $\cos x$, and e^x—are analytic for all x, while the natural logarithm $\ln x$ is analytic for all $x > 0$. The reader will remember that typical power series representations for these functions are

$$e^x = 1 + x + \frac{x^2}{2!} + \frac{x^3}{3!} + \cdots \qquad (-\infty < x < \infty) \tag{13a}$$

$$\sin x = x - \frac{x^3}{3!} + \frac{x^5}{5!} - \cdots \qquad (-\infty < x < \infty) \tag{13b}$$

$$\cos x = 1 - \frac{x^2}{2!} + \frac{x^4}{4!} - \cdots \qquad (-\infty < x < \infty) \tag{13c}$$

$$\ln x = (x - 1) - \frac{1}{2}(x - 1)^2 + \frac{1}{3}(x - 1)^3 - \cdots \qquad (0 < x \leq 2) \tag{13d}$$

A formula for the coefficients in the power series of an analytic function is given in Theorem 4.3.

THEOREM 4.3: Taylor and Maclaurin Series Representations

If f is an analytic function at x_0, then the coefficients a_n of the power series are given by $a_n = f^{(n)}(x_0)/n!$, giving the representation

$$f(x) = f(x_0) + f'(x_0)(x - x_0) + \frac{f''(x_0)}{2!}(x - x_0)^2 + \cdots$$

$$= \sum_{n=0}^{\infty} \frac{f^{(n)}(x_0)}{n!}(x - x_0)^n \tag{14}$$

This power series is called the **Taylor's series** of f about x_0. When $x_0 = 0$, it is often referred to as the **Maclaurin's series** for f.

SHIFTING INDICES

In this chapter our work with power series will involve some manipulation of power series. In this regard we will make use of the **shift-of-index theorem.**

> **THEOREM 4.4:** Shift-of-Index Theorem
>
> A power series remains unchanged if the *lower limit* of the summation is increased (decreased) by some constant, say L, while the index in the *summand* is decreased (increased) by L. For example,
>
> $$\sum_{n=c}^{\infty} a_n(x - x_0)^n = \sum_{n=c+L}^{\infty} a_{n-L}(x - x_0)^{n-L}$$
>
> are the same.

PROOF: Theorem 4.4 can be easily verified by writing out the first few terms of each series. For example, the two series

$$\sum_{n=0}^{\infty} (n + 2)(n + 3)x^{n+2} = \sum_{n=2}^{\infty} n(n + 1)x^n$$

are simply different ways of writing the same series, as can be seen by writing out the first few terms of each series, getting

$$2 \cdot 3x^2 + 3 \cdot 4x^3 + 4 \cdot 5x^4 + \cdots$$

PROBLEMS: Section 4.1

For Problems 1–8, determine the radius and interval of convergence of each power series.

1. $\displaystyle\sum_{n=0}^{\infty} \frac{x^n}{n!}$

2. $\displaystyle\sum_{n=0}^{\infty} \frac{(-1)^n x^{2n+1}}{(2n + 1)!}$

3. $\displaystyle\sum_{n=1}^{\infty} \frac{x^n}{n}$

4. $\displaystyle\sum_{n=1}^{\infty} (-1)^n x^n$

5. $\displaystyle\sum_{n=1}^{\infty} \frac{x^n}{3x + 1}$

6. $\displaystyle\sum_{n=1}^{\infty} (-1)^n n^2 x^n$

7. $\displaystyle\sum_{n=1}^{\infty} \frac{(-1)^n x^{2n}}{2n - 1}$

8. $\displaystyle\sum_{n=1}^{\infty} \frac{n^2 x^{2n}}{5^n}$

For Problems 9–10, find the power series expansion for the sum $f(x) + g(x)$, given the following expansions for $f(x)$ and $g(x)$.

9. $f(x) = \displaystyle\sum_{n=0}^{\infty} \frac{x^n}{n!}$ $g(x) = \displaystyle\sum_{n=0}^{\infty} \frac{(-1)^n}{n!} x^n$

10. $f(x) = \displaystyle\sum_{n=0}^{\infty} \frac{x^n}{n}$ $g(x) = \displaystyle\sum_{n=0}^{\infty} x^n$

11. Calculus Operations on a Power Series Given the power series

$$\frac{1}{1 - x} = 1 + x + x^2 + \cdots + x^n + \cdots \qquad (15)$$

find the interval of convergence and compute the following functions on the interval of convergence.
(a) $f'(x)$
(b) $f''(x)$
(c) $\int f(x)\, dx$

12. An Important Application of Power Series The error function

$$\operatorname{erf}(x) = \frac{2}{\sqrt{\pi}} \int_0^x e^{-u^2}\, du \qquad (16)$$

is an important function in the theory of heat conduction. Show that

$$\operatorname{erf}(x) = \frac{2}{\sqrt{\pi}}\left(x - \frac{x^3}{3} + \frac{x^5}{10} - \frac{x^7}{42} + \cdots\right) \qquad (17)$$

For Problems 13–16, verify the given identities.

13. $\sum_{n=0}^{\infty} a_n x^{n+1} = \sum_{n=1}^{\infty} a_{n-1} x^n$

14. $\sum_{n=2}^{\infty} n(n-1)a_n x^{n-2} = \sum_{n=0}^{\infty} (n+2)(n+1)a_{n+2} x^n$

15. $\sum_{n=0}^{\infty} (n+1)a_{n+1} x^n = \sum_{n=1}^{\infty} na_n x^{n-1}$

16. $\sum_{n=3}^{\infty} n(n+1)a_n x^{n-2} = \sum_{n=1}^{\infty} (n+2)(n+3)a_{n+2} x^n$

17. Useful Expansions Use the well-known formula from linear algebra

$$1 + x + x^2 + \cdots + x^n = \frac{1 - x^{n+1}}{1 - x} \qquad (x \neq 1)$$

to show that for $|x| < 1$,

$$\frac{1}{1-x} = 1 + x + x^2 + x^3 + \cdots$$

$$\frac{1}{1+x} = 1 - x + x^2 - x^3 + \cdots$$

For Problems 18–27, find the Taylor or Maclaurin series of the given function about the indicated point x_0. What is the interval of convergence of the series?

18. $f(x) = x^2 + 2x + 1 \quad x_0 = 0$
19. $f(x) = x^2 + 2x + 1 \quad x_0 = 1$
20. $f(x) = x^2 + 2x + 1 \quad x_0 = -1$
21. $f(x) = x \qquad\qquad x_0 = 1$
22. $f(x) = x^2 \qquad\qquad x_0 = 1$
23. $f(x) = \dfrac{1}{1+x} \qquad\quad x_0 = 0$
24. $f(x) = \dfrac{1}{1+x} \qquad\quad x_0 = 1$
25. $f(x) = \dfrac{1}{x} \qquad\qquad x_0 = 1$
26. $f(x) = \dfrac{x}{1-x} \qquad\quad x_0 = 0$
27. $f(x) = e^x \qquad\qquad x_0 = 1$

Taylor Series of Indefinite Integrals

The functions defined by the integrals in Problems 28–29 are important in applied mathematics. Find their Taylor series expansions about $x_0 = 0$.

28. $\text{erf}(x) = \dfrac{2}{\sqrt{\pi}} \displaystyle\int_0^x e^{-t^2}\,dt$

29. $\text{Si}(x) = \displaystyle\int_0^x \dfrac{\sin t}{t}\,dt$

Reversion of Power Series

The roles of x and y in a power series

$$y = y_0 + a_1(x - x_0) + a_2(x - x_0)^2 + \cdots \quad (18)$$

can be interchanged so that x may be expressed in terms of y in the form

$$x = x_0 + A_1(y - y_0) + A_2(y - y_0)^2 + \cdots \quad (19)$$

*by a process known as **reversion of power series**. We simply substitute the expression of $x - x_0$ given in Eq. (19) into Eq. (18), getting*

$$y - y_0 = a_1[A_1(y - y_0) + A_2(y - y_0)^2 + \cdots]$$
$$+ a_2[A_1(y - y_0) + A_2(y - y_0)^2 + \cdots] + \cdots$$

We then systematically solve for the coefficients A_i in terms of the coefficients a_i by equating the coefficients of powers of $y - y_0$.

30. Reversion of Power Series Show that the first four coefficients A_1, A_2, A_3, and A_4 of the **reversed power series** (19) are given by

$$A_1 = \frac{1}{a_1}$$

$$A_2 = -\frac{a_2}{a_1^2}$$

$$A_3 = \frac{2a_2^2 - a_1 a_3}{a_1^5}$$

$$A_4 = \frac{5a_1 a_2 a_3 - a_1^2 a_4 - 5a_2^3}{a_1^7}$$

31. Reversion of an Interesting Power Series Show that the reversion of the power series

$$y = 1 + x + \frac{1}{2!}x^2 + \frac{1}{3!}x^3 + \frac{1}{4!}x^4 + \cdots$$

is

$$x = (y - 1) - \frac{1}{2!}(y - 1)^2$$
$$+ \frac{1}{3!}(y - 1)^3 - \frac{1}{4!}(y - 1)^4 + \cdots$$

32. Computer Problem Use a computer to sketch the graphs of the functions $\text{erf}(x)$ and $\text{Si}(x)$ on the interval $[0, 10]$ as defined in Problems 28 and 29.

33. Journal Entry How would you explain the concept of an infinite series to someone who is unfamiliar with higher mathematics?

4.2   POWER SERIES EXPANSIONS ABOUT ORDINARY POINTS: PART i

PURPOSE
To provide the basic mechanics for finding power series solutions of first- and second-order linear homogeneous differential equations with *polynomial* coefficients. This section and Section 4.3 constitute a two-part discussion. In this section we stress the operational aspects for finding power series solutions, and in Section 4.3 we discuss the more theoretical aspects, such as the region of convergence of power series solutions and the linear independence of power series solutions.

We have seen that solutions of differential equations with constant coefficients can be expressed in terms of sines, cosines, and exponentials. However, since these functions are just *specific* power series, one wonders whether differential equations having *variable coefficients* can be found in terms of *general* power series. To illustrate the concept of finding power series solutions, let us find a power series solution of the form

$$y = a_0 + a_1 x + a_2 x^2 + a_3 x^3 + \cdots \tag{1}$$

for the simple equation

$$y' = y \tag{2}$$

The idea is simple: Find the coefficients of the power series (1) so that it satisfies the differential equation.* Differentiating the above series term by term, we get

$$y' = a_1 + 2a_2 x + 3a_3 x^2 + \cdots + na_n x^{n-1} + \cdots \tag{3}$$

Substituting y and y' into the differential equation gives

$$a_1 + 2a_2 x + 3a_3 x^2 + \cdots + na_n x^{n-1} = a_0 + a_1 x + a_2 x^2 + \cdots + a_n x^n + \cdots$$

Rearranging terms, we find

$$\left(a_1 - a_0\right) + \left(2a_2 - a_1\right)x + \left(3a_3 - a_2\right)x^2$$

$$+ \left((n + 1)a_{n+1} - a_n\right)x^n + \cdots = 0 \tag{4}$$

Now, in order for the differential equation (4) to hold for all x, it is necessary that all the coefficients of the powers of x be zero. Hence we have the equations in Table 4.1.

---

* This strategy is similar to the method of undetermined coefficients used for finding particular solutions of nonhomogeneous equations.

Table 4.1
Finding the Coefficients in
Terms of a_0

x^n	Coefficient of x^n		
1	$a_1 - a_0 = 0$	◊	$a_1 = a_0 = \dfrac{a_0}{0!}$
x	$2a_2 - a_1 = 0$	◊	$a_2 = \dfrac{a_1}{2} = \dfrac{a_0}{2!}$
x^2	$3a_3 - a_2 = 0$	◊	$a_3 = \dfrac{a_2}{3} = \dfrac{a_0}{3 \cdot 2} = \dfrac{a_0}{3!}$
x^3	$4a_4 - a_3 = 0$	◊	$a_4 = \dfrac{a_3}{4} = \dfrac{a_2}{4 \cdot 3} = \dfrac{a_0}{4 \cdot 3 \cdot 2} = \dfrac{a_0}{4!}$
$\vdots$	$\vdots$	$\vdots$	$\vdots$
x^n	$na_n - a_{n-1} = 0$	◊	$a_n = \dfrac{a_0}{n!}$

Note that we have determined all the coefficients a_1, a_2, ... in terms of the first coefficient a_0. Substituting the expressions for a_1, a_2, ... into the power series (1), we find

$$y = a_0 \left(1 + x + \frac{x^2}{2!} + \frac{x^3}{3!} + \cdots + \frac{x^n}{n!} + \cdots \right) = a_0 e^x \tag{5}$$

where a_0 is arbitrary. Hence we have obtained the same general solution ce^x that was found in Chapter 2.

SECOND-ORDER EQUATIONS WITH POLYNOMIAL COEFFICIENTS

Although the power series method works for first-order equations, our main objective is to find a power series solution to the second-order equation

$$P(x) \frac{d^2y}{dx^2} + Q(x) \frac{dy}{dx} + R(x)y = 0 \tag{6}$$

where the coefficients $P(x)$, $Q(x)$, and $R(x)$ are *polynomials*.* To do this, we look for solutions of the form

$$y = a_0 + a_1(x - x_0) + a_2(x - x_0)^2 + \cdots \tag{7}$$

The idea is to determine the coefficients so that the series satisfies the differential equation. But the question arises: How does one pick the center of expansion x_0 of the series? The choice of x_0 determines the region of convergence of the power series,† but for the time being we will pick x_0 in any way as long as $P(x_0) \neq 0$. We call a point

---

* The power series method developed here can also be used to solve linear equations with *analytic coefficients*. However, for simplicity we restrict ourselves to polynomial coefficients. Also, the restriction to polynomial coefficients does not seem so restrictive when one considers the important equations that have polynomial coefficients.

† We will study the more theoretical aspects of power series solutions in Section 4.3.

x_0 that satisfies $P(x_0) \neq 0$ an **ordinary point** of the differential equation (6) and a point x_0 that satisfies $P(x_0) = 0$ a **singular point** of the equation. It is possible to find a power series solution expanded about a singular point, but the series is not the usual type of power series. We will study series expansions about singular points in Section 4.4.

Example 1   Ordinary and Singular Points   Find the ordinary and singular points of the following equations.

(a) $y'' - 2xy' + \lambda y = 0$ (Hermite's equation)

(b) $x^2 y'' + xy' + (x^2 - v^2) y = 0$ (Bessel's equation)

(c) $(1 - x^2)y'' - 2xy' + \alpha(\alpha + 1) y = 0$ (Legendre's equation)

Solution   (a) Since $P(x) = 1$, all points are ordinary points.

(b) Since $P(x) = x^2$, the point $x = 0$ is a singular point; all other points are ordinary points.

(c) Since $P(x) = 1 - x^2$, the points $x = \pm 1$ are singular points; all other points are ordinary points.   ■

To gain more experience with power series, we find a power series solution of a well-known second-order equation with polynomial coefficients of degree *zero* (constants).

Example 2   Constant Coefficients   Find a power series solution of

$$y'' + y = 0 \qquad (-\infty < x < \infty) \tag{8}$$

Solution   Since all points are ordinary points, we pick (for simplicity) the center of expansion as $x_0 = 0$ and seek a solution of the form

$$y = a_0 + a_1 x + a_2 x^2 + \cdots + a_n x^n + \cdots = \sum_{n=0}^{\infty} a_n x^n \tag{9}$$

Differentiating the power series (9) term by term gives

$$y' = a_1 + 2a_2 x + 3a_3 x^2 + \cdots = \sum_{n=1}^{\infty} n a_n x^{n-1} \tag{10}$$

$$y'' = 2a_2 + 6a_3 x + \cdots = \sum_{n=2}^{\infty} n(n - 1)a_n x^{n-2} \tag{11}$$

Substituting y and y'' into Eq. (8) gives

$$\sum_{n=2}^{\infty} n(n - 1)a_n x^{n-2} + \sum_{n=0}^{\infty} a_n x^n = 0 \tag{12}$$

We now use the shift-of-index theorem (Theorem 4.4) from Section 4.1 and rewrite Eq. (12) so that the *exponent* of x in each series is n. We do this by decreasing the lower limit of the first summation from 2 to 0 while at the same time increasing the index in the summand from n to $n + 2$. This gives the equivalent equation

$$\sum_{n=0}^{\infty} (n + 2)(n + 1)a_{n+2}x^n + \sum_{n=0}^{\infty} a_n x^n = 0$$

or collecting terms, we get

$$\sum_{n=0}^{\infty} [(n + 2)(n + 1)a_{n+2} + a_n]x^n = 0 \tag{13}$$

For the power series on the right of Eq. (13) to be equal to 0 for all x, it is necessary that the coefficients of the powers of x be zero. Hence we have

$$(n + 2)(n + 1)a_{n+2} + a_n = 0$$

or

$$a_{n+2} = -\frac{a_n}{(n + 2)(n + 1)} \qquad (n = 0, 1, 2, ...) \tag{14}$$

Equation (14), which relates the coefficients of the power series, is called a **recurrence relation.** Using this relation for $n = 0, 1, 2, ...$, we can find successive coefficients of the power series in terms of earlier coefficients. Since the relation involves a_n and a_{n+2}, it is best to find the even and odd coefficients separately. We summarize these results in Table 4.2.

Table 4.2
Finding Coefficients Using the Recurrence Relation

Even indices	Even coefficients
$n = 0$	$2 \cdot 1 a_2 + a_0 = 0 \ \lozenge \ a_2 = -\dfrac{a_0}{2}$
$n = 2$	$4 \cdot 3 a_4 + a_2 = 0 \ \lozenge \ a_4 = -\dfrac{a_2}{4 \cdot 3} = +\dfrac{a_0}{4!}$
$n = 4$	$6 \cdot 5 a_6 + a_4 = 0 \ \lozenge \ a_6 = -\dfrac{a_4}{6 \cdot 5} = -\dfrac{a_0}{6!}$
$\vdots$	$\vdots \qquad \vdots \qquad \vdots$
	$a_{2k} = \dfrac{(-1)^k}{(2k)!} a_0 \qquad (k = 1, 2, 3, ...)$

Odd indices	Odd coefficients
$n = 1$	$3 \cdot 2 a_3 + a_1 = 0 \ \lozenge \ a_3 = -\dfrac{a_1}{3 \cdot 2}$
$n = 3$	$5 \cdot 4 a_5 + a_3 = 0 \ \lozenge \ a_5 = -\dfrac{a_3}{5 \cdot 4} = +\dfrac{a_1}{5!}$
$n = 5$	$7 \cdot 6 a_7 + a_5 = 0 \ \lozenge \ a_7 = -\dfrac{a_5}{7 \cdot 6} = -\dfrac{a_1}{7!}$
$\vdots$	$\vdots \qquad \vdots \qquad \vdots$
	$a_{2k+1} = \dfrac{(-1)^k}{(2k + 1)!} a_1 \qquad (k = 1, 2, 3, ...)$

Substituting the values in Table 4.2 into the power series (9) gives the solution

$$y = a_0 + a_1 x - \frac{a_0}{2!}x^2 - \frac{a_1}{3!}x^3 + \frac{a_0}{4!}x^4 + \frac{a_1}{5!}x^5 + \cdots$$

$$+ \frac{(-1)^n a_0}{(2n)!}x^{2n} + \frac{(-1)^{2n+1} a_0}{(2n+1)!}x^{2n+1} + \cdots$$

$$= a_0\left(1 - \frac{x^2}{2!} + \frac{x^4}{4!} - \cdots + \frac{(-1)^n}{(2n)!}x^{2n} + \cdots\right)$$

$$+ a_1\left(x - \frac{x^3}{3!} + \frac{x^5}{5!} - \cdots + \frac{(-1)^n}{(2n+1)!}x^{2n+1} + \cdots\right)$$

$$= a_0 y_1(x) + a_1 y_2(x) \tag{15}$$

Since a_0 and a_1 are arbitrary, we can pick $a_0 = 1$, $a_1 = 0$ (and conversely, $a_0 = 0$, $a_1 = 1$). In so doing, we see that *both* power series $y_1(x)$ and $y_2(x)$ are solutions of the differential equation. If we can verify that the power series $y_1(x)$ and $y_2(x)$ converge on some interval about $x_0 = 0$ and that they are linearly independent functions, then we have found a *fundamental set* of solutions. Of course, in this example we recognize the power series to be $y_1(x) = \cos x$ and $y_2(x) = \sin x$, and so we *know* that the series converge for all x and that they are linearly independent. ▪

In the next section we will learn how to determine the interval of convergence of a power series solution and whether the power series solutions are linearly independent.

Also, note in Eq. (15) that $y(0) = a_0$ and $y'(0) = a_1$. Hence if initial conditions $y(0)$ and $y'(0)$ are provided, we could substitute these values into the general solution and find the solution of the initial-value problem.

Although the respective power series in Eq. (15) converge to $\sin x$ and $\cos x$ for all x, it is instructive to sketch the graphs of the partial sums of these power series as more and more terms are included. Figure 4.1 shows the graphs of these partial sums.

We now solve an equation with variable coefficients: **Airy's equation.**

HISTORICAL NOTE

Sir George Airy (1801–1892) was an English astronomer and mathematician who was director of the Greenwich Observatory from 1835 to 1881. The equation that now bears his name has uses in the theory of diffraction. The solutions of Airy's equation, called Airy's functions, are interesting inasmuch as for negative x they resemble sines and cosines, and for positive x they resemble exponential functions.

Example 3   Airy's Equation   Find a series solution of Airy's equation

$$y'' = xy \qquad (-\infty < x < \infty) \tag{16}$$

Figure 4.1
Partial sums of power series
expansions for sin x and
cos x

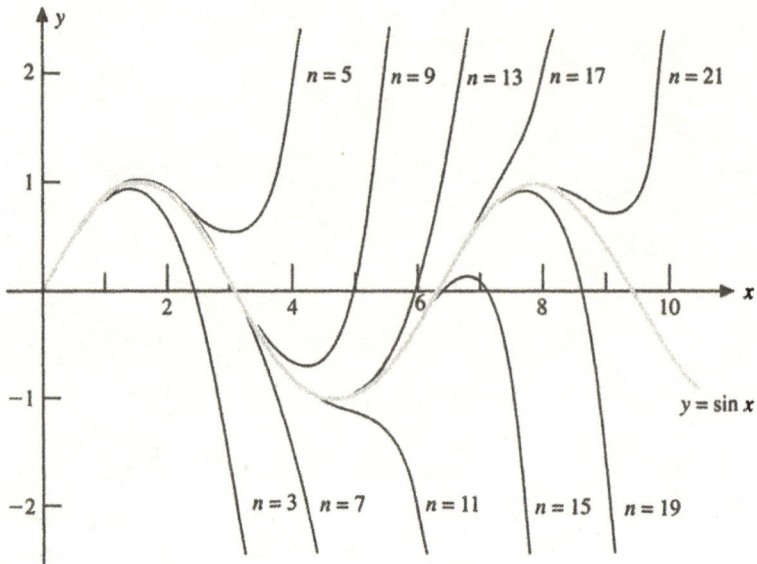

(a)   Polynomial approximations to the sine function. The value of n is the degree of the polynomial.

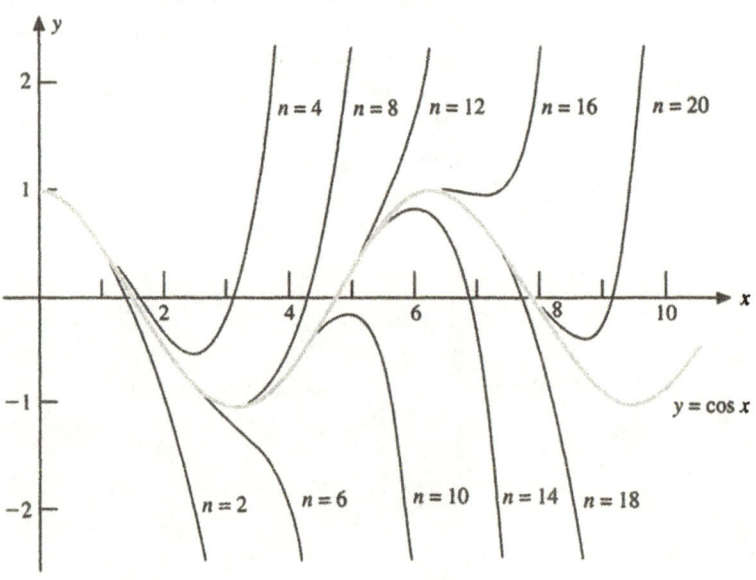

(b)   Polynomial approximations to the cosine function. The value of n is the degree of the polynomial.

Solution Since every real number is an ordinary point, we choose $x_0 = 0$ and seek a solution of the form

$$y = a_0 + a_1 x + a_2 x^2 + a_3 x^3 + \cdots + a_n x^n + \cdots \tag{17}$$

Differentiating term by term, we get

$$y' = a_1 + 2a_2 x + 3a_3 x^2 + \cdots = \sum_{n=1}^{\infty} n a_n x^{n-1} \tag{18}$$

$$y'' = 2a_2 + 6a_3 x + \cdots = \sum_{n=2}^{\infty} n(n-1) a_n x^{n-2} \tag{19}$$

Substituting y and y'' into Airy's equation gives

$$\sum_{n=2}^{\infty} n(n-1) a_n x^{n-2} = \sum_{n=0}^{\infty} a_n x^{n+1} \tag{20}$$

Again, the goal is to adjust the index n of all series so that the exponent of x is n. Doing this we get

$$\sum_{n=0}^{\infty} (n+2)(n+1) a_{n+2} x^n = \sum_{n=1}^{\infty} a_{n-1} x^n \tag{21}$$

Bringing out the first term of the series on the left so that both sums have the same lower limit of 1, after rearranging we get

$$2 \cdot 1 a_2 + \sum_{n=1}^{\infty} [(n+2)(n+1) a_{n+2} - a_{n-1}] x^n = 0 \tag{22}$$

In order for this power series to be satisfied for all x, it is necessary that all the coefficients of powers of x be zero. Hence we have $a_2 = 0$ and the recurrence relation

$$(n+2)(n+1) a_{n+2} - a_{n-1} = 0 \qquad (n = 1, 2, \ldots)$$

or

$$a_{n+2} = \frac{a_{n-1}}{(n+2)(n+1)} \qquad (n = 1, 2, \ldots) \tag{23}$$

Since the recurrence relation gives a_{n+2} in terms of a_{n-1} we have that the a's are determined in steps of three. In other words, starting at a_0, a_1, and a_2, we have

$$a_0 \rightarrow a_3 \rightarrow a_6 \rightarrow a_9 \rightarrow \cdots \rightarrow a_{3n} \rightarrow \cdots$$
$$a_1 \rightarrow a_4 \rightarrow a_7 \rightarrow a_{10} \rightarrow \cdots \rightarrow a_{3n+1} \rightarrow \cdots$$
$$a_2 \rightarrow a_5 \rightarrow a_8 \rightarrow a_{11} \rightarrow \cdots \rightarrow a_{3n+2} \rightarrow \cdots$$

Since $a_2 = 0$, we automatically have $a_5 = a_8 = \cdots = a_{3n+2} = 0$. The other two sequences are given in Table 4.3.

Table 4.3
Finding Coefficients Using
the Recurrence Relation

Odd indices	Odd coefficients
$n = 1$	$3 \cdot 2a_3 - a_0 = 0 \Rightarrow a_3 = \dfrac{a_0}{3 \cdot 2}$
$n = 4$	$6 \cdot 5a_6 - a_3 = 0 \Rightarrow a_6 = \dfrac{a_3}{6 \cdot 5} = \dfrac{a_0}{(6 \cdot 5)(3 \cdot 2)}$
$n = 7$	$9 \cdot 8a_9 - a_6 = 0 \Rightarrow a_9 = \dfrac{a_6}{9 \cdot 8} = \dfrac{a_0}{(9 \cdot 8)(6 \cdot 5)(3 \cdot 2)}$
$\vdots$	$\vdots \quad \vdots \quad \vdots$
$a_{3n} = \dfrac{1}{[(3n)(3n-1)][(3n-3)(3n-4)]\cdots[6\cdot5][3\cdot2]}\,a_0$	

Even indices	Even coefficients
$n = 2$	$4 \cdot 3a_4 - a_1 = 0 \Rightarrow a_4 = \dfrac{a_1}{4 \cdot 3}$
$n = 5$	$7 \cdot 6a_7 - a_4 = 0 \Rightarrow a_7 = \dfrac{a_4}{7 \cdot 6} = \dfrac{a_1}{(7 \cdot 6)(4 \cdot 3)}$
$n = 8$	$10 \cdot 9a_{10} - a_7 = 0 \Rightarrow a_{10} = \dfrac{a_7}{10 \cdot 9} = \dfrac{a_1}{(10 \cdot 9)(7 \cdot 6)(4 \cdot 3)}$
$\vdots$	$\vdots \quad \vdots \quad \vdots$
$a_{3n+1} = \dfrac{1}{[(3n+1)(3n)][(3n-2)(3n-3)]\cdots[7\cdot6][4\cdot3]}\,a_1$	

Hence a solution of Airy's equation is

$$
\begin{aligned}
y = a_0 & \left(1 + \frac{x^3}{3 \cdot 2} + \frac{x^6}{6 \cdot 5 \cdot 3 \cdot 2} + \cdots + \frac{x^{3n}}{(3n)(3n-1)\cdots 3 \cdot 2} + \cdots\right) \\
+ a_1 & \left(x + \frac{x^4}{4 \cdot 3} + \frac{x^7}{7 \cdot 6 \cdot 4 \cdot 3} + \cdots + \frac{x^{3n+1}}{(3n+1)(3n)(3n-3)\cdots 3 \cdot 2}\right) \\
= a_0 & y_1(x) + a_1 y_2(x) \hspace{4cm} (24)
\end{aligned}
$$

where a_0 and a_1 are arbitrary constants. Again, by picking $a_0 = 1$ and $a_1 = 0$ we see that the first power series $y_1(x)$ is a solution, and by picking $a_0 = 0$ and $a_1 = 1$ we see that $y_2(x)$ is a solution. Computing the Wronskian of y_1 and y_2 at $x = 0$, we find

$$W(y_1, y_2)(0) = y_1(0)\,y'_2(0) - y'_1(0)\,y_2(0) = (1)(1) - (0)(0) = 1 \neq 0 \quad (25)$$

Hence $y_1(x)$ and $y_2(x)$ are linearly independent solutions of Airy's equation, and so Eq. (23) defines the general solution of Airy's equation. To determine the region of convergence of the two proper series y_1 and y_2, known as **Airy's functions,** we will see in Section 4.3 that both series converge for all x. We would expect as much, inasmuch as the denominators in both series grow very quickly.

In Figure 4.2 we sketch the partial sums of Airy's functions.

Figure 4.2
Partial sums of Airy's
functions

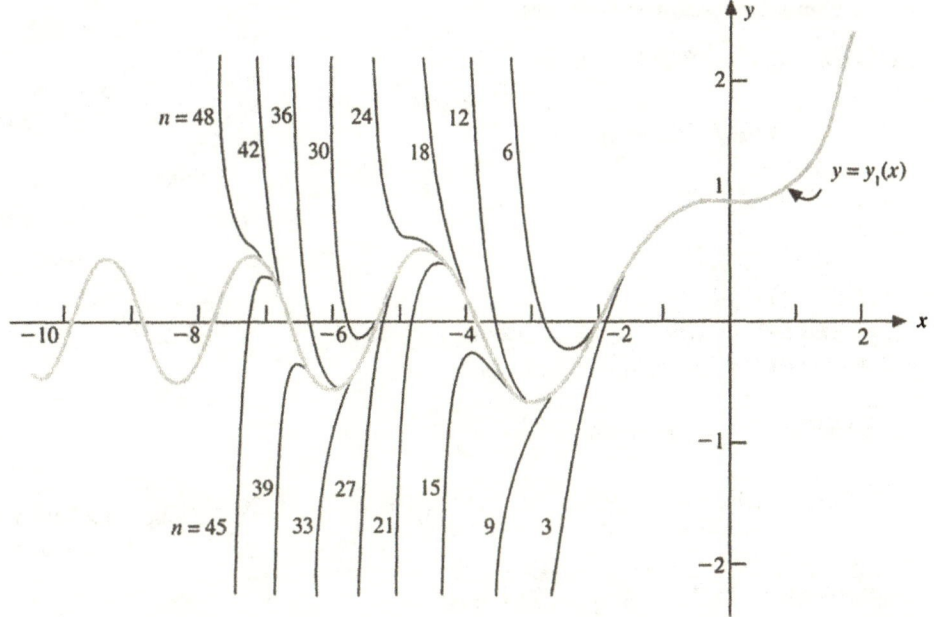

(a) Polynomial approximations to the solution $y_1(x)$ of Airy's equation. The value of n represents the degree of the polynomial.

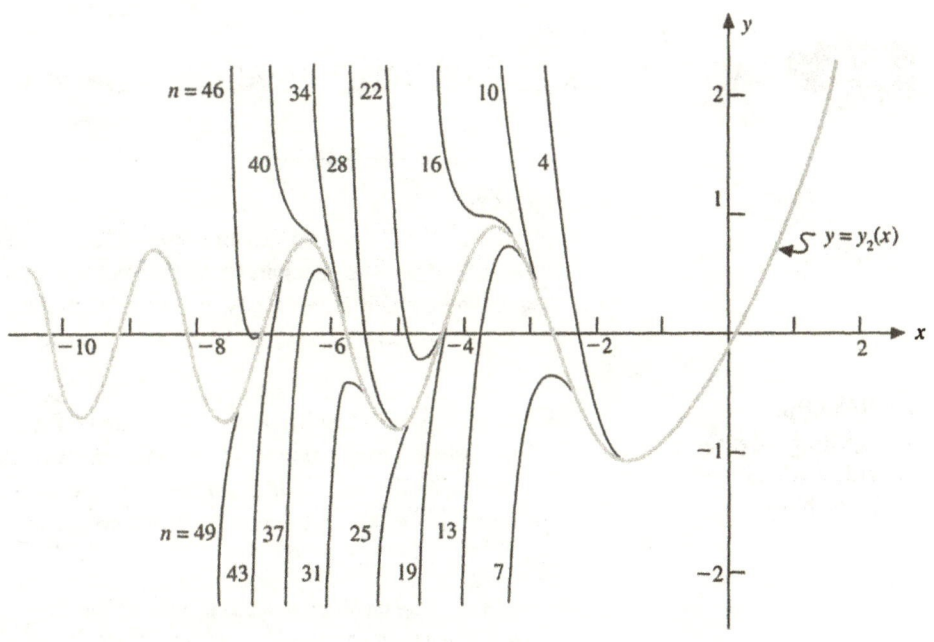

(b) Polynomial approximations to the solution $y_2(x)$ of Airy's equation. The value of n represents the degree of the polynomial.

PROBLEMS: Section 4.2

For Problems 1–5, determine the ordinary and singular points of the given differential equation.

1. $y'' + xy = 0$
2. $xy'' + y' + xy = 0$
3. $(1 - x^2)y'' + 2xy = 0$
4. $(1 + x^2)y'' + xy' + y = 0$
5. $x^2 y'' + 2xy' - y = 0$

For Problems 6–7, write the following expressions as a single power series and find the recurrence relation.

6. $\displaystyle\sum_{n=1}^{\infty} na_n x^{n-1} - \sum_{n=0}^{\infty} (n + 1)a_n x^n = 0$

7. $\displaystyle\sum_{n=2}^{\infty} 2n(n + 1)a_n x^{n-2} + \sum_{n=1}^{\infty} na_n x^n + \sum_{n=0}^{\infty} a_n x^n = 0$

For Problems 8–18, find the first few terms of the power series solution around $x_0 = 0$ of the given differential equation. If possible, determine the general series expansion and a closed form expression of the solution.

8. $y' = x$ (very easy)
9. $y' = x^2 + 2x + 1$ (very easy)
10. $y' = y + 1$ (pretty easy)
11. $y' - 2xy = 0$ 12. $y' + (x + 2)y = 0$
13. $y'' - y = 0$ 14. $y'' + xy = 0$

15. $y'' - x^2 y = 0$ 16. $y'' - xy' + 2y = 0$
17. $(1 + x^2)y'' - y' + y = 0$
18. $(1 + x^2)y'' - 4xy' + 6y = 0$
19. **Series Solutions and Initial Conditions** If initial conditions $y(0) = y_0$, $y'(0) = y'_0$ are specified with a second-order equation, show that the first two coefficients a_0 and a_1 in the power series solution

$$y = a_0 + a_1 x + a_2 x^2 + a_3 x^3 + \cdots$$

are $a_0 = y(0)$, $a_1 = y'(0)$. Use this result to solve the initial-value problem

$$y'' - xy' + 4y = 0$$
$$y(0) = 1$$
$$y'(0) = 0$$

20. **Comparing Power Series** Find the power series solution of the trivial equation

$$y' = 6x + 1$$

about the points $x_0 = 0$ and $x_0 = 1$, and verify that you get the same solution.

21. **Computer Problem** Use a computer to draw the graphs of some of the partial sums of the Airy's functions on the interval $[-1, 1]$.

22. **Journal Entry** Summarize in your own words the power series approach to finding solutions.

4.3   POWER SERIES EXPANSIONS ABOUT ORDINARY POINTS: PART II

PURPOSE

To continue the study of power series solutions expanded about ordinary points. We further illustrate the mechanics of the method by presenting more examples, and we discuss the convergence of power series solutions.

RADIUS OF CONVERGENCE OF POWER SERIES SOLUTIONS

Although the ratio test provides a useful tool for finding the radius of convergence of a power series, many power series solutions of differential equations are sufficiently unwieldy that the rule is difficult to apply. Fortunately, there is a general theorem that provides insight about the nature of all power series solutions. We state the theorem here without proof.*

---

* The details of the proof can be found in more advanced books in differential equations. For example, see *An Introduction to Ordinary Differential Equations* by Earl A. Coddington (Prentice-Hall, New York, 1961). The proof of the theorem depends on the important result from complex variables that states that if $P(x)$ and $Q(x)$ are polynomials with x_0 an ordinary point of $P(x)$, then the Taylor series expansion of the rational function $P(x)/Q(x)$ has a radius of convergence equal to the nearest root of $Q(z) = 0$ in the complex plane.

> **THEOREM 4.5:** Power Series Expansion About an Ordinary Point
>
> If x_0 is an ordinary point of the differential equation
>
> $$P(x)y'' + Q(x)y' + R(x)y = 0$$
>
> where $P(x)$, $Q(x)$, and $R(x)$ are polynomials, then the general solution of the equation can be written as
>
> $$y = \sum_{n=0}^{\infty} a_n(x - x_0)^n = a_0 y_1(x) + a_1 y_2(x)$$
>
> where a_0 and a_1 are arbitrary constants and $y_1(x)$ and $y_2(x)$ are linearly independent series solutions that are analytic at x_0. The radius of convergence of $y_1(x)$ and $y_2(x)$ is *at least as great* as the distance from x_0 to the nearest root of $P(x) = 0$.
>
> In computing the distance(s) between x_0 and the root(s) of $P(x) = 0$, one must consider possible complex roots, and if they occur, distances are computed in the complex plane.
>
>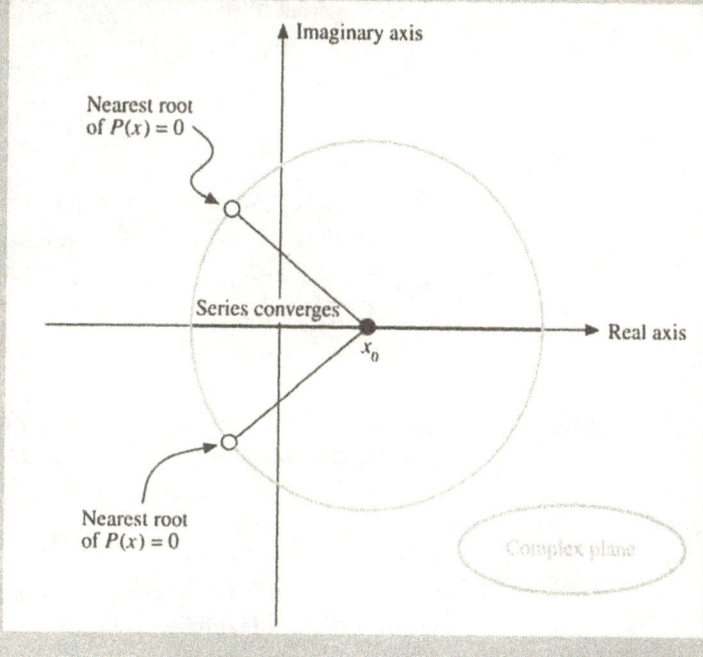

It should be noted that a series solution may converge for a wider range of x than is indicated by Theorem 4.5. The theorem gives a *lower bound* on the radius of convergence of the series solution.

We summarize the steps that should be carried out in solving differential equations using power series expansions about ordinary points.

Power Solutions About Ordinary Points

Suppose x_0 is an ordinary point of

$$P(x)y'' + Q(x)y' + R(x)y = 0 \qquad (1)$$

where P, Q, and R are polynomials that do not have a common factor. One can find a power series solution of the form

$$y = \sum_{n=0}^{\infty} a_n(x - x_0)^n \qquad (2)$$

by carrying out the following steps.

Step 1. Substitute the power series (2) into the differential equation (1).

Step 2. Rewrite the resulting power series as a single power series and set it equal to zero.

Step 3. Set the coefficients of the power series equal to zero. This gives a **recurrence relation** that relates later coefficients in the power series (2) to earlier ones.

Step 4. Find all the coefficients a_n in terms of the first two coefficients a_0 and a_1, thus writing the power series in the form

$$\sum_{n=0}^{\infty} a_n(x - x_0)^n = a_0 y_1(x) + a_1 y_2(x)$$

where y_1 and y_2 are two power series that define linearly independent solutions. The radius of convergence of the power series solutions y_1 and y_2 will converge on an interval at least as large as that determined by Theorem 4.5.

Example 1 **Power Series Solutions of Airy's Equation** Determine a lower bound for the radius of convergence of power series solutions about $x_0 = 1$ for Airy's equation

$$y'' = xy \qquad (3)$$

Solution Since $P(x) = 1$ has no zeros, the distance from $x_0 = 1$ to the nearest root of the equation $P(x) = 0$ is infinite. Hence the power series solution

$$y = a_0 + a_1(x - 1) + a_2(x - 1)^2 + \cdots$$

converges for all x. The rate of convergence will depend on the coefficients. ▨

Note that Example 1 illustrates the fact that if the coefficient of y'' is a constant, a power series solution expanded about any point can be found, and the power series will converge for all x.

Example 2   **Power Series Solutions of Chebyshev's Equation**   Determine a lower bound for the radius of convergence of power series solutions about $x_0 = 0$ of **Chebyshev's equation:**

$$(1 - x^2)\, y'' - xy' + \alpha^2 y = 0 \tag{4}$$

Solution   Here $P(x) = 1 - x^2$ has roots at ± 1. Since the distance from the center of expansion $x_0 = 0$ to both of these roots is 1, Theorem 4.5 says that a power series of the form $\sum_{n=0}^{\infty} a_n x^n$ will converge for *at least* $|x| < 1$ and possibly a larger interval of convergence. In fact, it can be shown that if α is a positive integer, say n, one solution of Chebyshev's equation is a *polynomial* of degree n, called Chebyshev's poynomial.   ◼

Example 3   **Lower Bound on the Radius of Convergence**   Determine a lower bound for the radius of convergence of power series solutions of the equation

$$(1 + x^2)y'' + xy' + 2y = 0 \tag{5}$$

about the ordinary points

(a) $x_0 = 0$

(b) $x_0 = 1$

(c) $x_0 = 2$

Solution   We identify $P(x) = 1 + x^2$, which has zeros at $x = \pm i$ in the complex plane.

(a) Since the distance (in the complex plane) from $x_0 = 0$ to both zeros is 1, any power series solution of the form $\sum_{n=0}^{\infty} a_n x^n$ will converge at least on the open interval $|x| < 1$. See Figure 4.3(a).

Figure 4.3
Finding the lower bound on the radius of convergence

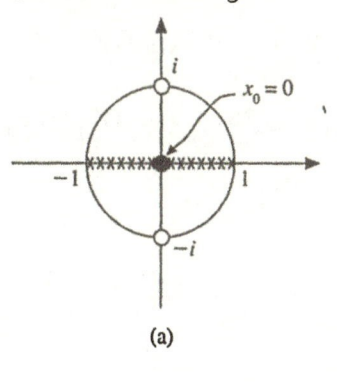

(a)

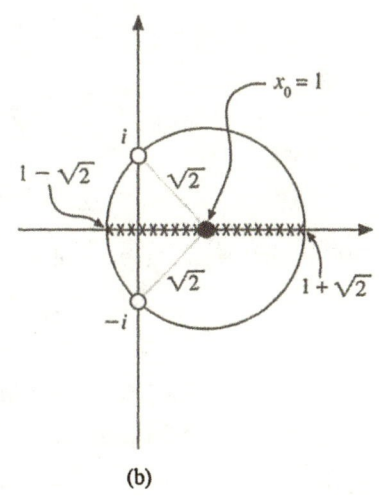

(b)

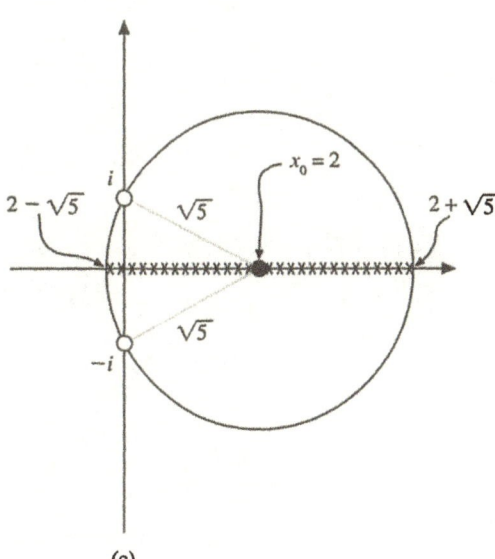

(c)

(b) Since the distance from $x_0 = 1$ to both zeros is $\sqrt{2}$, we conclude that any power series solution of the form $\sum\limits_{n=0}^{\infty} a_n(x - 1)^n$ will converge at least on the interval $|x - 1| < \sqrt{2}$. See Figure 4.3(b).

(c) Since the distance from $x_0 = 2$ to both zeros is $\sqrt{5}$, any power series of the form $\sum\limits_{n=0}^{\infty} a_n(x - 2)^n$ will converge at least on the interval $|x - 2| < \sqrt{5}$. See Figure 4.3(c).   ◼

Example 4   Airy's Equation Revisited   Find a solution of Airy's equation

$$y'' = xy \qquad (-\infty < x < \infty) \tag{6}$$

expanded about $x_0 = 1$. That is, find a solution of the form

$$y = a_0 + a_1(x - 1) + a_2(x - 1)^2 + \cdots \tag{7}$$

Solution   Since $x_0 = 1$ is an ordinary point of the differential equation and since $P(x) = 1$, any power series solution of the form of Eq. (7) exists and converges for all x. To find the power series, we compute

$$y' = \sum_{n=1}^{\infty} n a_n(x - 1)^{n-1} = \sum_{n=0}^{\infty} (n + 1)a_{n+1}(x - 1)^n \tag{8}$$

$$y'' = \sum_{n=2}^{\infty} n(n - 1)a_n(x - 1)^{n-2} = \sum_{n=0}^{\infty} (n + 2)(n + 1)a_{n+2}(x - 1)^n \tag{9}$$

Substituting the series for y, y', and y'' into the differential equation and rewriting x as $1 + (x - 1)$ so as to write the power series in powers of $x - 1$, we find

$$\sum_{n=0}^{\infty} (n + 2)(n + 1)a_{n+2}(x - 1)^n = x \sum_{n=0}^{\infty} a_n(x - 1)^n$$

$$= [1 + (x - 1)] \sum_{n=0}^{\infty} a_n(x - 1)^n$$

$$= \sum_{n=0}^{\infty} a_n(x - 1)^n + \sum_{n=0}^{\infty} a_n(x - 1)^{n+1}$$

$$= \sum_{n=0}^{\infty} a_n(x - 1)^n + \sum_{i=1}^{\infty} a_{n-1}(x - 1)^n \tag{10}$$

Finally, equating coefficients of like powers of x gives

$$2a_2 = a_0$$
$$3 \cdot 2a_3 = a_1 + a_0$$
$$4 \cdot 3a_4 = a_2 + a_1$$
$$5 \cdot 4a_5 = a_3 + a_2$$
$$\vdots$$

In general, we have

$$(n + 2)(n + 1)a_{n+2} = a_n + a_{n-1} \qquad (n \geq 1)$$

Solving for a few of these coefficients in terms of a_0 and a_1, we find

$$a_2 = \frac{a_0}{2}$$

$$a_3 = \frac{a_1}{6} + \frac{a_0}{6}$$

$$a_4 = \frac{a_2}{12} + \frac{a_1}{12} = \frac{a_0}{24} + \frac{a_1}{12}$$

$$a_5 = \frac{a_3}{20} + \frac{a_2}{20} = \frac{a_0}{30} + \frac{a_1}{120}$$

$$\vdots$$

Substituting these values into the power series (7), one finds

$$y = a_0 \left(1 + \frac{(x-1)^2}{2} + \frac{(x-1)^3}{6} + \frac{(x-1)^4}{24} + \frac{(x-1)^5}{30} + \cdots \right)$$

$$+ a_1 \left((x-1) + \frac{(x-1)^3}{6} + \frac{(x-1)^4}{12} + \frac{(x-1)^5}{120} + \cdots \right)$$

$$= a_0 y_1(x) + a_1 y_2(x) \tag{11}$$

The recurrent equation for this problem did not allow us to find a general expression for a_n in terms of a_0 and a_1, and for this reason it is difficult to use the ratio test to determine the radius of convergence for y_1 and y_2. However, Theorem 4.5 guarantees that both series converge for all x and that they are linearly independent. Hence we can conclude that Eq. (11) is the general solution of Airy's equation for $-\infty < x < \infty$. ∎

Example 5   Hermite's Example   Find the general solution of Hermite's equation*

$$y'' - 2xy' + 2py = 0 \qquad (-\infty < x < \infty) \tag{12}$$

of the form

$$y = \sum_{n=0}^{\infty} a_n x^n \tag{13}$$

Theorem 4.5 guarantees that a solution of the form of Eq. (13) converges for all x. Differentiating Eq. (13), we get

$$y' = \sum_{n=1}^{\infty} n a_n x^{n-1} \tag{14a}$$

$$y'' = \sum_{n=2}^{\infty} n(n-1) a_n x^{n-2} \tag{14b}$$

---

* Hermite's equation is one of the famous "classical" differential equations of physics. It is interesting inasmuch as when the parameter p that occurs in the equation is a positive integer, then one of the solutions of the equation is a polynomial of degree p, called *Hermite's polynomial*. These polynomials exhibit interesting and important properties.

Substituting these values into the differential equation, we find

$$\sum_{n=2}^{\infty} n(n-1)a_n x^{n-2} - 2x \sum_{n=1}^{\infty} n a_n x^{n-1} + 2p \sum_{n=0}^{\infty} a_n x^n = 0$$

Readjusting coefficients so that each summation contains x^n gives

$$\sum_{n=0}^{\infty} (n+2)(n+1)a_{n+2} x^n - \sum_{n=1}^{\infty} 2n a_n x^n + \sum_{n=0}^{\infty} 2p a_n x^n = 0$$

or

$$(2a_2 + 2pa_0) + \sum_{n=1}^{\infty} \Big((n+2)(n+1)a_{n+2} - 2n a_n + 2p a_n \Big) x^n = 0 \qquad (15)$$

From this we get $a_2 = -pa_0$ and the general recurrence relation

$$a_{n+2} = \frac{2(n-p)}{(n+2)(n+1)} a_n \qquad (n \geq 1) \qquad (16)$$

Using these equations, we can find from the two starting values a_0 and a_1 the two sequences

$$a_0 \to a_2 \to a_4 \to \cdots \to a_{2n} \to \cdots$$
$$a_1 \to a_3 \to a_5 \to \cdots \to a_{2n-1} \to \cdots$$

Substituting these values into the power series and simplifying, we find

$$y = a_0 \left(1 - \frac{2p}{2!}x^2 + \frac{2^2 p(p-2)}{4!}x^4 - \frac{2^3 p(p-2)(p-4)}{6!}x^6 + \cdots \right)$$
$$+ a_1 \left(x - \frac{2(p-1)}{3!}x^3 + \frac{2^2(p-1)(p-3)}{5!}x^5 - \cdots \right)$$
$$= a_0 y_1(x) + a_1 y_2(x) \qquad (17)$$

where a_0 and a_1 are arbitrary. Hence we have found two power series solutions $y_1(x)$ and $y_2(x)$. Theorem 4.5 guarantees that y_1 and y_2 are linearly independent functions. This can also be seen by computing the Wronskian of y_1 and y_2 at $x = 0$, getting.

$$W[y_1, y_2](0) = y_1(0)y_2'(0) - y_1'(0)y_2(0) = (1)(1) - (0)(0) = 1 \neq 0$$

Hence Eq. (13) is the general solution of Hermite's equation for $-\infty < x < \infty$. ∎

PROBLEMS: Section 4.3

For Problems 1–6, determine a lower bound for the radius of convergence of a power series solution about the given point x_0 of the indicated differential equation.

1. $y'' + 2y' + y = 0$ $x_0 = 0$
2. $xy'' + (1+x)y = 0$ $x_0 = -1$
3. $x^2 y'' + 2xy' - y = 0$ $x_0 = 2$

4. $(1 - x^2)y'' - xy' + y = 0$ $x_0 = 0$
5. $(1 + 2x^2)y'' + xy' - y = 0$ $x_0 = 1$
6. $(1 + x^3)y'' + xy' + y = 0$ $x_0 = 0$

The Classical Equations of Physics

Often a partial differential equation is solved by converting it to more than one differential equation. Many of these ordinary differential equations are known as the classical equations of physics, such as the equations of Bessel, Legendre, Hermite, Chebyshev, Laguerre, and others. Problems 7–9 study a few properties of these equations.

7. **Hermite's Equation** Use the power series solution (17) of Hermite's equation to show that when $p = n$ is a nonnegative integer, Hermite's equation has a polynomial solution of degree n, called **Hermite's polynomial.** Calling these polynomials $H_n(x)$, find $H_0(x)$, $H_1(x)$, $H_2(x)$, and $H_3(x)$.

8. **Chebyshev's Equation** The **Chebyshev** (Tchebichef) **polynomials** $T_n(x)$ are polynomial solutions of Chebyshev's equation:

$$(1 - x^2)y'' - xy' + n^2y = 0 \quad (-1 < x < 1) \quad (18)$$

where $n = 0, 1, 2, \dots$. The Chebyshev polynominals satisfy the recurrence relation

$$T_{n+1}(x) = 2xT_n(x) - T_{n-1}(x)$$

First use the power series method to determine $T_0(x) = 1$ and $T_1(x)$, and then use the recurrence relation to find the next four Chebyshev polynomials.

9. **Laguerre Polynomials** The differential equation

$$xy'' + (1 - x)y' + ny = 0 \quad (x > 0)$$

is called **Laguerre's equation** n, and for $n = 0, 1, 2, \dots$ the equation has polynomial solutions, called **Laguerre polynomials** $L_n(x)$. It can be shown that the Laguerre polynomials satisfy Rodrigues' formula,

$$L_n(x) = \frac{e^x}{n!} \frac{d^n}{dx^n}(x^n e^{-x}) \quad (19)$$

Use Rodrigues' formula to find the first four Laguerre polynomials.

Series Solutions for Analytic Coefficients

One can find series solutions of the form $\sum\limits_{n=0}^{\infty} a_n(x - x_0)^n$ to the equation

$$y'' + p(x)y' + q(x)y = 0$$

where $p(x)$ and $q(x)$ are analytic functions, by expanding $p(x)$ and $q(x)$ as power series about x_0 and proceeding as usual by treating the resulting equation as just an equation with polynomial coefficients of infinite degree. For Problems 10–11, use this idea to find the first three terms of either one or two linearly independent power series solutions about $x_0 = 0$ of the following equations.

10. $y'' + (\sin x)y = 0$

11. $y'' + e^{-x}y = 0$

12. **Nonhomogeneous Equation** Find the first three nonzero terms in a power series solution expansion about $x_0 = 0$ of the general solution of the equation

$$y' - xy = \sin x$$

13. **Computer Problem** Use a computer to sketch the graphs of some of the Chebyshev and Laguerre polynomials over appropriate intervals.

14. **Journal Entry** Summarize the results of this section.

4.4   SERIES SOLUTIONS ABOUT SINGULAR POINTS: THE METHOD OF FROBENIUS

PURPOSE

To present the method of Frobenius, which finds series solutions of linear differential equations with polynomial coefficients in a neighborhood of a regular singular point.

MOTIVATION FOR THE METHOD OF FROBENIUS

We have seen how linear differential equations with polynomial coefficients can be solved in terms of a power series expansion about an ordinary point. In many applications, however, it is the solution near a singular point that exhibits the most interesting behavior, and it is this behavior that we wish to study. For example, it may be necessary to know the solution of

$$xy' + (1 + x)y = 0 \quad (1)$$

near the singular point $x = 0$. However, substitution of the power series expansion

$$y = a_0 + a_1 x + a_2 x^2 + \cdots \tag{2}$$

into Eq. (1) results in the condition $a_0 = a_1 = \cdots = 0$, which only verifies the obvious fact that $y = 0$ is a solution of the equation. The reason the power series expansion fails to provide a nonzero solution of Eq. (1) can be understood by solving the equation by using the method of separation of variables. It can easily be shown that the general solution is

$$y = c\frac{e^{-x}}{x} \tag{3}$$

or in series form

$$
\begin{aligned}
y &= c\,\frac{1}{x}\left(1 - x + \frac{x^2}{2!} - \frac{x^3}{3!} + \frac{x^4}{4!} - \cdots\right) \\
&= c\left(\frac{1}{x} - 1 + \frac{x}{2!} - \frac{x^2}{3!} + \frac{x^3}{4!} - \cdots\right)
\end{aligned} \tag{4}
$$

Hence because of the term $1/x$, the differential equation (1) does not have a power series solution expanded about $x_0 = 0$. Note that (except when $c = 0$) the solution (3) "blows up" (grows without bound) when x approaches the singular point of zero. See Figure 4.4.

Figure 4.4
The solution looks like $\frac{1}{x} - 1$ near $x = 0$.

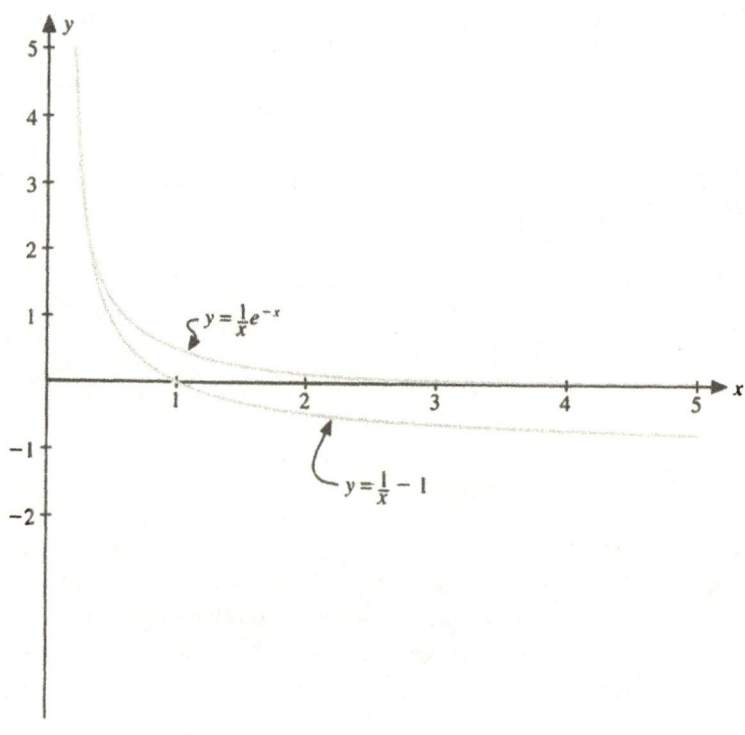

This example indicates that although the usual power series expansion about a singular point might not be found, it may be possible to find a series solution that contains negative powers of x. The problem of finding series solutions expanded about singular points was first studied by a mathematician named Ferdinand Georg Frobenius in the later part of the nineteenth century. He concluded that it is possible to find series solutions expanded about certain kinds of singular points, called regular singular points. Before studying the work of Frobenius, we introduce the concept of regular singular points.

REGULAR SINGULAR POINTS

When Frobenius investigated series solutions expanded about singular points, he discovered that it is necessary to further classify singular points into further subclasses: *regular singular points* and *irregular singular points*.

DEFINITION: Regular Singular Point

A **regular singular point** x_0 of the differential equation

$$P(x)y'' + Q(x)y' + R(x)y = 0 \tag{5}$$

is a singular point, that is, $P(x_0) = 0$, that satisfies

$$\lim_{x \to x_0} (x - x_0) \frac{Q(x)}{P(x)} < \infty$$

$$\lim_{x \to x_0} (x - x_0)^2 \frac{R(x)}{P(x)} < \infty$$

If either limit does not exist, a singular point x_0 is called an **irregular singular point.**

Note: If the coefficients $P(x)$, $Q(x)$, and $R(x)$ are polynomials and x_0 is a singular point $[P(x_0) = 0]$, then x_0 is a regular singular point if the factor $(x - x_0)$ appears *at most* to the first power in the denominator of $Q(x)/P(x)$, and *at most** to the second power in the denominator of $R(x)/P(x)$.

Example 1 **Classification of Points** For the following differential equations, classify points as ordinary or singular points. Classify singular points further as regular singular or irregular singular points.

(a) $x^2 y'' + xy' + y = 0$

(b) $(1 - x^2)y'' + y' + xy = 0$

---

* We assume that the quotients $Q(x)/P(x)$ and $R(x)/P(x)$ are reduced to lowest terms.

Solution   (a) We identify $P(x) = x^2$, $Q(x) = x$, and $R(x) = 1$. Hence all points are ordinary except for the singular point $x_0 = 0$. To determine whether $x_0 = 0$ is a regular singular or irregular singular point, we compute the limits

$$\lim_{x \to x_0} \left((x - x_0) \frac{Q(x)}{P(x)} \right) = \lim_{x \to 0} \left(x \left(\frac{x}{x^2} \right) \right) = \lim_{x \to 0} (1) = 1 \qquad (6a)$$

$$\lim_{x \to x_0} \left((x - x_0)^2 \frac{R(x)}{P(x)} \right) = \lim_{x \to 0} \left(x^2 \left(\frac{1}{x^2} \right) \right) = \lim_{x \to 0} (1) = 1 \qquad (6b)$$

Since both limits exist, the point $x_0 = 0$ is a regular singular point.

(b) We identify $P(x) = 1 - x^2$, $Q(x) = 1$, and $R(x) = x$. Hence the equation has singular points at $x_0 = \pm 1$, all other points being ordinary points. The singular point at $x_0 = -1$ is a regular singular point, since

$$\lim_{x \to x_0} \left((x - x_0) \frac{Q(x)}{P(x)} \right) = \lim_{x \to -1} \left((x + 1) \frac{1}{(1 - x^2)} \right) = \lim_{x \to -1} \left(\frac{1}{1 - x} \right) = \frac{1}{2}$$

$$\lim_{x \to x_0} \left((x - x_0)^2 \frac{R(x)}{P(x)} \right) = \lim_{x \to -1} \left((x + 1)^2 \frac{x}{(1 - x^2)} \right) = \lim_{x \to -1} \left(\frac{x(1 + x)}{1 - x} \right) = 0$$

Also, the singular point $x_0 = 1$ is a regular singular point, since

$$\lim_{x \to x_0} \left((x - x_0) \frac{Q(x)}{P(x)} \right) = \lim_{x \to 1} \left((x - 1) \frac{1}{(1 - x^2)} \right) = \lim_{x \to 1} \left(\frac{-1}{1 + x} \right) = -\frac{1}{2}$$

$$\lim_{x \to x_0} \left((x - x_0)^2 \frac{R(x)}{P(x)} \right) = \lim_{x \to 1} \left((x - 1)^2 \frac{x}{(1 - x^2)} \right) = \lim_{x \to 1} \left(\frac{x(1 - x)}{1 + x} \right) = 0$$

Note: Although any point can be a singular point of a differential equation, we consider only singular points at $x_0 = 0$. This condition is not restrictive, since if a differential equation has nonzero singular point(s) x_0, one can always introduce a new independent variable $z = x - x_0$ and transform the original equation to a new equation with a singular point at $z_0 = 0$. One can then solve this new equation in terms of z and then substitute $z = x - x_0$ to obtain the solution in terms of x.

THE METHOD OF FROBENIUS

Suppose $x_0 = 0$ is a regular singular point of the differential equation (5). The **method of Frobenius** refers to the procedure for finding a series solution of the form

$$y = x^r (a_0 + a_1 x + a_2 x^2 + \cdots) \qquad (7)$$

where the goal is to determine the coefficients a_2, a_3, ... in terms of the first two coefficients a_0 and a_1 and the exponent r. A series of the form of Eq. (7) is called a **Frobenius series,** and a solution of the type of Eq. (7) is called a **Frobenius-type solution.** Note that when r is a positive integer, the series becomes the usual power series. However, for negative values of r or even noninteger positive values, the series (7) is not the usual power series. Hence the Frobenius series includes the power series as a special case.

HISTORICAL NOTE

Ferdinand Georg Frobenius (1849–1917) was one of the eminent mathematicians of the nineteenth century. Although he made several important contributions to elliptic functions and differential equations, he is best remembered for his work in algebra as one of the founders of group theory.

Inasmuch as the general method of Frobenius is quite difficult and involved, we illustrate the method only by solving a specific equation. Consider

$$2xy'' + y' - y = 0 \tag{8}$$

which has a regular singular point at $x_0 = 0$. The method of Frobenius seeks a series solution of the form*

$$y = x^r \sum_{n=0}^{\infty} a_n x^n = \sum_{n=0}^{\infty} a_n x^{n+r}$$

where the coefficients $a_0, a_1, \ldots$ and r are to be determined so that the series satisfies the differential equation. The only assumption is that $a_0 \neq 0$. Differentiating, we find

$$y' = \sum_{n=0}^{\infty} (n + r)a_n x^{n+r-1} \tag{9a}$$

$$y'' = \sum_{n=0}^{\infty} (n + r)(n + r - 1)a_n x^{n+r-2} \tag{9b}$$

Substituting y, y', and y'' into the differential equation gives

$$2 \sum_{n=0}^{\infty} (n + r)(n + r - 1)a_n x^{n+r-1} + \sum_{n=0}^{\infty} (n + r)a_n x^{n+r-1} - \sum_{n=0}^{\infty} a_n x^{n+r} = 0 \tag{10}$$

which can be rewritten as

$$r(2r - 1)a_0 x^{r-1} + \sum_{n=0}^{\infty} [(n + r + 1)(2n + 2r + 1)a_{n+1} - a_n] x^{n+r} = 0 \tag{11}$$

For this equation to hold for all x, all the coefficients must be zero. Setting the coefficient of x^r to zero gives (remembering that $a_0 \neq 0$)

$$r(2r - 1) = 0 \tag{12}$$

---

* If series notation causes some confusion, it is always possible to write out the first few terms of the series to ''see'' what is going on.

This equation determines the value(s) of r and is called the **indicial equation.*** In this case the roots of the indicial equation are $r = 0, 1/2$. If we set the remaining coefficients of Eq. (11) equal to zero, we find

$$(n + r + 1)(2n + 2r + 1)a_{n+1} - a_n = 0 \tag{13}$$

The general idea is to systematically set r equal to 0 and 1/2 and use Eq. (13) to determine the coefficients $a_0, a_1, \ldots$. For each value of r we solve for a_{n+1}, getting

$$r = 0 \, \triangleright \, a_{n+1} = \frac{1}{(n + 1)(2n + 1)}a_n \qquad (n \geq 0) \tag{14a}$$

$$r = \frac{1}{2} \, \triangleright \, a_{n+1} = \frac{1}{(n + 1)(2n + 3)}a_n \qquad (n \geq 0) \tag{14b}$$

Using $r = 0$, the recurrence relation (14a) gives

$$a_1 = \frac{1}{1 \cdot 1}a_0, \qquad a_2 = \frac{1}{2 \cdot 3}a_1 = \frac{1}{(1 \cdot 2)(1 \cdot 3)}a_0$$

In general, we find

$$a_n = \frac{1}{n! \, [1 \cdot 3 \cdot 5 \cdots (2n - 1)]}a_0 = \frac{2^n}{(2n)!}a_0 \qquad (n \geq 0) \tag{15}$$

This gives one Frobenius-type solution

$$y_1(x) = x^r \sum_{n=0}^{\infty} a_n x^n = \sum_{n=0}^{\infty} \frac{2^n}{(2n)!}x^n \tag{16}$$

Using $r = 1/2$, the recurrence relation (14b) gives

$$a_1 = \frac{1}{1 \cdot 3}a_0, \qquad a_2 = \frac{1}{2 \cdot 5}a_1 = \frac{1}{(1 \cdot 2)(3 \cdot 5)}a_0$$

and in general

$$a_n = \frac{1}{n! \, [1 \cdot 3 \cdot 5 \cdot 7 \cdots (2n + 1)]}a_0 = \frac{2^n}{(2n + 1)!}a_0 \qquad (n \geq 0) \tag{17}$$

Hence we have found a second Frobenius-type solution

$$y_2(x) = x^r \sum_{n=0}^{\infty} a_n x^n = x^{1/2} \sum_{n=0}^{\infty} \frac{2^n}{(2n + 1)!}x^n \tag{18}$$

By using the ratio test, it can be shown that both Frobenius-type solutions (16) and (18) converge for all x. It can also be shown that y_1 and y_2 are linearly independent on $(0, \infty)$ by setting

$$c_1 y_1(x) + c_2 y_2(x) = 0 \qquad (0 < x < \infty) \tag{19}$$

---

* The equation is called *indicial* because it "indicates" the form of the series solution.

To show that $c_1 = c_2 = 0$, we let $x \to 0+$, getting $c_1 = 0$. But if c_1 is zero, c_2 must also be zero, since y_2 is not identically zero.

The graphs of the first several terms of y_1 and y_2 are shown in Figure 4.5.

The method of Frobenius for singular points at $x_0 = 0$ can be summarized as follows.

The Method of Frobenius

Assume that the equation

$$P(x)y'' + Q(x)y' + R(x)y = 0 \tag{20}$$

has a regular singular point at $x_0 = 0$. The method of Frobenius seeks a solution of the form

$$y = x^r \sum_{n=0}^{\infty} a_n x^n = a_0 x^r + a_1 x^{r+1} + a_2 x^{r+2} + \cdots$$

Step 1.   Substituting y, y', and y'' into Eq. (20), one gets the indicial equation (which yields roots r_1 and r_2) and the recurrence relation.

Step 2.   Using the recurrence relation and one root r_1 leads to a solution

$$y_1(x) = x^{r_1} \sum_{n=0}^{\infty} a_n x^n$$

Step 3.   A second linearly independent solution y_2 is obtained as follows.

(a) If $r_1 - r_2$ is not an integer, then

$$y_2(x) = x^{r_2} \sum_{n=0}^{\infty} b_n x^n$$

where the coefficients b_n are found by substituting $y_2(x)$ into Eq. (20).

(b) If $r_1 = r_2$, then

$$y_2(x) = x^{r_1} \sum_{n=0}^{\infty} b_n x^n + (\ln x) y_1(x)$$

where the coefficients b_n are found by substituting $y_2(x)$ into Eq. (20).

(c) If $r_1 - r_2 = m$ is a positive integer, then

$$y_2(x) = x^{r_2} \sum_{n=0}^{\infty} b_n x^n + c (\ln x) y_1(x)$$

where the constant c and the coefficients b_n are found by substituting y_2 into Eq. (20). It may happen that the constant c is zero, in which case no logarithm term occurs in y_2.

Figure 4.5
Graphs of $y_1(x)$ and $y_2(x)$ in a neighborhood of the regular singular point $x_0 = 0$

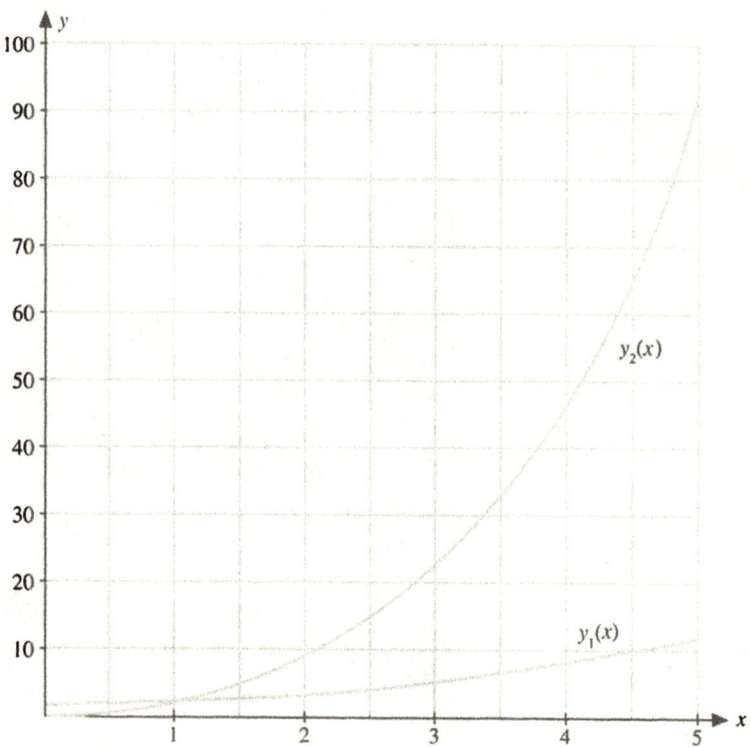

The following theorem summarizes much of the theory behind the method of Frobenius.

> **THEOREM 4.6:** The Theorem of Frobenius
>
> If x_0 is a regular singular point of
> $$P(x)y'' + Q(x)y' + R(x)y = 0 \tag{21}$$
> then Eq. (21) has *at least one* Frobenius-type solution
> $$y = (x - x_0)^r \left(a_0 + a_1(x - x_0) + a_2(x - x_0)^2 + \cdots \right)$$
> where apart from the factor $(x - x_0)^r$ the series converges for all x in the interval $|x - x_0| < R$, where R is the distance from x_0 to the nearest other singularity.

Example 2   Theorem of Frobenius   What does the theorem of Frobenius guarantee about a series solution expanded about $x_0 = 0$ of the equation

$$x^2 y'' - xy' + (1 - x)y = 0 \tag{22}$$

Solution   One can easily verify that $x_0 = 0$ is a regular singular point of Eq. (22). The theorem of Frobenius guarantees that Eq. (22) has a Frobenius-type solution of the form

$$y = x^r(a_0 + a_1 x + a_2 x^2 + \cdots)$$

which apart from the factor x^r converges for all x, since there is no other singular point. Carrying out the steps of the method of Frobenius, one finds the solution

$$y = x + x^2 + \frac{1}{4}x^3 + \frac{1}{36}x^4 + \cdots$$

The method of Frobenius can be used to find the other linearly independent solution (the indicial equation has a double root $r_1 = r_2 = 1$), or the other linearly independent solution can be obtained by using the reduction of order method.   ■

PROBLEMS: Section 4.4

For Problems 1–10, find and classify each singular point (real or complex) of the given equation as regular or irregular.

1. $y'' + xy = 0$
2. $xy'' + 2y' - xy = 0$
3. $x^3 y'' + 4y' - xy = 0$
4. $2x^2 y'' + 3xy' + y = 0$
5. $x^2 y'' + y' + y = 0$
6. $x^3 y'' + y = 0$
7. $x^4 y'' + 2y = 0$
8. $x^2(1 - x)^2(1 + x)y'' + 2y' + y = 0$
9. $x^2 y'' - xy' + (1 + x)y = 0$
10. $x^2(1 + x^2)y'' + x^2 y' + y = 0$

For Problems 11–14, find the form of two linearly independent solutions near the point $x_0 = 0$.

11. $x^2 y'' + xy' + \left(x^2 - \frac{1}{9}\right)y = 0$

12. $xy'' + \frac{1}{2}y' - x^2 y = 0$

13. $(2x^2 + x)y'' + 3xy' - 4y = 0$
14. $4x^2 y'' + 4xy' + (x^2 - 1)\, y = 0$

For Problems 15–21, find Frobenius-type solution of the given equation about $x_0 = 0$. If the general expression for the series cannot be found, find the first three terms of each linearly independent series.

15. $xy' - y = 0$
16. $xy' + 2(1 - x^2)y = 0$
17. $4xy'' + 2y' + y = 0$
18. $xy'' + 3y' - x^2 y = 0$

19. $2xy'' + (3 - x)y' - y = 0$
20. $x^2 y'' + xy' + (x^2 - 1)y = 0$
21. $x^2 y'' + 2xy' - 12y = 0$

Alternate Solutions to Equations with Variable Coefficients

There are other ways of solving differential equations with variable coefficients than by seeking series solutions. Problems 22–23 illustrate two common methods that apply to certain equations.

22. **Cauchy-Euler Equation**   Solve the **Cauchy-Euler equation**

$$ax^2 \frac{d^2 y}{dx^2} + bx \frac{dy}{dx} + cy = 0$$

by carrying out the following steps:

(a) Transform the Cauchy-Euler equation into the equation with constant coefficients

$$a\frac{d^2 y}{dt^2} + (b - a)\frac{dy}{dt} + cy = 0$$

by making the change of independent variable from x to t using $t = \ln x$. *Hint:* Show that $x\, dy/dx = dy/dt$ and $x^2 d^2 y/dx^2 = d^2 y/dt^2 - dy/dt$.

(b) Use the result of part (a) to solve

$$x^2 \frac{d^2 y}{dx^2} + 2x\frac{dy}{dx} - 6y = 0$$

23. Series by Successive Differentiation Solve the initial-value problem

$$y'' - (x + 1)y' + x^2y = 0$$
$$y(0) = 1$$
$$y'(0) = 1$$

by using the differential equation to compute the higher derivatives in the Taylor series

$$y(x) = y(0) + y'(0)x + \frac{y''(0)}{2!}x^2 + \frac{y'''(0)}{3!}x^3 + \cdots$$

The Theorem of Frobenius

For Problems 24–28, tell what the theorem of Frobenius (Theorem 4.6) guarantees about the convergence of Frobenius-type solutions of the stated equations about the regular singular point $x_0 = 0$.

24. $xy'' + y' - y = 0$
25. $x^2y'' + xy' + (x^2 - 1)y = 0$
26. $x^2(1 - x)y'' + xy' + x^2y = 0$
27. $x(2 - x)y'' + y' + xy = 0$
28. $x(1 + x^2)y'' + y' + y = 0$

The Amazing Hypergeometric Series

*One of the most famous differential equations is **Gauss's hypergeometric equation****

$$x(x - 1)y'' + [(\alpha + \beta + 1)x - \gamma]y' + \alpha\beta y = 0 \quad (23)$$

whose power series solution about the regular singular point at $x = 0$, which we denote as $F(\alpha, \beta, \gamma, x)$, is

$$F(\alpha, \beta, \gamma, x) = 1 + \frac{\alpha\beta}{1 \cdot \gamma}x + \frac{\alpha(\alpha + 1)\beta(\beta + 1)}{1 \cdot 2\gamma(\gamma + 1)}x^2$$
$$+ \frac{\alpha(\alpha + 1)(\alpha + \beta)\beta(\beta + 1)(\beta + 2)}{1 \cdot 2 \cdot 3\gamma(\gamma + 1)(\gamma + 2)}x^3 + \cdots (24)$$

*The interesting thing about this solution $F(\alpha, \beta, \gamma, x)$, called the **hypergeometric series** (or **hypergeometric function**), is the wide variety of important functions that can be attained by simply changing the constants α, β, and γ. The function acts as a unifying influence, since it represents many of the important functions in analysis. For Problems 29–34, verify the given identities by examining the series expansions of the functions on the left-hand side.*

29. $\dfrac{1}{1 - x} = F(1, \beta, \beta, x)$

30. $(1 - x)^p = F(-p, \beta, \beta, x)$

31. $\ln(1 - x) = -xF(1, 1, 2, x)$

32. $e^x = \lim_{\beta \to \infty} F\left(\alpha, \beta, \alpha, \dfrac{x}{\beta}\right)$

33. $\sin x = x\left(\lim_{\alpha \to \infty} F\left(\alpha, \alpha, \dfrac{3}{2}, \dfrac{-x^2}{4\alpha^2}\right)\right)$

34. $\cos x = \lim_{\alpha \to \infty} F\left(\alpha, \alpha, \dfrac{1}{2}, \dfrac{-x^2}{4\alpha^2}\right)$

35. Journal Entry Summarize how we solved the homogeneous linear equation with variable coefficients.

4.5   BESSEL FUNCTIONS

PURPOSE
To find a series solution of Bessel's equation using the method of Frobenius and to show how this famous equation can be used to predict the vibrations of a circular drumhead.

---

* Karl Friedrich Gauss (1777–1855) was the greatest mathematician of all time. Stories abound as to his genius: At the age of three he found an error in his father's payroll accounts, at the age of five he composed minuets, at the age of 18 he solved one of the most difficult problems in mathematics—how a regular polygon can be constructed with a ruler and compass.

BESSEL'S EQUATION

One of the most important differential equations in mathematical physics is **Bessel's equation of order p**

$$x^2 y'' + xy' + (x^2 - p^2)y = 0 \qquad (0 < x < \infty) \tag{1}$$

where $p \geq 0$ is a specified real number. Since the equation has a regular singular point at $x = 0$ and has no other singular points in the complex plane, it has a series solution of the form

$$y = x^r \sum_{k=0}^{\infty} a_k x^k \qquad (a_0 \neq 0) \tag{2}$$

which converges for $0 < x < \infty$. Substituting this series into the Eq. (1), we arrive at the indicial equation

$$r(r-1) + r - p^2 = (r - p)(r + p) = 0 \tag{3}$$

which has roots $r_1 = p$ and $r_2 = -p$. If p is not an integer, then the method of Frobenius yields the two linearly independent solutions

$$y_1(x) = a_0 x^p \sum_{k=0}^{\infty} \frac{(-1)^k x^{2k}}{2^{2k} k! \, (1+p)(2+p) \cdots (k+p)} \tag{4a}$$

$$y_2(x) = b_0 x^{-p} \sum_{k=0}^{\infty} \frac{(-1)^k x^{2k}}{2^{2k} k! \, (1-p)(2-p) \cdots (k-p)} \tag{4b}$$

where a_0 and b_0 are arbitrary constants. For reasons that will become apparent shortly, it is convenient to choose these constants as

$$a_0 = \frac{1}{2^p \Gamma(1+p)} \qquad \text{and} \qquad b_0 = \frac{1}{2^{-p} \Gamma(1-p)}$$

where $\Gamma(x)$ is the **gamma function** defined by

$$\Gamma(x) = \int_0^{\infty} t^{x-1} e^{-t} \, dt \tag{5}$$

We are not interested in the gamma function here per se, but only as a way to streamline our notation. For completeness, however, the graph of the gamma function is shown in Figure 4.6.

HISTORICAL NOTE

Friedrich Wilhelm Bessel (1784–1846) was a German astronomer who in 1840 predicted the existence of a planet beyond Uranus. He was also the first person to compute the orbit of Halley's comet. Although Bessel studied the equation that now bears his name in conjunction with his study of planetary orbits, it was Daniel Bernoulli who first studied the equation in his research on the displacement of an oscillating chain.

Figure 4.6
The gamma function. Note
that it "blows up" at zero
and at the negative integers.

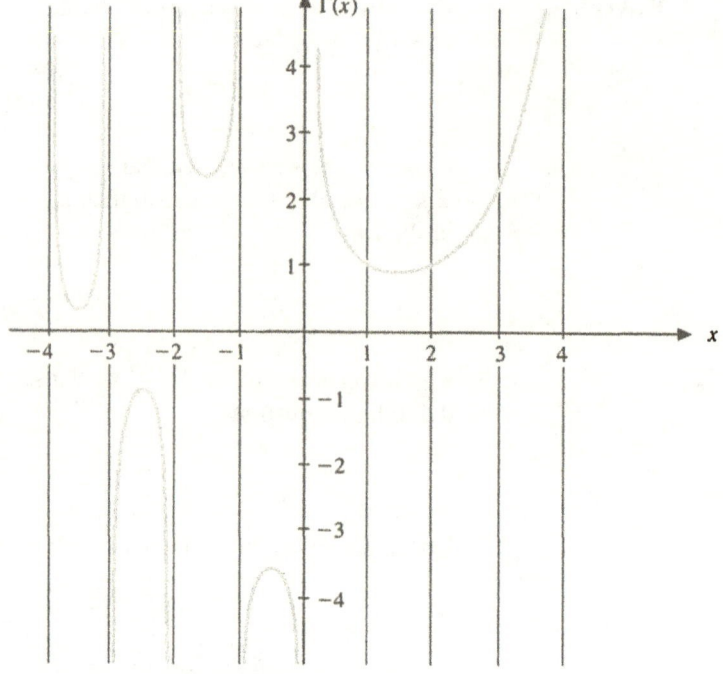

For our purposes the most important property of the gamma function is the property $\Gamma(p + 1) = p\Gamma(p)$, from which we can prove

$$\Gamma(1 + k + p) = (k + p) \cdots (2 + p)(1 + p)\,\Gamma(1 + p)$$
$$\Gamma(1 + k - p) = (k - p) \cdots (2 - p)(1 - p)\,\Gamma(1 - p)$$

and rewrite the two solutions (4a) and (4b) in the more compact form

$$J_p(x) = \sum_{k=0}^{\infty} \frac{(-1)^k}{k!\,\Gamma(1 + k + p)} \left(\frac{x}{2}\right)^{2k+p} \tag{6a}$$

$$J_{-p}(x) = \sum_{k=0}^{\infty} \frac{(-1)^k}{k!\,\Gamma(1 + k - p)} \left(\frac{x}{2}\right)^{2k-p} \tag{6b}$$

which are called **Bessel functions of the first kind** of order p and $-p$, respectively. Graphs of the two particular Bessel functions $J_0(x)$ and $J_1(x)$ are drawn in Figure 4.7.

When p is a *nonnegative integer*, which we denote by n, the Bessel function can be rewritten in the familiar *factorial* form

$$J_n(x) = \sum_{k=0}^{\infty} \frac{(-1)^k}{k!\,(k + n)!} \left(\frac{x}{2}\right)^{2k+n} \qquad (n = 0, 1, 2, \ldots) \tag{7}$$

Figure 4.7
The two important Bessel functions $J_0(x)$ and $J_1(x)$

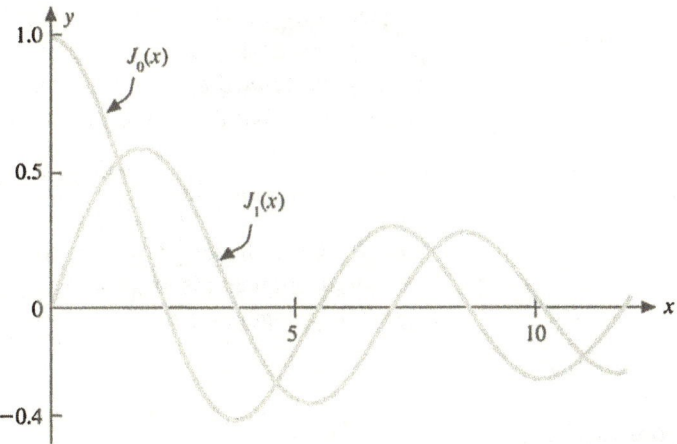

LINEARLY INDEPENDENT SOLUTIONS

We know from the method of Frobenius that when the difference between the roots r_1 and r_2 of the indicial equation ($r_1 - r_2 = 2p$ in this case) is *not* an integer, then the two solutions $J_p(x)$ and $J_{-p}(x)$ are linearly independent. Moreover, as long as p is *not* an integer, then even if $r_1 - r_2 = 2p$ *is* an integer, functions $J_p(x)$ and $J_{-p}(x)$ are *still* linearly independent. Hence the only remaining case to consider is when the parameter p is a nonnegative integer, which we denote $p = n$. Unfortunately, in this case, one can show that

$$J_n(x) = (-1)^n J_{-n}(x) \tag{8}$$

which means that the Bessel functions $J_n(x)$ and $J_{-n}(x)$ are linearly dependent. So we are left with the problem of finding a second linearly independent solution for the values $p = n = 0, 1, 2, \dots$.

FINDING A SECOND SOLUTION WHEN $p = 0, 1, 2, \dots$

Assume for a moment that $p \neq 0, 1, 2, \dots$ and consider the linear combination of solutions $J_p(x)$ and $J_{-p}(x)$ defined by

$$Y_p(x) = \frac{\cos(p\pi) J_p(x) - J_{-p}(x)}{\sin(p\pi)} \qquad (p \neq 0, 1, 2, \dots) \tag{9}$$

Clearly, this linear combination, called the **Bessel function of the second kind*** of order p, is also a solution of Bessel's equation, and in addition it can be shown to be linearly independent of $J_p(x)$. In other words, when $p \neq 0, 1, 2, \dots$, we have the two sets $\{J_p(x), J_{-p}(x)\}$ and $\{J_p(x), Y_p(x)\}$, which form fundamental sets of solutions of Bessel's equation. However, the purpose of introducing the new function $Y_p(x)$ is not to find a second fundamental set when $p \neq 0, 1, 2, \dots$ but to find a second linearly independent solution when $p = 0, 1, 2, \dots$. This goal seems hopeless at first glance,

---

* In the literature the function $Y_n(x)$ is often called the **Neumann function** or **Weber function** and is sometimes denoted by $N_n(x)$.

however, since the denominator of $Y_p(x)$ is zero when $p = 0, 1, 2, \ldots$, and so $Y_p(x)$ is not defined. However, the numerator of $Y_p(x)$ is *also* zero, and so there is hope that $Y_p(x)$ can be defined as a limiting value. This, in fact, is the case, and using l'Hôpital's rule, we can *define* $Y_p(x)$ when $p = n = 0, 1, 2, \ldots$ by

$$Y_n(x) = \lim_{p \to n} \left(\frac{\cos(p\pi) J_p(x) - J_{-p}(x)}{\sin p\pi} \right) \qquad (n = 0, 1, 2, \ldots) \qquad (10)$$

where we have changed notation to $Y_n(x)$. It can also be shown that $Y_n(x)$ satisfies Bessel's equation and is linearly independent of $J_n(x)$. The graphs of $Y_0(x)$ and $Y_1(x)$ are shown in Figure 4.8.

Figure 4.8
Bessel functions $Y_0(x)$ and $Y_1(x)$ of the second kind of orders 0 and 1

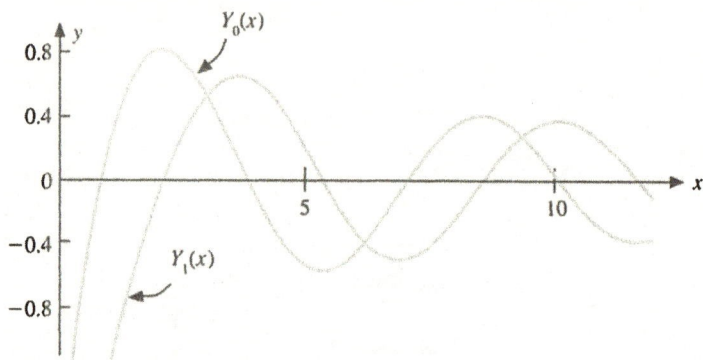

We summarize the general solution of Bessel's equation.

Solution of Bessel's Equation
The general solution of Bessel's equation of order $p \geq 0$

$$x^2 y'' + xy' + \left(x^2 - p^2\right)y = 0$$

depends on p. We have

(a) If $p \neq 0, 1, 2, \ldots$, then the general solution is
$$y = c_1 J_p(x) + c_2 J_{-p}(x)$$

(b) If $p = n = 0, 1, 2, \ldots$, then the general solution is
$$\boxed{y = c_1 J_n(x) + c_2 Y_n(x)}$$
where c_1 and c_2 are arbitrary constants.

Example 1   Bessel's Equation of Order Zero   Find the general solution of
$$x^2 y'' + xy' + x^2 y = 0 \qquad (11)$$

Solution   Since the equation is a Bessel's equation of order $p = 0$, we have

$$y = c_1 J_0(x) + c_2 Y_0(x) \tag{12}$$

where c_1 and c_2 are arbitrary constants.

Example 2   Bessel's Equation of Order One-Half   Find the general solution of

$$x^2 y'' + xy' + \left(x^2 - \frac{1}{4}\right)y = 0 \qquad (0 < x < \infty) \tag{13}$$

Solution   The equation is a Bessel's equation with $p^2 = 1/4$ or order $p = 1/2$. Hence the general solution is

$$y = c_1 J_{1/2}(x) + c_2 J_{-1/2}(x)$$

COMMENTS ON BESSEL FUNCTIONS

A wealth of knowledge is known about Bessel functions because of their importance in applied mathematics.* For example, the graphs of the functions $J_0(x)$ and $J_1(x)$ look like damped cosine and sine curves, respectively. It is because of such similarities that they belong to the class of functions called **almost periodic functions.** Also, each Bessel function $J_n(x)$ has an infinite number of roots, and the distance between consecutive roots of each Bessel function approaches π as $x \to \infty$. A listing of a few roots of some Bessel functions $J_n(x)$ is given in Table 4.4.

Table 4.4
A Few Roots of $J_n(x)$

	$J_0(x)$	$J_1(x)$	$J_2(x)$	$J_3(x)$	⋯
First root	2.40483	3.83171	5.13562	6.38016	
Second root	5.52008	7.01559	8.41724	9.76102	
Third root	8.65374	10.17347	11.61984	13.01520	
Fourth root	11.79153	13.32369	14.79595	16.22347	
⋮	⋮	⋮	⋮	⋮	

Example 3   A Few Bessel Functions   Write out the first few terms of the zeroth and first-order Bessel functions $J_0(x)$ and $J_1(x)$.

Solution

$$J_0(x) = 1 - \frac{x^2}{2^2(1!)^2} + \frac{x^4}{2^4(2!)^2} - \frac{x^6}{2^6(3!)^2} + \cdots$$

$$J_1(x) = \frac{x}{2} - \frac{x^3}{2^3 \cdot 1!2!} + \frac{x^5}{2^5 \cdot 2!3!} - \frac{x^7}{2^7 \cdot 3!4!} + \cdots$$

* The seminal work on Bessel functions is the classic treatise by G. N. Watson, *A Treatise on the Theory of Bessel Functions,* Second Edition (Cambridge University Press, London, 1944). This gargantuan work of 752 pages is packed full of equations and facts related to Bessel functions proved by dozens of researchers over the past three centuries.

Example 4 **Half-Order Bessel Function** Verify the identity

$$J_{1/2}(x) = \sqrt{\frac{2}{\pi x}} \sin x \tag{14}$$

Solution From the definition of $J_p(x)$ with $p = 1/2$ we have

$$J_{1/2}(x) = \sum_{k=0}^{\infty} \frac{(-1)^k}{k!\,\Gamma\left(1 + \dfrac{1}{2} + k\right)} \left(\frac{x}{2}\right)^{2k+1/2}$$

Now, using the property that $\Gamma\left(\dfrac{1}{2}\right) = \sqrt{\pi}$,* the gamma function in the first few terms is

$$k = 0 \ \lozenge \ \Gamma\left(1 + \frac{1}{2}\right) = \frac{1}{2}\Gamma\left(\frac{1}{2}\right) = \frac{1}{2}\sqrt{\pi}$$

$$k = 1 \ \lozenge \ \Gamma\left(1 + \frac{3}{2}\right) = \frac{3}{2}\Gamma\left(\frac{3}{2}\right) = \frac{3}{2^2}\sqrt{\pi}$$

$$k = 2 \ \lozenge \ \Gamma\left(1 + \frac{5}{2}\right) = \frac{5}{2}\Gamma\left(\frac{5}{2}\right) = \frac{5 \cdot 3}{2^3}\sqrt{\pi} = \frac{5!}{2^5 2!}\sqrt{\pi}$$

In general, we have

$$\Gamma\left(1 + \frac{1}{2} + k\right) = \frac{(2k+1)!}{2^{2k+1}k!}\sqrt{\pi}$$

Hence we have, after some simplification,

$$J_{1/2}(x) = \sqrt{\frac{2}{\pi x}} \sum_{k=0}^{\infty} \frac{(-1)^k}{(2k+1)!} x^{2k+1}$$

$$= \sqrt{\frac{2}{\pi x}} \sin x \tag{15}$$

By a similar argument it is possible to show

$$J_{1/2}(x) = \sqrt{\frac{2}{\pi x}} \cos x \tag{16}$$

---

* The integral $\Gamma\left(\dfrac{1}{2}\right) = \sqrt{\pi}$ can be verified by making the appropriate substitution.

THE CIRCULAR DRUM

One of the uses of Bessel's equation lies in the analysis of the vibrations of a circular drum. If you strike a drumhead, its motion, although irregular, is a sum of simple oscillations, called **standing waves.** The most important property of these standing waves is the fact that every point on a given standing wave oscillates with the same frequency. Such motion is called simple harmonic motion. Points on a standing wave that remain motionless are called **nodes** of the standing wave. Although we are getting ahead of ourselves,* the shapes $u(r, \theta)$ of the standing waves satisfy the partial differential equation

$$\frac{\partial^2 u}{\partial r^2} + \frac{1}{r}\frac{\partial u}{\partial r} + \frac{1}{r^2}\frac{\partial^2 u}{\partial \theta^2} + \lambda u = 0 \tag{17}$$

where u denotes the displacement of the standing wave from equilibrium and λ represents the frequency of oscillation of the standing wave. See Figure 4.9.

Figure 4.9
Shapes of a few standing waves of a circular drumhead

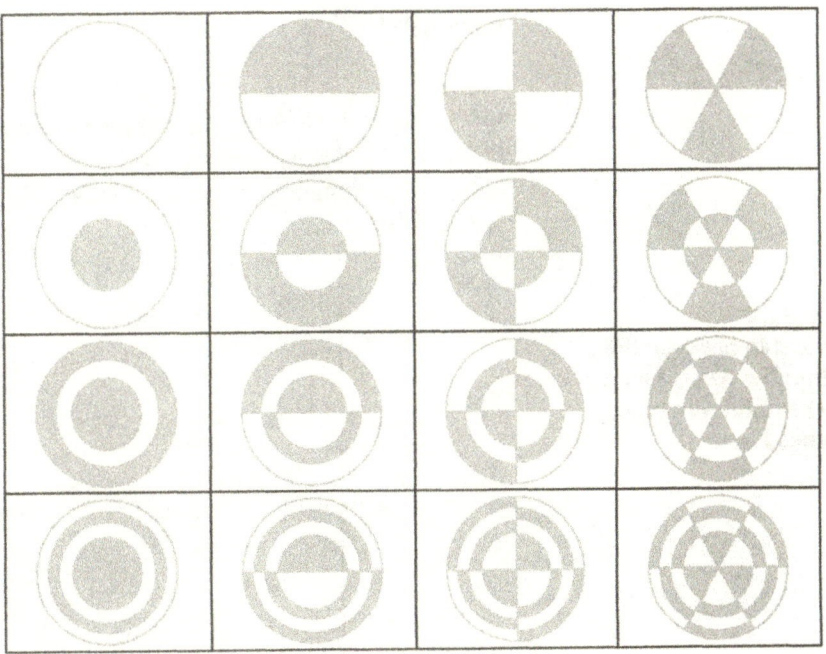

Now, if the standing waves are **radially symmetric,** that is, they depend only on r and not θ, the standing waves $R = R(r)$ are governed by the **parametric Bessel's equation** with boundary conditions

$$r^2 R''(r) + rR' + \lambda^2 r^2 R = 0 \qquad (0 \leq r \leq 1) \tag{18}$$

$$R(0) < \infty$$

$$R(1) = 0$$

* We will introduce standing waves in Section 9.4 in our study of partial differential equations.

To solve Eq. (18), we change the independent variable to $\xi = \lambda r$ and rewrite Eq. (18) as Bessel's equation:

$$\xi^2 R'' + \xi R' + \xi^2 R = 0 \tag{19}$$

This Bessel's equation has order zero and has the general solution

$$R(\xi) = c_1 J_0(\xi) + c_2 Y_0(\xi)$$

Hence the solution of the parametric Bessel's equation is

$$R(r) = c_1 J_0(\lambda r) + c_2 Y_0(\lambda r) \tag{20}$$

If we now impose the boundary condition that $R(0) < \infty$, we must have $c_2 = 0$. Hence we have

$$R(r) = c_1 J_0(\lambda r)$$

To determine the frequencies λ, we use the boundary condition $R(1) = 0$, which implies that

$$c_1 J_0(\lambda) = 0 \,\diamond\, J_0(\lambda) = 0$$

In other words, the frequencies of the standing waves are the *roots* of the zeroth-order Bessel function $J_0(x)$, the first few being $\lambda = 2.4, 5.5, 8.65, 11.8\ 14.9, \ldots$. Hence, the shapes of the standing waves are $J_0(2.4r)$, $J_0(5.5r)$, $J_0(8.65r)$, ..., which are shown in Figure 4.10.

Figure 4.10
Relationship between the zeroth-order Bessel function $J_0(\lambda r)$ and the vibrating drumhead

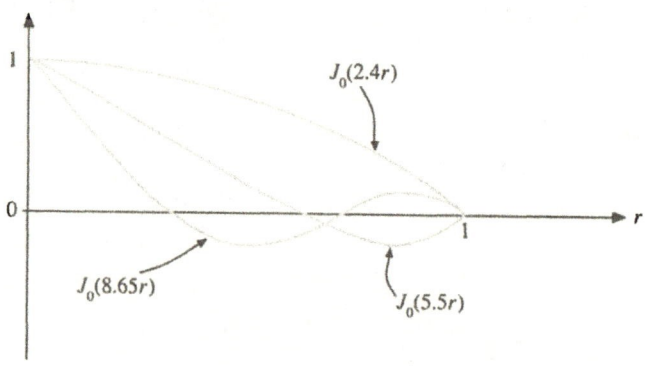

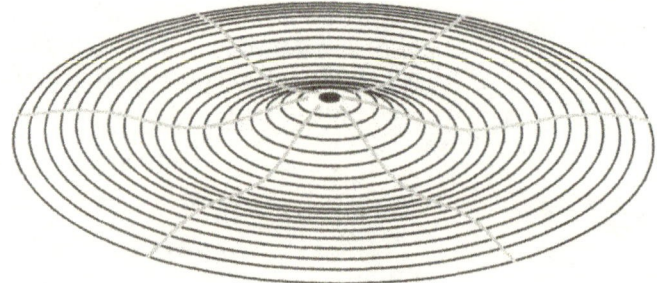

COMMENTS OF THE VIBRATION OF A DRUMHEAD

If you strike a circular drum exactly in the center, the resulting motion will be the sum of radially symmetric standing waves. For example, if a drumhead had an initial displacement described by $u(r, \theta) = J_0(2.4r)$, then subsequent vibration of the drumhead will be described by the simple harmonic motion $u(r, \theta, t) = J_0(2.4r) \cos (2.4\omega t)$, where ω is the frequency of vibration that depends on the specific drum.

PROBLEMS: Section 4.5

For Problems 1–6, express the general solution of the Bessel's equation in terms of Bessel functions.

1. $4x^2y'' + 4xy' + (4x^2 - 1)y = 0$

2. $x^2y'' + xy' + (x^2 - 1)y = 0$

3. $x^2y'' + xy' + (x^2 - 9)y = 0$

4. $9x^2y'' + 9xy' + (9x^2 - 4)y = 0$

5. $x^2y'' + xy' + (x^2 - \pi^2)y = 0$

6. $x^2y'' + xy' + (x^2 - 25)y = 0$

7. **Finding a Second Solution** When a Bessel's equation has a nonnegative integer index, $p = m = 0, 1, 2, ...$, one solution is $J_m(x)$. To find a second linearly independent solution, it is possible to use the reduction of order method to find a solution of the form

$$y_2(x) = u(x) J_m(x) \qquad (21)$$

where u is to be determined.

(a) Substitute Eq. (21) into Bessel's equation and arrive at the equation

$$u'' + \left\{ \frac{2 J_m'}{J_m} + \frac{1}{x} \right\} u' = 0 \qquad (22)$$

(b) Let $w = u'$ and solve Eq. (22) by first finding w and then u, getting

$$u(x) = c_1 \int \frac{1}{t J_m(t)^2} \, dt + c_2$$

where c_1 and c_2 are arbitrary constants.

(c) Show that a second solution is

$$y_2(x) = c_1 J_m(x) \int \frac{1}{t J_m(t)^2} \, dt + c_2 J_m(x) \qquad (23)$$

which we call $Y_m(x)$ and which is also called Bessel's function of the second kind of order m. Why can you say that this function is linearly independent of $J_m(x)$?

Hint: By looking at the integrand in Eq. (23), see that $Y_m(x)$ approaches infinity as $x \to 0$, whereas $J_m(x)$ remains finite as x approaches zero.

8. **A Familiar Equation** Can you write the general solution of

$$y'' + y = 0$$

in terms of Bessel functions?

9. **Parametric Bessel's Equation** Given the parametric Bessel's equation

$$x^2y'' + xy' + (\lambda^2 x^2 - p^2)y = 0 \qquad (24)$$

(a) Show that the parametric Bessel's equation can be transformed to the regular Bessel's equation

$$\xi^2 y'' + \xi y' + (\xi^2 - p^2)y = 0 \qquad (25)$$

by the substitution $\xi = \lambda x$.

(b) Show that the general solution to the parametric Bessel's equation is

$$y = \begin{cases} c_1 J_p(\lambda x) + c_2 J_{-p}(\lambda x) & (p \neq 0, 1, 2, ...) \\ c_1 J_p(\lambda x) + c_2 Y_p(\lambda x) & (p = 0, 1, 2, ...) \end{cases}$$
$$(26)$$

10. **Linear Dependence of $J_n(x)$ and $J_{-n}(x)$** Use Eq. (6b) with $p = n$, where n is a nonnegative integer, and the fact that $1/\Gamma(N) = 0$, where N is a negative integer, to show that

$$J_n(x) = (-1)^n J_{-n}(x)$$

11. **$J_n(x)$ Even and Odd Function** Use Eq. (6) with $p = n$, where n is a nonnegative integer, to show that

$$J_n(-x) = (-1)^n J_n(x)$$

12. **Parametric Bessel's Equation** Use the results from Problem 9 to find the general solution of

$$x^2y'' + xy' + (2x^2 - 1) y = 0$$

13. **Half-Order Identity** When p is half an odd integer, the Bessel function $J_p(x)$ is called a **half-order** or **spherical Bessel function.** Verify the half-order identity

$$J_{3/2}(x) = \sqrt{\frac{2}{\pi x}}\left(\frac{\sin x}{x} - \cos x\right)$$

Cauchy-Euler Equation

An equation that is closely related to Bessel's equation that is important in partial differential equations is the Cauchy-Euler equation

$$x^2 y'' + \alpha x y' + \beta y = 0 \quad (0 < x < \infty) \quad (27)$$

where α and β are given constants. By trying a solution of the form $y = x^r$, one arrives at the characteristic equation

$$r^2 + (\alpha - 1)r + \beta = 0$$

One can show that

   (a) If the roots r_1 and r_2 of the characteristic equation are real and unequal, then the general solution is

$$y = c_1 x^{r_1} + c_2 x^{r_2}$$

   (b) If the roots r_1 and r_2 are real and equal, then

$$y = (c_1 + c_2 \ln x)x^{r_1}$$

   (c) If the roots r_1 and r_2 are complex, then

$$y = x^\lambda\left(c_1 \cos (\mu \ln x) + c_2 \sin (\mu \ln x)\right)$$

   where $r_1, r_2 = \lambda \pm i\mu$.

For Problems 14–16, solve the given Cauchy-Euler equation.

14. $x^2 y'' + 4xy' + 2y = 0$
15. $x^2 y'' - 3xy' + 4y = 0$
16. $x^2 y'' + 3xy' + 5y = 0$

17. **The Gamma Function** Show that the gamma function $\Gamma(x)$ satisfies the following properties.
    (a) $\Gamma(x + 1) = x\Gamma(x)$
    (b) Use the property in part (a) to show that when n is a positive integer, $\Gamma(n + 1) = n!$ For this reason the gamma function is sometimes called the **generalized factorial** for those cases in which n is not a positive integer.

Basic Identities

Bessel functions are related by an amazing array of identities. Problems 18–22 ask the reader to verify a few of the most important of these identities.

18. Prove the derivative property

$$\frac{d[x^{n+1}J_{n+1}(x)]}{dx} = x^{n+1} J_n(x)$$

19. Prove the derivative property

$$\frac{d[x^{-n}J_{n+1}(x)]}{dx} = -x^{-n}J_{n+1}(x)$$

20. Prove the integral identity

$$\int x^{n+1}J_n(x)\, dx = x^{n+1}J_{n+1}(x) + c$$

21. Prove the integral identity

$$\int x^{-n}J_{n+1}(x)\, dx = -x^{-n}J_n(x) + c$$

22. Prove the recurrence relation

$$J_{n+1}(x) = \frac{2n}{x}J_n(x) - J_{n-1}(x)$$

23. **Computer Problem** Use a graphing calculator or computer to sketch the graph of $J_{1/2}(x)$ on the interval $(0, 2\pi)$.

24. **Journal Entry** Summarize the results of this section.

5

The Laplace Transform

5.1   DEFINITION OF THE LAPLACE TRANSFORM
5.2   PROPERTIES OF THE LAPLACE TRANSFORM
5.3   THE INVERSE LAPLACE TRANSFORM
5.4   INITIAL-VALUE PROBLEMS
5.5   STEP FUNCTIONS AND DELAYED FUNCTIONS
5.6   DIFFERENTIAL EQUATIONS WITH DISCONTINUOUS
        FORCING FUNCTIONS
5.7   IMPULSE FORCING FUNCTIONS
5.8   THE CONVOLUTION INTEGRAL

5.1   DEFINITION OF THE LAPLACE TRANSFORM

PURPOSE
To define the Laplace transform

$$F(s) = \int_0^\infty e^{-st} f(t)\, dt$$

and see how it assigns to each piecewise continuous function $f(t)$ of exponential order a new function $F(s)$. We also show that the Laplace transform is a linear transformation and evaluate some simple transforms.

One of the most powerful methods for solving initial-value problems in differential equations is by the use of the Laplace transform. Although some of the equations that we study in this chapter can be solved by using methods learned earlier, the Laplace transform often brings new insight to the equations, and for many equations, especially those having *discontinuous* and *impulse* forcing terms, the Laplace transform is essentially indispensable. However, before actually showing how the Laplace transform can be used to solve initial-value problems (Section 5.4), we spend Sections 5.1, 5.2, and 5.3 defining the Laplace transform and its inverse transform and studying its properties.

**THE CONCEPT OF
AN INTEGRAL
TRANSFORMATION**

Let $f(t)$ be a function defined on a finite or infinite interval $a < t < b$, and let $K(s, t)$ be a function of the variable t and a parameter s. Then

$$F(s) = \int_a^b K(s, t) f(t)\, dt \tag{1}$$

is called an **integral transform** of $f(t)$, where the function $K(s, t)$ is called the **kernel** of the transform. The function $F(s)$ is the **image,** or **transform,** of the function $f(t)$. We assume that the kernel $K(s, t)$ of the transform and the constants of integration, a and b, are known. Hence if one substitutes a function $f(t)$ into Eq. (1) and evaluates the definite integral, one obtains (if the integral exists) a new function $F(s)$. See Figure 5.1.

The Laplace transform is one of many integral transforms that are useful in solving differential equations because they transform *differential equations* into *algebraic equations,* thereby turning hard problems into easy ones. Figure 5.2 illustrates the general nature of this philosophy.

Figure 5.1
The Laplace transform transforms a function of t into a function of s.

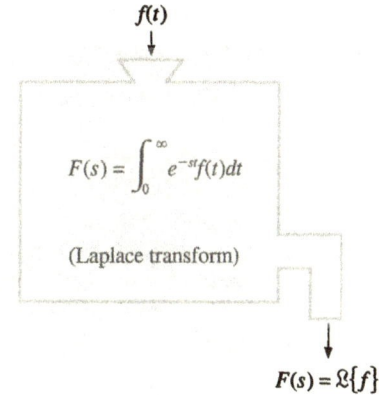

Figure 5.2
Transforming hard problems into easy problems

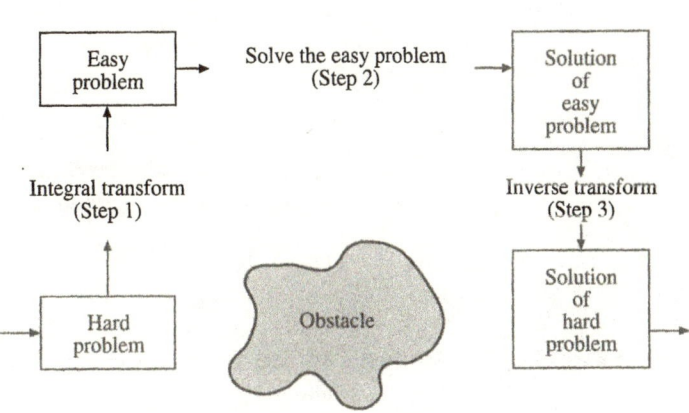

THE LAPLACE TRANSFORM

If the kernel $K(s, t)$ in the integral transform formula (1) is chosen to be the function $K(t, s) = e^{-st}$ and the limits of integration are chosen to be $a = 0$ and $b = \infty$, then the integral transform is called the Laplace transform.*

> **DEFINITION: Laplace Transform**
>
> Let $f(t)$ be a function defined on $(0, \infty)$. The **Laplace transform** of $f(t)$ is defined to be the function $F(s)$ given by the integral
>
> $$F(s) = \int_0^\infty e^{-st} f(t)\, dt$$
>
> The domain of the transform $F(s)$ is taken to be all values of s for which the integral exists.[†] The Laplace transform of $f(t)$ is denoted both by $F(s)$ and the alternate notation $\mathcal{L}\{f\}$.

* Readers who are interested in a more complete study of integral transformations can consult the standard reference, R. V. Churchill, *Operational Mathematics,* McGraw-Hill (New York) Third Edition, 1972.

[†] The variable s can be allowed to be a complex number, but for our purposes of solving differential equations, we will think of it as a real variable.

HISTORICAL NOTE
The Laplace transform is named after the French mathematician Pierre Simon de Laplace (1749–1827), who used the transform in his seminal treatise on probability, *Mécanique Céleste.* However, it was the English engineer Oliver Heaviside (1850–1925) who used the *operational calculus* in the 1800's to solve differential equations using techniques vaguely similar to the methods studied in this chapter. Heaviside was one of the first industrial mathematicians. While working for the Great Northern Telegraph Company, he made numerous contributions to the mathematical study of transmission lines and cables. However, he did not receive wide acceptance in the mathematics community, since he did not justify his techniques and because of his aversion to mathematical "rigor." His reply to mathematicians was "Should I refuse my dinner because I do not understand the process of digestion." It was only in this century that Heaviside's methods were verified by using the Laplace transform.

Note: It should be understood that the integral that defines the Laplace transform is an *improper* integral, defined by

$$F(s) = \int_0^\infty e^{-st} f(t) \, dt = \lim_{b \to \infty} \int_0^b e^{-st} f(t) \, dt \tag{2}$$

Example 1   Simplest Laplace Transform   Find the Laplace transform of the constant function $f(t) = 1$, where $t \geq 0$.

Solution   From the definition of the Laplace transform, we have

$$\mathcal{L}\{1\} = \int_0^\infty e^{-st} \cdot 1 \, dt$$

$$= \lim_{b \to \infty} \int_0^b e^{-st} \, dt$$

$$= \lim_{b \to \infty} \left(\frac{-e^{-st}}{s} \bigg|_{t=0}^{t=b} \right)$$

$$= \lim_{b \to \infty} \left(\frac{1}{s} - \frac{e^{-sb}}{s} \right)$$

$$= \frac{1}{s} \tag{3}$$

Note that the improper integral diverges when $s \leq 0$, and hence the Laplace transform of $f(t) = 1$ is $F(s) = 1/s$ for $s > 0$. See Figure 5.3.   ■

Example 2   Laplace Transform of Exponential Functions   Find the Laplace transform of $f(t) = e^{kt}$, where k is any constant.

Solution   We did not state the domain of $f(t)$, but for most problems dealing with the Laplace transform, it is taken to be $(0, \infty)$. To find the Laplace transform of $f(t)$, we simply use the definition, getting

$$\mathcal{L}\{e^{kt}\} = \int_0^\infty e^{-st} e^{kt}\, dt$$

$$= \int_0^\infty e^{-(s-k)t}\, dt$$

$$= \lim_{b\to\infty} \left(\int_0^b e^{-(s-k)t}\, dt \right)$$

$$= \lim_{b\to\infty} \left(-\frac{e^{-(s-k)t}}{s-k} \Big|_{t=0}^{t=b} \right)$$

$$= -\frac{1}{s-k} \lim_{b\to\infty} \left(e^{-(s-k)b} - 1 \right)$$

$$= \frac{1}{s-k} \qquad \text{for} \quad s > k$$

Figure 5.3
The Laplace transform of
$f(t) = 1$ is $F(s) = 1/s$.

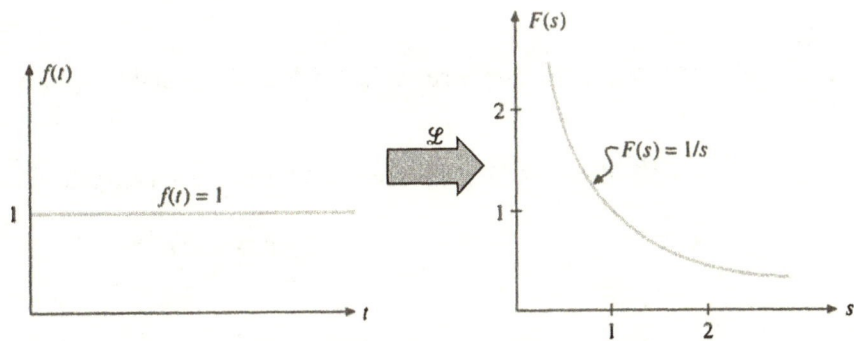

Again, note that the integral diverges when $s \leq k$; hence the domain of $F(s)$ is taken to be all real numbers $s > k$. See Figure 5.4.   ∎

Figure 5.4
The Laplace transform of
$f(t) = e^{kt}$ is $F(s) = 1/(s - k)$.

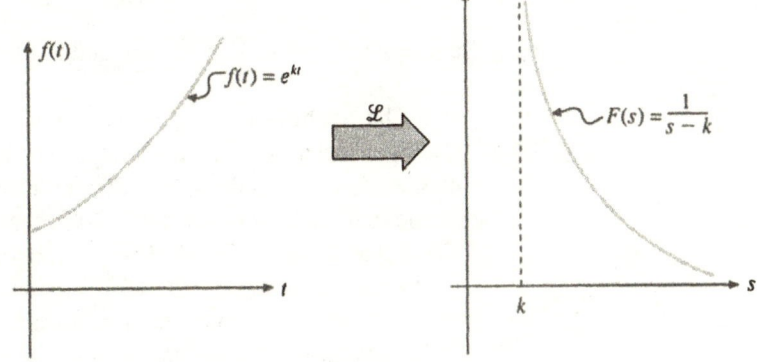

A useful property of the Laplace transform $\mathfrak{L}$ is the fact that it is a linear transformation.

> **THEOREM 5.1:** Laplace Transform, a Linear
>
> Let f and g be functions whose Laplace transform exists on a common domain. The Laplace transform satisfies the two properties of being a **linear transformation**:
>
> - $\mathfrak{L}\{f + g\} = \mathfrak{L}\{f\} + \mathfrak{L}\{g\}$
> - $\mathfrak{L}\{cf\} = c\mathfrak{L}\{f\}$
>
> where c is an arbitrary constant.

The proof of the linearity of the Laplace transform is a direct consequence of the linearity of the integral and is left as an exercise for the reader. (See Problem 43 at the end of this section.)

Example 3 **Importance of Linearity** Find the Laplace transform $\mathfrak{L}\{5 - 3e^{-2t}\}$.

Solution Using the fact that the Laplace transform is a linear transform, we write

$$\mathfrak{L}\{5 - 3e^{-2t}\} = \mathfrak{L}\{5\} + \mathfrak{L}\{-3e^{-2t}\}$$

$$= 5\mathfrak{L}\{1\} - 3\mathfrak{L}\{e^{-2t}\}$$

$$= 5\left(\frac{1}{s}\right) - 3\left(\frac{1}{s+2}\right)$$

$$= \frac{2(s+5)}{s(s+2)} \qquad (s > 0)$$

Since *both* $\mathfrak{L}\{1\}$ and $\mathfrak{L}\{e^{-2t}\}$ are defined for $s > 0$, so is $\mathfrak{L}\{5 - 3e^{-2t}\}$.

So far we have avoided asking whether *all* functions have Laplace transforms or only "well-behaved" ones. An important feature of the Laplace transform is that the functions that arise in the study of linear differential equations with constant coefficients (polynomials, exponentials, sines, and cosines) all have Laplace transforms. Later, we will see that all *piecewise continuous* functions that don't "grow too fast" have Laplace transforms. Of course, by "having" a Laplace transform, we mean that the improper integral in the definition of the transform exists.

To examine the exact conditions required for a function to have a Laplace transform, we define the concepts of piecewise continuity and exponential order $e^{\alpha t}$.

DEFINITION: Piecewise Continuity

A function $f(t)$ is called **piecewise continuous** on a finite interval if it is continuous at every point of the interval except possibly at a finite number of points, where the function has **jump discontinuities**. A jump discontinuity occurs at a point t_0 if the function is discontinuous at t_0 and if both the left- and right-handed limits exist as different numbers.

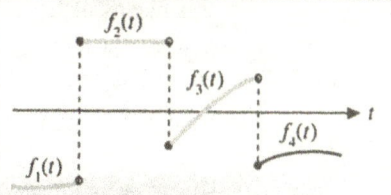

A function is piecewise continuous on $(0, \infty)$ if it is piecewise continuous on any finite interval.

The common functions such as polynomials, exponentials, and sines and cosines are all continuous and hence piecewise continuous. Later, we will study step functions that are not continuous but are piecewise continuous.

The second condition required for the existence of a Laplace transform is the condition of being of exponential order $e^{\alpha t}$.

DEFINITION: Exponential Order $e^{\alpha t}$

A function $f(t)$ is said to be of **exponential order** $e^{\alpha t}$ if there exist positive constants M and T such that $f(t)$ satisfies

$$-Me^{\alpha t} \leq f(t) \leq Me^{\alpha t} \tag{4}$$

for all $t \geq T$.

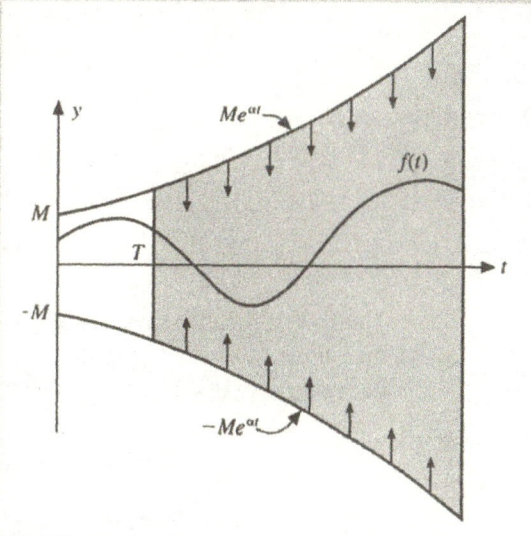

Intuitively, this says that $f(t)$ does not grow too "awfully" fast as $t \to \infty$. For example, $f(t) = e^{3t} \sin t$ is of exponential order e^{3t} ($\alpha = 3$), since

$$-e^{3t} \le e^{3t} \sin t \le e^{3t} \qquad (t \ge 0)$$

and hence Eq. (4) holds with $M = 1$ and T any positive constant. By a similar analysis, one can show that the familiar functions $\sin kt$, t^n, and e^{kt} are all of exponential order for some $e^{\alpha t}$. One function that is *not* of exponential order $e^{\alpha t}$ for any α is the function e^{t^2}. What this means is that e^{t^2} will eventually ($t \ge T$) grow faster than *any* exponential of the form $Me^{\alpha t}$ for any M and α.

We can now say more precisely which functions have Laplace transforms and how to determine the domain of the transform.

THEOREM 5.2: Existence of the Laplace Transform

> If $f(t)$ is a piecewise continuous function on $[0, \infty)$ and of exponential order $e^{\alpha t}$, then $f(t)$ has a Laplace transform $\mathcal{L}\{f\} = F(s)$ for $s > \alpha$.

PROOF: We split the Laplace transform in two parts:

$$F(s) = \int_0^\infty e^{-st} f(t)\, dt = \int_0^T e^{-st} f(t)\, dt + \int_T^\infty e^{-st} f(t)\, dt \qquad (5)$$

where T is chosen so that $-Me^{\alpha t} \le f(t) \le Me^{\alpha t}$ for $t \ge T$. The first integral on the right-hand side of Eq. (5) exists, since $f(t)e^{-st}$ is piecewise continuous on $[0, T]$. To verify that the second integral exists, we use the **comparison test** for improper integrals. We have

$$\left| \int_T^\infty e^{-st} f(t)\, dt \right| \le \int_T^\infty |e^{-st} f(t)|\, dt$$

$$\le \int_T^\infty Me^{-st} e^{\alpha t}\, dt \qquad \text{(by exponential order)}$$

$$= M \int_T^\infty e^{-(s-\alpha)t}\, dt$$

$$= M \frac{e^{-(s-\alpha)T}}{s - \alpha} \qquad (s > \alpha)$$

This last inequality says that the last integral in Eq. (5) is bounded above and below by a finite quantity. Using the comparison test for improper integrals* guarantees that the integral exists. Hence $F(s)$ exists for $s > \alpha$.

Several useful Laplace transforms are listed in Table 5.1.

---

* The comparison test for improper integrals can be found in any calculus text.

Table 5.1
Laplace Transforms

$f(t)$	$F(s) = \mathcal{L}\{f\}$	Domain of $F(s)$
1	$1/s$	$s > 0$
t^n (n positive integer)	$\dfrac{n!}{s^{n+1}}$	$s > 0$
t^p $(p > -1)$	$\dfrac{\Gamma(p+1)}{s^{p+1}}$	$s > 0$
e^{at}	$\dfrac{1}{s-a}$	$s > a$
$e^{at}\,t^n,\, n = 1, 2, \dots$	$\dfrac{n!}{(s-a)^{n+1}}$	$s > a$
$\sin bt$	$\dfrac{b}{s^2 + b^2}$	$s > 0$
$\cos bt$	$\dfrac{s}{s^2 + b^2}$	$s > 0$
$\sinh bt$	$\dfrac{b}{s^2 - b^2}$	$s > \lvert b \rvert$
$\cosh bt$	$\dfrac{s}{s^2 - b^2}$	$s > \lvert b \rvert$
$e^{at} \sin bt$	$\dfrac{b}{(s-a)^2 + b^2}$	$s > a$
$e^{at} \cos bt$	$\dfrac{s-a}{(s-a)^2 + b^2}$	$s > a$
$u(t - c)$	$\dfrac{e^{-cs}}{s}$	$s > 0$
$u(t - c)f(t - c)$	$e^{-cs} F(s)$	
$\displaystyle\int_0^t f(t - \tau)g(\tau)\,d\tau$	$F(s)G(s)$	
$\delta(t - c)$	e^{-cs}	
$\dfrac{d^n}{dt^n} f(t)$	$s^n F(s) - s^{n-1}f(0) - \cdots - f^{(n-1)}(0)$	
$t^n f(t)$	$(-1)^n F^{(n)}(s)$	

PROBLEMS: Section 5.1

For Problems 1–12, find the Laplace transform of the given function using the definition of the Laplace transform.

1. 5

2. t

3. e^{2t}

4. e^{-t}

5. $\sin 2t$

6. $\cos 3t$

7. $5 \sinh 2t$

8. $e^t \cos 2t$

9. $f(t) = \begin{cases} 0 & (0 < t < 1) \\ 1 & (1 \le t) \end{cases}$

10. $f(t) = \begin{cases} 1 & (0 < t < 1) \\ 0 & (1 \le t) \end{cases}$

11. $f(t) = \begin{cases} \sin t & (0 < t < \pi) \\ 0 & (\pi \le t) \end{cases}$

12. $f(t) = \begin{cases} t & (0 < t < 1) \\ 1 & (1 \leq t) \end{cases}$

For Problems 13–16, use integration by parts to find the Laplace transform of the given function.

13. te^{at}

14. $t^n e^{at}$

15. $t \sin at$

16. $t \cosh at$

For Problems 17–24, use Table 5.1 and the property of linearity to find the Laplace transform of the given function.

17. $\mathcal{L}\{a + bt + ct^2\}$

18. $\mathcal{L}\{1 + e^{-t}\}$

19. $\mathcal{L}\{e^{2t} + e^{-2t}\}$

20. $\mathcal{L}\{3 + t + e^{-t} \sin 2t\}$

21. $\mathcal{L}\{te^{-2t} + 3t^2 e^{-t}\}$

22. $\mathcal{L}\{t^3 e^{-3t} + 4e^{-t} \cos 3t\}$

23. $\mathcal{L}\{2e^{at} - e^{-at}\}$

24. $\mathcal{L}\{te^{-3t} + 2 \sin t\}$

For Problems 25–31, determine which of the "branched" functions is continuous on the interval $0 \leq t \leq 3$. Sketch the graphs of the functions.

25. $f(t) = \begin{cases} 0 & (0 \leq t < 1) \\ t - 1 & (1 \leq t \leq 3) \end{cases}$

26. $f(t) = \begin{cases} t & (0 \leq t < 2) \\ 2 - t & (2 \leq t \leq 3) \end{cases}$

27. $f(t) = \begin{cases} 0 & (1 \leq t < 1) \\ t & (1 \leq t \leq 3) \end{cases}$

28. $f(t) = \begin{cases} 1 & (0 \leq t < 1) \\ (t - 1)^2 & (1 \leq t \leq 3) \end{cases}$

29. $f(t) = \begin{cases} t^2 & (0 \leq t < 1) \\ t + 2 & (1 \leq t < 2) \\ 4 - t & (2 \leq t \leq 3) \end{cases}$

30. $f(t) = \begin{cases} t & (0 \leq t < 1) \\ 1 - t & (1 \leq t < 2) \\ 1 & (2 \leq t \leq 3) \end{cases}$

31. $f(t) = \begin{cases} 0 & (0 \leq t < 1) \\ \sin \pi t & (1 \leq t < 2) \\ t^2 & (2 \leq t \leq 3) \end{cases}$

For Problems 32–39, determine which of the functions are of exponential order $e^{\alpha t}$.

32. $f(t) = 3$

33. $f(t) = 3e^{2t}$

34. $f(t) = 3e^{-2t}$

35. $f(t) = t \ln t$

36. $f(t) = e^{-3t} e^{t^2}$

37. $f(t) = $ any bounded function

38. $f(t) = \dfrac{1}{t^2 + 1}$

39. $f(t) = \sin(e^{t^2})$ (Is the derivative of this function of exponential order?)

40. The Laplace Transform of t^n

(a) Show $\mathcal{L}\{t^n\} = \dfrac{n}{s} \mathcal{L}\{t^{n-1}\}$ by integrating by parts.

(b) Use the result from part (a) repeatedly to show the transform $\mathcal{L}\{t^n\} = \dfrac{n!}{s^{n+1}}$.

41. Laplace Transform of the Hyperbolic Sine and Cosine Use the identities

$$\sinh at = \frac{e^{at} - e^{-at}}{2}, \qquad \cosh at = \frac{e^{at} + e^{-at}}{2} \qquad (6)$$

to find the Laplace transform of the following functions:
(a) $f(t) = \sinh at$
(b) $f(t) = \cosh at$
(c) $f(t) = e^{at} \sinh bt$
(d) $f(t) = e^{at} \cosh bt$

42. Laplace Transform of the Circular Sine and Cosine Use the identities

$$\sin at = \frac{e^{iat} - e^{-iat}}{2i}, \qquad \cos at = \frac{e^{iat} + e^{-iat}}{2} \qquad (7)$$

to find the following Laplace transforms. Integrations involving the complex constant $i = \sqrt{-1}$ can be carried out in the same manner as real constants.
(a) $f(t) = \sin at$
(b) $f(t) = \cos at$
(c) $f(t) = e^{at} \sin bt$
(d) $f(t) = e^{at} \cos bt$

43. Linearity of the Laplace Transform Prove that the Laplace transform $\mathcal{L}$ is a linear transformation by showing that for all f and g whose transforms exist, one has

$$\mathcal{L}\{f + g\} = \mathcal{L}\{f\} + \mathcal{L}g$$
$$\mathcal{L}\{cf\} = c\mathcal{L}\{f\}$$

where c is an arbitrary constant.

44. Is There a Product Rule? We have seen that the Laplace transform of a sum is the sum of the transforms. Is there a product rule that says that $\mathcal{L}\{fg\}$ is equal to $\mathcal{L}\{f\}\mathcal{L}\{g\}$? *Hint:* Try some examples.

45. Laplace Transform of Circular Sine and Cosine One of the most important equations in mathematics is *Euler's formula:*

$$e^{ikt} = \cos kt + i \sin kt \qquad (8)$$

where i is the complex constant $i = \sqrt{-1}$, which can be treated the same as a real constant.

(a) Show that $\mathcal{L}\{e^{ikt}\} = \dfrac{1}{s - ik}$ from the definition of the Laplace transform.

(b) Verify the algebraic statement

$$\frac{1}{s - ik} = \frac{s}{s^2 + k^2} + i\frac{k}{s^2 + k^2}$$

(c) Use the results of parts (a) and (b) and the linearity of the Laplace transform to show that

$$\mathcal{L}\{\cos kt\} = \frac{s}{s^2 + k^2}, \qquad \mathcal{L}\{\sin kt\} = \frac{k}{s^2 + k^2}$$

46. Laplace Transform of Damped Sine and Cosine Functions Given the identity

$$e^{(a+ik)t} = e^{at}e^{ikt} = e^{at}(\cos kt + i \sin kt) \qquad (9)$$

(a) Show that

$$\mathcal{L}\{e^{(a+ik)t}\} = \frac{1}{1 - (a + ik)}$$

$$= \frac{s - a}{(s - a)^2 + k^2} + i\frac{k}{(s - a)^2 + k^2}$$

(b) Use the result from part (a) and the fact that the Laplace transform is a linear transformation to show that

$$\mathcal{L}\{e^{at} \cos kt\} = \frac{s - a}{(s - a)^2 + k^2} \qquad (10)$$

$$\mathcal{L}\{e^{at} \sin kt\} = \frac{k}{(s - a)^2 + k^2} \qquad (11)$$

47. The Laplace Transform of f' It is possible to find the Laplace transform of a derivative f' in terms of the transform of f. Use the definition of the Laplace transform to find $\mathcal{L}\{f'\}$ in terms of $\mathcal{L}\{f\}$. *Hint:* Integrate by parts, "shifting" the derivative from $f'(t)$ to e^{-st}.

48. Functions Without Laplace Transforms There are two properties, piecewise continuity and being of exponential order, that are sufficient for a function to have a Laplace transform. If either of these properties fails to hold, the function *may* fail to have a transform.

(a) Why isn't $f(t) = 1/t$ a piecewise continuous function on $[0, \infty)$, and why doesn't this function have a Laplace transform?

(b) Why isn't $f(t) = t^{-1/2}$ piecewise continuous on $[0, \infty)$? If one makes a change of variable $x = \sqrt{st}$ in the definition of the Laplace transform, one can show that the transform actually exists and is $\mathcal{L}\{t^{-1/2}\} = \sqrt{\pi/s}$. In other words, the condition of being piecewise continuous is *sufficient* for the transform to exist but is not *necessary*.

(c) Show that e^{t^2} is not of exponential order $e^{\alpha t}$ for any α and does not have a Laplace transform.

49. Fascinating Formula

(a) Use repeated integration by parts to show that if $P(t)$ is a polynomial, then apart from an additive constant, one has

$$\int e^{-st} P(t)\, dt =$$

$$-\frac{1}{s}e^{-st}\left(P(t) + \frac{1}{s}P'(t) + \frac{1}{s^2}P''(t) + \cdots\right) \qquad (12)$$

(b) Apply the formula from part (a) to find

$$\mathcal{L}\{1 + 2t - t^2 + 2t^3\}$$

(c) Can you think of a way in which Eq. (12) might be useful to approximate Laplace transforms of functions that are not polynomials?

50. The Gamma Function The Gamma function, denoted by $\Gamma(x)$, is defined by the integral

$$\Gamma(x) = \int_0^\infty t^{x-1} e^{-t}\, dt \qquad (13)$$

(a) Show that for $x > 0$, one has $\Gamma(x + 1) = x\,\Gamma(x)$.

(b) Given that $\Gamma(1) = 1$, show that $\Gamma(n + 1) = n!$ for n a positive integer.

(c) Given that $\Gamma\left(\frac{1}{2}\right) = \sqrt{\pi}$, find $\Gamma\left(\frac{3}{2}\right)$ and $\Gamma\left(\frac{5}{2}\right)$.

The graph of $\Gamma(x)$ for $x > 0$ is shown in Figure 5.5.

Figure 5.5
Graph of the gamma function

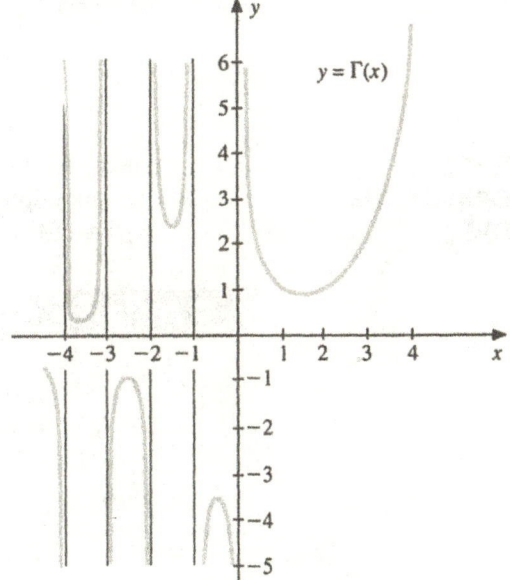

51. Bessel Function of Order Zero When a function has a Taylor's series expansion, it is generally possible to obtain the Laplace transform of the function by taking the term-by-term Laplace transform of the series. Use this method to find the Laplace transform of the *Bessel function of order* 0, denoted by $J_0(t)$, where its Taylor series expansion about zero is

$$J_0(t) = 1 - \frac{t^2}{2^2} + \frac{t^4}{2^2 \cdot 4^2} - \frac{t^6}{2^2 \cdot 4^2 \cdot 6^2} + \cdots \quad (14)$$

52. Journal Entry Summarize the results of this section.

5.2 PROPERTIES OF THE LAPLACE TRANSFORM

PURPOSE

To introduce several important properties of the Laplace transform. Three of the important ones are

- $\mathcal{L}\{f'\} = s\mathcal{L}\{f\} - f\{0\}$ (derivative property)
- $\mathcal{L}\{e^{at}f(t)\} = F(s - a)$ (translation property)
- $\mathcal{L}\{t^n f(t)\} = (-1)^n \dfrac{d^n}{ds^n} F(s)$ (derivative of transform)

GENERAL RULES FOR TRANSFORMS

One of the most important properties of the Laplace transform is one that relates the Laplace transform of the derivative of a function to the transform of the function. This rule is fundamental in the solution of differential equations.

THE LAPLACE TRANSFORM OF THE DERIVATIVE

In order that the Laplace transform be useful for solving differential equations, there must be a relationship between the Laplace transform of a function and the transform of the derivative of the function.

> **THEOREM 5.3:** Laplace Transform of the Derivative
>
> If f is a continuous function whose derivative f' is piecewise continuous on $[0, \infty)$, and if both functions are of exponential order $e^{\alpha t}$, then both $\mathcal{L}\{f\}$ and $\mathcal{L}\{f'\}$ exist for $s > \alpha$; moreover,
>
> $$\mathcal{L}\{f'\} = s\mathcal{L}\{f\} - f(0)$$

PROOF: Clearly the transforms of f and f' exist, since they are both assumed to be piecewise continuous on $[0, \infty)$ and of exponential order. We can also write the transform $\mathcal{L}\{f'\}$ as

$$\mathcal{L}\{f'\} = \int_0^\infty e^{-st}f'(t)\,dt = \lim_{b\to\infty}\int_0^b e^{-st}f'(t)\,dt$$

But this last integral can be integrated by parts ($u = e^{-st}$, $dv = f'(t)\,dt$), getting

$$\int_0^b e^{-st}f'(t)\,dt = e^{-st}f(t)\Big|_0^b + s\int_0^b e^{-st}f(t)\,dt$$

$$= s\int_0^b e^{-st}f(t)\,dt - f(0) + e^{-sb}f(b) \tag{1}$$

The theorem will now be proven if we can show that the last term in Eq. (1) approaches zero, or

$$\lim_{b\to\infty} e^{-sb}f(b) = 0$$

Since $f(t)$ is of exponential order $e^{\alpha t}$, there exist constants M and T that satisfy

$$-Me^{\alpha t} \le f(t) \le Me^{\alpha t}$$

or

$$e^{-st}\left(-Me^{\alpha t}\right) \le e^{-st}f(t) \le e^{-st}\left(Me^{\alpha t}\right)$$

for $t > T$. Rewriting this inequality as

$$-Me^{-(s-\alpha)t} \le e^{-st}f(t) \le Me^{-(s-\alpha)t}$$

we see that for t sufficiently large ($t \ge T$), the function $e^{-st}f(t)$ is bounded between two functions that both (for $s > \alpha$) approach 0 as $t \to \infty$. Hence $e^{-st}f(t)$ also approaches zero at $t \to \infty$. This proves the theorem.

In general we have the following result.

THEOREM 5.4: Laplace Transform of Higher Derivatives

If the functions $f, f', f'', \ldots, f^{(n-1)}$ are continuous on $[0, \infty)$ and $f^{(n)}$ is piecewise continuous, and if all these functions are of exponential order $e^{\alpha t}$, then the Laplace transforms of all these functions exist for $s > \alpha$; moreover,

$$\mathcal{L}\{f^{(n)}\} = s^n\mathcal{L}\{f\} - s^{n-1}f(0) - s^{n-2}f'(0) - s^{n-2}f''(0) - \cdots - f^{(n-1)}(0)$$

Roughly, the derivative property says that the Laplace transform transforms *differentiation* with respect to *t* into *multiplication* times *s*. This is the basic property that allows us to solve differential equations using the Laplace transform.

LAPLACE TRANSFORMS OF $\mathcal{L}\{e^{at}\, f(t)\}$ AND $\mathcal{L}\{t^n f(t)\}$

Using the definition of the Laplace transform to find Laplace transforms is about as impractical as using the definition of the derivative to find derivatives. In calculus, one develops rules of operation (product rule, power rule, chain rule, etc.) for finding derivatives without having to resort to the definition. We now introduce two properties that can be used to help find transforms. The first property is called the **translation property,** although a better name might be ''multiplication by e^{at} equals a translation in *s*.''

THEOREM 5.5: Translation Property

If the Laplace transform $F(s) = \mathcal{L}\{f\}$ exists for $s > \alpha$, then

$$\mathcal{L}\{e^{at}f(t)\} = F(s - a)$$

for $s > \alpha + a$.

PROOF: The translation property is simply a restatement of

$$\mathcal{L}\{e^{at}f(t)\} = \int_0^\infty e^{-st}e^{at}f(t)\, dt = \int_0^\infty e^{-(s-a)t}f(t)\, dt = F(s - a)$$

Note: In plain English, Theorem 5.5 says that the transform of e^{at} times a function can be obtained by taking the transform of the function and then replacing *s* in the transform by $s - a$.

Example 1 Translation Property Find $\mathcal{L}\{e^{at}\cos bt\}$.

Solution From Section 5.1 we know that

$$\mathcal{L}\{\cos bt\} = F(s) = \frac{s}{s^2 + b^2} \tag{2}$$

Using the translation property, we have

$$\mathcal{L}\{e^{at}\cos bt\} = F(s - a) = \frac{s - a}{(s - a)^2 + b^2} \tag{3}$$

∎

A rule that is useful for solving differential equations with variable coefficients is the following.

THEOREM 5.6: Laplace Transform $\mathcal{L}\{t^n f(t)\}$

If $f(t)$ is a piecewise continuous function on $[0, \infty)$ and of exponential order $e^{\alpha t}$, then for $s > \alpha$,

$$\mathcal{L}\{t^n f(t)\} = (-1)^n \frac{d^n F}{ds^n}(s)$$

where n is a positive integer.

PROOF: We can prove the result for $n = 1$ by using the fact that the derivative and integral can be interchanged. We have

$$\frac{d}{ds} F(s) = \frac{d}{ds} \int_0^\infty e^{-st} f(t)\, dt$$
$$= \int_0^\infty \frac{d}{ds} e^{-st} f(t)\, dt$$
$$= -\int_0^\infty e^{-st} t f(t)\, dt$$
$$= -\mathcal{L}\{t f(t)\}$$

The general result for arbitrary n can be obtained by induction.

Note: In plain English, Theorem 5.6 says that the transform of t^n times a function can be obtained by taking the nth derivative of the transform and then multiplying by -1 if n is odd.

Example 2 **Multiplication Times t** Find $\mathcal{L}\{t \cos bt\}$.

Solution Since

$$\mathcal{L}\{\cos bt\} = F(s) = \frac{s}{s^2 + b^2} \tag{4}$$

we have

$$\mathcal{L}\{t \cos bt\} = -\frac{d}{ds}\left(\frac{s}{s^2 + b^2}\right) = \frac{s^2 - b^2}{(s^2 + b^2)^2} \tag{5}$$

■

In Table 5.2 we list without proof several important properties of the Laplace transform.

Table 5.2
Properties of the Laplace
Transform, $\mathcal{L}\{f\} = F(s)$

1. $\mathcal{L}\{f + g\} = \mathcal{L}\{f\} + \mathcal{L}\{g\}$
2. $\mathcal{L}\{cf\} = c\mathcal{L}\{f\}$
3. $\mathcal{L}\{f'\} = s\mathcal{L}\{f\} - f(0)$
4. $\mathcal{L}\{f''\} = s^2\mathcal{L}\{f\} - sf(0) - f'(0)$
5. $\mathcal{L}\{f^{(n)}\} = s^n\mathcal{L}\{f\} - s^{n-1}f(0) - s^{n-2}f'(0) - \cdots - f^{(n-1)}(0)$
6. $\mathcal{L}\{e^{at}f(t)\} = F(s - a)$
7. $\mathcal{L}\{t^n f(t)\} = (-1)^n \dfrac{d^n}{ds^n} F(s)$
8. $\mathcal{L}\{f(at)\} = \dfrac{1}{a} F\left(\dfrac{s}{a}\right)$
9. $\mathcal{L}\left\{\displaystyle\int_0^t f(\tau)\, d\tau\right\} = \dfrac{1}{s} F(s)$
10. $\mathcal{L}\left\{\dfrac{f(t)}{t}\right\} = \displaystyle\int_s^\infty F(\xi)\, d\xi$
11. $\displaystyle\lim_{s\to\infty} sF(s) = f(0)$ (initial-value theorem)
12. $\displaystyle\lim_{s\to 0} sF(s) = f(\infty)*$ (final-value theorem)

* The notation $f(\infty)$ means the limiting value of $f(t)$ as $t \to \infty$.

Example 3   **Combination of Rules**   Find the transform

$$\mathcal{L}\left\{e^{-2t}\int_0^t \tau \sin \tau\, d\tau\right\}$$

Solution   Starting from the "inside," we compute the following Laplace transforms in succession, using rules from Table 5.2.

Step 1. $\mathcal{L}\{\sin \tau\} = \dfrac{1}{s^2 + 1}$

Step 2. $\mathcal{L}\{\tau \sin \tau\} = -\dfrac{d}{ds}\left\{\dfrac{1}{s^2 + 1}\right\} = \dfrac{2s}{(s^2 + 1)^2}$ (Property 7)

Step 3. $\mathcal{L}\left\{\displaystyle\int_0^t \tau \sin \tau\, d\tau\right\} = \dfrac{1}{s}\left(\dfrac{2s}{(s^2 + 1)^2}\right) = \dfrac{2}{(s^2 + 1)^2}$ (Property 9)

Step 4. $\mathcal{L}\left\{e^{-2t}\displaystyle\int_0^t \tau \sin \tau\, d\tau\right\} = \dfrac{2}{[(s + 2)^2 + 1]^2}$ (Property 6)

PROBLEMS: Section 5.2

For Problems 1–20, find the Laplace transform of the given function, using Table 5.1 from Section 5.1 and properties of the Laplace transform given in Table 5.2.

1. $at^2 + bt + c$
2. $t^2 + e^{2t} - 2$
3. $(t - 9)^2$
4. e^{2t-1}
5. $(1 + e^t)^2$
6. $3t \sin t$
7. $t^2 \sin 2t$
8. $e^{-2t} \sin 3t$
9. $5e^{5t} \cos 2t$
10. $t^2 e^{-3t}$
11. $te^t \cos t$
12. $t^2 e^t \sin t$

13. $\sin 2t \sinh 2t$
14. $\sin^2 t$
15. $\sinh 3t$
16. $\cos^3 t$
17. $t \sin^2 t$
18. $\cos mt \sin nt \quad (m \neq n)$
19. $\displaystyle\int_0^t \cos \tau \, d\tau$
20. $\displaystyle\int_0^t \sin 3\tau \, d\tau$

Multiplication by e^t

For Problems 21–24, use the "multiplication by an exponential" or translation property to find the given transform.

21. $\mathcal{L}\{e^{-t} \sin bt\}$
22. $\mathcal{L}\{e^{-2t} t^2\}$
23. $\mathcal{L}\{e^{2t} t\}$
24. $\mathcal{L}\{e^{-t} t^2\}$

Multiplication by t^n

For Problems 25–28, use the "multiplication by t^n property" to find the given transform.

25. $\mathcal{L}\{t \sin bt\}$
26. $\mathcal{L}\{t^2 \sin 2t\}$
27. $\mathcal{L}\{te^{-2t}\}$
28. $\mathcal{L}\{t^2 e^t\}$

29. **Multiplication by e^{at}** Use the transform of the sine function $\mathcal{L}\{\sin bt\} = b/(s^2 + b^2)$ to find $\mathcal{L}\{e^{at} \sin bt\}$.

30. **Using Two Rules** Show that $\mathcal{L}\{e^{at} t^n\} = n!/(s - a)^{n+1}$ in two different ways.

31. **Interesting Transform** Use the fact that $\mathcal{L}\{1\} = 1/s$ and one of the rules developed in this section to show that $\mathcal{L}\{t^n\} = n!/s^{n+1}$.

32. **Multiplication by cosh at** Prove the following property:

    (a) $\mathcal{L}\{\cosh at \cdot f(t)\} = \dfrac{1}{2}(F(s - a) + F(s + a))$ (6)

    (b) Use the property in part (a) to find $\mathcal{L}\{t^3 \cosh 2t\}$.

    (c) Use the property in part (a) to find $\mathcal{L}\{\sin t \cdot \cosh t\}$.

33. **Multiplication by sinh at** Prove the following property:

    (a) $\mathcal{L}\{\sinh at \cdot f(t)\} = \dfrac{1}{2}(F(s - a) - F(s + a))$

    (b) Use the property in part (a) to find $\mathcal{L}\{t^2 \sinh 3t\}$.

    (c) Use the property in part (a) to find $\mathcal{L}\{\cos t \cdot \sinh t\}$.

34. **Combining Properties** Use the given properties

$\mathcal{L}\{tf(t)\} = -\dfrac{dF(s)}{ds}$ and $\mathcal{L}\{f'(t)\} = s\,\mathcal{L}\{f\} - f(0)$ to show that

$$\mathcal{L}\{tf'(t)\} = -s\frac{dF(s)}{ds} - F(s)$$

35. **Change of Scale Property**

    (a) Verify Property 8, the **change of scale property** from Table 5.2:

    $$\mathcal{L}\{f(at)\} = \frac{1}{a}F\left(\frac{s}{a}\right)$$

    by replacing s by s/a in the Laplace transform integral. Use this rule to find the following transforms:

    (b) $\mathcal{L}\{\sin 2t\}$
    (c) $\mathcal{L}\{e^{5t}\}$

36. **Transform of an Integral** Use Property 9 from Table 5.2 to find the following transforms. Note that Property 9 says loosely that integration in the "t-domain" is the same as division by s in the "s-domain."

    (a) Find $\mathcal{L}\left\{\displaystyle\int_0^t \sin \tau \, d\tau\right\}$.

    (b) Find $\mathcal{L}\left\{\displaystyle\int_0^t e^{2\tau} \, d\tau\right\}$.

37. **Division by t** Use Property 10 from Table 5.2 to find

    $$\mathcal{L}\left\{\frac{\sin t}{t}\right\}$$

    Note that Property 10 says loosely that division by t in the "t-domain" is the same as integration in the "s-domain."

Some Harder Transforms

38. $\mathcal{L}\left\{e^t \displaystyle\int_0^t \tau e^{-2\tau} \, d\tau\right\}$

39. $\mathcal{L}\left\{te^{-2t} \displaystyle\int_0^t \cos 4\tau \, d\tau\right\}$

40. $\mathcal{L}\left\{\displaystyle\int_0^t \tau e^{-\tau} \sin \tau \, d\tau\right\}$

41. $\mathcal{L}\left\{\displaystyle\int_0^t \tau^2 e^{-\tau} \sin \tau \, d\tau\right\}$

42. $\mathcal{L}\left\{\displaystyle\int_0^t \left(\displaystyle\int_0^u \xi e^{-\xi} \, d\xi\right) d\tau\right\}$

43. $\mathcal{L}\left\{e^t \displaystyle\int_0^t e^{-\tau} \, d\tau\right\}$

44. **Journal Entry** Generally, operations in the "t-domain" are analogous to operations in the "s-domain." For instance, differentiation of a function with respect to t in the t-domain (time domain) is "almost" analogous to multiplication of the Laplace transform by s in the s-domain (frequency domain). Spend some time summarizing these "analogous" relations.

5.3 THE INVERSE LAPLACE TRANSFORM

PURPOSE
To introduce the inverse Laplace transform $\mathcal{L}^{-1}\{F(s)\}$ and show how it can be found by using tables with the help of some algebraic manipulations, such as the partial fraction decomposition.

DEFINITION OF THE INVERSE LAPLACE TRANSFORM

Until now, we have been concerned with finding the Laplace transform $F(s)$ of a given function $f(t)$. We now turn the problem around; that is, we find the function $f(t)$ whose transform is a given function $F(s)$. If there is a unique $f(t)$ for a given $F(s)$ that satisfies $\mathcal{L}\{f\} = F(s)$, then $f(t)$ is called the inverse Laplace transform of $F(s)$, and we write

$$f(t) = \mathcal{L}^{-1}\{F(s)\} \tag{1}$$

See Figure 5.6.

Figure 5.6
The inverse Laplace transform

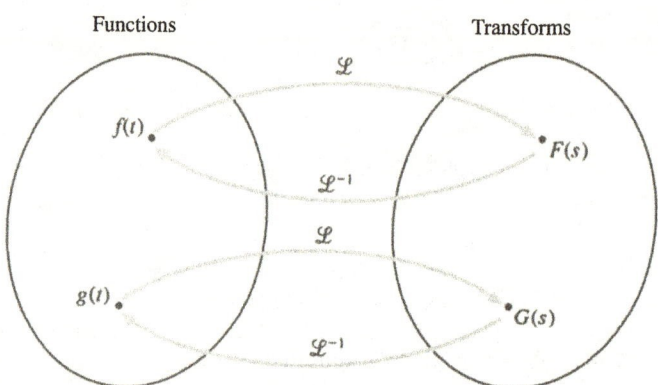

There is good news and bad news when it comes to the inverse Laplace transform. The bad news is that the inverse Laplace transform might not be unique. In other words, it may be possible that $\mathcal{L}\{f(t)\} = \mathcal{L}\{g(t)\}$ but yet $f \neq g$. (See Problem 27 at the end of this section.) However, the good news is that if f and g are *piecewise continuous functions* of *exponential order* and if $\mathcal{L}\{f(t)\} = \mathcal{L}\{g(t)\}$, then f and g are *for all practical purposes* the same; that is, they can differ only at points of discontinuity. So we simply select any one of those functions as the inverse Laplace transform, and if one of them is continuous, we select the continuous one. (See Problem 28 at the end of this section.) The more common situation occurs when f and g are continuous. In this case, if $\mathcal{L}\{f(t)\} = \mathcal{L}\{g(t)\}$, then $f = g$. These ideas are summarized in the following definition.

DEFINITION: Inverse Laplace Transform

The **inverse Laplace transform** of a function $F(s)$ is the unique continuous function f that satisfies

$$\mathcal{L}\{f\} = F(s)$$

which is denoted by $\mathcal{L}^{-1}\{F(s)\} = f(t)$. In case *all* functions f that satisfy $\mathcal{L}\{f\} = F(s)$ are discontinuous on $[0, \infty)$, select *any* piecewise continuous function that satisfies this condition to be $\mathcal{L}^{-1}\{F(s)\}$.

FINDING THE INVERSE LAPLACE TRANSFORM

We now get down to the operational problem of actually finding the inverse Laplace transform of a given function. Can we solve for $f(t)$ in the definition of the Laplace transform as a function of $F(s)$? If we could turn this integral formula around, we would have a formula that goes "backward." Although there *is* an inverse formula for finding $f(t)$ in terms of $F(s)$, the formula involves a *contour integral* over a region in the complex plane.* However, it is possible to avoid complex variable theory by simply using a table of Laplace transforms, such as Table 5.1 in Section 5.1. To find the inverse $f(t) = \mathcal{L}^{-1}\{F(s)\}$, simply locate $F(s)$ under the transform column and read off the corresponding function in the $f(t)$ column. Table 5.3 shows the results of using this strategy to find inverse transforms.

Table 5.3
Table of Inverse Laplace Transforms

Common inverse Laplace transforms

1. $\mathcal{L}^{-1}\left\{\dfrac{1}{s}\right\} = 1$

2. $\mathcal{L}^{-1}\left\{\dfrac{n!}{s^{n+1}}\right\} = t^n$

3. $\mathcal{L}^{-1}\left\{\dfrac{1}{s-a}\right\} = e^{at}$

4. $\mathcal{L}^{-1}\left\{\dfrac{b}{s^2 + b^2}\right\} = \sin bt$

5. $\mathcal{L}^{-1}\left\{\dfrac{s}{s^2 + b^2}\right\} = \cos bt$

6. $\mathcal{L}^{-1}\left\{\dfrac{b}{s^2 - b^2}\right\} = \sinh bt$

7. $\mathcal{L}^{-1}\left\{\dfrac{s}{s^2 - b^2}\right\} = \cosh bt$

---

* Interested readers can consult *Complex Variables and Applications by Mathematics* (Third Edition) by R. C. Churchill, J. W. Brown, and R. S. Verhey (McGraw-Hill, New York, 1974), for an in-depth discussion on the inverse Laplace transform. The contour integral formula that defines the inverse Laplace transform is $f(t) = \dfrac{1}{2\pi i} \displaystyle\int F(s)e^{st}\, ds$, where s is taken to be a complex variable and the integral is a contour integral in the complex plane.

Example 1 **Simple Inverses** Find the inverse Laplace transforms of the following functions.

(a) $F(s) = \dfrac{6}{s^4}$

(b) $F(s) = \dfrac{4}{s^2 + 16}$

(c) $F(s) = \dfrac{1}{s^2 + 6s + 10}$

Solution Using Table 5.3, we obtain

(a) $\mathcal{L}^{-1}\left\{\dfrac{6}{s^4}\right\} = t^3$

(b) $\mathcal{L}^{-1}\left\{\dfrac{4}{s^2 + 16}\right\} = \sin 4t$

(c) $\mathcal{L}^{-1}\left\{\dfrac{1}{s^2 + 6s + 10}\right\} = \mathcal{L}^{-1}\left\{\dfrac{1}{(s + 3)^2 + 1}\right\} = e^{-3t}\sin t$

Often, the inverse transform $F(s)$ cannot be found in any table. In these cases some preliminary algebraic manipulations are required. One property that is particularly useful in this regard is the property of linearity. It is stated here without proof.

> **THEOREM 5.7:** Linearity of the Inverse Laplace Transform
>
> If two inverse Laplace transforms $\mathcal{L}^{-1}\{F\}$ and $\mathcal{L}^{-1}\{G\}$ exist, then
>
> - $\mathcal{L}^{-1}\{F + G\} = \mathcal{L}^{-1}\{F\} + \mathcal{L}^{-1}\{G\}$
> - $\mathcal{L}^{-1}\{cF\} = c\mathcal{L}^{-1}\{F\}$
>
> These conditions state that the inverse Laplace transform is a linear transformation

It is often necessary to adjust the constants so that the expressions correspond to the ones in the table. The following example illustrates this idea.

Example 2 **Preliminary Algebraic Manipulation** Find the inverse transform

$$\mathcal{L}^{-1}\left\{\dfrac{2}{s - 1} - \dfrac{3}{s^2 + 5}\right\} \tag{2}$$

Solution Using the property of linearity, we have

$$\mathscr{L}^{-1}\left\{\frac{2}{s-1} - \frac{3}{s^2+5}\right\} = 2\mathscr{L}^{-1}\left\{\frac{1}{s-1}\right\} - \frac{3}{\sqrt{5}}\mathscr{L}^{-1}\left\{\frac{\sqrt{5}}{s^2+5}\right\}$$

$$= 2e^t - \frac{3}{\sqrt{5}}\sin\sqrt{5}t \qquad (3)$$

■

PARTIAL FRACTION DECOMPOSITION

If $P(s)$ and $Q(s)$ are polynomials in s, and if the degree of $P(s)$ is less than the degree of $Q(s)$, then $\mathscr{L}^{-1}\{P(s)/Q(s)\}$ exists and can be found by first writing $P(s)/Q(s)$ as its **partial fraction decomposition.** Since most readers have already encountered this technique in a course in calculus, we simply state the important facts related to the subject.

Rules for Partial Fraction Decomposition

$$\frac{P(s)}{Q(s)} = \text{terms of the partial fraction decomposition}$$

1. *Linear Factor:* For each factor $(as + b)$ in the denominator of $Q(s)$, there corresponds a term of the form

$$\frac{A}{as + b} \qquad (4)$$

in the partial fraction decomposition.

2. *Power of Linear Factor:* For each power $(as + b)^n$ of a linear factor in the denominator of $Q(s)$, there correspond terms of the form

$$\frac{A_1}{as + b} + \frac{A_2}{(as + b)^2} + \cdots + \frac{A_n}{(as + b)^n} \qquad (5)$$

in the partial fraction decomposition.

3. *Quadratic Factor:* For each quadratic factor $(as^2 + bs + c)$ in the denominator $Q(s)$, there corresponds term(s) of the form

$$\frac{As + B}{as^2 + bs + c} \qquad (6)$$

in the partial fraction decomposition.

4. *Power of Quadratic Factor:* For each power $(as^2 + bs + c)^n$ of a quadratic factor in the denominator $Q(s)$, there correspond terms of the form

$$\frac{A_1 s + B_1}{as^2 + bs + c} + \frac{A_2 s + B_2}{(as^2 + bs + c)^2} + \cdots + \frac{A_n s + B_n}{(as^2 + bs + c)^n} \qquad (7)$$

in the partial fraction decomposition.

Example 3 **Two Linear Factors** Find

$$\mathscr{L}^{-1}\left\{\frac{1}{s(s + 1)}\right\} \tag{8}$$

Solution Since the denominator contains two linear factors, we write

$$\frac{1}{s(s + 1)} = \frac{A}{s} + \frac{B}{s + 1} \tag{9}$$

Multiplying by $s(s + 1)$ gives

$$A(s + 1) + Bs = 1 \tag{10}$$

We find A and B by setting $s = 0$ and $s = -1$ one at a time and solving the resulting equation for A and B. Doing this, we find that $A = 1$ and $B = -1$. Hence the partial fraction decomposition is

$$\frac{1}{s(s + 1)} = \frac{1}{s} - \frac{1}{s + 1} \tag{11}$$

Hence

$$\begin{aligned}
\mathscr{L}^{-1}\left\{\frac{1}{s(s + 1)}\right\} &= \mathscr{L}^{-1}\left\{\frac{1}{s} - \frac{1}{s + 1}\right\} \\
&= \mathscr{L}^{-1}\left\{\frac{1}{s}\right\} - \mathscr{L}^{-1}\left\{\frac{1}{s + 1}\right\} \\
&= 1 - e^{-t}
\end{aligned} \tag{12}$$

Example 4 **One Linear and One Quadratic Factor** Find

$$\mathscr{L}^{-1}\left\{\frac{1}{s(s^2 + 1)}\right\} \tag{13}$$

Solution Since the denominator contains one linear factor and one quadratic factor, we try the decomposition

$$\frac{1}{s(s^2 + 1)} = \frac{A}{s} + \frac{Bs + C}{s^2 + 1} \tag{14}$$

Multiplying by $s(s^2 + 1)$, we find

$$A(s^2 + 1) + (Bs + C)s = 1$$

or

$$(A + B)s^2 + Cs + A = 1$$

We now use the fact that this expression is an identity only if the corresponding coefficients of 1, s, and s^2 are equal.* Hence we have

$$s^2: \quad A + B = 0$$
$$s: \quad C = 0$$
$$1: \quad A = 1$$

or $A = 1$, $B = -1$, and $C = 0$. Hence we have

$$\mathcal{L}^{-1}\left\{\frac{1}{s(s^2 + 1)}\right\} = \mathcal{L}^{-1}\left\{\frac{1}{s} + \frac{-s}{s^2 + 1}\right\}$$
$$= \mathcal{L}^{-1}\left\{\frac{1}{s}\right\} - \mathcal{L}^{-1}\left\{\frac{s}{s^2 + 1}\right\}$$
$$= 1 - \cos t \tag{15}$$

∎

Example 5 **Power of Linear Factor** Find

$$\mathcal{L}^{-1}\left\{\frac{s^2}{(s - 1)^3}\right\} \tag{16}$$

Solution Since the denominator contains a repeated linear factor, we try the partial fraction decomposition

$$\frac{s^2}{(s - 1)^3} = \frac{A}{s - 1} + \frac{B}{(s - 1)^2} + \frac{C}{(s - 1)^3} \tag{17}$$

Hence

$$A(s - 1)^2 + B(s - 1) + C = s^2 \tag{18}$$

As before, we can collect terms involving 1, s, and s^2 and equate coefficients. However, it is more convenient to substitute $s = 1$ into the equation and solve for $C = 1$. We now seem to have run out of convenient values of s to substitute. But remember that whenever two functions are equal, so are their derivatives. Hence we differentiate Eq. (18) with respect to s, getting

$$2A(s - 1) + B = 2s \tag{19}$$

Letting $s = 1$ now gives $B = 2$. Differentiating Eq. (19) again gives the equation $2A = 2$ or $A = 1$. Hence we have the partial fraction decomposition

$$\frac{s^2}{(s - 1)^3} = \frac{1}{s - 1} + \frac{2}{(s - 1)^2} + \frac{1}{(s - 1)^3}$$

---

* This statement is a consequence of the fact that the functions 1, s, and s^2 are *linearly independent* on the interval where the equation holds.

Hence

$$\mathcal{L}^{-1}\left\{\frac{s^2}{(s-1)^3}\right\} = \mathcal{L}^{-1}\left\{\frac{1}{s-1}\right\} + \mathcal{L}^{-1}\left\{\frac{2}{(s-1)^2}\right\} + \mathcal{L}^{-1}\left\{\frac{1}{(s-1)^3}\right\}$$

$$= e^t + 2te^t + \frac{1}{2}t^2e^t \tag{20}$$

Note: Differentiating works well with repeating linear factors.

Example 6   Completing the Square   Find

$$\mathcal{L}^{-1}\left\{\frac{1}{s^2 + 2s + 5}\right\} \tag{21}$$

Solution   Since the discriminant of the quadratic expression is negative, it cannot be factored into real linear factors. Hence we complete the square, getting

$$s^2 + 2s + 5 = s^2 + 2s + 1 + 4 = (s+1)^2 + 4$$

Hence

$$\mathcal{L}^{-1}\left\{\frac{1}{s^2 + 2s + 5}\right\} = \mathcal{L}^{-1}\left\{\frac{1}{(s+1)^2 + 4}\right\} = \frac{1}{2}e^{-t}\sin 2t \tag{22}$$

We summarize some of the properties of the inverse Laplace transform in Table 5.4.

Table 5.4
Properties of the Inverse
Laplace Transform

1. $\mathcal{L}^{-1}\{F + G\} = \mathcal{L}^{-1}\{F\} + \mathcal{L}^{-1}\{G\}$

2. $\mathcal{L}^{-1}\{cF\} = c\mathcal{L}^{-1}\{F\}$

3. $\mathcal{L}^{-1}\{F(s)\} = \dfrac{(-1)^n}{t^n} \mathcal{L}^{-1}\left\{\dfrac{d^n F(s)}{ds^n}\right\}$

4. $\mathcal{L}^{-1}\left\{\dfrac{F(s)}{s}\right\} = \displaystyle\int_0^t f(\tau)\, d\tau$

5. $\mathcal{L}^{-1}\left(F(s-a)\right) = e^{at}f(t)$

PROBLEMS: Section 5.3

For Problems 1–20, find the inverse Laplace transform of the given function.

1. $\dfrac{1}{s^3}$

2. $\dfrac{2}{s} + \dfrac{3}{s-1} + \dfrac{7}{s^3}$

3. $\dfrac{5}{s^2 + 3}$

4. $\dfrac{3}{s-3} + \dfrac{4}{s+3}$

5. $\dfrac{1}{s^2 + 3s}$

6. $\dfrac{s+1}{s^2 + 2s + 10}$

7. $\dfrac{1}{s^2 + 4s + 4}$

8. $\dfrac{3s+5}{s^2 - 6s + 25}$

9. $\dfrac{s+1}{s^2+s-2}$

10. $\dfrac{5}{s^2+s-6}$

11. $\dfrac{2s+4}{s^2-1}$

12. $\dfrac{7}{(s+2)^2+3}$

13. $\dfrac{2s+16}{s^2+4s+13}$

14. $\dfrac{6}{(s+2)^4}$

15. $\dfrac{7s^2+23s+30}{(s-2)(s^2+2s+5)}$

16. $\dfrac{4}{s^2(s^2+4)}$

17. $\dfrac{3}{(s^2+1)(s^2+4)}$

18. $\dfrac{7s-1}{(s+1)(s+2)(s-3)}$

19. $\dfrac{s^2-2}{s^3+8s^2+7s}$

20. $\dfrac{s^2+9s+2}{(s-1)^2(s+3)}$

The Inverse of Derivatives

For Problems 21–22, use Property 3 from Table 5.4 to find the given inverse Laplace transforms.

21. $\mathscr{L}^{-1}\left\{\tan^{-1}\left(\dfrac{1}{s}\right)\right\}$

22. $\mathscr{L}^{-1}\left\{\ln\left(\dfrac{s-a}{s-b}\right)\right\}$

Division by s Equals Integration in t

For Problems 23–26, use Property 4 from Table 5.4 to find the inverse transform of the given function.

23. $F(s)=\dfrac{1}{s^2}$

24. $F(s)=\dfrac{1}{s(s-1)}$

25. $F(s)=\dfrac{1}{s(s^2+1)}$

26. $F(s)=\dfrac{1}{s(s-1)^2}$

27. Nonunique Inverse Laplace Transform Show that both functions

$$f(t)=e^t \qquad (0\le t<\infty)$$

$$g(t)=\begin{cases}e^t & (t\ne 1)\\ 5 & (t=1)\end{cases}$$

have the same Laplace transform of $1/(s-1)$. Hence the function $1/(s-1)$ does not have a unique inverse. According to the definition of the inverse Laplace transform in the text, which inverse would you pick as the "official" inverse?

28. No Continuous Inverse Transform Show that both functions

$$LC(t)=\begin{cases}0 & (0\le t\le 1)\\ 1 & (1<t)\end{cases}$$

$$RC(t)=\begin{cases}0 & (0\le t<1)\\ 1 & (1\le t)\end{cases}$$

have the same Laplace transform e^{-s}. This particular function $F(s)=e^{-s}$ is a function that has no continuous inverse Laplace transform.

  (a) Which function, *LC* or *RC*, would you pick as the "official" Laplace transform?

  (b) Does it really make any difference which is chosen as the official Laplace transform?

  (c) Are there any more discontinuous inverse transforms of e^{-s}? That is, are there any more functions $h(t)$ that have e^{-s} as its transform?

29. Heaviside Expansion Formula It is possible to find a formula for the inverse transform of a given rational function $P(s)/Q(s)$, where the degree of $P(s)$ is less than the degree of $Q(s)$. Calling n the degree of $Q(s)$, we assume that the roots $r_1, r_2, \dots, r_n$ of $Q(s)$ are real and distinct. Expanding $P(s)/Q(s)$ in terms of partial fractions, we have

$$\frac{P(s)}{Q(s)}=\frac{A_1}{s-r_1}+\frac{A_2}{s-r_2}+\cdots+\frac{A_n}{s-r_n} \qquad (23)$$

where $A_1, A_2, \dots$ are undetermined coefficients that can be found.

  (a) Show that $A_k=P(r_k)/R(r_k)$ where $R(s)=Q(s)/(s-r_k)$. *Hint:* Multiply Eq. (23) by $s-r_k$ and let $s\to r_k$.

  (b) Prove the **Heaviside expansion formula**

$$\mathscr{L}^{-1}\{F(s)\}=\sum_{k=1}^{n}\frac{P(r_k)}{Q'(r_k)}e^{r_k t} \qquad (24)$$

30. Using the Heaviside Expansion Formula Use the Heaviside expansion formula (24) to find the inverse Laplace transform of

$$F(s)=\frac{2s^2-2s+6}{s^3+6s^2+11s+6}$$

31. Inverse Laplace Transform Using Residue Theory Suppose $P(s)/Q(s)$ is a rational function with the degree of $P(s)$ less than the degree of $Q(s)$. If r_0 is a nonrepeated root of $Q(s)$, then it can easily be shown that $P(s)/Q(s)$ has a partial fraction expansion of the form

$$\frac{P(s)}{Q(s)}=\frac{A}{s-r_0}+\cdots \qquad (25)$$

The important quantity in (25) is the constant A, which is called the **residue** of $P(s)/Q(s)$ at r_0.

  (a) Show that the residue A can be found from the formula

$$A=\lim_{s\to r_0}\left(\frac{(s-r_0)\,P(s)}{Q(s)}\right) \qquad (26)$$

  (b) Use the residue formula (26) to find the partial fraction decomposition for

$$F(s)=\frac{s+7}{(s+1)(s-2)}$$

32. Journal Entry Summarize the concept of the inverse Laplace transform.

5.4 INITIAL-VALUE PROBLEMS

PURPOSE
To show how the initial-value problem

$$ay'' + by' + cy = f(t)$$
$$y(0) = y_0$$
$$y'(0) = y_0'$$

can be solved by transforming the differential equation into an algebraic equation in $Y(s) = \mathcal{L}\{y\}$, which can then be solved by using simple algebra. We then show how the inverse Laplace transform $\mathcal{L}^{-1}\{Y(s)\}$ can be found, obtaining the solution $y(t)$.

Until now we have solved differential equations using a variety of techniques: integrating factor methods, trial-and-error methods, variation of parameters, transformations to easier problems, reduction of order methods, and so on. We now show how the Laplace transform can be used to transform differential equations into algebraic equations, thereby transforming equations that are difficult to solve into equations that are easy to solve. The general idea consists of three basic steps.

DEFINITION: Steps in Solving Differential Equations

Step 1. Take the Laplace transform of each side of the given differential equation, obtaining an algebraic equation in the transform of the solution $\mathcal{L}\{y\}$. The initial conditions for the problem will be included in the algebraic equations.

Step 2. Solve the algebraic equation obtained in Step 1 for the transform $\mathcal{L}\{y\}$ of the unknown solution.

Step 3. Find the inverse transform to find the solution $y(t)$.

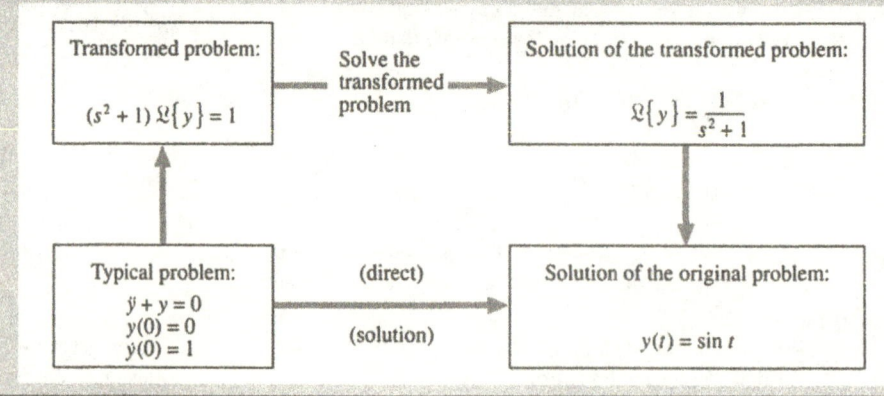

Example 1 **First-Order Equation** Solve the initial-value problem

$$y' + 3y = e^{-t}$$
$$y(0) = 1 \tag{1}$$

Solution Since the differential equation is an identity between two functions of t, their transforms are also equal. Hence

$$\mathcal{L}\{y' + 3y\} = \mathcal{L}\{e^{-t}\}$$

By the linearity of the Laplace transform, we have

$$\mathcal{L}\{y'\} + 3\mathcal{L}\{y\} = \mathcal{L}\{e^{-t}\}$$

Using the derivative property $\mathcal{L}\{y'\} = s\mathcal{L}\{y\} - y(0)$ and substituting the initial condition $y(0) = 1$ into the above equation, we find

$$\left(s\mathcal{L}\{y\} - 1\right) + 3\mathcal{L}\{y\} = \frac{1}{s+1}$$

Solving for $\mathcal{L}\{y\}$ gives

$$\mathcal{L}\{y\} = \frac{s+2}{(s+1)(s+3)}$$
$$= \frac{1}{s+3} + \frac{1}{(s+1)(s+3)}$$
$$= \frac{1}{s+3} + \left(\frac{1}{2(s+1)} - \frac{1}{2(s+3)}\right)$$
$$= \mathcal{L}\{e^{-3t}\} + \frac{1}{2}\mathcal{L}\{e^{-t}\} - \frac{1}{2}\mathcal{L}\{e^{-3t}\}$$

Hence we have the inverse

$$y(t) = e^{-3t} + \frac{1}{2}e^{-t} - \frac{1}{2}e^{-3t}$$
$$= \frac{1}{2}\left(e^{-t} + e^{-3t}\right) \tag{2}$$

$\blacksquare$

Example 2 **Use of the Laplace Transform** Solve the initial-value problem

$$y'' + 4y = \sin t$$
$$y(0) = 0 \tag{3}$$
$$y'(0) = 1$$

Solution Taking the Laplace transform of both sides of the differential equation gives

$$\mathcal{L}\{y'' + 4y\} = \mathcal{L}\{\sin t\}$$
$$\mathcal{L}\{y''\} + 4\mathcal{L}\{y\} = \mathcal{L}\{\sin t\}$$

Using the derivative formulas

$$\mathcal{L}\{y'\} = s\mathcal{L}\{y\} - y(0)$$
$$\mathcal{L}\{y''\} = s^2\mathcal{L}\{y\} - sy(0) - y'(0)$$

we obtain

$$s^2\mathcal{L}\{y\} - sy(0) - y'(0) + 4\mathcal{L}\{y\} = \frac{1}{s^2 + 1} \tag{4}$$

Substituting the initial conditions $y(0) = 0$, $y'(0) = 1$ into Eq. (4) and using the more suggestive notation $Y(s) = \mathcal{L}\{y\}$, we obtain

$$s^2 Y(s) - 1 + 4Y(s) = \frac{1}{s^2 + 1}$$

Solving for $Y(s)$ gives

$$Y(s) = \frac{s^2 + 2}{(s^2 + 1)(s^2 + 4)} = \frac{As + B}{s^2 + 1} + \frac{Cs + D}{s^2 + 4} \tag{5}$$

To find A, B, C, and D, we multiply by $(s^2 + 1)(s^2 + 4)$ and collect powers of s, getting

$$(A + C)s^3 + (B + D)s^2 + (4A + C)s + (4B + D) = s^2 + 2 \tag{6}$$

Equating like powers of s gives

$$A + C = 0$$
$$B + D = 1$$
$$4A + C = 0$$
$$4B + D = 2$$

or $A = C = 0$, $B = \frac{1}{3}$, $D = \frac{2}{3}$. Hence we have

$$Y(s) = \frac{1}{3(s^2 + 1)} + \frac{2}{3(s^2 + 4)}$$
$$= \frac{1}{3}\mathcal{L}\{\sin t\} + \frac{2}{3}\mathcal{L}\{\sin 2t\}$$

Hence the solution is

$$y(t) = \frac{1}{3}\sin t + \frac{2}{3}\sin 2t \tag{7}$$

The graph of $y(t)$ is shown in Figure 5.7.

Figure 5.7
Graph of $y(t)$

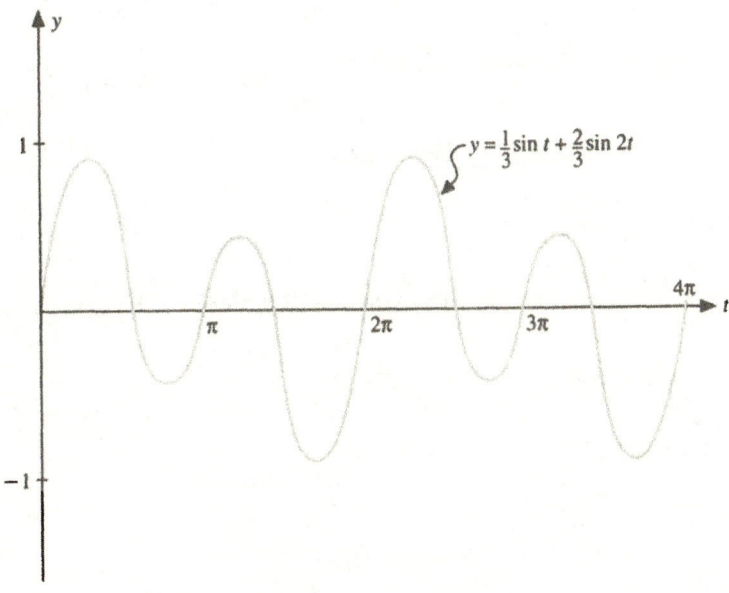

$$y = \tfrac{1}{3}\sin t + \tfrac{2}{3}\sin 2t$$

Example 3   Differential Equation with Damping   Solve the initial-value problem

$$y'' + 3y' + 2y = e^{-3t}$$
$$y(0) = 0 \qquad\qquad (8)$$
$$y'(0) = 1$$

Solution   Taking the Laplace transform, we have

$$\left(s^2\mathscr{L}\{y\} - sy(0) - y'(0)\right) + 3\left(\mathscr{L}\{y\} - y(0)\right) + 2\mathscr{L}\{y\} = \frac{1}{s+3} \qquad (9)$$

Substituting the initial values and solving for $\mathscr{L}\{y\}$ give

$$\mathscr{L}\{y\} = \frac{s+4}{(s+3)(s^2+3s+2)}$$
$$= \frac{s+4}{(s+3)(s+2)(s+1)}$$
$$= \frac{A}{s+3} + \frac{B}{s+2} + \frac{C}{s+1} \qquad (10)$$

Finding the undetermined coefficients gives $A = \frac{1}{2}$, $B = -2$, $C = \frac{3}{2}$. Hence we have

$$\mathcal{L}\{y\} = \frac{1}{2(s + 3)} - \frac{2}{s + 2} + \frac{3}{2(s + 1)}$$

$$= \frac{1}{2}\mathcal{L}\{e^{-3t}\} - 2\mathcal{L}\{e^{-2t}\} + \frac{3}{2}\mathcal{L}\{e^{-t}\}$$

Hence we have

$$y(t) = \frac{1}{2}e^{-3t} - 2e^{-2t} + \frac{3}{2}e^{-t} \tag{11}$$

The graph of $y(t)$ is shown in Figure 5.8.

Figure 5.8
Graph of $y(t)$

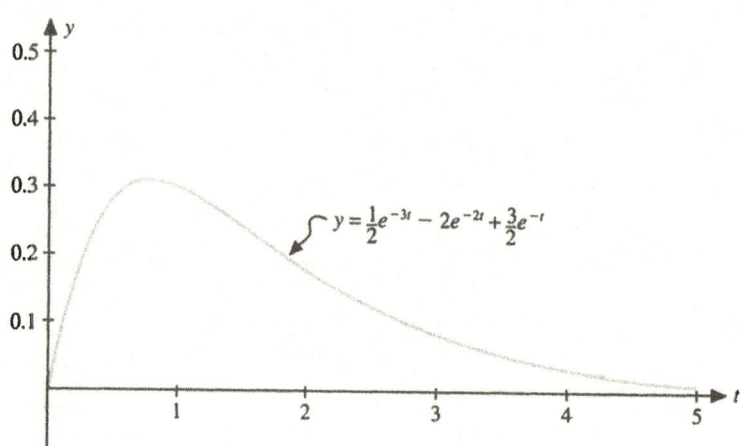

Although the Laplace transform is used primarily to solve differential equations with constant coefficients, there are some equations with variable coefficients that can be solved by using transform methods.

Example 4 **Differential Equations with Variable Coefficients** Solve the initial-value problem

$$\begin{aligned} y'' + ty' - y &= 0 \\ y(0) &= 0 \\ y'(0) &= 1 \end{aligned} \tag{12}$$

Solution Taking the transform of each side of Eq. (12), we have

$$\mathcal{L}\{y''\} + \mathcal{L}\{ty'\} - \mathcal{L}\{y\} = 0 \tag{13}$$

By using rules for the Laplace transform, the first two terms become

$$\mathcal{L}\{y''\} = s^2\mathcal{L}\{y\} - sy(0) - y'(0)$$
$$= s^2\mathcal{L}\{y\} - 1$$

$$\mathcal{L}\{ty'\} = -\frac{d}{ds}\mathcal{L}\{y'\}$$

$$= -\frac{d}{ds}\Big(sY(s) - y(0)\Big)$$

$$= -sY'(s) - Y(s) \tag{14}$$

Substituting these expressions into Eq. (13), we find

$$\Big(s^2Y(s) - 1\Big) + \Big(-sY'(s) - Y(s)\Big) - Y(s) = 0$$

or

$$Y'(s) + \Big(\frac{2}{s} - s\Big)Y(s) = -\frac{1}{s} \tag{15}$$

Equation (15) is a first-order linear equation that can be solved by finding an integrating factor. In this case the integrating factor is

$$\mu(t) = \exp\left\{\int\Big(\frac{2}{s} - s\Big)ds\right\} = \exp\Big(2\ln s - \frac{1}{2}s^2\Big) = s^2 e^{-s^2/2} \tag{16}$$

Hence Eq. (15) becomes

$$\frac{d}{ds}\left\{s^2 e^{-s^2/2}Y(s)\right\} = -se^{-s^2/2}$$

Integrating, we get

$$Y(s) = \frac{1}{s^2} + \frac{C}{s^2}e^{-s^2/2} \tag{17}$$

where C is an arbitrary constant. However, from the initial-value theorem (Property 11, Table 5.2), we have that $sY(s)$ tends toward zero as $s \to \infty$. However, the only way that this is possible is for $C = 0$. Hence we have $Y(s) = 1/s^2$, and so the solution is

$$y(t) = t \tag{18}$$

∎

Note: Any linear differential equation in $y(t)$ with polynomial coefficients in t can be transformed into a linear differential equation in $Y(s)$ whose coefficients are polynomials in s. If the coefficients of the given differential equation are first-order polynomials in t, then the differential equation in $Y(s)$ will be a *first-order* equation. Since we can solve a first-order equation, the major difficulty will be in finding the inverse transform.

PROBLEMS: Section 5.4

For Problems 1–14, use the Laplace transform to solve the given initial-value problem.

1. $y' = 1$ — $y(0) = 1$
2. $y' - y = 0$ — $y(0) = 1$
3. $y' - y = e^t$ — $y(0) = 1$
4. $y' + y = e^{-t}$ — $y(0) = 1$
5. $y'' = e^t$ — $y(0) = 1$ $y'(0) = 0$
6. $y'' - 3y' + 2y = 0$ — $y(0) = 1$ $y'(0) = 0$
7. $y'' + 2y' = 4$ — $y(0) = 1$ $y'(0) = -4$
8. $y'' + 9y = 20e^{-t}$ — $y(0) = 0$ $y'(0) = 1$
9. $y'' + 9y = \cos 3t$ — $y(0) = 1$ $y'(0) = -1$
10. $y'' + 4y = 4$ — $y(0) = 1$ $y'(0) = 0$
11. $y'' - 2y' + 5y = 0$ — $y(0) = 2$ $y'(0) = 4$
12. $y'' + 10y' + 25y = 0$ — $y(0) = 0$ $y'(0) = 10$
13. $y'' + 3y' + 2y = 6$ — $y(0) = 0$ $y'(0) = 2$
14. $y'' + y = \sin t$ — $y(0) = 2$ $y'(0) = -1$

For Problems 15–17, solve the given higher-order initial-value problem using the method of Laplace transforms.

15. $y''' - y'' - y' + y = 6e^t$
    $y(0) = 0$ $y'(0) = 0$ $y''(0) = 0$

16. $y''' + 2y'' + y' + 2y = 2$
    $y(0) = 3$ $y'(0) = -2$ $y''(0) = 3$

17. $y^{(4)} - y = 0$
    $y(0) = 1$ $y'(0) = 0$ $y''(0) = -1$ $y'''(0) = 0$

Finding Laplace Transforms by Taylor Series

Often a Laplace transform can be found by taking the Laplace transform of each term in the Taylor series of the function and then writing the sum of the transforms in closed algebraic form. For Problems 18–20, verify the indicated Laplace transform using the given Taylor series representation of the function.

18. $\sin t = t - \dfrac{t^3}{3!} + \dfrac{t^5}{5!} - \dfrac{t^7}{7!} + \cdots,$ $\mathcal{L}\{\sin t\} = \dfrac{1}{s^2 + 1}$

19. $\cos t = 1 - \dfrac{t^2}{2!} + \dfrac{t^4}{4!} - \dfrac{t^6}{6!} + \cdots,$ $\mathcal{L}\{\cos t\} = \dfrac{s}{s^2 + 1}$

20. $e^t = 1 + t + \dfrac{t^2}{2!} + \dfrac{t^3}{3!} + \cdots,$ $\mathcal{L}\{e^t\} = \dfrac{1}{s - 1}$

21. **Interesting Property** Use Property 3 from Table 5.4 to prove that

$$\mathcal{L}\{t^2 y'(t)\} = sY(s) + 2Y'(s) \tag{19}$$

22. **Differential Equation with Variable Coefficients**
Sometimes a differential equation with variable coefficients

can be solved by use of the Laplace transform. For example, the differential equation

$$ty'' + y' + ty = 0 \tag{20}$$

with a regular singular point at zero is called **Bessel's equation** of order zero.

(a) Show that if there are any solutions that remain bounded at $t = 0$ and whose derivatives remain bounded, then $Y(s) = \mathcal{L}\{y(t)\}$ must satisfy

$$(s^2 + 1)Y'(s) + sY(s) = 0 \tag{21}$$

(b) Solve the differential equation in $Y(s)$ to get

$$Y(s) = \frac{c}{\sqrt{1 + s^2}} \tag{22}$$

where c is an arbitrary constant.

(c) Expand $(1 + s^2)^{-1/2}$ as its binomial series

$$Y(s) = \frac{c}{\sqrt{1 + s^2}} = \frac{c}{s}\left(1 - \frac{1}{2}\cdot\frac{1}{s^2} + \frac{1}{2!}\cdot\frac{3}{2}\cdot\frac{1}{s^4}\right.$$
$$\left. + \cdots + \frac{1\cdot 3\cdot 5 \cdots (2n - 1)}{2^n n!}\cdot\frac{(-1)^n}{s^{2n}} + \cdots\right)$$

and, assuming that it is possible to take the inverse Laplace transform by taking the inverse transform term by term, show that the solution is

$$y(t) = \mathcal{L}^{-1}\{Y(s)\}$$
$$= c\left(1 - \frac{t^2}{2^2} + \frac{t^4}{2^2\cdot 4^2} - \frac{t^6}{2^2\cdot 4^2\cdot 6^2} + \cdots\right)$$

This power series is called Bessel's function of order zero and denoted by $J_0(t)$. It is the solution that satisfies $y(0) = 1$ and is shown in Figure 5.9.

23. **Airy's Equation** Find the differential equation satisfied by $Y(s) = \mathcal{L}\{y(t)\}$, where $y(t)$ is the solution of **Airy's equation:**

$$y'' = ty$$
$$y(0) = 0 \tag{23}$$
$$y'(0) = 1$$

24. **General Solutions of Differential Equations** One might be tempted to think that the Laplace transform is used only in solving initial-value problems. This is not true. Use the Laplace transform method to find the *general solution* of

$$y'' + y = 1$$

Figure 5.9
Graph of Bessel's function
$J_0(t)$

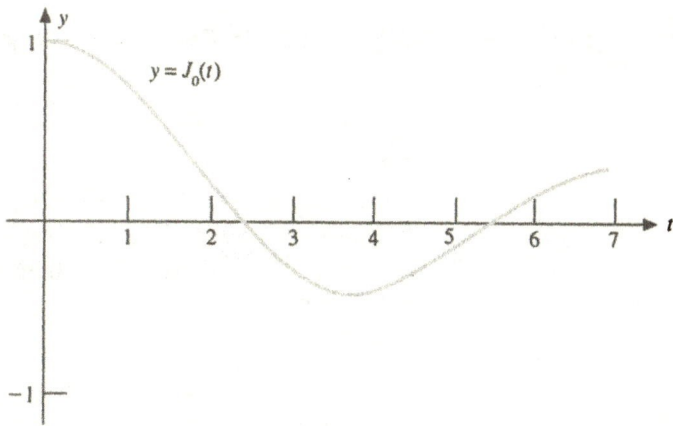

Hint: Carry along the initial conditions y_0 and y_0'. When the final solution is obtained, replace them by the more conventional arbitrary constants c_1 and c_2.

25. Which Grows Faster? Which curve grows faster: the curve that grows proportionally to its *value* or the curve that grows proportionally to its accumulated *area?* In other words, which of the following solutions ($k > 0$) is the larger?

$$y' = ky \qquad y(0) = 1$$

$$y' = k \int_0^t y(t)\, dt \qquad y(0) = 1$$

Operator Methods

Symbolic methods refer to methods for solving differential equations in which the derivatives are replaced by **symbolic operators** $D = d/dx$, $D^2 = d^2/dx^2$, ... Hence the differential equation $y'' + y' - 2y = 0$ would be written as the algebraic equation $D^2 y + Dy - 2y = 0$, or $(D^2 + D - 2)y = 0$, or even in factored form $(D - 1)(D + 2)y = 0$. The main advantage of symbolic methods over traditional methods lies in the fact that it is now possible to think of D as an algebraic quantity, thus ''turning'' a differential equation into an algebraic equation. Of course, one must verify that all these interpretations are mathematically sound, but this can be done. Without worrying much about the mathematical niceties, Prob-*

---

* Operator methods were first introduced to differential equations by the English engineer Oliver Heaviside (1850–1925). The reader will note the similarity between operator methods and the Laplace transform.

lems 26–27 ask the reader to use his or her imagination to carry out the requested steps.

26. Second-Order Operator as a Product of First-Order Operators Write

$$y'' + 3y' + 2y = 1 \qquad (24)$$

in factored form

$$(D + 1)(D + 2)y = 1 \qquad (25)$$

and then rewrite this equation as the two first-order equations

$$(D + 1)\, v = 1 \qquad (26a)$$
$$(D + 1)\, y = v \qquad (26b)$$

Find $y(x)$ by first solving Eq. (26a) for $v(x)$ and then substituting this value into Eq. (26b) to find $y(x)$. Show that you get the same result as if you had solved Eq. (24) directly.

27. Particular Solutions Using Operators Find a particular solution of

$$y'' + y = x^6 \qquad (27)$$

using symbolic operators. *Hint:* Write Eq. (27) in symbolic form, solve for y by dividing by $D^2 + 1$, divide $D^2 + 1$ into 1, getting a power series in D, and let the power series ''operate'' on x^6. This will give a particular solution. Verify your result.

28. Journal Entry Summarize the philosophy of using an integral transform in solving an initial-value problem.

5.5 · STEP FUNCTIONS AND DELAYED FUNCTIONS

PURPOSE
To show how the class of functions, called delayed functions, can be expressed in terms of step functions. Once this is done, we will find their Laplace transform. This will allow us to solve differential equations with delayed functions, which we will do in the next section.

THE UNIT STEP FUNCTION

Until now we have solved differential equations with continuous forcing functions. However, in electrical engineering, in which electronic components are often controlled by on-off switches, discontinuous forcing functions are the norm. Also, in physics, forces often change suddenly and are best described by discontinuous functions. A useful way for representing discontinuous functions is in terms of the unit step function.

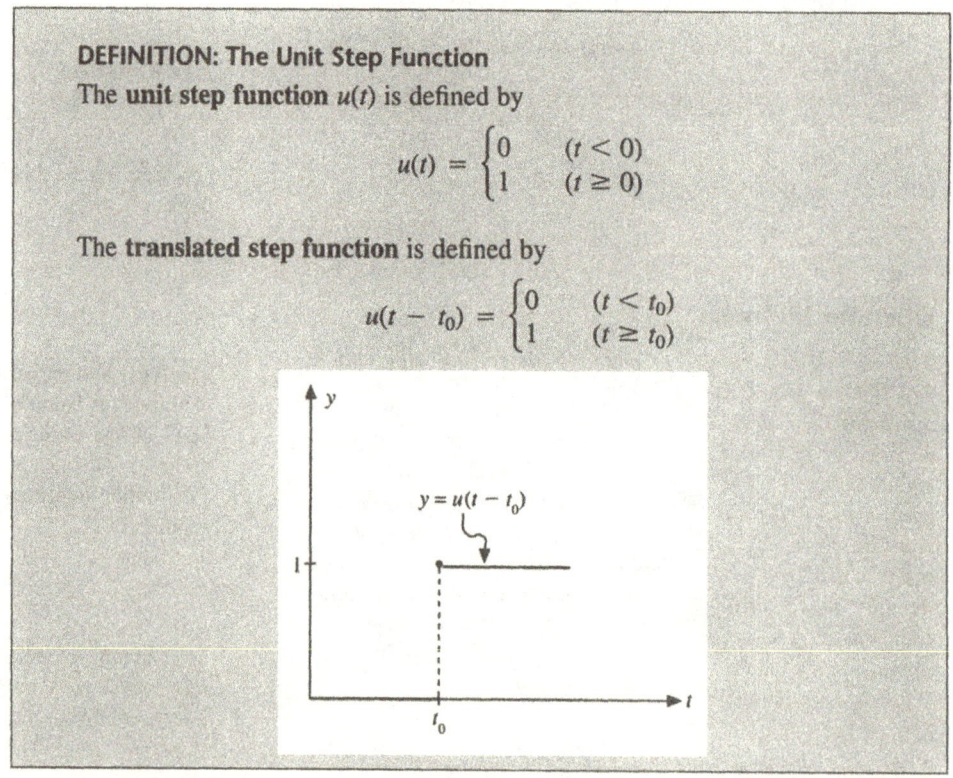

DEFINITION: The Unit Step Function
The **unit step function** $u(t)$ is defined by

$$u(t) = \begin{cases} 0 & (t < 0) \\ 1 & (t \geq 0) \end{cases}$$

The **translated step function** is defined by

$$u(t - t_0) = \begin{cases} 0 & (t < t_0) \\ 1 & (t \geq t_0) \end{cases}$$

$$y = u(t - t_0)$$

We often call the unit step function and the translated step function simply the *step function*.

Example 1   Graphing Step Functions   Graph the following functions on the interval $[0, \infty)$.

(a) $u(t - 2)$

(b) $1 - u(t - 1)$

(c) $1 - 2u(t - 1) + u(t - 2)$

Solution

(a) $u(t - 2) = \begin{cases} 0 & (t < 2) \\ 1 & (t \geq 2) \end{cases}$

See Figure 5.10(a).

(b) $1 - u(t - 1) = \begin{cases} 1 - 0 = 1 & (t < 1) \\ 1 - 1 = 0 & (t \geq 1) \end{cases}$

See Figure 5.10(b).

(c) $1 - 2u(t - 1) + u(t - 2) = \begin{cases} 1 - 2 \cdot 0 + 0 = 1 & (t < 1) \\ 1 - 2 \cdot 1 + 0 = -1 & (1 \leq t < 2) \\ 1 - 2 \cdot 1 + 1 = 0 & (2 \leq t) \end{cases}$

See Figure 5.10(c)   ∎

Figure 5.10
Combination
of step
functions

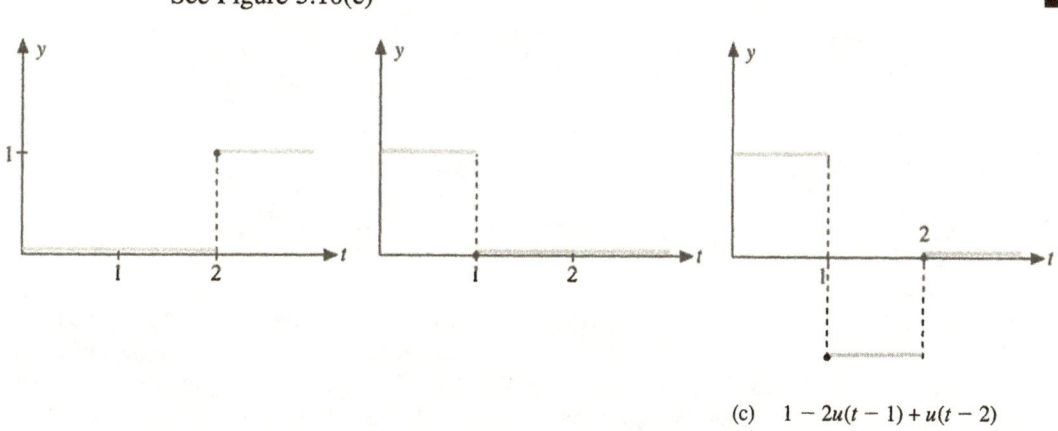

(c) $1 - 2u(t - 1) + u(t - 2)$

By means of the step function it is possible to write **branched functions** as a single equation. The following example illustrates this idea.

Example 2   Writing Branched Functions   Rewrite the branched function

$$f(t) = \begin{cases} 2 & (0 \leq t < 3) \\ -4 & (3 \leq t < 4) \\ 1 & (4 \leq t) \end{cases} \tag{1}$$

as a "single equation." See Figure 5.11.

Solution   The value of $f(t)$ is 2 until t reaches 3, at which time it jumps to -4. This can be expressed by $2 - 6u(t - 3)$, since $u(t - 3)$ is zero until t reaches 3, after which it becomes 1. When t reaches 4, the function jumps from -4 to 1, which can be expressed by adding on the term $5u(t-4)$. Hence

$$f(t) = 2 - 6u(t-3) + 5u(t-4) \qquad (2)$$

Figure 5.11
Graph of $f(t)$

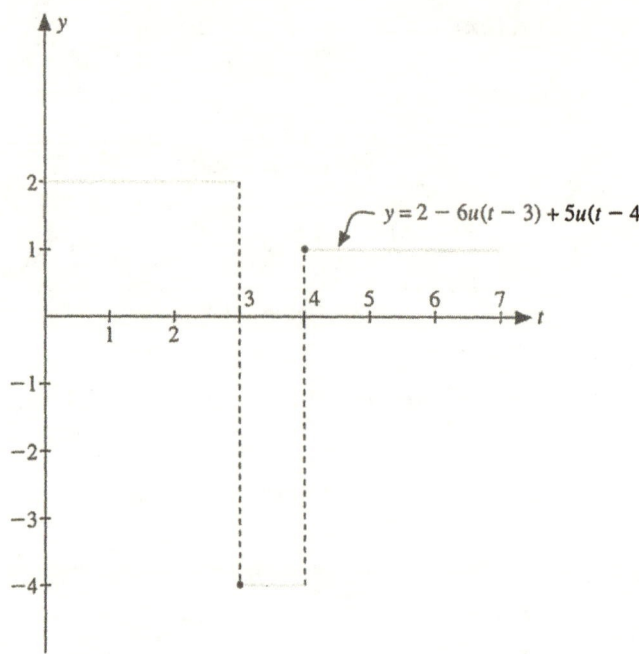

DELAYED FUNCTIONS

We are often interested in translating or *shifting* a given function $g(t)$ to the right c units and then replacing the function by zero to the left of c. For example, if we translate the function $g(t)$ shown in Figure 5.12(a) to the right c units and then replace the function by zero to the left of c, the graph of the new function $f(t)$, or **delayed $g(t)$,** would look like the graph in Figure 5.12(b). Mathematically, we could represent the delayed $g(t)$ as

$$f(t) = \begin{cases} 0 & (t < c) \\ g(t-c) & (t \ge c) \end{cases} \qquad (3)$$

Figure 5.12
Graph of a delayed function

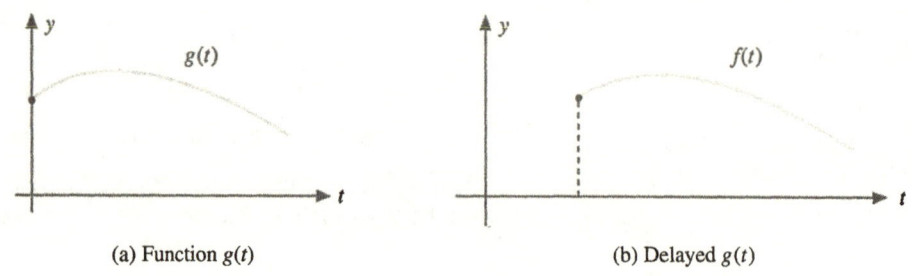

(a) Function $g(t)$ (b) Delayed $g(t)$

Example 3   **Representing Delayed Functions**   Represent mathematically the delayed function obtained when the graph of $g(t) = t^2$ is translated 1 unit to the right and then truncated to the left of 1.

Solution   The graph of $g(t - 1) = (t - 1)^2$ is the graph of $g(t) = t^2$ translated one unit to the right. Hence the graph of the delayed t^2 is

$$f(t) = u(t - 1)(t - 1)^2 = \begin{cases} 0 & (t < 1) \\ (t - 1)^2 & (t \geq 1) \end{cases} \tag{4}$$

See Figure 5.13.

Figure 5.13
Graph of t^2 delayed until $t = 1$

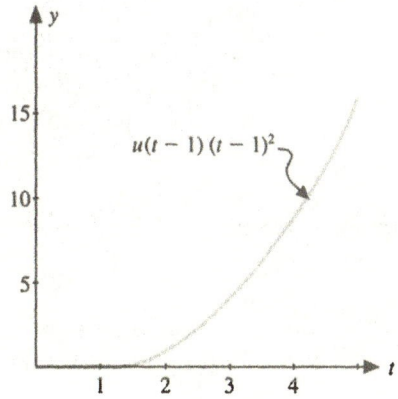

THE LAPLACE TRANSFORM OF DELAYED FUNCTIONS

For many problems it is necessary to find the Laplace transform of the delayed function $u(t - c)f(t - c)$ shown in Figure 5.12. We can find this transform directly from the definition of the Laplace transform. We have

$$\mathcal{L}\{u(t - c)f(t - c)\} = \int_0^\infty e^{-st} u(t - c)f(t - c)\, dt$$

$$= \int_c^\infty e^{-st} f(t - c)\, dt \tag{5}$$

Note that this last integral was taken over the interval $[c, \infty)$, since $u(t - c) = 0$ for $t < c$. Note too that $u(t - c) = 1$ on the interval $[c, \infty)$. If we now make the change of variables $\xi = t - c$, this last integral becomes

$$\mathcal{L}\{u(t - c)f(t - c)\} = \int_0^\infty e^{-(\xi + c)s} f(\xi)\, d\xi$$

$$= e^{-cs} \int_0^\infty e^{-s\xi} f(\xi)\, d\xi$$

$$= e^{-cs} F(s) \tag{6}$$

We summarize these results in the following theorem.

> **THEOREM 5.8:** Laplace Transform of a Delayed Function
>
> Assume that the Laplace transform $F(s) = \mathcal{L}\{f(t)\}$ exists for $s > \alpha \geq 0$. If c is a positive constant, then
>
> $$\mathcal{L}\{u(t - c)f(t - c)\} = e^{-cs}\mathcal{L}\{f(t)\} = e^{-cs}F(s) \qquad (s > \alpha)$$
>
> Conversely, if $f(t) = \mathcal{L}^{-1}\{F(s)\}$, then
>
> $$\mathcal{L}^{-1}\{e^{-cs}F(s)\} = u(t - c)f(t - c) \tag{7}$$

Note: In plain English, Theorem 5.8 basically says that to find the Laplace transform of the delayed function $u(t - c)f(t - c)$, take the transform of $f(t)$ and then multiply the transform by e^{-cs}. On the other hand, to find the inverse of the function $e^{-cs}F(s)$, just suppress the factor e^{-cs} and take the inverse of $F(s)$, getting $f(t)$; then replace t by $t - c$, getting $f(t - c)$ and then multiply by $u(t - c)$.

Example 4 **Transform of a Delayed Function** Find the Laplace transform of the function $f(t) = u(t - 1)t^2$.

Solution To apply Theorem 5.8, it is necessary to write $f(t) = u(t - 1)t^2$ in the proper form $u(t - 1)f(t - 1)$. What this means is that we must write t^2 as a polynomial in the variable "$t - 1$." We can do this by working backward and expanding $(t - 1)^2$ as

$$(t - 1)^2 = t^2 - 2t + 1$$

Solving for t^2, we can write the outcome in the proper form:

$$t^2 = (t - 1)^2 + 2t - 1$$
$$= (t - 1)^2 + 2(t - 1) + 1$$

Hence we have managed to write $f(t)$ as

$$f(t) = u(t - 1)t^2 = u(t - 1)[(t - 1)^2 + 2(t - 1) + 1]$$

From Theorem 5.8 the Laplace transform of $f(t)$ is

$$
\begin{aligned}
\mathcal{L}\{u(t - 1)t^2\} \\
&= \mathcal{L}\{u(t - 1)\,[(t - 1)^2 + 2(t - 1) + 1]\} \\
&= \mathcal{L}\{u(t - 1)(t - 1)^2\} + 2\mathcal{L}\{u(t - 1)(t - 1)\} + \mathcal{L}\{u(t - 1)\} \\
&= \frac{2e^{-s}}{s^3} + \frac{2e^{-s}}{s^2} + \frac{e^{-s}}{s} \\
&= e^{-s}\left(\frac{2}{s^3} + \frac{2}{s^2} + \frac{1}{s}\right) \tag{8}
\end{aligned}
$$

Example 5   Delayed Trigonometric Functions   Find the Laplace transform of the function

$$f(t) = \begin{cases} 0 & (0 \le t < \pi) \\ -\sin t & (\pi \le t < 2\pi) \\ 0 & (2\pi \le t) \end{cases} \tag{9}$$

The graph of $f(t)$ is shown in Figure 5.14.

Figure 5.14
Can you represent this function in terms of the unit step function?

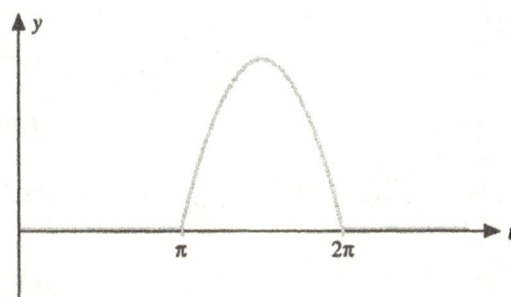

Solution   Using the unit step function, we can write

$$f(t) = -u(t - \pi) \sin t + u(t - 2\pi) \sin t \tag{10}$$

However, to use Theorem 5.8, we must rewrite *sin t* in such a way that the first term involves a "$t - \pi$" and the second term involves a "$t - 2\pi$." But from the graphs of the sine and cosine functions, it is easy to see

$$\sin t = -\sin (t - \pi)$$
$$\sin t = \sin (t - 2\pi)$$

Hence we can rewrite Eq. (10) in the proper form:

$$f(t) = u(t - \pi) \sin (t - \pi) + u(t - 2\pi) \sin (t - 2\pi) \tag{11}$$

Thus the transform is

$$\mathcal{L}\{f(t)\} = \mathcal{L}\{u(t - \pi) \sin (t - \pi) + u(t - 2\pi) \sin (t - 2\pi)\}$$
$$= \frac{e^{-\pi s}}{s^2 + 1} + \frac{e^{-2\pi s}}{s^2 + 1}$$
$$= \frac{e^{-\pi s}}{s^2 + 1} \left(1 + e^{-\pi s}\right) \tag{12}$$

∎

Example 6   Inverse of a Function involving e^{-cs}   Find the inverse of

$$F(s) = \frac{1 - e^{-3s}}{s^2}$$

Solution   Using the linearity of the inverse transform and Theorem 5.8, we have

$$\mathcal{L}^{-1}\left\{\frac{1 - e^{-3s}}{s^2}\right\} = \mathcal{L}^{-1}\left\{\frac{1}{s^2}\right\} - \mathcal{L}^{-1}\left\{\frac{e^{-3s}}{s^2}\right\}$$
$$= t - u(t - 3)(t - 3) \tag{13}$$

∎

PROBLEMS: Section 5.5

For Problems 1–4, write each of the given functions in terms of the unit step function and sketch their graphs.

1. $f(t) = \begin{cases} a & (0 \le t < 1) \\ b & (1 \le t < 2) \\ c & (2 \le t) \end{cases}$

2. $f(t) = \begin{cases} 1 & (0 \le t < 1) \\ e^t & (1 \le t < 2) \\ 2 & (2 \le t) \end{cases}$

3. $f(t) = \begin{cases} 1 & (0 \le t < 1) \\ 4t - t^2 & (1 \le t < 2) \\ 1 & (2 \le t) \end{cases}$

4. $f(t) = \begin{cases} 0 & (0 \le t < 1) \\ \sin \pi t & (1 \le t < 2) \\ 0 & (2 \le t) \end{cases}$

For Problems 5–14, determine the Laplace transform of the given function.

5. $1 - u(t - 1)$

6. $1 - 2u(t - 1) + u(t - 2)$

7. $u(t - 1)(t - 1)$

8. $u(t - 2)(t - 2)^2$

9. $u(t - \pi) \sin (t - \pi)$

10. $u(t - 3)e^t$

11. $u(t)$

12. $u(t - 2)t^2$

13. $u(t - \pi) \cos t$

14. $f(t) = \begin{cases} \sin t & (0 \le t < \pi) \\ 0 & (\pi \le t) \end{cases}$

For Problems 15–24, find the inverse transform of the given function and sketch their graphs.

15. $\dfrac{e^{-s}}{s}$

16. $\dfrac{e^{-s}}{s^2}$

17. $\dfrac{e^{-2s}}{s - 3}$

18. $\dfrac{e^{-3s}}{s^2 + 4}$

19. $\dfrac{e^{-4s}}{s + 4}$

20. $\dfrac{e^{-4s}}{(s + 4)^2}$

21. $\dfrac{e^{-s}}{s(s + 1)}$

22. $\dfrac{s + e^{-\pi s}}{s^2 + 1}$

23. $\dfrac{e^{-s} - 2e^{-2s} + 2e^{-3s} - e^{-4s}}{s}$

24. $\dfrac{e^{-s} - 2e^{-2s} + 2e^{-3s} - e^{-4s}}{s^2}$

Gates and Filters

*The function $G_a^b(t) = u(t - a) - u(t - b)$ is called a **gate function** (or filter). For Problems 25–28, sketch the graph of the given function and find the Laplace transform of the function.*

25. $G_a^b(t)$

26. $G_1^2(t)$

27. $G_1^3(t)t$

28. $G_\pi^{2\pi}(t) \sin t$

29. Laplace Transform of Periodic Functions If $f(t)$ satisfies $f(t + T) = f(t)$ for all $t \ge 0$, where T is some fixed positive number, then $f(t)$ is called a *periodic function* with period T. Show that for a periodic function with period T, one has

$$\mathcal{L}\{f\} = \frac{\displaystyle\int_0^T e^{-st}f(t)\, dt}{1 - e^{-sT}} \tag{14}$$

Transforms of Some Important Periodic Functions

For Problems 30–34, find the Laplace transform of the given periodic function using Eq. (14). In addition to using Eq. (14), find the Laplace transform by finding an expression for the periodic function $f(t)$ as an infinite series of translated truncated functions and then taking the term by term Laplace transform of this infinite series. Finally, if possible, write this infinite series in closed form, using the geometric series formula

$$\frac{1}{1 - r} = 1 + r + r^2 + \cdots$$

30. Square Wave See Figure 5.15.

$$f(t) = \begin{cases} 1 & (0 \le t < 1) \\ 0 & (1 \le t < 2) \end{cases}$$
$$f(t + 2) = f(t)$$

Figure 5.15
Square wave

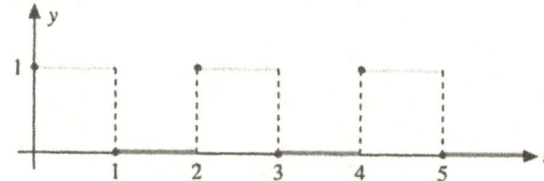

31. Square Cosine Wave See Figure 5.16.

$$f(t) = \begin{cases} 1 & (0 \le t < 1) \\ -1 & (1 \le t < 2) \end{cases}$$

$$f(t + 2) = f(t)$$

Figure 5.16
Square cosine wave

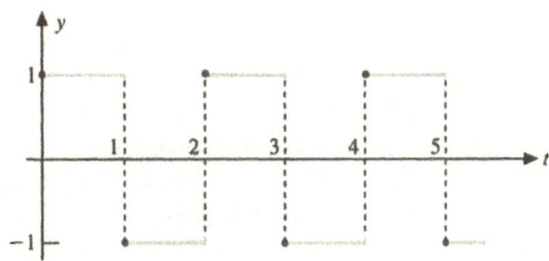

32. Sawtooth Wave See Figure 5.17.

$$f(t) = t \qquad (0 \le t < 1)$$

$$f(t + 1) = f(t)$$

Figure 5.17
Sawtooth wave

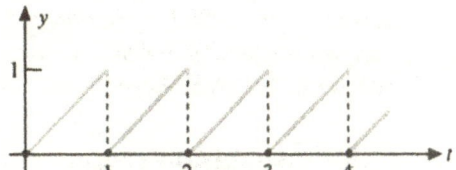

33. Triangular Wave See Figure 5.18.

$$f(t) = \begin{cases} t & (0 \le t < 1) \\ 2 - t & (1 \le t < 2) \end{cases}$$

$$f(t + 2) = f(t)$$

Figure 5.18
Triangular wave

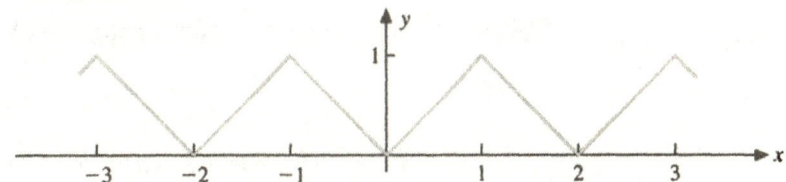

34. Rectified Sine Wave See Figure 5.19.

$$f(t) = \sin t \qquad (0 \le t < \pi)$$

$$f(t + \pi) = f(t)$$

Figure 5.19
Rectified sine wave

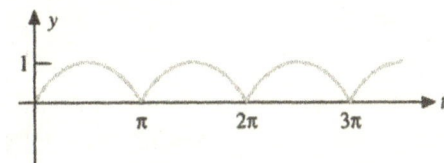

35. Computer Problem Use a computer to sketch the graphs of the step functions from Example 1.

36. Journal Entry Summarize your thoughts about this section.

5.6 DIFFERENTIAL EQUATIONS WITH DISCONTINUOUS FORCING FUNCTIONS

PURPOSE

To show how differential equations with constant coefficients and discontinuous forcing terms can be solved by using the Laplace transform. We will present examples and interpret their solutions.

As a tool for solving differential equations, the Laplace transform excels over other methods when it comes to solving equations with discontinuous forcing functions. If one used the method of variation of parameters to find a particular solution of a differential equation with a discontinuous forcing function, it would be necessary to find an expression of the solution *on each subinterval* where the forcing function was continuous. However, by rewriting discontinuous forcing functions in terms of the unit step function, it is possible to solve problems of this type quite easily using the Laplace transform. The following examples illustrate this idea.

Example 1   On-Off Forcing Term   Solve the initial-value problem

$$y'' + y = f(t) = \begin{cases} 1 & (0 \leq t < \pi) \\ 0 & (\pi \leq t) \end{cases}$$

$$y(0) = 0$$

$$y'(0) = 0 \tag{1}$$

Solution   We rewrite the on-off forcing function in terms of the unit step function $u(t)$, getting

$$y'' + y = 1 - u(t - \pi) = \begin{cases} 1 & (0 \leq t < \pi) \\ 0 & (\pi \leq t) \end{cases} \tag{2}$$

Taking the Laplace transform of this equation, we have

$$\left(s^2 Y(s) - sy(0) - y'(0)\right) + Y(s) = \mathcal{L}\{1\} - \mathcal{L}\{u(t - 1)\} = \frac{1 - e^{-\pi s}}{s}$$

Substituting the initial conditions into this equation and solving for $Y(s)$, we obtain

$$Y(s) = \frac{1 - e^{-\pi s}}{s(s^2 + 1)}$$

$$= \left(1 - e^{-\pi s}\right)\left\{\frac{1}{s} - \frac{s}{s^2 + 1}\right\}$$

$$= \left(1 - e^{-\pi s}\right)\left\{\mathcal{L}\{1\} - \mathcal{L}\{\cos t\}\right\}$$

$$= \left(\mathcal{L}\{1\} - \mathcal{L}\{\cos t\}\right) - e^{-\pi s}\left(\mathcal{L}\{1\} - \mathcal{L}\{\cos t\}\right) \tag{3}$$

Finally, using the translation property $\mathcal{L}^{-1}\{e^{-cs}F(s)\} = u(t - c)f(t - c)$, we get

$$
\begin{aligned}
y(t) &= \mathcal{L}^{-1}\{Y(s)\} \\
&= (1 - \cos t) - u(t - \pi)\,[1 - \cos(t - \pi)] \\
&= \begin{cases} 1 - \cos t & (0 \le t < \pi) \\ -2 \cos t & (\pi \le t) \end{cases}
\end{aligned}
\tag{4}
$$

The graph of $y(t)$ is drawn in Figure 5.20.   ■

Figure 5.20
The input f and output y of the system described by
$y'' + y = f(t)$,
$y(0) = y'(0) = 0$

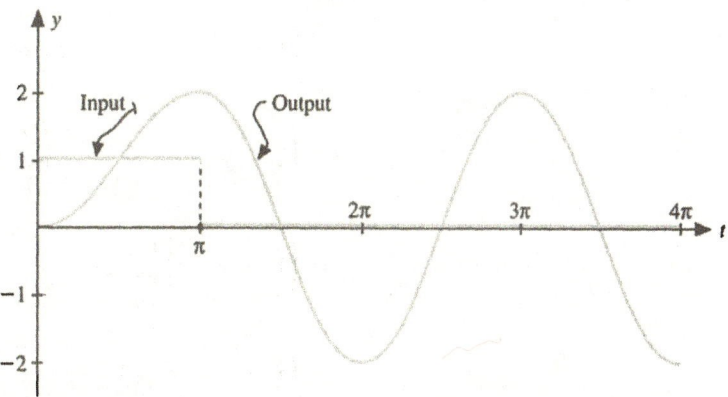

Note: The initial-value problem (1) can be interpreted as describing the motion of a weight attached to a spring, which is initially at rest, to which a constant external force is applied for π units of time.* Possibly, the force is caused by an electromagnet attracting the mass, or even an additional weight was added to the original weight but taken off at $t = \pi$. In any case, the solution $y(t)$ describes the net result of this action. It is interesting to observe the motion of the mass during the time interval $0 \le t \le \pi$ during the time when the force acts on the mass, and when time $t > \pi$, after the force is turned off. The graph of $y(t)$ in Figure 5.20 shows this behavior.

We now consider a discontinuous forcing term acting on a system that contains damping.

Example 2   On-Off Forcing Term on a Damped System   Find the solution of the damped differential equation

$$
\begin{aligned}
y'' + 2y' + 2y &= \begin{cases} 1 & (0 \le t < \pi) \\ 0 & (\pi \le t) \end{cases} \\
y(0) &= 0 \\
y'(0) &= 0
\end{aligned}
\tag{5}
$$

* This problem can also be interpreted as describing the charge on a condenser in a simple circuit containing a capacitor, an inductor, and an impressed voltage source. In this case the initial charge on the capacitor is zero farads ($y(0) = 0$), the initial current in the circuit is zero amperes ($y'(0) = 0$), and the driving voltage is 1 volt for π seconds and zero thereafter.

Solution Taking the Laplace transform of Eq. (5), we get

$$\left(s^2 Y(s) - sy(0) - y'(0)\right) + 2\left(sY(s) - y(0)\right) + 2Y(s) = \frac{1 - e^{-\pi s}}{s}$$

Substituting the initial conditions into the above equations are solving for $Y(s)$, we find

$$Y(s) = \frac{1 - e^{-\pi s}}{s(s^2 + 2s + 2)}$$

$$= \left(1 - e^{-\pi s}\right)\left\{\frac{1}{2s} - \frac{1}{2}\left(\frac{s + 2}{s^2 + 2s + 2}\right)\right\}$$

$$= \left(1 - e^{-\pi s}\right)\left\{\frac{1}{2s} - \frac{1}{2}\left(\frac{(s + 1) + 1}{(s + 1)^2 + 1}\right)\right\}$$

$$= \left(1 - e^{-\pi s}\right)\left\{\frac{1}{2s} - \frac{s + 1}{2[(s + 1)^2 + 1]} - \frac{1}{2[(s + 1)^2 + 1]}\right\}$$

$$= \left(1 - e^{-\pi s}\right)\left\{\frac{1}{2}\mathcal{L}\{1\} - \frac{1}{2}\mathcal{L}\{e^{-t}\cos t\} - \frac{1}{2}\mathcal{L}\{e^{-t}\sin t\}\right\} \tag{6}$$

Hence the solution is

$$y(t) = \mathcal{L}^{-1}\{Y(s)\}$$

$$= \frac{1}{2} - \frac{1}{2}e^{-t}\cos t - \frac{1}{2}e^{-t}\sin t$$

$$- u(t - \pi)\left(\frac{1}{2} - \frac{1}{2}e^{-(t - \pi)}\cos (t - \pi) - \frac{1}{2}e^{-(t - \pi)}\sin (t - \pi)\right)$$

or

$$y(t) = \begin{cases} \dfrac{1}{2}(1 - e^{-t}\cos t - e^{t}\sin t) & (0 \le t < \pi) \\[2mm] \dfrac{1}{2}e^{-t}(e^{\pi} + 1)(\sin t + \cos t) & (\pi \le t) \end{cases} \tag{7}$$

The graphs of the input and output $y(t)$ are shown in Figure 5.21. ∎

Example 3   Off-On-Off Forcing Function   Solve

$$y'' + 3y' + 2y = \begin{cases} t & (0 \le t < \pi) \\ 1 & (\pi \le t) \end{cases}$$

$$y(0) = 0$$

$$y'(0) = 1 \tag{8}$$

Figure 5.21
The input and output of a
damped linear system

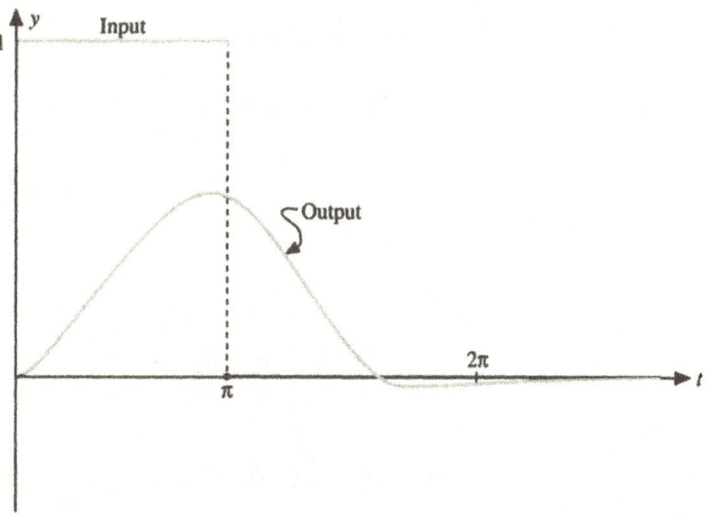

Solution   We begin by rewriting the forcing function in terms of the unit step function $u(t)$, or

$$y'' + 3y' + 2y = t - u(t - \pi)(t - \pi) \tag{9}$$

Taking the Laplace transform of this equation gives

$$\left(s^2 Y(s) - 1\right) + 3\left(s Y(s)\right) + 2\left(Y(s)\right) = \frac{1 - e^{-\pi s}}{s^2}$$

Solving for $Y(s)$ gives

$$Y(s) = \left(\frac{1}{s^2(s^2 + 3s + 2)}\right) + \left(\frac{1}{s^2 + 3s + 2}\right) - \left(\frac{e^{-\pi s}}{s^2(s^2 + 3s + 2)}\right) \tag{10}$$

We now write the two *distinct* fractions in Eq. (10) as partial fraction decompositions:

$$\frac{1}{s^2(s^2 + 3s + 2)} = -\frac{3}{4s} + \frac{1}{2s^2} - \frac{1}{4(s + 2)} + \frac{1}{s + 1}$$

$$\frac{1}{s^2 + 3s + 2} = \frac{1}{s + 1} - \frac{1}{s + 2}$$

Hence the inverse transforms of the three terms in Eq. (10) are

$$\mathcal{L}^{-1}\left\{\frac{1}{s^2(s^2 + 3s + 2)}\right\} = -\frac{3}{4} + \frac{t}{2} - \frac{1}{4}e^{-2t} + e^{-t} \tag{11}$$

$$\mathcal{L}^{-1}\left\{\frac{e^{-\pi s}}{s^2(s^2 + 3s + 2)}\right\} = u(t - \pi)\left(-\frac{3}{4} + \frac{t - \pi}{2} - \frac{1}{4}e^{-2(t-\pi)} + e^{-(t-\pi)}\right) \tag{12}$$

$$\mathcal{L}^{-1}\left\{\frac{1}{s^2 + 3s + 2}\right\} = e^{-t} - e^{-2t} \tag{13}$$

Here the solution is

$$y(t) = \mathcal{L}^{-1}\{Y(s)\} = \left(-\frac{3}{4} + \frac{t}{2} - \frac{1}{4}e^{-2t} + e^{-t}\right) + \left(e^{-t} - e^{-2t}\right)$$

$$- u(t - \pi)\left(-\frac{3}{4} + \frac{t - \pi}{2} - \frac{1}{4}e^{-2(t-\pi)} + e^{-(t-\pi)}\right)$$

$$y(t) = \begin{cases} 2e^{-t} - \dfrac{5}{4}e^{-2t} + \dfrac{t}{2} - \dfrac{3}{4} & (0 \le t < \pi) \\[2mm] \dfrac{\pi}{2} + \left(2 - e^{\pi}\right)e^{-t} + \dfrac{1}{4}\left(-5 + e^{2\pi}\right)e^{-2t} & (\pi \le t) \end{cases} \tag{14}$$

The graph of $y(t)$ is shown in Figure 5.22.

Figure 5.22
The input and output of a
linear system

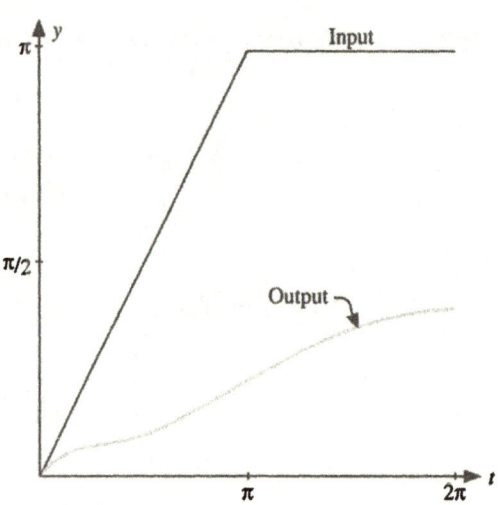

The Laplace transform is extremely valuable for solving differential equations in which the forcing function has an *infinite number* of discontinuities.

Example 4   Square Wave Excitation of a Damped System   Solve

$$y'' + 3y' + 2y = f(t)$$
$$y(0) = 0$$
$$y'(0) = 0 \tag{15}$$

where $f(t)$ is the square wave shown in Figure 5.23.

Solution Rewriting the square wave in terms of the unit step function, we have

$$f(t) = 1 - u(t - 1) + u(t - 2) - u(t - 3) + \cdots \tag{16}$$

Figure 5.23
Input square wave

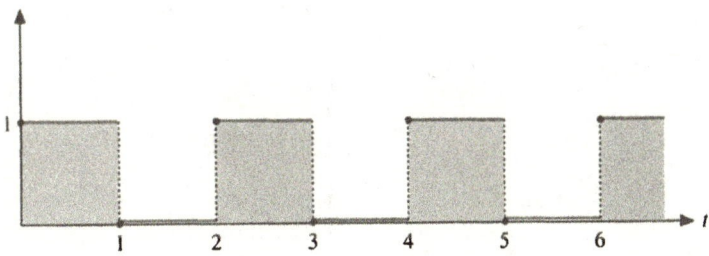

One should realize that although $f(t)$ is denoted as an infinite series, only a finite number of terms are nonzero for any specific value of t. Taking the Laplace transform of the differential equation (15), we have

$$(s^2 + 3s + 2)Y(s) = \frac{1}{s}\left(1 - e^{-s} + e^{-2s} - e^{-3s} + \cdots\right) \tag{17}$$

Solving for $Y(s)$ gives

$$
\begin{aligned}
Y(s) &= \left(1 - e^{-s} + e^{-2s} - e^{-3s} + \cdots\right)\left\{\frac{1}{s(s^2 + 3s + 2)}\right\} \\
&= \left(1 - e^{-s} + e^{-2s} - e^{-3s} + \cdots\right)\left\{\frac{1}{2s} - \frac{1}{s+1} + \frac{1}{2(s+2)}\right\} \\
&= \left(1 - e^{-s} + e^{-2s} - e^{-3s} + \cdots\right)\left(\frac{1}{2}\mathcal{L}\{1\} - \mathcal{L}\{e^{-t}\} + \frac{1}{2}\mathcal{L}\{e^{-2t}\}\right) \quad (18)
\end{aligned}
$$

Hence the solution is

$$
\begin{aligned}
y(t) = {} & \frac{1}{2}\left(1 - 2e^{-t} + e^{-2t}\right) - \frac{u(t - 1)}{2}\left(1 - 2e^{-(t-1)} + e^{-2(t-1)}\right) \\
& + \frac{u(t - 2)}{2}\left(1 - 2e^{-(t-2)} + e^{-2(t-2)}\right) \\
& - \frac{u(t - 3)}{2}\left(1 - 2e^{-(t-3)} + e^{-2(t-3)}\right) \\
& + \cdots \tag{19}
\end{aligned}
$$

Again, as in the earlier series, for any value of t there are only a finite number of terms in this series that are nonzero. The graph of $y(t)$ is shown in Figure 5.24. ■

Figure 5.24
Input and output of a
damped linear system

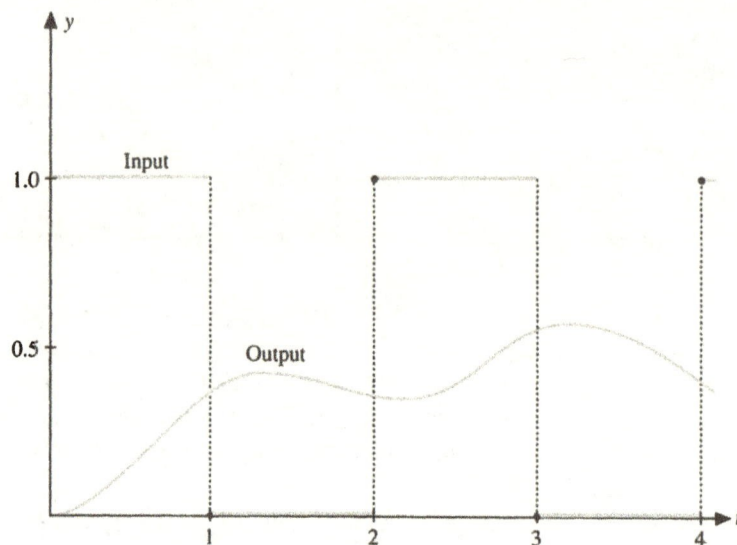

PROBLEMS: Section 5.6

For Problems 1–12, solve the given initial-value problem.
Sketch the graph of both the forcing function and the solution.

1. $y' = 1 - u(t - 1)$
   $y(0) = 0$

2. $y' = 1 - 2u(t - 1) + u(t - 2)$
   $y(0) = 0$

3. $y' + y = u(t - 1)$
   $y(0) = 0$

4. $y'' + y = t - u(t - 1)t$
   $y(0) = 0$
   $y'(0) = 0$

5. $y'' = 1 - u(t - 1)$
   $y(0) = 0$
   $y'(0) = 0$

6. $y'' + y = u(t - 3)$
   $y(0) = 0$
   $y'(0) = 1$

7. $y'' + y = u(t - \pi) - u(t - 2\pi)$
   $y(0) = 0$
   $y'(0) = 1$

8. $y'' + 4y = t - u\left(t - \dfrac{\pi}{2}\right)\left(t - \dfrac{\pi}{2}\right)$
   $y(0) = 0$
   $y'(0) = 0$

9. $y'' + 4y = u(t - 2\pi) \sin t$
   $y(0) = 1$
   $y'(0) = 0$

10. $y'' + 4y = \sin t - u(t - 2\pi) \sin (t - 2\pi)$
   $y(0) = 0$
   $y'(0) = 0$

11. $y'' + 3y' + 3y = u(t - 2)$
   $y(0) = 0$
   $y'(0) = 1$

12. $y'' + 2y' + 5y = 10 - 10u(t - \pi)$
   $y(0) = 0$
   $y'(0) = 0$

For Problems 13–15, solve the given initial-value problem with the indicated forcing function. Sketch the graph of the solution if you have access to a computer with graphing capabilities.

13. See Figure 5.25.

$$y' + y = f(t) \qquad y(0) = 0$$

Figure 5.25
Triangular sawtooth wave:
$$f(t) = t - u(t - 1) - u(t - 2) - \cdots$$

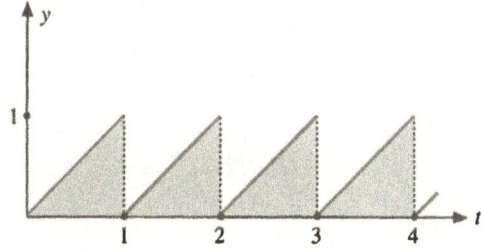

14. See Figure 5.26.

$$y'' + y = f(t) \qquad y(0) = 0 \quad y'(0) = 0$$

Figure 5.26
Rectangular wave:
$$f(t) = 1 - 2u(t - \pi) + 2u(t - 2\pi) - \cdots$$

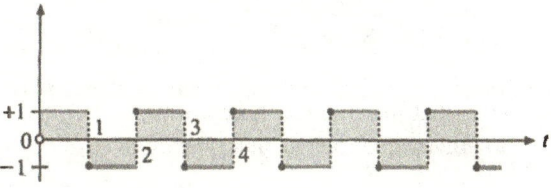

15. See Figure 5.27.

$$y'' + y = f(t) \qquad y(0) = 1 \quad y'(0) = 0$$

Figure 5.27
Rectangular sawtooth wave:
$$f(t) = 1 - u(t - \pi) + u(t - 2\pi) - \cdots$$

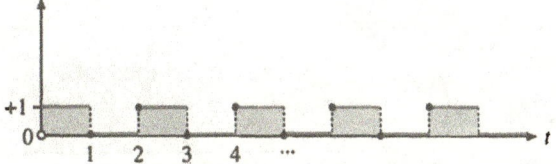

16. Hot Rod Problem The hot rod problem is a well-known problem in control theory* that seeks the *time optimal control function* $U(t)$ that "drives" the solution of the equation $\ddot{x} = U(t)$ from the initial state $x(0) = \dot{x}(0) = 0$ to the final state $x(t_f) = 1$, $\dot{x}(t_f) = 0$ in (unknown) *minimum time* t_f. See Figure 5.28(a) on page 292. The control function $U(t)$ represents a driving force when $U(t) > 0$ and a braking force when $U(t) < 0$. Its only restriction is the condition $-1 \leq U(t) \leq 1$. It can be shown that the *time optimal control* $U(t)$ for this problem is *bang-bang*, which means it is always $+1$ or -1. In this problem the bang-bang control is

$$U(t) = \begin{cases} 1 & (0 \leq t < 1) \\ -1 & (1 \leq t < 2) \\ 0 & (2 \leq t) \end{cases}$$
$$= 1 - 2u(t - 1) + u(t - 2)$$
$$(0 \leq t \leq 2) \qquad (20)$$

See Figure 5.28(b). Find the path of the hot rod using this optimal control strategy. Note that Eq. (20) says that the hot rod driver "puts the pedal to the metal" for 1 unit of time and then hits the brakes for the next 1 unit of time, thereby coming to a perfect stop at $x(2) = 0$. Sketch the graphs of $x(t)$ and $\dot{x}(t)$.

17. Unusual Application of the Laplace Transform Evaluate the integral

$$f(t) = \int_0^\infty \frac{\sin tx}{x} \, dx \qquad (21)$$

by showing that $\mathcal{L}\{f\} = 2/\pi s$ and hence f is the *constant* function $f(t) = 2/\pi$. Does this result agree with your intuition based on the graph of the integrand?

18. Computer Problem Use a computer to sketch the graph of the solution of Example 3.

19. Journal Entry It would be a good idea to write a brief paragraph outlining the advantages and disadvantages of solving the constant coefficient differential equation

$$ay'' + by' + cy = f(t)$$

by the Laplace transform. Contrast the Laplace transform method with the methods learned in Chapter 3.

---

* Optimal control theory is an important branch of differential equations that has significant engineering applications. Interested readers can consult the beginning text *Automatic Control Systems* by B.C. Kuo (Prentice-Hall, Englewood Cliffs, N.J., 1975).

Figure 5.28
Hot rod problem. Finding the
fastest way to get from one
stop sign to the next one.

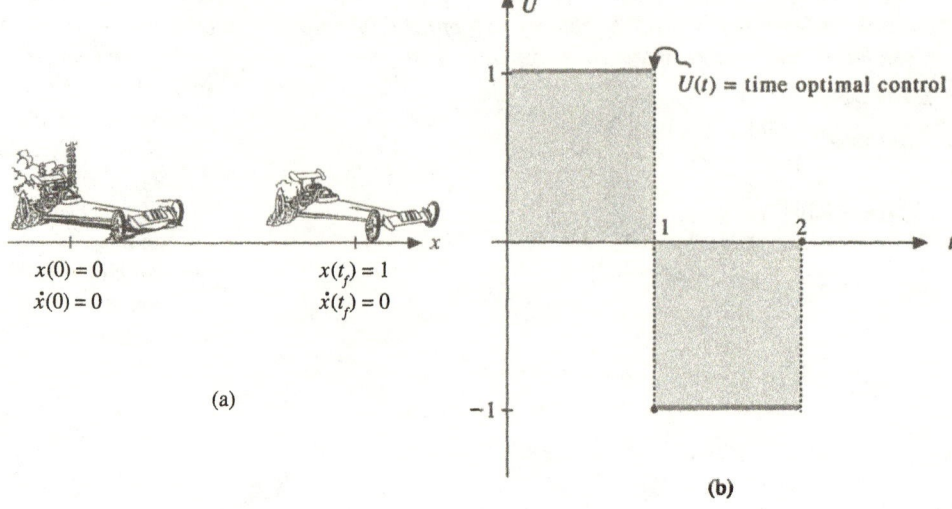

$x(0) = 0$
$\dot{x}(0) = 0$

$x(t_f) = 1$
$\dot{x}(t_f) = 0$

(a)

$U(t)$ = time optimal control

(b)

5.7   IMPULSE FORCING FUNCTIONS

PURPOSE
To show how the Laplace transform can be used to find the solution of a differential
equation of the form

$$ay'' + by' + cy = \delta(t - t_0)$$

where the forcing term $\delta(t - t_0)$, called an impulse function, is an idealization of an
impulsive force acting at t_0. We will show that the impulse function $\delta(t - t_0)$, has a
simple Laplace transform, allowing us to solve equations of this type quite easily.

INTRODUCTION TO THE
IMPULSE RESPONSE
FUNCTION

In physics and engineering, one often encounters impulsive forces that act over a short
period of time. A bat hitting a ball and two billiard balls colliding are impulsive forces
acting at nearly an instant of time. To deal with these types of forces mathematically,
physicist Paul A. M. Dirac defined an "idealized impulse," called the **impulse function,** or **Dirac delta function.**

HISTORICAL NOTE

Paul Dirac (1902–1984) was an English physicist who was awarded the Nobel
Prize at the age of 31 for his work in quantum theory.

> **DEFINITION: Intuitive Concept of the Unit Impulse Function**
> Loosely speaking, the **unit impulse function** or **Dirac delta function**, denoted $\delta(t)$, is defined as the "function" that satisfies the two (inconsistent) mathematical properties
>
> $$\delta(t) = \begin{cases} 0 & (t \neq 0) \\ \text{infinite} & (t = 0) \end{cases}$$
>
> $$\int_{-\infty}^{\infty} \delta(t) \, dt = 1$$
>

Note: Although the variable t in an impulse generally represents time, it sometimes represents distance. For example, the delta function can be used to represent a point load acting on a support beam, in which case t represents the distance measured along the beam.

A MORE PRECISE APPROACH TO THE IMPULSE FUNCTION

Although the above definition of the unit impulse function* provides an intuitive understanding of an impulsive force, it is too imprecise to be of much value mathematically. However, it is possible to approach problems involving impulsive forces indirectly, "through the back door," as the following spaceship example illustrates.

A spaceship of mass m sits motionless somewhere far off in space, away from any gravitational field. Suddenly, its thrusters are turned on for h seconds, exerting a thrust

---

* We often refer to the impulse function as the *unit* impulse function because the "area" under the curve is 1. The area under the curve has units of (force × time), which has units of energy. Hence an area of 1 refers to the fact that 1 unit of energy is applied by the unit impulse function.

Figure 5.29
A short impulse thrust of a rocket can be interpreted as a delta function

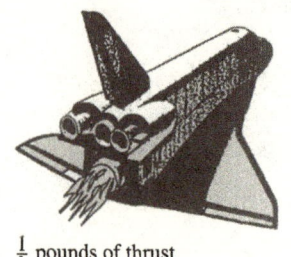

$\frac{1}{h}$ pounds of thrust

of $1/h$ pounds. See Figure 5.29. The position $y(t)$ of the spaceship along its straight line path can be found by solving the initial-value problem

$$my'' = \delta_h(t) = \frac{1}{h}[1 - u(t - h)]$$

$$y(0) = 0$$

$$y'(0) = 0 \tag{1}$$

See Figure 5.30.

Taking the Laplace transform of the differential equation gives

$$s^2 Y_h(s) = \frac{1 - e^{-hs}}{hs}$$

$$Y_h(s) = \frac{1 - e^{-hs}}{mhs^3} \tag{2}$$

Figure 5.30
Forcing function for the spaceship

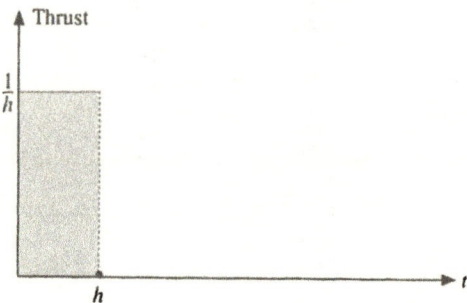

Note that we have denoted this transform by $Y_h(s)$ to reflect the fact that it depends on the *length h* of firing of the thrusters and the *magnitude* of the thrust ($1/h$). Taking the inverse of $Y_h(s)$ gives the desired response:

$$y_h(t) = \frac{1}{2mh}\left(t^2 - u(t - h)(t - h)^2\right)$$

$$= \begin{cases} \dfrac{1}{2mh}t^2 & (0 \le t < h) \\[2ex] \dfrac{1}{m}\left(t - \dfrac{h}{2}\right) & (h \le t) \end{cases} \tag{3}$$

The graphs of the thrusts $u_{0.5}$, u_1, and u_2 and the corresponding responses $y_{0.5}$, y_1, and y_2 are shown in Figure 5.31.

Figure 5.31
Three thrusts $u_{0.5}$, u_1, u_2, and their respective responses $y_{0.5}$, y_1, and y_2

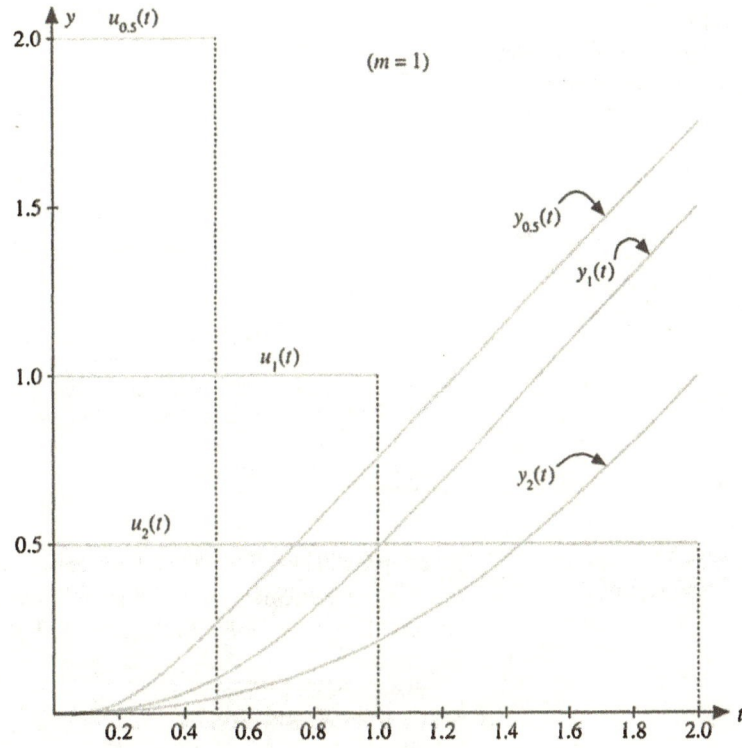

If we now let $h \to 0$ in Eq. (3) and use the "bottom" expression for $h \leq t$ (since it is the one that ultimately applies for $t > 0$), we find

$$y(t) = \lim_{h \to 0} y_h(t) = \frac{1}{m} \lim_{h \to 0} \left(t - \frac{h}{2}\right) = \frac{t}{m}$$

In other words, the limiting function $y(t) = t/m$ can be interpreted as the position of the spaceship at time t resulting from the delta function. Hence the delta function can be defined as the "input" that gives rise to this response $y(t)$. That is, we can define the delta function by "what it does" and not "what it is."*

A common interpretation of the delta function is that it is the "limiting" function of a sequence of functions that are getting "taller and taller" while at the same time getting "narrower and narrower." See Figure 5.32.

Also, since $\delta(t)$ represents an impulse of unit energy acting at $t = 0$, it follows that the translation $\delta(t - t_0)$ represents a **unit impulse acting at $t = t_0$**. We can also multiply the unit impulse $\delta(t - t_0)$ by a positive or negative constant A to obtain impulses of varying magnitudes acting in different directions.

---

* The delta function is like the Cheshire cat. It is the smile and not the cat on which we focus.

Figure 5.32
The delta function as the limit
of a certain sequence of
functions

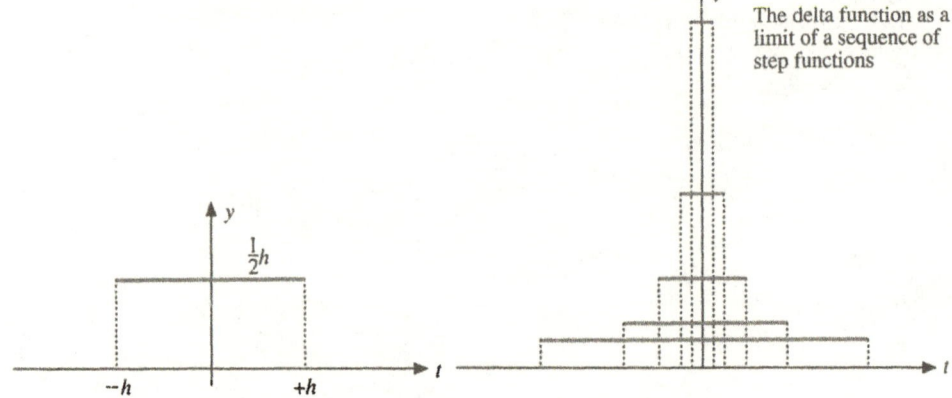

**THE LAPLACE
TRANSFORM OF $\delta(t)$**

In the spaceship problem the response $y(t)$ to the unit impulse function $\delta(t)$ was interpreted *indirectly* by taking the limiting response of $y_h(t)$ as $h \to 0$. If we *knew* the Laplace transform of $\delta(t)$, we could find $y(t)$ directly by solving the equation $my'' = \delta(t)$.

HISTORICAL NOTE

For many years, engineers and mathematicians alike interpreted the delta functions as a ''spike'' of *infinite* height and *infinitesimal* width that has unit area. Although this interpretation is physically appealing, it is not a genuine function. It was, however, legalized as a **distribution** or **generalized function** by the French mathematician Laurent Schwartz in the 1950's, and although this theory is extremely important in the study of differential equations (especially partial differential equations), it is outside the scope of this book.

To find the Laplace transform of $\delta(t)$, we first find the transform of the approximate impulse function $\delta_h(t) = 1 - u(t - h)$ and then find the limit of $\delta_h(t)$ as $h \to 0$. To find the transform of $\delta_h(t)$, we evaluate

$$
\begin{aligned}
\mathcal{L}\{\delta_h(t)\} &= \int_0^\infty e^{-st}\delta_h(t)\, dt \\
&= \frac{1}{h} \int_0^h e^{-st}\, dt && \text{(definition of } \delta_h(t)) \\
&= \frac{1}{h} \left(-\frac{1}{s} e^{-st} \bigg|_0^h \right) \\
&= \frac{1 - e^{hs}}{hs} && (4)
\end{aligned}
$$

Since both the numerator and denominator in Eq. (4) approach zero as $h \to 0$, we use L'Hôpital's rule to find the limit

$$\mathscr{L}\{\delta(t)\} = \lim_{h \to 0}\left(\frac{1 - e^{-hs}}{hs}\right) = \frac{1}{s}\lim_{h \to 0}\left(\frac{e^{-hs}}{1}\right) = 1 \tag{5}$$

By also using the shifting property $\mathscr{L}\{u(t - c)f(t - c)\} = e^{-cs}F(s)$, we find the transform of the unit impulse function acting at $t = t_0$ to be

$$\mathscr{L}\{\delta(t - t_0)\} = e^{-st_0} \tag{6}$$

The previous discussion leads to the following result.

Laplace Transform of the Delta Function

The Laplace transform of the **unit impulse function** or **Dirac delta function** $\delta(t)$ is defined to be

$$\mathscr{L}\{\delta(t)\} = 1$$

If the impulse function occurs at $t = t_0$, the Laplace transform of the shifted impulse function $\delta(t - t_0)$ is

$$\mathscr{L}\{\delta(t - t_0)\} = e^{-st_0}$$

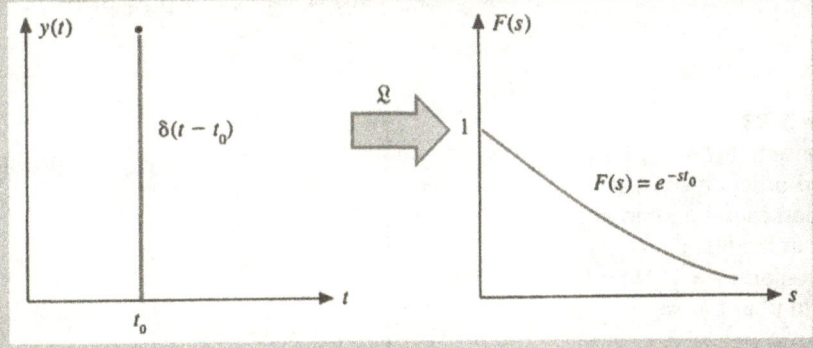

Note: To solve a differential equation with forcing term $\delta(t)$, the more precise mathematical approach is to first replace $\delta(t)$ by the step function

$$\delta_h(t) = \frac{1}{h}[1 - u(1 - h)] \tag{7}$$

and find the solution $y_h(t)$ corresponding to $\delta_h(t)$, and take the limit $y_h(t) \to y(t)$ as $h \to 0$. However, the same solution $y(t)$ will be obtained by using the Laplace transform to solve the differential equation directly with $\delta(t)$ as the forcing term.

Example 1   Impulse Forcing Function   A mass is attached to a spring and released from rest 1 ft below its equilibrium position. After the mass vibrates for π seconds, it is struck by a hammer in the downward direction, exerting unit energy on the mass. Assuming that the system is governed by the initial-value problem

$$y'' + y = \delta(t - \pi) \qquad y(0) = 1 \quad y'(0) = 0 \tag{8}$$

where $y(t)$ represents the downward displacement from equilibrium at time t, determine the subsequent motion of the mass.

Solution   Taking the Laplace transform of each side of Eq. (8), we obtain

$$s^2 Y(s) - s + Y(s) = e^{-\pi s}$$

Solving for $Y(s)$ gives

$$Y(s) = \frac{s}{s^2 + 1} + \frac{e^{-\pi s}}{s^2 + 1}$$

$$= \mathcal{L}\{\cos t\} + e^{-\pi s} \mathcal{L}\{\sin t\} \tag{9}$$

Using the shifting property, we find

$$y(t) = \cos t + u(t - \pi) \sin (t - \pi)$$

$$= \begin{cases} \cos t & (0 \le t < \pi) \\ \cos t - \sin t & (\pi \le t) \end{cases} \tag{10}$$

The graph of $y(t)$ is shown in Figure 5.33.   ∎

Figure 5.33
Response to $\delta(t - \pi)$ for a second-order differential equation causes a jump at $t = \pi$ in the first derivative of the solution by $+1$. Note the jump in y' at $t = \pi$.

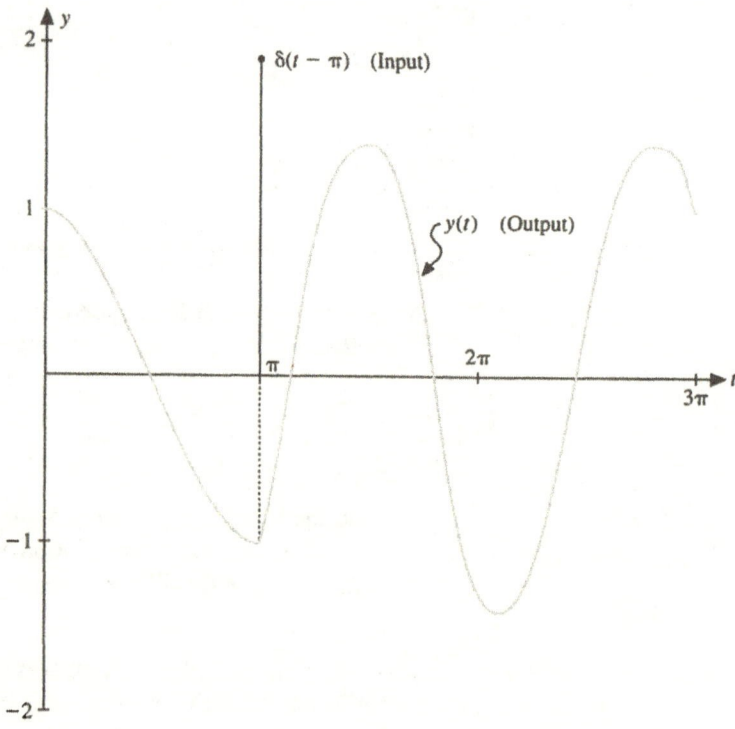

Example 2 **Delayed Impulse in a Damped System** Solve

$$y'' + 2y' + 5y = \delta(t - \pi) \qquad y(0) = 0 \quad y'(0) = 0 \tag{11}$$

Solution Taking the Laplace transform of the differential equation, we have

$$(s^2 + 2s + 5)Y(s) = e^{-\pi s}$$

Hence

$$\begin{aligned} Y(s) &= e^{-\pi s} \left\{ \frac{1}{s^2 + 2s + 5} \right\} \\ &= e^{-\pi s} \left\{ \frac{1}{(s + 1)^2 + 4} \right\} \\ &= \tfrac{1}{2} e^{-\pi s} \mathcal{L}\{ e^{-t} \sin 2t \} \end{aligned} \tag{12}$$

Again, using the shifting property, we find

$$y(t) = \frac{1}{2} u(t - \pi) e^{-(t-\pi)} \sin 2(t - \pi)$$

$$= \begin{cases} 0 & (0 \le t < \pi) \\ \frac{1}{2} e^{-(t-\pi)} \sin 2(t - \pi) & (\pi \le t) \end{cases} \tag{13}$$

The graph of $y(t)$ is shown in Figure 5.34. ∎

Figure 5.34
Notice the jump in y' from 0 to 1 at $t = \pi$ caused by the impulse function $\delta(t - \pi)$.

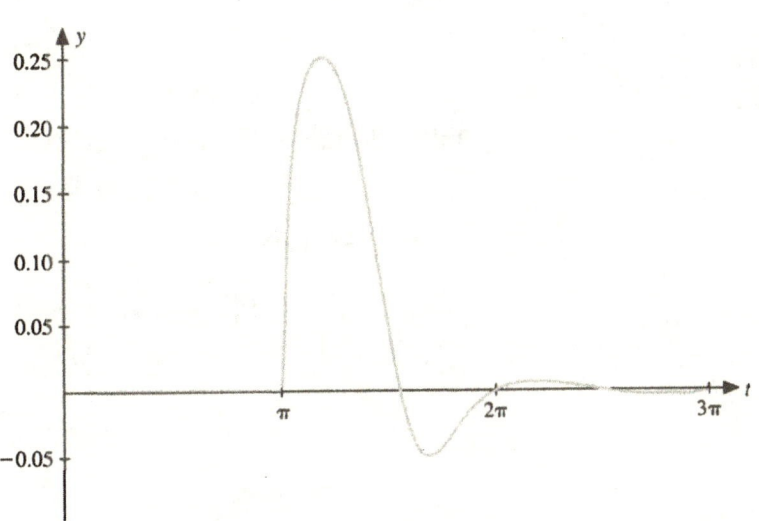

Note: The reader might have noticed that the delta function input in Example 2 did not give rise to a *discontinuity* in the solution, but it did make the *derivative*

of the solution jump. In general, for a second-order differential equation the derivative of the solution will have a jump discontinuity of magnitude A each time an impulse $A\delta(t - t_0)$ is applied. For a first-order equation a given delta input of $A\delta(t - t_0)$ will give rise to a jump discontinuity in the *function* of magnitude A.

Example 3   **Drug Therapy**   Sally's physician prescribes that she take 100 mg of Butalbital (stress medicine) tomorrow morning and another 100 mg on the following morning to relieve anxiety for an upcoming differential equations exam. After checking the *Physician's Desk Reference,* Sally discovers that the half-life of Butalbital in the body is 12 hours. How can she determine the amount of Butalbital in her body over the next several days? Will any stress medicine be left in her body by the time of the exam in four days?

Solution   Since drugs are usually eliminated from the body at a rate proportional to the amount present, the differential equation that describes the quantity of drug in her body at time t is

$$y' + ay = f(t) \tag{14}$$

where $f(t)$ represents the rate at which the drug is administered into the body. Since the half-life of Butalbital is $t_f = 12$ hr (1/2 day), we can find that the decay constant a in Eq. (14) is $a = \ln 2/t_f = 2 \ln 2 \doteq 1.4$. If Sally takes the drug in tablet form, the input function $f(t)$ can be taken to be an impulse function with magnitude equal to the quantity of drug ingested. Since Sally takes 100 mg of the drug on two consecutive days, the initial-value problem that describes the number of milligrams of Butalbital in her body after t days, starting from the time of the first ingestion, is

$$\begin{aligned} y' + 1.4y &= 100\delta(t) + 100\delta(t - 1) \\ y(0) &= 0 \end{aligned} \tag{15}$$

Taking the Laplace transform of Eq. (15), we get

$$s\mathcal{L}\{y\} + 1.4\mathcal{L}\{y\} = 100(1 + e^{-s})$$

Solving for $\mathcal{L}\{y\}$ gives

$$\begin{aligned} \mathcal{L}\{y\} &= 100 \left\{ \frac{1 + e^{-s}}{s + 1.4} \right\} \\ &= 100 \left\{ \frac{1}{s + 1.4} \right\} + 100 \left\{ \frac{e^{-s}}{s + 1.4} \right\} \\ &= 100\mathcal{L}\{e^{-1.4t}\} + 100e^{-s}\mathcal{L}\{e^{-1.4t}\} \end{aligned} \tag{16}$$

Hence

$$\begin{aligned} y(t) &= 100e^{-1.4t} + 100u(t - 1)e^{-1.4(t-1)} \\ &= \begin{cases} 100e^{-1.4t} & (0 \le t < 1) \\ 100(1 + e^{1.4})e^{-1.4t} & (1 \le t) \end{cases} \end{aligned} \tag{17}$$

See Figure 5.35. The quantity of drug in Sally's bloodstream four days after the therapy was begun has fallen drastically low to $y(4) \doteq 2$ mg. She hopes that it will be an easy exam.

Figure 5.35
Notice the *jumps* in the solution at $t = 0$ and $t = \pi$ caused by the impulses at those points. An impulse of magnitude $A\, \delta(t - t_0)$ will give rise to a jump in the solution y of a first-order equation by the amount A.

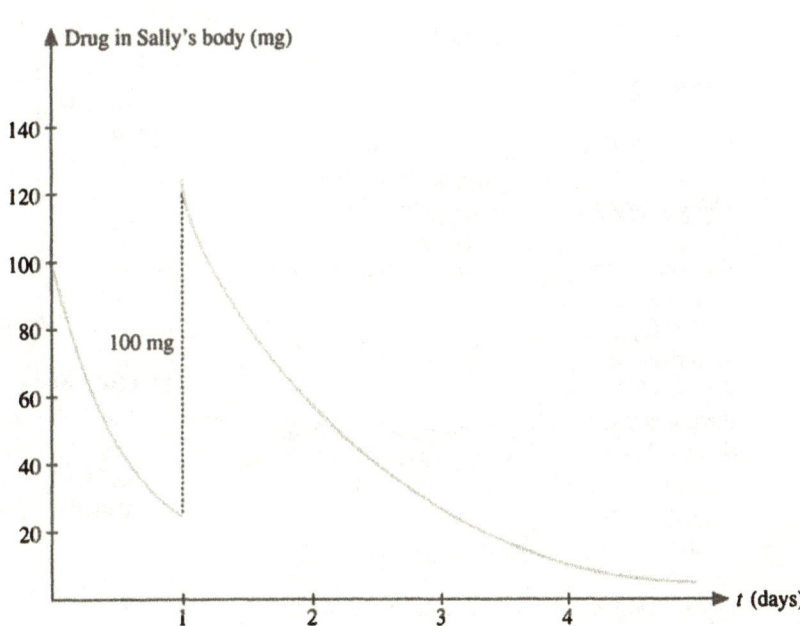

PROBLEMS: Section 5.7

For Problems 1–4, "plot" the given function and find the Laplace transform of the given function.

1. $\delta(t - 1) + 2\delta(t - 2) + 3\delta(t - 3)$
2. $\delta(t) - 2\delta(t - \pi) + \delta(t - 2\pi)$
3. $\delta(t) - \delta(t - 1) + \delta(t - 2) - \cdots$
4. $\delta(t) + \delta(t - \pi) + \delta(t - 2\pi) + \cdots$

For Problems 5–13, find the solution of the given initial-value problem. If you have access to a computer with a graphing package, sketch the graph of the solution.

5. $y' = \delta(t)$
   $y(0) = 0$
6. $y' = \delta(t) - \delta(t - 1)$
   $y(0) = 0$
7. $y'' + y = \delta(t - 2\pi)$
   $y(0) = 0 \quad y'(0) = 0$

8. $y'' + y = -\delta(t - \pi) + \delta(t - 2\pi)$
   $y(0) = 0 \quad y'(0) = 1$
9. $y'' + y = \delta(t - 2\pi)$
   $y(0) = 1 \quad y'(0) = 0$
10. $y'' + 4y' + 5y = \delta(t)$
    $y(0) = 0 \quad y'(0) = 0$
11. $y'' + 2y' + 2y = \delta(t - \pi)$
    $y(0) = 1 \quad y'(0) = 0$
12. $y'' + 2y' + 2y = \delta(t - \pi)$
    $y(0) = 0 \quad y'(0) = 0$
13. $y'' + 4y' + 5y = \delta(t - \pi) + \delta(t - 2\pi)$
    $y(0) = 0 \quad y'(0) = 2$
14. **Spaceship Driven by a Series of Thrusts** A spaceship with mass m is moving through frictionless space, driven only by a rocket that fires in short bursts of magnitude A at times $a, 2a, 3a, \ldots$. If the initial position and velocity of

the spaceship are $y(0) = y'(0) = 0$, find the position of the spaceship at any time. Sketch the graph of the solution.

15. **Differentiation in t Equals Multiplication by s** We start with the **integro-differential equation**

$$y' + 3y + 2\int_0^t y(t)\, dt = u(t - 1) \tag{18}$$

$$y(0) = 0$$

which is an equation that contains both a derivative of the unknown variable and an integral involving the unknown variable. The following three short problems illustrate an important fact.

(a) Take the Laplace transform of Eq. (18) and then multiply the resulting equation by s.

(b) Differentiate Eq. (18) with respect to t. (Accept the fact that the derivative of the step function $u(t - t_0)$ is the delta function $\delta(t - t_0)$.)

(c) Take the Laplace transform of the differential equation found in part (b) and observe that it is the same equation as is found in part (a). You have verified that differentiating a differential equation with respect to t is the same as multiplying the transform of the equation by s.

16. **The Soldier's Problem** In 1831 a column of soldiers marched over the Broughton suspension bridge near Manchester, England. Unfortunately, their cadence set up a forced vibration that had the same frequency as the natural frequency of the bridge. The bridge collapsed. Since that disaster, soldiers have always broken step when crossing a bridge. Can you tell what the solution of

$$y'' + y = \sum_{k=0}^{\infty} \delta(t - 2\pi k) \qquad y(0) = 0 \quad y'(0) = 0$$

has to do with this historical decision?

17. **Effect of an Impulse** Show that the solution $y(t)$ of

$$y'' + y = \delta(t) \qquad y(0) = 0 \quad y'(0) = 0$$

with a delta function input is the same as the solution of the unforced system

$$y'' + y = 0 \qquad y(0) = 0 \quad y'(0) = 1$$

18. **Impulse Functions Change Direction of the Solution** Show that the solution of

$$y'' + y = A\delta(t - t_0) \qquad y(0) = 0 \quad y'(0) = 0$$

is continuous as t_0 but the derivative jumps at t_0 by the amount A.

Drug Therapy

Problems 19–24 describe the quantity of drug in a person's body using various drug therapies and decay rates. For each problem, find $y(t)$ and sketch its graph.

19. $y' + y = \delta(t)$
  $y(0) = 0$ (single injection)

20. $y' + y = 1$
  $y(0) = 0$ (intravenous therapy)

21. $y' + y = 1 - u(t - 1)$
  $y(0) = 0$ (intravenous therapy)

22. $y' + y = \delta(t) + \delta(t - 1)$
  $y(0) = 0$ (two injections)

23. $y' + y = \delta(t) + \delta(t - 1) + \delta(t - 2)$
  $y(0) = 0$ (three injections)

24. $y' + y = \delta(t) + \delta(t - 1) + \delta(t - 2) + \cdots$
  $y(0) = 0$ (infinite injections)

25. **Steady State Drug Problem** If a drug is taken periodically, then depending on the dosage and the rate at which the drug is expelled from the body, a steady state accumulation of the drug will occur in the body. We can find the limiting amount by examining the long-term solution of

$$y' + ky = A\delta(t) + A\delta(t - 1) + A\delta(t - 2) + \cdots$$
$$y(0) = 0 \tag{19}$$

(a) Show that

$$\mathcal{L}\{y\} = \left(\frac{A}{s + k}\right)(1 + e^{-s} + e^{-2s} + e^{-3s} + \cdots)$$

(b) Show that

$$y(t) = Ae^{-kt}\left(1 + u(t - 1)e^k + u(t - 2)e^{2k} + \cdots\right)$$

(c) Show that at the end of day $t = N$ the amount of drug in the body is

$$y(N) = A\left(\frac{e^{-aN} - e^k}{1 - e^k}\right)$$

(d) Show that limiting amount of drug in the body is

$$A_L = \frac{Ae^k}{e^k - 1}$$

The graph of $y(t)$ and its limiting value are shown in Figure 5.36.

26. **Computer Problem** Use a computer to sketch the graph of the solution of Problem 19.

27. **Journal Entry** Summarize for yourself the concept of the delta function.

Figure 5.36
Drug accumulation curve due to a periodic dosage of size A with decay constant k

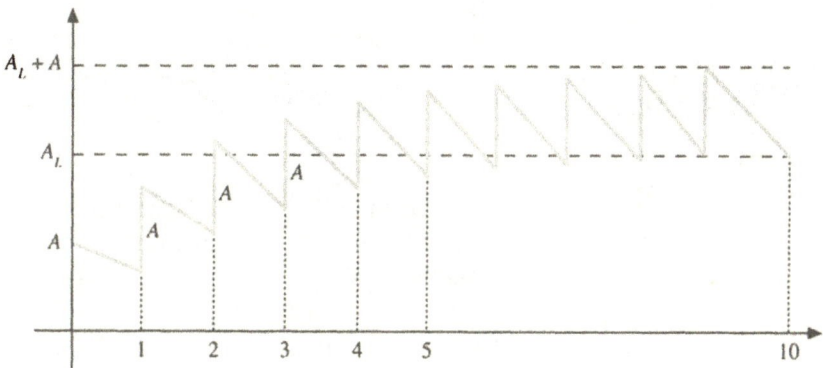

5.8   THE CONVOLUTION INTEGRAL

PURPOSE
To define the convolution of two functions $f(t)$ and $g(t)$ as

$$(f * g)(t) = \int_0^t f(t - \tau)g(\tau)\,d\tau$$

and prove the important convolution property

$$\mathcal{L}\{f * g\} = \mathcal{L}\{f\}\mathcal{L}\{g\}$$

INTRODUCTION TO THE CONVOLUTION

We have seen that the Laplace transform of a sum of two functions is the sum of the transforms. This leads one to wonder whether it is also true that the transform of a *product* of two functions is always equal to the *product* of their transforms. The answer is no, but there is a sort of "generalized multiplication" of two functions, known as the *convolution*, whose transform *is* the product of the transforms.

DEFINITION: Convolution of Two Functions
Let $f(t)$ and $g(t)$ be piecewise continuous functions defined on $(0, \infty)$. The **convolution** of $f(t)$ and $g(t)$, denoted $(f * g)(t)$, is defined by either (equivalent) integral

$$(f * g)(t) = \int_0^t f(t - \tau)g(\tau)\,d\tau$$

$$= \int_0^t f(\tau)\,g(t - \tau)\,d\tau \qquad (t \geq 0)$$

GEOMETRIC INTERPRETATION OF THE CONVOLUTION

The operation of computing the convolution of two functions $f(t)$ and $g(t)$ to obtain a third function $h(t) = (f * g)(t)$ has many of the properties of a multiplication. To better understand the convolution integral, consider the two simple functions

$$f(\tau) = 1 - u(\tau - 1) \qquad \text{and} \qquad g(\tau) = e^{-\tau}$$

whose graphs are drawn in Figure 5.37(a).

We use the variable τ because it is the variable of integration in the convolution integral. We also sketch the graph of $g(\tau - t)$, which is the graph of $g(\tau)$ moved to the right t units, and the graph of $g(t - \tau)$, which is the graph of $g(\tau - t)$ reflected about the point $\tau = t$. See Figure 5.37(b).

Figure 5.37
The convolution is somewhat reminiscent of the dot product of two vectors except that here we are working with functions and the arguments of the function go in "opposite directions."

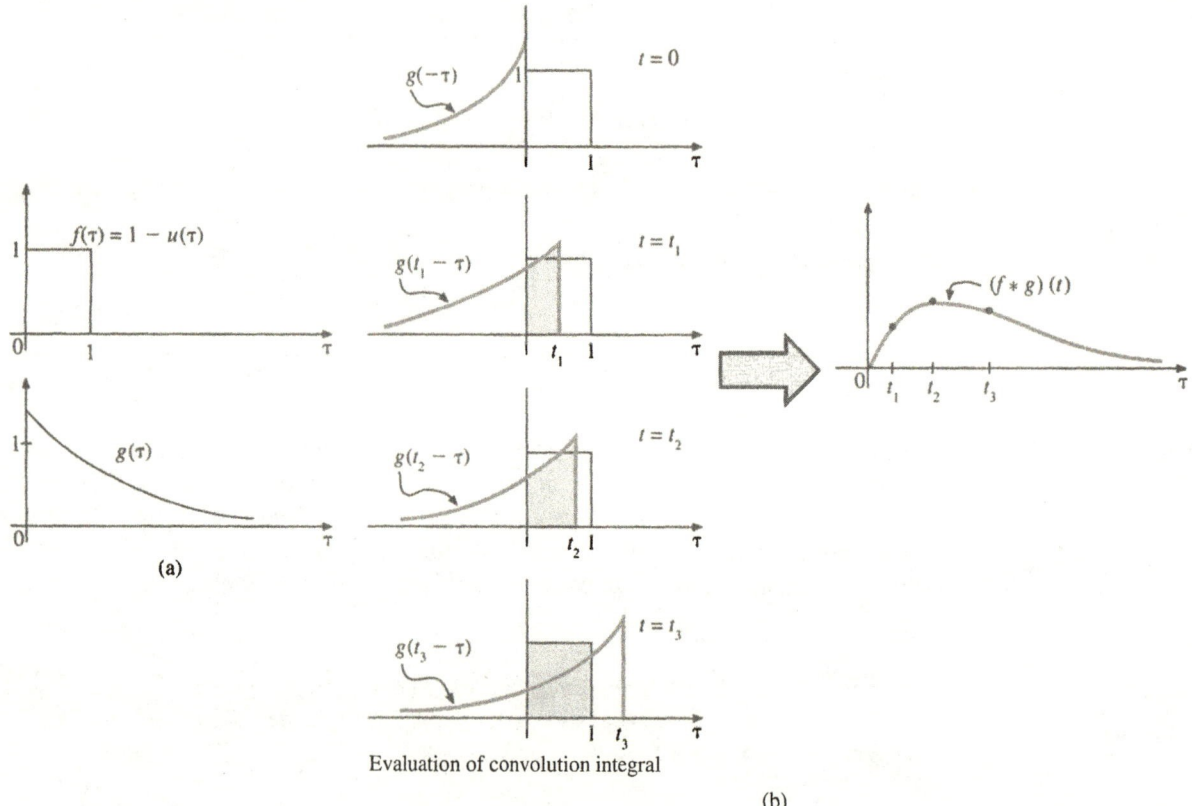

Evaluation of convolution integral

(b)

Now observe that the integrand $g(t - \tau)f(\tau)$ that appears in the convolution integral will be different from zero only when the nonzero portions of the functions $g(t - \tau)$ and $f(\tau)$ "overlap." In Figure 5.37(b) the graphs of the functions $g(t - \tau)$ and $f(\tau)$ are drawn (as functions of τ) for fixed values of t. For *each* of these values of t, the value of the convolution integral is equal to the shaded area. By plotting the values of these shaded areas as t changes, it is possible to sketch the graph of $(f * g)(t)$.

Although the convolution of two functions is not the same as function multiplication, it shares many of the same properties.

THEOREM 5.9: Properties of the Convolution

If $f(t)$ and $g(t)$ are piecewise continuous on $(0, \infty)$, then

- $f * g = g * f$ (commutative property)
- $f * (g * h) = (f * g) * h$ (associative property)
- $f * (g + h) = f * g + f * h$ (distributive property)
- $f * 0 = 0$ (zero multiplication)

PROOF: We will verify the commutative property. The other properties are left to the reader. (See Problems 1–3 at the end of this section.) The commutative property is a simple consequence of making a change of variable from τ to $\xi = t - \tau$ in the convolution integral. We have

$$
\begin{aligned}
(f * g)(t) &= \int_0^t f(t - \tau) g(\tau)\, d\tau && \text{(definition of } f * g) \\
&= \int_t^0 f(\xi) g(t - \xi)(-d\xi) && \text{(changing variable } \xi = t - \tau) \\
&= \int_0^t g(t - \xi) f(\xi)\, d\xi && \text{(simple calculus)} \\
&= (g * f)(t) && \text{(definition of } g * f) \quad\quad (1)
\end{aligned}
$$

This completes the proof.

Note: We often denote the convolution by both $f * g$ and $(f * g)(t)$. This admittedly "back-and-forth" use of function notation is analogous to interchanging of f and $f(t)$, as mathematicians often do. When we wish to emphasize the convolution as an operation, we use $f * g$; when we wish to emphasize the value of the convolution at t, we use $(f * g)(t)$.

There are several reasons for the importance of the convolution of two functions. The following theorem describes one of the most important reasons.

THEOREM 5.10: Transform of a Convolution

If $f(t)$ and $g(t)$ are piecewise continuous functions of exponential order $e^{\alpha t}$, then

$$\mathcal{L}\{f * g\} = \mathcal{L}\{f\}\mathcal{L}\{g\} = F(s)G(s) \quad (s > a)$$

where $F(s) = \mathcal{L}\{f\}$ and $G(s) = \mathcal{L}\{g\}$. Stated in inverse form,

$$\mathcal{L}^{-1}\{F(s)G(s)\} = (f * g)(t)$$

PROOF: Starting from the definition of the Laplace transform, we write

$$F(s)G(s) = \left(\int_0^\infty e^{-su}f(u)\,du\right)\left(\int_0^\infty e^{-sv}g(v)\,dv\right)$$

$$= \int_0^\infty \int_0^\infty e^{-s(u+v)}f(u)g(v)\,du\,dv$$

$$= \int_0^\infty g(v)\left\{\int_0^\infty e^{-s(u+v)}f(u)\,du\right\}dv$$

The goal is to show that the above expression is $\mathscr{L}\{f * g\}$. To do this, we first simplify the integral in u by replacing u by $t = u + v$ for fixed v. This gives

$$F(s)G(s) = \int_0^\infty g(v)\left\{\int_v^\infty e^{-st}f(t-v)\,dt\right\}dv$$

and replacing v by τ, we obtain

$$F(s)G(s) = \int_0^\infty g(\tau)\left\{\int_\tau^\infty e^{-st}f(t-\tau)\,dt\right\}d\tau \tag{2}$$

The integration in Eq. (2) is carried out over the shaded wedge-shaped region shown in Figure 5.38. Note that if the first integration on t (from τ to ∞) is changed to an integration on τ first (from 0 to t), then we get*

$$F(s)G(s) = \int_0^\infty e^{-st}\left\{\int_0^t f(t-\tau)g(\tau)\,d\tau\right\}dt = \mathscr{L}\{f * g\}$$

Figure 5.38
Region of integration

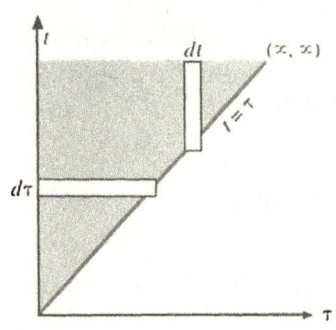

Example 1   Use of the Convolution Property   Express the solution of

$$
\begin{aligned}
y'' + y &= f(t)\\
y(0) &= 0\\
y'(0) &= 0
\end{aligned} \tag{3}
$$

in terms of the external forcing function $f(t)$.

Solution   Taking the Laplace transform of each side of Eq. (3), we have

$$s^2 Y(s) + Y(s) = F(s) \tag{4}$$

where $F(s) = \mathscr{L}\{f\}$. Solving for $Y(s)$ gives

$$Y(s) = \frac{F(s)}{s^2 + 1}$$

---

* This interchange of integrals is valid if the functions $f(t)$ and $g(t)$ are piecewise continuous and of exponential order.

Using the convolution property, we have

$$y(t) = \mathcal{L}^{-1}\left\{\frac{F(s)}{s^2 + 1}\right\}$$

$$= \sin t * f(t)$$

$$= \int_0^t \sin (t - \tau) f(\tau)\, d\tau \tag{5}$$

This gives a useful formula for $y(t)$ in terms of the forcing function $f(t)$.   ◼

The convolution property has other applications. One principal use is to find inverses of transforms.

Example 2   Application of the Convolution Property   Find the inverse transform

$$\mathcal{L}^{-1}\left\{\frac{1}{(s^2 + 1)^2}\right\}$$

Solution

$$\mathcal{L}^{-1}\left\{\frac{1}{(s^2 + 1)^2}\right\} = \mathcal{L}^{-1}\left\{\frac{1}{s^2 + 1}\right\} * \mathcal{L}^{-1}\left\{\frac{1}{s^2 + 1}\right\}$$

$$= \sin t * \sin t$$

$$= \int_0^t \sin (t - \tau) \sin \tau\, d\tau$$

$$= \frac{1}{2}\int_0^t [\cos (2\tau - t) - \cos t]\, d\tau *$$

$$= \frac{1}{2}\left(\frac{\sin (2\tau - t)}{2}\bigg|_0^t\right) - \frac{1}{2}t \cos t$$

$$= \frac{1}{2}\left(\frac{\sin t}{2} - \frac{\sin (-t)}{2}\right) - \frac{1}{2}t \cos t$$

$$= \frac{1}{2}\left(\sin t - t \cos t\right) \tag{6}$$

◼

Another application of the convolution theorem lies in the solution of *integral equations* in which the unknown function appears under the integral sign.

Example 3   Solving Integral Equations   Solve for $y(t)$ in the integral equation

$$y(t) = 1 + \int_0^t \cos (t - \tau)\, y(\tau)\, d\tau \tag{7}$$

---

* We have used the identity $\sin A \sin B = \frac{1}{2}[\cos (B - A) - \cos (B + A)]$.

Solution Taking the Laplace transform of each side of the equation and using the convolution theorem, we have

$$Y(s) = \frac{1}{s} + \left(\frac{s}{s^2 + 1}\right) Y(s) \tag{8}$$

Solving for $Y(s)$ gives

$$Y(s) = \frac{s^2 + 1}{s(s^2 - s + 1)}$$

$$= \frac{1}{s} + \frac{1}{s^2 - s + 1}$$

$$= \frac{1}{s} + \frac{2}{\sqrt{3}} \frac{\sqrt{3}/2}{\left[(s - \frac{1}{2})^2 + \frac{3}{4}\right]}$$

Hence

$$y(t) = 1 + \frac{2\sqrt{3}}{3} e^{t/2} \sin\left(\frac{\sqrt{3}}{2}t\right) \tag{9}$$

Two of the most important concepts in engineering system theory are the concepts of the transfer function and the impulse response function.

THE TRANSFER FUNCTION AND THE IMPULSE RESPONSE FUNCTION

Impulse input $\delta(T)$	General input $f(t)$
$aI'' + bI' + cI = \delta(t)$	$ay'' + by' + cy = f(t)$
$I(0) = 0$	$y(0) = 0$
$I'(0) = 0$	$y'(0) = 0$ \hfill (10)

For the impulse input problem the system is driven by a unit impulse at $t = 0$, and in the general input problem the system is driven by a general forcing function $f(t)$. The solution $I(t)$ to the impulse input problem is called the **impulse response function.** If we now take the Laplace transform of both differential equations (10), we find

$$\mathcal{L}\{I(t)\} = \frac{1}{as^2 + bs + c} \qquad \text{(impulse input)} \tag{11}$$

$$\mathcal{L}\{y(t)\} = \left\{\frac{1}{as^2 + bs + c}\right\} \mathcal{L}\{f\} \qquad \text{(general input)} \tag{12}$$

where $\mathcal{L}\{I(t)\}$ is called the **transfer function** for the linear system. From Eqs. (11) and (12) we obtain the relationship

$$\mathcal{L}\{y(t)\} = \mathcal{L}\{I(t)\}\mathcal{L}\{f(t)\} \tag{13}$$

and using the convolution property, we have

$$y(t) = \mathfrak{L}^{-1}\left\{\mathfrak{L}\{I\}\mathfrak{L}\{f\}\right\} = I(t) * f(t) = \int_0^t I(t - \tau) f(\tau)\, d\tau \qquad (14)$$

The above discussion can be summarized in the following important theorem.

THEOREM 5.11: Linear System Representation as a Convolution

The solution $y(t)$ to the general linear system

$$ay'' + by' + cy = f(t)$$
$$y(0) = 0$$
$$y'(0) = 0$$

can be written as the convolution of the impulse response function $I(t)$ and the forcing function $f(t)$. That is,

$$y(t) = I(t) * f(t) = \int_0^t I(t - \tau) f(\tau)\, d\tau$$

INTERPRETATION OF THE CONVOLUTION PROPERTY

Consider the linear system*

$$ay'' + by' + cy = f(t)$$
$$y(0) = 0 \qquad\qquad (15)$$
$$y'(0) = 0$$

To interpret the convolution representation property stated in Theorem 5.11, think of breaking the input $f(t)$ into a succession of small rectangular pulses. See Figure 5.39. If the time interval Δh is small, then each rectangular pulse will produce a response roughly the same as that produced by an impulse function of magnitude equal to the area under the rectangular pulse. For example, the first pulse of area $f(0)\,\Delta h$ can be approximated by the impulse function, $f(0)\,\Delta h\,\delta(t)$, whose response would be the impulse response function, $f(0)\,\Delta h\,I(t)$. Likewise, the second rectangular pulse can be approximated by the impulse function $f(\Delta h)\,\Delta h\,\delta(t - \Delta h)$, whose response would be $f(\Delta h)\,\Delta h\,I(t - \Delta h)$. In general, the nth rectangular pulse can be approximated by the impulse function $f(n\,\Delta h)\,\Delta h\,\delta(t - n\,\Delta h)$, whose response would be

$$\text{Response to the } n\text{th impulse} = f(n\,\Delta h)\,\Delta h\,I(t - n\,\Delta h) \qquad (16)$$

By using the superposition principle for linear differential equations, the response $y_{\Delta h}(t)$ to the succession of N rectangular pulses of width Δh can be found by summing the N individual responses. In other words,

$$y_{\Delta h}(t) = \sum_{n=0}^{N} f(n\,\Delta h)\,\Delta h\,I(t - n\,\Delta h) \qquad (17)$$

* It is not necessary that the differential equation have constant coefficients, only that it be linear. We write the equation with constant coefficients only because we have stressed equations with constant coefficients in this chapter.

Figure 5.39
The representation of the solution
of a linear system as the convolution
of the impulse response function.
The input can be interpreted
as a "superposition" or "summing"
of an infinite number of responses to
delta functions.

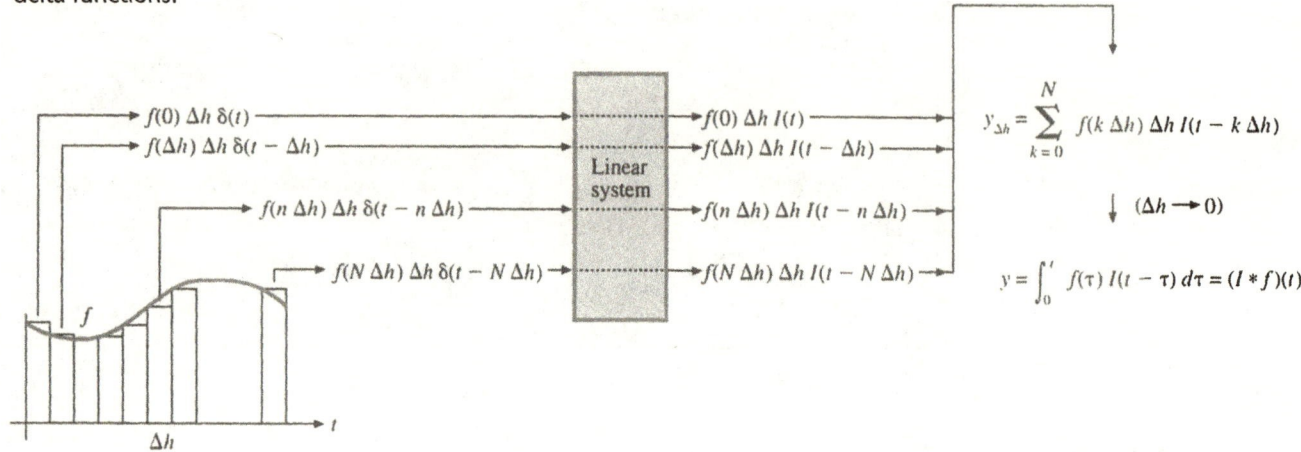

But as the number of pulses n increases without bound ($n \to \infty$) and the width Δh of each pulse approaches zero ($\Delta h \to 0$), the system response $y(t)$ to the general system (15) will approach

$$y(t) = \int_0^t I(t - \tau) f(\tau) \, d\tau = I * f \tag{18}$$

In other words, the output $y(t)$ of a linear differential equation can be interpreted as "breaking" the input $f(t)$ into an infinite number (a *continuum*) of impulses, finding the response to each impulse, and then adding up (integrating) the impulse responses. Of course, communications engineers have known this for years, but the convolution integral provides a mathematical justification of this important process.

PROBLEMS: Section 5.8

For Problems 1–3, verify the properties of the convolution.

1. $f * (g * h) = (f * g) * h$
2. $f * (g + h) = (f * g) + (f * h)$
3. $f * 0 = 0$

For Problems 4–10, find the given convolution.

4. $1 * 1$
5. $1 * 1 * 1$
6. $1 * t$
7. $t * t$
8. $t * t * t * \cdots * t$ (k factors)
9. $e^t * e^{-t}$
10. $e^{at} * e^{-at}$
11. **First-Order Linear Convolution Equation** Inasmuch as the convolution is sometimes called the "generalized product," is the solution of the equation

$$a * t = b \qquad (a \neq 0)$$

equal to $t = b/a$ as it is for $at = b$?

For Problems 12–19, find the solution y of the given initial-value problem in terms of a convolution involving f.

12. $y' = f(t)$ $\qquad\qquad$ $y(0) = 0$
13. $y' = f(t)$ $\qquad\qquad$ $y(0) = 1$
14. $y' + y = f(t)$ $\qquad$ $y(0) = 0$
15. $y' + y = f(t)$ $\qquad$ $y(0) = 1$
16. $y'' + y = f(t)$ $\qquad$ $y(0) = 1$ $\quad$ $y'(0) = 0$
17. $y'' + 4y = f(t)$ $\qquad$ $y(0) = 0$ $\quad$ $y'(0) = 1$
18. $y'' + 3y' + 2y = f(t)$ $\quad$ $y(0) = 0$ $\quad$ $y'(0) = 0$
19. $y'' + 4y' + 5y = f(t)$ $\quad$ $y(0) = 1$ $\quad$ $y'(0) = 0$

For Problems 20–25, find the given inverse transform in terms of a convolution and evaluate the convolution.

20. $\mathcal{L}^{-1}\left\{\dfrac{1}{s^2}\right\}$ $\qquad$ **23.** $\mathcal{L}^{-1}\left\{\dfrac{4}{s^2(s-2)}\right\}$

21. $\mathcal{L}^{-1}\left\{\dfrac{1}{s^3}\right\}$ $\qquad$ **24.** $\mathcal{L}^{-1}\left\{\dfrac{1}{s^2(s^2+1)}\right\}$

22. $\mathcal{L}^{-1}\left\{\dfrac{1}{s(s+1)}\right\}$ $\qquad$ **25.** $\mathcal{L}^{-1}\left\{\dfrac{1}{(s^2+1)^2}\right\}$

Fractional Calculus

*Although most students of calculus learn only how to integrate and differentiate functions once, twice, three times, and so on, there is a theory of calculus, called **fractional calculus**, that defines integrals and derivatives of any real (or even complex) order. Problems 26–28 deal with this subject.*

26. Fractional Integral The **one-half integral** of a function $f(t)$ is defined by

$$I_{1/2}\{f\} = \frac{1}{\sqrt{\pi}}\left[t^{-1/2} * f\right] \qquad (19)$$

Find the following one-half integrals and compare them with the first integrals (antiderivatives).
(a) $I_{1/2}\{1\}$ $\qquad$ (b) $I_{1/2}\{t\}$
(c) $I_{1/2}\{at^2 + bt + c\}$

27. Something That Must Hold Verify the property

$$I_{1/2}\{I_{1/2}(f)\}(t) = \int_0^t f(\tau)\,d\tau \qquad (20)$$

Hint: $\mathcal{L}\{t^{-1/2}\} = \sqrt{\pi/s}.$

28. Fractional Derivatives Using the definition of the one-half integral in Problem 26, the **one-half derivative** of a function $f(t)$ is defined by

$$\frac{d^{-1/2}f(t)}{dt^{-1/2}} = \frac{d}{dt}I_{-1/2}\{f\} \qquad (21)$$

As with whole derivatives, fractional derivatives may or may not exist. Find the following one-half derivatives and compare them with the first derivatives.

(a) $\dfrac{d^{-1/2}}{dt^{-1/2}}\{1\}$

(b) $\dfrac{d^{-1/2}}{dt^{-1/2}}\{t\}$

(c) $\dfrac{d^{-1/2}}{dt^{-1/2}}\{at^2 + bt + c\}$

Volterra Integral Equations

An equation of the type

$$y(t) = g(t) + \int_0^t k(t - \tau)\,y(\tau)\,d\tau \qquad (22)$$

*where g and k are known functions and the unknown function y appears under the integral sign is called a **Volterra integral equation**. Since the integral in this equation is the convolution $k * y$, it is possible to solve this equation by use of the Laplace transform. For Problems 29–33, solve the given Volterra integral equation.*

29. $y(t) = 1 - \displaystyle\int_0^t y(\tau)\,d\tau$

30. $y(t) = t - \displaystyle\int_0^t (t - \tau)\,y(\tau)\,d\tau$

31. $y(t) = t^3 + \displaystyle\int_0^t \sin(t - \tau)y(\tau)\,d\tau$

32. $y(t) = e^t\left\{1 + \displaystyle\int_0^t e^{-\tau}\,y(\tau)\,d\tau\right\}$

33. $y(t) = \cos t + \displaystyle\int_0^t \sin(t - \tau)y(\tau)\,d\tau$

34. General Solution of Volterra's Equation Show that the general solution $y(t)$ of the Volterra integral equation

$$y(t) = f(t) + \int_0^t g(t - \tau)y(\tau)\,d\tau$$

is given by

$$y(t) = \mathcal{L}^{-1}\left\{\frac{F(s)}{1 - G(s)}\right\}$$

where $F(s)$ and $G(s)$ are the Laplace transforms of $f(t)$ and $g(t)$, respectively.

35. The Equivalence of Integral and Differential Equations Show that the initial-value problem

$$\frac{dy}{dt} = p(t)y(t) + f(t) \qquad y(0) = y_0 \qquad (23)$$

is equivalent to the integral equation

$$y(t) = y_0 + f(t) + \int_0^t p(\tau)y(\tau)\,d\tau \qquad (24)$$

Differo-Integral Equations

*Equations in which the unknown function both is differentiated and appears in an integral are called **differo-integral equations**. For Problems 36–38, solve the given differo-integral equation with initial condition.*

36. $y' + \int_0^t y(\tau)\, d\tau = 1$ $\qquad\qquad y(0) = 0$

37. $y' = 1 - \sin t - \int_0^t y(\tau)\, d\tau$ $\qquad y(0) = 0$

38. $y' - \int_0^t \cos(t - \tau)y(\tau)\, d\tau = \sin t$ $\quad y(0) = 0$

39. Interesting Integral Equation Solve the integral equation

$$\int_0^t y(\tau)\, d\tau = y(t) * y(t)$$

and verify your result.

Transfer Function and the Impulse Response Function

For Problems 40–43, find the transfer function and the impulse response function of the given input-output system. Write the output of the system as a convolution of the impulse response function and the input function. The initial conditions are always taken to be zero.

40. $y' = f(t)$

41. $y' + ay = f(t)$

42. $y'' + y = f(t)$

43. $y'' + 4y' + 5y = f(t)$

44. Duhamel's Formula Let $y(t)$ be a solution of the initial-value problem

$$ay'' + by' + cy = f(t) \qquad y(0) = y'(0) = 0$$

where we assume that $f(t)$ is piecewise continuous and of exponential order.

(a) Show that $y(t)$ can be expressed as the Duhamel integral

$$y(t) = \int_0^t z'(\tau)f(t - \tau)\, d\tau$$

where $z(t)$ is a solution of the initial-value problem

$$az'' + bz' + cz = 1 \qquad z(0) = z'(0) = 0$$

(b) Use the Duhamel integral formula to find the solution of the initial-value problem

$$y'' - y = f(t) \qquad y(0) = y'(0) = 0$$

45. Journal Entry Summarize briefly the Laplace transform and its applications.

6

Systems of Differential Equations

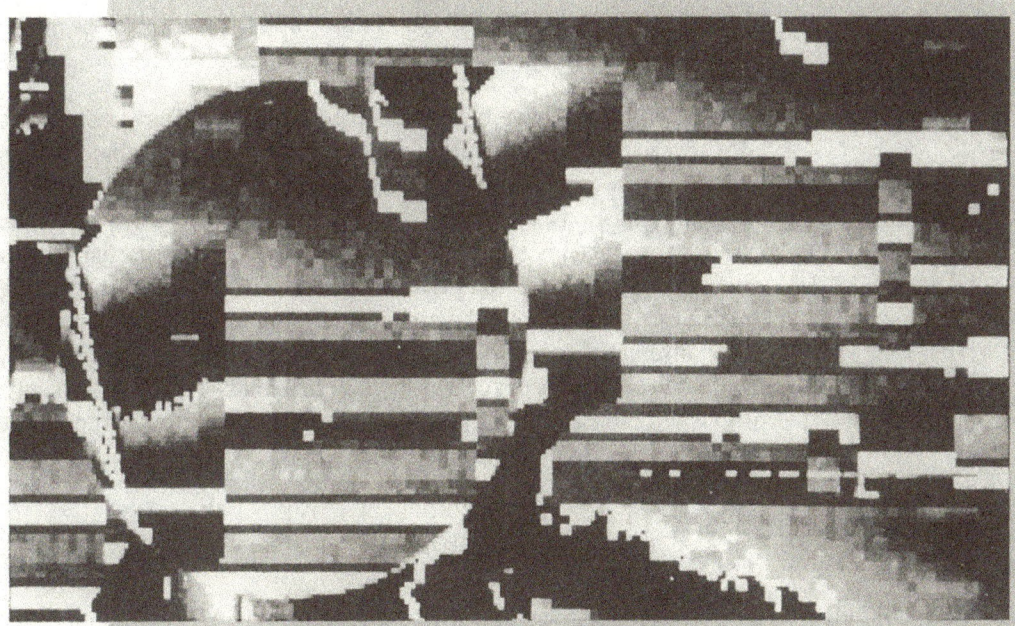

6.1 Introduction to Linear Systems: The Method of Elimination

6.2 Review of Matrices

6.3 Basic Theory of First-Order Linear Systems

6.4 Homogeneous Linear Systems with Real Eigenvalues

6.5 Homogeneous Linear Systems with Complex Eigenvalues

6.6 Nonhomogeneous Linear Systems

6.7 Nonhomogeneous Linear Systems: Laplace Transform (Optional)

6.8 Applications of Linear Systems

6.9 Numerical Solution of Systems of Differential Equations

6.1 INTRODUCTION TO LINEAR SYSTEMS: THE METHOD OF ELIMINATION

PURPOSE

To introduce the concept of a system of linear differential equations and present the simplest method of solution, the method of elimination, which works well for small systems of two or three equations. We will also show how a single differential equation of order n can be written as a system of n first-order differential equations.

Until now, we have been concerned with *one* differential equation with *one* unknown variable. Now, however, we will study *systems* of *linear* differential equations that involve two or more unknown variables. This is analogous to a system of linear *algebraic* equations in which the equations must be solved simultaneously. Linear systems arise in many fields of science, such as vibrating mechanical systems, multiloop electrical networks, and compartmental analysis in chemistry and biology. For example, the study of the motion of the nine planets around the sun requires a system of 54 first-order differential equations (six first-order equations for each planet). In this section we introduce the concept of a system of differential equations and solve linear systems of two and three equations. We will leave the 54 first-order (nonlinear) equations that describe planetary motion for the computer.

GENERAL FIRST-ORDER LINEAR SYSTEM

By a **first-order system of linear differential equations** we mean the set of simultaneous differential equations of the form

$$
\begin{aligned}
\frac{dx_1(t)}{dt} &= a_{11}(t)x_1 + a_{12}(t)x_2 + \cdots + a_{1n}(t)x_n + f_1(t) \\
\frac{dx_2(t)}{dt} &= a_{21}(t)x_1 + a_{22}(t)x_2 + \cdots + a_{2n}(t)x_n + f_2(t) \\
&\ \ \vdots \qquad\quad\ \vdots \qquad\qquad\ \vdots \qquad\qquad\quad \vdots \qquad\qquad \vdots \\
\frac{dx_n(t)}{dt} &= a_{n1}(t)x_1 + a_{n2}(t)x_2 + \cdots + a_{nn}(t)x_n + f_n(t)
\end{aligned}
\tag{1}
$$

where the *coefficients* $a_{ij}(t)$ and the *nonhomogeneous terms* $f_i(t)$ are continuous functions on the interval of interest. If the coefficients $a_{ij}(t)$ are constants, then the linear system (1) is said to have **constant coefficients.** Also, if the functions $f_i(t)$ are zero for all t on the interval of interest, then the system (1) is called **homogeneous;** otherwise, the system is said to be **nonhomogeneous.** By a **solution** of system (1) we mean a collection of functions $x_1(t), x_2(t), ..., x_n(t)$ that satisfy system (1) on some interval. In addition to the linear system (1) there may also be specified **initial conditions**

$$
x_1(t_0) = A_1, \ x_2(t_0) = A_2, \ ..., \ x_n(t_0) = A_n
\tag{2}
$$

where t_0 is a specified value of t and $A_1, A_2, ..., A_n$ are prescribed constant values. The problem of finding a solution to the linear system (1) that also satisfies initial conditions (2) is called the **initial-value problem** for the linear system.

Example 1   Solution of a Linear System   Verify that the pair of functions

$$x_1(t) = e^{2t} + e^{6t} + 1$$

$$x_2(t) = 2e^{2t} - 2e^{6t} + 1$$

$$(3)$$

is a solution of the system

$$\frac{dx_1}{dt} = 4x_1 - x_2 - 3 \tag{4a}$$

$$\frac{dx_2}{dt} = -4x_1 + 4x_2 \tag{4b}$$

Solution   Substituting $x_1(t)$ and $x_2(t)$ into Eqs. (4a) and (4b), we check whether

$$\underbrace{2e^{2t} + 6e^{6t}}_{\uparrow \atop x_1{}'} \overset{?}{=} 4\underbrace{(e^{2t} + e^{6t} + 1)}_{\uparrow \atop x_1} - \underbrace{(2e^{2t} - 2e^{6t} + 1)}_{\uparrow \atop x_2} - 3$$

$$\underbrace{4e^{2t} - 12e^{6t}}_{\uparrow \atop x_2{}'} \overset{?}{=} -4\underbrace{(e^{2t} + e^{6t} + 1)}_{\uparrow \atop x_1} + 4\underbrace{(2e^{2t} - 2e^{6t} + 1)}_{\uparrow \atop x_2}$$

Upon closer inspection we see that these equations hold. The reader should realize that $x_1(t)$ by itself or $x_2(t)$ by itself is not a solution, but the *pair* $x_1(t)$ and $x_2(t)$ constitute a solution. We will see later that there are many more solutions (pairs) of this linear system.   ∎

TRANSFORMING HIGHER-ORDER EQUATIONS TO FIRST-ORDER SYSTEMS

There are several reasons for studying systems of differential equations. One reason lies in the fact that all higher-order differential equations can be written as an *equivalent* system of first-order equations. Being able to write higher-order equations as systems of first-order equations allows one to combine much of the subject matter of differential equations under a single unifying theory. And then from a numerical point of view, most computer programs are written to approximate solutions of first-order systems of equations, and so users who wish to approximate solutions of higher-order equations must rewrite these equations as systems of equations.

Reduction of an *n*th-Order Equation to a System of *n* First-Order Equations

The *n*th-order linear differential equation

$$y^{(n)} + a_1(t)y^{(n-1)} + \cdots + a_{n-1}(t)y' + a_n(t)y = f(t) \qquad (5)$$

with initial conditions

$$y(x_0) = A_1, \quad y'(x_0) = A_2, \quad y''(x_0) = A_3, \quad \ldots, \quad y^{(n-1)}(x_0) = A_n$$

can be rewritten as a system of *n* first-order linear* equations by defining

$$x_1 = y, \quad x_2 = y', \quad x_3 = y'', \quad \ldots, \quad x_n = y^{(n-1)}$$

Differentiating these equations and rewriting the derivatives in terms of the new variables $x_1, x_2, \ldots, x_n$, we arrive at the following system of first-order equations:

$$
\begin{aligned}
x_1' &= x_2 \\
x_2' &= x_3 \\
x_3' &= x_4 \\
&\vdots \\
x_{n-1}' &= x_n \\
x_n' &= -a_n(t)x_1 - \cdots - a_1(t)x_n + f(t)
\end{aligned}
\qquad
\begin{aligned}
x_1(x_0) &= A_1 \\
x_2(x_0) &= A_2 \\
x_3(x_0) &= A_3 \\
&\vdots \\
x_{n-1}(x_0) &= A_{n-1} \\
x_n(x_0) &= A_n
\end{aligned}
$$

Note: In other words, when we transform an *n*th-order differential equation to a system of first-order differential equations, the solution of the *n*th-order equation is the *first component* of the first-order system, the first derivative of the *n*th-order equation is the *second component* of the system, the second derivative of the *n*th-order equation is the *third component* of the system, and so on.

Example 2 **The Vibrating Spring Equation as a Linear System** Write the initial-value problem

$$y'' + y = 0 \qquad y(0) = 0 \qquad y'(0) = 1 \qquad (6)$$

as a system of two first-order equations.

Solution Making the substitution $x_1 = y$ and $x_2 = y'$ and differentiating these equations, we get the linear system

$$
\begin{aligned}
x_1' &= x_2 & x_1(0) &= 0 \\
x_2' &= -x_1 & x_2(0) &= 1
\end{aligned}
\qquad (7)
$$

Note that $y(t) = \sin t$ is the solution of the second-order equation (6) and that the pair of functions $x_1(t) = \sin t$, $x_2(t) = \cos t$ is a solution of the first-order linear system (7). ∎

---

* Equivalence between higher-order differential equations and first-order systems is also valid for *nonlinear* differential equations. However, we restrict ourselves here to linear systems.

Example 3   **Third-Order Equation as a Linear System**   Rewrite the third-order equation

$$y''' + 3y'' + 2y' + y = \sin t \tag{8}$$

as a system of three first-order equations.

Solution   Introduce the new variables

$$\begin{aligned} x_1 &= y \\ x_2 &= y' \\ x_3 &= y'' \end{aligned} \tag{9}$$

Now differentiating these equations and rewriting the derivatives in terms of x_1, x_2, and x_3, we get

$$\begin{aligned} x_1' &= x_2 \\ x_2' &= x_3 \\ x_3' &= -x_1 - 2x_2 - 3x_3 + \sin t \end{aligned} \tag{10}$$ ■

SOLVING LINEAR SYSTEMS

Most beginning students of mathematics are familiar with the *method of elimination* for solving two simultaneous algebraic equations, whereby one eliminates one of the variables and solves for the remaining one. This method also works to a limited degree for solving linear systems of differential equations when two or possibly three equations are involved. Here we seek to reduce a given system of first-order differential equations to a *single* higher-order differential equation in one unknown. Although we will study more effective methods for solving similar systems later in this chapter,* the method of elimination, facilitated by the **D operator,** illustrates how differential equations can be ''turned into'' algebraic problems. Before presenting this method, we introduce the D operator.

THE *D* OPERATOR

Let D denote differentiation with respect to t, and let D^2 denote the second derivative with respect to t. That is, $D = d/dt$ and $D^2 = d^2/dt^2$. In general, we denote the quantity $D^n = d^n/dt^n$, where $D^0 = I$ is the identity operator. By using the D operator, it is possible to form more general **differential operators,** such as $D^2 + 2D - 3$, $(D - 1)(D + 3)$, and so on.

Example 4   **Operating with the *D* Operator**   Evaluate

(a) $D(\sin t + t^2 + 3)$

(b) $D(5)$

(c) $(D^2 + 2D - 3)y(t)$

---

* Later in this chapter we will use the Laplace transform and the method of eigenvalues/eigenvectors to solve the same types of problems in a more systematic manner.

Solution Keeping in mind that D is simply the derivative d/dt, we get

(a) $D(\sin t + t^2 + 3) = \dfrac{d}{dt}(\sin t + t^2 + 3) = \cos t + 2t$

(b) $D(5) = 0$ (The derivative of a constant is zero.)

(c) $(D^2 + 2D - 3)y(t) = \left(\dfrac{d^2}{dt^2} + 2\dfrac{d}{dt} - 3\right)y(t) = \dfrac{d^2y}{dt^2} + 2\dfrac{dy}{dt} - 3y$ ∎

Example 5 **Factoring Differential Operators**   Show that the operator $D^2 + 5D + 6$ is the same as the product $(D + 2)(D + 3)$.

Solution For a twice differentiable function $y(t)$ and using properties of the derivative, we can write

$$(D + 2)(D + 3)y = (D + 2)(y' + 3y)$$
$$= D(y' + 3y) + 2(y' + 3y)$$
$$= (y'' + 3y') + (2y' + 6y)$$
$$= y'' + 5y' + 6y$$
$$= (D^2 + 5D + 6)y$$

Hence

$$(D + 2)(D + 3) = D^2 + 5D + 6$$ ∎

> *Note:* The reader should realize that the above factorization is not an ''algebraic'' factorization as learned in elementary algebra although it can be treated as such; thus in a way we can treat homogeneous linear differential equations with constant coefficients as algebraic equations.

THE METHOD OF ELIMINATION USING THE D OPERATOR

Consider solving the linear homogeneous system

$$x_1' = x_1 + x_2 \tag{11}$$
$$x_2' = 4x_1 + x_2$$

by eliminating one of the two variables x_1 or x_2, thereby arriving at a single differential equation in the remaining variable. This is the **method of elimination** applied to differential equations. Unfortunately, the steps required for eliminating one of the unknown functions and its derivatives are not as obvious as when one solves systems of algebraic equations. However, if one replaces the derivative d/dt by D and treats Dx_1 and Dx_2 as *multiplication* of x_1 and x_2 by D, the two equations (11) ''look and act'' like a system of two algebraic equations.* Doing this, we get

$$Dx_1 = x_1 + x_2 \tag{12}$$
$$Dx_2 = 4x_1 + x_2$$

---

* The reader should realize that although Dx_1 is really the derivative dx_1/dt, nevertheless it behaves like multiplication of a variable D times x_1.

Of course, it's one thing to "think" of Dx_1 and Dx_2 as a multiplication and another to verify that the procedure is mathematically valid. However, the procedure can be verified for linear systems with constant coefficients, which illustrates the power of proper symbolism. Thinking now of Eqs. (12) as a system of two algebraic equations in x_1 and x_2, we rewrite it as

$$(D - 1)x_1 - x_2 = 0 \tag{13}$$
$$-4x_1 + (D - 1)x_2 = 0$$

Multiplying the first equation in system (13) by $D - 1$ and adding the two equations, we eliminate x_2, getting

$$(D - 1)[(D - 1)x_1 - x_2] = (D - 1)0$$
$$\underline{-4x_1 + (D - 1)\,x_2 = 0}$$
$$[(D - 1)(D - 1) - 4]\,x_1 = 0$$

or

$$(D^2 - 2D - 3)x_1 = 0 \tag{14}$$

It is now time to replace the D operator by the derivative d/dt and write Eq. (14) as the differential equation

$$x_1'' - 2x_1' - 3x_1 = 0 \tag{15}$$

We find the general solution of this equation to be

$$x_1(t) = c_1 e^{3t} + c_2 e^{-t} \tag{16}$$

We can now find $x_2(t)$ by simply substituting $x_1(t)$ into the first equation of system (11), getting

$$3c_1 e^{3t} - c_2 e^{-t} = x_2 c_1 e^{3t} + c_2 e^{-t} + x_2$$

or

$$x_2(t) = 2c_1 e^{3t} - 2c_2 e^{-t} \tag{17}$$

Note: Since c_1 and c_2 are arbitrary constants, we have found an infinite number of solutions. However, if x_1 and x_2 are required to satisfy initial conditions, say, $x_1(0) = 1$, $x_2(0) = 0$, then the constants c_1 and c_2 must be chosen to satisfy

$$x_1(0) = c_1 + c_2 = 1$$
$$x_2(0) = 2c_1 - 2c_2 = 0$$

or $c_1 = c_2 = 1/2$. Hence

$$x_1(t) = \frac{1}{2}\left(e^{3t} + e^{-t}\right)$$
$$x_2(t) = e^{3t} - e^{-t} \tag{18}$$

See Figure 6.1

Figure 6.1
Graphs of $x_1(t)$ and $x_2(t)$

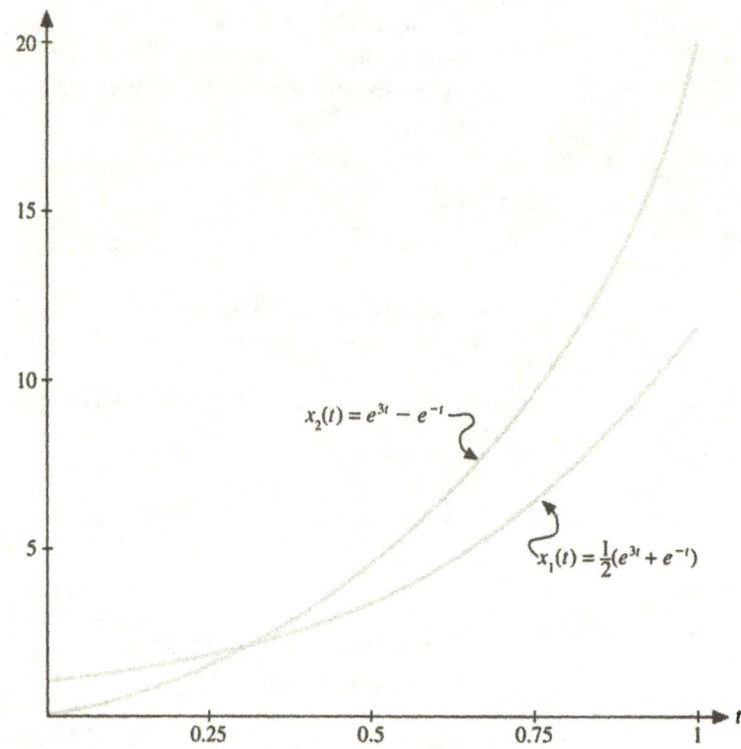

Example 6   Nonhomogeneous Linear System   Solve the initial-value problem

$$x' = y + t \qquad\qquad x(0) = 0 \tag{19}$$
$$y' = -2x + 3y + 5 \qquad y(0) = 0$$

Solution   Introducing $D = d/dt$, we rewrite the differential equations as

$$Dx - y = t \tag{20}$$
$$2x + (D - 3)y = 5$$

We can eliminate y in these equations by multiplying the first equation by $(D - 3)$ and summing the equations. Doing this, we get

$$
\begin{array}{r}
(D - 3)(Dx - y) = (D - 3)\,(t) \\
+ \qquad 2x + (D - 3)y = 5 \\
\hline
(D - 3)Dx + 2x = (D - 3)(t) + 5
\end{array}
\tag{21}
$$

or

$$(D^2 - 3D + 2)x = -3t + 6$$

or

$$x'' - 3x' + 2x = -3t + 6 \tag{22}$$

Solving Eq. (22), we get

$$x(t) = c_1 e^t + c_2 e^{2t} - \frac{3}{2}t + \frac{3}{4}$$

Substituting $y(t)$ into

$$x' = y + t$$

we get

$$y(t) = c_1 e^t + 2c_2 e^{2t} - t - \frac{3}{2} \tag{23}$$

To find the solution of the initial-value problem, we substitute $x(t)$ and $y(t)$ into the initial conditions $x(0) = y(0) = 0$ and solve for c_1 and c_2. This gives

$$c_1 + c_2 + \frac{3}{4} = 0$$

$$c_1 + 2c_2 - \frac{3}{2} = 0 \tag{24}$$

or $c_1 = -3$, $c_2 = 9/4$. This solution is graphed in Figure 6.2.   ■

Figure 6.2
Solution of the initial-value problem

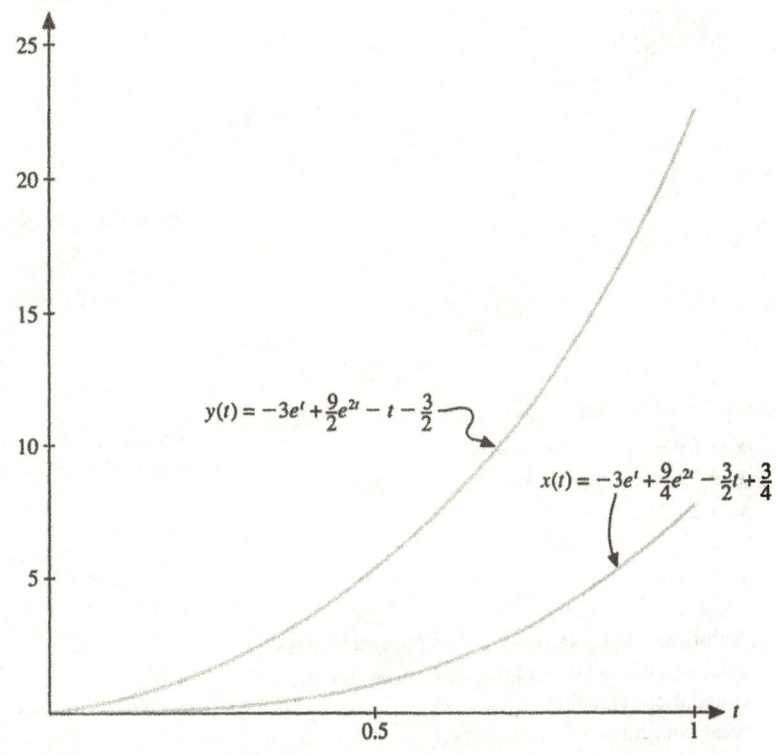

PROBLEMS: Section 6.1

For Problems 1–6, transform the given differential equation into a system of first-order equations. If initial conditions are given, transform the initial conditions as well and solve the initial-value problem.

1. $y'' = 1$
2. $y'' - y = 1$
3. $y'' + y = t$ $\qquad$ $y(0) = 1$ $\quad y'(0) = 0$
4. $y'' + 3y' + 2y = 1$ $\qquad y(0) = 1$ $\quad y'(0) = 0$
5. $y''' + py'' + qy' + ry = f(x)$
6. $y^{(4)} - y = 0$

For Problems 7–15, use the D operator and the method of elimination to find a general solution of the given linear system, where x', y', and z' refer to differentiation with respect to t.

7. $x' = x$
   $y' = y$
8. $x' = y$
   $y' = x$
9. $x' = y$
   $y' = -2x + 3y$
10. $x' = -3x + 4y$
    $y' = -2x + 3y$
11. $x' = y + t$
    $y' = -2x + 3y + 1$
12. $x' = 5x - 6y + 1$
    $y' = 6x - 7y + 1$
13. $x' = 2y - 2$
    $y' = 2x + 2$
14. $x' = 3x - y - 4$
    $y' = x + y + e^t$
15. $x' = y$
    $y' = z$
    $z' = w$
    $w' = x$

For Problems 16–18, solve the given initial-value problem.

16. $x' = 5x - 6y + 1$ $\qquad x(0) = 0$
    $y' = 6x - 7y + 1$ $\qquad y(0) = 0$
17. $x' = 2y - 2$ $\qquad\qquad x(0) = 0$
    $y' = 2x + 2$ $\qquad\qquad y(0) = 0$
18. $x' = 3x - y - 4$ $\qquad x(0) = 1$
    $y' = x + y + e^t$ $\qquad y(0) = 0$

19. **When Solutions Are Periodic, Try Polar Coordinates** Differential equations sometimes have solutions that are periodic, and those equations can often be solved by changing to polar coordinates. For example, the harmonic oscillator, which is described by

$$\dot{x}_1 = x_2 \qquad\qquad (25)$$
$$\dot{x}_2 = -x_1$$

is equivalent to $\ddot{x}_1 + x_1 = 0$, whose periodic solution can be written in the form $x_1(t) = R \sin(t + \delta)$, where R and δ are arbitrary constants. Solve Eq. (25) by transforming x_1 and x_2 to polar coordinates r and θ. *Hint:* Compute $d(r^2)/dt$ and $d\theta/dt$. The form of the solution is not exactly what you have seen before, but it still represents the same class of functions.

20. **Transforming to Constant Coefficients** Given the linear system

$$t\frac{dx}{dt} = a_{11}x + a_{12}y$$
$$t\frac{dy}{dt} = a_{21}x + a_{22}y$$

where a_{11}, a_{12}, a_{21}, and a_{22} are real constants, show that the transformation to the new independent variable defined by $w = \ln t$ ($t = e^w$) transforms Eq. (25) to a new system with constant coefficients.

21. **Transforming to Constant Coefficients** Use the results from Problem 20 to solve the system

$$t\frac{dx}{dt} = -3x + 4y$$
$$t\frac{dy}{dt} = -2x + 3y \qquad\qquad (26)$$

Solving More General Linear Systems

It is possible to solve a system of two possibly higher-order differential equations of the form

$$P_{11}(D)x(t) + P_{12}(D)y(t) = f_1(t) \qquad (27)$$
$$P_{21}(D)x(t) + P_{22}(D)y(t) = f_2(t)$$

where P_{11}, P_{12}, P_{21}, and P_{22} are polynomials in $D = d/dt$ and the determinant satisfies the nondegenerate condition

$$|P_{11}(D)P_{22}(D) - P_{12}(D)P_{21}(D)| \neq 0 \qquad (28)$$

For Problems 22–23, verify the nondegenerate condition (28) and find the general solution by eliminating one of the two variables x or y.

22. $x' - 3x + 2y = 0$
    $2x - y' - 2y = 0$
23. $x' - 2x - y = 0$
    $x + y' - 2y = 0$

24. An Old Story* By a stretch of the imagination the system of two differential equations

$$\frac{dr}{dt} = -aj \qquad (a > 0)$$

$$\frac{dj}{dt} = br \qquad (b > 0)$$

(29)

might represent the affections of two lovers, Romeo and Juliet, where the positive values r and j measure their respective love for the other. Note that the more Juliet loves him, the more Romeo begins to dislike her, but when she loses interest, his feelings for her warm up. She, on the other hand, has the opposite experience. Her love for him grows when he loves her and turns to hate when he hates her. Under these conditions, what do you think the solution of this system of equations would look like?

25. Romeo and Juliet—More Mature Love The relationship between Romeo and Juliet described in Problem 24 has matured into one modeled by the linear system

$$\frac{dr}{dt} = a_{11}r + a_{12}j$$

$$\frac{dj}{dt} = a_{21}r + a_{22}j$$

(30)

where the "love/hate" parameters a_{ij} can be either positive (love) or negative (hate) and measure the affection each lover has for the other. Note that when we have $a_{11} > 0$ and $a_{12} > 0$, Romeo's love for Juliet is spurred on by his own feelings and by Juliet's love for him. In this case we might call Romeo an "eager beaver lover." In parts (a)–(c), can you describe roughly the other three types of lovers modeled by the following parameters?

(a) $a_{11} > 0$, $a_{12} < 0$
(b) $a_{11} < 0$, $a_{12} > 0$
(c) $a_{11} < 0$, $a_{12} < 0$

26. Journal Entry Try summarizing the concept of a system of differential equations to yourself. What do you know about a system of two algebraic equations, and how might they be the same as a system of two differential equations?

6.2 REVIEW OF MATRICES

PURPOSE

To review the basic concepts of matrices and matrix algebra that will be needed to solve systems of differential equations by the method of eigenvalues and eigenvectors.

Most readers of this section have probably been introduced to matrices and determinants at some time in the past. However, to ensure an adequate background for the important sections that follow, we review the relevant ideas.

BASIC TERMINOLOGY AND MATRIX OPERATIONS

We start with the definition of a **matrix.** An $m \times n$ (pronounced "m by n") matrix A is simply a rectangular array of mn numbers, called **elements** of the matrix, arranged in m **rows** and n **columns,** denoted by

$$A = \begin{bmatrix} a_{11} & a_{12} & \cdots & a_{1n} \\ a_{21} & a_{22} & \cdots & a_{2n} \\ \vdots & \vdots & \cdots & \vdots \\ a_{m1} & a_{m2} & \cdots & a_{mn} \end{bmatrix}$$

(1)

* This problem is based on the note "Love Affairs and Differential Equations" by Steven H. Strogatz, *Mathematics Magazine,* Vol. 61, No. 1, February 1988.

where m and n define the **order** of the matrix. Unless otherwise indicated, matrices will be denoted by capital letters in **bold type**. Sometimes, we will use the matrix notation $A = (a_{ij})$ to denote that a_{ij} is the element in the ith row and jth column of A. Associated with each matrix A is the **transpose matrix A^T**, which is the matrix obtained by interchanging the rows and columns of A. Two special matrices are the $m \times n$ **zero matrix**, denoted by $O_{m \times n}$ or simply O, which is the $m \times n$ matrix with all zero elements, and the **identity matrix** of order n, denoted by $I_{n \times n}$ or simply I, which is the square $n \times n$ matrix having 1's down the diagonal and 0's elsewhere. That is,

$$O_{m \times n} = \begin{bmatrix} 0 & 0 & \cdots & 0 \\ 0 & 0 & \cdots & 0 \\ \vdots & \vdots & \cdots & \vdots \\ 0 & 0 & \cdots & 0 \end{bmatrix} \tag{2}$$

$$I_{n \times n} = \begin{bmatrix} 1 & 0 & \cdots & 0 \\ 0 & 1 & \cdots & 0 \\ \vdots & \vdots & \vdots & \vdots \\ 0 & 0 & \cdots & 1 \end{bmatrix} \tag{3}$$

Finally, we say that two $m \times n$ matrices $A = (a_{ij})$ and $B = (b_{ij})$ are *equal* if their corresponding elements are equal; that is, $a_{ij} = b_{ij}$ for $1 \le i \le m$ and $1 \le j \le n$.

ARITHMETIC OF MATRICES

One often thinks of matrices as *generalized* numbers, since it is possible to add, subtract, multiply, and in a certain sense divide matrices in much the same way as real numbers.

1. **Addition of Matrices.** The sum of the two $m \times n$ matrices $A = (a_{ij})$ and $B = (b_{ij})$ is defined as the matrix formed by adding the corresponding entries of A and B. That is,

$$A + B = (a_{ij}) + (b_{ij}) = (a_{ij} + b_{ij}) \tag{4}$$

2. **Multiplication by a Constant.** To find the product of a matrix A and a real or complex constant c, we simply multiply each element of A by c. That is,

$$cA = Ac = (ca_{ij}) \tag{5}$$

Sometimes constants such as c are called **scalars** in this context; we would call this **scalar multiplication.**

3. **Subtraction of Matrices.** We denote $(-1)A$ by $-A$ and then define the difference $A - B$ as

$$A - B = A + (-B) \qquad (6)$$

That is,

$$A - B = (a_{ij}) + (-b_{ij}) = (a_{ij} - b_{ij}) \qquad (7)$$

4. **Multiplication of Vectors.** A $1 \times n$ matrix A is called a **row vector,** and an $n \times 1$ matrix B is called a **column vector.** Denoting the row vector by x and the column vector by y, we define the product of these special matrices, sometimes called the **dot** or **scalar product,** by

$$xy = [x_1 \quad x_2 \quad \cdots \quad x_n] \begin{bmatrix} y_1 \\ y_2 \\ \vdots \\ y_n \end{bmatrix} = \sum_{i=1}^{n} x_i y_i \qquad (8)$$

We use the letters x, y, and z to denote both row and column vectors. Whether x denotes a row or column vector should be determined from the context of the problem. Also, to save space, we often denote a column vector

$$x = \begin{bmatrix} x_1 \\ \vdots \\ x_n \end{bmatrix}$$

by writing $x = \text{col}\,(x_1, \ldots, x_n)$.

5. **Multiplication of General Matrices.** The product AB of two matrices is defined whenever the number of *columns* of the first matrix is the same as the number of *rows* of the second matrix. If A is an $m \times r$ matrix and B is an $r \times n$ matrix, then the product $C = AB$ is an $m \times n$ matrix whose element c_{ij} in the ith row and jth column is defined as the dot or scalar product of the ith row of A and the jth column of B. That is,

$$c_{ij} = a_{i1}b_{1j} + a_{i2}b_{2j} + \cdots + a_{ir}b_{rj} = \sum_{k=1}^{r} a_{ik}b_{kj} \qquad (9)$$

We illustrate the choice of row i from A and column j from B to find the element c_{ij} of the product $C = AB$ by the equation

$$\begin{bmatrix} c_{11} & c_{12} & \cdots & \cdots & c_{1n} \\ c_{21} & \cdots & \cdots & \cdots & c_{2n} \\ \vdots & \cdots & c_{ij} & \cdots & \vdots \\ \vdots & \cdots & \cdots & \cdots & \vdots \\ c_{m1} & \cdots & \cdots & \cdots & c_{mn} \end{bmatrix} = \begin{bmatrix} a_{11} & \cdots & \cdots & a_{1r} \\ \vdots & \vdots & \cdots & \vdots \\ a_{i1} & a_{i2} & \cdots & a_{ir} \\ \vdots & \vdots & \cdots & \vdots \\ a_{m1} & \cdots & \cdots & a_{mr} \end{bmatrix} \begin{bmatrix} b_{11} & \cdots & b_{1j} & \cdots & b_{1n} \\ \vdots & \cdots & b_{2j} & \cdots & \cdots \\ \vdots & \cdots & \vdots & \cdots & \vdots \\ b_{r1} & \cdots & b_{rj} & \cdots & b_{rn} \end{bmatrix}$$

It is easy to show that matrix multiplication satisfies the **associative law**

$$(AB)C = A(BC) \qquad (10)$$

and the *distributive law*

$$A(B + C) = AB + AC \tag{11}$$

However, matrix multiplication in general is not **commutative.** That is,

$$AB \neq BA \tag{12}$$

In fact, in order that *both* products AB and BA exist, it is necessary that both A and B are square matrices of the same order. Even then, the two products AB and BA need not be the same.

Example 1 **Product of Matrices** Find the matrix product of the following matrices, and show that $AB \neq BA$.

$$A = \begin{bmatrix} 1 & 1 & 3 \\ 0 & 4 & 2 \\ 2 & 0 & 4 \end{bmatrix} \qquad B = \begin{bmatrix} 3 & 1 & 0 \\ 2 & 4 & 2 \\ 1 & 0 & 5 \end{bmatrix}$$

Solution

$$AB = \begin{bmatrix} 1 \cdot 3 + 1 \cdot 2 + 3 \cdot 1 & 1 \cdot 1 + 1 \cdot 4 + 3 \cdot 0 & 1 \cdot 0 + 1 \cdot 2 + 3 \cdot 5 \\ 0 \cdot 3 + 4 \cdot 2 + 2 \cdot 1 & 0 \cdot 1 + 4 \cdot 4 + 2 \cdot 0 & 0 \cdot 0 + 4 \cdot 2 + 2 \cdot 5 \\ 2 \cdot 3 + 0 \cdot 2 + 4 \cdot 1 & 2 \cdot 1 + 0 \cdot 4 + 4 \cdot 0 & 2 \cdot 0 + 0 \cdot 2 + 4 \cdot 5 \end{bmatrix}$$

$$= \begin{bmatrix} 8 & 5 & 17 \\ 10 & 16 & 18 \\ 10 & 2 & 20 \end{bmatrix}$$

$$BA = \begin{bmatrix} 3 \cdot 1 + 1 \cdot 0 + 0 \cdot 2 & 3 \cdot 1 + 1 \cdot 4 + 0 \cdot 0 & 3 \cdot 3 + 1 \cdot 2 + 0 \cdot 4 \\ 2 \cdot 1 + 4 \cdot 0 + 2 \cdot 2 & 2 \cdot 1 + 4 \cdot 4 + 2 \cdot 0 & 2 \cdot 3 + 4 \cdot 2 + 2 \cdot 4 \\ 1 \cdot 1 + 0 \cdot 0 + 5 \cdot 2 & 1 \cdot 1 + 0 \cdot 4 + 5 \cdot 0 & 1 \cdot 3 + 0 \cdot 2 + 5 \cdot 4 \end{bmatrix}$$

$$= \begin{bmatrix} 3 & 7 & 11 \\ 6 & 18 & 22 \\ 11 & 1 & 23 \end{bmatrix}$$

Clearly, $AB \neq BA$.

6. **The Determinant of a Matrix.** Most readers have probably been introduced to the determinant of a matrix. As a brief review, if $A = (a_{ij})$ is a 2×2 matrix, then the determinant of A is defined by

$$|A| = \begin{vmatrix} a_{11} & a_{12} \\ a_{21} & a_{22} \end{vmatrix} = a_{11}a_{22} - a_{12}a_{21} \tag{13}$$

Determinants of higher-order matrices can be determined by induction as follows. Let $A = (a_{ij})$ be an $n \times n$ matrix. Associated with each element a_{ij} of A is a determinant M_{ij}, called the **minor** of a_{ij}, which is the determinant of the $(n - 1) \times (n - 1)$ matrix obtained by deleting the ith row and jth column of A. The determinant $|A|$ can then be found by computing the "expansion" of $|A|$ along any row i of A from the formula

$$|A| = \sum_{j=1}^{n} (-1)^{i+j} a_{ij} M_{ij} \tag{14a}$$

Equivalently, $|A|$ can be found by "expanding" along any column j of A from the formula

$$|A| = \sum_{i=1}^{n} (-1)^{i+j} a_{ij} M_{ij} \tag{14b}$$

HISTORICAL NOTE

The subject of determinants grew from the efforts of many people in the late 1700's and early 1800's. The most complete work on determinants was written by the French mathematician Augustin Louis Cauchy (1789–1857), who first coined the word "determinant" in 1812. It was Cauchy who first showed that the determinant of an array can be found by "expanding by minors" down a row or column.

Example 2   Computing Determinants   Find the determinant of the matrix

$$A = \begin{bmatrix} 3 & 1 & -1 \\ 2 & 1 & 3 \\ 0 & 1 & 2 \end{bmatrix}$$

Solution   Since it is possible to expand along any row or column, we expand down the first column, getting

$$|A| = a_{11} M_{11} - a_{21} M_{21} + a_{31} M_{31}$$

$$= + 3 \begin{vmatrix} 1 & 3 \\ 1 & 2 \end{vmatrix} - 2 \begin{vmatrix} 1 & -1 \\ 1 & 2 \end{vmatrix} + 0 \begin{vmatrix} 1 & -1 \\ 1 & 3 \end{vmatrix}$$

$$= 3(-1) - 2(3) + 0(4)$$

$$= -9 \qquad \blacksquare$$

QUICK DETERMINANT FOR THE 3 $\times$ 3 MATRIX   There is a streamlined way to find the determinant of a 3×3 matrix. If one evaluates the determinant of the matrix

$$A = \begin{bmatrix} a_{11} & a_{12} & a_{13} \\ a_{21} & a_{22} & a_{23} \\ a_{31} & a_{32} & a_{33} \end{bmatrix}$$

using a minor expansion, one finds

$$|A| = (a_{11} a_{22} a_{33} + a_{21} a_{32} a_{13} + a_{12} a_{23} a_{31})$$
$$- (a_{31} a_{22} a_{13} + a_{21} a_{12} a_{33} + a_{32} a_{23} a_{11}) \tag{15}$$

7. Inverse of a Matrix. If A is an $n \times n$ matrix, then the **inverse** of A (if it exists) is a square matrix, denoted by A^{-1}, that satisfies

$$AA^{-1} = A^{-1}A = I \tag{16}$$

It is not difficult to show that if A has an inverse A^{-1}, then the inverse is *unique*. Hence we speak of *the* inverse of A. It is also shown in courses in linear algebra that A^{-1} exists if and only if the **determinant** of the square matrix A, denoted by $|A|$, is nonzero, in which case the matrix A is called **nonsingular.** When $|A| = 0$, the square matrix A is called **singular,** and the matrix A does not have an inverse. For a 2×2 nonsingular matrix.

$$A = \begin{bmatrix} a & b \\ c & d \end{bmatrix}$$

the inverse can easily be found to be

$$A^{-1} = \frac{1}{|A|} \begin{bmatrix} d & -b \\ -c & a \end{bmatrix} \tag{17}$$

where $|A| = ad - bc$ is the determinant of A. The reader can easily verify that $AA^{-1} = A^{-1}A = I$ for these matrices.

MATRIX FUNCTIONS

Later in this chapter we will consider vectors and matrices whose elements are functions rather than constants. We write

$$x(t) = \begin{bmatrix} x_1(t) \\ \vdots \\ x_n(t) \end{bmatrix} \qquad A(t) = \begin{bmatrix} a_{11}(t) & \cdots & a_{1n}(t) \\ \vdots & \vdots & \vdots \\ a_{n1}(t) & \cdots & a_{nn}(t) \end{bmatrix}$$

We say that $x(t)$ and $A(t)$ are differentiable on an interval if their respective elements are differentiable on the interval, and we define the derivative of a vector or matrix as the vector or matrix of its derivatives. That is,

$$x'(t) = \frac{dx(t)}{dt} = \left(\frac{dx_i}{dt} \right) \qquad A'(t) = \frac{dA(t)}{dt} = \left(\frac{da_{ij}(t)}{dt} \right) \tag{18}$$

For instance, if

$$A(t) = \begin{bmatrix} e^t & t^2 \\ \sin t & 2t \end{bmatrix}$$

then

$$A'(t) = \begin{bmatrix} e^t & 2t \\ \cos t & 2 \end{bmatrix}$$

Most of the rules of calculus for matrix-valued functions appear almost the same as for real-valued functions. The reader can verify the following general rules:

1. $\dfrac{d}{dt}\left(A(t) + B(t)\right) = \dfrac{d}{dt}A(t) + \dfrac{d}{dt}B(t)$

2. $\dfrac{d}{dt}\left(CA(t)\right) = C\dfrac{d}{dt}A(t)$ (where C is a constant matrix)

3. $\dfrac{d}{dt}\left(A(t)B(t)\right) = A(t)\dfrac{d}{dt}[B(t)] + \dfrac{d}{dt}[A(t)]B(t)$

HISTORICAL NOTE

The concept of the inverse of a matrix was first introduced into mathematics in 1855 by the English mathematician Arthur Cayley (1821–1895). Cayley and his close friend, James Sylvester, are primarily responsible for the development of the subject of matrices.

THE MATRIX EXPONENTIAL e^{At}

Recall from calculus that many functions can be expressed as a Maclaurin series. This leads us to define functions of matrices in an analogous manner. For example, the Maclaurin series of e^t is

$$e^t = 1 + t + \frac{1}{2!}t^2 + \frac{1}{3!}t^3 + \cdots$$

Hence we define the **matrix exponential*** as

$$e^{At} = I + tA + \frac{t^2}{2!}A^2 + \frac{t^3}{3!}A^3 + \cdots \tag{19}$$

where A is a constant square matrix and t is a scalar variable. It can be shown that for any square matrix A and for any real number t, all the elements of the matrix defined by the series (19) converges, and so e^{At} is well defined.

Example 3 **Matrix Exponential** Find the matrix exponential e^{At} of the matrix

$$A = \begin{bmatrix} 3 & 0 \\ 0 & 2 \end{bmatrix}$$

Solution By direct computation we find

$$A^n = \begin{bmatrix} 3^n & 0 \\ 0 & 2^n \end{bmatrix} \qquad (n = 1, 2, 3, \ldots)$$

---

* The matrix exponential of a matrix A is defined as the *constant* matrix $e^A = I + A + A^2/2! + \cdots$. However, for our purposes we are interested in the matrix *function* e^{At}.

Hence we get

$$e^{At} = I + tA + \frac{t^2}{2!}A^2 + \frac{t^3}{3!}A^3 + \cdots$$

$$= \begin{bmatrix} 1 & 0 \\ 0 & 1 \end{bmatrix} + t \begin{bmatrix} 3 & 0 \\ 0 & 2 \end{bmatrix} + \frac{t^2}{2!} \begin{bmatrix} 3^2 & 0 \\ 0 & 2^2 \end{bmatrix} + \frac{t^3}{3!} \begin{bmatrix} 3^3 & 0 \\ 0 & 2^3 \end{bmatrix} + \cdots$$

$$= \begin{bmatrix} \sum_{k=0}^{\infty} (3t)^k/k! & 0 \\ 0 & \sum_{k=0}^{\infty} (2t)^k/k! \end{bmatrix}$$

$$= \begin{bmatrix} e^{3t} & 0 \\ 0 & e^{2t} \end{bmatrix}$$

■

REPLACING DIFFERENTIAL EQUATIONS WITH MATRICES

It is sometimes possible to differentiate a function by matrix multiplication. For example, the derivative of the quadratic polynomial

$$\frac{d}{dx}(at^2 + bt + c) = 2at + b \tag{20}$$

can be found by multiplying the column vector of *coefficients* (*a, b, c*) by a suitable matrix. In this case the matrix multiplication

$$\begin{bmatrix} 0 & 0 & 0 \\ 2 & 0 & 0 \\ 0 & 1 & 0 \end{bmatrix} \begin{bmatrix} a \\ b \\ c \end{bmatrix} = \begin{bmatrix} 0 \\ 2a \\ b \end{bmatrix} \begin{matrix} \leftarrow & t^2 \\ \leftarrow & t \\ \leftarrow & 1 \end{matrix} \tag{21}$$

gives the new vector (0, 2*a*, *b*), whose elements are simply the coefficients of the derivative $0 \cdot t^2 + 2at + b$. That is, instead of differentiating a polynomial, we can replace the polynomial by a vector of its coefficients and then multiply this coefficient vector by a suitable matrix.* In fact, it is possible to differentiate any order polynomial by matrix multiplication. For example, the derivative of the general cubic polynomial is

$$\frac{d}{dt}\left(at^3 + bt^2 + ct + d\right) = 3at^2 + 2bt + c \tag{22}$$

This differentiation can also be carried out by multiplying the vector of coefficients column (*a, b, c, d*) by the matrix

$$\begin{bmatrix} 0 & 0 & 0 & 0 \\ 3 & 0 & 0 & 0 \\ 0 & 2 & 0 & 0 \\ 0 & 0 & 1 & 0 \end{bmatrix} \begin{bmatrix} a \\ b \\ c \\ d \end{bmatrix} = \begin{bmatrix} 0 \\ 3a \\ 2b \\ c \end{bmatrix} \begin{matrix} \leftarrow & t^3 \\ \leftarrow & t^2 \\ \leftarrow & t \\ \leftarrow & 1 \end{matrix} \tag{23}$$

* The reader who has taken a course in linear algebra will recognize this technique as finding the matrix representation of a linear transformation (derivative) and multiplying the coordinates of the vector we wish to transform (polynomial) in order to find the coordinates of the transformed vector (derivative of the polynomial).

The product is $(0, 3a, 2b, c)$, whose elements are the coefficients of the derivative. We call the matrix in Eq. (23) the **matrix representation** of the derivative acting on cubic polynomials.

The reader can verify that the following matrix multiplications carry out the indicated analytical calculations.

Matrix Representations of Analytical Operations

The following matrices perform the same operations at the analytical operations.

1. First derivative of a cubic polynomial:

$$\frac{d}{dt}[at^3 + bt^2 + ct + d] = 3at^2 + 2bt + c$$

Matrix representation: $\begin{bmatrix} 0 & 0 & 0 & 0 \\ 3 & 0 & 0 & 0 \\ 0 & 2 & 0 & 0 \\ 0 & 0 & 1 & 0 \end{bmatrix} \begin{bmatrix} a \\ b \\ c \\ d \end{bmatrix} = \begin{bmatrix} 0 \\ 3a \\ 2b \\ c \end{bmatrix}$

2. Second derivative of a cubic polynomial:

$$\frac{d^2}{dt^2}[at^3 + bt^3 + ct + d] = 6at + 2b$$

Matrix representation: $\begin{bmatrix} 0 & 0 & 0 & 0 \\ 0 & 0 & 0 & 0 \\ 6 & 0 & 0 & 0 \\ 0 & 2 & 0 & 0 \end{bmatrix} \begin{bmatrix} a \\ b \\ c \\ d \end{bmatrix} = \begin{bmatrix} 0 \\ 0 \\ 6a \\ 2b \end{bmatrix}$

3. Derivative of a sine and cosine combination:

$$\frac{d}{dt}[c_1 \sin t + c_2 \cos t] = -c_2 \sin t + c_1 \cos t$$

Matrix representation: $\begin{bmatrix} 0 & -1 \\ 1 & 0 \end{bmatrix} \begin{bmatrix} c_1 \\ c_2 \end{bmatrix} = \begin{bmatrix} -c_2 \\ c_1 \end{bmatrix}$

It is also possible to transform some linear differential equations with constant coefficients to systems of algebraic equations. In so doing, it is often possible to find the general solution of a differential equation by solving a system of algebraic equations. The following example illustrates this idea.

Example 4 Solving Differential Equations with Matrices   Find the general solution of

$$y'' + y = t \tag{24}$$

Solution Although we do not know the precise solution, we know that it is a linear combination of functions*

$$\mathfrak{B} = \{\cos t, \sin t, t, 1\} \tag{25}$$

which we call the **basis functions.** Letting y be an *unknown* linear combination of these functions, we write $y = a \cos t + b \sin t + ct + d$. To find the matrix representation of the **differential operator** $L(y) = y'' + y$ we observe the identity

$$\frac{d^2}{dt^2}[a \cos t + b \sin t + ct + d] + [a \cos t + b \sin t + ct + d] = ct + d \tag{26}$$

Since the matrix multiplication

$$\begin{bmatrix} 0 & 0 & 0 & 0 \\ 0 & 0 & 0 & 0 \\ 0 & 0 & 1 & 0 \\ 0 & 0 & 0 & 1 \end{bmatrix} \begin{bmatrix} a \\ b \\ c \\ d \end{bmatrix} = \begin{bmatrix} 0 \\ 0 \\ c \\ d \end{bmatrix} \tag{27}$$

performs the identical operation on the *coefficients*, we have that the matrix representation of the differential operator $L(y) = y'' + y$ is

$$A = \begin{bmatrix} 0 & 0 & 0 & 0 \\ 0 & 0 & 0 & 0 \\ 0 & 0 & 1 & 0 \\ 0 & 0 & 0 & 1 \end{bmatrix}$$

Hence the differential equation (25) can be rewritten in matrix form as

$$\begin{bmatrix} 0 & 0 & 0 & 0 \\ 0 & 0 & 0 & 0 \\ 0 & 0 & 1 & 0 \\ 0 & 0 & 0 & 1 \end{bmatrix} \begin{bmatrix} a \\ b \\ c \\ d \end{bmatrix} = \begin{bmatrix} 0 \\ 0 \\ 1 \\ 0 \end{bmatrix} \begin{matrix} \leftarrow \cos t \\ \leftarrow \sin t \\ \leftarrow t \\ \leftarrow 1 \end{matrix} \tag{28}$$

$$\underset{\dfrac{d^2}{dt^2} + 1}{\uparrow} \qquad \underset{y}{\uparrow} \qquad \underset{t}{\uparrow}$$

Solving this system of equations, we see that a and b are arbitrary, $c = 1$, and $d = 0$. Hence the general solution of the differential equation (25) is

$$\begin{aligned} y(t) &= a \cos t + b \sin t + ct + d \\ &= a \cos t + b \sin t + t \end{aligned} \tag{29}$$

where a and b are arbitrary constants.

---

* Although this method seems contrived, since we need to know the general form of the solution before we can find the matrix representation, it is possible to use a computer and allow many common functions (hundreds) in the basic set. Hence we can solve general classes of equations, such as equations of the form $ay'' + by' + cy = f(t)$, where $f(t)$ is of the type considered when we solved nonhomogeneous equations by the method of undetermined coefficients.

PROBLEMS: Section 6.2

For Problems 1–14, let

$$A = \begin{bmatrix} -1 & 0 & 3 \\ 2 & 1 & 2 \\ -1 & 0 & 1 \end{bmatrix} \quad B = \begin{bmatrix} 1 & 3 & 0 \\ 0 & 1 & 0 \\ 0 & 0 & 1 \end{bmatrix}$$

$$C = \begin{bmatrix} 1 & 0 \\ 2 & 1 \\ 1 & 3 \end{bmatrix} \quad D = \begin{bmatrix} 3 & -1 & 0 \\ 2 & 1 & 2 \end{bmatrix}$$

and compute the indicated quantity, if defined.

1. $2A$
2. $A + 2B$
3. $2C - D$
4. AB
5. BA
6. CD
7. DC
8. $(DC)^T$
9. $C^T D$
10. $D^T C$
11. A^2
12. $|A|$
13. $|C|$
14. $|D|$

True or False?

For Problems 15–23, say whether you think the given statement is true or false. If you think the statement is true, prove it for general 2×2 matrices. If you think the statement is false, find an example for which the statement fails. The quantities a, b, and c represent scalars.

15. $(A + B)^T = A^T + B^T$
16. $(AB)^T = A^T B^T$
17. $(a + b)A = aA + bA$
18. $a(A + B) = aA + aB$
19. $A(B + C) = AB + AC$
20. $(cA)^T = cA^T$
21. $(AB)^{-1} = A^{-1}B^{-1}$
22. $(AB)^{-1} = B^{-1}A^{-1}$
23. $(A^{-1})^{-1} = A$

Complex Matrices

Matrix operations for complex-valued matrices obey the same rules as do real matrices. For Problems 24–26, compute the indicated matrix, given

$$A = \begin{bmatrix} 1 + i & 2i \\ 2 & 2 - 3i \end{bmatrix} \quad B = \begin{bmatrix} 1 & -i \\ 2i & 1 + i \end{bmatrix}$$

24. $A + 2B$
25. AB
26. BA

Matrix Derivatives

For Problems 27–30, let

$$A = \begin{bmatrix} 1 & \sin t \\ t & 0 \end{bmatrix} \quad B = \begin{bmatrix} t & 1 \\ 0 & t^2 \end{bmatrix} \quad C = \begin{bmatrix} 1 & 2 \\ 2 & 1 \end{bmatrix}$$

and compute the indicated matrix.

27. $A'(t)$
28. $B'(t)$
29. $(AB)'$
30. $(A + B)'$
31. **Verification of the Product Law** For the matrices A and B listed before Problems 27–30, verify the product law for the derivative

$$(AB)' = A'B + AB'$$

32. **Real and Complex Parts of Matrices** Write the complex-valued matrix

$$A = \begin{bmatrix} 1 + i & 2i \\ 2 & 2 - 3i \end{bmatrix}$$

in the form $A = B + iC$, where B and C are real matrices.

33. **Finding Determinant Using Minors** Find the determinant of the matrix

$$A = \begin{bmatrix} 1 & 3 & 2 \\ 2 & -1 & 3 \\ 2 & 4 & 5 \end{bmatrix}$$

by first expanding by minors across the top row, then expanding down the first column.

34. **Inverse of a 2×2 Square Matrix** Verify that the inverse of the 2×2 matrix

$$A = \begin{bmatrix} a & b \\ c & d \end{bmatrix}$$

is

$$A^{-1} = \frac{1}{ad - bc} \begin{bmatrix} d & -b \\ -c & a \end{bmatrix}$$

provided that $|A| \neq 0$.

35. **Matrix Inverse** Use the result from Eq. (17) to find the inverse of

$$A = \begin{bmatrix} 1 & 2 \\ 3 & 5 \end{bmatrix}$$

36. **Matrix Exponential** For the matrix

$$A = \begin{bmatrix} 0 & 1 & 0 & 0 \\ 0 & 0 & 1 & 0 \\ 0 & 0 & 0 & 1 \\ 0 & 0 & 0 & 0 \end{bmatrix}$$

find A^2, A^3, and A^4. Use these matrices to find the matrix exponential

$$e^A = I + A + \frac{1}{2}A^2 + \frac{1}{3!}A^3 + \cdots$$

37. Properties of the Matrix Exponential Verify the following properties of the matrix exponential e^{At}.

(a) $(e^{-At})(e^{At}) = I$

(b) $\dfrac{d}{dt} e^{At} = e^{At} \cdot A$

Differentiation by Matrix Multiplication

For Problems 38–41, find the matrix that carries out the indicated differentiation on the given class of functions.

	Calculus operation	Function	Basis functions
38.	First derivative	$at^2 + bt + c$	$\{t^2, t, 1\}$
39.	Second derivative	$at^2 + bt + c$	$\{t^2, t, 1\}$
40.	Third derivative	$at^2 + bt + c$	$\{t^2, t, 1\}$
41.	First derivative	$a \sin t + b \cos t$	$\{\sin t, \cos t\}$

Matrix Representations of Differential Equations

For Problems 42–47, the solution of the given differential equation can be written as a linear combination of the given basis functions. Use this information to rewrite the differential equation in matrix form, and solve the system of equations to find the general solution.

	Differential equation	Basis functions
42.	$y' = t + 2$	$\{t^2, t, 1\}$
43.	$y' + y = t$	$\{e^{-t}, t, 1\}$
44.	$y'' = t$	$\{t^3, t^2, t, 1\}$
45.	$y'' + y = 1$	$\{\sin t, \cos t, 1\}$
46.	$y'' + y = t$	$\{\sin t, \cos t, t, 1\}$
47.	$y'' + y = \sin t$	$\{t \sin t, t \cos t, \sin t, \cos t, t, 1\}$

48. Finding Your Own Basis Find the general solution of

$$y'' + y = \sin t$$

using matrices.

49. Differo-Integral Equation Find the general solution of

$$y' - \int_0^1 y(s)\, ds = t$$

using matrices. *Hint:* Try a solution having the general form $y = a + bt + \frac{1}{2}t^2$. This general idea can be extended to higher derivatives and more complicated integrals as long as the problem is linear in y and its derivatives and as long as one knows the form of the solution.

50. Journal Entry Is it possible to find *antiderivatives* by using matrix operations?

6.3 BASIC THEORY OF FIRST-ORDER LINEAR SYSTEMS

PURPOSE

To introduce some of the basic concepts necessary for the study of systems of linear differential equations such as existence and uniqueness theory, superposition of solutions, linearly independent solutions, fundamental sets of solutions, and the general solution.

In Section 6.1 we introduced the system of n first-order linear equations

$$
\begin{aligned}
x_1' &= a_{11}(t)x_1 + a_{12}(t)x_2 + \cdots + a_{1n}(t)x_n + f_1(t) \\
x_2' &= a_{21}(t)x_1 + a_{22}(t)x_2 + \cdots + a_{2n}(t)x_n + f_2(t) \\
&\ \vdots\ \ \vdots\qquad\ \ \vdots \\
x_n' &= a_{n1}(t)x_1 + a_{n2}(t)x_2 + \cdots + a_{nn}(t)x_n + f_n(t)
\end{aligned}
\tag{1}
$$

Using matrix notation

$$
\mathbf{x}(t) = \begin{bmatrix} x_1(t) \\ \vdots \\ x_n(t) \end{bmatrix}
\qquad
\mathbf{A}(t) = \begin{bmatrix} a_{11}(t) & \cdots & a_{1n}(t) \\ \vdots & & \vdots \\ a_{n1}(t) & \cdots & a_{nn}(t) \end{bmatrix}
\qquad
\mathbf{f}(t) = \begin{bmatrix} f_1(t) \\ \vdots \\ f_n(t) \end{bmatrix}
$$

we can write this system in matrix form as

$$x'(t) = A(t)x + f(t) \tag{2}$$

We recall from Section 6.1 that a solution of the linear system (2) on an open interval I is a *vector* of functions $x(t) = (x_i(t))$ such that the components $x_i(t)$ satisfy the linear system on the given interval. We will see that the general theory for the linear system (2) closely parallels that of a single differential equation studied in Chapter 3. We state without proof the basic existence and uniqueness theorem regarding linear systems.

THEOREM 6.1: Existence and Uniqueness of Solutions to Linear Systems

If the elements of the matrix $A = (a_{ij}(t))$ and column vector $f = (f_i(t))$ are continuous on an open interval I that contains the point t_0, then for any given vector x_0, the initial-value problem

$$x'(t) = A(t)x(t) + f(t) \qquad x(t_0) = x_0$$

has a unique solution on I.

THE HOMOGENEOUS LINEAR SYSTEM

We consider the linear homogeneous system of n equations

$$x'(t) = A(t)x(t) \tag{3}$$

obtained from Eq. (3) by setting $f(t) = 0$. The reader will recall the **principle of superposition** from the discussion of linear second-order equations in Chapter 3, which stated that linear combinations of solutions of linear homogeneous equations were again solutions. We now state the systems version of this important principle.

THEOREM 6.2: Principle of Superposition for Linear Systems

If the two vector functions $x^{(1)}(t)$ and $x^{(2)}(t)$ are solutions of $x' = A(t)x$, then so is any linear combination $c_1 x^{(1)}(t) + c_2 x^{(2)}(t)$, where c_1 and c_2 are arbitrary constants.

PROOF: One simply verifies that $c_1 x^{(1)}(t) + c_2 x^{(2)}(t)$ satisfies $x' = Ax,$ assuming that $x^{(1)}(t)$ and $x^{(2)}(t)$ satisfy $x' = Ax.$ (See Problem 20 at the end of this section.)

LINEAR INDEPENDENCE AND FUNDAMENTAL SETS

The concept of linearly independent functions played a crucial role in the general theory of a single linear differential equation. We now discuss the generalization of this concept for vector functions.

> **DEFINITION: Linearly Dependent and Independent Vector Functions**
>
> The n vector functions $x^{(1)}(t)$, $x^{(2)}(t)$, ..., $x^{(n)}(t)$ are **linearly dependent** on an interval I if there exist constants c_1, c_2, ..., c_n not all zero for which
>
> $$c_1 x^{(1)}(t) + c_2 x^{(2)}(t) + \cdots + c_n x^{(n)}(t) = 0$$
>
> for all t in I. If the vectors are not linearly dependent, they are said to be **linearly independent** on I.

Example 1   **Linearly Independent Vector Functions**   Show that the vectors

$$x^{(1)}(t) = \begin{bmatrix} e^t \\ 0 \\ 2e^t \end{bmatrix} \qquad x^{(2)}(t) = \begin{bmatrix} e^{-t} \\ 3e^{-t} \\ 0 \end{bmatrix} \qquad x^{(3)}(t) = \begin{bmatrix} e^{2t} \\ e^{2t} \\ e^{2t} \end{bmatrix}$$

are linearly independent vector functions on $(-\infty, \infty)$.

Solution   We show that the only constants c_1, c_2, and c_3 for which

$$c_1 x^{(1)}(t) + c_2 x^{(2)}(t) + c_3 x^{(3)}(t) = 0 \tag{4}$$

for all $-\infty < t < \infty$ are $c_1 = c_2 = c_3 = 0$. Since Eq. (4) holds for all t, it must hold for $t = 0$. Hence we have

$$c_1 \begin{bmatrix} 1 \\ 0 \\ 2 \end{bmatrix} + c_2 \begin{bmatrix} 1 \\ 3 \\ 0 \end{bmatrix} + c_3 \begin{bmatrix} 1 \\ 1 \\ 1 \end{bmatrix} = \begin{bmatrix} 0 \\ 0 \\ 0 \end{bmatrix} \tag{5}$$

Writing this matrix equation in scalar form, we have

$$\begin{aligned} c_1 + c_2 + c_3 &= 0 \\ 3c_2 + c_3 &= 0 \\ 2c_1 + c_3 &= 0 \end{aligned} \tag{6}$$

Since the determinant of the coefficient matrix can easily be seen to be nonzero, a basic result from the theory of equations says that $c_1 = c_2 = c_3 = 0$. If the reader is unfamiliar with this result, simply solve (6) by elimination to show $c_1 = c_2 = c_3 = 0$. ∎

Example 2   **Linearly Dependent Vector Functions**   Show that the vector functions

$$x^{(1)}(t) = \begin{bmatrix} e^{3t} \\ 2e^{3t} \end{bmatrix} \qquad x^{(2)}(t) = \begin{bmatrix} 2e^{3t} \\ 4e^{3t} \end{bmatrix}$$

are linearly dependent on $(-\infty, \infty)$.

Solution   Since we can write

$$2x^{(1)}(t) - x^{(2)}(t) = \mathbf{0}$$

the vector functions are linearly dependent on any interval. Here we have chosen the constants $c_1 = 2$ and $c_2 = -1$.

We are now led to the general representation theorem for linear systems.

THEOREM 6.3: Fundamental Set of Solutions

If $x^{(1)}(t)$, $x^{(2)}(t)$, ..., $x^{(n)}(t)$ are n linearly independent solutions of the homogeneous linear system

$$x' = A(t)x \tag{7}$$

on an interval I where the elements of $A(t)$ are continuous, then any solution $x(t)$ of Eq. (7) can be written in the form

$$x(t) = c_1 x^{(1)}(t) + c_2 x^{(2)}(t) + \cdots + c_n x^{(n)}(t) \tag{8}$$

where $c_1, c_2, ..., c_n$ are constants.

We call a set $\{x^{(1)}(t), ..., x^{(n)}(t)\}$ of n linearly independent solutions of Eq. (7) a **fundamental set of solutions,** and the collection of all linear combinations (8) of such solutions is called the **general solution** of the linear system. The reader will recall from second-order equations that the Wronskian played an important role in determining whether two solutions were linearly independent. The Wronskian can be extended to the vector Wronskian, and it plays a similar role here in the theory of linear systems of differential equations.

THE FUNDAMENTAL MATRIX AND THE WRONSKIAN

If we write the equation

$$c_1 x^{(1)}(t) + c_2 x^{(2)}(t) + \cdots + c_n x^{(n)}(t) = \mathbf{0} \tag{9}$$

in matrix form, we have

$$\begin{bmatrix} x_{11}(t) & \cdots & x_{1n}(t) \\ \vdots & & \vdots \\ x_{n1}(t) & \cdots & x_{nn}(t) \end{bmatrix} \begin{bmatrix} c_1 \\ \vdots \\ c_n \end{bmatrix} = \begin{bmatrix} 0 \\ \vdots \\ 0 \end{bmatrix} \tag{10}$$

where

$$x^{(1)}(t) = \begin{bmatrix} x_{11}(t) \\ \vdots \\ x_{n1}(t) \end{bmatrix}, \quad ..., \quad x^{(k)}(t) = \begin{bmatrix} x_{1k}(t) \\ \vdots \\ x_{nk}(t) \end{bmatrix}, \quad ..., \quad x^{(n)}(t) = \begin{bmatrix} x_{1n}(t) \\ \vdots \\ x_{nn}(t) \end{bmatrix}$$

From this we conclude that the vectors $x^{(1)}(t)$, $x^{(2)}(t)$, ..., $x^{(n)}(t)$ are linearly independent at t if and only if the determinant of the coefficient matrix $X(t) = (x_{ij}(x))$ in Eq. (10) is nonzero at t. This leads us to the definition of the fundamental matrix.

DEFINITION: The Fundamental Matrix

An $n \times n$ matrix

$$X(t) = \begin{bmatrix} x_{11}(t) & \cdots & x_{1n}(t) \\ \vdots & & \vdots \\ x_{n1}(t) & \cdots & x_{nn}(t) \end{bmatrix}$$

whose columns are n linearly independent solutions of $x' = Ax$ on an interval I is called a **fundamental matrix** for $x' = Ax$ on I.

Note that the general solution (8) of $x' = Ax$ can be written in terms of the fundamental matrix as $x(t) = X(t)c$, where c is an arbitrary column vector of constants. (See Problem 21 at the end of this section.) Also, note that since the columns of the fundamental matrix are linearly independent on I, then the determinant of the fundamental matrix will be nonzero on I. Critical to these ideas is the concept of the Wronskian.

DEFINITION: The Wronskian

The **Wronskian** of n vector functions

$$x^{(1)} = \text{col}\,(x_{11}, x_{21}, ..., x_{n1})$$
$$x^{(2)} = \text{col}\,(x_{12}, x_{22}, ..., x_{n2})$$
$$\vdots$$
$$x^{(n)} = \text{col}\,(x_{1n}, x_{2n}, ..., x_{nn})$$

is the real-valued function determined by the $(n \times n)$ determinant

$$W[x^{(1)}(t), \cdots, x^{(n)}(t)] = \begin{vmatrix} x_{11}(t) & \cdots & x_{1n}(t) \\ \vdots & & \vdots \\ x_{n1}(t) & \cdots & x_{nn}(t) \end{vmatrix}$$

The importance of the Wronskian lies in two facts. First, the Wronskian of solutions of $x' = Ax$ is either *identically zero or never zero*. Also, a set $\{x^{(1)}, x^{(2)}, ..., x^{(n)}\}$ of solutions of $x' = Ax$ is linearly independent on I *if and only if* their Wronskian is never zero on I. Hence if the Wronskian of a set of solutions of $x' = Ax$ is nonzero at a single point, then the solutions form a fundamental set.

Example 3   **General Solution of a Linear System**   Verify that the pair

$$x^{(1)}(t) = \begin{bmatrix} e^{3t} \\ e^{3t} \end{bmatrix} \qquad x^{(2)}(t) = \begin{bmatrix} e^{-t} \\ -e^{-t} \end{bmatrix}$$

form a fundamental set of solutions of

$$x'(t) = \begin{bmatrix} 1 & 2 \\ 2 & 1 \end{bmatrix} x(t) \tag{11}$$

on the interval $(-\infty, \infty)$. Write the solution in the form $x(t) = X(t)c$, where $X(t)$ is the fundamental matrix and c is an arbitrary vector.

Solution   First, we can easily verify that $x^{(1)}(t)$ and $x^{(2)}(t)$ satisfy Eq. (11) for all t. To show that $\{x^{(1)}(t), x^{(2)}(t)\}$ form a fundamental set of solutions, we compute the vector Wronskian, getting

$$W(t) = \begin{vmatrix} e^{3t} & e^{-t} \\ e^{3t} & -e^{-t} \end{vmatrix} = -e^{2t} - e^{2t} = -2e^{2t}$$

Since the Wronskian is nonzero, we conclude that $x^{(1)}(t)$ and $x^{(2)}(t)$ are linearly independent solutions. The general solution can be written

$$x(t) = Xc = \begin{bmatrix} e^{3t} & e^{-t} \\ e^{3t} & -e^{-t} \end{bmatrix} \begin{bmatrix} c_1 \\ c_2 \end{bmatrix} = c_1 \begin{bmatrix} e^{3t} \\ e^{3t} \end{bmatrix} + c_2 \begin{bmatrix} e^{-t} \\ -e^{-t} \end{bmatrix} \qquad \blacksquare$$

Recall from Chapter 3 that the general solution of a linear nonhomogeneous differential equation is the sum of a *particular solution* plus the *general solution* of the corresponding homogeneous equation. An analogous theorem holds for linear systems.

THEOREM 6.4:   General Solution of a Nonhomogeneous Linear System

If $x_p(t)$ is a particular solution of the nonhomogeneous linear system

$$x'(t) = A(t)x(t) + f(t) \tag{12}$$

on an interval I, and if $\{x^{(1)}(t), x^{(2)}(t), ..., x^{(n)}(t)\}$ is a fundamental set of solutions of the corresponding homogeneous system $x'(t) = A(t)x(t)$, then *any* solution of Eq. (12) can be written in the form

$$x(t) = c_1 x^{(1)}(t) + c_2 x^{(2)}(t) + \cdots + c_2 x^{(n)}(t) + x_p(t)$$

where $c_1, c_2, ..., c_n$ are constants.

The proof of Theorem 6.4 is essentially the same as that of Theorem 3.6 in Chapter 3, and we leave it as an exercise.

The general solution of a nonhomogeneous linear system can be found by carrying out the following steps.

Finding the General Solution of a Nonhomogeneous Linear System

To find the general solution of the *nonhomogeneous* linear system of n differential equations, $x' = A(t)x + f$, we carry out the following steps.

Step 1. Find n linearly independent solutions (fundamental set) $x^{(1)}, x^{(2)}, ...,$ $x^{(n)}$ of the corresponding *homogeneous* linear system $x' = A(t)x$. The general solution of the homogeneous linear system is

$$x_h(t) = c_1 x^{(1)}(t) + c_2 x^{(2)}(t) + \cdots + c_n x^{(n)}(t)$$

where $c_1, c_2, ..., c_n$ are arbitrary constants.

Step 2. Find a particular solution $x_p(t)$ of the nonhomogeneous linear system $x' = A(t)x + f$. The general solution of the nonhomogeneous system is then

$$x(t) = c_1 x^{(1)}(t) + c_2 x^{(2)}(t) + \cdots + c_n x^{(n)}(t) + x_p(t)$$

Note: In Sections 6.4 and 6.5 we will find the general solution of the homogeneous linear system $x' = Ax$ using the method of eigenvalues and eigenvectors. The methods of variation of parameters and undetermined coefficients that were used in Chapter 3 to find particular solutions of nonhomogeneous equations can be extended to linear systems, and we will touch upon those methods in Section 6.6. We will also solve the nonhomogeneous equation $x' = Ax + f$ in Section 6.7 using the matrix version of the Laplace transform.

PROBLEMS: Section 6.3

For Problems 1–3, determine whether each set of functions is linearly independent for $-\infty < t < \infty$.

1. $x^{(1)}(t) = \begin{bmatrix} e^t \\ e^t \end{bmatrix}$ $x^{(2)}(t) = \begin{bmatrix} 2e^{2t} \\ e^t \end{bmatrix}$

2. $x^{(1)}(t) = \begin{bmatrix} \sin t \\ \cos t \end{bmatrix}$ $x^{(2)}(t) = \begin{bmatrix} \cos t \\ -\sin t \end{bmatrix}$

3. $x^{(1)}(t) = \begin{bmatrix} e^t \\ 2e^t \\ e^t \end{bmatrix}$ $x^{(2)}(t) = \begin{bmatrix} e^{-t} \\ 2e^{-t} \\ e^t \end{bmatrix}$ $x^{(3)}(t) = \begin{bmatrix} e^{2t} \\ 3e^{2t} \\ e^{2t} \end{bmatrix}$

For Problems 4–7, write the given linear system in matrix form $x' = Ax + f$. The independent variable is taken to be t.

4. $x' = y$
$y' = -x$

5. $x' = 2x + 3y$
$y' = x - y$

6. $x' = 2x + y + 1$
$y' = 4x + y + \sin t$

7. $x' = 2x + y + e^t$
$y' = x - y + 1$

For Problems 8–11, write the single differential equation in the form $x' = Ax + f$.

8. $y'' + y = 0$

9. $y'' + 2y' + y = 1$

10. $y''' + y'' - y' + 2y = e^t$

11. $y^{(4)} + y = 1$

For Problems 12–15, write the linear system as a set of equations.

12. $x'(t) = \begin{bmatrix} 1 & 2 \\ 4 & -1 \end{bmatrix} x(t)$

13. $x'(t) = \begin{bmatrix} 1 & 0 \\ 0 & -1 \end{bmatrix} x(t) + \begin{bmatrix} 0 \\ 1 \end{bmatrix}$

14. $x'(t) = \begin{bmatrix} 4 & 3 \\ -1 & -1 \end{bmatrix} x(t) + e^{-t} \begin{bmatrix} 1 \\ 0 \end{bmatrix}$

15. $x'(t) = \begin{bmatrix} 0 & 1 & 0 \\ 0 & 0 & 1 \\ -2 & 1 & 3 \end{bmatrix} x(t) + \sin t \begin{bmatrix} 0 \\ 0 \\ 1 \end{bmatrix}$

For Problems 16–19, the given vector functions are solutions of a linear system $x' = Ax$. Tell whether they form a funda-mental set of solutions of the linear system by (a) first determining whether they are linearly independent and (b) computing their vector Wronskian.

16. $x^{(1)}(t) = e^{2t} \begin{bmatrix} 2 \\ 1 \end{bmatrix}$ $x^{(2)}(t) = e^t \begin{bmatrix} 1 \\ 0 \end{bmatrix}$

17. $x^{(1)}(t) = e^{3t} \begin{bmatrix} 1 \\ 1 \end{bmatrix}$ $x^{(2)}(t) = e^{-t} \begin{bmatrix} 2 \\ -3 \end{bmatrix}$

18. $x^{(1)}(t) = e^t \begin{bmatrix} 2 \\ 1 \end{bmatrix}$ $x^{(2)}(t) = e^t \begin{bmatrix} 1 \\ 0 \end{bmatrix}$

19. $x^{(1)}(t) = e^{4t} \begin{bmatrix} 3 \\ 1 \end{bmatrix}$ $x^{(2)}(t) = e^{4t} \begin{bmatrix} 1 \\ 1 \end{bmatrix}$

20. Principle of Superposition of Linear Systems Show that if $x^{(1)}$ and $x^{(2)}$ are solutions of $x' = Ax$, then the linear combination $c_1 x^{(1)} + c_2 x^{(2)}$ is a solution, where c_1 and c_2 are arbitrary constants.

21. Fundamental Matrix Show that the general solution of $x' = A(t)x$ can be written in the form $x = X(t)c$, where $X(t)$ is a fundamental matrix of $x' = A(t)x$ and c is a column vector of arbitrary constants.

22. Journal Entry Summarize the results of this section.

6.4   HOMOGENEOUS LINEAR SYSTEMS WITH REAL EIGENVALUES

PURPOSE

To find the general solution of the linear homogeneous system $x' = Ax$ in terms of the eigenvalues and eigenvectors of the matrix A for those cases in which the eigenvalues are real numbers. We will see that if the matrix A has n distinct eigenvalues, then A has n linearly independent eigenvectors and the general solution can easily be found. For cases in which some of the eigenvalues of A are repeated, we see that the situation becomes more involved, depending on whether or not A has n linearly independent eigenvectors.

EIGENVALUE AND EIGENVECTOR SOLUTIONS

Our goal in this section is to find the general solution of the homogeneous linear system

$$x' = Ax \tag{1}$$

where the elements of the matrix A are real numbers. To find the general solution of this system, we recall that the single equation $y' = ay$ has a general solution of the form $y(t) = ce^{at}$. This suggests that we seek a general solution to Eq. (1) having the similar form

$$x(t) = e^{\lambda t}\xi \tag{2}$$

where λ is an unknown constant and $\xi = \text{col}(\xi_1, \xi_2, ..., \xi_n)$ is a nonzero* constant unknown vector. The idea is to substitute $x(t)$ into the linear system (1) and determine λ and ξ. Doing this, we get

$$\lambda e^{\lambda t}\xi = Ae^{\lambda t}\xi = e^{\lambda t}A\xi \tag{3}$$

Canceling the nonzero factor $e^{\lambda t}$ and rearranging terms, we arrive at the system of algebraic equations

$$(A - \lambda I)\xi = 0 \tag{4}$$

where λI denotes multiplying the $n \times n$ identity matrix I by the constant λ—in other words, a matrix with λ's located down the main diagonal and zeros elsewhere. Hence the coefficient matrix $A - \lambda I$ in Eq. (4) is constructed by simply subtracting λ's from the main diagonal of A.

Summarizing what we have done thus far, remember that our ultimate goal is to find the constant λ and the vector ξ so that $x(t) = e^{\lambda t}\xi$ satisfies $x' = Ax$. What we have shown is that $x(t)$ will satisfy $x' = Ax$, provided that λ and ξ satisfy the system of algebraic equations (4). Since the zero solution $\xi = 0$ will not help in finding a set of linearly independent solutions, we require that $\xi \neq 0$. However, a system of linear equations such as $(A - \lambda I)\xi = 0$ will have *nonzero* solution ξ if and only if the determinant of its coefficient matrix is zero. Hence a necessary and sufficient condition for the linear system $(A - \lambda I)\xi = 0$ to have a nonzero solution is that

$$p(\lambda) \equiv |A - \lambda I| = 0 \tag{5}$$

or, in expanded form,

$$\begin{vmatrix} a_{11} - \lambda & a_{12} & \cdots & a_{1n} \\ a_{21} & a_{22} - \lambda & \cdots & a_{2n} \\ \vdots & \vdots & \vdots & \vdots \\ a_{n1} & a_{n2} & \cdots & a_{nn} - \lambda \end{vmatrix} = 0 \tag{6}$$

By expanding this determinant this equation can be seen to be an nth-order polynomial equation in λ called the *characteristic equation,* and the roots of this equation are called the *eigenvalues* of A. The polynomial $p(\lambda)$ is called the *characteristic polynomial.* Substituting any eigenvalue λ into the system of equations $(A - \lambda I)\xi = 0$, we can find a nonzero solution ξ. Nonzero solutions ξ that correspond to a given eigenvalue λ are called the *eigenvectors* corresponding to λ. Now that we have found λ and ξ, we have found a solution $x(t) = e^{\lambda t}\xi$. The above discussion leads to the following definition.

---

* The zero vector $\xi = 0$ has no value to us, since we eventually hope to find n linearly independent solutions.

> **DEFINITION: Eigenvalues and Eigenvectors of a Matrix**
>
> Let $A = (a_{ij})$ be a constant $n \times n$ matrix. The **eigenvalues** of A are the values of λ (possibly complex) for which the system of equations
>
> $$(A - \lambda I)\xi = 0 \qquad (7)$$
>
> has *at least one* nonzero solution ξ. They can be found by solving the **characteristic equation**
>
> $$p(\lambda) = |A - \lambda I| = 0$$
>
> The corresponding nonzero solutions ξ are called the **eigenvectors** of A corresponding to λ.
>
> If any root $\lambda = \mu$ of the characteristic equation is repeated r times, the eigenvalue μ is said to have **algebraic multiplicity** of r. An eigenvalue is said to have **geometric multiplicity** k if it has associated with it k linearly independent eigenvectors.*

Example 1   **Distinct Eigenvalues of a 2 × 2 Matrix**   Find the characteristic equation, eigenvalues, and eigenvectors of

$$A = \begin{bmatrix} 1 & 1 \\ 4 & 1 \end{bmatrix} \qquad (8)$$

Solution   The characteristic equation is

$$|A - \lambda I| = \begin{vmatrix} 1 - \lambda & 1 \\ 4 & 1 - \lambda \end{vmatrix} = (1 - \lambda)^2 - 4 = 0 \qquad (9)$$

which simplifies to $\lambda^2 - 2\lambda - 3 = 0$. Solving this equation, we find two eigenvalues $\lambda_1 = -1$, $\lambda_2 = 3$. To find the eigenvector(s) corresponding to the eigenvalue $\lambda_1 = -1$, we solve the system of equations $(A - \lambda_1 I)\xi = 0$ for ξ. Substituting for A and λ_1, we get

$$\begin{bmatrix} 1 - (-1) & 1 \\ 4 & 1 - (-1) \end{bmatrix} \begin{bmatrix} \xi_1 \\ \xi_2 \end{bmatrix} = \begin{bmatrix} 0 \\ 0 \end{bmatrix} \qquad (10)$$

The reader will observe after simplification the two scalar equations in Eq. (10) are the same,[†] namely, $2\xi_1 + \xi_2 = 0$. A geometrical interpretation of the solutions of this equation are shown in Figure 6.3. Hence we conclude that there are an infinite number

---

* It can be proven that the algebraic multiplicity of an eigenvalue $\lambda = \mu$ is always *greater than or equal to* the geometric multiplicity of μ. Interested readers can consult any textbook in linear algebra, such as *Introductory Linear Algebra with Applications*, Third Edition, by Bernard Kolman (Macmillan, New York, 1980).

[†] The fact that the two equations are the same when $\lambda = \lambda_1$ is substituted into $(A - \lambda_1 I)\xi = 0$ is not surprising, since $|A - \lambda_1 I| = 0$, which means that the system of equations is degenerate.

of eigenvectors $\boldsymbol{\xi} = \text{col}\,(\xi_1, \xi_2)$. Letting $\xi_1 = c$, we get $\xi_2 = -2c$, where we require $c \neq 0$. Hence the eigenvectors corresponding to $\lambda_1 = -1$ are

$$\boldsymbol{\xi}^{(1)} = c\begin{bmatrix} 1 \\ -2 \end{bmatrix} \qquad (c \neq 0) \tag{11}$$

Figure 6.3
The geometrical interpretation of an eigenvector

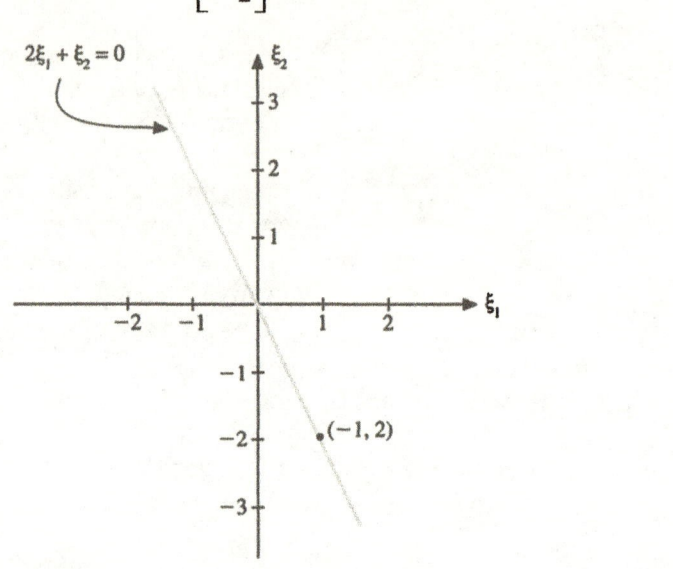

To find the eigenvectors corresponding to $\lambda_2 = 3$, we solve

$$\begin{bmatrix} 1 - 3 & 1 \\ 4 & 1 - 3 \end{bmatrix} \begin{bmatrix} \xi_1 \\ \xi_2 \end{bmatrix} = \begin{bmatrix} 0 \\ 0 \end{bmatrix} \tag{12}$$

Again, both equations in Eq. (12) yield a single equation, this time $-2\xi_1 + \xi_2 = 0$; by letting $\xi_1 = c$, we get $\xi_2 = 2c$, where $c \neq 0$. Hence the eigenvectors corresponding to $\lambda_2 = 3$ are

$$\boldsymbol{\xi}^{(2)} = c\begin{bmatrix} 1 \\ 2 \end{bmatrix} \qquad (c \neq 0) \tag{13}$$

Since the constant c in $\boldsymbol{\xi}^{(1)}$ and $\boldsymbol{\xi}^{(2)}$ can be any nonzero real number, we have found an infinite number of eigenvectors corresponding to each eigenvalue. However, for each eigenvalue we have found only one *linearly independent* eigenvector.* See Figure 6.4. ∎

Example 2 **Distinct Eigenvalues of a 3×3 Matrix** Find the eigenvalues and eigenvectors of the matrix

$$A = \begin{bmatrix} 1 & 1 & -2 \\ -1 & 2 & 1 \\ 0 & 1 & -1 \end{bmatrix}$$

---

* In this problem, each eigenvalue has one linearly independent eigenvector. Later, we will see that some eigenvalues have more than one linearly independent eigenvector.

Figure 6.4
The two linearly independent
eigenvectors of **A**

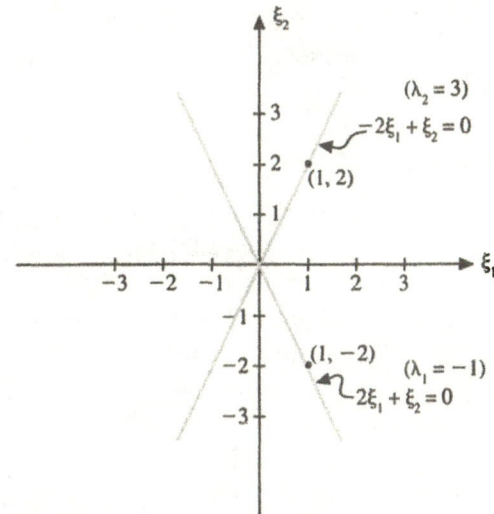

Solution   The characteristic equation for A is

$$|A - \lambda I| = \begin{vmatrix} 1 - \lambda & 1 & -2 \\ -1 & 2 - \lambda & 1 \\ 0 & 1 & -1 - \lambda \end{vmatrix} = 0 \tag{14}$$

or

$$\lambda^3 - 2\lambda^2 - \lambda + 2 = 0 \tag{15}$$

which simplifies* to $(\lambda + 1)(\lambda - 1)(\lambda - 2) = 0$. Hence the three eigenvalues of A are $\lambda_1 = -1$, $\lambda_2 = 1$, and $\lambda_3 = 2$. To find the corresponding eigenvectors, we systematically substitute λ_1, λ_2, and λ_3 into $(A - \lambda I)\xi = \mathbf{0}$ for λ and solve for $\boldsymbol{\xi}$. Doing this, we find

$$\lambda_1 = -1 \ \lozenge \ \begin{cases} 2\xi_1 + \xi_2 - 2\xi_3 = 0 \\ -\xi_1 + 3\xi_2 + \xi_3 = 0 \\ \xi_2 = 0 \end{cases} \tag{16a}$$

$$\lambda_2 = \ \ 1 \ \lozenge \ \begin{cases} \xi_2 - 2\xi_3 = 0 \\ -\xi_1 + \xi_2 + \xi_3 = 0 \\ \xi_2 - 2\xi_3 = 0 \end{cases} \tag{16b}$$

$$\lambda_3 = \ \ 2 \ \lozenge \ \begin{cases} -\xi_1 + \xi_2 - 2\xi_3 = 0 \\ -\xi_1 + \xi_3 = 0 \\ \xi_2 - 3\xi_3 = 0 \end{cases} \tag{16c}$$

* One generally approximates the roots of a polynomial equation of order greater than two by using a computer.

The system (16a) reduces* to $\xi_2 = 0$, $\xi_1 = \xi_3$. Hence to find a nonzero solution, we set $\xi_1 = \xi_3 = c$. Hence the eigenvectors associated with $\lambda_1 = -1$ are

$$\xi^{(1)} = c \begin{bmatrix} 1 \\ 0 \\ 1 \end{bmatrix} \qquad (c \neq 0) \tag{17}$$

which represents a line in three dimensions.

The system (16b) corresponding to $\lambda_2 = 1$ is equivalent to the pair of equations $\xi_1 = 3\xi_3$ and $\xi_2 = 2\xi_3$. Since we have an **underdetermined system** of three unknowns and two equations, we can let $\xi_3 = c$ and solve for $\xi_1 = 3c$ and $\xi_2 = 2c$. Hence the eigenvectors corresponding to $\lambda_2 = 1$ are

$$\xi^{(2)} = c \begin{bmatrix} 3 \\ 2 \\ 1 \end{bmatrix} \qquad (c \neq 0) \tag{18}$$

which represents a line in three dimensions.

Finally, the system (16c) corresponding to $\lambda_3 = 2$ is equivalent to the two equations $-\xi_1 + \xi_3 = 0$ and $\xi_2 - 3\xi_3 = 0$. Letting $\xi_3 = c$, we then find $\xi_1 = c$ and $\xi_2 = 3c$. Hence the eigenvectors corresponding to $\lambda_3 = 2$ are

$$\xi^{(3)} = c \begin{bmatrix} 1 \\ 3 \\ 1 \end{bmatrix} \qquad (c \neq 0) \tag{19}$$

which represents a line in three dimensions. ◼

Example 3   **Geometric Multiplicity Equal to Algebraic Multiplicity**   Find the eigenvalues and eigenvectors of

$$A = \begin{bmatrix} -2 & 1 & 1 \\ 1 & -2 & 1 \\ 1 & 1 & -2 \end{bmatrix}$$

Solution   The characteristic equation is

$$|A - \lambda I| = \begin{vmatrix} -2-\lambda & 1 & 1 \\ 1 & -2-\lambda & 1 \\ 1 & 1 & -2-\lambda \end{vmatrix} = 0 \tag{20}$$

Evaluating this determinant, we find $\lambda(\lambda + 3)^2 = 0$, and so the three eigenvalues are $\lambda_1 = 0$, $\lambda_2 = -3$, $\lambda_3 = -3$. The eigenvalue 0 has an algebraic multiplicity of 1, and the eigenvalue -3 has an algebraic multiplicity of 2.

---

* The fact that the three algebraic equations in ξ_1, ξ_2, and ξ_3 reduce to two equations is not surprising, since by substituting λ_1 into the linear system $(A - \lambda_1 I)\xi = 0$ results in a degenerate system of equations. In fact, to save yourself a little work, just drop one of the three equations and solve for ξ_1, ξ_2, and ξ_3 from the other two equations.

To find the eigenvectors, we first consider $\lambda_1 = 0$. Substituting this value into $(A - \lambda_1 I)\xi = 0$, this system reduces to the equations $\xi_1 - \xi_3 = 0$ and $\xi_2 - \xi_3 = 0$. Hence we get

$$\xi^{(1)} = c \begin{bmatrix} 1 \\ 1 \\ 1 \end{bmatrix} \qquad (c \neq 0) \tag{21}$$

We now consider the case of the multiple eigenvalue $\lambda_2 = \lambda_3 = -3$. Substituting -3 for λ in the system $(A - \lambda I)\xi = 0$, we get

$$\begin{bmatrix} 1 & 1 & 1 \\ 1 & 1 & 1 \\ 1 & 1 & 1 \end{bmatrix} \begin{bmatrix} \xi_1 \\ \xi_2 \\ \xi_3 \end{bmatrix} = \begin{bmatrix} 0 \\ 0 \\ 0 \end{bmatrix} \tag{22}$$

Since the three equations in Eq. (22) are equivalent to the single equation

$$\xi_1 + \xi_2 + \xi_3 = 0$$

we can find *two* linearly independent solutions by first picking $\xi_1 = 1$, $\xi_2 = 0$ and computing $\xi_3 = -1$ and then choosing $\xi_1 = 0$, $\xi_2 = 1$ and computing $\xi_3 = -1$. In so doing, we obtain the *two* linearly independent eigenvectors

$$\xi^{(2)} = c \begin{bmatrix} 1 \\ 0 \\ -1 \end{bmatrix} \qquad (c \neq 0)$$

$$\xi^{(3)} = c \begin{bmatrix} 0 \\ 1 \\ -1 \end{bmatrix} \qquad (c \neq 0)$$

corresponding to $\lambda_2 = \lambda_3 = -3$. Hence the eigenvalue of -3 has a geometric multiplicity of 2. ∎

Example 4 **Geometric Multiplicity Less Than Algebraic Multiplicity** Find the eigenvalues and eigenvectors of

$$A = \begin{bmatrix} 5 & 10 & 7 \\ 0 & -3 & -3 \\ 0 & 3 & 3 \end{bmatrix} \tag{23}$$

Solution The characteristic equation of A is

$$\lambda^2(\lambda - 5) = 0$$

and so the eigenvalues are $\lambda_1 = 5$, $\lambda_2 = 0$, $\lambda_3 = 0$. Hence $\lambda_1 = 5$ has an algebraic multiplicity of 1, and $\lambda_2 = \lambda_3 = 0$ has an algebraic multiplicity of 2.

We first find the eigenvector corresponding to $\lambda_1 = 5$ by finding the solution(s) of the linear system $(A - 5I)\boldsymbol{\xi} = \mathbf{0}$. This gives

$$\begin{bmatrix} 0 & 10 & 7 \\ 0 & -8 & -3 \\ 0 & 3 & -2 \end{bmatrix} \begin{bmatrix} \xi_1 \\ \xi_2 \\ \xi_3 \end{bmatrix} = \begin{bmatrix} 0 \\ 0 \\ 0 \end{bmatrix}$$

This system is equivalent to the equations

$$10\xi_2 + 7\xi_3 = 0$$
$$-8\xi_2 - 3\xi_3 = 0$$
$$3\xi_2 - 2\xi_3 = 0$$

These equations impose no condition on ξ_1 but require $\xi_2 = \xi_3 = 0$. Hence we have the nonzero eigenvectors

$$\boldsymbol{\xi}^{(1)} = c \begin{bmatrix} 1 \\ 0 \\ 0 \end{bmatrix} \qquad (c \neq 0)$$

To find the eigenvectors corresponding to $\lambda_2 = \lambda_3 = 0$, we solve the system

$$(A - 0I)\boldsymbol{\xi} = \begin{bmatrix} 5 & 10 & 7 \\ 0 & -3 & -3 \\ 0 & 3 & 3 \end{bmatrix} \begin{bmatrix} \xi_1 \\ \xi_2 \\ \xi_3 \end{bmatrix} = \begin{bmatrix} 0 \\ 0 \\ 0 \end{bmatrix}$$

or

$$5\xi_1 + 10\xi_2 + 7\xi_3 = 0$$
$$-3\xi_2 - 3\xi_3 = 0$$
$$3\xi_2 + 3\xi_3 = 0$$

Noting that the second and third equations are similar, we can arbitrarily set $\xi_3 = 5c$ and solve for $\xi_1 = -3\xi_3$ and $\xi_2 = -5\xi_3$. Hence we arrive at the eigenvectors

$$\boldsymbol{\xi}^{(2)} = c \begin{bmatrix} 3 \\ -5 \\ 5 \end{bmatrix} \qquad (c \neq 0)$$

In other words, the eigenvalue $\lambda = 0$ has a geometric multiplicity of 1. ◪

THE GENERAL SOLUTION OF THE HOMOGENEOUS LINEAR SYSTEM

We have found that the homogeneous linear system $x' = Ax$ has a solution of the form $e^{\lambda t}\boldsymbol{\xi}$, provided that λ is an eigenvalue of A and $\boldsymbol{\xi}$ is an eigenvector of A corresponding to λ. We now ask whether a system of n linear equations has n linearly independent solutions from which the general solution would be found. The answer is given in the following theorem.

THEOREM 6.5: General Solution of $x' = Ax$

If A is an $n \times n$ constant matrix with n linearly independent eigenvectors $\xi^{(1)}$, $\xi^{(2)}$, ..., $\xi^{(n)}$ corresponding to the real eigenvalues* $\lambda_1, \lambda_2, ..., \lambda_n$, then

$$\{e^{\lambda_1 t}\, \xi^{(1)},\ e^{\lambda_2 t}\xi^{(2)},\ ...,\ e^{\lambda_n t}\xi^{(n)}\} \tag{24}$$

constitutes a set of n linearly independent solutions, called a **fundamental set**, of the linear homogeneous system $x' = Ax$. The general solution of $x' = Ax$ is

$$x(t) = c_1 e^{\lambda_1 t}\xi^{(1)} + c_2 e^{\lambda_2 t}\xi^{(2)} + \cdots + c_n e^{\lambda_n t}\xi^{(n)} \tag{25}$$

where $c_1, c_2, ..., c_n$ are arbitrary constants.

PROOF: We have seen that each vector in the fundamental set (24) is a solution of the homogeneous system $x' = Ax$. To show that they are linearly independent, we compute the vector Wronskian, given by

$$\begin{aligned} W(t) &= \det\, [e^{\lambda_1 t}\xi^{(1)},\ e^{\lambda_2 t}\xi^{(2)},\ ...,\ e^{\lambda_n t}\xi^{(n)}] \\ &= e^{(\lambda_1 + \cdots + \lambda_n)t} \det\, [\xi^{(1)},\ \xi^{(2)},\ ...,\ \xi^{(n)}] \end{aligned} \tag{26}$$

Since the eigenvectors have been assumed to be linearly independent, the determinant $\det\, [\xi^{(1)}, \xi^{(2)}, ..., \xi^{(n)}]$ is nonzero. Hence the Wronskian $W(t)$ is never zero, and thus the n solutions are linearly independent.

Note: It can also be shown that if an $n \times n$ matrix A has n distinct eigenvalues, then the matrix has n linearly independent eigenvectors. Hence the solution of such a system $x' = Ax$ has the form of Eq. (25).

Example 5   Fundamental Set of Solutions for Distinct Eigenvalues   Find two linearly independent solutions of the system

$$x' = Ax \quad \text{where} \quad A = \begin{bmatrix} 1 & 1 \\ 4 & 1 \end{bmatrix} \tag{27}$$

Solution   In Example 1 we found the eigenvalues of A to be $\lambda_1 = -1, \lambda_2 = 3$ with corresponding eigenvectors

$$\xi^{(1)} = c\begin{bmatrix} 1 \\ 2 \end{bmatrix} \qquad \xi^{(2)} = c\begin{bmatrix} 1 \\ -2 \end{bmatrix} \tag{28}$$

---

* As we saw in Example 3, some of the eigenvalues may have algebraic multiplicity greater than 1.

Hence two linearly independent solutions* are

$$x^{(1)}(t) = e^{\lambda_1 t}\xi^{(1)} = e^{-t}\begin{bmatrix} 1 \\ 2 \end{bmatrix} \tag{29a}$$

$$x^{(2)}(t) = e^{\lambda_2 t}\xi^{(2)} = e^{3t}\begin{bmatrix} 1 \\ -2 \end{bmatrix} \tag{29b}$$

The general solution is

$$x(t) = c_1 e^{-t}\begin{bmatrix} 1 \\ 2 \end{bmatrix} + c_2 e^{3t}\begin{bmatrix} 1 \\ -2 \end{bmatrix} \tag{30}$$

where c_1 and c_2 are arbitrary constants. ∎

Example 6 **Fundamental Set of Solutions for Distinct Eigenvalues** Find the general solution of

$$x' = Ax \qquad \text{where} \qquad A = \begin{bmatrix} 1 & 1 & -2 \\ -1 & 2 & 1 \\ 0 & 1 & -1 \end{bmatrix} \tag{31}$$

Solution In Example 2 we found the eigenvalues and eigenvectors of A to be

$$\lambda_1 = -1 \qquad \xi^{(1)} = \begin{bmatrix} 1 \\ 0 \\ 1 \end{bmatrix} \tag{32a}$$

$$\lambda_2 = 1 \qquad \xi^{(2)} = \begin{bmatrix} 3 \\ 2 \\ 1 \end{bmatrix} \tag{32b}$$

$$\lambda_3 = 2 \qquad \xi^{(3)} = \begin{bmatrix} 1 \\ 3 \\ 1 \end{bmatrix} \tag{32c}$$

Hence a general solution is

$$x(t) = c_1 e^{-t}\begin{bmatrix} 1 \\ 0 \\ 1 \end{bmatrix} + c_2 e^{t}\begin{bmatrix} 3 \\ 2 \\ 1 \end{bmatrix} + c_3 e^{2t}\begin{bmatrix} 1 \\ 3 \\ 1 \end{bmatrix} \tag{33}$$

where c_1, c_2, and c_3 are arbitrary constants. ▣

Example 7 **Fundamental Set of Solutions for Repeated Eigenvalues** Find the general solution of

$$x' = Ax \qquad \text{where} \qquad A = \begin{bmatrix} -2 & 1 & 1 \\ 1 & -2 & 1 \\ 1 & 1 & -2 \end{bmatrix} \tag{34}$$

* Since we need only one eigenvector to obtain a solution, we select the constant c in the eigenvector to be $c = 1$.

Solution   In Example 3 we found the eigenvalues and eigenvectors to be

$$\lambda_1 = 0 \qquad \boldsymbol{\xi}^{(1)} = \begin{bmatrix} 1 \\ 1 \\ 1 \end{bmatrix} \tag{35a}$$

$$\lambda_2 = -3 \qquad \boldsymbol{\xi}^{(2)} = \begin{bmatrix} 1 \\ 0 \\ -1 \end{bmatrix} \tag{35b}$$

$$\lambda_3 = -3 \qquad \boldsymbol{\xi}^{(3)} = \begin{bmatrix} 0 \\ 1 \\ -1 \end{bmatrix} \tag{35c}$$

Hence a general solution is

$$\boldsymbol{x}(t) = c_1 \begin{bmatrix} 1 \\ 1 \\ 1 \end{bmatrix} + c_2 e^{-3t} \begin{bmatrix} 1 \\ 0 \\ -1 \end{bmatrix} + c_3 e^{-3t} \begin{bmatrix} 0 \\ 1 \\ -1 \end{bmatrix} \tag{36}$$

where c_1, c_2, and c_3 are arbitrary constants.*   ∎

PROBLEMS: Section 6.4

For Problems 1–8, find the eigenvalues and eigenvectors of the indicated matrix. For the 2×2 matrices, plot the eigenvectors.

1. $\begin{bmatrix} 2 & 0 \\ 0 & 1 \end{bmatrix}$

2. $\begin{bmatrix} 3 & 2 \\ 2 & 0 \end{bmatrix}$

3. $\begin{bmatrix} 1 & 0 \\ 0 & 1 \end{bmatrix}$ (repeated eigenvalues)

4. $\begin{bmatrix} 1 & 0 \\ 1 & 2 \end{bmatrix}$

5. $\begin{bmatrix} 2 & 0 & 0 \\ 1 & -1 & -2 \\ -1 & 0 & 1 \end{bmatrix}$

6. $\begin{bmatrix} 1 & 2 & -1 \\ 1 & 0 & 1 \\ 4 & -4 & 5 \end{bmatrix}$

7. $\begin{bmatrix} 1 & 2 & 2 \\ 2 & 0 & 3 \\ 2 & 3 & 0 \end{bmatrix}$

8. $\begin{bmatrix} 0 & 1 & 1 \\ 1 & 0 & 1 \\ 1 & 1 & 0 \end{bmatrix}$ (repeated roots)

For Problems 9–15, find the general solution of the homogeneous linear system $\boldsymbol{x}' = A\boldsymbol{x}$ for the given matrix A.

9. $A = \begin{bmatrix} -4 & 2 \\ 2 & -1 \end{bmatrix}$

* In this problem, although the eigenvalue $\lambda_2 = \lambda_3 = -3$ is repeated (algebraic multiplicity 2), there are *two linearly independent* eigenvectors $\boldsymbol{\xi}^{(2)}$ and $\boldsymbol{\xi}^{(3)}$ associated with it (geometric multiplicity 2). Hence we have the two linearly independent solutions $e^{-3t}\boldsymbol{\xi}^{(2)}$ and $e^{-3t}\boldsymbol{\xi}^{(3)}$. Sometimes, however, the geometric multiplicity of an eigenvalue is *less* than the algebraic multiplicity, and in those cases a *new strategy* must be found for finding enough linearly independent solutions. The strategy for finding n linearly independent solutions of an $n \times n$ system that contains *fewer* than n linearly independent eigenvectors is (partially) described in Problems 24–29.

10. $A = \begin{bmatrix} 1 & -1 \\ 2 & 4 \end{bmatrix}$

11. $A = \begin{bmatrix} 5 & -1 \\ 3 & 1 \end{bmatrix}$

12. $A = \begin{bmatrix} 1 & 0 \\ -2 & 2 \end{bmatrix}$

13. $A = \begin{bmatrix} 3 & 2 & 2 \\ 1 & 4 & 1 \\ -2 & -4 & -1 \end{bmatrix}$

14. $A = \begin{bmatrix} -1 & 1 & 0 \\ 1 & 2 & 1 \\ 0 & 3 & -1 \end{bmatrix}$

15. $A = \begin{bmatrix} 1 & -2 & 2 \\ -2 & 1 & 2 \\ 2 & 2 & 1 \end{bmatrix}$ (repeated eigenvalues)

For Problems 16–21, solve the given initial-value problem.

16. $x' = \begin{bmatrix} 2 & 0 \\ 0 & 3 \end{bmatrix} x$ $\qquad x(0) = \begin{bmatrix} 5 \\ 4 \end{bmatrix}$

17. $x' = \begin{bmatrix} 1 & 1 \\ 1 & 1 \end{bmatrix} x$ $\qquad x(0) = \begin{bmatrix} 2 \\ 3 \end{bmatrix}$

18. $x' = \begin{bmatrix} 1 & 2 \\ 2 & 1 \end{bmatrix} x$ $\qquad x(0) = \begin{bmatrix} 1 \\ 3 \end{bmatrix}$

19. $x' = \begin{bmatrix} 1 & -1 \\ 2 & 4 \end{bmatrix} x$ $\qquad x(0) = \begin{bmatrix} 1 \\ 0 \end{bmatrix}$

20. $x' = \begin{bmatrix} 1 & -1 & 0 \\ 0 & -1 & 3 \\ -1 & 1 & 0 \end{bmatrix} x$ $\quad x(0) = \begin{bmatrix} 0 \\ 0 \\ 1 \end{bmatrix}$

21. $x' = \begin{bmatrix} 1 & 1 & 0 \\ 1 & 1 & 0 \\ 0 & 0 & -1 \end{bmatrix} x$ $\quad x(0) = \begin{bmatrix} 2 \\ 4 \\ 2 \end{bmatrix}$

Converting Higher-Order Equations to Systems

For Problems 22–23, convert the given differential equation to a first-order linear system $x' = Ax$.

(a) Show that the characteristic equation of the first-order system $x' = Ax$ is the same as the characteristic equation of the given differential equation.

(b) Solve both the linear system $x' = Ax$ and the given equation and compare the results.

22. $y'' - y = 0$

23. $y'' + 3y' + 2y = 0$

Fewer than n Linearly Independent Eigenvectors

We have seen that if the $n \times n$ matrix A in the linear system $x' = Ax$ has n distinct eigenvalues, then its eigenvectors are linearly independent, and the vectors

$$\{e^{\lambda_1 t}\xi^{(1)}, e^{\lambda_2 t}\xi^{(2)}, ..., e^{\lambda_n t}\xi^{(n)}\} \tag{37}$$

form a fundamental set of solutions. However, if A has eigenvalues with algebraic multiplicity greater than 1, then in some instances the solutions in Eq. (37) will not be linearly independent. Problems 24–29 show how a fundamental set of solutions can be found when the matrix A does not have n linearly independent eigenvectors.

24. One Independent Eigenvector Carry out Steps (a)–(c) to solve the linear system

$$x' = \begin{bmatrix} 3 & -4 \\ 1 & -1 \end{bmatrix} x \tag{38}$$

(a) Show that the characteristic equation of the matrix

$$A = \begin{bmatrix} 3 & -4 \\ 1 & -1 \end{bmatrix}$$

has the repeated root (eigenvalue) $\lambda = 1$ and that *all* eigenvectors have the form

$$\xi = c \begin{bmatrix} 2 \\ 1 \end{bmatrix}$$

(b) Use the results from Step (a) to find a single solution $x^{(1)}(t)$ of $x' = Ax$.

(c) Find a second linearly independent solution by trying a solution of the form

$$x^{(2)}(t) = te^{\lambda t}\xi + e^{\lambda t}\eta = te^t \begin{bmatrix} 2 \\ 1 \end{bmatrix} + e^t \begin{bmatrix} \eta_1 \\ \eta_2 \end{bmatrix} \tag{39}$$

Substitute $x^{(2)}(t)$ into $x' = Ax$ and derive the equations

$$\begin{aligned} (A - \lambda I)\xi &= 0 & \text{(40a)} \\ (A - \lambda I)\eta &= \xi & \text{(40b)} \end{aligned}$$

Using Eq. (40b), we can substitute the known values of λ and ξ and solve for η.

25. One Independent Eigenvector Use the strategy discussed in Problem 24 to find the general solution of

$$x' = \begin{bmatrix} 7 & 1 \\ -4 & 3 \end{bmatrix} x \tag{41}$$

26. One Linearly Independent Eigenvector Consider the linear system $x' = Ax$, where

$$A = \begin{bmatrix} 0 & 0 & 1 \\ 1 & 0 & -3 \\ 0 & 1 & 3 \end{bmatrix}$$

(a) Show that A has a repeated eigenvalue $\lambda = 1$ with algebraic multiplicity 3, all eigenvectors being scalar multiples of $\xi = \text{col}(1, -2, 1)$.

(b) Use the result from part (a) to find a solution of the system $x' = Ax$ of the form $x^{(1)}(t) = e^t \xi$.

(c) Find a second solution corresponding to $\lambda = 1$ by looking for a solution of the form

$$x^{(2)}(t) = te^t \xi + e^t \eta$$

where the vector η is to be determined. *Hint:* Show that η must satisfy $(A - I)\eta = \xi$.

(d) Use η found in part (c) to find a third solution corresponding to $\lambda = 1$ by looking for a solution of the form

$$x^{(3)}(t) = \frac{t^2}{2} e^t \xi + te^t \eta + e^t \zeta$$

Hint: Show that ζ must satisfy $(A - I)\zeta = \eta$.

27. General Repeated Eigenvalue Problem Given the homogeneous linear system

$$x'(t) = \begin{bmatrix} a & b \\ c & d \end{bmatrix} x(t)$$

(a) Show that the characteristic equation has a repeated root only if $(a - d)^2 + 4bc = 0$.

(b) Show that if $a \neq d$ and $(a - d)^2 = 0$, then the general solution of the linear system is

$$x(t) = c_1 \begin{bmatrix} 2b \\ d - a \end{bmatrix} e^{(a+d)t/2}$$

$$+ c_2 \left\{ \begin{bmatrix} 2b \\ d - a \end{bmatrix} t + \begin{bmatrix} 0 \\ 2 \end{bmatrix} \right\} e^{(a+d)t/2}$$

28. Cauchy-Euler System Show that the **Cauchy-Euler system**

$$tx'(t) = Ax(t) \qquad (t > 0) \qquad (42)$$

where A is a constant matrix, has a solution of the form $x(t) = t^\lambda \xi$, where λ is an eigenvalue of A and ξ is a corresponding eigenvector.

29. Typical Cauchy-Euler System Use the results from Problem 28 to solve

$$tx'(t) = \begin{bmatrix} 3 & -2 \\ 2 & -2 \end{bmatrix} x(t) \qquad (t > 0)$$

30. Adjoint Systems The linear system

$$\dot{x} = Ax \qquad (43)$$

has a "cousin" system

$$\dot{y} = -A^T y \qquad (44)$$

called the **adjoint system** to Eq. (43). Since finding the adjoint system of a given system consists of taking the negative transpose of the coefficient matrix A, we conclude that Eq. (43) is *also* the adjoint system to Eq. (44). Hence the systems (43) and (44) are simply referred to as **adjoint systems.**

(a) Find the adjoint system to

$$\dot{x} = \begin{bmatrix} 0 & 1 \\ 1 & 0 \end{bmatrix} x \qquad (45)$$

(b) Show the general relationship

$$\frac{d}{dt}\{y^T x\} = \dot{y}^T x + y^T \dot{x} = 0$$

between solutions x and y of adjoint systems. This implies that $y^T x \equiv$ constant. *Hint:* $(AB)^T = B^T A^T$.

(c) Find the solution of Eq. (45) with initial conditions

$$x(0) = \begin{bmatrix} 1 \\ 0 \end{bmatrix}$$

(d) Find the solution of the adjoint system of Eq. (45) with initial conditions

$$y(0) = \begin{bmatrix} 0 \\ 1 \end{bmatrix}$$

(e) Since $y^T(0)x(0) = 0$ for the initial conditions given in (c) and (d), what is true about the paths $x = x(t)$ and $y = y(t)$ in the xy-plane?

31. Systems of Equations in Radioactive Decay The radioactive isotope of iodine, ^{135}I, decays into the radioactive isotope of xenon, ^{135}Xe, which in turn decays into something else (not important). The half-lives of iodine and xenon are 6.7 hr and 9.2 hr, respectively.

(a) Write the system of differential equations that describes the amount of ^{135}I and ^{135}Xe present.

(b) What is the general solution of the system of equations found in part (a)?

32. Journal Entry Explain the concept of an eigenvalue and eigenvector of a matrix A.

6.5   HOMOGENEOUS LINEAR SYSTEMS WITH COMPLEX EIGENVALUES

PURPOSE

To solve the linear homogeneous system $x' = Ax$ when the coefficient matrix A has at least one pair of complex conjugate eigenvalues.

SOLUTIONS CORRESPONDING TO COMPLEX EIGENVALUES

In Section 6.4 we found the general solution of the linear homogeneous system

$$x' = Ax \tag{1}$$

when A was a constant real matrix with real eigenvalues. We now consider the case in which A has at least one pair of complex eigenvalues,* $\alpha + i\beta$, $\alpha - i\beta$. We will find a fundamental set of solutions, provided that the complex eigenvalues are not repeated. We will see that when the eigenvalues of A are complex, the corresponding eigenvectors are complex and hence the solution is complex. It is possible, however, to use Euler's equation to obtain real-valued solutions in much the same way as we did for second-order equations.

We begin by assuming that A has a complex eigenvalue $\lambda_1 = \alpha + i\beta$ with a corresponding eigenvector $a + ib$, where α and β are constant real numbers and a and b are constant real vectors. Since complex eigenvalues occur in pairs for real matrices, we also know that $\lambda_2 = \alpha - i\beta$ is an eigenvalue of A. It can also be shown that the eigenvector corresponding to $\lambda_2 = \alpha - i\beta$ is the conjugate eigenvector $a - ib$. In short, corresponding to the complex eigenvalues $\lambda_1 = \alpha + i\beta$ and $\lambda_2 = \alpha - i\beta$ and their corresponding eigenvectors $a + ib$ and $a - ib$, we have the two linearly independent solutions

$$w^{(1)}(t) = e^{(\alpha + i\beta)t}(a + ib) \tag{2a}$$
$$w^{(2)}(t) = e^{(\alpha - i\beta)t}(a - ib) \tag{2b}$$

These complex exponentials are similar to the ones studied in Chapter 3 when the roots of the characteristic equation were complex. There we used Euler's formula,

$$e^{ix} = \cos x + i \sin x \tag{3}$$

to obtain two real solutions. Using Euler's formula again, we can rewrite the first complex solution (2a) as

$$
\begin{aligned}
w^{(1)}(t) &= e^{(\alpha + i\beta)t}(a + ib) \\
&= e^{\alpha t} e^{i\beta t}(a + ib) \\
&= e^{\alpha t}(\cos \beta t + i \sin \beta t)(a + ib) \\
&= e^{\alpha t}(\cos \beta t\, a - \sin \beta t\, b) + i e^{\alpha t}(\sin \beta t\, a + \cos \beta t\, b) \tag{4}
\end{aligned}
$$

---

* Since the elements of the matrix A are real, the coefficients of the characteristic polynomial are also real, and hence the roots of the characteristic equation occur in *complex conjugate pairs* $\lambda_1 = \alpha + i\beta$, and $\lambda_2 = \alpha - i\beta$.

Hence we have rewritten the complex solution as $w^{(1)}(t) = x^{(1)}(t) + ix^{(2)}(t)$, where

$$x^{(1)}(t) = e^{\alpha t} \cos \beta t\, a - e^{\alpha t} \sin \beta t\, b \qquad (5a)$$
$$x^{(2)}(t) = e^{\alpha t} \sin \beta t\, a + e^{\alpha t} \cos \beta t\, b \qquad (5b)$$

Now, since $w^{(1)}(t) = x^{(1)}(t) + ix^{(2)}(t)$ is a solution of $x' = Ax$, we have

$$\frac{d}{dt}\left(x^{(1)}(t) + ix^{(2)}(t)\right) = A\left(x^{(1)}(t) + ix^{(2)}(t)\right) = Ax^{(1)}(t) + iAx^{(2)}(t) \qquad (6)$$

But by equating the real and imaginary parts, we find

$$\frac{d}{dt}x^{(1)}(t) = Ax^{(1)}(t) \qquad (7a)$$

$$\frac{d}{dt}x^{(2)}(t) = Ax^{(2)}(t) \qquad (7b)$$

Hence we have proven that $x^{(1)}(t)$ and $x^{(2)}(t)$ are *real* vector solutions of $x' = Ax$. If we substitute the second complex solution $w^{(2)}(t)$ into $x' = Ax$ we will obtain the same real solutions $x^{(1)}(t)$ and $x^{(2)}(t)$. Finally, it can be shown (see Problem 12 at the end of this section) that $x^{(1)}(t)$ and $x^{(2)}(t)$ are linearly independent vectors on the real line $(-\infty, \infty)$; hence they form a set of two linearly independent solutions. We summarize these results.

THEOREM 6.6: Solutions Corresponding to Complex Eigenvalues

If $\alpha \pm i\beta$ are complex conjugate eigenvalues of the real matrix A with corresponding eigenvectors $a \pm ib$, then

$$x^{(1)}(t) = e^{\alpha t} \cos \beta t\, a - e^{\alpha t} \sin \beta t\, b \qquad (8a)$$
$$x^{(2)}(t) = e^{\alpha t} \sin \beta t\, a + e^{\alpha t} \cos \beta t\, b \qquad (8b)$$

are linearly independent vector solutions of $x' = Ax$, and the general real-valued vector solution of Eqs. (8) is given by

$$x(t) = c_1 x^{(1)}(t) + c_2 x^{(2)}(t)$$

where c_1 and c_2 are arbitrary real constants.

Example 1   Complex Eigenvalues Find a general solution of

$$x'(t) = \begin{bmatrix} -1 & -1 \\ 4 & -1 \end{bmatrix} x(t) \qquad (9)$$

Solution   The characteristic equation is

$$|A - \lambda I| = \begin{vmatrix} -1 - \lambda & -1 \\ 4 & -1 - \lambda \end{vmatrix} = \lambda^2 + 2\lambda + 5 = 0 \qquad (10)$$

which has complex conjugate roots (eigenvalues) $\lambda = -1 \pm 2i$. From these eigenvalues we identify the real and complex parts $\alpha = -1$ and $\beta = 2$. To find the general solution, we need only find the eigenvector associated with one of the eigenvalues, say, $-1 + 2i$. We do this by substituting $-1 + 2i$ into the system of algebraic equations $(A - \lambda I)\,\xi = 0$ and solving for ξ. Doing this, we find

$$\begin{bmatrix} -2i & -1 \\ 4 & -2i \end{bmatrix} \begin{bmatrix} \xi_1 \\ \xi_2 \end{bmatrix} = \begin{bmatrix} 0 \\ 0 \end{bmatrix} \tag{11}$$

Writing out these equations, we see that they both simplify to $2\xi_1 - i\xi_2 = 0$. If we now let $\xi_2 = 2$, we get $\xi_1 = i$. Hence the eigenvector corresponding to $-1 + 2i$ is $(i, 2)$, which we can write as

$$\begin{bmatrix} i \\ 2 \end{bmatrix} = \begin{bmatrix} 0 \\ 2 \end{bmatrix} + i \begin{bmatrix} 1 \\ 0 \end{bmatrix} \tag{12}$$

We can now identify the vectors $a = \operatorname{col}(0, 2)$ and $b = \operatorname{col}(1, 0)$. Finally, substituting $\alpha = -1$, $\beta = 2$, and a and b into Eqs. (8), we obtain the two linearly independent solutions

$$\begin{aligned} x^{(1)}(t) &= e^{\alpha t} \cos \beta t\, a - e^{\alpha t} \sin \beta t\, b \\ &= e^{-t} \cos 2t \begin{bmatrix} 0 \\ 2 \end{bmatrix} - e^{-t} \sin 2t \begin{bmatrix} 1 \\ 0 \end{bmatrix} \\ &= \begin{bmatrix} -e^{-t} \sin 2t \\ 2e^{-t} \cos 2t \end{bmatrix} \end{aligned}$$

$$\begin{aligned} x^{(2)}(t) &= e^{\alpha t} \sin \beta t\, a + e^{\alpha t} \cos \beta t\, b \\ &= e^{-t} \sin 2t \begin{bmatrix} 0 \\ 2 \end{bmatrix} + e^{-t} \cos 2t \begin{bmatrix} 1 \\ 0 \end{bmatrix} \\ &= \begin{bmatrix} e^{-t} \cos 2t \\ 2e^{-t} \sin 2t \end{bmatrix} \end{aligned}$$

Hence the general solution is

$$x(t) = c_1 \begin{bmatrix} -e^{-t} \sin 2t \\ 2e^{-t} \cos 2t \end{bmatrix} + c_2 \begin{bmatrix} e^{-t} \cos 2t \\ 2e^{-t} \sin 2t \end{bmatrix} \qquad \blacksquare$$

Example 2 **One Real, Two Complex Eigenvalues** Find the general solution of

$$x'(t) = \begin{bmatrix} 1 & 0 & 0 \\ 0 & 0 & 2 \\ 0 & -2 & 0 \end{bmatrix} x(t) \tag{13}$$

Solution   The characteristic equation is

$$|A - \lambda I| = \begin{vmatrix} 1 - \lambda & 0 & 0 \\ 0 & -\lambda & 2 \\ 0 & -2 & -\lambda \end{vmatrix} = (1 - \lambda)(\lambda^2 + 2) = 0 \tag{14}$$

which has the roots (eigenvalues) $\lambda_1 = 1$, $\lambda_2 = 2i$, and $\lambda_3 = -2i$. Substituting the real eigenvalue $\lambda_1 = 1$ into the equation $(A - \lambda I)\xi = 0$ gives the real eigenvector $\xi = (1, 0, 0)$. Hence one solution is

$$x^{(1)}(t) = e^t \begin{bmatrix} 1 \\ 0 \\ 0 \end{bmatrix} \tag{15}$$

Substituting $\lambda_2 = 2i$ into $(A - \lambda I)\xi = 0$ and solving for ξ, we get $\xi = (0, 1, i)$. Writing this complex eigenvector in the form $a + ib$ gives

$$\begin{bmatrix} 0 \\ 1 \\ i \end{bmatrix} = \begin{bmatrix} 0 \\ 1 \\ 0 \end{bmatrix} + i \begin{bmatrix} 0 \\ 0 \\ 1 \end{bmatrix} \tag{16}$$

Hence we have $a = (0, 1, 0)$ and $b = (0, 0, 1)$. Substituting these values and $\alpha = 0$, $\beta = 2$ into the general solution (8), we have

$$\begin{aligned} x^{(2)}(t) &= e^{\alpha t} \cos \beta t\, a - e^{\alpha t} \sin \beta t\, b \\ &= \cos 2t \begin{bmatrix} 0 \\ 1 \\ 0 \end{bmatrix} - \sin 2t \begin{bmatrix} 0 \\ 0 \\ 1 \end{bmatrix} \\ &= \begin{bmatrix} 0 \\ \cos 2t \\ -\sin 2t \end{bmatrix} \end{aligned}$$

$$\begin{aligned} x^{(3)}(t) &= e^{\alpha t} \sin \beta t\, a + e^{\alpha t} \cos \beta t\, b \\ &= \sin 2t \begin{bmatrix} 0 \\ 1 \\ 0 \end{bmatrix} + \cos 2t \begin{bmatrix} 0 \\ 0 \\ 1 \end{bmatrix} \\ &= \begin{bmatrix} 0 \\ \sin 2t \\ \cos 2t \end{bmatrix} \end{aligned}$$

Hence the general solution is

$$x(t) = c_1 e^t \begin{bmatrix} 1 \\ 0 \\ 0 \end{bmatrix} + c_2 \begin{bmatrix} 0 \\ \cos 2t \\ -\sin 2t \end{bmatrix} + c_3 \begin{bmatrix} 0 \\ \sin 2t \\ \cos 2t \end{bmatrix} \tag{17}$$

■

PROBLEMS: Section 6.5

For Problems 1–5, find a general solution of the linear system $x' = Ax$ *for the given matrix A.*

1. $\begin{bmatrix} 0 & 1 \\ -1 & 0 \end{bmatrix}$

2. $\begin{bmatrix} -1 & 1 \\ -4 & -1 \end{bmatrix}$

3. $\begin{bmatrix} 1 & 2 \\ -2 & 1 \end{bmatrix}$

4. $\begin{bmatrix} 1 & 0 & -1 \\ 0 & 2 & 0 \\ 1 & 0 & 1 \end{bmatrix}$

5. $\begin{bmatrix} 1 & 0 & 0 \\ 2 & 1 & -2 \\ 3 & 2 & 1 \end{bmatrix}$

For Problems 6–7, find the solution of the given initial-value problem.

6. $x'(t) = \begin{bmatrix} 2 & 1 \\ -1 & 2 \end{bmatrix} x(t) \quad x(0) = \begin{bmatrix} 2 \\ -3 \end{bmatrix}$

7. $x'(t) = \begin{bmatrix} 2 & -1 & 1 \\ 1 & 0 & 1 \\ -2 & 0 & -1 \end{bmatrix} x(t) \quad x(0) = \begin{bmatrix} -1 \\ 2 \\ 7 \end{bmatrix}$

Writing Higher-Order Equations as a Linear System

For Problems 8–11, convert the given differential equation to a first-order system $x' = Ax$.

(a) *Show that the characteristic equation* $|A - \lambda I| = 0$ *for the first-order system is the same as the characteristic equation of the differential equation.*

(b) *Solve both the first-order linear system and the higher-order differential equation and compare the results.*

8. $y'' + y = 0$

9. $y'' + 2y' + 5y = 0$

10. $y''' + y = 0$

11. $y^{(4)} - y = 0$

12. Linearly Independent Solutions Show that the two solutions

$$x^{(1)}(t) = e^{\alpha t} \cos \beta t \, a - e^{\alpha t} \sin \beta t \, b$$
$$x^{(2)}(t) = e^{\alpha t} \sin \beta t \, a + e^{\alpha t} \cos \beta t \, b$$

of $x' = Ax$ are linearly independent on $(-\infty, \infty)$, provided that $\beta \neq 0$ and a and b are not both the zero vector. *Hint:* Compute the Wronskian at $t = 0$.

The Matrix Exponential

Since the general solution of the first-order equation $x' = ax$ *is* $x(t) = ce^{at}$, *where c is an arbitrary constant, it may not be surprising to learn that the general solution of the general linear system* $x' = Ax$, *where A is an n × n constant matrix, can be written in the form*

$$x(t) = e^{At}c \qquad (18)$$

where c is an arbitrary column vector with n components, and e^{At} *is an n × n matrix, called the* **matrix exponential** *and defined by*

$$e^{At} = I + tA + \frac{t^2}{2}A^2 + \frac{t^3}{3!}A^3 + \cdots \qquad (19)$$

For Problems 13–14, find the general solution using Eq. (18).

13. $x'(t) = \begin{bmatrix} 0 & 1 \\ 1 & 0 \end{bmatrix} x(t)$

14. $x'(t) = \begin{bmatrix} 0 & 1 \\ -1 & 0 \end{bmatrix} x(t)$

The Fundamental Matrix

If the columns of a matrix $X(t)$ *are linearly independent solutions of the linear system* $x' = Ax$, *then the matrix* $X(t)$ *is called a* **fundamental matrix** *for the system. For example, the linear system*

$$x_1' = x_1 + x_2$$
$$x_2' = 4x_1 + x_2$$

has two linearly independent solutions

$$x^{(1)}(t) = \begin{bmatrix} e^{-t} \\ 2e^{-t} \end{bmatrix} \qquad x^{(2)}(t) = \begin{bmatrix} e^{3t} \\ -2e^{3t} \end{bmatrix}$$

Hence a fundamental matrix for the system is

$$X(t) = \begin{bmatrix} e^{-t} & e^{3t} \\ 2e^{-t} & -2e^{3t} \end{bmatrix}$$

Problems 15–19 allow the reader to gain some experience with the fundamental matrix.

15. The General Solution as a Fundamental Matrix Show that the general solution of the $n \times n$ system $x' = Ax$ can be written in the form $x(t) = X(t)c$, where $X(t)$ is a fundamental matrix for the system and c is an arbitrary $n \times 1$ constant vector.

16. Initial-Value Problem and the Fundamental Matrix Let $X(t)$ be a fundamental matrix for the general linear system $x' = Ax$. Show that $x(t) = X(t)X^{-1}(0)x_0$ is the solution for the initial-value problem $x' = Ax$, $x(0) = x_0$.

17. Verifying a Fundamental Matrix Verify that

$$X(t) = \begin{bmatrix} \sin t & \cos t \\ \cos t & -\sin t \end{bmatrix}$$

is a fundamental matrix for the system

$$x'(t) = \begin{bmatrix} 0 & 1 \\ -1 & 0 \end{bmatrix} x(t)$$

18. Finding a Fundamental Matrix Find a fundamental matrix for the linear system

$$x'(t) = \begin{bmatrix} 2 & -1 \\ 3 & -2 \end{bmatrix} x(t) \qquad (20)$$

and write the general solution as $x(t) = X(t)c$, where c is an arbitrary $n \times 1$ vector of constants.

19. Initial-Value Problem Write the solution of the initial-value problem

$$x'(t) = \begin{bmatrix} 2 & -1 \\ 3 & -2 \end{bmatrix} x(t) \qquad x(0) = \begin{bmatrix} 1 \\ 0 \end{bmatrix}$$

in the form $x(t) = X(t)X^{-1}(0)x(0)$.

20. Self-Adjoint Systems The linear system $\dot{x} = Ax$ where the constant matrix A is **skew-symmetric**, in other words, $A = -A^T$, is called a **self-adjoint system**. Self-adjoint systems have the special property that the length $\|x\| = \sqrt{x^T x}$ of any solution $x(t)$ remains *constant over time*.

(a) Show that the length of any solution $x(t)$ of a self-adjoint system remains constant over time by proving that

$$\frac{d}{dt}\left(x^T x\right) = 0$$

(b) Verify that $\|x\|^2 = x^T x$ is a constant function of t for any solution $x(t)$ of the self-adjoint system

$$\dot{x} = \begin{bmatrix} 0 & 1 \\ -1 & 0 \end{bmatrix} x$$

21. Journal Entry Summarize the results of this section.

6.6 NONHOMOGENEOUS LINEAR SYSTEMS

PURPOSE

We first find the general solution of the linear autonomous system $\dot{x} = Ax$, where A is a constant matrix, in terms of the matrix exponential e^{At}. We then use a matrix version of the method of variation of parameters to find a particular solution of the more general nonhomogeneous system $\dot{x} = A(t) + f(t)$, where the elements of $A(t)$ are allowed to be continuous functions of t.

THE MATRIX EXPONENTIAL

The reader will recall from Chapter 3 that the general solution of the first-order equation $dy/dt = ay$ is $y(t) = ce^{at}$. By analogy the general solution of the $n \times n$ linear homogeneous system $\dot{x} = Ax$, where A is a constant matrix, is

$$x(t) = e^{At}c$$

where e^{At} is the $n \times n$ matrix exponential

$$e^{At} = I + tA + \frac{t^2}{2!}A^2 + \frac{t^n}{n!}A^n + \cdots \qquad (1)$$

and c is an arbitrary $n \times 1$ vector of constants.* By direct substitution it can be shown that $x(t) = e^{At}c$ reduces $\dot{x} = Ax$ to an identity. This discussion leads us to the following theorem.

---

* In this section we denote the vector derivative by $\dot{x}$ since the independent variable is t.

> **THEOREM 6.7:** Homogeneous Solution for Systems with Constant Coefficients
>
> If A is an $n \times n$ constant matrix, then the solution of the initial-value problem
>
> $$\dot{x} = Ax \qquad x(0) = x_0$$
>
> is
>
> $$x(t) = e^{At}x_0$$

PROOF: By straightforward differentiation of the matrix exponential (see Problem 11 at the end of this section), one can show that $d(e^{At})/dt = Ae^{At}$. Hence $x(t) = e^{At}x_0$ reduces the differential equation to an identity. To verify $x(0) = x_0$, we observe the identity $e^{A \cdot 0} = e^0 = I$.

GENERALIZED INTEGRATING FACTOR AND THE NONHOMOGENEOUS SOLUTION

We now show how the matrix exponential e^{At} acts as a *matrix integrating factor* for solving the nonhomogeneous linear system

$$\dot{x} = Ax + f(t) \qquad x(0) = x_0 \tag{2}$$

in the case in which A is a constant matrix. We first write the system in the more convenient form

$$\dot{x} - Ax = f(t) \tag{3}$$

and multiply each side of the equation by the matrix exponential e^{-At}, getting

$$e^{-At}[\dot{x} - Ax] = e^{-At} f(t) \tag{4}$$

By matrix differentiation we find

$$\frac{d}{dt}[e^{-At}x(t)] = e^{-At}(-A)x(t) + e^{-At}\dot{x}(t)$$
$$= e^{-At}[\dot{x}(t) - Ax(t)] \tag{5}$$

Hence Eq. (4) can be written

$$\frac{d}{dt}[e^{-At}x(t)] = e^{-At}f(t)$$

Integrating, we have

$$e^{-At}x(t) = \int e^{-As}f(s)\,ds + c$$

and using the property (see Problem 10 at the end of this section) of the matrix exponential $e^{At}e^{-At} = I$, we obtain the general solution

$$x(t) = e^{At}c + e^{At}\int e^{-As}f(s)\,ds \tag{6}$$

This leads us to the following theorem.

> **THEOREM 6.8:** General Solution of the Nonhomogeneous System with Constant Coefficients
>
> If A is an $n \times n$ matrix with constant elements and $f(t)$ is a given vector whose elements are continuous functions, then the general solution of the nonhomogeneous linear system
>
> $$\dot{x} = Ax + f(t)$$
>
> is
>
> $$x(t) = e^{At}c + e^{At} \int e^{-As}f(s)\, ds$$
>
> The solution of the nonhomogeneous equation with the initial conditions $x(0) = x_0$ is
>
> $$x(t) = e^{At}x_0 + e^{At} \int_0^t e^{-As}f(s)\, ds$$

Example 1   Nonhomogeneous System   Find the solution of the initial-value problem

$$\dot{x} = \begin{bmatrix} 0 & 1 \\ -1 & 0 \end{bmatrix} x + \begin{bmatrix} 0 \\ 2 \end{bmatrix} \qquad x(0) = \begin{bmatrix} 1 \\ 0 \end{bmatrix}$$

Solution   We first observe that the powers of

$$A = \begin{bmatrix} 0 & 1 \\ -1 & 0 \end{bmatrix}$$

repeat in the sense that $A^2 = -I$, $A^3 = -A$, $A^4 = A$, ... Substituting these powers of A into the matrix exponential gives

$$
\begin{aligned}
e^{At} &= I + t\overset{A}{\begin{bmatrix} 0 & 1 \\ -1 & 0 \end{bmatrix}} + \frac{t^2}{2!}\overset{A^2}{\begin{bmatrix} -1 & 0 \\ 0 & -1 \end{bmatrix}} + \frac{t^3}{3!}\overset{A^3}{\begin{bmatrix} 0 & -1 \\ 1 & 0 \end{bmatrix}} + \frac{t^4}{4!}\overset{A^4}{\begin{bmatrix} 0 & 1 \\ -1 & 0 \end{bmatrix}} + \cdots \\[2mm]
&= \begin{bmatrix} 1 - \dfrac{t^2}{2!} + \dfrac{t^4}{4!} - \cdots & t - \dfrac{t^3}{3!} + \dfrac{t^5}{5!} - \cdots \\[3mm] -t + \dfrac{t^3}{3!} - \dfrac{t^5}{5!} + \cdots & 1 - \dfrac{t^2}{2!} + \dfrac{t^4}{4!} + \cdots \end{bmatrix} \\[2mm]
&= \begin{bmatrix} \cos t & \sin t \\ -\sin t & \cos t \end{bmatrix}
\end{aligned}
$$

$$(7)$$

Thus the solution is

$$x(t) = e^{At}x_0 + e^{At} \int_0^t e^{-As}f(s)\, ds$$

$$= \begin{bmatrix} \cos t & \sin t \\ -\sin t & \cos t \end{bmatrix} \begin{bmatrix} 1 \\ 0 \end{bmatrix} + \begin{bmatrix} \cos t & \sin t \\ -\sin t & \cos t \end{bmatrix} \int_0^t \begin{bmatrix} \cos s & -\sin s \\ \sin s & \cos s \end{bmatrix} \begin{bmatrix} 0 \\ 2 \end{bmatrix} ds$$

$$= \begin{bmatrix} \cos t \\ -\sin t \end{bmatrix} + 2 \begin{bmatrix} \cos t & \sin t \\ -\sin t & \cos t \end{bmatrix} \begin{bmatrix} \cos t - 1 \\ \sin t \end{bmatrix}$$

$$= \begin{bmatrix} -\cos t + 2 \\ \sin t \end{bmatrix}$$

GENERALIZED VARIATION OF PARAMETERS

In Chapter 3 we used the methods of undetermined coefficients and variation of parameters to find particular solutions of nonhomogeneous linear equations. Although both of these methods can be adapted to nonhomogeneous linear systems, the method of variation of parameters is the more effective procedure. The method of undetermined coefficients becomes bogged down in a maze of equations except for the simplest problems. The method of variation of parameters gives a useful form of the solution that can be used in theoretical work.

We now solve the general solution of the nonhomogeneous linear system

$$\dot{x} = A(t)x(t) + f(t) \tag{8}$$

We know that the general solution has the form $x = x_h + x_p$, where x_h is the general solution of the corresponding homogeneous system and x_p is any solution of the nonhomogeneous equation (8). In Section 6.3 we saw that the general solution of the corresponding homogeneous system $\dot{x} = A(t)x$ is

$$x(t) = X(t)c \tag{9}$$

where $X(t)$ is a fundamental matrix for the system and c is a column vector of arbitrary constants. Assuming that we have already found a fundamental matrix of the homogeneous system $\dot{x} = A(t)x$, our goal now is to find a particular solution of Eq. (8). Recalling the method of variation of parameters for a single nonhomogeneous equation, we replace the constant vector c in Eq. (9) by a variable vector u. Hence we seek a solution of the form

$$x_p(t) = X(t)u(t) \tag{10}$$

where

$$u(t) = \begin{bmatrix} u_1(t) \\ u_2(t) \\ \vdots \\ u_n(t) \end{bmatrix}$$

To find $u(t)$, we differentiate Eq. (10) using the product rule for matrices, getting

$$\dot{x}_p(t) = X(t)\dot{u}(t) + \dot{X}(t)u(t) \tag{11}$$

Substituting Eqs. (10) and (11) into Eq. (8) we get

$$X(t)\dot{u}(t) + \dot{X}(t)\,u(t) = A(t)X(t)u(t) + f(t) \tag{12}$$

However, since the fundamental matrix satisfies $\dot{X}(t) = A(t)X(t)$, Eq. (12) becomes

$$X(t)\dot{u}(t) + A(t)X(t)u(t) = A(t)X(t)u(t) + f(t)$$

or

$$X(t)\dot{u}(t) = f(t)$$

Multiplying each side of this equation by $X^{-1}(t)$ gives

$$\dot{u}(t) = X^{-1}(t)f(t)$$

or

$$u(t) = \int X^{-1}(s)f(s)\,ds \tag{13}$$

Substituting this value back into Eq. (10) gives the particular solution

$$x_p(t) = X(t)u(t)$$

$$= X(t)\int X^{-1}(s)f(s)\,ds \tag{14}$$

Hence the general solution of Eq. (8) is

$$x = X(t)c + X(t)\int X^{-1}(s)f(s)\,ds \tag{15}$$

There will be arbitrary constants in the indefinite integral in Eq. (15). However, the terms involving these constants can be "absorbed" by the vector $X(t)c$, and so they can be picked to be zero.

We can also determine the arbitrary vector c so that Eq. (15) satisfies an initial condition $x(t_0) = x_0$. To do this, we rewrite Eq. (15) in the equivalent form*

$$x(t) = X(t)c + X(t)\int_{t_0}^{t} X^{-1}(s)f(s)\,ds \tag{16}$$

Substituting Eq. (16) into $x(t_0) = x_0$ results in

$$c = X^{-1}(t_0)x_0$$

---

* The indefinite integral in Eq. (15) contains an arbitrary constant vector. This vector can be chosen so that the integral is zero when $t = t_0$, which is the integral in Eq. (16).

Thus we have

$$x(t) = X(t)X^{-1}(t_0)x_0 + X(t)\int_{t_0}^{t} X^{-1}(s)f(s)\,ds \qquad (17)$$

These ideas lead to the following theorem.

THEOREM 6.9: General Solution of $\dot{x} = A(t)x + f(t)$

Let $A(t)$ be an $n \times n$ matrix whose elements are continuous functions on the interval under consideration, and let $f(t)$ be an $n \times 1$ vector with continuous elements. If $X(t)$ is a fundamental matrix for $\dot{x} = A(t)x$, then the general solution of the nonhomogeneous linear system

$$\dot{x} = A(t)x + f(t)$$

is

$$\dot{x} = X(t)c + X(t)\int X^{-1}(s)f(s)\,ds$$

where c is a vector of arbitrary constants. The solution of the initial-value problem

$$\dot{x} = A(t)x + f(t)$$
$$x(t_0) = x_0$$

is

$$x = X(t)X^{-1}(t_0)x_0 + X(t)\int_{t_0}^{t} X^{-1}(s)f(s)\,ds$$

Example 2   Example   Use variation of parameters to find a general solution of

$$\dot{x} = \begin{bmatrix} 1 & 1 \\ 4 & 1 \end{bmatrix} \begin{bmatrix} x_1 \\ x_2 \end{bmatrix} + \begin{bmatrix} 2e^{3t} \\ 0 \end{bmatrix} \qquad (18)$$

Solution   We could use the formula given in Theorem 6.9, but we find the following methodology more insightful. In Section 6.4 we found the fundamental matrix for Eq. (18) to be

$$X(t) = \begin{bmatrix} e^{3t} & e^{-t} \\ 2e^{3t} & -2e^{-t} \end{bmatrix}$$

Hence a particular solution of Eq. (18) has the form $x = X(t)u(t)$, where $u(t)$ satisfies $X(t)\dot{u}(t) = f(t)$, or

$$\begin{bmatrix} e^{3t} & e^{-t} \\ 2e^{3t} & -2e^{-t} \end{bmatrix} \begin{bmatrix} \dot{u}_1 \\ \dot{u}_2 \end{bmatrix} = \begin{bmatrix} 2e^{3t} \\ 0 \end{bmatrix}$$

Solving for $\dot{u}_1$ and $\dot{u}_2$, we find

$$\dot{u}_1 = 1 \,\triangleright\, u_1(t) = t + c_1$$

$$\dot{u}_2 = e^{4t} \,\triangleright\, u_2(t) = \frac{1}{4}e^{4t} + c_2$$

Hence

$$
\begin{aligned}
x &= X(t)u(t) \\
&= \begin{bmatrix} e^{3t} & e^{-t} \\ 2e^{3t} & -2e^{-t} \end{bmatrix} \begin{bmatrix} t + c_1 \\ \frac{1}{4}e^{4t} + c_2 \end{bmatrix} \\
&= c_1 \begin{bmatrix} 1 \\ 2 \end{bmatrix} e^{3t} + c_2 \begin{bmatrix} 1 \\ -2 \end{bmatrix} e^{-t} + \begin{bmatrix} t + \frac{1}{4} \\ 2t - \frac{1}{2} \end{bmatrix} e^{3t}
\end{aligned}
\tag{19}
$$

Note that by including the arbitrary constants c_1 and c_2 we have actually found the general solution of Eq. (18) and not just a single solution.

Example 3   Initial-Value Problem   Use variation of parameters to solve

$$\dot{x} = \begin{bmatrix} 1 & 1 \\ 4 & 1 \end{bmatrix} \begin{bmatrix} x_1 \\ x_2 \end{bmatrix} + \begin{bmatrix} 2e^{3t} \\ 0 \end{bmatrix} \qquad x(0) = \begin{bmatrix} 1 \\ 0 \end{bmatrix}$$

For demonstration purposes we will use the formula given in Theorem 6.9. However, we first recall from Section 6.4 that a fundamental matrix of the corresponding homogeneous system is

$$X(t) = \begin{bmatrix} e^{3t} & e^{-t} \\ 2e^{3t} & -2e^{-t} \end{bmatrix} \qquad \text{and} \qquad X^{-1}(t) = \frac{1}{4} \begin{bmatrix} 2e^{-3t} & e^{-3t} \\ 2e^{t} & -e^{t} \end{bmatrix}$$

Hence using the formula for a particular solution, we get

$$
\begin{aligned}
x_p &= X(t) \int_0^t X^{-1}(s)f(s)\,ds \\
&= \begin{bmatrix} e^{3t} & e^{-t} \\ 2e^{3t} & -2e^{-t} \end{bmatrix} \int_0^t \frac{1}{4} \begin{bmatrix} 2e^{-3s} & e^{-3s} \\ 2e^{s} & -e^{s} \end{bmatrix} \begin{bmatrix} 2e^{3s} \\ 0 \end{bmatrix} ds \\
&= \begin{bmatrix} e^{3t} & e^{-t} \\ 2e^{3t} & -2e^{-t} \end{bmatrix} \int_0^t \begin{bmatrix} 1 \\ e^{4s} \end{bmatrix} ds \\
&= \frac{1}{4} \begin{bmatrix} e^{3t} & e^{-t} \\ 2e^{3t} & -2e^{-t} \end{bmatrix} \begin{bmatrix} 4t \\ e^{4t} - 1 \end{bmatrix} \\
&= \frac{1}{4} \begin{bmatrix} -e^{-t} + (4t + 1)e^{3t} \\ 2e^{-t} + (8t - 2)e^{3t} \end{bmatrix}
\end{aligned}
$$

Hence the solution of the initial-value problem is

$$x(t) = X(t)X^{-1}(0)x_0 + X(t)\int_{t_0}^{t} X^{-1}(s)f(s)\, ds$$

$$= \frac{1}{4}\begin{bmatrix} e^{3t} & e^{-t} \\ 2e^{3t} & -2e^{-t} \end{bmatrix}\begin{bmatrix} 2 & 1 \\ 2 & -1 \end{bmatrix}\begin{bmatrix} 1 \\ 0 \end{bmatrix} + \frac{1}{4}\begin{bmatrix} -e^{-t} + (4t + 1)e^{3t} \\ 2e^{-t} + (8t - 2)e^{3t} \end{bmatrix}$$

$$= \frac{1}{4}\begin{bmatrix} e^{-t} + (4t + 3)e^{3t} \\ -2e^{-t} + (8t + 2)e^{3t} \end{bmatrix}$$

PROBLEMS: Section 6.6

For Problems 1–4, find the matrix exponential e^{At} and use this matrix to find the general solution of the given system.

1. $\dot{x} = \begin{bmatrix} -1 & 0 \\ 0 & 2 \end{bmatrix}\begin{bmatrix} x_1 \\ x_2 \end{bmatrix} + \begin{bmatrix} 1 \\ 0 \end{bmatrix}$

2. $\dot{x} = \begin{bmatrix} 2 & 0 \\ 0 & 3 \end{bmatrix}\begin{bmatrix} x_1 \\ x_2 \end{bmatrix} + \begin{bmatrix} 0 \\ 6 \end{bmatrix}$

3. $\dot{x} = \begin{bmatrix} 0 & 1 \\ 1 & 0 \end{bmatrix}\begin{bmatrix} x_1 \\ x_2 \end{bmatrix} + \begin{bmatrix} 1 \\ 1 \end{bmatrix}$

4. $\dot{x} = \begin{bmatrix} 0 & 1 \\ -1 & 0 \end{bmatrix}\begin{bmatrix} x_1 \\ x_2 \end{bmatrix} + \begin{bmatrix} 1 \\ 0 \end{bmatrix}$ $x(0) = \begin{bmatrix} 1 \\ 1 \end{bmatrix}$

For Problems 5–8, use the generalized version of variation of parameters to find the general solution.

5. $\dot{x} = \begin{bmatrix} 1 & 4 \\ 1 & 1 \end{bmatrix}\begin{bmatrix} x_1 \\ x_2 \end{bmatrix} + \begin{bmatrix} -10 \\ 1 \end{bmatrix}$

6. $\dot{x} = \begin{bmatrix} 3 & -3 \\ 2 & -2 \end{bmatrix}\begin{bmatrix} x_1 \\ x_2 \end{bmatrix} + \begin{bmatrix} 4 \\ -1 \end{bmatrix}$

7. $\dot{x} = \begin{bmatrix} 2 & 1 \\ -3 & 6 \end{bmatrix}\begin{bmatrix} x_1 \\ x_2 \end{bmatrix} + \begin{bmatrix} e^{5t} \\ e^{5t} \end{bmatrix}$

8. $\dot{x} = \begin{bmatrix} 0 & -1 \\ 3 & 4 \end{bmatrix}\begin{bmatrix} x_1 \\ x_2 \end{bmatrix} + \begin{bmatrix} t \\ -4t - 2 \end{bmatrix}$

9. **Matrix Exponential** Let

$$A = \begin{bmatrix} 0 & 1 \\ 1 & 0 \end{bmatrix}$$

   (a) Find the matrix exponential e^{At}.
   (b) Show that e^{-At} is the inverse of e^{At}.
   (c) Use the matrix exponential to find the general solution of $\dot{x} = Ax$.

10. **Properties of the Matrix Exponential** Verify the following properties for the matrix

$$A = \begin{bmatrix} 1 & 0 \\ 0 & 2 \end{bmatrix}$$

   (a) $(e^{At})^{-1} = e^{-At}$
   (b) $e^{A(t+s)} = e^{At}e^{As}$ (*s, t* are scalars)
   (c) $e^{-At}e^{At} = I$

Although we verify these properties for this specific matrix, they hold for the general matrix exponential. Property (a) shows that for any square matrix A the matrix exponential has an inverse for all t. Property (c) can be easily verified for an arbitrary matrix exponential.

11. **Differentiating the Matrix Exponential** Show that

$$\frac{d}{dt}e^{-At} = e^{At}(-A)$$

The Method of Undetermined Coefficients

In Chapter 3 we found particular solutions of nonhomogeneous differential equations using the method of undetermined coefficients. This method can be generalized to nonhomogeneous linear systems. The method is similar to the method applied to single equations; we make an intelligent guess and substitute the guess into the system and determine the coefficients. The choice of the particular solution is essentially the same as that given in Table 3.2 in Section 3.7 except that now we use vectors. Problems 12–14 illustrate this method.

12. **Undetermined Coefficients** Given the nonhomogeneous linear system

$$\begin{aligned} \dot{x} &= x + y - 3 \\ \dot{y} &= 4x + y - 6 \end{aligned} \tag{20}$$

   (a) Find the general solution of the corresponding homogeneous system.

(b) Find a particular solution of the form

$$\begin{bmatrix} x_p(t) \\ y_p(t) \end{bmatrix} = K = \begin{bmatrix} k_1 \\ k_2 \end{bmatrix}$$

(c) Find the general solution of the nonhomogeneous system (20).

13. Undetermined Coefficients Given the nonhomogeneous linear system

$$\begin{aligned} \dot{x} &= -y + t \\ \dot{y} &= 3x + 4y - 2 - 4t \end{aligned} \tag{21}$$

(a) Find the general solution of the corresponding homogeneous system.
(b) Find a particular solution of the form

$$\begin{bmatrix} x_p(t) \\ y_p(t) \end{bmatrix} = at + b = \begin{bmatrix} a_1 \\ a_2 \end{bmatrix} t + \begin{bmatrix} b_1 \\ b_2 \end{bmatrix}$$

(c) Find the general solution of the nonhomogeneous system (21).

14. Undetermined Coefficients Given the nonhomogeneous system

$$\begin{aligned} \dot{x} &= 4x + 2y \\ \dot{y} &= 3x - y + 7e^{-2t} \end{aligned} \tag{22}$$

(a) Find the general solution of the corresponding homogeneous system.
(b) Find a particular solution of the form

$$\begin{bmatrix} x_p(t) \\ y_p(t) \end{bmatrix} = ae^t = \begin{bmatrix} a_1 \\ a_2 \end{bmatrix} e^t$$

(c) Find the general solution of the nonhomogeneous system (22).

15. Journal Entry Summarize the results of this section.

6.7 NONHOMOGENEOUS LINEAR SYSTEMS: LAPLACE TRANSFORM (Optional)

PURPOSE
To solve the nonhomogeneous linear system

$$\dot{x} = Ax + f$$

by means of the Laplace transform. We will show how the Laplace transform can be used to transform a linear nonhomogeneous system of differential equations into a system of linear algebraic equations in the transformed variables. We then solve the system of algebraic equations to obtain the transforms and then compute the inverse transforms to obtain the solution of the original system of differential equations.

Earlier we saw how the Laplace transform could be used to solve a single nonhomogeneous differential equation with constant coefficients by transforming the equation to an algebraic equation. By solving the algebraic equation for the transformed variable and then finding its inverse, it is possible to solve the original problem. In this section we use the Laplace transform to solve systems of nonhomogeneous differential equations with constant coefficients by transforming those systems to systems of linear algebraic equations. The process works exactly as it did for a single equation, except that now we are working with systems of equations. The following system of equations is a *homogeneous* linear system, but it illustrates the basic idea.

Example 1 **Homogeneous Linear System** Solve the initial-value problem*

$$\dot{x}_1 = x_1 + x_2 \qquad x_1(0) = 1$$
$$\dot{x}_2 = 4x_1 + x_2 \qquad x_2(0) = 0 \tag{1}$$

Solution Taking the Laplace transform of both sides of the equations in system (1) and using the fact the transform is a linear operator, we have

$$\mathcal{L}\{\dot{x}_1\} = \mathcal{L}\{x_1\} + \mathcal{L}\{x_2\}$$
$$\mathcal{L}\{\dot{x}_2\} = 4\mathcal{L}\{x_1\} + \mathcal{L}\{x_2\} \tag{2}$$

We now let $X_1(x) = \mathcal{L}\{x_1\}$ and $X_2(s) = \mathcal{L}\{x_2\}$ and use the derivative property of the transform to get[†]

$$sX_1(s) - x_1(0) = X_1(s) + X_2(s)$$
$$sX_2(s) - x_2(0) = 4X_1(s) + X_2(s) \tag{3}$$

Substituting the initial conditions $x_1(0) = 1$, $x_2(0) = 0$ into system (3) and rewriting the system in standard form, we have

$$(s - 1)X_1(s) - X_2(s) = 1$$
$$-4X_1(s) + (s - 1)X_2(s) = 0 \tag{4}$$

Solving these algebraic equations for $X_1(s)$ and $X_2(s)$ gives

$$X_1(s) = \frac{s - 1}{s^2 - 2s - 3} = \frac{1}{2(s - 3)} + \frac{1}{2(s + 1)} \tag{5a}$$

$$X_2(s) = \frac{4}{s^2 - 2s - 3} = \frac{1}{s - 3} - \frac{1}{s + 1} \tag{5b}$$

Taking the inverse transform of these expressions, we get

$$x_1(t) = \frac{1}{2}(e^{3t} + e^{-t}) \tag{6a}$$

$$x_2(t) = e^{3t} - e^{-t} \tag{6b}$$

This solution is graphed in Figure 6.5. ∎

The following example illustrates that nonlinear equations are handled in the same way.

---

* We solved this initial-value problem in Section 6.1 using the method of elimination in conjunction with the D method. The reason we solve this problem now is to compare the D method with the Laplace transform.

[†] The reader should compare the system of Eqs. (3) with the analogous system found in Section 6.1 as a result of replacing d/dt by D in the same initial-value problem. The two systems are almost the same; whereas the D method replaces d/dt by D, the Laplace transform replaces d/dt by s. However, the Laplace transform method has the advantage over the D method in that the Laplace transform will yield the general solution if the initial conditions are not specified.

Figure 6.5
Graph of the solution

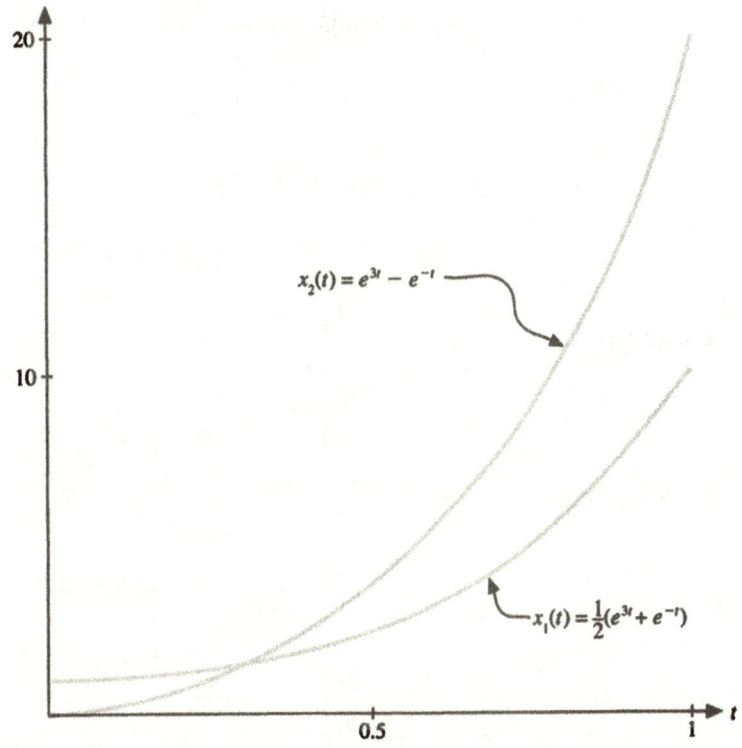

Example 2 Nonhomogeneous Linear System Solve the initial-value problem

$$\dot{x}_1 = -2x_1 + x_2 + 1 \qquad x_1(0) = 0$$
$$\dot{x}_2 = x_1 - 2x_2 \qquad\qquad x_2(0) = 1 \tag{7}$$

Solution Taking the Laplace transform of each question, we have

$$sX_1(s) = -2X_1(s) + X_2(s) + \frac{1}{s}$$

$$sX_2(s) - 1 = X_1(s) - 2X_2(s)$$

Rewriting this system of equations, we find

$$(s + 2)X_1(s) - X_2(s) = \frac{1}{s} \tag{8}$$

$$-X_1(x) + (s + 2)X_2(s) = 1$$

Solving this linear system, we get

$$X_1(s) = \frac{2}{s(s + 3)}$$

$$X_2(s) = \frac{s + 1}{s(s + 3)} \tag{9}$$

Finally, finding the inverse transforms, we obtain the solution

$$x_1(t) = \mathcal{L}^{-1}\left\{\frac{2}{s(s+3)}\right\} = \mathcal{L}^{-1}\left\{\frac{2}{3s} - \frac{2}{3(s+3)}\right\} = \frac{2}{3} - \frac{2}{3}e^{-3t} \quad \text{(10a)}$$

$$x_2(t) = \mathcal{L}^{-1}\left\{\frac{s+1}{s(s+3)}\right\} = \mathcal{L}^{-1}\left\{\frac{1}{3s} + \frac{2}{3(s+3)}\right\} = \frac{1}{3} + \frac{2}{3}e^{-3t} \quad \text{(10b)}$$

This solution is graphed in Figure 6.6.   ■

Figure 6.6
Graph of the solution

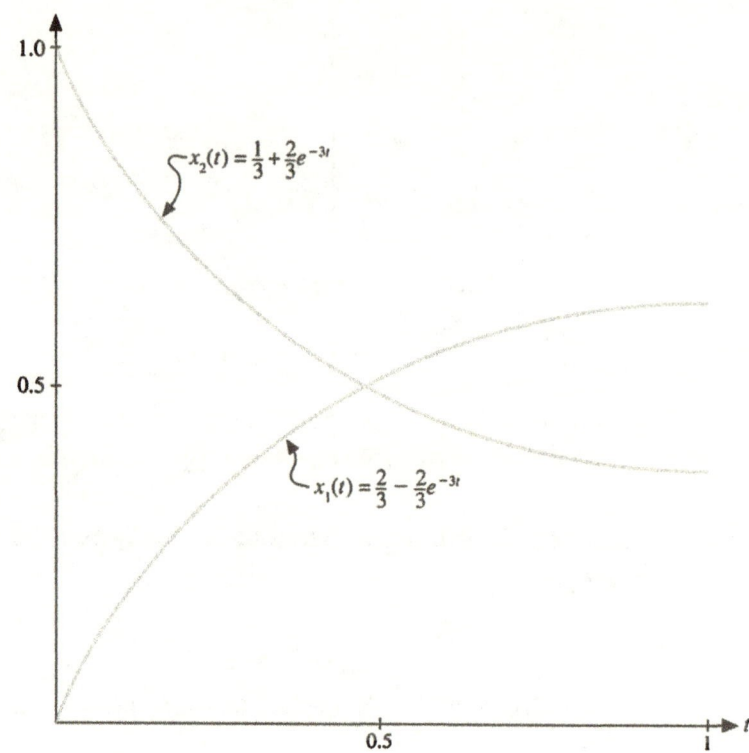

SIMPLE TWO-COMPARTMENT MODEL

Compartmental models have proven extremely useful in predicting drug concentration levels in organs and in estimating rates at which a drug is eliminated from the body. A simple two-compartment model for describing the kinetics of a metabolic process is illustrated in Figure 6.7.

Figure 6.7
Two-compartment model

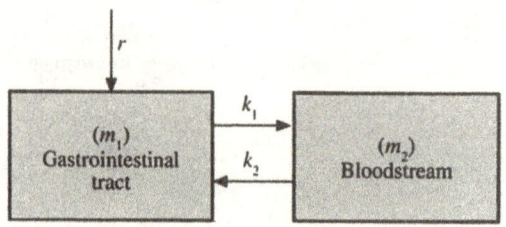

We assume that a drug is ingested (orally or intravenously) at a given rate $u(t)$. Units of u are normally chosen to be mg/hr, mg/day, ... We denote

$$m_1(t) = \text{mass of the drug in the gastrointestinal tract}$$

$$m_2(t) = \text{mass of the drug in the bloodstream}$$

The rate of change of the drug in the gastrointestinal tract is equal to the input rate at which the drug is ingested, minus an output rate, which is generally proportional to the amount present in the gastrointestinal tract. That is,

$$\dot{m}_1 = -k_1 m_1 + u \tag{11}$$

where k_1 is a constant of proportionality* that varies from person to person.

The rate of change of the drug in the bloodstream is proportional to the input amount coming from the gastrointestinal tract, minus an output rate lost to metabolism, which is proportional to the mass present in the bloodstream. That is,

$$\dot{m}_2 = k_1 m_1 - k_2 m_2 \tag{12}$$

where k_2 is a constant of proportionality that characterizes the metabolic and excretory processes of the individual. Hence a simple two-compartment model used to describe drug metabolism is

$$\begin{aligned} \dot{m}_1 &= -k_1 m_1 + u & \text{(gastrointestinal compartment)} \\ \dot{m}_2 &= k_1 m_1 - k_2 m_2 & \text{(bloodstream compartment)} \end{aligned} \tag{13}$$

The following example illustrates the use of the metabolism model.

Example 3   Lidocaine Metabolism   The drug Lidocaine is commonly used in the treatment of *ventricular arrhythmias* (irregular heartbeat). The schematic model shown in Figure 6.8 is a widely used model of Lidocaine kinetics.

We assume that 2 mg of Lidocaine are injected into the bloodstream and then move into the tissues of the heart. Since an initial injection of a drug is often represented mathematically by a delta function,[†] the initial-value problem that describes the amount (in milligrams) of Lidocaine in the bloodstream (m_1) and in the tissues (m_2) is

$$\begin{aligned} \dot{m}_1 &= -m_1 + 2\delta(t) & m_1(0) &= 0 \\ \dot{m}_2 &= -m_1 - m_2 & m_2(0) &= 0 \end{aligned} \tag{14}$$

where $\delta(t)$ is the Dirac delta function. Find the future amount of Lidocaine in the bloodstream and heart tissue.

---

* The constant of proportionality k_1 has units of hr^{-1} and measures how fast a given drug goes from one compartment to another. A rough rule of thumb interpretation of the reciprocal $1/k_1$ (hr) is that it is the time required for two-thirds of a drug to go from compartment A to compartment B.

[†] Drug injections are sometimes modeled by delta functions or exponential functions describing the rate at which the drug enters the first compartment. Sometimes an initial drug injection is described by the initial conditions of the problem.

Figure 6.8
A two-compartment model predicting the amount of Lidocaine in the blood (m_1) and in tissue (m_2) resulting from a 2-mg injection of Lidocaine into the bloodstream

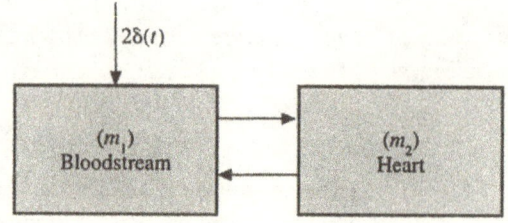

Solution Calling $M_1(s) = \mathcal{L}\{m_1\}$ and $M_2(s) = \mathcal{L}\{m_2\}$ and taking the Laplace transform of Eqs. (14), we get

$$sM_1(s) = -M_1(s) + 2$$
$$sM_2(s) = M_1(s) - M_2(s) \tag{15}$$

Solving for $M_1(s)$ and $M_2(s)$ gives

$$M_1(s) = \frac{2}{s + 1}$$

$$M_2(s) = \frac{2}{(s + 1)^2} \tag{16}$$

Hence

$$m_1(t) = \mathcal{L}^{-1}\left\{\frac{2}{s + 1}\right\} = 2e^{-t} \tag{17a}$$

$$m_2(t) = \mathcal{L}^{-1}\left\{\frac{2}{(s + 1)^2}\right\} = 2te^{-t} \tag{17b}$$

These functions are graphed in Figure 6.9. ■

Figure 6.9
Metabolism of Lidocaine in blood (m_1) and tissue (m_2)

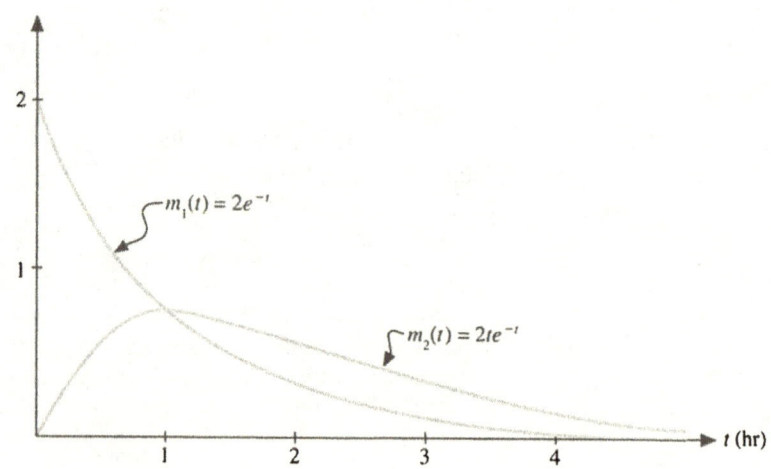

PROBLEMS: Section 6.7

For Problems 1–6, solve the given initial-value problem.

1. $\dot{x} = y$ $x(0) = 0$
$\dot{y} = -x$ $y(0) = 1$

2. $\dot{x} = x - y$ $x(0) = -1$
$\dot{y} = 2x + 4y$ $y(0) = 0$

3. $\dot{x} = y$ $x(0) = 1$
$\dot{y} = -2x + 3y + 12e^{4t}$ $y(0) = 0$

4. $\dot{x} = y$ $x(0) = 0$
$\dot{y} = -x + 2\cos t$ $y(0) = 0$

5. $\dot{x} = y + e^{3t}$ $x(0) = 0$
$\dot{y} = -2x + 3y$ $y(0) = 0$

6. $\dot{x} = -y + t$ $x(0) = 0$
$\dot{y} = 3x + 4y - 2 - 4t$ $y(0) = 0$

General Solutions of Linear Systems

It is possible to find the general solution of a linear system of differential equations by simply letting the initial conditions be arbitrary constants. For Problems 7–8, find the general solution of the given linear system.

7. $\dot{x} = x + y$
$\dot{y} = 4x + y$

8. $\dot{x} = -x - 4y$
$\dot{y} = x - y$

9. More General Linear System Use the Laplace transform to solve the system

$$\frac{dx}{dt} + 4x + \frac{dy}{dt} = 1 \qquad x(0) = 2$$

$$\frac{dx}{dt} - 2x + y = t^2 \qquad y(0) = -1$$

10. Higher-Order System Use the Laplace transform to solve the nonhomogeneous linear system

$$\dot{x}_1 = x_2 + t^2 \qquad x_1(0) = 1$$
$$\dot{x}_2 = x_3 \qquad x_2(0) = 1$$
$$\dot{x}_3 = x_4 \qquad x_3(0) = 1$$
$$\dot{x}_4 = x_1 \qquad x_4(0) = -1$$

11. Finding General Solutions Consider the linear system

$$\dot{x} = a_{11}x + a_{12}y + f_1(t) \qquad x(0) = c_1$$
$$\dot{y} = a_{21}x + a_{22}y + f_2(t) \qquad y(0) = c_2$$

where the a's and c's are constants. Use the Laplace transform to show that the solution can be written in the form

$$x(t) = x_h(t) + x_p(t)$$
$$y(t) = y_h(t) + y_p(t)$$

where $x_h(t)$ and $y_h(t)$ depend on c_1 and c_2, while $x_p(t)$ and $y_p(t)$ depend on $f_1(t)$ and $f_2(t)$.

12. Solving Linear Systems in Matrix Form Show that the Laplace transform $X(s) = \mathcal{L}\{x\}$ of the solution of the linear system of n differential equations

$$\dot{x} = Ax + f(t) \qquad x(0) = x_0$$

where x, $\dot{x}$, and f are vectors and A is an $n \times n$ constant matrix, is

$$X(s) = -(A - sI)^{-1}[F(s) + x_0] \qquad (18)$$

where $F(s) = \mathcal{L}\{f\}$ and I is the $n \times n$ identity matrix. *Hint:* First solve the problem for the general 2×2 linear system. The general case will then be much easier.

Solving Old Problems with a New Method

*In Chapter 5 we found particular solutions of single nonhomogeneous differential equations using the Laplace transform. This method can be generalized to nonhomogeneous linear systems. Problems 13–15 illustrate this method.** *

13. Undetermined Coefficients Find the general solution of the nonhomogeneous linear system

$$\dot{x} = x + y - 3$$
$$\dot{y} = 4x + y - 6 \qquad (19)$$

using the Laplace transform. *Hint:* Let the initial conditions be arbitrary constants.

14. Undetermined Coefficients Find the general solution of the nonhomogeneous linear system

$$\dot{x} = -y + t$$
$$\dot{y} = 3x + 4y - 2 - 4t \qquad (20)$$

using the Laplace transform. *Hint:* Let the initial conditions be arbitrary constants.

---

* This set of problems (13–15) is identical to Problems 12–14 in Section 6.6. This should illustrate how the same equations can be solved using two different methods.

15. Undetermined Coefficients Find the general solution of the nonhomogeneous linear system

$$\dot{x} = 4x + 2y$$
$$\dot{y} = 3x - y + 7e^{-2t} \tag{21}$$

using the Laplace transform. *Hint:* Let the initial conditions be arbitrary constants

16. Drug Metabolism* A drug is taken orally at a constant rate r. If the mass of the drug in the gastrointestinal tract and bloodstream is denoted m_1 and m_2, respectively, then

the rates of change m_1' and m_2' can be shown to be

$$\frac{dm_1}{dt} = k_1 m_1 + r$$
$$\frac{dm_2}{dt} = k_1 m_1 - k_2 m_2 \tag{22}$$

The positive constants k_1 and k_2 are rate constants that vary from person to person. Taking the initial conditions to be $m_1(0) = 1$ and $m_2(0) = 0$, use the Laplace transform to solve this initial-value problem for $k_1 = k_2 = 1$ and $r = 0$.

17. Journal Entry Summarize the results of this section.

6.8 APPLICATIONS OF LINEAR SYSTEMS

PURPOSE

To show how systems of linear differential equations can be used to describe phenomena involving more than one unknown variable. We do this by presenting more general versions of some of the phenomena discussed in Chapters 2 and 3.

In Chapter 2 we saw how heat flow and mixing problems can be described with first-order differential equations. In Chapter 3 we saw how a vibrating mass-spring system can be described with a second-order equation. In this section we generalize those problems to *systems* of differential equations. We start with a two-room heating problem.

Example 1 **The Two-Room Heating Problem** Consider a building that contains two rooms A and B. See Figure 6.10. Room A is heated with a furnace that generates 100,000 Btu of heat per hour and has a heat capacity[†] of 0.20°F per thousand Btu. The time constants that specify the rate of heat flow between the rooms are as follows: Between room A and the outside is 5 hr, between room B and the outside is 10 hr, and between room A and room B is 2 hr. If the outside temperature is fixed at 0°F, how cold will it eventually get in the unheated room B?

Solution We let $x(t)$ and $y(t)$ denote the temperatures in rooms A and B, respectively. Newton's law of cooling states that the rate of change in the temperature in a room is proportional to the difference between the temperature of the room and the temperature of the

---

* Taken from N. H. McClamrock, *State Models of Dynamic Systems* (Springer-Verlag, New York, 1980).

[†] The heat capacity of a room (deg/Btu) is a measure of the increase in temperature of a sealed and insulated room as a function of the energy added to the room. It can be determined by a simple experiment.

surrounding medium. In this problem the rate at which the temperature in room A changes depends on the addition of heat from the furnace and heat gain (or loss) from room B and/or the outside. Also, the rate of change of temperature in room B depends on the heat gain (or loss) from room A and/or the outside.

The heat capacity of 0.20°F per thousand Btu for room A means that if room A is sealed and insulated, the temperature in room A increases by 0.20°F for every thousand Btu of heat generated by the furnace. Since the furnace generates 100,000 Btu per hour, the temperature in an insulated room A will increase by 20°F every hour the furnace is running. Applying Newton's law of cooling to both rooms A and B, we have

$$\frac{dx}{dt} = 20 - \frac{1}{5}[x(t) - 0] - \frac{1}{2}[x(t) - y(t)]$$

$$= -\frac{7}{10}x(t) + \frac{1}{2}y(t) + 20 \tag{1a}$$

$$\frac{dy}{dt} = -\frac{1}{2}[y(t) - x(t)] - \frac{1}{10}[y(t) - 0]$$

$$= \frac{1}{2}x(t) - \frac{3}{5}y(t) \tag{1b}$$

Since we are interested only in finding the steady state or limiting solution, it is not necessary to solve these equations. We simply set dx/dt and dy/dt equal to zero and solve for x and y. Doing this gives $x = 1200/17$ (71°) and $y = 1000/17$ (59°). Hence the limiting temperature in room B is 59°. ∎

Figure 6.10
Flow of heat in a two-room house

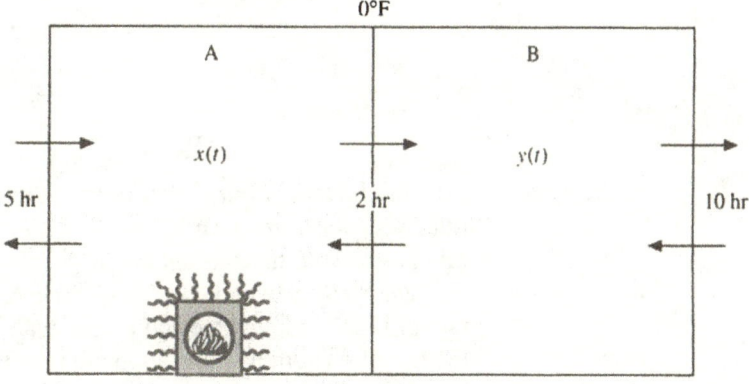

If we also wanted to find the *transient* temperature $x(t)$ and $y(t)$, we could easily solve this system by eliminating one of the two variables x and y. Doing this, we find*

$$\begin{bmatrix} x(t) \\ y(t) \end{bmatrix} \doteq c_1 e^{-0.15t} \begin{bmatrix} 0.9 \\ 1 \end{bmatrix} + c_2 e^{-1.15t} \begin{bmatrix} -1.1 \\ 1 \end{bmatrix} + \begin{bmatrix} 71 \\ 59 \end{bmatrix} \tag{2}$$

* We use the approximation sign because all constants have been rounded to two-place accuracy. Often in applied work we dispense with the approximation sign and simply write equality.

If the initial temperature in each room were chosen to be $x(0) = y(0) = 70°$, then we would find $c_1 \doteq 5.85$ and $c_2 \doteq 5.32$. These transient temperatures are graphed in Figure 6.11.

We now see how mixing problems extend to systems of differential equations.

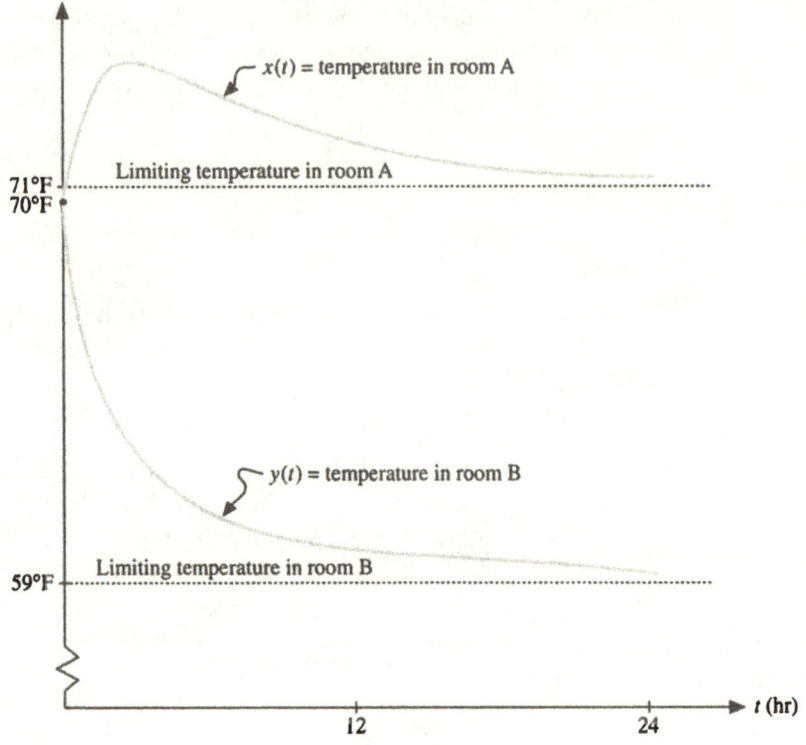

Figure 6.11
Temperature in rooms A and B

Example 2   **The Two-Tank Mixing Problem**   Consider two 100-gal tanks. Tank A is initially filled with water in which 25 lb of salt are dissolved. A 0.5-lb/gal salt mixture is poured into this tank at the constant rate of 4 gal/min. See Figure 6.12.

The well-mixed solution from tank A is constantly being pumped to tank B at a rate of 6 gal/min, and the solution in tank B is constantly being pumped to tank A at the rate of 2 gal/min. The solution in tank B also exits the tank at the rate of 4 gal/min. Find the amount of salt in each tank at any time.

Solution   If we let

$$x_1(t) = \text{amount of salt (lb) in tank A at time } t$$
$$x_2(t) = \text{amount of salt (lb) in tank B at time } t$$

the equations that describe the rate of change $\dot{x}_1$ and $\dot{x}_2$ of salt (lb/min) in tanks A and B are

$$\dot{x}_1 = \left\{ \begin{array}{c} \text{rate of change} \\ \text{in tank A} \end{array} \right\} = \left\{ \begin{array}{c} \text{rate into A} \\ \text{from outside} \end{array} \right\} + \left\{ \begin{array}{c} \text{rate into A} \\ \text{from B} \end{array} \right\} - \left\{ \begin{array}{c} \text{rate out of} \\ \text{A into B} \end{array} \right\}$$

$$= \left(0.5 \text{ lb/gal} \right)\left(4 \text{ gal/min} \right) + \left(\frac{x_2}{100} \text{ lb/gal} \right)\left(2 \text{ gal/min} \right) - \left(\frac{x_1}{100} \text{ lb/gal} \right)\left(6 \text{ gal/min} \right)$$

$$\dot{x}_2 = \left\{ \begin{array}{c} \text{rate of change} \\ \text{in tank B} \end{array} \right\} = \left\{ \begin{array}{c} \text{rate into B} \\ \text{from A} \end{array} \right\} - \left\{ \begin{array}{c} \text{rate out of} \\ \text{B to A} \end{array} \right\} - \left\{ \begin{array}{c} \text{rate out of} \\ \text{B to outside} \end{array} \right\}$$

$$= \left(\frac{x_1}{100} \text{ lb/gal} \right)\left(6 \text{ gal/min} \right) - \left(\frac{x_2}{100} \text{ lb/gal} \right)\left(2 \text{ gal/min} \right) - \left(\frac{x_2}{100} \text{ lb/gal} \right)\left(4 \text{ gal/min} \right)$$

Hence the initial-value problem that describes the amount of salt in tanks A and B is

$$\begin{aligned} \dot{x}_1 &= -0.06x_1 + 0.02x_2 + 2 & x_1(0) &= 25 \\ \dot{x}_2 &= 0.06x_1 - 0.06x_2 & x_2(0) &= 0 \end{aligned} \tag{3}$$

Although we can solve this nonhomogeneous linear system either by the method of elimination or by the Laplace transform, we will use a third method for the purpose of illustration. We first find the solution of the corresponding homogeneous linear system $\dot{x} = Ax$, where

$$x = \begin{bmatrix} x_1 \\ x_2 \end{bmatrix} \qquad A = \begin{bmatrix} -0.06 & 0.02 \\ 0.06 & -0.06 \end{bmatrix}$$

Finding the eigenvalues and eigenvectors of A, we find

$$x_h(t) = \begin{bmatrix} x_1(t) \\ x_2(t) \end{bmatrix} = c_1 e^{-0.025t} \begin{bmatrix} 1 \\ 0.706 \end{bmatrix} + c_2 e^{-0.095t} \begin{bmatrix} 1 \\ 7.750 \end{bmatrix} \tag{4}$$

Figure 6.12
Two-tank problem

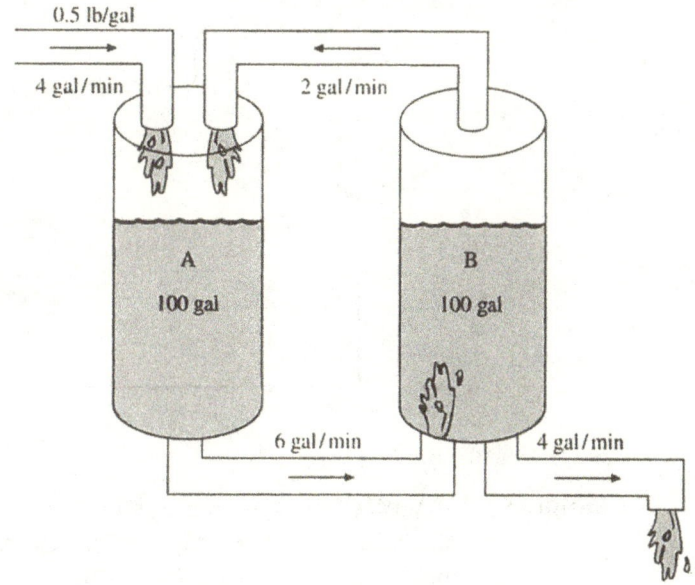

0.5 lb/gal

4 gal/min   2 gal/min

A   B

100 gal   100 gal

6 gal/min   4 gal/min

To find a particular solution

$$x_p(t) = \begin{bmatrix} x_{1p}(t) \\ x_{2p}(t) \end{bmatrix}$$

we use a generalized version of the method of undetermined coefficients and try unknown constants $x_{1p}(t) = A$ and $x_{2p}(t) = B$. Substituting these values for x_1 and x_2 in system (3), we get

$$0 = -0.06A + 0.02B + 2$$
$$0 = 0.06A - 0.06B$$

which gives $A = B = 50$. Hence the general solution is

$$x(t) = c_1 e^{-0.025t} \begin{bmatrix} 1 \\ 0.706 \end{bmatrix} + c_2 e^{-0.095t} \begin{bmatrix} 1 \\ 7.750 \end{bmatrix} + \begin{bmatrix} 50 \\ 50 \end{bmatrix} \tag{5}$$

Substituting this general solution into the initial conditions $x_1(0) = 25$, $x_2(0) = 0$ gives $c_1 = -20.4$ and $c_2 = -4.6$. This solution is graphed in Figure 6.13. ∎

Figure 6.13
Salt in two tanks

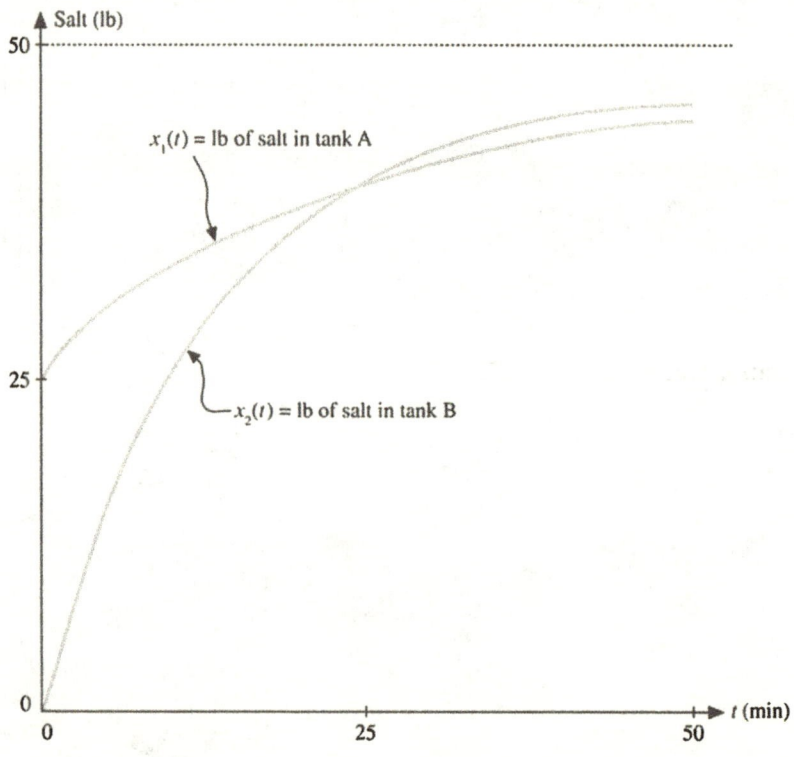

Example 3   **Coupled Mass-Spring System**   Two equal masses $m_1 = m_2 = m$ are attached to three springs, each having the same spring constant $k_1 = k_2 = k_3 = k$, where the two outside springs are attached to walls. The masses slide in a straight line on a frictionless

surface. See Figure 6.14. The system is set in motion by holding the left mass in its equilibrium position while at the same time pulling the right mass to the right of its equilibrium a distance of d. Find the subsequent motion of the masses.

Figure 6.14
Coupled vibrating system

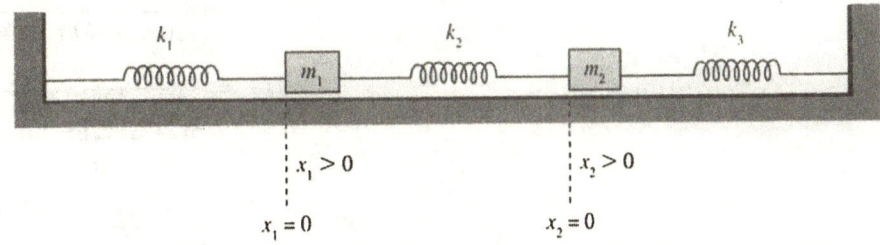

Solution   We denote by $x_1(t)$ and $x_2(t)$ the positions of the respective masses m_1 and m_2 from their respective equilibrium positions. Since the only forces acting on the masses are the forces due to the connecting springs, Hooke's law says the following:

- Force on m_1 due to the left spring $= -k_1 x_1$

- Force on m_1 due to the middle spring $= +k_2(x_2 - x_1)$

- Force on m_2 due to the middle spring $= -k_2(x_2 - x_1)$

- Force on m_2 due to the right spring $= -k_3 x_2$

Hence we have the following initial-value problem:

$$m_1 \ddot{x}_1 = -k_1 x_1 + k_2(x_2 - x_1) \qquad x_1(0) = 0 \quad \dot{x}_1(0) = 0 \tag{6}$$
$$m_2 \ddot{x}_2 = -k_2(x_2 - x_1) - k_3 x_2 \qquad x_2(0) = d \quad \dot{x}_2(0) = 0 \tag{7}$$

Although we have not used the Laplace transform to solve a system of second-order equations, the method is identical to that of first-order equations. Taking the Laplace transform of these equations, we get

$$m_1\left(s^2 X_1(s) - sx_1(0) - \dot{x}_1(0)\right) = -k_1 X_1(s) + k_2\left(X_2(s) - X_1(s)\right)$$
$$m_1\left(s^2 X_2(s) - sx_2(0) - \dot{x}_2(0)\right) = -k_2\left(X_2(s) - X_1(s)\right) - k_3 X_2(s) \tag{8}$$

Letting $m_1 = m_2 = m$ and $k_1 = k_2 = k_3 = k$ and substituting the initial conditions, we get

$$\left(ms^2 + 2k\right)X_1(s) - kX_2(s) = 0$$
$$-kX_1(s) + \left(ms^2 + 2k\right)X_2(s) = msd \tag{9}$$

Solving for $X_1(s)$ and $X_2(s)$ and expanding the expressions as continued fractions, we find

$$X_1(s) = \frac{kmds}{(ms^2 + k)(ms^2 + 3k)}$$

$$= \frac{mds}{2(ms^2 + k)} - \frac{mds}{2(ms^2 + 3k)}$$

$$= \frac{ds}{2\left(s^2 + \dfrac{k}{m}\right)} - \frac{ds}{2\left(s^2 + \dfrac{3k}{m}\right)} \tag{10a}$$

$$X_2(s) = \frac{mds(ms^2 + 2k)}{(ms^2 + k)(ms^2 + 3k)}$$

$$= \frac{mds}{2(ms^2 + k)} + \frac{mds}{2(ms^2 + 3k)}$$

$$= \frac{ds}{2\left(s^2 + \dfrac{k}{m}\right)} + \frac{ds}{2\left(s^2 + \dfrac{3k}{m}\right)} \tag{10b}$$

Hence the solution is

$$x_1(t) = \frac{d}{2}\left(\cos\sqrt{\frac{k}{m}}t - \cos\sqrt{\frac{3k}{m}}t\right) \tag{11a}$$

$$x_2(t) = \frac{d}{2}\left(\cos\sqrt{\frac{k}{m}}t + \cos\sqrt{\frac{3k}{m}}t\right) \tag{11b}$$

Figure 6.15 shows this solution when $d = 2$ and $k = m$. ∎

Figure 6.15
Motion of vibrating masses
for $d = 2$, $k = m$

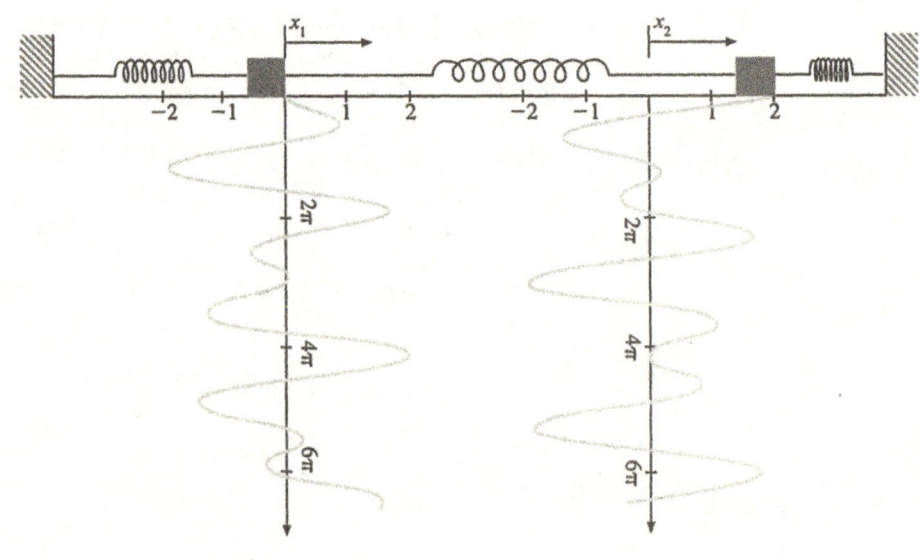

PROBLEMS: Section 6.8

Heating, Cooling, and Mixing Problems

1. **Two-Room Heating Problem** A building consists of two heating regions in which region A is heated by a 80,000-Btu furnace. See Figure 6.16. The heat capacity of region A is 0.25°F per thousand Btu. The time constants for heat transfer are as follows: Between region A and the outside is 4 hr, between region A and room B is 2 hr, and between region B and the outside is 5 hr. Assume that the temperature outside is steady at 0°F.
   (a) What is the system of differential equations that describes the temperature in regions A and B?
   (b) What are the steady state temperatures in the two regions?
   (c) Use a computer program to approximate the future temperature in the two regions if the initial temperatures in the two rooms are 70°F.

Figure 6.16
Two-room heating problem

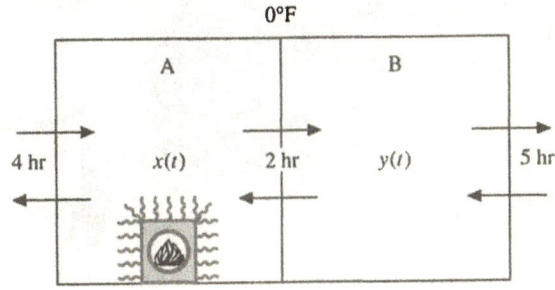

2. **Air-Conditioning Problem** A house consists of two areas, the living area and the attic. See Figure 6.17. A 24,000-Btu per hour air-conditioner cools the living area, which has a heat capacity of 0.50°F per thousand Btu. The time constants between the regions are as follows: Between the living area and the outside is 4 hr, between the living area and the attic is 4 hr, and between the attic and the outside is 2 hr. Suppose the temperature outside is a constant 100°F.
   (a) What is the system of differential equations that describes the future temperature in the living area and the attic?
   (b) How warm will it get in the attic?
   (c) Use a computer to approximate the future temperatures in the two regions if the initial temperatures are 70°F.

3. **Heating Professor Snarf's Den** The building described in Example 1 is Professor Snarf's home. Professor Snarf now adds an electric heater in his den (room B) that generates 10,000 Btu/hr of heat. Under the same conditions as in Example 1, what is the coldest his den will get if the heating capacity of his den is 0.5°F per thousand Btu?

4. **Cooling Professor Snarf's Den** Professor Snarf keeps an air-conditioner in his home to keep him cool in the summer. For cooling purposes his home consists of two regions, region A, where the air conditioner is located, and region B, where his den is located. Assume that the air-conditioner removes 24,000 Btu/hr and that the heat capacity of room A is 0.50°F per thousand Btu. The time constants measuring the rate of heat transfer between the relevant regions are as follows: Between room A and room B is 2 hr, between room A and the outside is 5 hr, and between room B and the outside is 10 hr. If the outside temperature is fixed at 100°F, how cool will it get in Professor Snarf's den (region B)? Professor Snarf likes to keep it rather cool.

5. **Two-Tank Mixing Problem** Two tanks, each of capacity 100 gal, are initially filled with fresh water. Brine containing 1 lb of salt per gallon flows into the first tank at a rate of 3 gal/min, and the dissolved mixture flows into the second tank at a rate of 5 gal/min. The resulting stirred mixture is simultaneously pumped back into the first tank at the rate of 2 gal/min and out of the second tank at the

Figure 6.17
Air-conditioned building

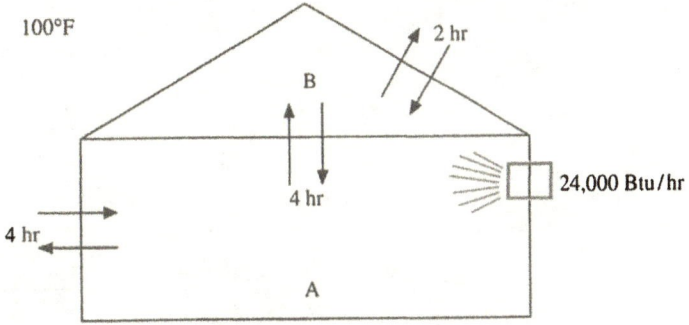

rate of 3 gal/min. See Figure 6.18. What is the initial-value problem that describes the future amount of salt in the two tanks?

Compartmental Analysis

An important problem in biological research is finding the distribution of a given substance throughout the body. A common approach used in studying such systems is **compartmental analysis,** *in which the body is divided into several interconnected* **compartments** *or* **pools,** *which exchange the substance being studied. Compartmental analysis is also playing an important role in ecology for researchers who wish to determine the flow of nutrients throughout an ecological system. Figure 6.19 shows a three-compartment system in which $x_i(t)$ is the amount of material in compartment i at time t and r_{ij} denotes the rate at which material passes to i from j. Typical units for r_{ij} might be grams per minute. The ratio $a_{ij}(t) = r_{ij}/x_i(t)$ is called the* **transfer coefficient** *from j to i and is a fractional rate of change. Typical units for a_{ij} might be 1/min. Hence if $a_{12}(t) = 0.2$ per minute, this means that at time t, material passes from compartment 2 to compartment 1 at the rate of $0.2x_2(t)$ g/min. If the rate of transfer from one compartment to another compartment is proportional to the amount present in the former, then the rate of change of the material in the compartment can be described by the system of equations*

$$\begin{bmatrix} \dot{x}_1 \\ \dot{x}_2 \\ \dot{x}_3 \end{bmatrix} = \begin{bmatrix} a_{11} & a_{12} & a_{13} \\ a_{21} & a_{22} & a_{23} \\ a_{31} & a_{32} & a_{33} \end{bmatrix} \begin{bmatrix} x_1 \\ x_2 \\ x_3 \end{bmatrix} \qquad (12)$$

The following problems deal with compartmental systems in biological and ecological systems.

Figure 6.18
Two-tank mixing problem

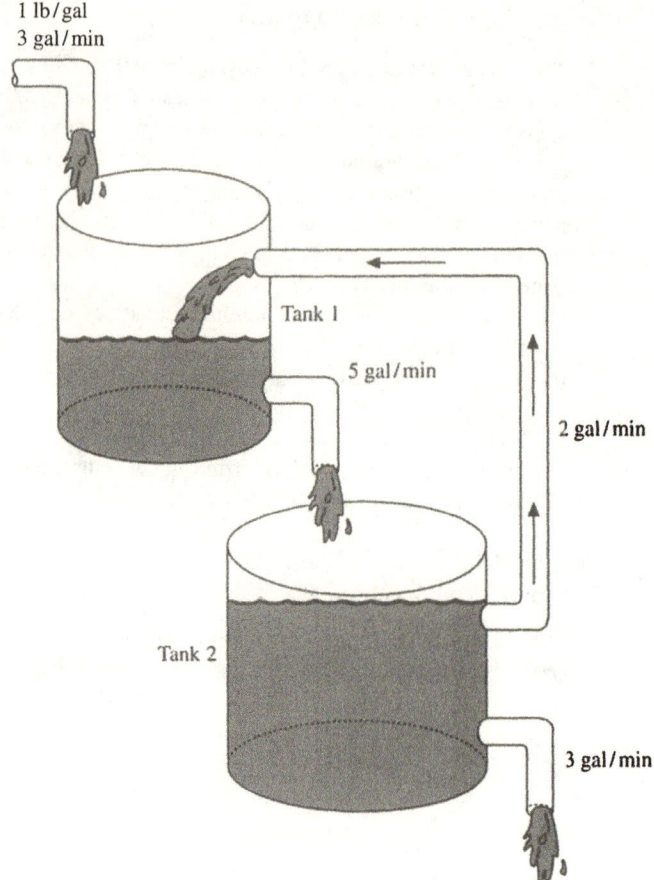

1 lb/gal
3 gal/min

Tank 1

5 gal/min

2 gal/min

Tank 2

3 gal/min

Figure 6.19
General three-compartment system

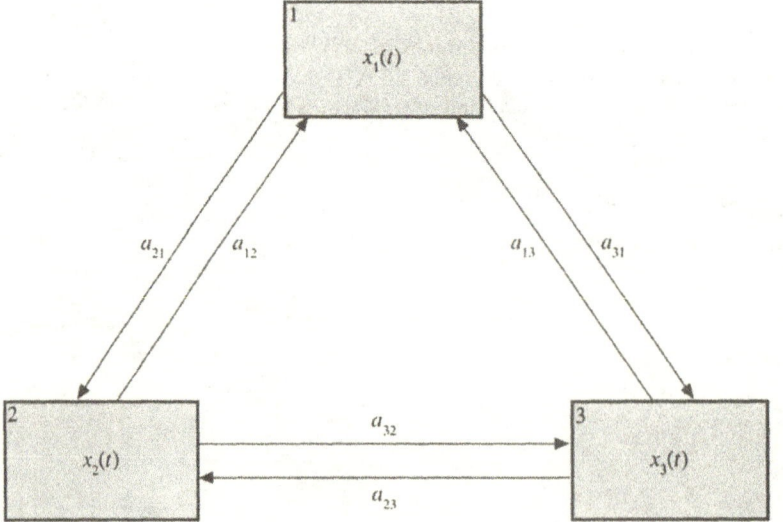

6. **Aquatic Compartmental System** A simple three-compartment model that describes nutrients in a food chain has been studied by M. R. Cullen.* See Figure 6.20. As an example, the constant $a_{31} = 0.06$/hr alongside the arrow connecting compartment 1 (phytoplankton) to compartment 3 (zooplankton) means that at any given time, nutrients pass from the phytoplankton compartment to the zooplankton compartment at the rate of $0.06 x_1$ per hour.
   (a) Find the linear system $\dot{x} = Ax$ that describes the amount of nutrients in each compartment.

(b) Why does the linear system of three equations $\dot{x} = Ax$ describe the amount of nutrients $x_1(t)$, $x_2(t)$, and $x_3(t)$ in the three compartments?

7. **Ecological Compartmental System**[†] The three-compartment system shown in Figure 6.21 illustrates the recovery of a field destroyed by fire. The variables x_1, x_2, and x_3 represent the size (in acres) of each of the three compartments: bare space, grasses, and small shrubs. As an illustration, the transfer coefficient $a_{13} = 0.15$/year means that at any time the small shrub acreage is changing to bare

Figure 6.20
A compartmental model used to predict the amount of a given type of nutrient in the food chain. This three-compartment model describes nutrients in three compartments: phytoplankton, zooplankton, and water.

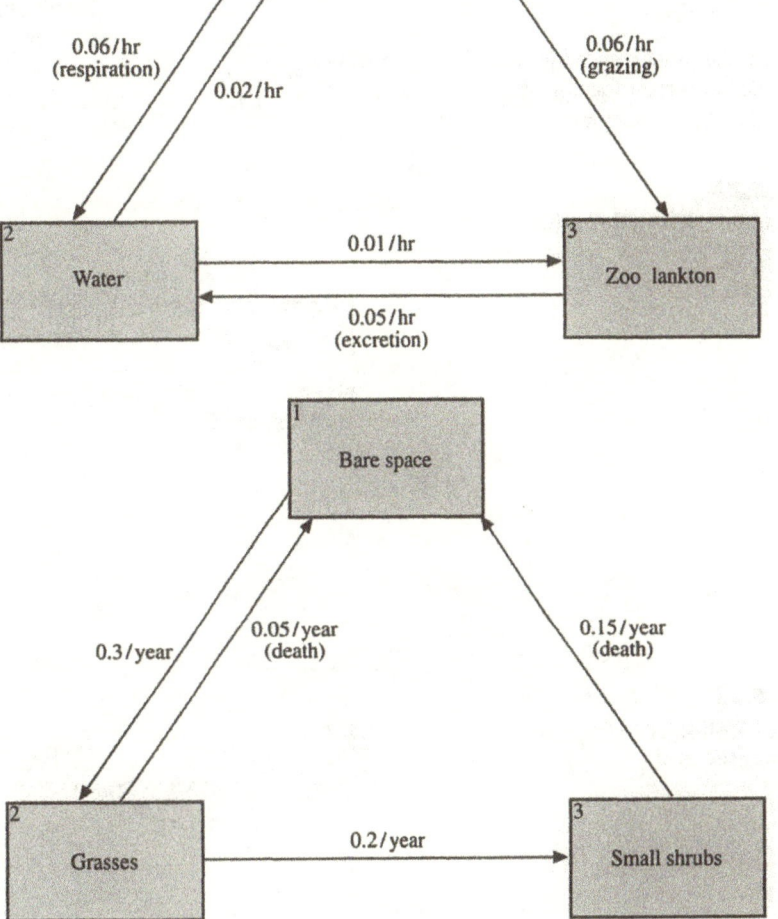

Figure 6.21
A three-compartment model describing the size (acres) of the three compartments after a fire

* Adapted from M. R. Cullen, *Mathematics for the Biosciences* (PWS Publishers, 1983).

† This problem is based on work carried out by M. R. Cullen, *Mathematics for the Biosciences*, (PWS Publishers, 1983).

space acreage at the rate of $0.15x_3$ acre/year. Determine the linear system $\dot{x} = Ax$ that describes the number of acres in each compartment. What are the initial conditions if there are initially 500 acres of bare space, 100 acres of grasses, and 250 acres of small shrubs?

8. **Glucose-Insulin Kinetics** Compartmental analysis can be used to help understand the physiological control of blood glucose by the pancreatic hormone insulin, whose dysfunction can cause the disorder known as diabetes mellitus. In 1982 a two-compartment model for glucose-insulin (g, i) hormonal system was suggested by Bollemo et al.* with the equations

$$\frac{di}{dt} = -k_{11}i + k_{12}(g - g_d) + di_r(t)$$

$$\frac{dg}{dt} = k_{21}g - k_{22}gi$$

where the exact values for the positive constants k_{ij} and d are not important for this problem. Suggest an interpretation for these equations.

9. **A Three-Compartment Model** A person ingests a semi-toxic chemical, which enters the bloodstream at the constant rate R, whereupon it distributes itself in the bloodstream, tissue, and bone. It is excreted in urine and sweat at rates u and s, respectively. The variables x_1, x_2, and x_3 define the *concentrations* of the chemical in the three pools. See Figure 6.22. The equation for x_1 is given by

$$\frac{dx_1}{dt} = R - ux_1 - k_{12}x_1 + k_{21}x_2$$

Find the equations for x_2 and x_3.

Vibration Problems

10. **Vibrations with a Free End** Two springs and two masses vibrate on a frictionless surface as illustrated in Figure 6.23. The system is set in motion by holding the mass m_1 in its equilibrium position and pulling the mass m_2 to the right of its equilibrium position 1 ft and then releasing both masses. Assume that $m_1 = 2$, $m_2 = 1$ slug, $k_1 = 4$, and $k_2 = 2$ lb/ft.

Figure 6.22
A three-compartment model

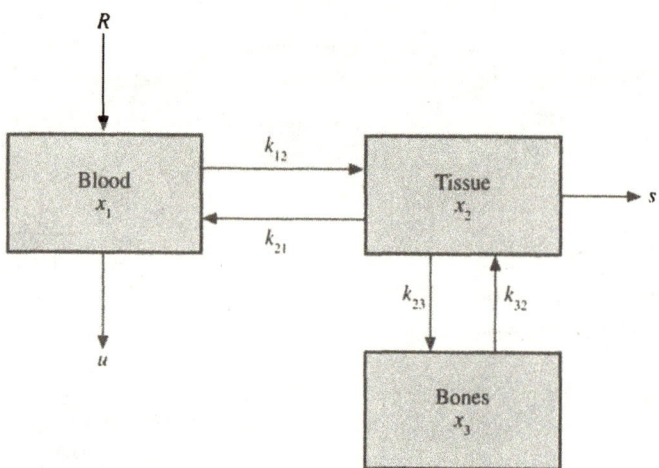

Figure 6.23
Coupled spring-mass system with one free end

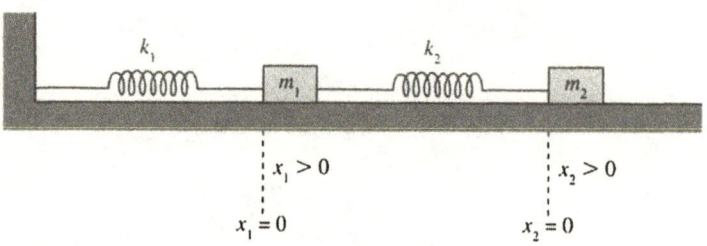

* Optimal Feedback of Giycemia Regulation in Diabetics'' by V. W. Bollomo et al., *Med. Biol. Eng. Comp.*, *20*, 329–335.

(a) Determine the equations of motion of the masses.

(b) Use the Laplace transform to solve the equations found in part (a) by using the initial conditions $x_1(0) = 1$, $\dot{x}_1(0) = 0$, $x_2(0) = 5$, and $\dot{x}_2(0) = 0$.

11. Automobile Spring System The mechanical system illustrated in Figure 6.24 is an idealization of an automobile suspension system. The chassis of the car is represented by the mass m_2, the springs and shock absorbers by the spring constant k_2 and damping constant d, the wheel and axle by mass m_1, and the flexible tire by the spring constant k_1. The irregularities of the road surface give rise to the forcing function $f(t)$ acting through the tire. Convince yourself that the differential equations that describe the positions from equilibrium $x_1(t)$ and $x_2(t)$ of the two masses m_1 and m_2 are

$$m_1\ddot{x}_1 + k_1(x_1 - f(t)) + k_2(x_1 - x_2) + d(\dot{x}_1 - \dot{x}_2) = 0$$
$$m_2\ddot{x}_2 + d(\dot{x}_1 - \dot{x}_2) + k_2(x_2 - x_1) = 0 \tag{13}$$

Figure 6.24
A schematic drawing of the suspension system of an automobile

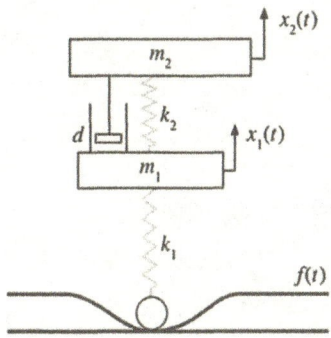

12. Interesting Problem In determining the ratio e/m of the charge to the mass of an electron, Thompson discovered that the planar path $x = x(t)$, $y = y(t)$ is governed by the equations

$$m\ddot{x} + He\dot{y} = Ee$$
$$m\ddot{y} - He\dot{x} = 0 \tag{14}$$

where E and H are the intensities of the electric and magnetic fields, respectively. If the initial conditions are given by $x(0) = \dot{x}(0) = y(0) = \dot{y}(0) = 0$, show that the path of the electron is a cycloid.

13. Computer Problem Use a computer to sketch the graphs of the temperature functions $x_1(t)$ and $x_2(t)$ of the two rooms described in Problem 1.

14. Journal Entry Summarize some of the applications of systems of differential equations.

6.9   NUMERICAL SOLUTION OF SYSTEMS OF DIFFERENTIAL EQUATIONS

PURPOSE

To show how numerical methods, such as higher-order Taylor series methods and the Runge-Kutta method, can be extended to systems of differential equations.

THE TAYLOR SERIES METHOD

In Chapter 2 we saw how a Taylor series expansion could be used to approximate the solution of a first-order differential equation. We now show how a Taylor series expansion can be used to approximate the solution of a higher-order differential equation or a system of differential equations. For example, consider the system of two equations

$$\begin{aligned} x' &= x - y + t^2 + 1 & x(0) &= 1 \\ y' &= x + y + t^3 + t^2 & y(0) &= 0 \end{aligned} \tag{1}$$

where x' and y' denote differentiation with respect to t. Differentiating both equations with respect to t, we get

$$\begin{cases} x'' = x' - y' + 2t \\ y'' = x' + y' + 3t^2 + 2t \end{cases} \tag{2}$$

$$\begin{cases} x''' = x'' - y'' + 2 \\ y''' = x'' + y'' + 6t + 2 \end{cases} \tag{3}$$

$$\begin{cases} x^{(iv)} = x''' - y''' \\ y^{(iv)} = x''' + y''' + 6 \end{cases} \tag{4}$$

$$\vdots$$

If we now substitute these derivatives into the Taylor series expansions

$$x(t + h) = x(t) + hx' + \frac{h^2}{2}x'' + \frac{h^3}{6}x''' + \frac{h^4}{24}x^{(iv)} + \cdots \tag{5a}$$

$$y(t + h) = y(t) + hy' + \frac{h^2}{2}y'' + \frac{h^3}{6}y''' + \frac{h^4}{24}y^{(iv)} + \cdots \tag{5b}$$

we obtain formulas for $x(t + h)$ and $y(t + h)$ in terms of $x(t)$ and $y(t)$. By using formulas (5) it is possible to find an approximate solution of the initial-value problem (1) at the points $t = 0, h, 2h, \ldots, T$.

Although the above example illustrates only two equations and two unknowns, the Taylor series method can easily be extended to any number of equations. Today, by using computer programs that perform analytic differentiation, such approximations are feasible.

Example 1   **Numerical Solution of a Second-Order Equation**   Use the first four terms of the Taylor series to approximate the solution of the (nonlinear) initial-value problem

$$y'' + y^2 = 0 \qquad y(0) = 1 \quad y'(0) = 0$$

at the point $t = 0.1$.

Solution   Defining the new variables $x_1 = y$ and $x_2 = y'$, we can rewrite the second-order equation as the first-order system

$$\begin{aligned} x_1' &= x_2 & x_1(0) &= 1 \\ x_2' &= - x_1^2 & x_2(0) &= 0 \end{aligned} \tag{6}$$

Differentiating and writing all derivatives in terms of x_1 and x_2, we get

$$\begin{cases} x_1'' = x_2' = -x_1^2 \\ x_2'' = -2x_1x_1' = -2x_1x_2 \end{cases} \tag{7}$$

$$\begin{cases} x_1''' = -2x_1x_1' = -2x_1x_2 \\ x_2''' = -2x_1x_2' - 2x_1'x_2 = 2x_1{}^3 - 2x_2^2 \end{cases} \tag{8}$$

Using a Taylor series including the third derivative, we can approximate $x_1(0.1)$ and $x_2(0.1)$ by

$$x_1(0.1) \doteq x_1(0) + 0.1\,x_1'(0) + \frac{0.01}{2}x_1''(0) + \frac{0.001}{6}x_1'''(0)$$

$$= 1 + 0.1x_2(0) - \frac{0.01}{2}x_1^2(0) - \frac{0.001}{3}x_1(0)x_2(0)$$

$$= 1 + 0.1(0) - 0.005(1) - \frac{0.001}{3}(1)(0)$$

$$= 0.995 \tag{9a}$$

$$x_2(0.1) \doteq x_2(0) + 0.1x_2'(0) + \frac{0.01}{2}x_2''(0) + \frac{0.001}{6}x_2'''(0)$$

$$= 0 - 0.1x_1^2(0) - \frac{0.02}{2}x_1(0)x_2(0) + \frac{0.001}{6}\left(2x_1^3(0) - 2x_2^2(0)\right)$$

$$= 0 - 0.1(1) + \frac{0.02}{2}(1)(0) + \frac{0.001}{6}(1)$$

$$= -0.1 \tag{9b}$$

The approximate values generated by a computer are given in Table 6.1.

Table 6.1

Taylor Series Approximation of $y'' + y^2 = 0$

t	$x_1(t)$	$x_2(t)$
0	1.0	0
0.1	0.995	-0.0996666
0.2	0.9800830	-0.1999938
0.3	0.9553461	-0.2979942
0.4	0.9210781	-0.3920713
0.5	0.8777497	-0.4804497
0.6	0.8259928	-0.5615974
0.7	0.7665763	-0.6342983
0.8	0.7003703	-0.6977055
0.9	0.6283100	-0.7513684
1.0	0.5513567	-0.7952371

THE RUNGE-KUTTA METHOD FOR SYSTEMS

In Chapter 2 we learned that the Runge-Kutta method allows one to obtain the accuracy of "higher-order methods" while requiring knowledge only of first-order derivatives. Consider the initial-value problem

$$x' = f(t, x) \qquad x(t_0) = x_0 \tag{10}$$

We recall that the Runge-Kutta formulas for approximating the solution of Eq. (10) at the points $t_0, t_0 + h, t_0 + 2h, \dots$ are

$$t_{n+1} = t_n + h \qquad (n = 0, 1, \dots)$$
$$x_{n+1} = x_n + \frac{h}{6}\left(k_1 + 2k_2 + 2k_3 + k_4\right) \tag{11a}$$

where

$$k_1 = f(t_n, x_n)$$
$$k_2 = f\left(t_n + \frac{h}{2}, x_n + \frac{1}{2}k_1\right)$$
$$k_3 = f\left(t_n + \frac{h}{2}, x_n + \frac{1}{2}k_2\right) \tag{11b}$$
$$k_4 = f(t_n + h, x_n + k_3)$$

To extend the Runge-Kutta method to systems of first-order equations, consider the system of two* first-order equations:

$$x' = f(t, x, y) \qquad x(t_0) = x_0$$
$$y' = g(t, x, y) \qquad y(t_0) = y_0 \tag{12}$$

If we denote by $x_n = x(t_n)$ and $y_n = y(t_n)$ the approximate solution at $t_n = t_0 + nh$, $n = 0, 1, \dots$, the generalization of the Runge-Kutta formula (11) to two equations is

$$x_{n+1} = x_n + \frac{h}{6}\left(k_1 + 2k_2 + 2k_3 + k_4\right) \tag{13a}$$

$$y_{n+1} = y_n + \frac{h}{6}\left(j_1 + 2j_2 + 2j_3 + j_4\right) \tag{13b}$$

where

$$\begin{cases} k_1 = f(t_n, x_n, y_n) \\ j_1 = g(t_n, x_n, y_n) \end{cases} \tag{14a}$$

---

* The generalization to more than two first-order equations is obvious, so for notational simplicity we restrict ourselves to two equations.

$$\begin{cases} k_2 = f\left(t_n + \dfrac{h}{2}, x_n + \dfrac{h}{2}k_1, y_n + \dfrac{h}{2}j_1\right) \\[2mm] j_2 = g\left(t_n + \dfrac{h}{2}, x_n + \dfrac{h}{2}k_1, y_n + \dfrac{h}{2}j_1\right) \end{cases} \tag{14b}$$

$$\begin{cases} k_3 = f\left(t_n + \dfrac{h}{2}, x_n + \dfrac{h}{2}k_2, y_n + \dfrac{h}{2}j_2\right) \\[2mm] j_3 = g\left(t_n + \dfrac{h}{2}, x_n + \dfrac{h}{2}k_2, y_n + \dfrac{h}{2}j_2\right) \end{cases} \tag{14c}$$

$$\begin{cases} k_4 = f(t_n + h, x_n + hk_3, y_n + hj_3) \\ j_4 = g(t_n + h, x_n + hk_3, y_n + hj_3) \end{cases} \tag{14d}$$

The four constants k_1, k_2, k_3, and k_4 represent slopes* at different points used in moving from x_n to x_{n+1}, and the constants j_1, j_2, j_3, and j_4 represent slopes used in moving from y_n to y_{n+1}.

Example 2   Runge-Kutta Method   Approximate the solution of the initial-value problem

$$\begin{array}{ll} x' = x - xy & x(0) = 2 \\ y' = -y + xy & y(0) = 1 \end{array} \tag{15}$$

Solution   First, we identify the derivatives $f(t, x, y) = x - xy$, $g(t, x, y) = -y + xy$ and step size $h = 0.1$. Then, starting with initial values $n = 0$, $t_0 = 0$, $x_0 = 1$, $y_0 = 0$, we compute

$$k_1 = f(t_0, x_0, y_0) = x_0 - x_0 y_0 = 2 - (2)(1) = 0$$

$$j_1 = g(t_0, x_0, y_0) = -y_0 + x_0 y_0 = -1 + (2)(1) = 1$$

$$k_2 = f\left(t_0 + \frac{h}{2}, x_0 + \frac{h}{2}k_1, y_0 + \frac{h}{2}j_1\right)$$

$$= \left(x_0 + \frac{h}{2}k_1\right) - \left(x_0 + \frac{h}{2}k_1\right)\left(y_0 + \frac{h}{2}j_1\right)$$

$$= \left(2 + \frac{0.1}{2}(0)\right) - \left(2 + \frac{0.1}{2}(0)\right)\left(1 + \frac{0.1}{2}(1)\right)$$

$$= -0.10$$

---

* The reader should realize that although the notation does not indicate this fact, the slopes k_1, k_2, k_3, and k_4 depend on n and are often denoted by k_{n1}, k_{n2}, k_{n3}, and k_{n4}. However, for simplicity we use the streamlined notation.

$$j_2 = g\left(t_0 + \frac{h}{2}, x_0 + \frac{h}{2}k_1, y_0 + \frac{h}{2}j_1\right)$$

$$= -\left(y_0 + \frac{h}{2}j_1\right) + \left(x_0 + \frac{h}{2}k_1\right)\left(y_0 + \frac{h}{2}j_1\right)$$

$$= -\left(1 + \frac{0.1}{2}(1)\right) + \left(2 + \frac{0.1}{2}(0)\right)\left(1 + \frac{0.1}{2}(1)\right)$$

$$= 1.05$$

Similarly, we compute

$$k_3 = -0.105 \qquad k_4 = -0.208$$
$$j_3 = 1.047 \qquad j_4 = 1.093$$

Hence we get

$$x_1 = x_0 + \frac{h}{6}\left(k_1 + 2k_2 + 2k_3 + k_4\right) = 1.9897$$

$$y_1 = y_0 + \frac{h}{6}\left(j_1 + 2j_2 + 2j_3 + j_4\right) = 1.1048 \qquad \blacksquare$$

Table 6.2 gives the approximate values for $t = 0, 0.1, 0.2, ..., 0.9, 1.0$.

Table 6.2
Runge-Kutta Method for
Systems*

t	$x(t)$	$y(t)$
0	2.0	1.0
0.1	1.9897	1.1048
0.2	1.9580	1.2180
0.3	1.9044	1.3371
0.4	1.8302	1.4585
0.5	1.7377	1.5777
0.6	1.6309	1.6896
0.7	1.5144	1.7892
0.8	1.3935	1.8724
0.9	1.2728	1.9357
1.0	1.1565	1.9777

* Although the numbers are displayed to four places, the calculations were carried out to eight places.

REWRITING HIGHER-ORDER SYSTEMS AS FIRST-ORDER SYSTEMS

We have seen that a higher-order differential equation can be replaced by an equivalent system of first-order systems. We can also write higher-order systems of equations as equivalent systems of first-order equations. It is often necessary to carry out this manipulation in order to use computer programs written only for first-order systems.

For instance, consider the following system of two second-order equations:

$$x'' = f(t, x, y, x', y') \qquad x(0) = x_0 \qquad x'(0) = x_0' \tag{16}$$
$$y'' = g(t, x, y, x', y') \qquad y(0) = y_0 \qquad y'(0) = y_0'$$

From these two second-order equations, we can obtain an equivalent system of four first-order equations by introducing the new variables

$$x_1 = x$$
$$x_2 = y$$
$$x_3 = x' \tag{17}$$
$$x_4 = y'$$

Differentiating these equations with respect to t, we obtain the equivalent system of four first-order equations:

$$x_1' = x_3 \qquad\qquad x_1(0) = x_0$$
$$x_2' = x_4 \qquad\qquad x_2(0) = y_0$$
$$x_3' = f(t, x_1, x_2, x_3, x_4) \qquad x_3(0) = x_0' \tag{18}$$
$$x_4' = g(t, x_1, x_2, x_3, x_4) \qquad x_4(0) = y_0'$$

It is often useful to rewrite a system of n second-order equations (which often occur in mechanics) as an equivalent system of $2n$ first-order equations inasmuch as most computer packages are designed to solve systems of first-order equations.

PROBLEMS: Section 6.9

For Problems 1–3, use the Taylor series method with the four terms (using the first three derivatives) to find the approximate solution of the given initial-value problem at $t = 0.1$ using the step size $h = 0.1$.

1. $x' = 1 \qquad\qquad x(0) = 1$
   $y' = t \qquad\qquad y(0) = 1$
2. $x' = ty \qquad\qquad x(0) = 1$
   $y' = xy \qquad\qquad y(0) = 1$
3. $x' = x - y \qquad\quad x(0) = 1$
   $y' = -4x + y \quad y(0) = 1$

Numerical Solution of Higher-Order Equations

For Problems 4–8, rewrite the given higher-order equation as a system of first-order equations and find the approximate solution at $t = 0.1$ using the Runge-Kutta method. If you have access to a microcomputer with a differential equations package, evaluate the solution on the interval $[0, 1]$ with step size $h = 0.1$.

4. $y'' + y^2 = 0$
   $y(0) = 1 \quad y'(0) = 0$
5. $y'' + ty' + y = 0$
   $y(0) = 1 \quad y'(0) = 0$
6. $y'' + \sin y = 0$
   $y(0) = 1 \quad y'(0) = 0$
7. $(1 + t^2)y'' + y^2 = t + 1$
   $y(0) = 1 \quad y'(0) = 0$
8. $y''' + y^2 = 0$
   $y(0) = 1 \quad y'(0) = 0 \quad y''(0) = 0$

Transforming Higher-Order Equations to Autonomous Systems

*If a system of differential equations does not explicitly contain the independent variable t, then the system is called an **autonomous system**. For Problems 9–12, show that the given nth-order equation can be written as an equivalent autonomous system of $n + 1$ first-order equations by defining the first n variables as we have done before and a new variable given by $x_{n+1} = t$ by defining $x_{n+1}' = 1$, $x_{n+1}(0) = 0$.*

9. $y'' + y^2 = t^2$
$y(0) = 1 \quad y'(0) = 0$

10. $y'' + ty' + y = \sin t$
$y(0) = 0 \quad y'(0) = 0$

11. $y'' + (1 - y^2)y' + y = t$
$y(0) = 2 \quad y'(0) = -1$

12. $y''' + 2y'' + ty = \cos t$
$y(0) = 1 \quad y'(0) = 0 \quad y''(0) = 2$

Computer Projects

Problems 13–21 describe phenomena described by differential equations. Use a computer with computer package to perform the indicated experiments. For second-order equations it might be necessary to rewrite the equation as an equivalent system of first-order equations to be compatible with existing software.

13. Pendulum Equation Without Damping The undamped pendulum equation

$$\ddot{\theta} + \frac{g}{L} \sin \theta = 0 \qquad (19)$$

describes the oscillations of a frictionless pendulum of length L, where θ measures the angle (radians) of the bob from equilibrium. See Figure 6.25. Take $g = 32$ ft/sec² and $L = 8$ ft.

(a) Does the period of oscillation depend on the initial conditions? Try the initial position $\theta(0) = \theta_0$ and $\dot{\theta}(0) = 0$ for $\theta_0 = \pi/6$, $\pi/4$, and $\pi/3$.

(b) If the bob is initially located in its equilibrium position, what must the initial velocity be in order that the bob go ''over the top'' (i.e., reach $\theta = \pi$ at some time)? When this happens, does the bob ever stop?

Figure 6.25
Oscillations of a pendulum

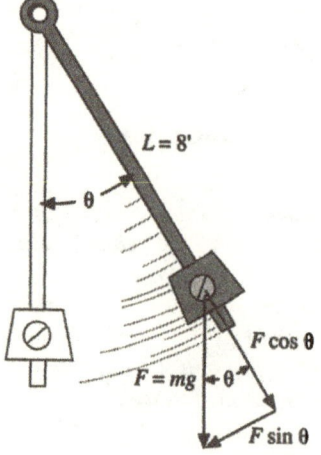

$L = 8'$

θ

$F \cos \theta$

$F = mg$ θ

$F \sin \theta$

14. The Almost Pendulum Equation If the sine function is approximated by the first two terms of its Maclaurin series expansion, the pendulum equation (19) becomes

$$\ddot{\theta} + \frac{g}{L}\left(\theta - \frac{\theta^3}{6}\right) = 0 \qquad (20)$$

Compare the solution of Eq. (20) with the solution of the pendulum equation (19). Try the initial conditions $\theta(0) = \pi/4$ and $\dot{\theta}(0) = 0$.

15. Damped Pendulum Equation The damped pendulum equation

$$\ddot{\theta} + \frac{k}{L}\dot{\theta} + \frac{g}{L} \sin \theta = 0 \qquad (21)$$

describes the oscillations of a pendulum of length L about equilibrium, where $\theta(t)$ measures the angle (radians) of the bob from equilibrium. We take $g = 32$ ft/sec² and $L = 8$ ft.

(a) Find the oscillations of the bob starting at rest 45° from equilibrium with the three different damping constants $k = 1$, 2, and 3.

(b) If the bob is initially located in its equilibrium position, what must the initial velocity be in order that the bob go ''over the top'' (i.e., reach $y = \pi$ at some time)? Use the damping constant $k = 1$.

16. The Nonlinear Spring Duffing's equation

$$\ddot{y} + y + ky^3 = 0 \qquad (22)$$

where k is a constant that describes the vibrations of a mass-spring system in which the restoring force of the spring is the nonlinear function $F(y) = -y - ky^3$. Use the Runge-Kutta method to approximate the solution with $h = 0.1$ with initial conditions $y(0) = y_0$, $\dot{y}(0) = 0$ for $y_0 = 1, 2$, and 3. The constant k describes the strength of the spring. Determine the period of oscillation of the vibrating mass as a function of k. Try values $k = 0.5, 1.0, 1.5$.

17. Bombs Away An airplane flying horizontally at v mph drops a bomb weighing m lb. The equations that describe the motion of the bomb have been estimated to be

$$\frac{d^2x}{dt^2} = -\frac{a}{m}\left(\frac{dx}{dt}\right)^2 \qquad x(0) = 0 \quad \dot{x}(0) = v$$
$$\frac{d^2y}{dt^2} = g - \frac{a}{m}\left(\frac{dy}{dt}\right)^2 \qquad y(0) = 0 \quad \dot{y}(0) = 0 \qquad (23)$$

where a is the coefficient of friction and $g = 32$ ft/sec² is the acceleration due to gravity. The origin $x = 0$, $y = 0$ is taken from the drop point, with the positive y-axis pointing downward. See Figure 6.26.

Figure 6.26
Bombs away

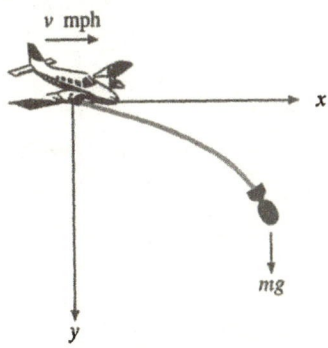

(a) Rewrite the two second-order equations as an equivalent system of four first-order equations.

(b) If the velocity of the airplane is 1000 mph and the bomb weighs 500 lb, find the trajectory of the bomb for coefficients of friction $a = 1, 2$, and 3. Sketch the graph of y versus x if you have access to the necessary graphing facilities.

18. Projectile Problem A projectile is fired at an angle of $\theta = \pi/4$ radians (45°) from the horizontal. The origin $x = 0, y = 0$ is taken as the point of firing, assuming a flat earth. Assume a constant acceleration $g = 32 \text{ ft/sec}^2$ due to gravity and a drag force that is proportional to the magnitude of the projectile's velocity and acts opposite to the direction of the motion of the projectile. Under these assumptions the equations that describe the motion of the projectile are

$$\ddot{x} = -D \cos \alpha$$
$$\ddot{y} = -g - D \sin \alpha \qquad (24)$$

where

$$D = 0.01 \sqrt{(\dot{x})^2 + (\dot{y})^2}$$

and α is the angle (radians) between the direction of the projectile and the horizontal. That is, $\tan \alpha = \dot{y}/\dot{x}$. See Figure 6.27.

(a) Rewrite the system (24) as an equivalent system of four first-order equations.

(b) If the muzzle velocity of the projectile is 1000 ft/sec, determine the resulting trajectory. How far will the projectile travel?

19. Harmonic Oscillator The initial-value problem

$$\dot{x}_1 = x_2 \qquad x_1(0) = 0$$
$$\dot{x}_2 = -x_2 \qquad x_2(0) = 1$$

has the solution $x_1(t) = \sin t, x_2(t) = \cos t$. Approximate the solution of this problem for $0 \leq t \leq 2\pi$ and plot the solution. How accurate is your solution? You may want to try different step sizes and compare the results.

20. Hmmmmm Approximate the solution of the initial-value problem

$$\dot{x} = y \qquad x(0) = 1$$
$$\dot{y} = -4x + x^2 \qquad y(0) = 0$$

on some interval. Is the motion periodic? Can you think of an interpretation of these equations?

21. Predator-Prey Equations The two nonlinear differential equations

$$\dot{x}_1 = 2x_1 - 2x_1 x_2 \qquad x_1(0) = 2$$
$$\dot{x}_2 = -x_2 + x_1 x_2 \qquad x_2(0) = 4 \qquad (25)$$

are called **predator-prey equations,** where x_1 is the population size (in thousands) of the prey and x_2 is the population size (in thousands) of the predator. A typical predator and prey problem might involve a fox/rabbit ecosystem in which the fox survives by eating the rabbit. Of course, most ecosystems involve more variables, which could be described by more differential equations. Approximate the solution of Eq. (25). Plot the graphs of $x_1(t)$ and $x_2(t)$ on your computer video screen for $0 \leq t \leq 20$.

22. Journal Entry Summarize what you know about systems of differential equations.

Figure 6.27
The path of a projectile with air friction

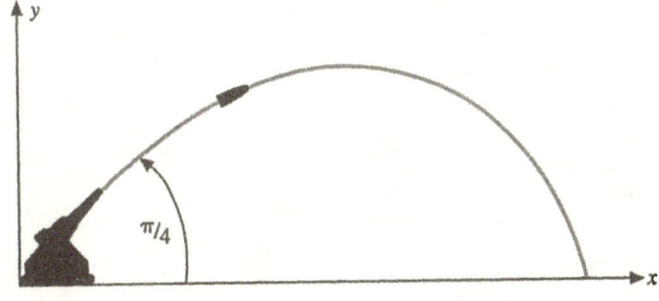

7

Difference Equations

7.1 INTRODUCTION TO DIFFERENCE EQUATIONS

7.2 HOMOGENEOUS EQUATIONS

7.3 NONHOMOGENEOUS EQUATIONS

7.4 APPLICATIONS OF DIFFERENCE EQUATIONS

7.5 THE LOGISTIC EQUATION AND THE PATH TO CHAOS

7.6 ITERATIVE SYSTEMS: JULIA SETS AND THE MANDELBROT SET (OPTIONAL)

7.1   INTRODUCTION TO DIFFERENCE EQUATIONS

PURPOSE
To introduce the basic definitions, ideas and theory of first- and second-order linear difference equations. This section provides much of the foundation of the remainder of the chapter.

In the previous chapters we studied differential equations in which an unknown variable y depends on a *continuous* variable t. In many applications, however, change takes place only at *discrete* instances of time, a feature that leads to the study of *difference equations*. During recent years, widespread use of computer simulation, sampled data systems, digital control systems, and numerical methods has led to a rediscovery of this old and honorable subject.

A **difference equation** is an equation that relates different members of a sequence of numbers $\{y_0, y_1, y_2, ..., y_n, ...\}$, where the values y_n of the sequence are unknown quantities that we wish to find. Examples of difference equations are

$$y_{n+1} - 0.08y_n = 1 \tag{1a}$$

$$y_{n+1} - 0.5y_n^2 = 1 \tag{1b}$$

$$y_{n+2} - y_{n+1} + y_n = 2n + 1 \tag{1c}$$

$$y_{n+2} + y_n = n^2 \tag{1d}$$

We often denote a sequence by $\{y_n\}_{n=0}^{n=\infty}$ or simply by $\{y_n\}$ or even by y_n.* Unless otherwise indicated, we always assume that the subscript n is an integer 0, 1, 2, ...

The order of a difference equation is defined as the difference between the largest and smallest subscripts of the members of the sequence that appear in the equation. For instance, Eqs. (1a) and (1b) are first-order equations, and Eqs. (1c) and (1d) are of second order.

SOLUTION OF A DIFFERENCE EQUATION

A solution of a difference equation is simply a sequence that satisfies the difference equation for all $n = 0, 1, 2, ...$

Example 1   Solutions of Difference Equations   Verify that the given sequences $\{y_n\}$ are solutions of the given difference equations.

---

* The reader may object to the use of the brace notation, $\{\cdots\}$, for denoting sequences, since braces are generally used to denote sets. To avoid this ambiguity, a few authors find it preferable to denote sequences by $\langle\cdots\rangle$. We join the majority, however, and use the ambiguous brace notation with the understanding that the reader will realize that we are referring to sequences.

Difference Equation	**Solution**
(a) $y_{n+1} - \frac{1}{2}y_n = 0$	$\{y_n\} = \{2^{-n}\} = \{1, \frac{1}{2}, \frac{1}{4}, \frac{1}{8}, ...\}$
(b) $y_{n+1} - 3y_{n+1} + 2y_n = 0$	$\{y_n\} = \{2^n\} = \{1, 2, 4, 8, ...\}$
(c) $y_{n+2} + 6y_{n+1} + 9y_n = 0$	$\{y_n\} = \{-3^n\} = \{1, -3, 9, -27, ...\}$

Solution   Substituting each sequence into the respective difference equation gives

(a) $y_{n+1} - \frac{1}{2}y_n = 2^{-(n+1)} - \frac{1}{2}2^{-n} = 2^{-(n+1)} - 2^{-(n+1)} = 0$

(b) $y_{n+2} - 3y_{n+1} + 2y_n = 2^{n+2} - 3 \cdot 2^{n+1} + 2 \cdot 2^n = 0$

(c) $y_{n+2} + 6y_{n+1} + 9y_n = (-3)^{n+2} + 6(-3)^{n+1} + 9(-3)^n = 0$ ∎

LINEAR DIFFERENCE EQUATIONS

The general first- and second-order linear difference equations are difference equations having the form

$$a_n y_{n+1} + b_n y_n = f_n \qquad \text{(first-order linear equation)} \qquad (2a)$$
$$a_n y_{n+2} + b_n y_{n+1} + c_n y_n = f_n \qquad \text{(second-order linear equation)} \qquad (2b)$$

The sequences $\{a_n\}$, $\{b_n\}$, and $\{c_n\}$ represent given *coefficients* and either are constant sequences or depend on n. When all the members of these sequences are constant, the equation is said to have **constant coefficients**; otherwise, they are said to have **variable coefficients.**

The sequence $\{f_n\}$ is called the **nonhomogeneous sequence,** or **forcing sequence** of the equation. If all the members of $\{f_n\}$ are zero, then the linear difference equation is called a **homogeneous equation.**

Example 2   Classifying Difference Equations   Classify the following difference equations.

(a) $y_{n+2} + ny_{n+1} + y_n = 0$ $\qquad\qquad$ $(n = 0, 1, 2, ...)$

(b) $y_{n+1} = y_n + 0.05y_n^2$ $\qquad\qquad$ $(n = 0, 1, 2, ...)$

(c) $y_{n+1} - y_n = 0.08y_n + 1000(1 + 0.05n)$ $\qquad$ $(n = 0, 1, 2, ...)$

(d) $y_{n+2} + 3y_{n+1} + y_n = \sin(n\pi)$ $\qquad\qquad$ $(n = 0, 1, 2, ...)$

Solution   (a) Second-order, linear, homogeneous with variable coefficients.

(b) First-order, nonlinear.

(c) First-order, linear, nonhomogeneous with constant coefficients.

(d) Second-order, linear, nonhomogeneous with constant coefficients.   ∎

It is possible to write a difference equation in several ways. For example, the difference equations (3a) and (3b),

$$y_{n+2} + ny_{n+1} - y_n = n^2 \qquad\qquad (n = 0, 1, 2, ...) \qquad (3a)$$
$$y_{m+1} + (m - 1)y_m - y_{m-1} = (m - 1)^2 \qquad (m = 1, 2, 3, ...) \qquad (3b)$$

define exactly the same relations between the members of the sequence $\{y_n\}$. This is easy to see by simply comparing a few of the equations. The convention in this book, unless otherwise indicated, is to use the notation of Eq. (3a), where equations are defined for integers $n \geq 0$.

INITIAL-VALUE PROBLEM

When modeling problems using a first- or second-order difference equation, we often assume that the first member y_0 or the first two members y_0 and y_1 of the sequence are known. For a first-order equation, since one can solve for y_{n+1} in terms of y_n, it is easy to determine the successive values y_1, y_2, ... of the sequence given a *single initial condition* y_0. In the case of a second-order difference equation, one can solve for y_{n+2} in terms of y_{n+1} and y_n, and hence it is possible to find y_2, y_3, ... from *two initial conditions* y_0 and y_1. The problem of finding a solution to a difference equation that satisfies given initial conditions is called an **initial-value problem.** See Figure 7.1.

Figure 7.1
For a first-order difference equation, specifying y_0 allows one to find the remaining members of the sequence; for a second-order equation, one must specify y_0 and y_1 in order to find the remaining members.

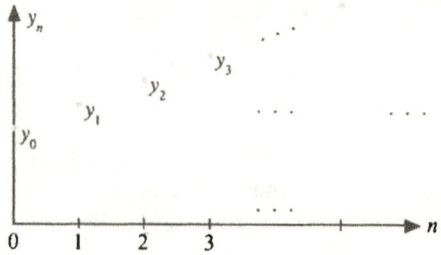

Example 3 **Initial-Value Problem** Find the first five terms of the solution of the following initial-value problems:

(a) $y_{n+1} - 2y_n = 1$ $\qquad\qquad y_0 = 1$
(b) $y_{n+2} + ny_{n+1} - y_n = 1$ $\quad y_0 = 3 \quad y_1 = 0$

Solution We solve for the term y_n with the largest subscript and compute *recursively** the members of the sequence.

(a) Solving for y_{n+1}, we have $y_{n+1} = 1 + 2y_n$. Hence we compute y_1, y_2, y_3, y_4, and y_5 recursively, getting

$$y_1 = 1 + 2y_0 = 1 + 2 \cdot 1 = 3$$
$$y_2 = 1 + 2y_1 = 1 + 2 \cdot 3 = 7$$
$$y_3 = 1 + 2y_2 = 1 + 2 \cdot 7 = 15$$
$$y_4 = 1 + 2y_3 = 1 + 2 \cdot 15 = 31$$
$$y_5 = 1 + 2y_4 = 1 + 2 \cdot 31 = 63$$

---

* The term *recursively* refers to the fact that the difference equation is being used recurrently to find the terms of the sequence. In fact, the equations that we call difference equations are sometimes called *recurrence equations*.

See Figure 7.2(a).

(b) Solving for y_{n+2} in terms of y_{n+1} and y_n gives $y_{n+2} = 1 - ny_{n+1} + y_n$. Hence we compute y_2, y_3, y_4, and y_5 recursively, getting

$$y_2 = 1 - 0 \cdot y_1 + y_0 = 1 - 0 \cdot 0 + 3 = 4$$
$$y_3 = 1 - 1 \cdot y_2 + y_1 = 1 - 1 \cdot 4 + 0 = -3$$
$$y_4 = 1 - 2 \cdot y_3 + y_2 = 1 - 2 \cdot (-3) + 4 = 11$$
$$y_5 = 1 - 3 \cdot y_4 + y_3 = 1 - 3 \cdot 11 - 3 = -35$$

See Figure 7.2(b). ∎

Figure 7.2
The first few members of the solution of the initial-value problem

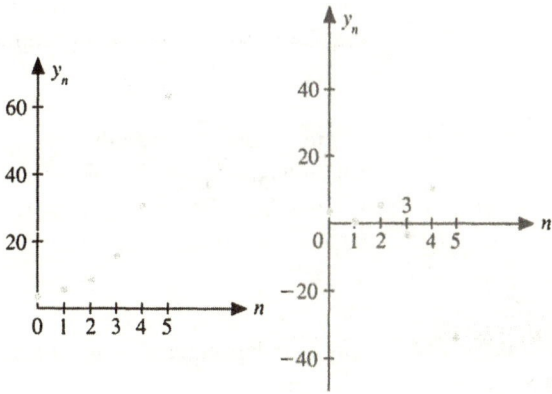

SOME BASIC SOLUTION THEORY

The verification of the existence and uniqueness of a solution to the general first-order differential equation $y' = f(x, y)$ is quite difficult and is almost never given in beginning differential equations texts. On the other hand, verifying the existence and uniqueness of solutions of difference equations is trivial. For example, in Example 3 we found the first five members of the solution to both a first- and a second-order difference equation by repeated use of the equation. By continuing this process again and again, it is possible to find *any* member of the solution. This leads us to a general existence and uniqueness theorem for difference equations, which we state for first- and second-order equations.

> **THEOREM 7.1:** Existence and Uniqueness of Solutions of Difference Equations
>
> Let a_n, b_n, c_n, and f_n be given sequences for $n \geq 0$, and let A and B be given constants. Each of the first- and second-order initial-value problems
>
> $$a_n y_{n+1} + b_n y_n = f_n \qquad\qquad y_0 = A$$
> $$a_n y_{n+2} + b_n y_{n+1} + c_n y_n = f_n \qquad y_0 = A \quad y_1 = B$$
>
> has a unique solution, provided that $a_n \neq 0$ for all $n = 0, 1, 2, \dots$.

The reader should note that although we state Theorem 7.1 for first- and second-order equations, analogous results hold for higher-order equations.

GENERAL THEORY OF LINEAR HOMOGENEOUS EQUATIONS

In Chapter 3 we saw that if $y_1(x)$ and $y_2(x)$ were solutions of a linear homogeneous differential equation, then so was any linear combination of these solutions. It is possible to show (see Problem 31 at the end of this section) that if two *sequences* $\{u_n\}$ and $\{v_n\}$ are solutions of a linear homogeneous *difference* equation

$$A_n y_{n+2} + B_n y_{n+1} + C_n y_n = 0 \tag{4}$$

then so is any linear combination of these sequences. By a linear combination of two sequences we mean a sequence having the form

$$c_1\{u_n\} + c_2\{v_n\} = \{c_1 u_n + c_2 v_n\}$$

where c_1 and c_2 are arbitrary constants. The next question is: Can we find *all* the solutions of Eq. (4) by taking all linear combinations of two "different" solutions of Eq. (4). The answer to this crucial question is based on the concept of linearly independent sequences.

LINEAR INDEPENDENCE OF SEQUENCES

Linear independence of sequences is analogous to linear independence of functions.

> **DEFINITION: Linear Independence of Sequences**
> Two sequences $\{u_n\}$ and $\{v_n\}$, $n \geq 0$, are **linearly independent sequences** if
> $$Au_n + Bv_n = 0 \tag{5}$$
> for all $n = 0, 1, 2, \ldots$ implies that $A = B = 0$. If, on the other hand, there exist constants B and A, not both zero, such that Eq. (5) holds for all $n \geq 0$, then the sequences are **linearly dependent sequences.**

The concept of linear independence and dependence is very simple for only two sequences. Two sequences are linearly independent if one is not a constant multiple of the other. If one is a constant multiple of the other, they are linearly dependent.

Example 4 **Checking Independence or Dependence** Determine whether the following sequences are linearly independent.

(a) $\{2^n\}$ and $\{3^n\}$

(b) $\{3n + 1\}$ and $\{4n + \frac{4}{3}\}$

Solution (a) If

$$A(2^n) + B(3^n) = 0 \tag{6}$$

for all $n = 0, 1, 2, \ldots$, then the relationship holds for the specific values $n = 0$

and 1. Substituting these values into Eq. (6), we get $A + B = 0, 2A + 3B = 0$. But since the only solution of these equations is $A = B = 0$, we have that $\{2^n\}$ and $\{3^n\}$ are linearly independent sequences.

(b) Since

$$4(3n + 1) - 3\left(4n + \tfrac{4}{3}\right) = 0$$

for $n = 0, 1, 2, ...$, the two sequences are linearly dependent.   ∎

THE CASORATIAN

In linear differential equation theory the Wronskian provided a useful test of linear independence of two solutions. In difference equations the Casoratian* provides an easy test of linear independence of two sequence solutions.

DEFINITION: Casoratian of Two Sequences

The **Casoratian** $C(u_n, v_n)$ of two sequences $\{u_n\}$ and $\{v_n\}$ is the sequence defined by the determinant

$$C(u_n, v_n) = \begin{vmatrix} u_n & v_n \\ u_{n+1} & v_{n+1} \end{vmatrix} = u_n v_{n+1} - u_{n+1} v_n \qquad (7)$$

for $n = 0, 1, 2, \ldots$.

It can be shown (see Problem 32 at the end of this section) that if $\{u_n\}$ and $\{v_n\}$ are solutions of a linear homogeneous equation, then their Casoratian $C(u_n, v_n)$ is either *always* zero for all n or *never* zero. In the case in which the Casoratian is never zero, the two solutions $\{u_n\}$ and $\{v_n\}$ are linearly independent; otherwise, they are linearly dependent.

The previous discussion leads us to the following important theorem.

THEOREM 7.2: Representation of Solutions of the Homogeneous Equation

If $\{u_n\}$ and $\{v_n\}$ are any two linear independent solutions of

$$a_n y_{n+2} + b_n y_{n+1} + c_n y_n = 0 \qquad (8)$$

then *every* solution $\{w_n\}$ of Eq. (8) can be expressed as

$$\{w_n\} = c_1\{u_n\} + c_2\{v_n\}$$

where c_1 and c_2 are constants.

The two linear independent solutions $\{u_n\}$ and $\{v_n\}$ are called a **fundamental set of solutions,** and the set of all linear combinations of these solutions is called the **general solution** of Eq. (8).

---

* The *Casoratian sequence* is analogous to the *Wronskian* of linear differential equation theory. It is named after the Italian mathematician Felice Casorati (1835–1890).

PROOF: Since $\{u_n\}$ and $\{v_n\}$ are solutions of Eq. (8), then so is the linear combination $c_1\{u_n\} + c_2\{v_n\}$. The goal is to find c_1 and c_2 so that $\{w_n\} = c_1\{u_n\} + c_2\{v_n\}$, where $\{w_n\}$ is an arbitrary solution of Eq. (8). Certainly, it is no problem to find c_1 and c_2 that satisfy the *first two terms*

$$w_0 = c_1 u_0 + c_2 v_0$$
$$w_1 = c_1 u_1 + c_2 v_1 \tag{9}$$

since by the linear independence hypothesis, *no* term of the Casoratian sequence $\{u_n v_{n+1} - u_{n+1} v_n\}$ is zero, and hence $u_0 v_1 - u_1 v_0 \neq 0$. But this says that both sequences $\{w_n\}$ and $c_1\{u_n\} + c_2\{v_n\}$ satisfy the *same* difference equation and initial conditions. Hence by the existence and uniqueness conclusion of Theorem 7.1, they must be the same sequence. This completes the proof.

Example 5   **Test for Fundamental Set**   Determine whether the sequences $\{u_n\} = \{1^n\}$ and $\{v_n\} = \{(-1)^n\}$ form a fundamental set of solutions of

$$y_{n+2} - y_n = 0 \tag{10}$$

If so, find the general solution of Eq. (10).

Solution   Since the first member of the Casoratian sequence of $\{u_n\}$ and $\{v_n\}$ is

$$u_0 v_1 - u_1 v_0 = 1 \cdot (-1) - 1 \cdot 1 = -2 \neq 0$$

the sequences $\{1^n\} = \{1\}$ and $\{(-1)^n\}$ are linearly independent solutions and hence form a fundamental set of solutions. The general solution is given by the sequence $y_n = c_1 1^n + c_2(-1)^n = c_1 + c_2(-1)^n$. ∎

PROBLEMS: Section 7.1

For Problems 1–5, write out the first five terms of the sequences. If you have access to a computer, you can find more terms.

1. $y_n = 3 \cdot \left(\dfrac{1}{2}\right)^n$ $(n \geq 0)$

2. $y_n = 2^n + (-1)^n$ $(n \geq 0)$

3. $y_n = 1^n + (-1)^n$ $(n \geq 0)$

4. $y_n = \cos\left(\dfrac{n\pi}{2}\right) + \sin\left(\dfrac{n\pi}{2}\right)$ $(n \geq 0)$

5. $y_n = \left(\dfrac{1}{2}\right)^n \left\{\cos(n\pi) + \sin(n\pi)\right\}$ $(n \geq 0)$

For Problems 6–15, write out the first five terms of the sequences and tell whether the sequences are linearly independent or linearly dependent.

6. $\{1^n\}$ $\{(-1)^n\}$

7. $\{1\}$ $\{2^n\}$

8. $\{2^n\}$ $\{3^n\}$

9. $\{2n\}$ $\{n + 1\}$

10. $\{\cos n\pi\}$ $\{1\}$

11. $\{5n + 1\}$ $\{2n + 4\}$

12. $\{5n + 1\}$ $\{an + b\}$

13. $\{(-1)^n\}$ $\{1^n\}$

14. $\{2^n \cos n\pi\}$ $\{\cos n\pi\}$

15. $\{2^n\}$ $\{n\, 2^n\}$

For Problems 16–25, write out the difference equation for values $n = 0$ through 5. Then find the first five terms of the solution of the given initial-value problem. If you have a computer, you can find more terms.

16. $y_{n+1} - y_n = 0$ $y_0 = 1$

17. $y_{n+1} - y_n = 1$ $y_0 = 2$

18. $y_{n+1} - y_n = 0.25 y_n$ $y_0 = 1$

19. $y_{n+1} = \frac{1}{2}\left(1 + \frac{2}{y_n}\right)$ $y_0 = 1$ (Newton's method)

20. $y_{n+1} = 1 + \frac{1}{y_n}$ $y_0 = 1$

21. $y_{n+1} - y_n = n + 1$ $y_0 = 0$

22. $y_{n+2} - y_{n+1} + y_n = 1$ $y_0 = 1$ $y_1 = 0$

23. $y_{n+2} + y_n = 0$ $y_0 = 1$ $y_1 = 0$

24. $y_{n+2} - y_n = 0$ $y_0 = 0$ $y_1 = 1$

25. $y_{n+2} = y_{n+1} + y_n$ $y_0 = 1$ $y_1 = 1$

For Problems 26–30, determine which pairs of sequences form a fundamental set of solutions of the given difference equation by computing the first member of the Casoratian sequence.

Difference equation		Solutions
26. $y_{n+2} - 2y_{n+1} + y_n = 0$	$\{1\}$	$\{n\}$
27. $y_{n+2} - 2y_{n+1} + y_n = 0$	$\{n\}$	$\{2n\}$
28. $2y_{n+2} + 5y_{n+1} - 3y_n = 0$	$\{2^{-n}\}$	$\{(-3)^n\}$
29. $y_{n+2} + 6y_{n+1} + 9y_n = 0$	$\{(-3)^n\}$	$\{n(-3)^n\}$
30. $y_{n+2} + y_n = 0$	$\{\cos(n\pi/2)\}$	$\{\sin(n\pi/2)\}$

31. Superposition of Solutions Let $\{u_n\}$ and $\{v_n\}$ be two solutions of

$$A_n y_{n+2} + B_n y_{n+1} + C_n y_n = 0$$

where A_n, B_n, and C_n are given sequences. Show that any linear combination $c_1 u_n + c_2 v_n$ is also a solution, where c_1 and c_2 are arbitrary constants.

32. Important Property of the Casoratian Let $\{u_n\}$ and $\{v_n\}$ be two solutions of the linear homogeneous equation written in the form

$$y_{n+2} + p y_{n+1} + q y_n = 0 \qquad (11)$$

where p and q and constants.

(a) Show that the Casoratian sequence defined by the equation $c_n = C(u_n, v_n) = u_n v_{n+1} - u_{n+1} v_n$ satisfies the difference equation $c_{n+1} = q c_n$.

(b) Solve the first-order difference equation found in part (a) for the Casoratian, getting $c_n = C q^n$, where C is an arbitrary constant.

This problem verifies the fact that the Casoratian sequence $\{c_n\}$ is either identically zero $(C = 0)$ or is never zero $(C \neq 0)$.*

33. Difference Equation with Variable Coefficients Show that the difference equation

$$y_n + n y_{n-1} = 1$$

is satisfied by the integral

$$y_n = \int_0^1 x^n e^{x-1} \, dx$$

34. Journal Entry Summarize to yourself the importance of difference equations as a model for describing physical systems.

7.2   Homogeneous Equations

PURPOSE

To find the general solution of the linear homogeneous equation with constant coefficients

$$a y_{n+2} + b y_{n+1} + c y_n = 0$$

Consider the linear second-order homogeneous difference equation with constant coefficients

$$a y_{n+2} + b y_{n+1} + c y_n = 0 \qquad (n = 0, 1, 2, ...) \qquad (1)$$

In Chapter 3 we solved the analogous differential equation $a y'' + b y' + c y = 0$ by trying solutions of the form $y(x) = e^{mx}$. The same methodology works here for Eq.

* This result also holds for linear homogeneous equations with *variable* coefficients, although the proof is more difficult.

(1), although it is more convenient to write the solution in the variant form $y_n = r^n$, where $r \neq 0$ is an unknown constant.* Substituting $y_n = r^n$ into Eq. (1) gives

$$ar^{n+2} + br^{n+1} + cr^n = r^n[ar^2 + br + c] = 0$$

But this equation holds for all $n = 0, 1, 2, ...$ if and only if r satisfies the **characteristic equation**

$$ar^2 + br + c = 0 \qquad (2)$$

Hence we are led to a situation similar to when we solved the second-order differential equation. We summarize the results in Theorem 7.3.

THEOREM 7.3: General Solution of $ay_{n+2} + by_{n+1} + cy_n = 0$

The general solution of the difference equation

$$ay_{n+2} + by_{n+1} + cy_n = 0 \qquad (n = 0, 1, 2, ...) \qquad (3)$$

depends on the two roots r_1 and r_2 of the characteristic equation

$$ar^2 + br + c = 0$$

Real and Distinct Roots $(r_1 \neq r_2)$. If r_1 and r_2 are distinct real roots, then the general solution of Eq. (3) is

$$y_n = c_1 r_1{}^n + c_2 r_2{}^n \qquad (n = 0, 1, 2, ...) \qquad (4a)$$

Real and Equal Roots $(r_1 = r_2)$. If $r_1 = r_2 = -b/2a$, then the general solution of Eq. (3) is

$$y_n = c_1 r^n + c_2 n r^n \qquad (n = 0, 1, 2, ...) \qquad (4b)$$

Complex Conjugate Roots $(r_1 = re^{i\theta}, r_2 = re^{-i\theta})$. If the roots are complex conjugates, $r_1 = p + iq$ and $r_2 = p - iq$, written in polar form as $r_1 = re^{i\theta}$ and $r_2 = re^{-i\theta}$, where $r = \sqrt{p^2 + q^2}$ and $\tan \theta = q/p$, then the general solution of Eq. (3) is

$$y_n = r^n [c_1 \cos n\theta + c_2 \sin n\theta] \qquad (n = 0, 1, 2, ...) \qquad (4c)$$

PROOF: When the characteristic equation has real roots, the solutions of the difference equation are basically the same as the solutions of the second-order differential equation, and we omit the proofs for these cases. In the case when the roots have complex conjugate values $r_1 = p + iq$, $r_2 = p - iq$, we use $p + iq = re^{i\theta}$ and $p - iq = e^{-i\theta}$ and rewrite the solution in polar form. We get

---

* The function r^n can be rewritten in the form e^{kn} by writing $r^n = e^{n \ln r} = e^{kn}$, where $k = \ln r$.

$$y_n = c_1(p + iq)^n + c_2(p - iq)^n$$

$$= c_1(re^{i\theta})^n + c_2(re^{-i\theta})^n$$

$$= c_1 r^n e^{in\theta} + c_2 r^n e^{-in\theta}$$

$$= c_1 r^n(\cos n\theta + i \sin n\theta) + c_2 r^n(\cos n\theta - i \sin n\theta)$$

$$= (c_1 + c_2)r^n \cos n\theta + (ic_1 - ic_2)r^n \sin n\theta$$

$$= C_1 r^n \cos n\theta + C_2 r^n \sin n\theta$$

where $C_1 = c_1 + c_2$, $C_2 = ic_1 - ic_2$. Since c_1 and c_2 are arbitrary constants, then so are C_1 and C_2. Hence we have found two new linearly independent solutions, $r^n \cos n\theta$ and $r^n \sin n\theta$. This proves the theorem.

We summarize these ideas in Table 7.1.

Table 7.1
Solutions of
$ay_{n+2} + by_{n+1} + cy_n = 0$

Nature of roots of the characteristic equation $ar^2 + br + c = 0$	General solution
Real and Unequal ($r_1 \neq r_2$)	$y_n = c_1 r_1^n + c_2 r_2^n$
Real and Equal ($r_1 = r_2 = r$)	$y_n = c_1 r^n + c_2 n r^n$
Conjugate Complex	$y_n = r^n[c_1 \cos n\theta + c_2 \sin n\theta]$
$\quad r_1 = p + iq$	
$\quad r_2 = p - iq$	$r = \sqrt{p^2 + q^2}$, $\tan \theta = q/p$

Example 1   Real Unequal Roots   Solve the initial-value problem

$$y_{n+2} + y_{n+1} - 6y_n = 0 \qquad y_0 = 1 \quad y_1 = 0 \tag{5}$$

Solution   We first find the general solution of the difference equation. The characteristic equation is

$$r^2 + r - 6 = (r - 2)(r + 3) = 0$$

which has roots $r_1 = 2$, $r_2 = -3$. Hence the general solution is

$$y_n = c_1 2^n + c_2(-3)^n \qquad (n = 0, 1, 2, \dots) \tag{6}$$

Substituting this expression into the initial conditions gives

$$y_0 = c_1 + c_2 = 1$$
$$y_1 = 2c_1 - 3c_2 = 0$$

which has the solution $c_1 = \dfrac{3}{5}$, $c_2 = \dfrac{2}{5}$. Hence the solution of the initial-value problem is

$$y_n = \frac{3}{5}2^n + \frac{2}{5}(-3)^n \qquad (n = 0, 1, 2, \ldots)$$

A few terms of this sequence are computed in Table 7.2.   ∎

Table 7.2
Solution of
$y_{n+2} + y_{n+1} - 6y_n = 0$
$y_0 = 1, y_1 = 0$

n	$y_n = \dfrac{3}{5}2^n + \dfrac{2}{5}(-3)^n$
0	1
1	0
2	6
3	-6
4	42
5	-78
⋮	⋮

Example 2   Real and Equal Roots   Solve the initial-value problem

$$y_{n+2} + 4y_{n+1} + 4y_n = 0 \qquad y_0 = 0 \quad y_1 = 1 \tag{7}$$

Solution   We first find the general solution of the difference equation. The characteristic equation is

$$r^2 + 4r + 4 = (r + 2)^2 = 0$$

which has the double real root $r_1 = r_2 = -2$. Hence the general solution is

$$y_n = c_1(-2)^n + c_2 n(-2)^n \qquad (n = 0, 1, 2, \ldots) \tag{8}$$

Substituting this expression into the initial conditions gives

$$y_0 = c_1 = 0$$
$$y_1 = -2c_1 - 2c_2 = 1$$

which has the solution $c_1 = 0$, $c_2 = -\dfrac{1}{2}$. Hence the solution of the initial-value problem is

$$y_n = -\frac{1}{2}n(-2)^n \qquad (n = 0, 1, 2, \ldots)$$

A few terms of this sequence are computed in Table 7.3.   ∎

Table 7.3
Solution of
$y_{n+2} + 4y_{n+1} + 4y_n = 0$
$y_0 = 0, y_1 = 1$

n	$y_n = -\dfrac{1}{2}n(-2)^n$
0	0
1	1
2	-4
3	12
4	-32
5	80
⋮	⋮

Example 3 **Complex Roots** Solve the initial-value problem

$$y_{n+2} + 2y_{n+1} + 4y_n = 0 \qquad y_0 = 1 \quad y_1 = 0 \tag{9}$$

Solution Again, we first find the general solution of the difference equation. The characteristic equation is

$$r^2 + 2r + 4 = 0$$

which has roots $r_1, r_2 = -1 \pm i\sqrt{3}$. Writing $-1 + i\sqrt{3}$ in polar form $r^{i\theta}$, we have $r = 2$ and $\theta = \tan^{-1}\left(-\sqrt{3}\right) = 2\pi/3$. Hence the general solution is

$$y_n = 2^n[c_1 \cos(2n\pi/3) + c_2 \sin(2n\pi/3)] \qquad (n = 0, 1, 2, ...) \tag{10}$$

Substituting this expression into the initial conditions gives

$$y_0 = c_1 = 1$$
$$y_1 = 2c_1 \cos(2\pi/3) + 2c_2 \sin(2\pi/3) = 0$$

or $c_1 = 1, c_2 = \sqrt{3}/3$. Hence the solution is

$$y_n = 2^n\left(\cos\left(\frac{2\pi}{3}n\right) + \frac{\sqrt{3}}{3}\sin\left(\frac{2\pi}{3}n\right)\right) \tag{11}$$

A few terms of this sequence are shown in Table 7.4. ∎

Table 7.4
Solution of
$y_{n+2} + 2y_{n+1} + 4y_n = 0$
$y_0 = 1, \quad y_1 = 0$

n	$y_n = 2^n\left(\cos(2n\pi/3) + \dfrac{\sqrt{3}}{3}\sin(2n\pi/3)\right)$
0	1
1	0
2	-4
3	8
4	0
5	-32
⋮	⋮

THE FIRST-ORDER DIFFERENCE EQUATION

The general solution of the first-order difference equation

$$ay_{n+1} + by_n = 0 \tag{12}$$

can be found by trying $y_n = r^n$, obtaining the characteristic equation $ar + b = 0$. Solving for r gives $r = -b/a$, which gives the general solution:

$$y_n = C(-b/a)^n \tag{13}$$

where C is an arbitrary constant. If the first member y_0 of the sequence y_n is given, then the solution of the resulting initial-value problem is

$$y_n = y_0(-b/a)^n \tag{14}$$

Example 4 **First-Order Homogeneous Equation** Find the solution of the initial-value problem

$$y_{n+1} - 0.08y_n = 0 \qquad y_0 = 100$$

Solution The characteristic equation is $r - 0.08 = 0$, which has the root $r = 0.08$. Hence the general solution of the difference equation is $y_n = C(0.08)^n$, and the solution of the initial-value problem is $y_n = 100(0.08)^n$. ∎

PROBLEMS: Section 7.2

For Problems 1–10, find the general solution of the homogeneous difference equation.

1. $y_{n+1} - 0.5y_n = 0$
2. $y_{n+2} - y_n = 0$
3. $y_{n+2} + y_{n+1} - 6y_n = 0$
4. $y_{n+2} + y_{n+1} - 2y_n = 0$
5. $y_{n+2} - 2y_{n+1} + y_n = 0$
6. $y_{n+2} - y_n - 6y_n = 0$
7. $y_{n+2} + 10y_{n+1} + 25y_n = 0$
8. $y_{n+2} - 2y_{n+1} + 4y_n = 0$
9. $y_{n+2} + 4y_n = 0$
10. $y_{n+2} + y_{n+1} + y_n = 0$

For Problems 11–21, find the solution of the given initial-value problem. Sketch the graph of the first five terms of the solution.

11. $y_{n+1} - 2y_n = 0$ $y_0 = 1$
12. $y_{n+1} - 0.5y_n = 0$ $y_0 = 1$
13. $y_{n+2} + y_n = 0$ $y_0 = 0$ $y_1 = 1$
14. $y_{n+2} + 4y_n = 0$ $y_0 = 0$ $y_1 = 1$
15. $y_{n+2} + \alpha^2 y_n = 0$ $y_0 = 1$ $y_1 = 0$
16. $y_{n+2} - 6y_{n+1} + 9y_n = 0$ $y_0 = 2$ $y_1 = 15$
17. $y_{n+2} - 2y_{n+1} + 2y_n = 0$ $y_0 = 1$ $y_1 = 0$
18. $y_{n+2} - 4y_{n+1} + 4y_n = 0$ $y_0 = 1$ $y_1 = 3$

19. $y_{n+2} - 2y_{n+1} + y_n = 0$ $y_0 = 1$ $y_1 = 3$
20. $y_{n+2} + y_{n+1} + y_n = 0$ $y_0 = 0$ $y_1 = 1$
21. $y_{n+2} - 2y_{n+1} + 4y_n = 0$ $y_0 = 0$ $y_1 = 1$

The ideas of first- and second-order difference equations extend readily to higher-order equations. For Problems 22–27, find the general solution to the given difference equation.

22. $y_{n+3} + y_{n+2} - 4y_{n+1} - 4y_n = 0$
23. $y_{n+3} - 3y_{n+2} + 4y_{n+1} - 2y_n = 0$
24. $y_{n+4} - 4y_{n+3} + 6y_{n+2} - 4y_{n+1} + y_n = 0$
25. $y_{n+4} - y_n = 0$
26. $y_{n+4} + 2y_{n+2} + y_n = 0$
27. $y_{n+4} - 8y_{n+2} + 16y_n = 0$

28. **Gambler's Ruin** Mary and John are flipping pennies; between the two of them, they have N pennies. We let n denote the number of pennies that Mary has at any time during play and p_n the probability that Mary will be ruined given that she has n pennies. In this case, p_n satisfies

$$p_n = \frac{1}{2}p_{n+1} + \frac{1}{2}p_{n-1} \qquad (n = 1, 2, \dots, N-1) \tag{15}$$

If $n = 0$, then Mary is ruined, and no further play takes place. Also, if $n = N$, then John is ruined, and no further play takes place. Hence we require the two boundary conditions $p_0 = 1$ and $p_N = 0$. Solve the resulting difference equation and boundary conditions for p_n.

Comparison Between Difference and Differential Equations

For Problems 29–33, first solve the given differential equation with its initial condition. Then approximate the differential equation by a difference equation by replacing the derivative by

$$\frac{dy}{dt} \doteq \frac{y_{n+1} - y_n}{h} \qquad (16)$$

where $y_n = y(nh)$, $n = 0, 1, 2, \ldots$. For example the initial-value problem

$$y' = y \qquad y(0) = 1 \qquad (17)$$

is approximated by the discrete initial-value problem

$$\frac{y_{n+1} - y_n}{h} = y_n \qquad y_0 = 1 \qquad (18)$$

or, equivalently,

$$y_{n+1} = (1 + h)y_n \qquad y_0 = 1 \qquad (19)$$

Solve the difference equation with $h = 0.1$, either analytically or numerically, and compare this with the solution of the differential equation.

29. $\dfrac{dy}{dt} = y \qquad\qquad y(0) = 1$

30. $\dfrac{dy}{dt} = -y \qquad\qquad y(0) = 1$

31. $\dfrac{dy}{dt} = y^2 \qquad\qquad y(0) = 1$

32. $\dfrac{dy}{dt} = y(1 - y) \quad y(0) = 1$

33. $\dfrac{dy}{dt} = e^{-y} \qquad\qquad y(0) = 1$

34. Three Difference Equations The decay equation

$$\frac{dy}{dt} = -y \qquad y(0) = y_0$$

can be approximated by many difference equations.* Two common ones are

$$\frac{y_{n+1} - y_n}{h} = -y_n \qquad \text{(forward Euler equation)}$$

* There exists a huge amount of literature on the approximate solution of differential equations by difference equations. The interested reader can consult *Computational Methods in Ordinary Differential Equations* by J. D. Lambert (Wiley, New York, 1973).

$$\frac{y_n - y_{n-1}}{h} = -y_n \qquad \text{(backward Euler equation)}$$

Solve these difference equations and compare the solution with the decay equation for $y_0 = 1$ and $h = 0.1$.

35. Fibonacci Sequence Possibly the most famous difference equation is

$$F_{n+2} = F_n + F_{n+1} \qquad F_0 = 1 \quad F_1 = 1 \qquad (20)$$

whose solution $\{F_n\} = \{1, 1, 2, 3, 5, 8, 13, \ldots\}$ yields what are called **Fibonacci numbers.**[†] Find the general expression for the nth Fibonacci number F_n. The limiting value of F_{n+1}/F_n as $n \to \infty$ is called the *golden mean* or ratio and, according to ancient Greek mathematicians, represents the ratio of the sides of a rectangle that is the "most pleasing" to the eye. Approximate the golden mean.

Difference Equations of Functions

Difference equations do not always represent the relation between members of a sequence of numbers, but of a sequence of functions $\{y_0(x), y_1(x), y_2(x), \ldots\}$. The following problem illustrates difference equations involving sequences of functions.

36. Chebyshev Polynomials An important class of functions is the Chebyshev polynomials, $\{T_0(x), T_1(x), T_2(x), \ldots\}$, where each $T_n(x)$ is defined on the interval $-1 \le x \le 1$ and satisfies the second-order difference equation

$$T_{n+2} - xT_{n+1} + \frac{1}{4}T_n = 0 \qquad (21)$$

(a) If $T_0(x) = 2$ and $T_1(x) = x$, show that

$$T_n(x) = \frac{1}{2^n}\left(\left(x + \sqrt{x^2 - 1}\right)^n + \left(x - \sqrt{x^2 - 1}\right)^n\right)$$

Hint: Treat the coefficient of T_{n+1} in Eq. (21) as a constant and proceed normally.

(b) Define the relationship $\tan \theta = \sqrt{x^2 - 1}/x$ and rewrite $T_n(x)$ in the form

$$T_n(x) = \frac{1}{2^{n-1}} \cos (n \cos^{-1} x) \qquad (n = 0, 1, 2, \ldots)$$

37. Computer Problem Use the equation in Problem 36(a) to plot the Chebyshev polynomial $T_2(x)$ at several points in the interval $-1 \le x \le 1$.

38. Journal Entry Summarize the results of this section.

[†] The Fibonacci numbers were introduced into mathematics by the Italian Leonardo Fibonacci (1170?–1250?), the most talented mathematician during the Middle Ages.

7.3 · NONHOMOGENEOUS EQUATIONS

PURPOSE

To find the general solution of the first- and second-order linear nonhomogeneous equations with constant coefficients

$$ay_{n+1} + by_n = f_n$$
$$ay_{n+2} + by_{n+1} + cy_n = f_n$$

by the method of undetermined coefficients. We also introduce the $\mathbb{Z}$-transform, which is the discrete version of the Laplace transform which is often used for solving difference equations.

GENERAL THEORY OF NONHOMOGENEOUS EQUATIONS

Consider the linear nonhomogeneous difference equation with constant coefficients

$$ay_{n+2} + by_{n+1} + cy_n = f_n \tag{1}$$

As in the case of differential equations, the general solution of Eq. (1) is the sum of the general solution of the corresponding homogeneous equation plus any particular solution of Eq. (1). We state this important result and leave the proof to the reader. (See Problem 14 at the end of this section.)

THEOREM 7.4: Representation of Nonhomogeneous Solutions

If $\{p_n\}$ is a particular solution of the nonhomogeneous equation

$$a_n y_{n+2} + b_n y_{n+1} + c_n y_n = f_n \qquad (n = 0, 1, 2, ...)$$

$(a_n \neq 0)$, then *any* solution of this equation* can be written in the form

$$\{y_n\} = \{h_n\} + \{p_n\} \qquad (n = 0, 1, 2, ...)$$

where $\{h_n\}$ is the general solution of the corresponding homogeneous equation.

Again, we are led to finding a particular solution of the nonhomogeneous equation. As we saw in Chapter 3, the two standard methods for finding a particular solution of a nonhomogeneous linear differential equation are the method of undetermined coefficients and the method of variation of parameters. Although there are "discrete ver-

---

* Although Theorem 7.4 is stated only for the second-order linear difference equation, the theorem holds for any order.

sions'' of both these techniques, we will study only the method of undetermined coefficients.*

DISCRETE METHOD OF UNDETERMINED COEFFICIENTS

As it was for differential equations, the method of undetermined coefficients allows one to find a particular solution of a difference equation when the nonhomogeneous term has a special form. Table 7.5 gives the necessary rules for finding a particular solution.

Table 7.5
Particular Solution of
$ay_{n+2} + by_{n+1} + cy_n = f_n$

f_n*	Form of the particular solution†
1. a (constant)	A
2. an^k (k a given integer)	$A_0 n^k + A_1 n^{k-1} + \cdots + A_{k-1}n + A_k$
3. ab^n	Ab^n
4. $a \cos(bn)$	$A \cos(bn) + B \sin(bn)$
5. $a \sin(bn)$	$A \cos(bn) + B \sin(bn)$

* When f_n consists of several terms, the particular solution is the sum of the particular solutions corresponding to each term in f_n.

† Whenever a term in the particular solution duplicates a term in the homogeneous solution, all terms in the particular solution must be multiplied by the lowest positive integer power of n to ensure that no duplications exist.

Example 1 **Polynomial Forcing Term** Find the general solution of

$$y_{n+2} - 4y_{n+1} + 4y_n = n + 1 \tag{2}$$

Solution The characteristic equation of the corresponding homogeneous equation is the quadratic equation $r^2 - 4r + 4 = 0$, which has the double root $r_1 = r_2 = 2$. Hence the general solution of the corresponding homogeneous equation is

$$h_n = c_1 2^n + c_2 n 2^n \qquad (n = 0, 1, 2, ...) \tag{3}$$

Using Table 7.5, we try a particular solution of the form $p_n = An + B$. Substituting p_n into Eq. (2), we find

$$An + (-2A + B) = n + 1 \qquad (n = 0, 1, 2, ...)$$

Since these equations must hold for all n, they hold for $n = 0, 1$. Hence we get the equations $A = 1$ and $-2A + B = 1$, which give $A = 1$, $B = 3$. This gives the particular solution $p_n = n + 3$, which gives the general solution of Eq. (2):

$$y_n = c_1 2^n + c_2 n 2^n + n + 3 \qquad (n = 0, 1, 2, ...) \tag{4}$$

■

Example 2 **Exponential Forcing Term** Find the general solution of

$$y_{n+2} - 3y_{n+1} + 2y_n = 2^n \qquad (n = 0, 1, 2, ...) \tag{5}$$

* An excellent treatment of the method of variation of parameters for difference equations can be found in *An Introduction to Differential Equations* by N. Finzio and G. Ladas (Wadsworth, Belmont, Calif., 1982).

Solution The characteristic equation of the corresponding homogeneous equation is the quadratic equation $r^2 - 3r + 2 = 0$, which has roots $r_1 = 1$, $r_2 = 2$. Hence the general solution of the homogeneous equation is

$$h_n = c_1 + c_2 2^n \qquad (n = 0, 1, 2, ...) \tag{6}$$

To find a particular solution, we are tempted to try $p_n = A2^n$. However, the sequence 2^n occurs in the homogeneous solution, and so Table 7.5 tells us to multiply by n, and we get $p_n = An \cdot 2^n$. Substituting this sequence into Eq. (5), we obtain

$$A(n + 2)2^{n+2} - 3A(n + 1)2^{n+1} + 2An2^n = 2^n \tag{7}$$

Simplifying this equation gives

$$[4(n + 2)A - 6(n + 1)A + 2nA]\, 2^n = 2^n$$

or

$$(4A - 6A + 2A)n + (8A - 6A) = 1$$

This equation simply says $2A = 1$, or $A = 1/2$. Hence $p_n = n2^{n-1}$ is a particular solution, and so the general solution is

$$y_n = c_1 + c_2 2^n + n2^{n-1} \qquad (n = 0, 1, 2, ...) \tag{8}$$

■

Example 3 **Trigonometric Forcing Term** Find the general solution of

$$y_{n+2} - y_n = \sin\left(\frac{n\pi}{2}\right) \qquad (n = 0, 1, 2, ...) \tag{9}$$

Solution The characteristic equation of the corresponding homogeneous equation is the quadratic equation $r^2 - 1 = 0$, which has roots of $r_1 = 1$, $r_2 = -1$. Hence the general solution of the homogeneous equation is

$$h_n = c_1 + c_2(-1)^n \qquad (n = 0, 1, 2, ...) \tag{10}$$

For a particular solution p_n we try $p_n = A \cos(n\pi/2) + B \sin(n\pi/2)$. Substituting p_n into Eq. (9), we have

$$A \cos\left(\frac{(n + 2)\pi}{2}\right) + B \sin\left(\frac{(n + 2)\pi}{2}\right) - A \cos\left(\frac{n\pi}{2}\right) - B \sin\left(\frac{n\pi}{2}\right) = \sin\left(\frac{n\pi}{2}\right)$$

or

$$-A \cos\left(\frac{n\pi}{2}\right) - B \sin\left(\frac{n\pi}{2}\right) - A \cos\left(\frac{n\pi}{2}\right) - B \sin\left(\frac{n\pi}{2}\right) = \sin\left(\frac{n\pi}{2}\right)$$

or

$$-2A \cos\left(\frac{n\pi}{2}\right) - 2B \sin\left(\frac{n\pi}{2}\right) = \sin\left(\frac{n\pi}{2}\right)$$

Setting the coefficients of the sine and cosine functions equal to each other, we get

$-2A = 0$ and $-2B = 1$, which give $A = 0$, $B = -1/2$. Hence a particular solution is

$$p_n = -\frac{1}{2} \sin\left(\frac{n\pi}{2}\right) \qquad (n = 0, 1, 2, ...) \tag{11}$$

and the general solution of Eq. (9) is

$$y_n = c_1 + c_2(-1)^n - \frac{1}{2} \sin\left(\frac{n\pi}{2}\right) \qquad (n = 0, 1, 2, ...) \tag{12}$$

■

We close this section by introducing the "discrete cousin" of the Laplace transform, the $\mathbb{Z}$-transform. It is an important tool for systems engineers, who are often interested in discrete time systems modeled by difference equations rather than continuous time systems modeled by differential equations.

The $\mathbb{Z}$-Transform

We have seen that the Laplace transform is a valuable tool for solving linear differential equations with constant coefficients and discontinuous inputs. Although the Laplace transform cannot be used to solve difference equations, there is an analogous transform, called the **$\mathbb{Z}$-transform,** that can be used to solve difference equations. To understand the $\mathbb{Z}$-transform, we begin with a sequence of numbers $\{y_0, y_1, y_2, ...\}$ and construct an infinite series of *descending* powers of a variable z as follows:

$$\mathbb{Z}\{y_n\} = \sum_{n=0}^{\infty} y_n z^{-n} = y_0 + \frac{y_1}{z} + \frac{y_2}{z^2} + \frac{y_3}{z^3} + \cdots \tag{13}$$

This series will converge for $|z| > R$ for some number R depending on $\{y_n\}$ and on this region defines a function of z.* This function of z is called the $\mathbb{Z}$-transform of the sequence $\{y_n\}$.

> **DEFINITION: The $\mathbb{Z}$-Transform**
> The $\mathbb{Z}$-transform of the sequence $\{y_0, y_1, y_2, ...\}$ is the function $F(z)$ defined by the infinite series
>
> $$F(z) = \mathbb{Z}\{y_n\} = y_0 + \frac{y_1}{z} + \frac{y_2}{z^2} + \frac{y_3}{z^3} + \cdots$$

Note: Since the $\mathbb{Z}$-transform is a *mapping* from a class of sequences to a class of functions, it is often denoted by

$$\mathbb{Z}: \{y_n\} \to F(z) \tag{14}$$

Also, since $\mathbb{Z}\{y_n\}$ is a function of z, it is often denoted as $F(z)$ in the same way that the Laplace transform $\mathfrak{L}\{f\}$ was denoted by $F(s)$.

---

* The variable z can be interpreted as either a real or a complex variable. However, our discussion is formal, and we will not dwell on this matter here.

Example 4 **Finding $\mathbb{Z}$-Transforms** Find the $\mathbb{Z}$-transform of the following sequences.

(a) $\{1\} = \{1, 1, 1, ..., 1, ...\}$ (unit sequence)

(b) $\{y_n\} = \{1, r, r^2, ..., r^n, ...\}$ (geometric sequence)

(c) $\{y_n\} = \{1, 2, 3, ..., n, ...\}$ (arithmetic sequence)

(d) $\{y_n\} = \{0, 1, 1, 1, ..., 1, ...\}$ (step sequence)

Solution

(a) $\mathbb{Z}\{1\} = 1 + \dfrac{1}{z} + \dfrac{1}{z^2} + \cdots = \dfrac{1}{1 - z^{-1}} = \dfrac{z}{z - 1}$ $|z| > 1$

(b) $\mathbb{Z}\{y_n\} = 1 + \dfrac{r}{z} + \dfrac{r^2}{z^2} + \cdots = \dfrac{1}{1 - rz^{-1}} = \dfrac{z}{z - r}$ $|z| > r$

(c) $\mathbb{Z}\{y_n\} = 1 + \dfrac{2}{z} + \dfrac{3}{z^2} + \cdots = \dfrac{z}{(1 - z)^2}$ $|z| > 1$

(d) $\mathbb{Z}\{y_n\} = 0 + \dfrac{1}{z} + \dfrac{1}{z^2} + \dfrac{1}{z^3} + \cdots = \dfrac{1}{z}\dfrac{z}{z - 1} = \dfrac{1}{z - 1}$ $|z| > 1$ ■

Note that the $\mathbb{Z}$-transforms in Example 4 have been written in closed form. The ability to write many $\mathbb{Z}$-transforms in closed form is crucial to its usefulness. Table 7.6 lists the $\mathbb{Z}$-transform for a few important sequences. The sequence $U_k(r)$, called the **shifted geometric sequence,** has the first k members 0's followed by the terms entries 1, r, r^2, ... For example $U_0(r) = \{1, r, r^2, ...\}$, $U_1(r) = \{0, 1, r, r^2, ...\}$, $U_2(r) = \{0, 0, 1, r, r^2, ...\}$, and so on.

Table 7.6
Typical $\mathbb{Z}$-Transforms

$\{y_n\}$	$F(z)$
1. $\{1\}$	$\dfrac{z}{z - 1}$
2. $\{r^n\}$	$\dfrac{z}{z - r}$
3. $\{n\}$	$\dfrac{z}{(z - 1)^2}$
4. $\{n^2\}$	$\dfrac{z(z + 1)}{(z - 1)^3}$
5. $U_k(r)$	$\dfrac{z}{z^k(z - r)}$
6. $\{\sin bn\}$	$\dfrac{z \sin b}{z^2 - 2z \cos b + 1}$
7. $\{\cos bn\}$	$\dfrac{z(z - \cos b)}{z^2 - 2z \cos b + 1}$

Although there are tables of inverse $\mathbb{Z}$-transforms,* one can often use the definition to find the inverse transform.

Example 5 **Inverse $\mathbb{Z}$-Transform** Find the inverse $\mathbb{Z}$-transform of the following functions.

(a) $F(z) = \dfrac{1}{z - 1}$

(b) $F(z) = \dfrac{z^2}{z^2 - 1}$

Solution (a) First dividing the numerator and denominator by z, we get

$$F(z) = \frac{1}{z - 1}$$

$$= \frac{1}{z} \cdot \frac{1}{1 - 1/z}$$

$$= \frac{1}{z}\left(1 + \frac{1}{z} + \frac{1}{z^2} + \cdots\right)$$

$$= 0 + \frac{1}{z} + \frac{1}{z^2} + \frac{1}{z^3} + \cdots$$

Hence the inverse transform is $\{0, 1, 1, 1, \ldots\}$.

(b) Using a partial fraction decomposition, we write

$$F(z) = \frac{z^2}{z^2 - 1}$$

$$= \frac{1}{2}\left(\frac{z}{z - 1}\right) + \frac{1}{2}\left(\frac{z}{z + 1}\right)$$

Hence

$$y_n = \mathbb{Z}^{-1}\left(\frac{1}{2}\left(\frac{z}{z - 1}\right) + \frac{1}{2}\left(\frac{z}{z + 1}\right)\right)$$

$$= \frac{1}{2}\left(1\right)^n + \frac{1}{2}\left(-1\right)^n$$

or

$$\{y_n\} = \{1, 0, 1, 0, 1, 0, \ldots\} \qquad \blacksquare$$

PROPERTIES OF THE $\mathbb{Z}$-TRANSFORM

Just as there were important properties of the Laplace transform, there are important properties of the $\mathbb{Z}$-transform.

* An extensive table of $\mathbb{Z}$-transforms can be found in the *Handbook of Mathematical Sciences* by William H. Beyer (CRC Press, Boca Raton, Fla., 1988)

Properties of the ℤ-Transform
The following properties hold for the Z-transform.

Linearity: $\mathbb{Z}\{ay_n + bz_n\} = a\mathbb{Z}\{y_n\} + b\mathbb{Z}\{z_n\}$

Shift Properties: $\mathbb{Z}\{y_{n+1}\} = z\mathbb{Z}\{y_n\} - zy_0$
$\mathbb{Z}\{y_{n+2}\} = z^2\,\mathbb{Z}\{y_n\} - z^2 y_1 - zy_0$

Derivative Property: $\mathbb{Z}\{ny_n\} = -z\dfrac{d}{dz}F(z)$

where

$$\{y_n\} = \{y_0, y_1, y_2, \ldots\}$$
$$\{y_{n+1}\} = \{y_1, y_2, y_3 \ldots\}$$
$$\{y_{n+2}\} = \{y_2, y_3, y_4, \ldots\}$$

PROOF: The first shift property $\mathbb{Z}\{y_{n+1}\} = z\mathbb{Z}\{y_n\} - zy_0$ can be proven by substituting the transforms

$$\mathbb{Z}\{y_n\} = y_0 + \frac{y_1}{z} + \frac{y_2}{z^2} + \cdots$$

$$\mathbb{Z}\{y_{n+1}\} = y_1 + \frac{y_2}{z} + \frac{y_3}{z^2} + \cdots$$

into the equation and observing that equality holds. The proofs of the other properties are left to the reader in the problems set.

The following example illustrates how the ℤ-transform can be used to solve a nonhomogeneous difference equation.

Example 6   **ℤ-Transform Solution of Difference Equations**   Solve the difference equation

$$y_{n+1} + y_n = 2 \qquad y_0 = 0 \tag{15}$$

Solution   Taking the ℤ-transform of each side of the equation and using the fact that it is a linear transform, we get

$$\mathbb{Z}\{y_{n+1} + y_n\} = \mathbb{Z}\{2\}$$
$$\mathbb{Z}\{y_{n+1}\} + \mathbb{Z}\{y_n\} = \mathbb{Z}\{2\}$$

or

$$z\,\mathbb{Z}\{y_n\} + \mathbb{Z}\{y_n\} = \frac{2z}{z-1}$$

Substituting the value $y_0 = 0$ and solving for $\mathbb{Z}\{y_n\}$, we arrive at

$$\mathbb{Z}\{y_n\} = \frac{2z}{(z-1)(z+1)} = \frac{1}{z-1} + \frac{1}{z+1} = \frac{1}{z}\left(\frac{z}{z-1}\right) + \frac{1}{z}\left(\frac{z}{z+1}\right) \quad (16)$$

Using Table 7.6 (inverse No. 5) we see that the factor of $1/z$ in the $\mathbb{Z}$-transform acts to shift the inverse transform to the right k terms, putting a zero in the first k positions. Hence we have

$$
\begin{aligned}
y_n &= \mathbb{Z}^{-1}\left(\frac{1}{z}\frac{z}{z-1}\right) + \mathbb{Z}^{-1}\left(\frac{1}{z}\frac{z}{z+1}\right) \\
&= \{0, 1, 1, 1, 1, ...\} + \{0, 1, -1, 1, -1, ...\} \quad (17) \\
&= \{0, 2, 0, 2, 0, ...\} \qquad\qquad\qquad\qquad \blacksquare
\end{aligned}
$$

PROBLEMS: Section 7.3

For Problems 1–8, find the general solution of the given equation. Find all particular solutions by the method of undetermined coefficients.

1. $y_{n+1} + y_n = 2$
2. $y_{n+1} - 2y_n = n^2$
3. $y_{n+2} + y_n = n^2$
4. $y_{n+2} + 2y_{n+1} + y_n = 1$
5. $y_{n+2} - 2y_{n+1} + y_n = 6n$
6. $y_{n+2} - y_{n+1} - 6y_n = 6n^2$
7. $y_{n+2} - 4y_{n+1} + 4y_n = 2^{n+3}$
8. $y_{n+2} + 4y_n = \cos n$

For Problems 9–12, find the solution of the given initial-value problem. Find all particular solutions by the method of undetermined coefficients.

9. $y_{n+1} - y_n = 2^n$ $\qquad y_0 = 0$
10. $y_{n+2} + y_n = n^2$ $\qquad y_0 = 0 \quad y_1 = 1$
11. $y_{n+2} + 2y_{n+1} + y_n = 1$ $\quad y_0 = 0 \quad y_1 = 0$
12. $y_{n+2} - y_{n+1} - 6y_n = 6n^2$ $\quad y_0 = 0 \quad y_1 = 0$

13. Superposition Principle Let $\{y_n\}$ be a solution of the nonhomogeneous equation (1) with nonhomogeneous term f_n, and let $\{z_n\}$ be a solution of Eq. (1) with nonhomogeneous term g_n. Show that the sequence $\{c_1 y_n + c_2 z_n\}$ is a solution of Eq. (1) with a nonhomogeneous term $f_n + g_n$.

14. General Solution of the Nonhomogeneous Equation Use the representation theorem for homogeneous equations (Theorem 7.2) and the superposition principle (see Problem 13) to prove the representation theorem for nonhomogeneous equations (Theorem 7.4).

15. Boundary-Value Problems for Difference Equations Show that the **boundary-value problem**

$$ay_{n+2} + by_{n+1} + cy_n = f_n \qquad y_0 = A \quad y_N = B$$

has a unique solution y_n for $n = 0, 1, 2, ..., N$ if the roots of the characteristic equation are either equal or distinct and $r_1{}^N \neq r_2{}^N$.

16. Homogeneous Equation with Variable Coefficients It is possible to find the general solution of the equation

$$y_{n+1} = b_n y_n \qquad (n = 0, 1, 2, ...) \quad (18)$$

by repeated use of the equation, writing y_n in terms of y_0. If we treat y_0 as an arbitrary constant, this expression gives the general solution.

(a) Use this strategy to find the general solution of Eq. (18).

(b) Use the solution found in part (a) to solve

$$y_{n+1} - (n+1)y_n = 0 \qquad y_0 = 1$$

17. Alternative Solution of the First-Order Equation Find the solution of the first-order initial-value problem

$$y_{n+1} = ay_n + b \qquad y_0 \text{ given} \quad (19)$$

by solving the following problems.

(a) By repeated use of Eq. (19), write y_n in terms of y_0.

(b) Use the formula

$$1 + a + a^2 + \cdots + a^n = \frac{1 - a^{n+1}}{1 - a}$$

to write the solution found in part (a) in the closed form

$$y_n = \left(y_0 - \frac{b}{1-a}\right)a^n + \frac{b}{1-a} \quad (20)$$

where $a \neq 1$.

18. **Missing Link** The solution (20) found in Problem 17 is invalid for $a = 1$. Find the solution of

$$y_{n+1} = y_n + b \qquad y_0 \text{ given} \qquad (21)$$

19. **Comparing Solutions** Solve the first-order initial-value problem (19) by first finding the general solution of the differential equation and then substituting the general solution into the initial condition. Show that solution (20) is obtained.

Odds and Ends

20. **Tower of Hanoi Puzzle** One can verify that the number of moves M_n required to transfer n disks from needle A to needle C (see Figure 7.3), provided that only one needle is moved at a time and a larger disk is *never* placed on a smaller disk, satisfies the equation $M_{n+1} = 2M_n + 1$. Given that $M_1 = 1$, find the general expression for M_n.

In Hindu mythology, God placed 64 gold disks on a diamond needle, and Hindu priests transferred these disks one at a time to a third diamond needle using the rules described above. When all the disks are transferred, the world will come to an end. If Hindu priests can transfer one disk each second, how long will it take the priests to transfer the disks?

21. **Triangular Numbers** Greek mathematicians attributed mystical properties to numbers. A number T_n is said to be

Figure 7.3
Tower of Hanoi problem

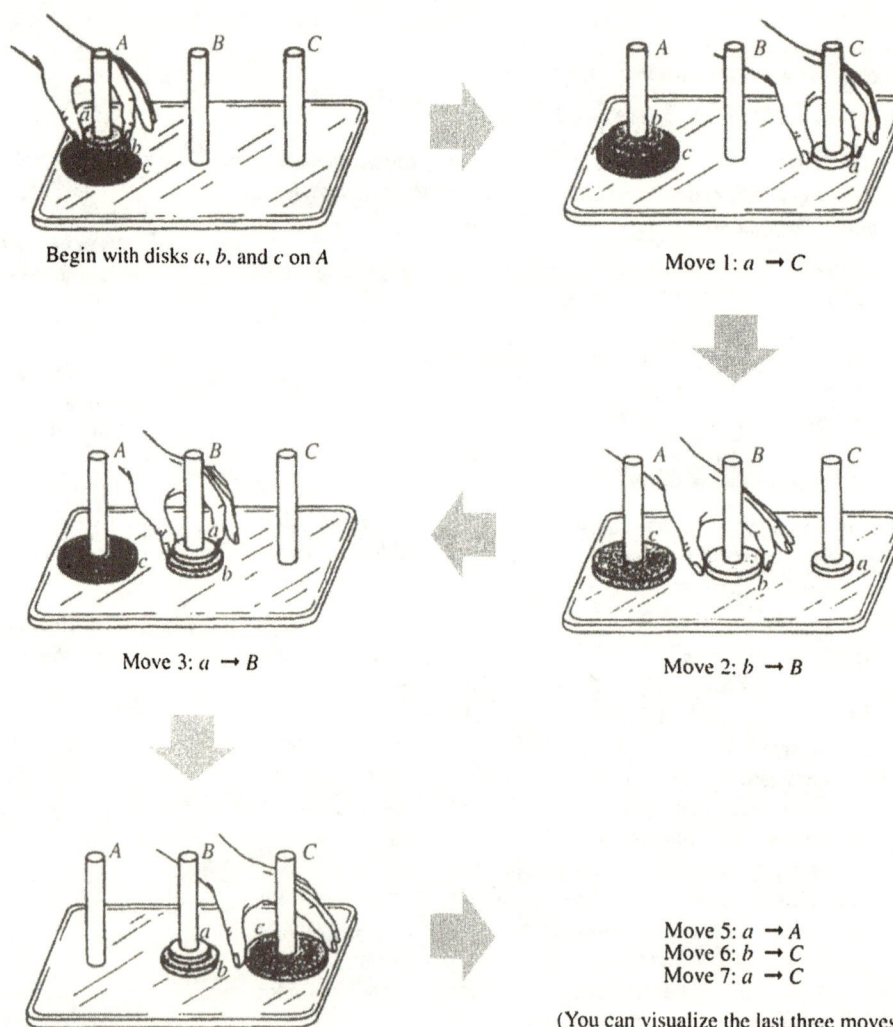

Begin with disks a, b, and c on A

Move 1: $a \rightarrow C$

Move 2: $b \rightarrow B$

Move 3: $a \rightarrow B$

Move 4: $c \rightarrow C$

Move 5: $a \rightarrow A$
Move 6: $b \rightarrow C$
Move 7: $a \rightarrow C$

(You can visualize the last three moves.)

a **triangular number** if it is the number of dots that form one of the triangles in the sequence illustrated in Figure 7.4.

(a) What is the initial-value problem that describes the triangular numbers?

(b) What is the expression for the nth triangular number T_n?

22. Square Numbers A number S_n is said to be a **square number** if it is the number of dots that form one of the squares in the sequence illustrated in Figure 7.5.

(a) What is the initial-value problem that describes the square numbers?

(b) What is the expression for the nth square number S_n?

23. Pentagonal Numbers A number is said to be a **pentagonal number** if it is the number of dots that form one of the pentagons in the sequence illustrated in Figure 7.6.

(a) What is the initial-value problem that describes the pentagonal numbers?

(b) What is the expression for the nth pentagonal number P_n?

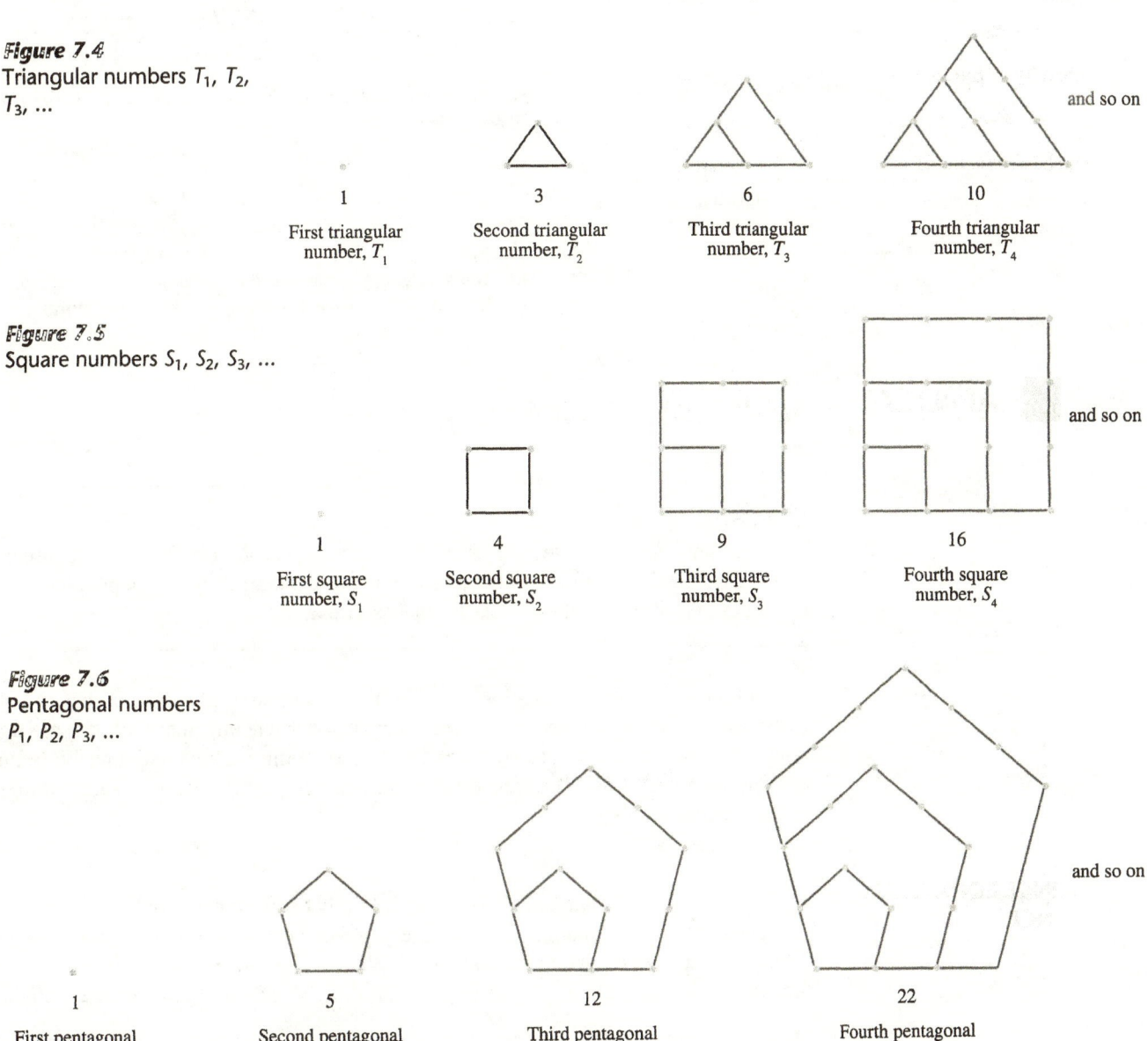

Figure 7.4
Triangular numbers T_1, T_2, T_3, ...

1
First triangular number, T_1

3
Second triangular number, T_2

6
Third triangular number, T_3

10
Fourth triangular number, T_4

and so on

Figure 7.5
Square numbers S_1, S_2, S_3, ...

1
First square number, S_1

4
Second square number, S_2

9
Third square number, S_3

16
Fourth square number, S_4

and so on

Figure 7.6
Pentagonal numbers P_1, P_2, P_3, ...

1
First pentagonal number, P_1

5
Second pentagonal number, P_2

12
Third pentagonal number, P_3

22
Fourth pentagonal number, P_4

and so on

Z-Transform Problems

For Problems 24–28, use the definition of the Z-transform to find the Z-transform of the given sequence. If possible, write the transform in closed form.

24. $\{0, 0, 1, 1, ..., 1, ...\}$
25. $\{0, 0, 1, 1, 0, 0, ..., 0, ...\}$
26. $\{1, -1, 1, -1, ...\}$
27. $\{a, ae^{-kt}, ae^{-2kt}, ae^{-3kt}, ...\}$
28. $\{0, 1, 2, 0, 0, 0, ...\}$
29. Analog of the Second Derivative Rule Verify the rule

$$\mathbb{Z}\{y_{n+2}\} = z^2 \mathbb{Z}\{y_n\} - z^2 y_1 - z y_0$$

30. Linearity of the Z-Transform Prove the relationship

$$\mathbb{Z}\{ay_n + bz_n\} = a\mathbb{Z}\{y_n\} + b\mathbb{Z}\{z_n\}$$

that shows that the Z-transform is a linear transformation.
31. Derivative Property of the Z-Transform Verify the property

$$\mathbb{Z}\{ny_n\} = -\frac{d}{dz} F(z)$$

of the Z-transform and use this property to find the Z-transform of $\{n^3\}$.

For Problems 32–37, find the inverse Z-transform of the given function F(z) using the definition.

32. $F(z) = \dfrac{1}{z}$

33. $F(z) = \dfrac{1}{z^2}$

34. $F(z) = \dfrac{1}{1 - 1/z}$

35. $F(z) = \dfrac{1}{1 - 2z}$

36. $F(z) = \dfrac{z}{z - 1}$

37. $F(z) = \dfrac{z}{z^2 - 1/4}$

For Problems 38–41, solve the difference equations using the Z-transform.

38. $y_{n+1} - y_n = 2n + 1 \qquad y_0 = 1$
39. $y_{n+2} - 4y_n = 0 \qquad y_0 = 1 \quad y_1 = -1$
40. $y_{n+2} + 2y_{n+1} + y_n = 0 \quad y_0 = 1 \quad y_1 = 0$
41. $y_{n+2} + y_n = n^2 \qquad y_0 = 0 \quad y_1 = 0$
42. Journal Entry Compare the similarities and differences between the Laplace transform and the Z-transform.

7.4 APPLICATIONS OF DIFFERENCE EQUATIONS

PURPOSE

To show how first- and second-order difference equations can be used to predict phenomena ranging from the value of annuities in banking to future population sizes in biology. We illustrate these ideas by means of examples.

Although differential equations are useful for predicting phenomena that change continuously over time, many phenomena change at discrete time intervals, like a watch hand that jerks forward second by second instead of gliding along continuously. In this section we will show how difference equations describe phenomena in finance, biology, and numerical analysis.

DIFFERENCE EQUATIONS IN FINANCE

Problems involving compound interest lead naturally to difference equations. To review the meaning of compound interest, suppose you deposit $1000 in a bank that pays 8% compound interest, compounded annually.* What this means is that at the end of one year the bank will make a deposit of $(0.08)(\$1000) = \$80$ to your account, bringing

---

* Most banks compound interest on a daily basis. Of course, 8% annual interest is the same interest as $8/365 \doteq 0.0219178\%$ *daily* interest.

the total to \$1080. At the end of the second year the bank will make another deposit into your account, this time $(0.08)(\$1080) = \$86.40$, bringing the total to \$1166.40, and so on. However, nowadays with the widespread use of computers, banks compound interest every day, allowing the investor to earn interest more often. Interest rates are quoted as annual interest rates, and so an 8% annual interest rate is the same as a 2% per quarter rate (three-month period). Table 7.7 compares the quarterly versus annual compounding if the bank pays 8% annual interest.

Table 7.7
Future Value of a \$1000 Deposit at 8% Annual Interest

Length of deposit	Value of account (quarterly compounding)	Value of account (annual compounding)
3 months	$\$1000 + (0.02)(\$1000) = \$1020$	\$1000
6 months	$\$1020 + (0.02)(\$1020) = \$1040.40$	\$1000
9 months	$\$1040.40 + (0.02)(\$1040.40) = \$1061.21$	\$1000
12 months	$\$1061.21 + (0.02)(\$1061.21) = \$1082.43$	\$1080

In other words, 8% annual interest compounded annually yields the final value of \$1080, whereas 8% annual interest compounded quarterly yields \$1082.43.

The following examples illustrate how difference equations can be used in finance.

Example 1 **A Million-Dollar Annuity** Ed has just entered college and has decided to quit smoking. He has decided to make daily deposits of the money he saves into a bank account that pays long-term annual interest of 10.95%. If the bank compounds interest daily, what will be the value of Ed's account when he retires in 50 years?

Solution If we call y_n the value of Ed's account at the end of day n after interest has been deposited into Ed's account, then $y_{n+1} - y_n$ is the *increase* in the account from day n to day $n + 1$. Since the bank pays 10.95% annual interest, then at the end of day n the bank makes a deposit of

$$\frac{0.1095}{365} y_n = 0.0003 y_n$$

into Ed's account. Hence the daily increase in Ed's account will come about from the daily interest payments made by the bank and Ed's own deposits of \$3. That is, the value y_n of Ed's account at the end of day n will satisfy the equation

$$\underset{\substack{\uparrow \\ \textit{Increase in} \\ \textit{account}}}{y_{n+1} - y_n} = \underset{\substack{\uparrow \\ \textit{Interest} \\ \textit{earned}}}{0.0003 y_n} + \underset{\substack{\uparrow \\ \textit{Daily} \\ \textit{deposit}}}{3} \qquad \underset{\substack{\uparrow \\ \textit{Initial} \\ \textit{value}}}{y_0 = 0} \tag{1}$$

Rewriting this equation, we obtain

$$y_{n+1} - 1.0003 y_n = 3 \qquad y_0 = 0 \tag{2}$$

The characteristic equation of the homogeneous equation is $r - 0.0003 = 0$, which has the root $r = 0.0003$. Hence the homogeneous solution is

$$h_n = C(1.0003)^n \qquad (n = 0, 1, 2, ...) \tag{3}$$

To find a particular solution, we try $p_n = A$. Substituting this constant into Eq. (2) gives $A = -10,000$. Hence a particular solution is $p_n = -10,000$, and the general solution of Eq. (2) is

$$y_n = C(1.0003)^n - 10,000 \tag{4}$$

Substituting this equation into the initial condition $y_0 = 0$ gives $C = 10,000$. Hence the value of Ed's account after day n will be

$$y_n = 10,000\{(1.0003)^n - 1\} \qquad (n = 0, 1, 2, ...) \tag{5}$$

The general trend of this sequence is illustrated in Figure 7.7.

Figure 7.7
Result of compound interest after n days

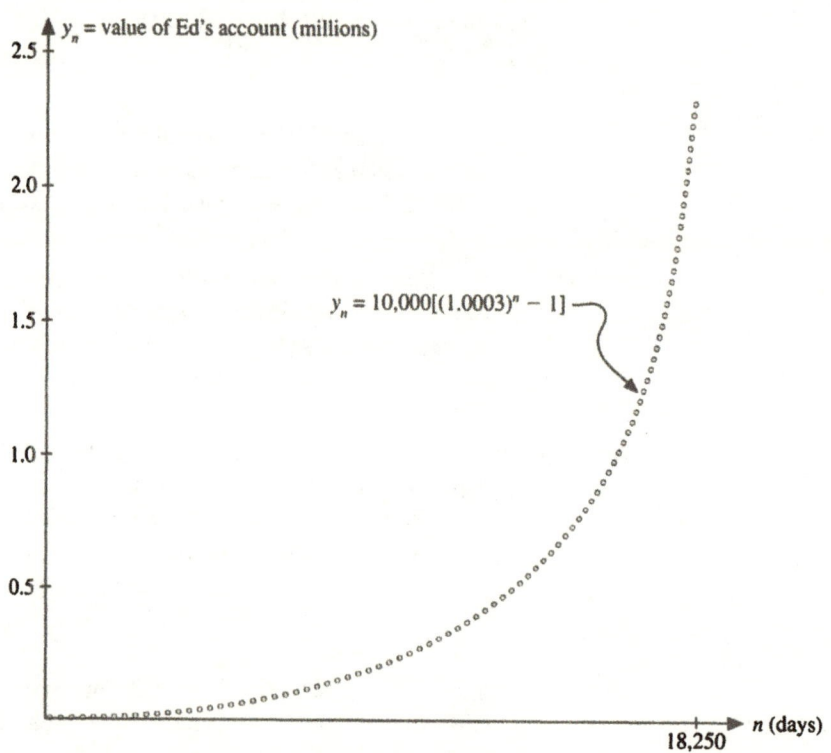

Hence the value of Ed's account after 50 years [50(365) = 18,250 days] will be

$$y_{18,250} = \$10,000\{(1.0003)^{18,250} - 1\}$$
$$= \$2,374,545.90 \qquad \blacksquare$$

Note: Although this large amount seems hard to believe, it is true. Keep in mind that in 50 years a million dollars will not buy what it buys today. However, keep in mind too that as cigarette prices increase over the next 50 years, Ed will make *larger and larger* deposits, not just the paltry $3 daily deposit that he makes now. In the problem set at the end of this section (Problem 9) the reader will have the opportunity to find the future value of Ed's account if he increases his deposits by 5% per year over the next 50 years. (Now we're talking *real money*.)

We now summarize the general payment problem.

Periodic Deposits Problem (Annuity Problem)

If periodic payments, each of amount d, are deposited in a bank account whose initial amount is S_0 that pays interest at a rate r per period, then the difference equation that describes the total amount S_n in the account after n periods is

$$S_{n+1} = (1 + r)S_n + d$$

whose solution is

$$S_n = S_0(1 + r)^n + \frac{d}{r}[(1 + r)^n - 1] \qquad (n = 0, 1, 2 \ldots)$$

BIOLOGICAL POPULATIONS

A difference equation is an ideal tool for studying some types of biological populations, since many species of plants and animals do what they do in neat one-year periods. For instance, the size of a deer herd changes essentially on a yearly basis because of mating and birthing habits. Then too, counts in population size of many wild animals are generally conducted only at yearly time intervals. For these and other reasons, difference equations are generally the equations of preference for studying population dynamics in biology. The following example illustrates a typical problem in population biology.

Example 2   **Controlled Fisheries**   A fisheries biologist is studying the effects of fishing on a given trout population. The fish population in a given lake is currently 1 million, and it grows at an annual rate of 4% per year in the absence of fishing. Fishing regulations allow for 80,000 fish per year to be caught. Under these conditions, what will be the future size of the fish population?

Solution   If we call y_n the size of the fish population at the end of the nth year, the initial-value problem that describes the future fish population is

$$\underset{\substack{\uparrow \\ \textit{Change in fish} \\ \textit{population}}}{y_{n+1} - y_n} = \underset{\substack{\uparrow \\ \textit{Nature} \\ \textit{growth}}}{0.04\,y_n} - \underset{\substack{\uparrow \\ \textit{Fish} \\ \textit{caught}}}{80{,}000} \qquad \underset{\substack{\uparrow \\ \textit{Initial population} \\ \textit{size}}}{y_0 = 1{,}000{,}000} \qquad (6)$$

This equation can easily be solved and has the solution

$$y_n = 1{,}000{,}000[2 - (1.04)^n] \qquad (n = 0, 1, 2, \ldots) \qquad (7)$$

The general trend of this sequence is illustrated in Figure 7.8.   ■

Figure 7.8
A difference equation model for the growth of a fish population in the presence of harvesting

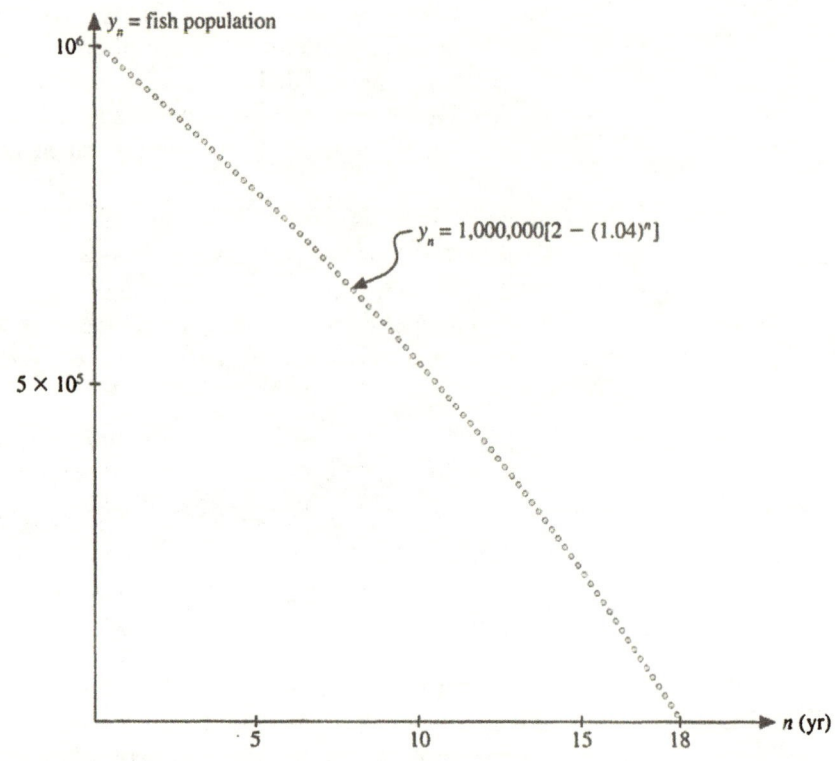

Biological Growth with Periodic Depletion

If a biological population grows at a constant rate of r per time period, and if the population is periodically depleted by the amount d, then the difference equation that describes the size y_n of the population at the end of the nth period is

$$y_{n+1} = (1 + r)y_n - d$$

whose solution is

$$y_n = y_0(1 + r)^n - \frac{d}{r}[(1 + r)^n - 1] \qquad (n = 0, 1, 2, \dots)$$

SECOND-ORDER EQUATIONS IN BIOLOGICAL GROWTH

The previous examples illustrate the importance of first-order difference equations in the description of phenomena in biology and banking. The next example illustrates how second-order difference equations arise in problems in epidemic modeling. The analysis of the spread of diseases (malaria, AIDS, etc.) is a major area of activity for many public health service researchers.

Example 3   **Spread of Disease**   The number of cases of the dreaded disease fungus amongus at Treadmore College is growing in such a way that its growth increase in any week is twice its growth rate during the previous week. If 20 people were sick this last week ($n = 0$) and 25 are sick this week ($n = 1$), how many will be sick in future weeks? If growth continues at this rate, how long will it be until 500 students have the disease?

Solution   Letting y_n denote the number of students having the disease during week n, we can write

$$y_{n+2} - y_{n+1} = 2(y_{n+1} - y_n) \qquad y_0 = 20 \quad y_1 = 25$$

Increase   Increase
this week   last week

(8)

Rewriting this equation, we obtain

$$y_{n+2} - 3y_{n+1} + 2y_n = 0 \qquad y_0 = 20 \quad y_1 = 25 \tag{9}$$

The characteristic equation of this homogeneous equation is $r^2 - 3r + 2 = 0$, which has real roots of $r_1 = 1$, $r_2 = 2$. Hence the general solution of the difference equation is

$$y_n = c_1 + c_2 2^n \qquad (n = 0, 1, 2, ...) \tag{10}$$

Substituting this expression into the intial conditions yields $c_1 = 15$, $c_2 = 5$. Hence the number of students sick with fungus amongus during week n will be

$$y_n = 15 + 5 \cdot 2^n \qquad (n = 0, 1, 2, ...) \tag{11}$$

Referring to Table 7.8, we see that sometimes during the seventh week, 500 students will have the disease. ∎

Table 7.8

Number of Students sick with Fungus Amongus

Week	Number sick	Weekly increase
0	$15 + 5 \cdot 2^0 = 20$	—
1	$15 + 5 \cdot 2^1 = 25$	5
2	$15 + 5 \cdot 2^2 = 35$	10
3	$15 + 5 \cdot 2^3 = 55$	20
4	$15 + 5 \cdot 2^4 = 95$	40
5	$15 + 5 \cdot 2^5 = 175$	80
6	$15 + 5 \cdot 2^6 = 335$	160
7	$15 + 5 \cdot 2^7 = 655$	320

Note: From Table 7.8 we see that the weekly increase in the number of sick students doubles every week, as was stated in the problem. Just how long the dynamics of this epidemic will obey the given difference equations depends on many factors such as the size of the student body and whether students are getting inoculated.

PROBLEMS: Section 7.4

Finding Closed-Form Expressions for Sums

*Difference equations can be used to find **closed-form expressions** to certain sums from open-form expressions. For Problems 1–4, find a difference equation for the given sum s_n using the open-form expression, and verify the given closed-form expression for s_n by solving the difference equation.*

Open form	Closed form
1. $s_n = 1 + 2 + 3 + \cdots + n$	$s_n = \dfrac{n(n + 1)}{2}$
2. $s_n = 1 + 2^2 + \cdots + n^2$	$s_n = \dfrac{n(n + 1)(2n + 1)}{6}$
3. $s_n = 1 + 2 + 2^2 + \cdots + 2^n$	$s_n = 2^{n+1} - 1$
4. $s_n = 1 + 3 + 3^2 + \cdots + 3^n$	$s_n = \dfrac{1}{2}(3^{n+1} - 1)$

Difference Equations in Finance

5. Check This with Your Banker John deposits $1000 in a bank that pays 7.3% annual interest, compounded daily. How much money will be in John's account after 1 day, 10 days, 365 days?

6. How Much Money is Enough? Suppose you have won $200,000 in the tri-state lottery, and you have decided to retire on your winnings. Suppose you deposit your winnings in a bank that pays 8% annual interest (compounded annually) and make yearly withdrawals of $30,000.
   (a) What are the difference equation and initial condition that describe the future value of your account?
   (b) Will you ever run out of money? If so, when?

7. How to Retire a Millionaire Ann wants to become a millionaire, so she takes a differential equations class and learns all about compound interest. Starting with no capital, she deposits d dollars per year in a bank that pays an annual interest rate of 8%. Assume that interest is compounded annually.*
   (a) How much money will Ann have in the bank at the end of n years?
   (b) How much will she have to deposit every year so that her account will be worth a million dollars in 50 years?

8. Sensitivity to the Interest Rate The effect of a small change in interest rates is substantial over long periods of time. To see this, suppose that John and Ann start with no capital and John makes annual deposits of $1000 into a bank that pays 8% annual interest, whereas Ann deposits the same amount in another bank that pays 8.5% annual

---

* Although interest is probably compounded on a daily basis, we will assume annual compounding for simplicity.

---

interest. If both banks compound interest annually, compare the sums accumulated by each person after 50 years.

9. Marboros for Millions Ed has just entered college and has decided to quit smoking. At his current rate of two packs a day, he figures he can save $1000 per year. Suppose Ed deposits $1000 into a bank at the end of the first year and then increases his annual deposit by 5% each year to reflect the increasing price of cigarettes. The bank pays Ed 10% annual interest. For simplicity we assume that the bank compounds interest annually, although most banks compound interest daily.
   (a) What is the difference equation that describes the amount of money Ed has in the bank at the end of n years?
   (b) How much money will be in Ed's account by the time he graduates in four years?
   (c) How much money will be in Ed's account when he retires in 50 years?

10. Amazing But True Amanda has just entered college and has decided to quit smoking. At her current rate of smoking, she figures she can save $25 per week. Suppose each week Amanda deposits this amount into a savings account that pays 10.4% annual interest, compounded weekly. Note that 10.4% annual interest is the same as $10.4/52 \doteq 0.20\%$ weekly interest.
   (a) What is the initial-value problem that describes the amount of money in Amanda's account after n weeks?
   (b) How much money will be in Amanda's account after n weeks?
   (c) How much money will be in Amanda's account when she graduates after 208 weeks (four years)?

11. The Amortization Problem An *amortization* is a method for repaying a loan by a series of equal payments, such as when a person buys a car or house. Each payment goes partially toward payment of interest and partially toward reducing the outstanding principal. If a person borrows S dollars to buy a house, and if p_n denotes the outstanding principal after the nth payment of d dollars, then p_n satisfies the difference equation

$$p_{n+1} = (1 + i)p_n - d \qquad p_0 = S$$

where i is the interest per payment period.
   (a) Find p_n.
   (b) Use the solution found in part (a) to find the payment d per period that must be made so as to pay back the debt in exactly N periods.
   (c) Suppose you take out a $100,000 mortgage on a house from a bank that charges monthly interest of 1%. If the loan is to be repaid in 360 monthly payments (30 years) of equal amounts, what will be the amount of each payment?

Difference Equations in Biology

12. Fisheries Management It is estimated that the population of haddock in the North Atlantic will increase each year by 2% per year when a total ban on fishing is imposed. Under new guidelines, fishermen can collectively harvest 1000 tons of haddock per year. Suppose the current tonnage of haddock in the North Atlantic is 100,000 tons.

(a) Determine the initial-value problem that describes the future tonnage of haddock in the North Atlantic. (Measure haddock in thousand of tons.)

(b) Find the tonnage in the future under these fishing guidelines.

(c) What would happen if fishing regulations allowed for 5000 tons of haddock to be caught each year?

13. Deer Population A Wyoming deer population has an initial size of 100,000. Owing to natural growth, the herd increases by 10% per year.

(a) Find the future size of the deer population if the state allows 15,000 deer to be hunted each year.

(b) Find the number of deer that can be killed each year that will allow the present population to remain constant.

14. Save the Whales Assume that the natural increase in the number of blue whales is 25% per year. Assume that the current population is 1000 and that some countries still harvest blue whales, taking a total of 300 per year.

(a) Find the whale population in the future.

(b) If this trend continues, what will be the fate of the whales?

15. Drug Therapy Betty has diabetes, which means that her body does not produce insulin. If she does not take insulin shots, the amount of insulin in her body decreases 25% per day. Currently, she has no insulin in her body but takes shots amounting to 100 grams per day.

(a) What will be the number of grams of insulin in her body after n days?

(b) What will be the long-term amount of insulin in her body?

16. Consequences of Periodic Drug Therapy Jerry takes 100 mg of a given drug to control his asthma. The drug is eliminated from his body at such a rate that it decreases by 25% each day. Let y_n be the amount of the drug present in Jerry's bloodstream on day n immediately after he takes the drug.

(a) Show that the difference equation and initial condition for y_n are

$$y_{n+1} = 0.75y_n + 100 \qquad y_0 = 100$$

(b) Solve this initial-value problem.

(c) What is the limiting amount of drug in Jerry's body?

(d) What should be Jerry's daily drug dosage so that the limiting amount of drug in his body is 800 mg?

17. Periodic Drug Therapy Jerry from Problem 16 takes d mg of a drug each day. The manufacturer of the drug says that the drug is eliminated from the body exponentially with a half-life of $t_f = 36$ hours. Let y_n be the amount of drug (in milligrams) in Jerry's body on day n immediately after he takes the drug.

(a) Find the difference equation and initial condition that describe the amount of drug in Jerry's body after day n. *Hint:* First convert the half-life to daily decay rate.

(b) Find the limiting amount of this drug in Jerry's body.

(c) Find the daily dosage d of drug that Jerry should take in order that the limiting amount of drug in his body by 350 mg.

18. General Drug Therapy Your doctor has prescribed that every t_d hours you take a tablet containing d mg of a certain drug. Suppose you know that the drug is eliminated from your body according to the law exponential e^{-kt} with a given half-life of $t_{1/2}$. Under these conditions the cumulative amount of drug in your body will follow a curve like the one shown in Figure 7.9.

(a) Show that the cumulated amount of drug, y_n, in your body immediately after taking the $(n + 1)$ tablet will satisfy the difference equation.

$$y_{n+1} = ay_n + d \qquad (n = 0, 1, 2, ...)$$
$$y_0 = d$$

where $a = e^r$, $r = -\left(\dfrac{t_d}{t_{1/2}}\right) \ln 2$.

Figure 7.9

Amount of drug accumulated in the body resulting from periodic dosages

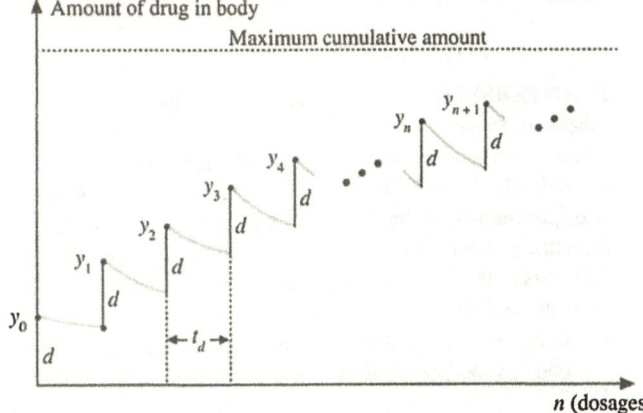

(b) Show that the maximum amount of drug cumulated in your body will be

$$\text{Maximum cumulated amount} = \left(\frac{1}{1 - e^r}\right)d$$

(c) A promising new antidepressant drug, Prozac, has a half-life between 2 and 3 days. Assuming a half-life of 60 hours (2.5 days), suppose you take one 50-mg tablet of Prozac each day. What will be the maximum cumulated amount of Prozac in your body?

(d) How many tablets (days in this case) will it take before the cumulated amount of Prozac reaches 90% of the maximum cumulated value as found in part (c)?

19. **Spruce Budworm Infestation** An attack of the spruce budworm beetle is killing trees in the eastern United States. Every year the increase in the number of acres destroyed is three times the increase during the previous year. Suppose 10,000 acres were destroyed last year and 12,000 acres are destroyed this year.

(a) How many acres will be destroyed after year n?

(b) If trees are destroyed at this same rate, how long will it be until 100,000 acres are being destroyed each year?

20. **General Growth Problem** Suppose the growth of a given type of bacteria is such that the increase in any given week is r times the increase on the previous week. If the number last week is y_0 and the number this week is y_1, how many bacteria will be present on week n?

Other Applications

21. **Very Interesting** If any collection of n nonparallel lines are drawn in the plane, how many distinct regions in the plane are formed? *Hint:* If y_n denotes the number of regions formed by n nonparallel lines, then the $(n + 1)$st line will intersect the previous n lines at n points and hence will divide $n + 1$ previously constructed regions into twice that number. Convince yourself that

$$y_{n+1} = y_n + n + 1 \qquad y_0 = 1$$

22. **PLANTERSPEANUTS Puzzle** In the early 1950's the makers of Planters Peanuts sponsored a contest and offered a prize to anyone who could determine the number of ways to spell PLANTERSPEANUTS by moving either downward, to the left, to the right, or any combination of these directions in the fifteen rowed pyramid shown in Figure 7.10. Solve the PLANTERSPEANUTS puzzle by working parts (a) and (b).

(a) If T_n: $n = 1, 2, \ldots$ denotes the number of ways in which one can spell an n-lettered word in an n-rowed pyramid, show that T_n satisfies

$$T_{n+1} = 2T_n + 1 \qquad T_1 = 1$$

(b) Solve the initial-value problem in part (a) and use this result to solve the PLANTERSPEANUTS puzzle.

Figure 7.10

In how many ways can you spell PLANTERSPEANUTS by moving downward, to the left, to the right, or any combination of these directions?

```
                      P
                    P L P
                  P L A L P
                P L A N A L P
              P L A N T N A L P
            P L A N T E T N A L P
          P L A N T E R E T N A L P
        P L A N T E R S R E T N A L P
      P L A N T E R S P S R E T N A L P
    P L A N T E R S P E P S R E T N A L P
  P L A N T E R S P E A E P S R E T N A L P
P L A N T E R S P E A N A E P S R E T N A L P
P L A N T E R S P E A N U N A E P S R E T N A L P
P L A N T E R S P E A N U T U N A E P S R E T N A L P
P L A N T E R S P E A N U T S T U N A E P S R E T N A L P
```

23. **Computer Problem** Use a computer to make a table giving the solution of the difference equation in Problem 9. Using this table, how much money will Ed have in his bank account after 10 years?

24. **Ask Marilyn*** In the "Ask Marilyn" column, a popular syndicated column run in *Parade* magazine, the following question was asked: You currently make $10,000 per year. You can have a $1000 raise at the end of each year or a $300 raise at the end of each six months. Which option do you choose? Letting

A_n = amount earned during year n if a $1000 annual raise is given

B_n = amount earned during year n if a $300 biannual raise is given

find and solve the initial-value problems for A_n and B_n. How much will you earn in the twentieth year using each strategy?

25. **Journal Entry** Ask yourself when difference equations are more appropriate for modeling physical phenomena and when differential equations are more appropriate.

* This problem appeared in the "Ask Marilyn" column in *Parade* magazine on June 7, 1991. The column is written by Marilyn Vos Savant. According to the column, most people made the incorrect choice.

7.5   THE LOGISTIC EQUATION AND THE PATH TO CHAOS

PURPOSE

To introduce the (discrete) logistic equation and study the complex behavior of its solution and how it changes from orderly behavior to chaotic behavior. This leads to the doubling sequence, the bifurcation diagram, and the Feigenbaum universal constant. We also introduce the concept of the cobweb diagram and show how it can be used to visualize the solution of an iterative system.

It has been discovered in recent years that a large class of nonlinear differential and difference equations exhibit a certain deterministic but seemingly randomlike behavior known nowadays as **chaotic motion.** By chaotic motion or behavior we simply mean "without apparent order"* Often, equations that exhibit chaotic behavior are those that describe phenomena whose output is fed back into the system, in much the same way as a loudspeaker's sound is fed back into a microphone. One interesting idea related to these *iterative* systems is the fact that their solutions often depend on a parameter, and simply changing the parameter can make the difference between orderly and chaotic behavior. Possibly the simplest (certainly the most famous) equation whose solution changes from an orderly to chaotic behavior is the (discrete) logistic equation. The logistic equation is an ideal vehicle in which to study chaotic behavior, first because the equation is very simple and second because many characteristics of chaotic motion are universal in nature and depend only on the presence of feedback in the equation and not on the particular equation. This is important because a mathematical model that provides insight into the transition from orderliness to chaotic behavior might provide an understanding of the onslaught of chaotic behavior in nature, such as in the complex phenomenon of turbulence.

THE LOGISTIC EQUATION

An important difference equation, sometimes called a **mapping** in the more general context of iterative systems, that is used to study simple biological populations is the **logistic equation:**[†]

$$x_{n+1} = \lambda x_n(1 - x_n) \qquad (n = 0, 1, 2, ...) \tag{1}$$

where the parameter λ, called the *growth parameter,* lies in the interval $0 < \lambda \le 4$.

---

* There are many intuitive definitions of *chaotic motion* in the current literature. It is often defined as "irregular behavior displaying extreme sensitive dependence on initial conditions" or "deterministic behavior that appears to be randomlike." There *are* precise mathematical definitions for chaos, and the interested reader can consult *An Introduction to Chaotic Dynamical Systems,* Second Edition, by R. L. Devaney (Addison-Wesley, Reading, Mass., 1989).

[†] The logistic equation was first studied in 1845 by population biologist P. F. Verhulst, and for that reason it sometimes called Verhulst's equation. The reader should realize that the logistic *difference* discussed here is slightly different from the logistic *differential* equation $\dot{x} = \lambda x(1 - x)$, since by letting $\dot{x}(nt) = x_{n+1} - x_n$ and $x(nt) = x_n$ the logistic differential equation becomes $x_{n+1} = x_n + \lambda x_n(1 - x_n)$. Each has its place in the study of biological populations.

By picking λ in this interval, one is assured that for any initial condition or **seed** lying in the interval $[0, 1]$, all future iterates $x_1, x_2, \ldots$ (often called the **orbit** of x_0) also lie in the interval $[0, 1]$. In biological problems the values x_n denote *fractional* populations, where 1 represents the maximum possible population, 0.5 represents half the maximum population, 0 represents zero population, and so on. The reason the logistic equation is a useful tool for predicting population size is the competing factors of x_n and $1 - x_n$ on the right-hand side of Eq. (1). As x_n gets close to the maximum value of 1, the quantity $(1 - x_n)$ diminishes. Hence these "competing" factors act in opposite directions, one trying to increase the population and the other trying to stop it. This, of course, is the way in which many biological populations behave.

Although the importance of the discrete logistic equation has been known by population biologists for over a hundred years, it was only recently that many of its most interesting properties were uncovered. The big surprise came when researchers, including biologist Robert May,[*] tried to understand what effect the growth rate λ had on the *long-term* or *ultimate* solution of the equation. This is an important question because in biological systems, λ corresponds to such things as temperature, nutrients, and all sorts of natural things.[†] May was interested not so much in the solution of the logistic equation for a specific λ, but more globally in the effects of λ on the *eventual* solution. To answer this question, he would try a given λ and then, starting with x_0, grind out the first several terms of the solution, sometimes even using his pocket calculator. After that, he then took another value of λ and repeated the process again. What he discovered was that when $0 \le \lambda < 1$, then no matter what the initial condition $0 \le x_0 \le 1$, the solution always approached the ultimate population of 0. And for λ between 1 and 3, then regardless of the initial condition $0 < x_0 < 1$, the resulting solution approached the nonzero final population of $(1 - \lambda)/\lambda$. Finally, for values of λ larger than 3, strange things began to happen. We hint at what he discovered in Figure 7.11.

But we're getting ahead of ourselves. Let's analyze this "path to chaos" in a little more detail.

THE COBWEB DIAGRAM

A useful geometric construction that allows one to *visualize* the iterates of the logistic equation is the **cobweb diagram**. The idea is to superimpose on the same graph paper the graphs of the two functions $f(x) = \lambda x(1 - x)$ and the 45° line $y = x$. Then, to visualize the iterates $x_0, x_1, x_2, \ldots$ starting from an arbitrary seed x_0, we move up the vertical line $x = x_0$ to the height $x_1 = \lambda x_0(1 - x_0)$; from there we make a dogleg turn, moving horizontally left or right until we hit the 45° line. Since $x = y$ on this line, the value of x where we hit the 45° line gives the new value x_1. See Figure 7.12.

---

[*] One of the people responsible for discovering the chaotic properties of the logistic equation was physicist-turned-biologist Robert May. It was May who in the early 1970's first investigated the fascinating character of the solution of the logistic equation for different values of λ.

[†] To understand why λ is called the *growth rate* parameter, suppose $\lambda = 2$. Neglecting the damping factor $(1 - x_n)$ on the right-hand side of the logistic equation, the equation becomes the *unrestricted* growth equation $x_{n+1} = 2x_n$ in which the population doubles in size every time period. The factor $(1 - x_n)$ only limits the growth when x_n is near 1.

We then draw another vertical line $x = x_1$ and proceed vertically along this line until we hit the logistic curve at $x_2 = \lambda x_1(1 - x_1)$. Again, making a dogleg horizontal turn, we move horizontally until we reach the 45° line, where the x-value will be x_2. Continuing on in this manner, we obtain the sequence of vertical and horizontal lines having the appearance of a cobweb and providing the iterates $x_0, x_1, x_2, \ldots$ of the logistic

Figure 7.11
The discrete logistic equation $x_{n+1} = \lambda x_n(1 - x_n)$ is one of the simplest nonlinear equations whose complex behavior depends on a single parameter.

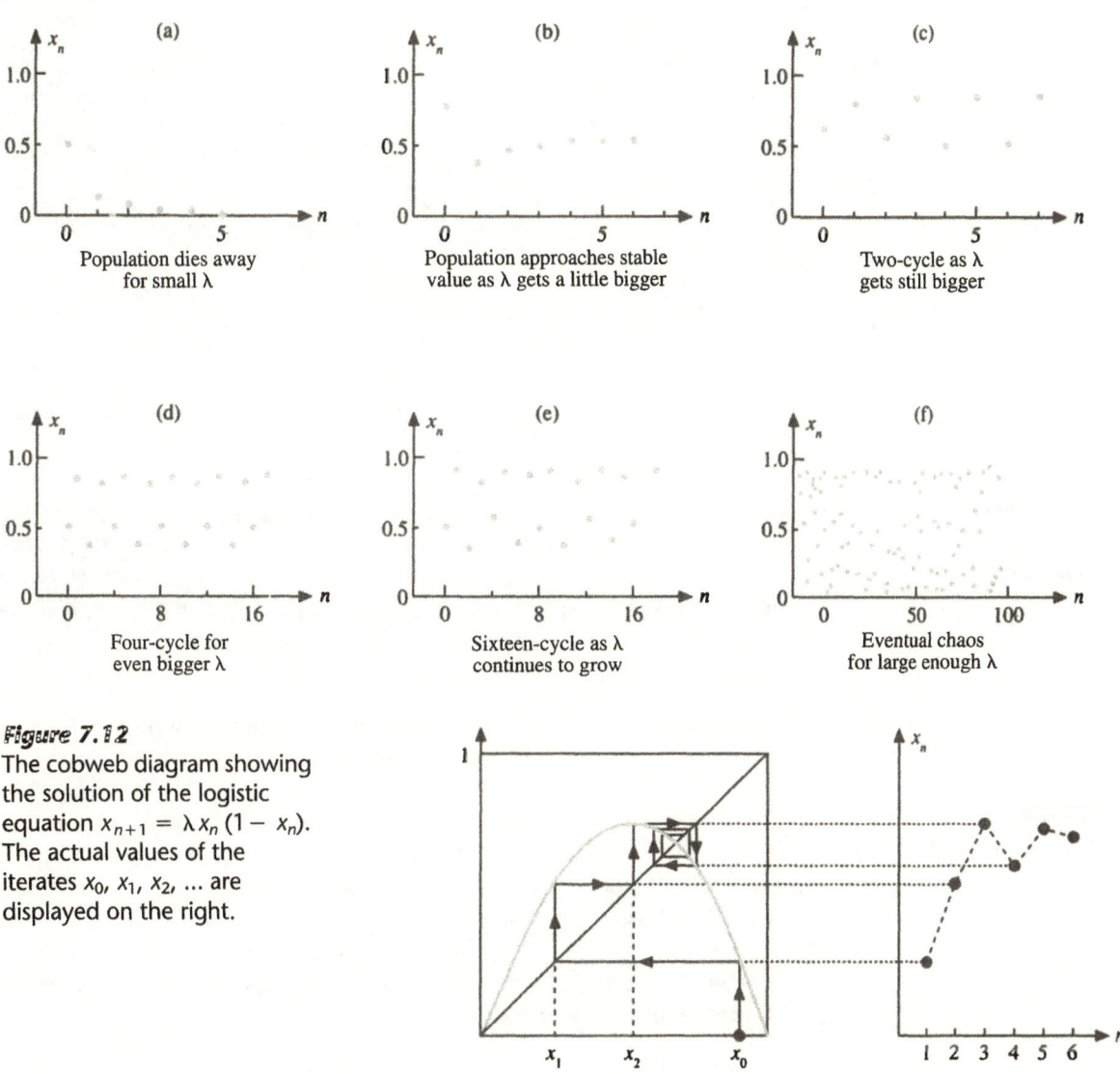

(a) Population dies away for small λ

(b) Population approaches stable value as λ gets a little bigger

(c) Two-cycle as λ gets still bigger

(d) Four-cycle for even bigger λ

(e) Sixteen-cycle as λ continues to grow

(f) Eventual chaos for large enough λ

Figure 7.12
The cobweb diagram showing the solution of the logistic equation $x_{n+1} = \lambda x_n(1 - x_n)$. The actual values of the iterates $x_0, x_1, x_2, \ldots$ are displayed on the right.

equation. The process of finding the solution x_0, x_1, x_2, ... by drawing a cobweb diagram is called the **cobweb diagram algorithm** and is described here.

The Cobweb Diagram Algorithm for Visualizing the Iterates of the Logistic Equation

Starting with x_0, perform the following steps.

Step 1.   Move vertically until one hits the graph of the function given by
$y = f(x, \lambda) = \lambda x(1 - x)$.

Step 2.   Move horizontally until one hits the 45° line; the x-value (or y-value) at this line will be the next x_n.

Step 3.   Repeat Steps 1 and 2 again and again.

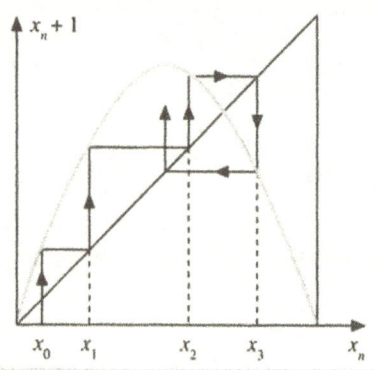

THE PATH TO CHAOS

Population dies out
$(0 < \lambda \le 1)$

From a purely experimental point of view the reader can use a computer or even a pocket calculator and observe that for any λ in the interval $0 < \lambda \le 1$ and *any* starting point $0 \le x_0 \le 1$ the resulting sequence $x_{n+1} = \lambda x_n(1 - x_n)$ approaches 0. If one analyzes the cobweb diagram in Figure 7.13(a), using the fact that the graph of the equation $y = \lambda x(1 - x)$ lies *below* the 45° line $y = x$ when $0 \le \lambda < 1$, one can convince oneself that all solutions converge to 0.

> *Note:* If a gypsy moth population satisfies the logistic equation, then for values $0 \le \lambda < 1$, the population will become extinct.

Population approaches limiting value $(1 < \lambda \le 3)$

Continuing on, for any λ in the interval $1 < \lambda \le 3$, the reader can again use a calculator to verify that for any initial population $0 < x_0 < 1$, the solution of the logistic equation will approach the value $(\lambda - 1)/\lambda$. We can also draw a cobweb diagram to develop a visual sense of why this is true. If we graph the logistic function $y = 2.8\,x(1 - x)$ for $\lambda = 2.8$ shown in Figure 7.13(b) and determine where it crosses the 45° line $y = x$, we get the point $(2.8 - 1)/2.8 = 9/14 \doteq 0.643$. This point is called a **fixed point** x_λ (or **equilibrium point**) and satisfies the equation

$$x_n = \lambda x_n(1 - x_n) \tag{2}$$

Figure 7.13
The long-term behavior of
the logistic equation does not
depend on the initial
condition but on the
parameter λ.

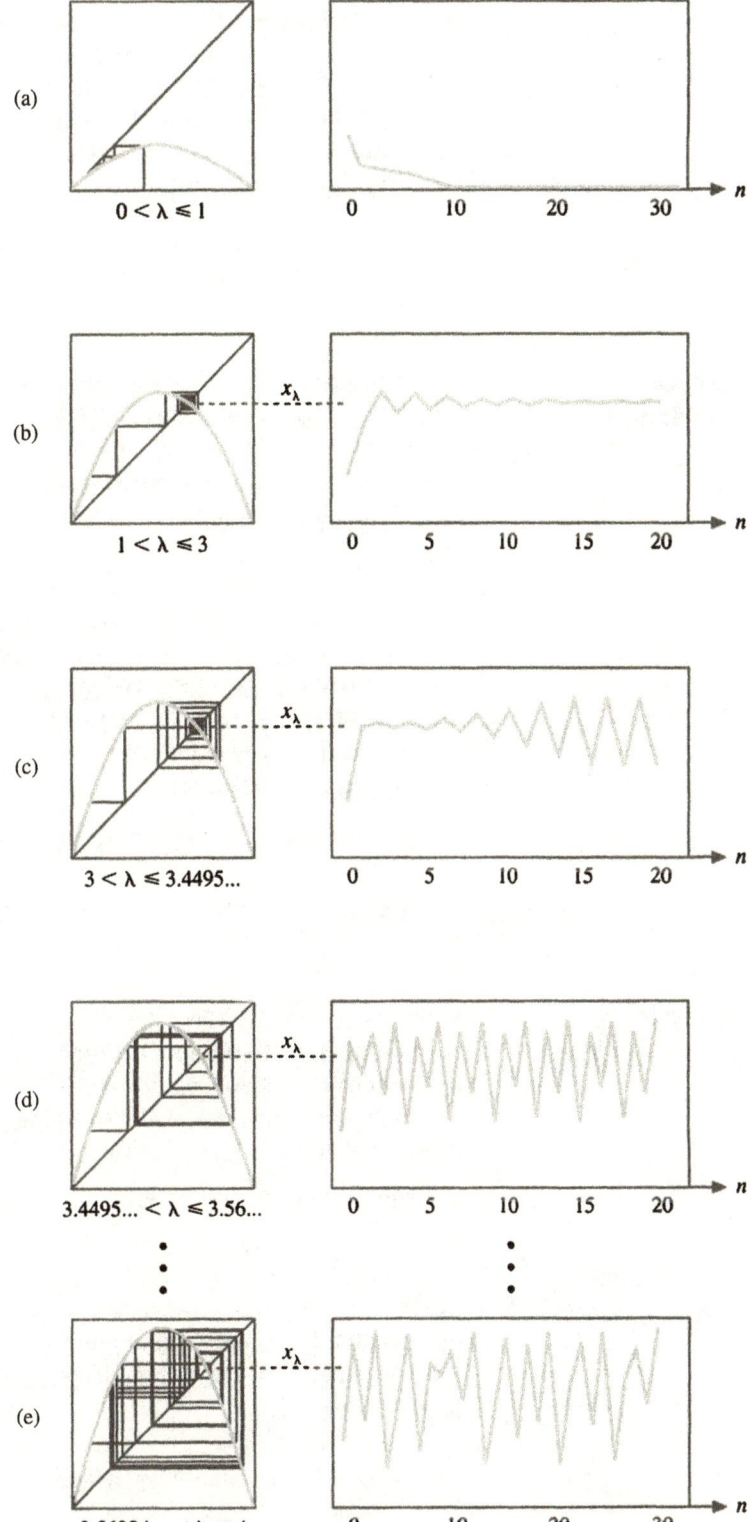

Note that 0 is *always* a fixed point (or equilibrium point) of the logistic equation and that $x_\lambda = (\lambda - 1)/\lambda$ is a fixed point for $1 \leq \lambda \leq 3$. If one examines carefully the paths of the cobweb diagrams starting with any point x_0, one can see that all paths (with the exception of the one starting at 0) lead to x_λ. For that reason it is called an **attracting fixed point.** On the other hand, since initial points near 0 move away from 0, it is called a **repelling fixed point.**

> *Note:* If the solution of the logistic equation represents a biological population, such as a gypsy moth population, this means that all nonzero initial populations ultimately settle down to a fixed positive value (that depends on λ).

Period two-cycle
$(3 < \lambda \leq 3.4495...)$

As λ is increased *beyond* 3 and in the interval $3 < \lambda \leq 3.4495...$, the solution makes an interesting *qualitative* change, called a **bifurcation.** Here the attracting fixed point, which is now $(3 - 1)/3 = 2/3$ at $\lambda = 3$, becomes *repelling* and splits or **bifurcates** into two points.* The new solution, instead of settling down to a single value, jumps back and forth between two points called an **attracting two-cycle** (the exact two values depending on λ). See Figure 7.13(c).

To understand why this phenomenon occurs, observe in Figure 7.13(c) that the seed $x_0 = 0.10$ first moves *toward* the fixed point x_λ but then gradually "spirals" away from it and settles down to jumping back and forth between two points around x_λ (although it never reaches these two points exactly). We also observe from the cobweb diagram that the *reason* for the repelling fixed point x_λ is that the slope of the curve $y = f(x)$ at x_λ is *less than* one. We would say that the two-cycle is attracting but that the fixed point x_λ is repelling.

> *Note:* In this case a gypsy moth population would experience a "boom-and-bust" cycle in alternating years. We have possibly witnessed situations like this in which fish, deer, and insect populations oscillate in two-year periods owing to a variety of causes.

Period four-cycle
$(3.4495... < \lambda \leq 3.56...)$

Moving on, as λ moves past 3.4495..., the two-cycle bifurcates once again, resulting in a (limiting) period four-cycle. In other words, for any nonzero seed x_0 the limiting population will eventually settle down to oscillating among *four* different values, called an **attracting four-cycle,** the exact values depending on the value of λ. This phenomenon is illustrated in Figure 7.13(d).

> *Note:* Here, the gypsy moth population undergoes a four-year cycle.

More period doublings
$(3.56... < \lambda \leq 3.56994...)$

As λ moves past 3.56, the period doubles again and again, faster and faster with periods (8, 16, 32, $\cdots$), until λ reaches a point of accumulation at $\lambda_\infty = 3.56994...$. For a biologist who didn't keep long-term records, the period would become so large that the yearly population of gypsy moths would appear to be changing in a random manner without any pattern.

---

* We have been a little lax with mathematical rigor. We simply state that in general a fixed point x_λ of $x_{n+1} = f(x_n, \lambda)$ is attracting if $|f'(x_\lambda)| < 1$ and repelling if $|f'(x_\lambda)| \geq 1$. A reader with keen insight might have guessed that this was the case.

Chaos
(3.56994... $< \lambda \leq 4$)

When λ reaches the "magical" point of accumulation λ_∞, called the **chaos number,** the limiting solution behaves in an interesting way. It does not approach an attracting fixed point, it does not approach a periodic cycle of any kind, but on the other hand it does not go to infinity. One would almost think that there is no way in which it can behave, but there is one more type of motion, known as **chaotic motion,** in which it appears to jump around aimlessly and have no orderliness. See Figure 7.13(e). In other words, the solution of the logistic equation becomes *chaotic*. To a biologist the population of gypsy moths would appear to change randomly with no underlying dynamical law.*

THE BIFURCATION DIAGRAM

To further understand this **period-doubling phenomenon,** May drew a diagram, known as a **bifurcation diagram** (or **orbit diagram**), which portrays the period-doubling route to chaos and what happens once you get there. The parameter λ is plotted on the horizontal axis, and the limiting values of the solution are plotted on the vertical axis. See Figure 7.14. One can use a computer to show that for certain values of λ in the

Figure 7.14
The cascade of bifurcations of the logistic equation showing the doubling path to chaos as λ increases from 0 to 4. The "chaotic" region beginning at $\lambda_\infty = 3.56994...$ shows an unexpectedly fine structure. Note the certain "windows" of order inside the *chaos region* $\lambda_\infty < \lambda \leq 4$, when the solution returns from chaos to order. Specifically, note the three-cycle in this region in the general vicinity of the value $\lambda = 3.85$.

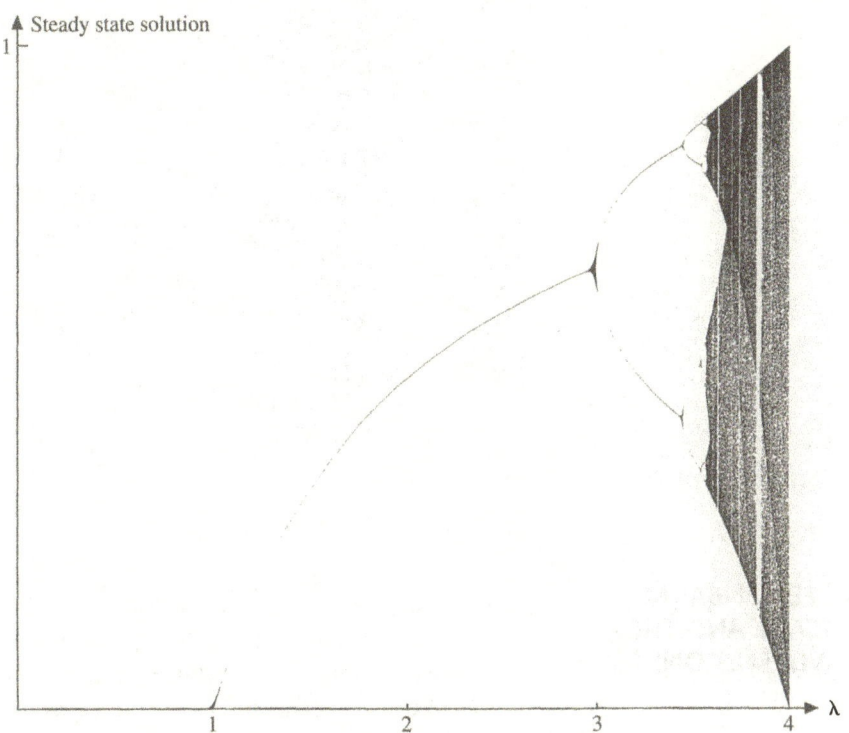

---

* In the 1970's, Robert May carried out experiments on insect populations that allowed him to increase or decrease the birthrate by altering the food supply. He discovered that the time required for the population to return to its starting value doubled at certain critical values of the birthrate. For larger values of the birthrate, the insect populations varied chaotically, as they generally do in the real world.

chaos region, $\lambda_\infty < \lambda \leq 4$, the limiting solution momentarily loses its chaotic behavior and has a limiting three-cycle motion.

A very simple BASIC computer program that will plot the points of the bifurcation diagram (no axes, no labels, nothing fancy) for the logistic equation is given in Table 7.9. This BASIC program can easily be modified to draw bifurcation diagrams for iterative equations other than the logistic equation. Note that the seed is always chosen to be 0.5 and that the first ten iterates are not plotted to allow the orbit to "settle down."

Table 7.9
BASIC Program for Drawing the Bifurcation Diagram

```
10    REM BIFURCATION DIAGRAM FOR THE LOGISTIC EQUATION
20    REM N = NUMBER OF ITERATIONS FOR EACH LAMBDA
30    REM LMIN = MINIMUM LAMBDA
40    REM LMAX = MAXIMUM LAMBDA
50    REM NSTEP = NUMBER OF LAMBDA STEPS
60    SCREEN 2
70    WINDOW (0,0) − (4,1)
80    N = 30
90    LMIN = 1
100   LMAX = 4
110   NSTEP = 101
120   FOR I = 1 TO NSTEP
130      LET X = 0.5
140      LET LAMBDA = LMIN + (I − 1)*(LMAX − LMIN)/(NSTEP − 1)
150      FOR J = 1 TO N
160         X = LAMBDA * X * (1 − X)
170         IF J < 10 THEN 190
180         CIRCLE (LAMBDA, X), 0
190      NEXT J
200   NEXT I
210   END
```

THE FEIGENBAUM CASCADE AND THE UNIVERSAL CONSTANT

In the summer of 1975, physicist Mitchell Feigenbaum of the Los Alamos National Laboratory, while studying the sequence of values $\lambda_1, = 3$, $\lambda_2 = 3.4495$, $\lambda_3 = 3.56$, ..., which constitute the **doubling sequence,** made a fascinating discovery. Using only a hand calculator, he observed that the *ratios* of the differences between members of the doubling sequence $\lambda_1, \lambda_2, \lambda_3, \ldots$ converged to a finite number, the number being 4.6692.... That is, what he discovered was

$$\lim_{n \to \infty} \delta_n = \lim_{n \to \infty} \left(\frac{\lambda_n - \lambda_{n-1}}{\lambda_{n+1} - \lambda_n} \right) = 4.6692\ldots \tag{3}$$

See Table 7.10.

Table 7.10
The Road to Chaos for
$x_{n+1} = \lambda x_n(1 - x_n)$

n	Period 2^n	λ_n	$\delta_n = \dfrac{\lambda_n - \lambda_{n-1}}{\lambda_{n+1} - \lambda_n}$
0	1	3.0	—
1	2	3.4495...	—
2	4	3.5441...	4.844...
3	8	3.5687...	4.327...
4	16	3.5698...	4.696...
5	32	3.5694...	4.636...
⋮	⋮	⋮	⋮
∞	∞	3.56994...	4.6692...

Feigenbaum then went on to study other difference equations arising in electrical circuit theory, optical systems, economic systems, and population problems that obeyed the period-doubling phenomenon. To his amazement the parameters in these difference equations also satisfied this *same property*. Feigenbaum's limiting ratio $\delta_\infty = 4.6692...$ is now called the **Feigenbaum number** or **universal constants;** the word "universal" refers to the fact that it is a universal property of the period-doubling route to chaos for mappings that have a "quadratic maximum" and not just the logistic equation.

The real importance of Feigenbaum's discovery was that it gave experimenters a criterion to determine whether a dynamical system is becoming chaotic by simply observing its "prechaotic" periodic behavior. It has been applied to electrical, chemical, and fluid systems. Scientists have always been interested in the transition from orderly systems to chaotic systems, such as the change from laminar flow to turbulence, and the period-doubling approach is just *one way* in which this occurs.* Much is still unknown about how physical phenomena change from orderly systems to chaotic systems, and research is being carried out at a frantic rate.

PROBLEMS: Section 7.5

For Problems 1–4, find the fixed points of the iterative systems and determine whether they are attractive or repelling. Use your calculator to verify your conclusions.

1. $x_{n+1} = x_n^2$
2. $x_{n+1} = x_n^2 - 1$
3. $x_{n+1} = x_n^3$
4. $x_{n+1} = \cos x_n$

Numerical Experiments in Chaos

For Problems 5–12, use a calculator to iterate each of the given functions (using an arbitrary initial point) and explain the results.

5. $L(x) = 0.5x(1 - x)$
6. $L(x) = 2.8x(1 - x)$
7. $L(x) = 4x(1 - x)$
8. $S(x) = \sin x$

* Other paths to chaos are discussed in *Chaotic Vibrations* by Francis Moon (John Wiley & Sons, New York, 1987).

9. $C(x) = \cos x$

10. $P(x) = x^2 + 0.1$

11. $P(x) = x^2 - 0.1$

12. $P(x) = x^2 - 2$

13. Bifurcation Diagram of Sine Mapping Modify the computer program given in the text (Table 7.9) to draw the difurcation diagram of the iterative system

$$x_{n+1} = \lambda \sin x_n \qquad x_0 = 0.5$$

for $0 \le \lambda \le 1$.

14. Bifurcation Diagram of the Tent Mapping Modify the computer program given in the text (Table 7.9) to draw the bifurcation diagram for the **tent mapping**

$$x_0 = 0.5$$
$$x_{n+1} = 2\lambda \begin{cases} x_n & 0 \le x_n < 0.5 \\ 1 - x_n & 0.5 \le x_n \le 1 \end{cases} \quad (n = 0, 1, 2, ...)$$

for different values of $0 \le \lambda \le 1$.

15. Windows in the Bifurcations Diagram In the bifurcation diagram of Figure 7.14, note the presence of dark bands or curves, which indicate regions where there is a high probability that the solution will occur. An unshaded vertical band occurs between $3.83 < \lambda < 3.85$, which indicates that in the case of gypsy moths, the population is no longer chaotic and is predictable, taking on only three different values. Write a computer program to compute the first 1000 points of the solution of the logistic equation for $\lambda = 3.84$ to verify that the limiting solution is an attracting three-cycle. Use the seed $x_0 = 0.5$.

16. Cobweb Diagrams Write a computer program that draws the cobweb diagram of the logistic equation with the given parameter λ and initial conditions:

(a) $\lambda = 1$ $\quad x_0 = 0.1$ $\qquad$ (c) $\lambda = 3$ $\quad x_0 = 0.5$

(b) $\lambda = 2.5$ $\quad x_0 = 0.3$ $\qquad$ (d) $\lambda = 4$ $\quad x_0 = 1.0$

What is the nature of the limiting solution? The computer program should compute and plot the first 250 iterates of the solution of the logistic equation. Are the cobweb diagram and the graphed solution consistent with one another?

17. The Cobweb Diagram and Cos x Enter any number into your calculator and start hitting the cosine key. No matter what number you enter, the sequence will converge to 0.73908... if your calculator is working in radians or 0.99984... if it is working in degrees. Draw a cobweb diagram to say why this is true. Are there any seeds that don't lead to one of these values?

18. Cobweb Analysis of the Baker Mapping Write a computer program to convince yourself that the **Baker map**

$$x_{n+1} = \begin{cases} 2x_n & 0 \le x_n < 0.5 \\ 2x_n - 1 & 0.5 \le x_n \le 1 \end{cases} \quad (n = 0, 1, 2, ...)$$

gives rise to chaotic behavior for any $0 \le x_0 \le 1$.

19. Chaotic Numerical Iterations Iteration is a common tool used in numerical analysis. For example, **Newton's method** for approximating a root of $f(x) = 0$ consists in picking a seed x_0 and then iterating the formula

$$x_{n+1} = x_n - \frac{f(x_n)}{f'(x_n)} \qquad (4)$$

The sequence of values $x_1, x_3, ...$ should converge to a root. Write a computer program that draws the bifurcation diagram for the Newton's equation formula (4) for finding a root of the equation

$$x^3 + \lambda x + 1 = 0$$

for $-1.25 \le \lambda \le -1.3$. *Hint:* As an aid in scaling the vertical axis of the bifurcation diagram, the computed values of x_n lie in the interval $-0.25 \le x_n \le 0.25$.

20. A Geometric Interpretation of Chaos The logistic mapping can be interpreted geometrically as a process in which the interval $[0, 1]$ is continually "folded" and "stretched." Note in Figure 7.15 that for $\lambda = 4$ the points near the ends of the interval $[0, 1]$ are "folded" over on top of themselves, while points near 0.5 get "stretched" into new points near the endpoint of the new interval $[0, 1]$. The repeated folding and stretching process gives rise to chaotic motion. (Basically, the process resembles the one used in

Figure 7.15
Many mappings whose motion is chaotic essentially "stretch" and "fold" the domain of the mapping. Here the interval $[0, 1]$ is stretched near the middle and folded at the endpoints for $\lambda = 4$.

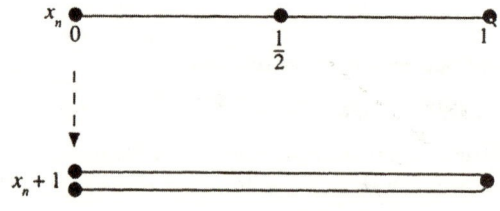

making taffy candy or kneading dough.) Which of the following mappings "stretch" and "fold" the interval $[0, 1]$ when $\lambda = 1$. Do you think the equation will give rise to a chaotic solution?

(a) $x_{n+1} = \lambda \sin \pi x_n$ $\quad 0 \le x_n \le 1$ $\quad 0 \le \lambda \le 1$

(b) $x_{n+1} = \begin{cases} 2\lambda x_n & 0 \le x_n < 0.5 \\ 2\lambda(1 - x_n) & 0.5 \le x_n \le 1 \end{cases} \quad 0 \le \lambda \le 1$

The Dynamical System $x_{n+1} = x_n{}^2 + c$

The simple squaring system, where both c and x_n are real numbers, is an example of a simple iterative or dynamical system whose long-run solution is not so simple. Problems 21–23 examine this dynamical system.

21. **Hmmmmm** Examine the long-run solution of the iterative systems
    (a) $x_{n+1} = x_n{}^2 - 1.3 \qquad x_0 = 0$
    (b) $x_{n+1} = x_n{}^2 - 1.755 \quad x_0 = 0$
    (c) $x_{n+1} = x_n{}^2 - 2 \qquad\quad x_0 = 0$

22. **Bifurcation Diagram for the Squaring System** Write a computer program to draw the bifurcation or orbit diagram for the iterative system

$$x_{n+1} = x_n{}^2 + c \qquad x_0 = 0$$

for $-2 \le c \le 0.25$. *Hint:* As an aid in setting up the range for the *y*-axis, the computed values of x_n always lie in the interval $-2 \le x_n \le 2$. Later, if you want to zoom in on a region of the entire bifurcation diagram, simply restrict both the *c*-interval and the x_n-interval.

23. **Zooming in on the Bifurcation Diagram** Write a computer program to zoom in on the bifurcation diagram computed in Problem 22 by restricting the range of values $-1.525 \le c \le -1.05$ and printing out the bifurcation diagram only for values of *x* in the interval $-0.2 \le x \le 0.2$. Can you see the three-cycle window in the diagram?

24. **Attracting and Repelling Fixed Points** We have seen that a fixed point x_f of an iterative system $x_{n+1} = f(x_n)$ is a point that satisfies $x_f = f(x_f)$. It can be shown that a fixed point is
    (a) attracting if $|f'(x_f)| < 1$
    (b) repelling if $|f'(x_f)| > 1$
    (c) neither if $|f'(x_f)| = 1$
    Determine whether the fixed points in Figure 7.16 are attracting, repelling, or neither, and draw a cobweb path to verify your answers.

25. **Journal Entry** Summarize to yourself the concept of the period-doubling sequence to chaos.

Figure 7.16
The classification of fixed points of an iterative system $x_{n+1} = f(x_n)$ as either *attracting* (nearby iterates converge to the point), *repelling* (nearby iterates move away from the point), or *neither* (nearby points move neither toward nor away from the point) depends on whether the absolute value of the derivative at the fixed point is less than, greater than, or equal to 1, respectively.

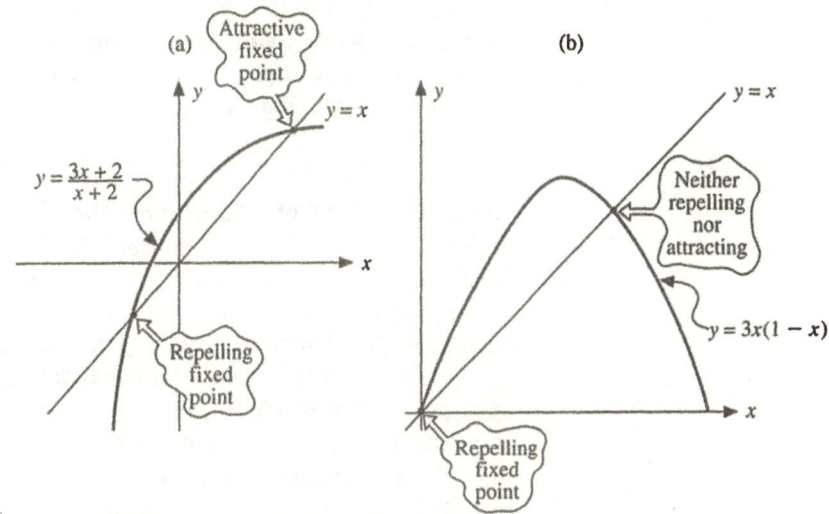

ADDITIONAL READING ON CHAOS

1. *Chaos: Making of a New Science* by James Gleick (Penguin Books, New York, 1987). This paperback is an excellent nontechnical introduction to the subject of chaos.

2. *Turbulent Mirror* by John Briggs and David Peat (Harper and Row, New York, 1989). Another excellent introduction to the subject of chaos. Slightly more technical than Gleick's book.

3. *An Introduction to Chaotic Dynamical Systems* by Robert L. Devaney (Addison-Wesley, Reading, Mass., 1989). One of the first textbooks intended to give an entire course in dynamical systems and chaos.

ITERATIVE SYSTEMS: JULIA SETS AND THE MANDELBROT SET (OPTIONAL)

PURPOSE

To show how the solution of a system of difference equations, which we call an iterative system, depends on the initial condition and how this gives rise to Julia sets and the Mandelbrot set for this iterative system.

Although the material in this section has not traditionally been part of the curriculum of differential equations, many of the concepts in this section are beginning to find their way into the thinking of the broader area of *dynamics*.* One of the surprises in recent times has been the realization that even the simplest dynamical systems can exhibit extremely complex behavior, leading to such newly discovered concepts as *fractals, chaotic behavior*, and *strange attractors*. We begin with our old friend, the first-order difference equation

$$x_{n+1} = f(x_n) \tag{1}$$

which, in today's world of computers and iteration, is usually referred to as an **iterative system** or a **discrete dynamical system.** As we have seen, the idea is to start with an initial condition, or *seed* x_0, and *iterate* the function f, using the previous output as input and computing the values

$$x_1 = f(x_0), \qquad x_2 = f(x_1), \qquad x_3 = f(x_2), \qquad x_4 = f(x_3), \qquad \cdots \tag{2}$$

of the solution, or **orbit.** Although we have already studied (linear) difference equations in which x_0, x_1, ... are real numbers, Eq. (2) could just as well define a sequence of complex numbers, vectors, matrices, sets, or anything.

It may come as a surprise to the reader, but many properties of iterative systems have been discovered only in the past few years. This contrasts with most of the material studied in a beginning course in differential equations, which was discovered during the eighteenth and nineteenth centuries. Although at one time the subject of *discrete* dynamical systems was considered the ''poor cousin'' of *continuous* dynamical systems (i.e., differential equations) and more limited in scope, researchers are starting to realize that this is not the case. This section shows how a very simple iterative system gives rise to interesting ideas that even now are not completely understood.

THE QUADRATIC ITERATIVE SYSTEM AND ITS JULIA SETS

In the previous section we were given the seed x_0 of the logistic equation and found the long-term behavior of the solution. In this section we are interested in the ''opposite'' problem: How does the value of the seed affect the behavior of the orbit? Although the answer to this equation depends on the specific dynamical system, we will see that

---

* The subject of *dynamics* is the study of motion and includes difference equations, differential equations, iterative systems, and so on.

even for the simplest dynamical system the answer can be quite complex. For example, consider the **complex quadratic** iterative system

$$z_{n+1} = z_n^2 + c \qquad (n = 0, 1, 2, \ldots) \tag{3}$$

where $c = c_1 + ic_2$ is a given complex number, and the sequence of complex numbers, $z_n = x_n + iy_n$, is determined once a seed $z_0 = x_0 + iy_0$ has been chosen.

To determine how the seed z_0 affects the orbit of Eq. (3), we must first pick a *specific* complex number c so that we are talking about a *specific* dynamical system.* For example, if we select $c = 1 + i$, then a seed $z_0 = 0$ results in the orbit

$$
\begin{aligned}
z_0 &= 0 \\
z_1 &= 0^2 + (1 + i) = 1 + i \\
z_2 &= (1 + i)^2 + (1 + i) = 1 + 3i \\
z_3 &= (1 + 3i)^2 + (1 + i) = 1 + 6i - 9 + 1 + i = -7 + 7i \\
z_4 &= (-7 + 7i)^2 + (1 + i) = 1 - 97i \\
&\;\vdots
\end{aligned}
\tag{4}
$$

Note: Although we have written the dynamical system (3) as a *single* complex equation, it can in fact be written as a system of *two* real difference equations by observing that

$$z^2 = (x + iy)^2 = x^2 - y^2 + 2xyi$$

Hence the dynamical system (3) can be rewritten as

$$
\begin{aligned}
x_{n+1} + iy_{n+1} &= (x_n + iy_n)^2 + (c_1 + ic_2) \\
&= (x_n^2 - y_n^2 + c_1) + i(2x_n y_n + c_2)
\end{aligned}
$$

Then, setting real and complex parts equal to each other gives the equivalent system of two real difference equations

$$
\begin{aligned}
x_{n+1} &= x_n^2 - y_n^2 + c_1 \\
y_{n+1} &= 2x_n y_n + c_2
\end{aligned}
$$

Our goal now is to distinguish seeds z_0 that have *bounded orbits*[†] (convergent, periodic, etc.) from seeds whose orbits are *unbounded* (go to infinity). Furthermore, if a seed *has* a bounded orbit, do nearby seeds have bounded orbits? These are important questions because experimental observations always contain some error, and as one of the pioneers of chaos theory, Jim Yorke, stated, ''The whole electrical power grid of the East Coast is an oscillatory (dynamical) system, most of the time stable, and one would like to know what happens when you disturb it (change the seed).''

The collection of starting points z_0 of the iterative system (3) whose orbits are *bounded* (remain inside some given circle in the complex plane) is called a **Julia set**

---

* For each c we have a different iterative system, so in effect $z_{n+1} = z_n^2 + c$ defines a *one-parameter family* of iterative systems.

[†] When we say that the orbit is bounded or unbounded, we are referring to whether the sequence of absolute values $|z_n| = \sqrt{x_n^2 + y_n^2}$ is bounded or unbounded, respectively.

and denoted by J_c. (Note that there is a separate Julia set for each parameter c.) This can be stated mathematically as

$$J_c = \{z_0: z_0, \, z_0^2 + c, \, (z_0^2 + c)^2 + c, \, \cdots \not\rightarrow \infty\} \tag{5}$$

To determine the starting points z_0 that have bounded orbits, we use a computer to examine the orbits of starting values z_0 in a square *grid* in the complex plane. For example, if $c = -1$, then taking each z_0 in a grid of points about the origin, we iterate z_0 a certain number of times (maybe 100 or 200) and determine whether the orbit is bounded or unbounded. If the orbit is bounded, we color z_0 *black;* if the orbit is unbounded, we color z_0 *white*. For example, the seed $z_0 = 0$ would give rise to the orbit

$$z_0 = 0$$
$$z_1 = 0^2 - 1 = -1$$
$$z_2 = (-1)^2 - 1 = 0$$
$$z_3 = 0^2 - 1 = -1$$
$$\vdots$$

Since this orbit is bounded (it alternates between 0 and -1), we color the seed value $z_0 = 0$ black. On the other hand, the seed $z_0 = 1 + i$ has the orbit

$$z_0 = 1 + i$$
$$z_1 = (1 + i)^2 - 1 = -1 + 2i$$
$$z_2 = (-1 + 2i)^2 - 1 = -4 - 4i$$
$$z_3 = (-4 - 4i)^2 - 1 = -1 + 32i$$
$$z_4 = (-1 + 32i)^2 - 1 = -1024 - 64i$$
$$z_5 = 1044479 + 131072i$$
$$\vdots$$

Although we haven't proven it, it seems clear that this orbit is unbounded. As a practical test, for the quadratic iterative system, $z_{n+1} = z_n^2 + c$, one can prove that if the absolute value of any iterate z_n satisfies

$$|z_n| = \sqrt{x_n^2 + y_n^2} > |c| + 1$$

then the orbit is unbounded. For example, if the parameter is $c = -1$, then if the absolute value of any iterate is greater than $|c| + 1 = 2$, one can stop the iteration and color the seed white. We now repeat this process for *each seed* in the grid. The Julia set J_{-1} was approximated with the help of a computer* and is shown in Figure 7.17. Other Julia sets, J_c, are shown[†] in Figure 7.18.

Some Julia sets exhibit exotic patterns shaped like those seen in a kaleidoscope, while others are shaped like dragons and others like disconnected specks of dust. One

---

* The Julia sets and the Mandelbrot set (Figure 7.20) in this section were originally drawn by Homer Smith of ART MATRIX. We didn't superimpose the complex coordinate system over these sets for aesthetic reasons.

[†] Figure 7.21 was taken with permission from the book, *Fractal Creations,* The Waite Group, Inc. (1991).

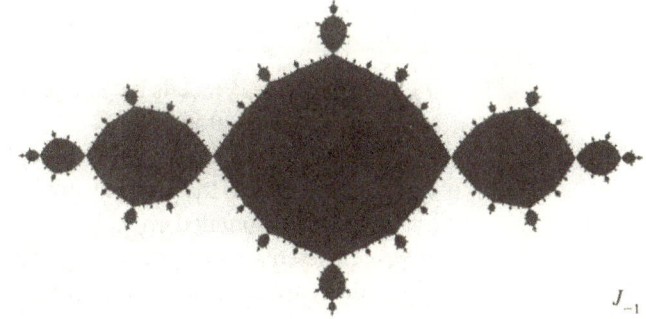

J_{-1}

Julia set even has the shape of a horse's tail, and another has the shape of a rabbit. Most Julia sets are **fractals**,* meaning a set that exhibits *increasing detail with increas-*

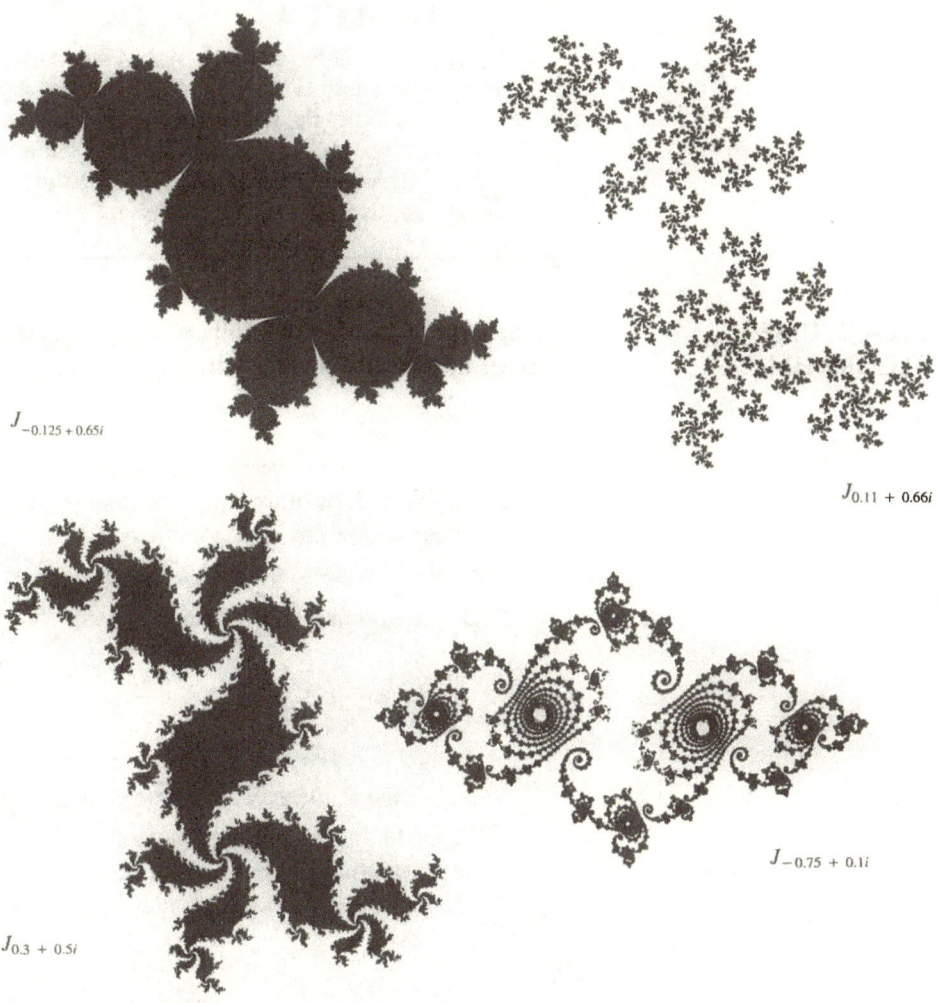

$J_{-0.125 + 0.65i}$

$J_{0.11 + 0.66i}$

$J_{-0.75 + 0.1i}$

$J_{0.3 + 0.5i}$

* Fractal sets are the creation of the mathematician Benoît Mandelbrot and are the embodiment of irregularity. Although a more precise definition of fractals can be given, they are geometric shapes that exhibit increasing detail with increasing magnification. Mandelbrot informally defines fractals as objects such that if a piece of the fractal is suitably magnified to become the same size as the whole, it should look like the whole. Another definition says that a fractal is a geometric shape of *fractional* dimension.

ing magnification. Fractals have begun to play an important role as a geometric means to describe chaotic motion.

The Julia sets illustrate the complex manner in which the orbits of dynamical systems depend on the starting point. Observe the complex way in which seeds of bounded and unbounded orbits interweave with one another. If orbits of most nonlinear physical systems are as sensitive to initial conditions as they are for this "trivial" quadratic system, it makes one wonder about the ability to predict the long-run behavior of many nonlinear phenomena. The prospect of a complex Julia set spreads terror among engineers and computer scientists who are involved in the computer simulation of such systems.

HISTORICAL NOTE

The Julia set is named in honor of the French mathematician Gaston Julia (1893–1978), who along with Pierre Fatou studied iterations of rational functions of complex variables during World War I. Normally, the set of points with bounded orbits is called the "filled-in" Julia set, and the *boundary* of this set is called the Julia set. However, we will not be discussing the boundary of the set, so we simply refer to the set as the "Julia set."

ESCAPE TIME ALGORITHM

A simple method for finding Julia sets is the **escape time algorithm.** Although there are other methods for finding Julia sets, the escape time algorithm is easy to program on a computer.

Escape Time Algorithm for Finding Julia Sets for $z_{n+1} = z_n^2 + c$

Start with a complex parameter $c = p + iq$ and a seed $z_0 = x_0 + iy_0$. Often the first seed* is chosen to be $(x_0, y_0) = (-1.5, -1.5)$.

Step 1. Beginning with the seed (x_0, y_0), compute successive values

$$x_{n+1} = x_n^2 - y_n^2 + p$$
$$y_{n+1} = 2x_ny_n + q \tag{6}$$

Step 2. If the sequence of absolute values $|z_n| = \sqrt{x_n^2 + y_n^2}$ grows without bound ("goes to infinity"), then the initial point (x_0, y_0) is *not* a member of the Julia set. Shade either points in the Julia set or points outside the Julia set (depending on which style looks better to you).

Step 3. Choose a nearby initial point (x_0, y_0) and repeat Steps 1 and 2 to determine whether this new point (x_0, y_0) belongs to the Julia set. Continue this process for all points of a grid in the square. Often the square with opposite corners $(-1.5, -1.5)$ and $(1.5, 1.5)$ is chosen.

---

* It is convenient to represent the complex numbers as pairs of real numbers in the plane. For example, we represent z_n by the pair (x_n, y_n) and the complex parameter c by (p, q).

A simple BASIC program is given in Table 7.11 to draw Julia sets of the mapping $z_{n+1} = z_n^2 + c$ on a computer. By changing statements 120 and 130 it is possible to draw the Julia sets of other iterative systems of complex functions. Figure 7.19 shows a Julia set generated on a computer by the escape time algorithm.

Table 7.11
BASIC Program JULIA to Find the Julia Sets of
$z_{n+1} = z_n^2 + c$

```
 10      REM PROGRAM JULIA COMPUTES (APPROXIMATION) OF A
 20      JULIA SET FOR SQUARING ITERATION. POINTS IN THE
 30      JULIA SET ARE PLOTTED IN STATEMENT 190
 40      SCREEN 2
 50      WINDOW (-1.5, -1.5) - (1.5, 1.5)
 60      INPUT "ENTER REAL AND COMPLEX PARTS FOR C"; P, Q
 70      FOR Y0 = -1.5 TO 1.5 STEP .01
 80          FOR X0 = -1.5 TO 1.5 STEP .01
 90              X = X0
100              X = Y0
110              FOR ITER = 1 TO 20
120                  XNP1 = X*X - Y*Y + P
130                  YNP1 = 2*X*Y + Q
140                  SS = XNP1*XNP1 + YNP1*YNP1
150                  IF SS > 4 THEN
160                      GOTO 200
170                  END IF
180              NEXT ITER
190              CIRCLE (X0, Y0), 0
200          NEXT XO
210      NEXT Y0
220      END
```

Figure 7.19
The "dendrite" Julia set J_i computed by the escape time algorithm. In this 90,000-point grid, the black region represents the Julia set. This set was computed in 30 seconds on a 80386-33-MHz microcomputer.

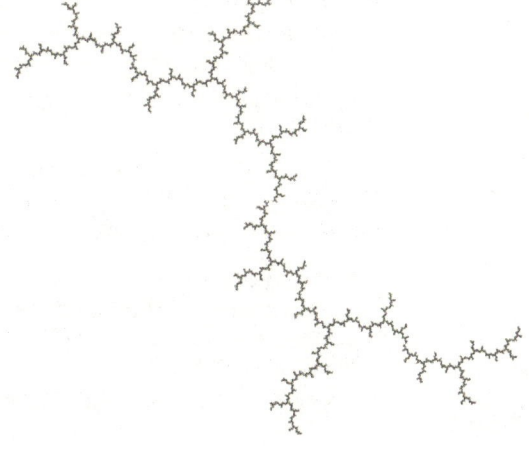

THE MANDELBROT SET: A CATALOG OF JULIA SETS

In 1980 the mathematician Benoît Mandelbrot discovered that he could construct a single set that would act as a sort of catalog for *all* Julia sets. The idea is simple: Find the values of the parameter c in the complex *c-plane* for which the orbit of $z_0 = 0$ remains bounded after a large number of iterations. The computer algorithm is essentially the same as it was when one found Julia sets, except that here one takes different values of the parameter c in the grid and iterates the equation $z^2 + c$, always starting at $z_0 = 0$. The complex numbers c for which the orbit of $z_0 = 0$ remains bounded is called the **Mandelbrot set,** denoted $\mathfrak{M}$. Stated another way,

$$\mathfrak{M} = \{c: c, c^2 + c, (c^2 + c)^2 + c, \ldots \nrightarrow \infty\} \tag{7}$$

In theory it can be approximated by using just a pocket calculator. See Figure 7.20.

Although impossible to tell from Figure 7.20, the *boundary* of the Mandelbrot set is a fractal.*

Figure 7.20
The Mandelbrot set has sometimes been called the most complex object in mathematics. It may or may not be, but it is one of the most popular in the nonmathematical community.

In the Mandelbrot set in Figure 7.20 the origin in the complex plane is located in the center of the largest "blob." We have not superimposed the complex coordinates on the Mandelbrot set for aesthetic reasons.

The Mandelbrot set, $\mathfrak{M}$, is useful in determining properties of Julia sets. For example, if c belongs to the Mandelbrot set, then one can show that the boundary of the Julia set, J_c, is *connected*.[†] If c does *not* belong to $\mathfrak{M}$, then the boundary of J_c is not connected, and it appears as disjoint pieces of dust, sometimes called "Fatau dust."

It should be understood that there is only one Mandelbrot set but a Julia set J_c for each value of c. Various Julia sets have been drawn for different values of c in Figure 7.21.

The Mandelbrot set has been dubbed by some people the "most complex" set in mathematics. At first sight, it appears to be like a molecule made up of bonded atoms, the largest being shaped like a cardioid and the smaller ones being nearly circular.

---

* Although there is no universally accepted definition of a fractal, one definition states that fractals are sets that exhibit increasing detail with increasing magnification (very jagged sets). A more mathematical definition states that fractals are sets of fractional dimension. A good beginning reference to fractals is *Chaotic Vibrations* by Francis C. Moon (John Wiley & Sons, New York, 1987).

[†] Roughly, a set is *connected* if it is in one piece.

Upon close inspection, however, it shows an infinite number of smaller molecules, all shaped like the bigger one. Figure 7.21 shows the amazing structure of this set as one zooms in closer and closer. Note how intertwined are the regions of bounded orbits (colored black) and the regions of unbounded orbits (colored white). With access to a large enough computer and a high-resolution computer monitor, one can "zoom in" (i.e., construct a finer grid in the c-plane) and examine the unearthly landscape in any region of the Mandelbrot set. As Cornell pioneer mathematician John Hubbard stated, "There are zillions of beautiful spots to visit."

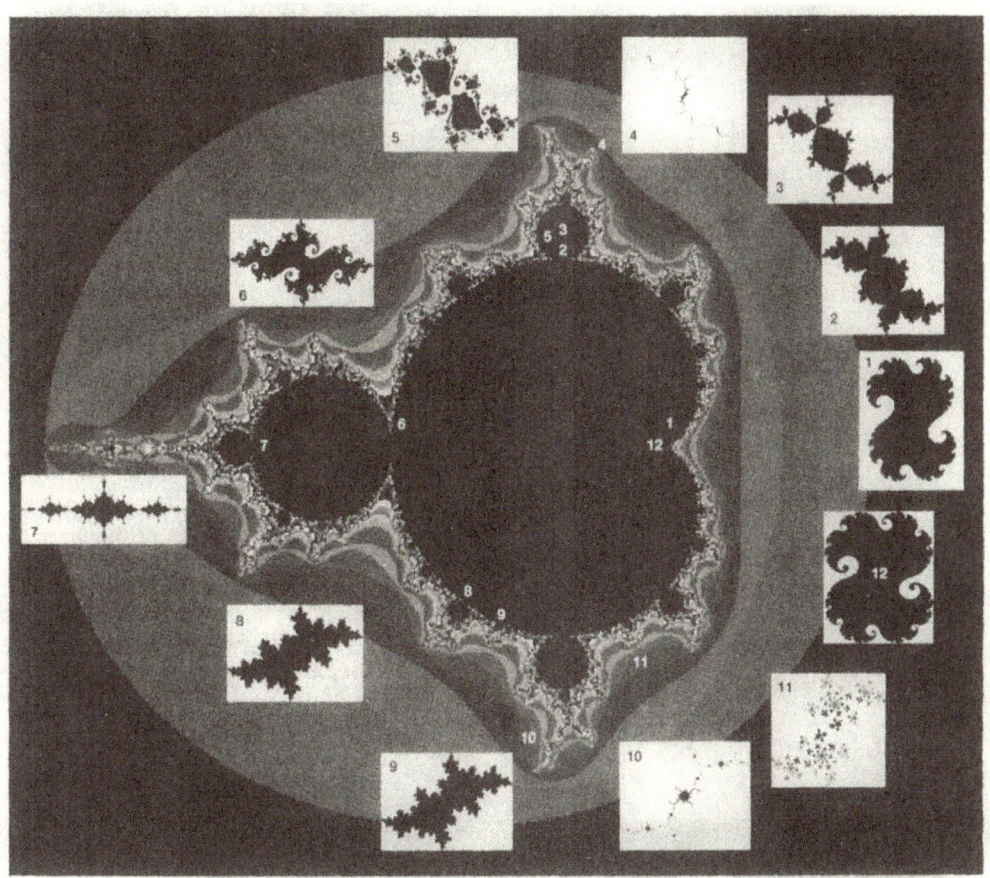

Figure 7.21
For complex numbers c in the Mandelbrot set, the corresponding Julia set J_c is connected, whereas for c not in the Mandelbrot set the Julia set J_c is not connected. The most interesting Julia sets J_c are those for which c lies near the boundary of the Mandelbrot set. (Reprinted by permission from The Waite Group's *Fractal Creations* by Timothy Wegner and Mark Peterson, © 1991 Waite Group Press, Corte Madera, California.)

There are several algorithms for drawing approximations of the Mandelbrot set. One of the simplest is the escape time algorithm. This algorithm is similar to the escape time algorithm for finding Julia sets except that in this case the initial point is always taken to be $z_0 = 0$ ($x_0 = y_0 = 0$), and the *parameter* "runs" over a given set of grid points. Table 7.12 lists a simple BASIC program that shades the Mandelbrot set using this algorithm.

Table 7.12
BASIC Program MANDEL to
Find the Mandelbrot Set

```
10      REM PROGRAM MANDEL DRAWS THE MANDELBROT SET
20      SCREEN 2
30      WINDOW (-2, -2) - (2, 2)
40      FOR Y0 = -2 TO 2 STEP .01
50          FOR X0 = -2 TO 2 STEP .01
60              X = 0
70              Y = 0
80              FOR ITER = 1 TO 20
90                  XNP1 = X*X - Y*Y + X0
100                 YNP1 = 2*X*Y + Y0
110                 IF XNP1 * XNP1 + YNP1 * YNP1 > 4 THEN
120                     X = XNP1
130                     Y = YNP1
140                     CIRCLE (X0, Y0), 0
150                     GOTO 180
160                 END IF
170             NEXT ITER
180         NEXT X0
190     NEXT Y0
200     END
```

Note: The study of dynamical systems (of which differential equations is a part) is undergoing nothing less than a revolution, motivated by the concept of chaotic motion. One of the principal facets of this new area of mathematics is the concept of *experimental mathematics.* To some, the phrase ''experimental mathematics'' may seem to be an oxymoron or contradiction in terms. Mathematics is supposed to be founded on logic in which theorems are proven from a given collection of theorems in which experiments have no place. However, in actual practice, math-

HISTORICAL NOTE

Benoit Mandelbrot (1924–   ) was born in Warsaw, Poland, and is the person responsible for giving the fractal its name. As Euclid is associated with plane geometry, it might be argued that Mandelbrot will forever be associated with the ''geometry of nature.'' Benoit Mandelbrot is an IBM Fellow at the Thomas J. Watson Research Center in Yorktown Heights, New York, and a Professor of Mathematics at Yale University. Often called the ''father of fractals,'' his research has concentrated on extreme and unpredictable irregularity in natural phenomena in the physical, social, and biological sciences. His seminal book, *The Fractal Geometry of Nature* (Freeman, New York, 1977), is one of the most influential mathematics books of this century.

ematicians carry out experiments in much the same way as other scientists: with paper and pencil, trial and error, pictures, diagrams, or whatever else is available. After one's intuition has been developed with experimentation, relationships are proven rigorously.

Computers and computer graphic systems have enabled one to investigate dynamical systems that have not been studied before.

PROBLEMS: Secton 7.6

For Problems 1–7, use program JULIA in Table 7.11 to draw pictures of the Julia set J_c for the quadratic iterative system $z_{n+1} = z_n^2 + c$.

1. $c = i$ (the "dendrite")
2. $c = -0.12256117 + 0.74486177i$ ("Doudady's Rabbit")
3. $c = -0.125 + 0.64919i$ (the "Fat Rabbit")
4. $c = -0.75$ (you name it)
5. $c = 0.3 + 0.5i$ (the "dragon")
6. $c = -0.9 + 0.12i$ (you name it)
7. $c = -0.3$ (boring)
8. **Only Simple Julia Sets** The only Julia sets of

$$z_{n+1} = z_n^2 + c$$

that have simple geometric shapes are

$J_0 = \{(x, y): x^2 + y^2 \leq 1\}$ (unit disk)
$J_{-2} = [-2, 2]$ (interval on the real line)

Use program JULIA in Table 7.11 (or paper and pencil) to verify these Julia sets.

Other Julia Sets

It is possible to draw Julia sets for other mappings other than the squaring map. For Problems 9–10, use program JULIA in Table 7.11 to draw the Julia sets of the following maps. Rewrite the complex equation as two real equations and then change Statements 120 and 130 in program JULIA. In all problems, c is a complex parameter.

9. $z_{n+1} = z_n^3 + c$ 10. $z_{n+1} = z_n^4 + c$
11. **Your Very Own Julia Set** The most interesting Julia sets occur for values of c that are near the boundary of the Mandelbrot set. Use program JULIA to find some interesting Julia sets.
12. **Mandelbrot Set** Use the program MANDEL in Table 7.12 to draw the Mandelbrot set.
13. **Other Mandelbrot Type Sets** You can modify MANDEL listed in Table 7.12 by changing statements 90 and 100 to find "Mandelbrot-type" sets for essentially any complex function you wish. Find the Mandelbrot set for the map $z_{n+1} = c(z_n - 1/z_n^2)$, where $c = p + iq$. *Hint:* Rewrite $z - 1/z^2$ in terms of its real and complex parts and make the approximate changes in statements 90 and 100.
14. **Journal Entry** Summarize to yourself Julia sets and the Mandelbrot set.

ADDITIONAL READING ON JULIA SETS AND THE MANDELBROT SET

1. *The Fractal Geometry Nature* by Benoit B. Mandelbrot (W. H. Freeman and Company, New York, 1977). The definitive overview of the subject of fractals, including Julia sets and the Mandelbrot set.

2. *The Science of Fractal Images* edited by Heinz-Otto Peitgen and Dietman Saupe (Springer-Verlag, New York, 1988). An excellent summary of fractals (including Julia sets and the Mandelbrot set). It includes beautiful pictures and computer programs for constructing these sets.

3. *Fractals Everywhere* by Michael Barnsley (Academic Press, New York, 1988). Another of the excellent new books dealing with fractals written by another pioneer in the field. The concept of the *iterated function system*, a procedure that allows one to construct fractals, is stressed throughout the book.

4. *Dynamic Systems and Fractals* by K. Becker and M. Dörfler (Cambridge University Press, New York, 1989). This book contains dozens of computer programs written in Pascal with which the reader can experiment.

Nonlinear Differential Equations and Chaos

8.1 PHASE PLANE ANALYSIS OF AUTONOMOUS SYSTEMS

8.2 EQUILIBRIUM POINTS AND STABILITY FOR LINEAR SYSTEMS

8.3 STABILITY: ALMOST LINEAR SYSTEMS

8.4 CHAOS, POINCARÉ SECTIONS, AND STRANGE ATTRACTORS

8.1   PHASE PLANE ANALYSIS OF AUTONOMOUS SYSTEMS

PURPOSE

To introduce the concepts of the state vector and the phase plane of a system of two autonomous differential equations. We then show how they can be used to illustrate descriptive properties of the system, even when the solution of the system is unknown.

QUALITATIVE ANALYSIS AND NONLINEAR EQUATIONS

Usually, it is not possible to find solutions in closed form for nonlinear differential equations. This is not because past investigators lacked ability, but because the collection of elementary functions, such as $p_n(x)$, e^x, $\sin x$, $\cos x$, $\ln x$, and so on, is too limited to express *all* the solutions of the wide variety of nonlinear equations that arise in practice. Even when a formula is found, it is often expressed as an infinite series or as an integral, so it is difficult to determine the relevant properties of the solution. For instance, solutions of the relatively simple nonlinear pendulum equation

$$\ddot{\theta} + \frac{g}{L} \sin \theta = 0 \tag{1}$$

are expressed in terms of a computationally awkward class of functions known as *Jacobi elliptic integrals*.

At the end of the nineteenth century the French mathematician Jules Henri Poincaré introduced a new approach to the study of differential equations. Realizing that analytic solutions were unattainable for most nonlinear equations, he focused his efforts on finding *descriptive* properties of solutions of differential equations. This area of differential equations is called the **qualitative theory*** of differential equations in contrast to the **quantitative theory** of differential equations, which is more concerned with actually finding solutions. For example, in qualitative theory, one would like to know the limiting behavior of all solutions of an equation as $t \to \pm \infty$. Does the solution approach a constant, a periodic cycle, infinity, or something else? These are important questions, since many phenomena ''live'' in their limiting behavior. In many systems, transient solutions approach zero so rapidly that only the limiting behavior is observed.

One should realize that although we are ''backing off'' from finding solutions, that does not mean that qualitative analysis is easier than quantitative analysis. Keep in mind that the basic goal of qualitative analysis is to analyze *nonlinear* differential equations, which are considerably more difficult than linear equations.

The cornerstone of the Poincaré qualitative approach is the *phase plane*, which often reveals many important properties of solutions of a differential equation, even when the solution is unknown. The use of the phase plane gives the qualitative theory of differential equations a more *geometric* flavor whereas the quantitative theory of differential equations has more of an *analytic* flavor. The following mass-spring system illustrates some of Poincaré's ideas.

---

* The term *qualitative* refers to ''big, small, tall, short, etc.,'' while the term *quantitative* refers to specific quantities, such as 2.3, π, and $\sin t$.

PHASE PLANE ANALYSIS OF THE MASS-SPRING SYSTEM

Consider the linear mass-spring equation

$$m\ddot{x} + kx = 0 \tag{2}$$

where until now we have concentrated on finding the dependent variable x in terms of the independent variable t, which is given by $x(t) = c_1 \sin \omega_0 t + c_2 \cos \omega_0 t$, where $\omega_0^2 = k/m$ and c_1 and c_2 are arbitrary constants. It is possible, however, to obtain useful information about the solution of the differential equation (2) without actually knowing the solution.* The idea behind phase plane analysis is to find an *algebraic relationship* between x (position) and $\dot{x}$ (velocity) and to plot this relationship in the Cartesian plane, where the horizontal axis is labeled x and the vertical axis is labeled $\dot{x}$. This $x\dot{x}$-plane is called the **phase plane.** If all this can be done, then as t increases, one obtains paths $(x, \dot{x}) = (x(t), \dot{x}(t))$ in the phase plane, called **orbits** or **trajectories.** Although these paths are *not* solutions of Eq. (2), they do provide information about the solution. In the case of the mass-spring equation, if we multiply Eq. (2) by $2\dot{x}$ and replace $\omega_0^2 = k/m$, we get

$$2\dot{x}\ddot{x} + 2\omega_0^2 \dot{x}x = 0 \tag{3}$$

or simply

$$\frac{d}{dt}\left(\dot{x}^2 + \omega_0^2 x^2\right) = 0 \tag{4}$$

Integrating gives

$$\dot{x}^2 + \omega_0^2 x^2 = c^2 \tag{5}$$

where c is an arbitrary constant. Equation (5) describes the trajectories of the mass-spring system. These trajectories consist of a family of ellipses centered at the origin, as shown in Figure 8.1. Depending on the initial conditions $(x_0, \dot{x}_0)$ of the mass, the point $(x, \dot{x})$ will move along a specific ellipse, giving the position and velocity of the mass at each instant of time.

Figure 8.1
Trajectories of the mass-spring equation in the phase plane

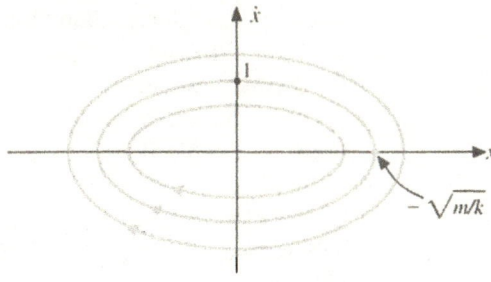

* Of course, for this linear equation, carrying out a phase plane analysis doesn't provide us with any new information, since we already know the solution. For many nonlinear equations, however, for which we cannot find an analytical solution, such an analysis is often very useful.

Of course, for Eq. (2) the trajectories in Figure 8.1 could have been found directly from the solution, but for many nonlinear equations it is possible to find its trajectories *without knowing the solution.*

The direction of motion of a trajectory in Figure 8.1 can be determined from the general fact that x always *increases* in the upper half-plane (since $\dot{x} > 0$) and *decreases* in the lower half-plane (since $\dot{x} < 0$). This fact is denoted by the arrowheads drawn on the trajectories. Note that all the trajectories in Figure 8.1 are *closed paths*, referring to the fact that the solutions of $m\ddot{x} + kx = 0$ are periodic. The trivial solution $x(t) \equiv 0$ corresponds to the origin $(x, \dot{x}) = (0, 0)$ in the phase plane, referring to the fact that if the system ever reaches this point, it remains there.

We now summarize many of the above ideas in more generality. Since any second-order equation can be written as a system of two first-order equations, our discussion focuses on first-order systems.

HISTORICAL NOTE

The qualitative theory of differential equations, which is the focus of this chapter, is the creation of the French mathematician Jules Henri Poincaré (1854–1912), who developed the theory in a series of papers dating from 1880 to 1886. Poincaré was considered the leading mathematician of his time and is often called the "last universalist," referring to the fact that he was the last mathematician to make major discoveries in every mathematical discipline. He made major contributions to ordinary and partial differential equations, number theory, real and complex function theory, celestial mechanics, and other fields.

THE GENERAL PHASE PLANE AND RELATED IDEAS

Consider the system of two equations

$$\dot{x} = P(x, y) \tag{6}$$
$$\dot{y} = Q(x, y)$$

where P and Q are continuous functions of x and y with continuous first partial derivatives. Systems of differential equations such as system (6) in which P and Q do not depend explicitly on t are called **autonomous systems.** The values of $x = x(t)$ and $y = y(t)$ defined by Eqs. (6) define a point (x, y) in the phase plane, called the **state vector.** As the independent variable t changes, the state vector moves about in the phase plane. An algebraic equation relating the two dependent variables x and y defines a path (or trajectory or orbit) in the phase plane. With this interpretation we can think of solutions of differential equations as state vectors moving along trajectories in the phase plane.

The simplest trajectories are those that settle down to a steady equilibrium. A point (x_e, y_e) in the phase plane is called an **equilibrium point** (or *steady state* point) if both derivatives $\dot{x}$ and $\dot{y}$ are zero. In other words,

$$\dot{x} = P(x_e, y_e) = 0 \tag{7}$$
$$\dot{y} = Q(x_e, y_e) = 0$$

Equilibrium points are points where the motion of the state vector is at rest. If an equilibrium point (x_e, y_e) of Eqs. (6) is the initial condition for Eqs. (6), then the resulting solution will be the constant solution $x(t) \equiv x_e, y(t) \equiv y_e$. Hence an equilibrium point is sometimes called an **equilibrium solution.** Points that are not equilibrium points are called **regular points.**

Often, all points near an equilibrium point settle down to the equilibrium point. Those equilibrium points are called **stable equilibrium points** or **attractors.** In addition to equilibrium *points*, there are other types of attractors, such as **limit cycles,** in which all nearby trajectories settle down or "converge" to a *closed loop*. Closed loops correspond to periodic solutions because the trajectory keeps tracing out the same points in the phase plane over and over.

Finally, a second-order equation, such as $\ddot{x} + kx = 0$, is called an **autonomous differential equation,** since it can be rewritten as an autonomous system by letting $y = \dot{x}$ and writing

$$
\begin{aligned}
\dot{x} &= y \\
\dot{y} &= -kx
\end{aligned}
\tag{8}
$$

FINDING TRAJECTORIES

Although there is no standard set of rules for finding the trajectories of the general system (6), several approaches are possible. We list a few here.

Three Ways to Find Trajectories

Elimination of *t* from the Solution

If an analytic solution $(x(t), y(t))$ can be found, and if the independent variable t can be eliminated from the equations $x = x(t), y = y(t)$, then one can obtain an algebraic relationship between x and y that describes the trajectories.

Changing the Independent Variable

If one divides $\dot{x} = P(x, y)$ by $\dot{y} = Q(x, y)$ and recalls from calculus that $dy/dx = \dot{y}/\dot{x}$, one obtains the first-order equation

$$
\frac{dy}{dx} = \frac{P(x, y)}{Q(x, y)}
$$

that can (one hopes) be solved for y in terms of x. Of course, one can use a computer to draw the direction field of this equation.

Direct Integration

If a differential equation has the form $\ddot{x} + f(x) = 0$, one can use the chain rule $\ddot{x} = d(\dot{x})/dt = (d\dot{x}/dx)\dot{x}$ and rewrite the equation in the form $(d\dot{x}/dx)\dot{x} + f(x) = 0$. This equation can be integrated directly, giving the trajectories

$$
\frac{1}{2}\dot{x}^2 + \int f(x)\, dx = c
$$

where $\int f(x)\, dx$ is an antiderivative of $f(x)$ and c is an arbitrary constant.

Example 1 **Finding Trajectories by Eliminating t** Find the trajectories of the linear autonomous system

$$\frac{dx}{dt} = x \qquad \frac{dy}{dt} = 2y \qquad (9)$$

Solution Here we can solve these equations separately, finding the general solution given by $x(t) = c_1 e^t$, $y(t) = c_2 e^{2t}$. Writing y in terms of x, we find the trajectories

$$y = c_2 (e^t)^2 = c_2(x/c_1)^2 = kx^2 \qquad (10)$$

See Figure 8.2. ∎

Figure 8.2
Parabolic trajectories for
$dx/dt = x$, $dy/dt = 2y$

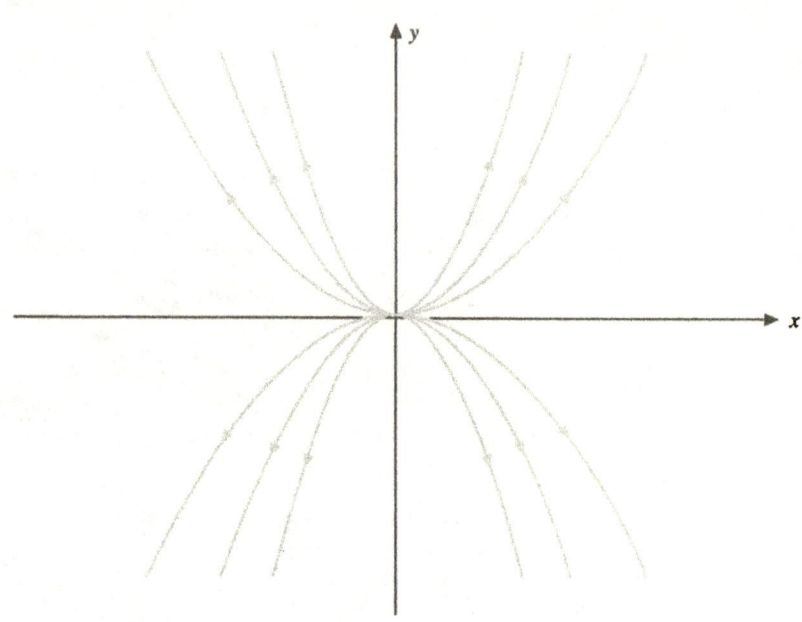

Example 2 **Finding Trajectories by Changing Variable** Find the trajectories of the nonlinear autonomous system

$$\frac{dx}{dt} = y(x^2 + 1) \qquad \frac{dy}{dt} = 2xy^2 \qquad (11)$$

Solution Although this system is nonlinear and cannot be solved by using the methods studied in Chapter 6, we can still find the trajectories of the system in the phase plane. For $y \neq 0$ the above equations imply the first-order (nonlinear) equation

$$\frac{dy}{dx} = \frac{dy/dt}{dx/dt} = \frac{2xy}{x^2 + 1}$$

whose general solution can be found by separation of variables and gives the trajectories

$$y = c(x^2 + 1) \qquad (12)$$

Hence although we do not know the relationships $x = x(t)$, $y = y(t)$, we do know the relationship between y and x, which may be very important. See Figure 8.3. ∎

Figure 8.3
Parabolic trajectories for
$dx/dt = y(x^2 + 1)$,
$dy/dt = 2xy^2$

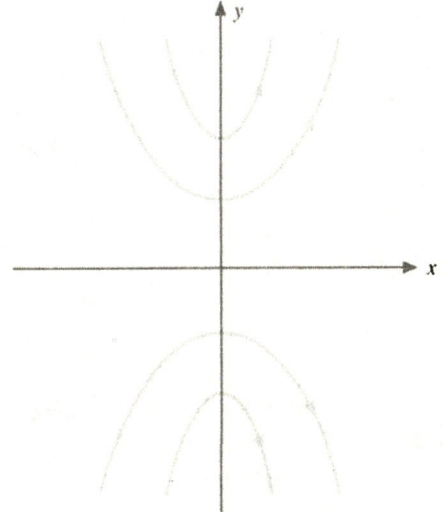

Example 3   **Trajectories by Direct Integration**   Determine the trajectories of the nonlinear autonomous equation

$$\ddot{x} + \frac{x}{1 + x^2} = 0 \tag{13}$$

Solution   Using the chain rule $\ddot{x} = d(\dot{x})/dt = d(\dot{x})/dx\, \dot{x}$, we rewrite the equation as

$$\frac{d\dot{x}}{dx}\dot{x} + \frac{x}{1 + x^2} = 0$$

Integrating with respect to x gives the trajectories

$$\dot{x}^2 + \ln\left(1 + x^2\right) = c$$

or

$$y^2 + \ln\left(1 + x^2\right) = c \tag{14}$$

where $y = \dot{x}$. A computer was used to draw these trajectories as shown in Figure 8.4. ∎

We now illustrate how concepts relating to the phase plane can be used to analyze two simple problems from control theory.

Figure 8.4
Trajectories for the
autonomous equation

$$\ddot{x} + \frac{x}{1 + x^2} = 0$$

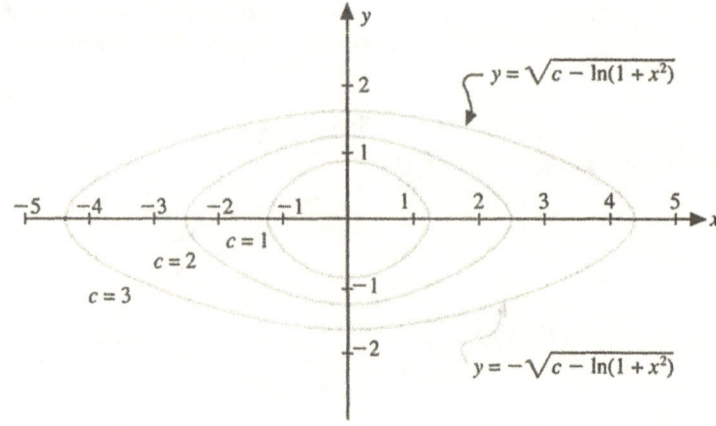

**HOW TO STOP A
SPACESHIP**

A spaceship moves along a straight line far from any gravitational field. Under this ideal situation the equation of motion that describes its position x along its path of motion is

$$m\ddot{x} = F(t) \tag{15}$$

where m is the mass of the spaceship and $F(t)$ is the *net* thrust or force of two opposing rockets. See Figure 8.5.

We assume that each rocket has only two states: off, which yields a thrust of 0 units, and on full blast, which exhibits a thrust of 1 unit.* This means that if the left rocket, shown in Figure 8.5, is turned on and the right rocket is turned off, then there is a net thrust of 1 unit pushing to the right, and so $F(t) = 1$. On the other hand, if the right rocket is turned on and the left rocket is turned off, then the net thrust pushes to the left, or $F(t) = -1$. Finally, if both rockets are turned off or if both rockets are turned on, the net thrust is zero.

Figure 8.5
What is the thrust strategy
that will stop a moving
spaceship in minimum time?

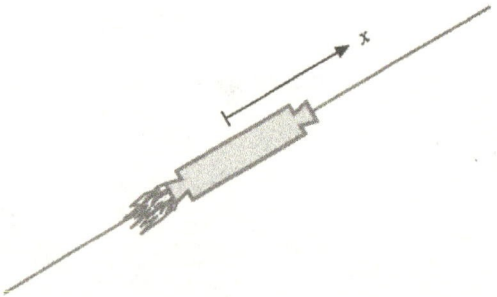

The goal is to determine when to switch the rockets on and off (possibly many times) so that the spaceship, with an arbitrary initial position and velocity $(x_0, \dot{x}_0)$, can be "driven" to a final state $(x, \dot{x})$, which for simplicity we pick as $(0, 0)$, in *minimum time*. See Figure 8.6.

* One unit can stand for anything: 1 pound, 10 pounds, 1000 pounds, etc.

Since the net thrust $F(t)$ must be either $+1$, 0, -1, we draw the trajectories of the system corresponding to the three constant functions $F(t) \equiv +1$, $F(t) \equiv 0$, and $F(t) \equiv -1$. Using the chain rule $\ddot{x} = d\dot{x}/dt = (d\dot{x}/dx)\dot{x}$, we rewrite $m\ddot{x} = F(t)$ as

$$m\dot{x}\frac{d\dot{x}}{dx} = F(t) \tag{16}$$

Letting $F(t) \equiv +1$, -1, and 0, we get

$$F(t) \equiv 1 \Rightarrow m\dot{x}\frac{d\dot{x}}{dx} = 1 \Rightarrow \frac{m}{2}\dot{x}^2 = x + c \tag{17a}$$

$$F(t) \equiv -1 \Rightarrow m\dot{x}\frac{d\dot{x}}{dx} = -1 \Rightarrow \frac{m}{2}\dot{x}^2 = -x + c \tag{17b}$$

$$F(t) \equiv 0 \Rightarrow m\dot{x}\frac{d\dot{x}}{dx} = 0 \Rightarrow \frac{m}{2}\dot{x}^2 = c \tag{17c}$$

Figure 8.6
How can the spaceship be driven from an initial state $(x_0, \dot{x}_0)$ to $(0, 0)$?

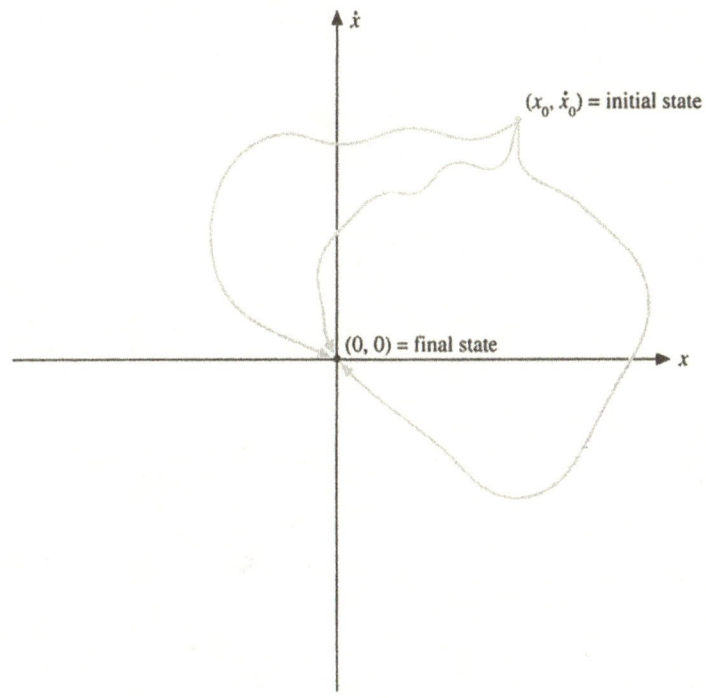

Each of the trajectories (17a) and (17b) defines a one-parameter family of parabolas in the phase plane. See Figures 8.7(a) and 8.7(b). When $F(t) \equiv 1$, the parabolas open to the right, and when $F(t) \equiv -1$, the parabolas open to the left. Finally, when we have $F(t) \equiv 0$, the trajectories define a family of horizontal lines. See Figure 8.7(c).

Figure 8.7
Trajectories of the spaceship
for thrusts $F(t) \equiv +1$,
$F(t) \equiv -1$, and $F(t) \equiv 0$

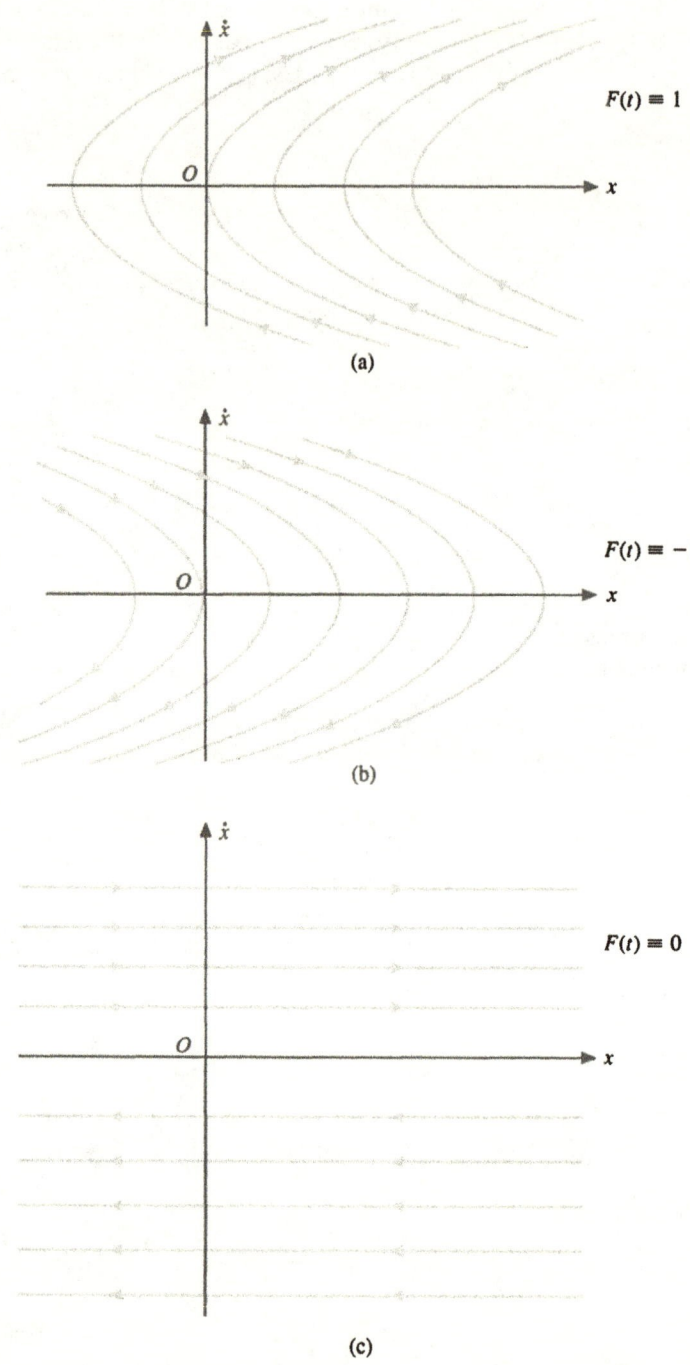

(a)

(b)

(c)

By observing the trajectories it is clear that to eventually reach the final state of $(0, 0)$, the state of the system must eventually move along one of the two **switching boundaries:**

$$\text{Switching Boundary I:} \qquad \frac{m}{2}\dot{x}^2 = -x \qquad (\dot{x} > 0)$$

$$\text{Switching Boundary II:} \quad \frac{m}{2}\dot{x}^2 = x \quad (\dot{x} < 0)$$

as shown in Figure 8.8.

For example, in order for a spaceship with an initial state A to reach the origin, the trajectories indicate that we should pick $F(t) = -1$ (right rocket on, left rocket off) until the state vector $(x, \dot{x})$ reaches B, which occurs when the condition $\frac{m}{2}\dot{x}^2 = x$ is met. At that time we should pick $F(t) = +1$ (left rocket on, right rocket off). This strategy will cause the spaceship to "move" along Switching Boundary II to the origin. In fact, our intuition is correct; it can be proven* that to drive the spaceship to the origin, the forcing function F exhibits at most one *switch* from $+1$ to -1 (or vice versa) along the trajectory. The forcing function F is called a **control function,** since it *controls* the movement of the spaceship. Problems such as this are called **control problems,** and their analysis is often facilitated by use of the phase plane.

Figure 8.8
Switching curves for stopping a spaceship

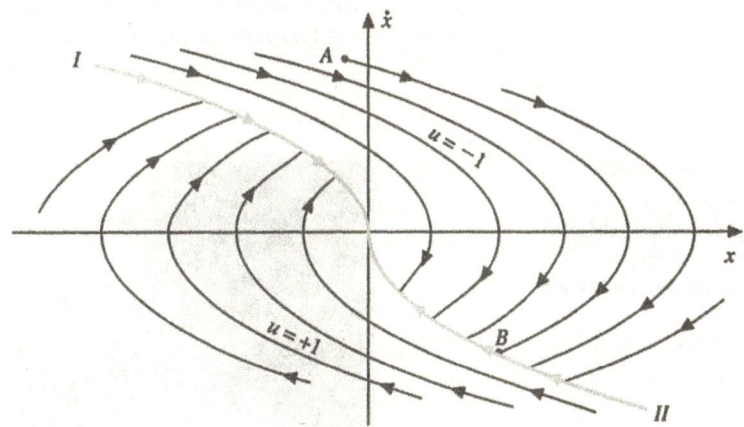

ANGLE CONTROL OF A MISSILE (OPTIONAL)

When a missile, such as a Patriot missile, is fired at a target, such as a Scud missile, a feedback control system guides the missile toward the target. This system constantly (or nearly so) monitors the angle θ between the direction of the missile and the target and, from this "error" angle, adjusts a set of fins, resulting in decreasing the angle θ. See Figure 8.9.

Figure 8.9
Feedback control of a missile

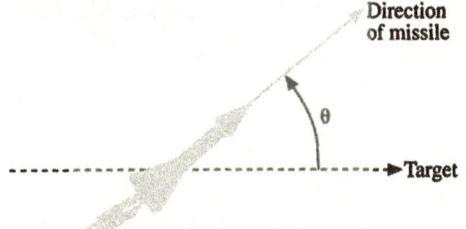

---

* The proof that the forcing function for this problem has at most one switch (from $+1$ to -1 or vice versa) can be found in almost any beginning text in optimal control theory, including *Introducing Systems and Control* by D. M. Auslander, Y. Takahashi, and M. J. Rabins (McGraw-Hill, New York, 1974).

If we assume that the controlling fins can be positioned in one of two fully deflected positions, then a simple model that governs the error angle θ is the second-order equation (Newton's equation of motion)

$$\ddot{\theta} = -FU(\gamma\theta + \dot{\theta}) = \begin{cases} -F & (\gamma\theta + \dot{\theta} \geq 0) \\ F & (\gamma\theta + \dot{\theta} < 0) \end{cases} \tag{18}$$

where U is the **unit switching function** defined by

$$U(\xi) = \begin{cases} -1 & (\xi < 0) \\ 1 & (\xi \geq 0) \end{cases}$$

The constant $F > 0$ is the angular force caused by the fully deflected fins and depends on the size and shape of the fins. The constant $\gamma > 0$ is a system parameter picked by the engineer who designs the control system.* The function on the right-hand side of system (18) is called the **control function** of the system, and one can observe that its value (F or $-F$) depends on θ, $\dot{\theta}$, and γ. See Figure 8.10.

Figure 8.10
The control function has the value of $+F$ or $-F$, depending on which side of the switching line $\dot{\theta} = -\gamma\theta$ the state vector $(\theta, \dot{\theta})$ lies on.

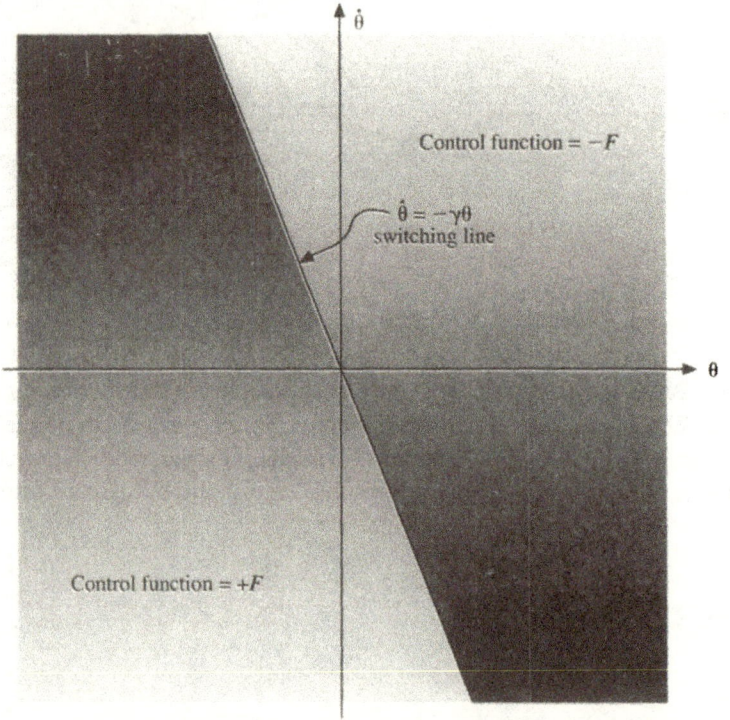

* The choice of the parameter $\gamma > 0$ affects the performance of the control system. The determination of a control function that will make a system behave in a desired manner is a problem in **control theory.** Often, one wishes to find a control function that maximizes (or minimizes) a performance criterion, which is a problem in **optimal control theory.**

To analyze the performance of the control function governed by (18), we solve the equation separately for $\gamma\theta + \dot{\theta} \geq 0$ and $\gamma\theta + \dot{\theta} < 0$, getting

Region	Equation	Implicit solution	
$\gamma\theta + \dot{\theta} \geq 0$	$\ddot{\theta} = -F$ $\Rightarrow$	$\frac{1}{2}\dot{\theta}^2 = -F\theta + c$	(19a)
$\gamma\theta + \dot{\theta} < 0$	$\ddot{\theta} = F$ $\Rightarrow$	$\frac{1}{2}\dot{\theta}^2 = F\theta + c$	(19b)

The trajectories in Eqs. (19a) and (19b) describe parabolas in the phase plane to the left and right of the line $\gamma\theta + \dot{\theta} = 0$. See Figure 8.11.

Figure 8.11
Trajectories of the equation $\ddot{\theta} = -FU(\gamma\theta + \dot{\theta})$ in the phase plane. To the right of the switching line $\dot{\theta} = -\gamma\theta$ the state vector $(\theta, \dot{\theta})$ follows the parabolas that open to the left, and to the left of the switching line the state vector follows the parabolas that open to the right.

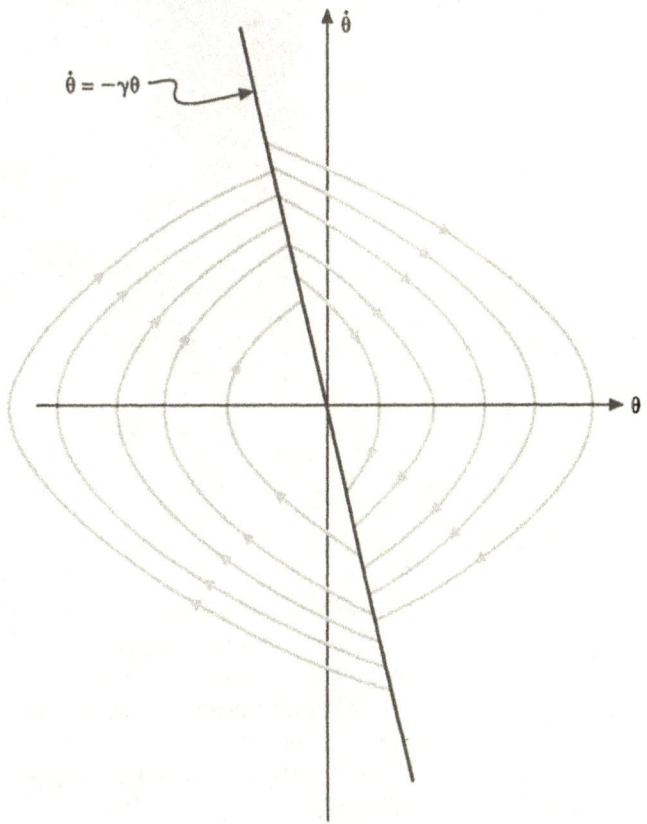

$\dot{\theta} = -\gamma\theta$

The straight line $\gamma\theta + \dot{\theta} = 0$ in the $\theta\dot{\theta}$-plane separating the two regions where the control is given by $-F$ and $+F$ is called the **switching line.** For example, suppose a missile is fired at a target with an initial error angle θ_0, resulting in the initial state of the system being $(\theta, \dot{\theta}) = (\theta_0, 0)$. For simplicity we assume that the design parameter is $\gamma = 3$. To determine the resulting angle of attack at which this control system "drives" the missile, we follow the state vector $(\theta, \dot{\theta})$ along the parabolic trajectory starting at $(\theta_0, 0)$ until we reach the switching line $3\theta + \dot{\theta} = 0$ at p_1. Note that the control function initially has the value $+F$.

When the state $(\theta, \dot{\theta})$ reaches the point P_1 as illustrated in Figure 8.12, the control function switches from $+F$ to $-F$. (The fins are rotated in the opposite direction.) The state vector then travels along a new parabolic trajectory until it returns to the switching line for the second time at P_2. At that time the control function switches from $-F$ back to $+F$. This process is continued again and again (unless the missile's motion is interrupted by hitting the target), each time getting closer to the final state (0, 0).

Figure 8.12
The missile's trajectory starting from the initial state $(\theta_0, 0)$ to the final point (0, 0)

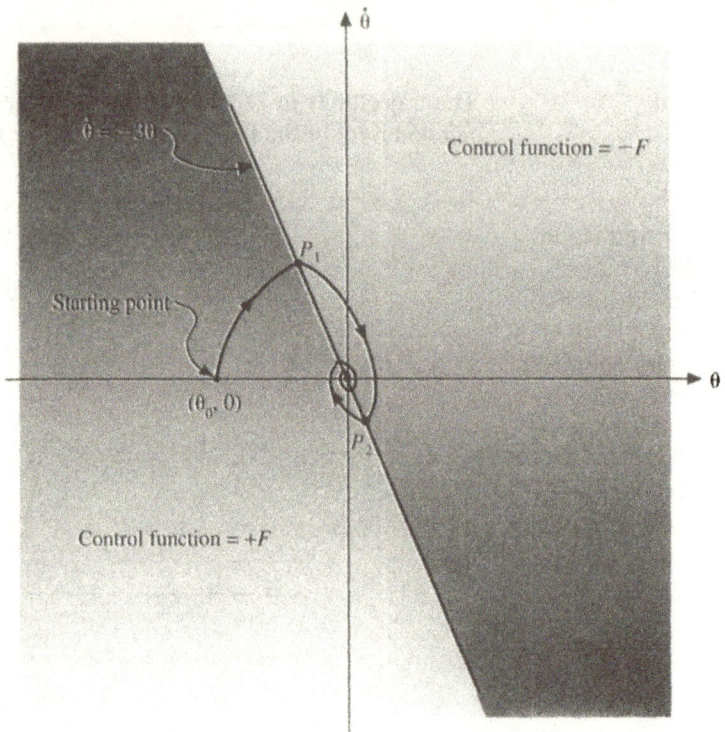

However, if we "zoom in" on the phase plane in a small region around the target point (0, 0), we will discover that *eventually* after some time t_1 the state vector becomes *trapped* on the switching line. We say "trapped" because the trajectories on both sides of the switching lines point *toward* the switching line. See Figure 8.13.

What this means is that we must now keep switching from $+F$ to $-F$ (and vice versa) at an *infinite rate* in order to drive the state vector down the switching line toward (0, 0). If we could do this, then we could find $\theta(t)$ by simply solving the initial-value problem

$$\dot{\theta} = -3\theta \qquad \theta(t_1) = \theta_1$$

where θ_1 is the error angle at time t_1. Solving the initial-value problem, we find

$$\theta(t) = \theta_1 \, e^{-(t-t_1)} \qquad (t \geq t_1) \tag{20}$$

From this we conclude that $\theta(t) \to 0$ (and $\dot{\theta} \to 0$), which means that the "continuous" switching strategy for $t > t_1$ results in the state vector $(\theta, \dot{\theta})$ moving toward the final point (0, 0) *on* the switching line. Of course, in practice we cannot *continuously* switch the control function back and forth between $-F$ and $+F$, but it would switch back and forth rapidly with the net result that the state vector $(\theta, \dot{\theta})$ would zigzag around the switching line on its way toward the final state (0, 0). See Figure 8.13.

Figure 3.13
Since the parabolic trajectories on both sides of the switching line point toward the switching line near the final point (0, 0), the state vector becomes trapped on this line. Hence after this point the state vector approaches the target point (0, 0) along this line.

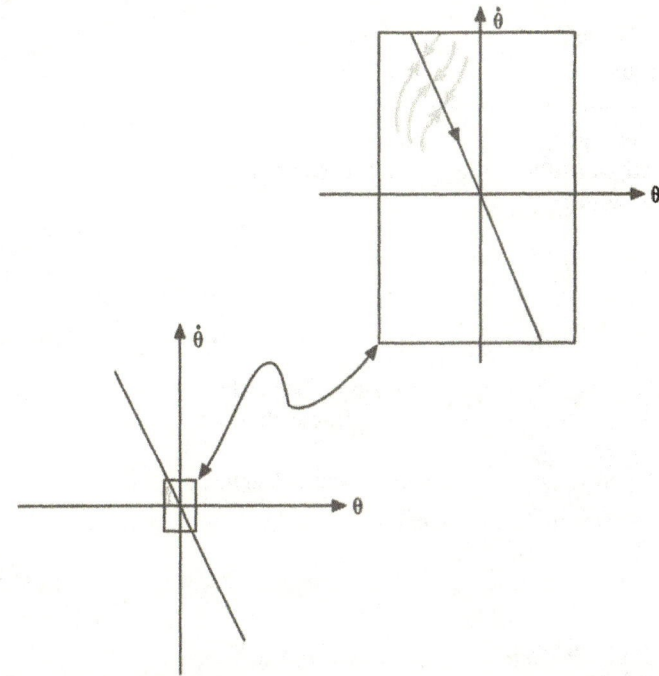

Note: The control function depends on θ and $\dot{\theta}$. Control functions that depend on the state of the system are called **closed loop controls** or **feedback controls** in contrast to control functions that depend only on time, which are called **open loop control functions.** Closed loop controls are usually the controls of preference, since they are better equipped to handle unexpected difficulties. However, closed loop controls require constant monitoring of the state of the system, which may be costly and difficult, whereas open loop controls require only the availability of a clock.

PROBLEMS: Section 8.1

For Problems 1–6, rewrite the given second-order equation as an equivalent system of first-order equations and tell whether the system is autonomous or nonautonomous.

1. $\ddot{x} + \dfrac{g}{L} \sin x = 0$
   (pendulum equation)

2. $\ddot{x} + \dfrac{g}{L} \sin x = \sin t$
   (forced pendulum equation)

3. $\ddot{x} + \omega^2 x + \beta x^3 = 0$
   (Duffing's equation)

4. $\ddot{x} + \omega^2 x + \beta x^3 = F_0 \cos \gamma t$
   (forced Duffing's equation)

5. $\ddot{x} + (\alpha + \beta \cos t)x = 0$
   (Mathieu's equation)

6. $\ddot{x} + \epsilon(x^2 - 1)\dot{x} + x = F_0 \cos \gamma t$
   (forced Van der Pol's equation)

For Problems 7–10, determine the equilibrium points (if any) of the given system.

7. $\dot{x} = y \qquad \dot{y} = -\sin x$
   (undamped pendulum)

8. $\dot{x} = y \qquad \dot{y} = -\dfrac{k}{m}x - \dfrac{d}{m}y; \quad m, d, k > 0$
   (vibrating spring)

9. $\dot{x} = y \qquad \dot{y} = -\dfrac{g}{L}\sin x - \dfrac{d}{mL}y; \quad g, m, d, L > 0$
   (pendulum equation)

10. $\dot{x} = y \qquad \dot{y} = -\epsilon(x^2 - 1)y - x; \quad \epsilon > 0$
    (Van der Pol equation)

11. **Phase Portrait of Duffing's Equation** The long-term behavior of the Duffing equation

$$\ddot{x} + 0.1\dot{x} + x^3 = 0$$
$$x(0) = 4$$
$$\dot{x}(0) = 0 \qquad (21)$$

is illustrated in Figure 8.14 both in the phase plane and by a graph of its solution. Verify to yourself that the two representations are compatible.

12. **Pendulum Equation** An undamped pendulum of length L is governed by the **pendulum equation**

$$\ddot{x} + \frac{g}{m}\sin x = 0 \qquad (22)$$

(a) Rewrite the pendulum equation as a system of two first-order equations.
(b) Find the equilibrium points of this system.
(c) Find the trajectories of the system.
(d) Convince yourself the trajectories look like the curves in Figure 8.15.
(e) Interpret the motion of the pendulum by analyzing the trajectories.

13. **Oscillations of a Hard Spring** Larger vibrations of an undamped mass-spring system are often described by the **hard spring equation**

$$\frac{d^2x}{dt^2} + x + x^3 = 0 \qquad (23)$$

where x represents the displacement from equilibrium.
(a) Write this equation as a first-order linear system.
(b) Find the equilibrium point(s) of this system.
(c) Find the trajectories of the system.

14. **Spaceship Revisited** Consider the spaceship described in this section, where we let $m = 1$. Describe the minimum-time control function $F(t)$ for driving the spaceship to the origin from the following points.

(a) $(-10, 10)$ (c) $(10, -10)$
(b) $(10, 10)$ (d) $(-10, 10)$

15. **Missile Attack** Suppose the error angle θ between the direction of a missile and its target is governed by the equation

$$\ddot{\theta} = -U(\theta) = \begin{cases} -1 & (\theta \geq 0) \\ 1 & (\theta < 0) \end{cases}$$

See Figure 8.16. Determine the motion of the attack angle with initial conditions $(\theta_0, \dot{\theta}_0) = (-\pi/2, 0)$.

16. **Qualitative Behavior of Two Competing Species** Two species of fish, neither of which preys on the other, occupy

Figure 8.14
Solution and trajectory for Duffing's equation

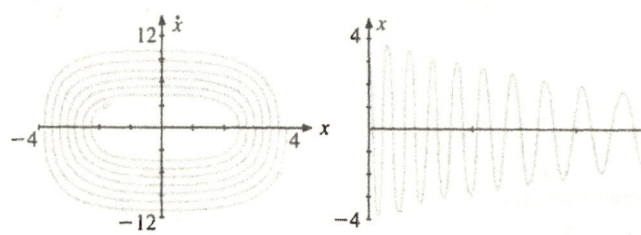

Figure 8.15
Trajectories of the pendulum equation

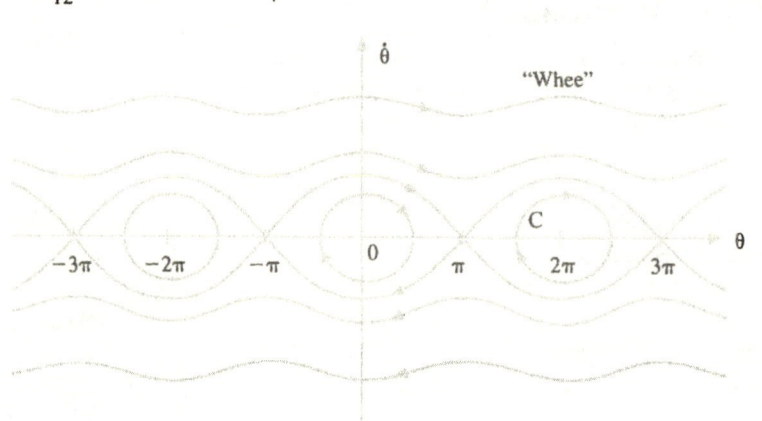

Figure 8.16
Missile attack using control
$\ddot{\theta} = -U(\theta)$

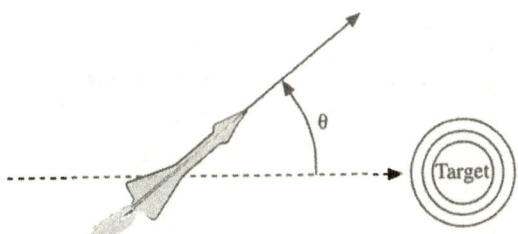

the same pond and compete for a limited food supply. If we let $x(t)$ and $y(t)$ denote the populations of the two species at time t, a typical model for predicting future population sizes (maybe measured in thousands of tons of fish) is

$$\dot{x} = x(1 - x - y) \qquad (24)$$
$$\dot{y} = y(2 - x - 4y)$$

(a) Find the equilibrium point of this system.
(b) The two lines $1 - x - y = 0$ and $2 - x - 4y = 0$ divide the phase plane into four regions. Find the sign of $\dot{x}$ and $\dot{y}$ in each of these regions to get a rough idea of the trajectories of the system about the equilibrium point.
(c) What can you say about the fish populations, knowing what you do about the trajectories shown in Figure 8.17?

Trajectories of Conservative Equations

A differential equation of the form

$$\ddot{x} + f(x) = 0 \qquad (25)$$

is called a **conservative equation** (no friction).* If one uses the chain rule and replaces $\ddot{x} = \dot{x}\,(d\dot{x}/dt)$, the equation can be rewritten as $\dot{x}\,(d\dot{x}/dx) + f(x) = 0$. Letting $y = \dot{x}$ and integrating with respect to x, one obtains the trajectories of Eq. (25) as

$$\frac{1}{2}y^2 + \int f(x)\,dx = C \qquad (26)$$

where $\int f(x)\,dx$ is an antiderivative of $f(x)$ and C is an arbitrary constant. Use this technique to find the trajectories of the conservative systems in Problems 17–19.

17. $\ddot{x} + \sin x = 0$ **18.** $\ddot{x} + x^2 = 0$
19. $\ddot{x} + x + x^3 = 0$

The Hamiltonian of a Conservative System

Problems 20–21 are concerned with the Hamiltonian of a conservative system.

20. Hamiltonian of a Conservative System The equation

$$H(x, y) = \frac{1}{2}y^2 + \int f(x)\,dx \qquad (27)$$

(which is found on the left-hand side of Eq. (26)) is called

---

* An equation that is *not* conservative is called **dissipative** (friction).

Figure 8.17
Some of the trajectories of competing fish populations

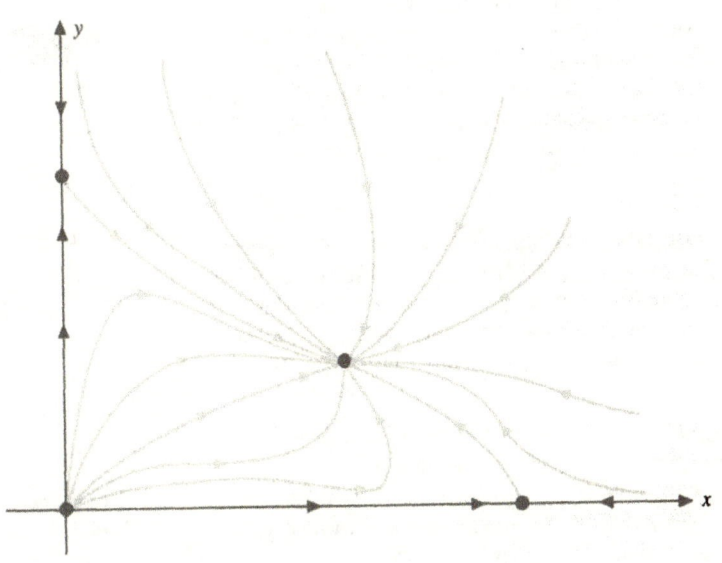

Figure 8.18
Areas are preserved in
conservative systems

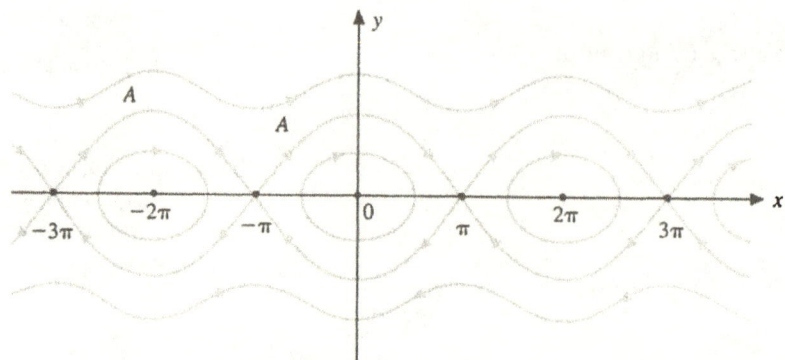

the **Hamiltonian** of the conservative system governed by $\ddot{x} + f(x) = 0$. It represents the total energy of the system (the sum of the potential and kinetic energy).

(a) Show that the Hamiltonian is a constant along trajectories in the phase plane.

(b) Show that the Hamiltonian satisfies the system of differential equations

$$\dot{x} = \frac{\partial H(x, y)}{\partial y}$$

$$\dot{y} = -\frac{\partial H(x, y)}{\partial x} \qquad (28)$$

21. **Preservation of Areas for Conservative Systems** In the previous problems we defined an equation of the general form $\ddot{x} + f(x) = 0$ as conservative, which means that the total energy of the system described by the equation remains constant over time. Another important property of conservative equations (or conservative systems) is the **preservation of area** (or volume in higher dimensions) **property,** which means that states or points in the phase plane move in such a way that the area of any given *set* of points remains constant over time.* See Figure 8.18. Show that areas are conserved in the following systems.

(a) $\ddot{x} = 0$

(b) $\ddot{x} + x = 0$

22. **The Phase Space in Three Dimensions** The concept of the phase plane can easily be extended to higher dimensions. Figure 8.19 illustrates the change in three variables (sharks, herring, haddock).[†] Here the number of any one of the three variables depends on the other two. From the

trajectories, analyze the dynamics of these three species of fish.

23. **Solving a Nonlinear Equation by Factoring** Once in a while,* a nonlinear equation can be solved by using a "once in a lifetime" strategy. Solve the nonlinear equation

$$y'(y' + y) = x(x + y)$$

by transferring all terms in the equation to the left-hand side, factoring the resulting expression into two factors, and setting each factor to zero, getting two linear equations.

24. **Journal Entry** Summarize the concept of the phase plane.

Figure 8.19
Three-dimensional phase
space

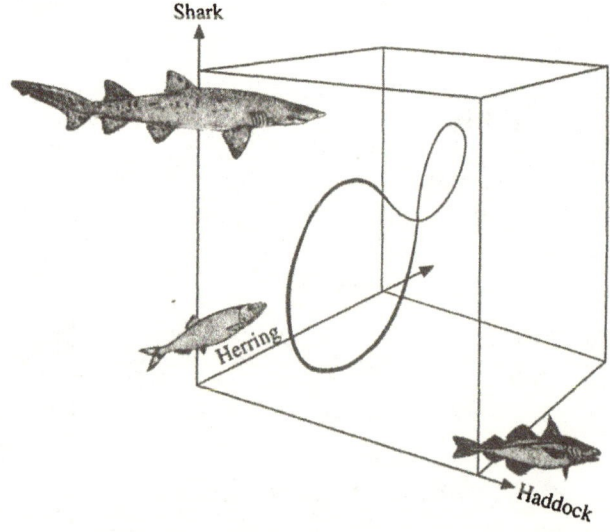

---

* For nonconservative systems, or dissipative systems, areas (and volumes in higher dimensions) decrease over time.

[†] This three-dimensional phase space was taken from the delightful little book *Turbulent Mirror* by John Briggs and F. David Peat (Perennial Library, Harper & Row, New York, 1990).

* This problem was taken from the interesting and useful reference book *Handbook of Differential Equations* by Daniel Zwillinger (Academic Press, New York, 1989).

8.2   EQUILIBRIUM POINTS AND STABILITY FOR LINEAR SYSTEMS

PURPOSE

To find equilibrium points and trajectories of the linear homogeneous system

$$\dot{x} = ax + by$$
$$\dot{y} = cx + dy$$

and to introduce the concept of stability.

A pin lying on its side is unaffected by a small disturbance, whereas a pin balanced on its tip will probably topple over even if a bug happens to flap its wings in the next room. See Figure 8.20. Roughly, this is the difference between stability and instability. A state is stable if it is unaffected by small disturbances in the system. If the balanced pin stands perfectly vertically, it will remain there. However, to determine whether this vertical state is stable, we must concern ourselves with *nearby* states as well.* Any small disturbance in the pin will result in its falling. Hence the vertical position of the pin is unstable.

Figure 8.20
A balanced pin is unstable, whereas one lying on its side is stable.

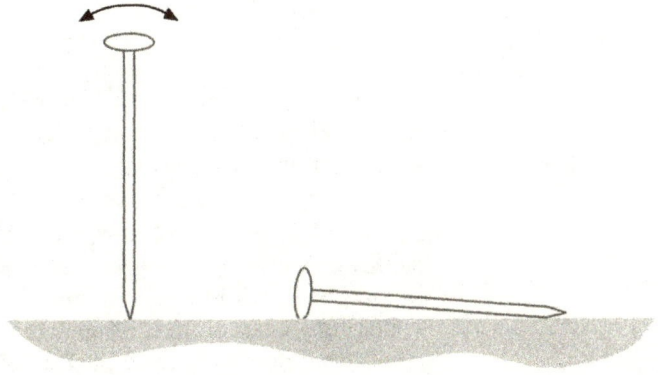

Stability is a very important concept in modern control and systems theory. The motion of an airplane must be stable. Today's airplanes are microguided by elaborate control systems so that any disturbance (wind shear, turbulance, etc.) will be quickly damped. The dynamics of an airplane under guidance of its control system must be stable in order to handle unexpected disturbances.

---

* We refer here to a *state* being stable, but we could also refer to any motion as being stable. Here, the pin standing on its end represents *constant* motion. A *motion* is stable if it is unaffected by small disturbances.

BEHAVIOR OF TRAJECTORIES OF $\dot{x} = Ax$

In this section we study the shapes of the trajectories of the two-dimensional linear autonomous system $\dot{x} = Ax$ in the phase plane. The shapes are fairly simple, since trajectories *cannot cross themselves* (other than forming closed loops). The reason for this phenomenon is that the vector $\dot{x}$ that determines the *direction* of a trajectory at a point, is determined uniquely by the point. This being the case, there are not very many different "types" of trajectories. For example, if a trajectory forms a *closed loop*, then the system has a *periodic solution*. All trajectories inside the closed loop are trapped there, and all the trajectories on the outside are excluded, so things are fairly simple. Now, in *three dimensions* the trajectories can move around each other, and the trajectories can get so complicated that they can almost never be sorted out. But fortunately, we are in two dimensions, where the situation is much simpler.

Consider then the linear homogeneous system

$$\dot{x} = ax + by$$
$$\dot{y} = cx + dy \tag{1}$$

where a, b, c, and d are constants. We recall from Chapter 6 that the general solution of this linear system depends on the eigenvalues and eigenvectors of the matrix

$$A = \begin{bmatrix} a & b \\ c & d \end{bmatrix} \tag{2}$$

Recall too that the eigenvalues λ_1 and λ_2 are the roots of the characteristic polynomial

$$\begin{vmatrix} a - \lambda & b \\ c & d - \lambda \end{vmatrix} = \lambda^2 - (a + d)\lambda + (ad - bc) = 0 \tag{3}$$

In this section we restrict ourselves to matrices A that are nonsingular (i.e., that satisfy $|A| \neq 0$). The reason for this assumption is that for nonsingular matrices the zero vector $x = 0$ is the only solution of the homogeneous linear system $Ax = 0$. This property is important for us in this chapter, since it means that the origin $(0, 0)$ is the *only equilibrium point* of $\dot{x} = Ax$. We should also note that if A is nonsingular, then neither eigenvalue of A is zero; if an eigenvalue λ *were zero*, then since the equation $|A - \lambda I| = 0$, it follows that $|A| = 0$, which we have assumed is not true.

Recalling again from Chapter 6 that the form of the general solution of $\dot{x} = Ax$ *depends on the eigenvalues of A*. Real and unequal eigenvalues give rise to one form, complex eigenvalues give rise to another form, and so on. For that reason we must consider several cases when finding the trajectories of $\dot{x} = Ax$.

TRAJECTORIES OF $\dot{x} = Ax$

Several distinct cases arise in discussing the trajectories of $\dot{x} = Ax$. It was shown in Chapter 6 that there are five distinct types of eigenvalue-eigenvector pairs that arise.

Type 1: Real unequal eigenvalues of the same sign (improper node)

Consider the case in which the eigenvalues λ_1 and λ_2 of A satisfy $\lambda_1 < \lambda_2 < 0$ or else $0 < \lambda_1 < \lambda_2$. We saw in Chapter 6 that the general solution is

$$x(t) = c_1 e^{\lambda_1 t} x^{(1)} + c_2 e^{\lambda_2 t} x^{(2)} \tag{4}$$

where $x^{(1)}$ and $x^{(2)}$ are the linearly independent* eigenvectors corresponding to λ_1 and λ_2, respectively, and c_1 and c_2 are arbitrary constants. In the case in which both eigenvalues are negative, it is clear that $x(t) \to 0$ as $t \to \infty$ regardless of the values of c_1 and c_2. Hence all solutions approach the equilibrium point located at the origin. See Figure 8.21.

If the solution starts at an initial point that lies on a line passing through one of the eigenvectors, say $x^{(1)}$, then it follows that $c_2 = 0$, and so the solution approaches the origin along that line. Similarly, if a solution starts at an initial point on the line through $x^{(2)}$, then the solution approached $(0, 0)$ along that line.

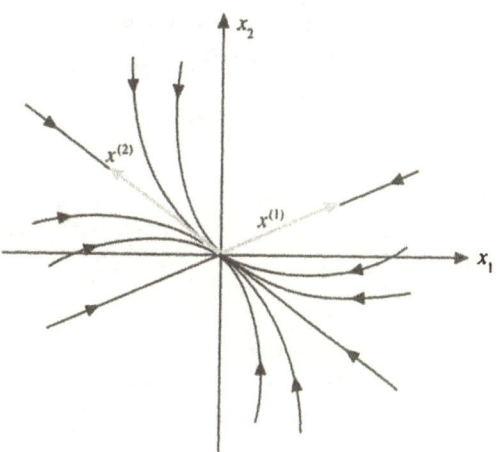

Figure 8.21
An improper node; $\lambda_1 < \lambda_2 < 0$

It is useful to rewrite the general solution (4) in the form

$$x(t) = e^{\lambda_2 t} \left(c_1 e^{(\lambda_1 - \lambda_2)t} x^{(1)} + c_2 x^{(2)} \right) \tag{5}$$

where as long as $c_2 \neq 0$, the term $c_1 e^{(\lambda_1 - \lambda_2)t} x^{(1)}$ is negligible in comparison to $c_2 x^{(2)}$. Hence with the exception of solutions starting on the line through $x^{(1)}$, all other solutions will approach the equilibrium point $(0, 0)$ *tangent* to $x^{(2)}$. See Figure 8.21. An equilibrium point of this type is called a **node** or sometimes an **improper node.** Note that in this case the improper node acts like a "magnet," pulling all the other points in the phase plane to it.

Finally, when both eigenvalues are positive ($0 < \lambda_2 < \lambda_1$), the type of the trajectories in Figure 8.21 is exactly the same with the exception that all arrowheads on the trajectories are reversed and motion is directed away from the origin.

Type 2: Real unequal eigenvalues of opposite sign (saddle point)

Consider the case in which one eigenvalue is positive and the other is negative. For convenience, $\lambda_2 < 0 < \lambda_1$. Here the general solution can be written in the form

$$x(t) = c_1 e^{\lambda_1 t} x^{(1)} + c_2 e^{\lambda_2 t} x^{(2)} \tag{6}$$

---

* It can be shown that eigenvectors corresponding to distinct eigenvalues are linearly independent.

where, as before, $x^{(1)}$ and $x^{(2)}$ are the linearly independent eigenvectors corresponding to λ_1 and λ_2. Typical eigenvectors are drawn in Figure 8.22.

If a solution starts at an initial point on the line through one of the eigenvectors, say $x^{(1)}$, then it follows that $c_2 = 0$, and hence the solution will remain on that line. Also, since $\lambda_1 > 0$, the solution $x(t)$ approaches infinity* as $t \to \infty$. Similarly, if a solution starts on the line through $x^{(2)}$, the solution remains on the line except that now the solution approaches the origin as $t \to \infty$. Solutions that start at other points in the phase plane follow trajectories shown in Figure 8.22. To analyze these trajectories, realize that the term involving the positive exponential λ_1 is the dominant term, so for large t the solution (6) will approach infinity in a direction tangent to the line determined by $x^{(1)}$. Also, when t is a large *negative number*, the term involving the *negative* eigenvalue λ_2 is the dominant term, so the solution asymptotically "comes from" a direction determined by $x^{(2)}$. In this case the equilibrium point $(0, 0)$ is called a **saddle point.**

Figure 8.22
Saddle point; $\lambda_1 > 0$, $\lambda_2 < 0$

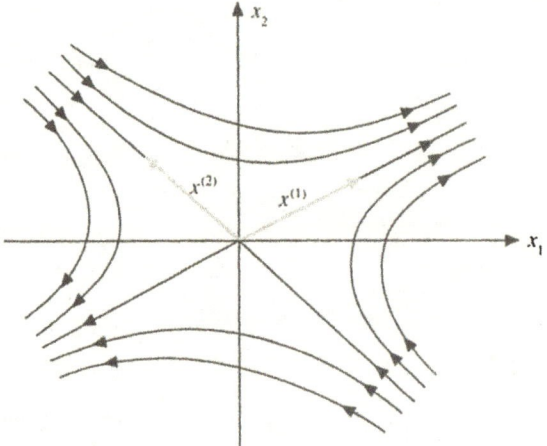

Type 3: Equal eigenvalues (proper and improper nodes)

Consider the case in which both eigenvalues are the same, that is, $\lambda_1 = \lambda_2 = \lambda$. We assume too that the eigenvalues are negative. (The phase portrait is the same for positive eigenvalues except that the direction of motion is reversed.) There are two subcases that we must consider: the case in which there are two linearly independent eigenvectors and the case in which there is only one linearly independent eigenvector.

Subtype 1: Two Independent Eigenvectors (Proper Node). Here the general solution is

$$x(t) = c_1 e^{\lambda t} x^{(1)} + c_2 e^{\lambda t} x^{(2)}$$
$$= e^{\lambda t} (c_1 x^{(1)} + c_2 x^{(2)}) \tag{7}$$

which is a constant vector multiplied times a scalar function of t. Hence all trajectories are straight lines passing through the origin. See Figure 8.23. This equilibrium point is called a **proper node.**

---

* By $x(t) \to \infty$ we mean that the distance from $x(t)$ to the origin approaches infinity.

Figure 8.23
A proper node, two linearly independent eigenvectors; $\lambda_1 = \lambda_2$

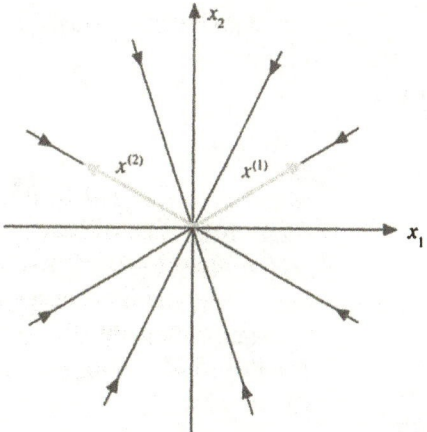

Subtype 2: One Linearly Independent Eigenvector (Improper Node). Although this particular subtype was not solved in Chapter 6, we asked the reader to solve a special case of this problem in the problem set in Section 6.4 (Problem 24). There we saw that the solution can be written in the form*

$$x(t) = c_1 e^{\lambda t}\boldsymbol{\xi} + c_2 e^{\lambda t}(t\boldsymbol{\xi} + \boldsymbol{\eta}) \tag{8}$$

where $\boldsymbol{\xi}$ is the only linearly independent eigenvector of A. For large t it is clear that the dominant term is $c_2 t e^{\lambda t}\boldsymbol{\xi}$, which means that as $t \rightarrow \infty$, all solutions approach the origin tangent to the line passing through $\boldsymbol{\xi}$. See Figure 8.24. In this case the equilibrium point $(0, 0)$ is called an **improper node.** When λ is negative, all solutions tend toward the improper node. If λ is positive, all solutions would move away from the improper node.

Figure 8.24
An improper node, one linearly independent eigenvector, $\lambda_1 = \lambda_2$

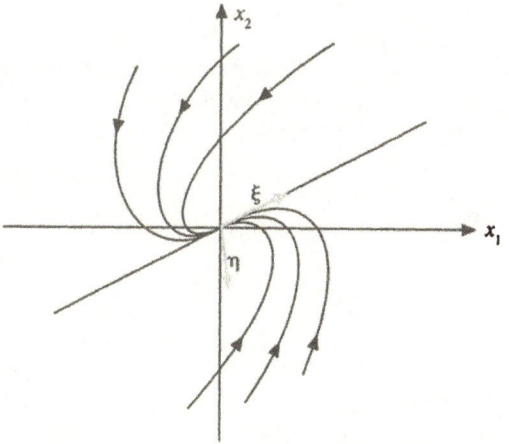

* Recall that $\boldsymbol{\xi}$ is the eigenvector of A and the vector $\boldsymbol{\eta}$ was determined to be the solution of the linear system $(A - \lambda I)\boldsymbol{\eta} = \boldsymbol{\xi}$, called the *generalized eigenvector* of A.

Type 4: Complex conjugate but not purely imaginary eigenvalues (spiral point)

For complex eigenvalues $\alpha \pm i\beta$ with $\beta \neq 0$, the general solution can be written in the form

$$x(t) = R_1 e^{\alpha t} \cos(\beta t - \delta_1)$$
$$y(t) = R_2 e^{\alpha t} \cos(\beta t - \delta_2)$$

$$(9)$$

where R_1, R_2, δ_1, and δ_2 are arbitrary constants. Hence the functions $x(t)$ and $y(t)$ oscillate between positive and negative values as t increases, which means that the state of the system $(x(t),\, y(t))$ spirals around the equilibrium point $(0, 0)$. This type of equilibrium point is called a **spiral point.** When $\alpha < 0$, the trajectories *spiral in* toward the equilibrium point $(0, 0)$. In this case the spiral point is called a **point attractor.** See Figure 8.25. On the other hand, if $\alpha > 0$, then the trajectories *spiral out* away from $(0, 0)$.

Figure 8.25
Spiral point; $\lambda = \alpha \pm i\beta$

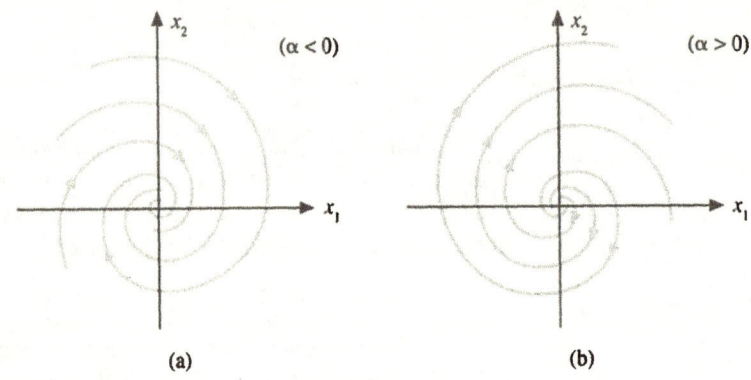

(a) (b)

Type 5: Pure imaginary roots (center)

If $\lambda_1 = i\beta$ and $\lambda_2 = -i\beta$ with $\beta \neq 0$, then the general solution can be written in the form

$$x(t) = R_1 \cos(\beta t - \delta_1)$$
$$y(t) = R_1 \cos(\beta t - \delta_2)$$

where R_1, R_2, δ_1, and δ_2 are arbitrary constants. Hence the trajectories are ellipses centered at the origin. This type of equilibrium point is called a **center point.** See Figure 8.26.

Figure 8.26
A center point; $\lambda_1 = i\beta$, $\lambda_2 = -i\beta$

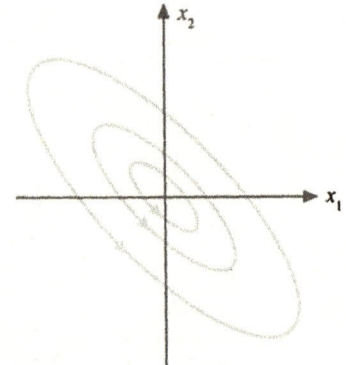

Table 8.1 summarizes the five types of equilibrium points. We distinguish between major and minor cases because although the borderline cases are important theoretically, they have little importance in applications, since the conditions for which they occur are unlikely.

Table 8.1
Types of Equilibrium Points

Roots of the characteristic equation	Type of equilibrium point
Major cases:	
Real, unequal, same sign	Improper node
Real, unequal, opposite sign	Saddle point
Complex conjugate but not pure imaginary	Spiral point
Minor cases:	
Real and equal	Proper or improper node
Pure imaginary	Center point

STABILITY OF EQUILIBRIUM POINTS

Equilibrium points can be interpreted as steady state solutions of a physical system and thus are often the focus in the study of natural phenomena. However, a steady state solution has little importance unless it has some degree of ''permanence,'' that is, unless it has the property of stability. We say that an equilibrium point (steady state solution) is **stable** if for any initial conditions sufficiently near the equilibrium point, the trajectory of the resulting solution remains close to the equilibrium point. Further, an equilibrium point is **asymptotically stable** if it is stable *and* the trajectory of nearby solutions approach the equilibrium point as $t \to \infty$. If an equilibrium point is not stable, it is called **unstable.** See Figure 8.27.

Figure 8.27
Concepts of stability and asymptotic stability

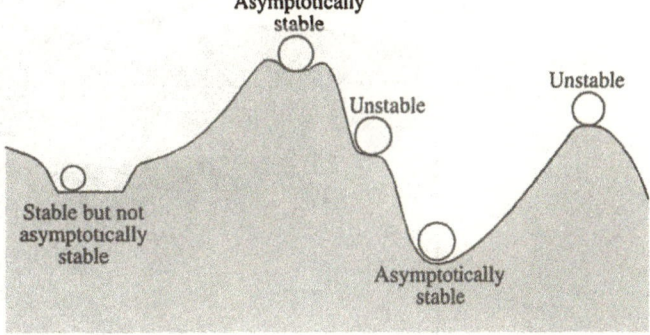

For example, it is not difficult to show that the pendulum equation

$$\ddot{\theta} + \frac{g}{L} \sin \theta = 0 \tag{10}$$

has equilibrium points (steady state solutions) at both $(\theta, \dot{\theta}) = (0, 0)$ and the point

$(\theta, \dot{\theta}) = (\pi, 0)$, corresponding to when the mass is at rest at the lowest and highest points. See Figure 8.28. Clearly, the low state is asymptotically stable, and the high state is unstable.

Although we are mostly interested in stability of nonlinear differential equations (and nonlinear systems), we will discuss stability in the context of the two-dimensional linear system $\dot{x} = Ax$. We have already seen that this system has a single equilibrium point at $x = (0, 0)$ when the matrix A is nonsingular. By examining the trajectories found earlier, we first make the observations summarized in Theorem 8.1.

THEOREM 8.1: Stability of the Linear System $\dot{x} = Ax$

Let λ_1 and λ_2 be the eigenvalues of the matrix A in the linear system $\dot{x} = Ax$. Then the equilibrium point $(0, 0)$ is

- asymptotically stable if the real parts of λ_1 and λ_2 are both negative;
- stable but not asymptotically stable if the real parts of λ_1 and λ_2 are both zero, that is, $\lambda_1, \lambda_2 = \pm i\beta$;
- unstable if either λ_1 or λ_2 has a positive real part.

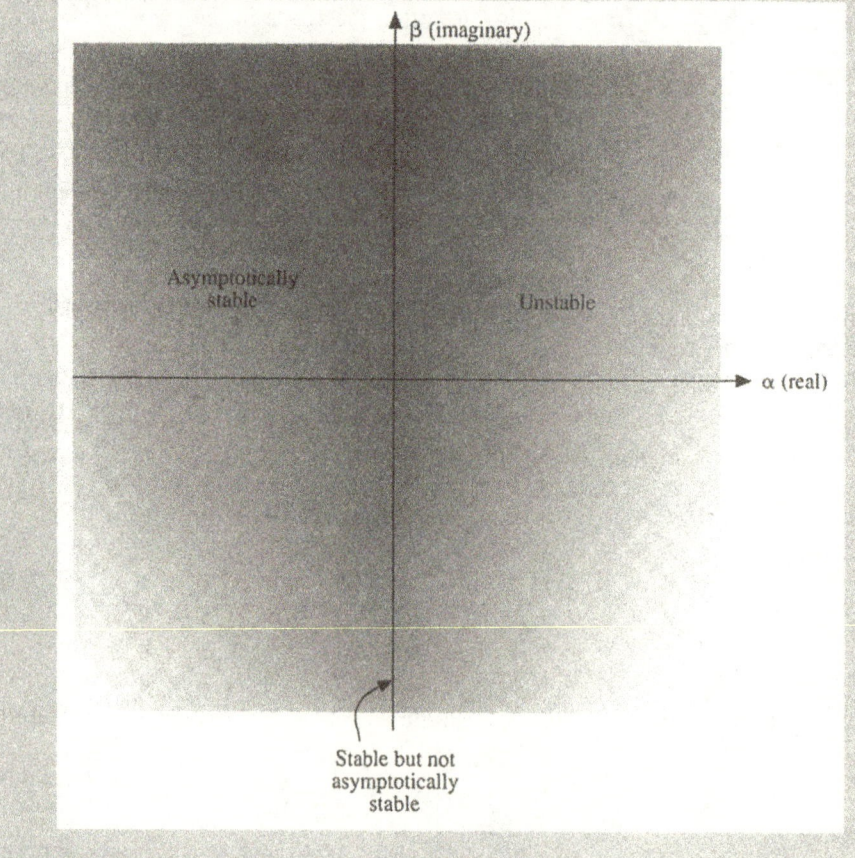

Next, observe that the eigenvalues λ_1 and λ_2 of the matrix

$$A = \begin{bmatrix} a & b \\ c & d \end{bmatrix}$$

can be determined from the two numbers, the **trace** of A, defined by $a + d$, and the determinant, $|A| = ad - bc$. In fact, one can easily show that the sum of the eigenvalues is the trace and the product is the determinant. Hence it should be possible to determine the stability of the equilibrium point $(0, 0)$ of $x' = Ax$ from the trace and determinant of A. It is possible, and the major result is illustrated in Figure 8.29. See Problem 22 at the end of this section.

Figure 8.28
Two equilibrium points of the
pendulum equation; one
stable, one unstable

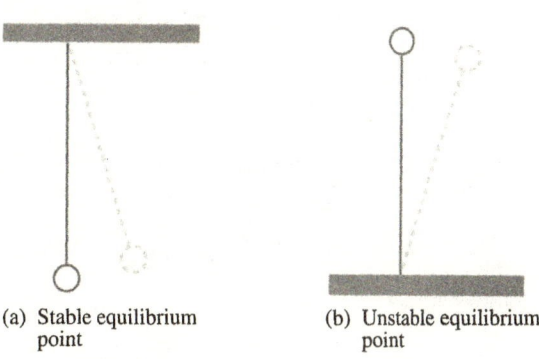

(a) Stable equilibrium
    point

(b) Unstable equilibrium
    point

Example 1   Stability of the Mass-Spring   Find the equilibrium point(s) and determine the stability of the frictionless mass-spring equation

$$\ddot{x} + \omega_0^2 x = 0 \tag{11}$$

Solution   Letting $y = \dot{x}$, we write this single equation as the linear system

$$\begin{bmatrix} \dot{x} \\ \dot{y} \end{bmatrix} = \begin{bmatrix} 0 & 1 \\ -\omega_0^2 & 0 \end{bmatrix} \begin{bmatrix} x \\ y \end{bmatrix} \tag{12}$$

Setting $\dot{x} = \dot{y} = 0$, we find the equilibrium point(s) by solving

$$\begin{bmatrix} 0 & 1 \\ -\omega_0^2 & 0 \end{bmatrix} \begin{bmatrix} x \\ y \end{bmatrix} = \begin{bmatrix} 0 \\ 0 \end{bmatrix} \tag{13}$$

for x and y, getting $x = y = 0$. Hence there is a single equilibrium point at $(0, 0)$. The eigenvalues of A can be found to be the purely imaginary numbers $\lambda_1 = \lambda_2 = \pm i\omega_0$. Hence the equilibrium point $(0, 0)$ is stable but not asymptotically stable. This reflects the fact that if a frictionless pendulum is moved slightly from the equilibrium position $\theta = 0$, $\dot{\theta} = 0$, it will not move far from this position, but it will also not return to equilibrium. ∎

Figure 8.29
Regions of stability and
instability of the equilibrium
point $(0, 0)$ of $\dot{x} = Ax$

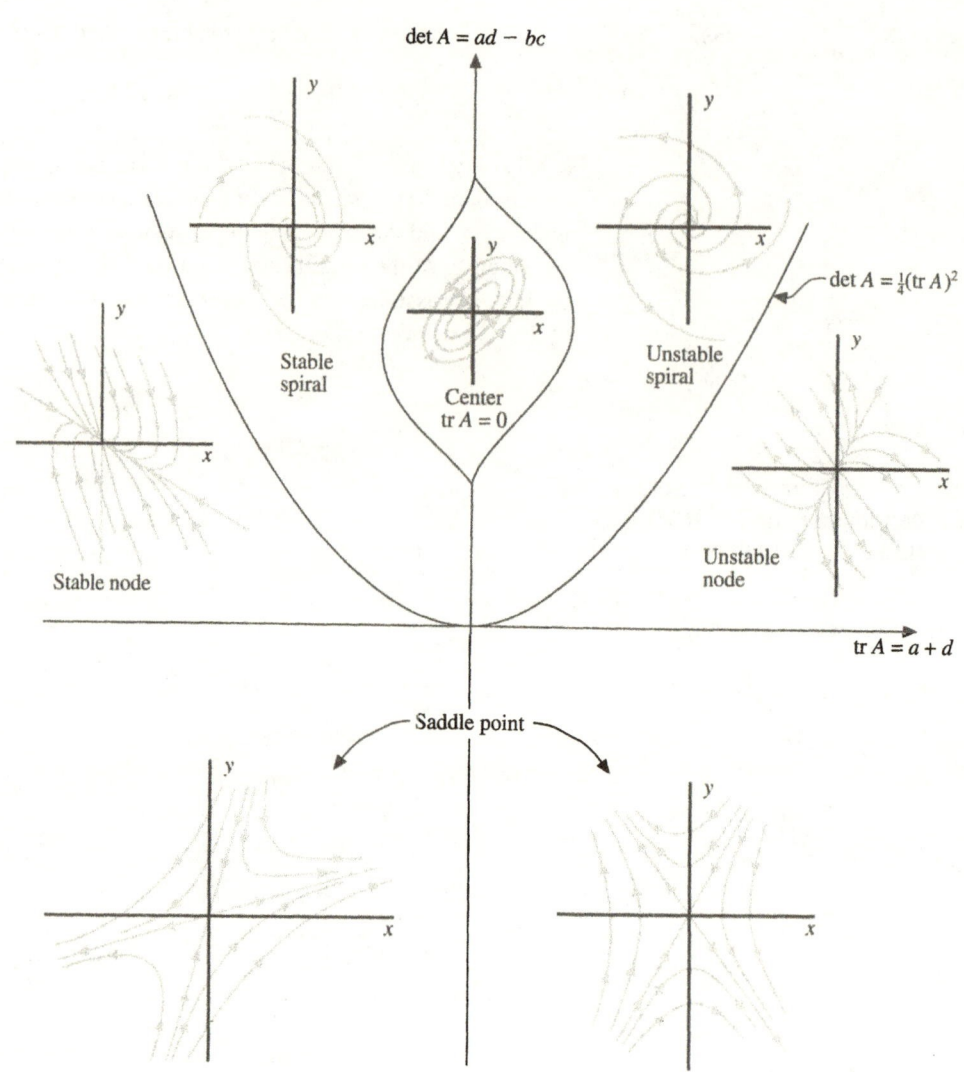

PROBLEMS: Section 8.2

For Problems 1–8, first determine whether the equilibrium point
(0, 0) is a node, saddle point, spiral point, or center point. Then
determine whether the equilibrium point is stable, asymptoti-
cally stable, or unstable.

1. $\dot{x} = x$
   $\dot{y} = 2y$

2. $\dot{x} = x$
   $\dot{y} = -y$

3. $\dot{x} = -x$
   $\dot{y} = -y$

4. $\dot{x} = -2x + y$
   $\dot{y} = -x - 2y$

5. $\dot{x} = x + 12y$
   $\dot{y} = 3x + y$

6. $\dot{x} = -8y$
   $\dot{y} = 2x$

7. $\dot{x} = -2x + y$
$\dot{y} = -4x + 3y$

8. $\dot{x} = x - 4y$
$\dot{y} = x + y$

In Problems 9–13, find the equilibrium point $(x_0\, y_0)$, and then classify the equilibrium point according to its type by making the substitution $u = x - x_0$, $v = y - y_0$ and finding the new system of differential equations in u and v. Determine whether the equilibrium point $(0, 0)$ of this new system is stable, asymptotically stable, or unstable.

9. $\dot{x} = x - 1$
$\dot{y} = y + 1$

10. $\dot{x} = -2x + y + 1$
$\dot{y} = -x - 2y + 3$

11. $\dot{x} = -8y + 16$
$\dot{y} = 2x + 4$

12. $\dot{x} = -2x + y - 2$
$\dot{y} = -4x + 3y - 6$

13. $\dot{x} = ax + by - h$
$\dot{y} = cx + dy - k$

14. Phase Plane Analysis of the Vibrating Spring The equation of motion of the vibrating mass-spring system is

$$m\ddot{u} + d\dot{u} + ku = 0 \qquad (14)$$

where m, d, and k are positive. Write the second-order equation as a system of two first-order equations by letting $x = u$, $y = \dot{u}$, and show that $x = y = 0$ is an equilibrium point of this system. Classify this type of equilibrium point as a function of m, d, and k. Determine whether the equilibrium point is stable, asymptotically stable, or unstable.

15. What Happens for Zero Eigenvalues Consider the system

$$\dot{x} = 0$$
$$\dot{y} = -x + y$$

(a) Show that the eigenvalues of this system are $\lambda_1 = 0$ and $\lambda_2 = 1$.

(b) Find the equilibrium points of this system.

(c) Find the general solution of the system.

(d) Show that the trajectories are straight lines.

16. One Eigenvalue Zero If $\lambda_1 = 0$ and $\lambda_2 \neq 0$, then show the following.
(a) There is a line of equilibrium points.

(b) All solutions that do not start on the line of equilibrium points tend toward this line if $\lambda_2 < 0$ and move away from this line if $\lambda_2 > 0$.

17. Both Eigenvalues Zero What is the nature of the solution of the 2×2 system $\dot{x} = Ax$ when both eigenvalues of A are zero?

Linearizing Nonlinear Systems

It is possible to approximate the trajectories of a nonlinear system

$$\dot{x} = f(x, y)$$
$$\dot{y} = g(x, y) \qquad (15)$$

near an equilibrium point (x_0, y_0) with trajectories of a linear system. To find the linear system that approximates the nonlinear system (15), make the substitution

$$x = \eta + x_0 \qquad y = \xi + y_0$$

from the original variables (x, y) to the new variables (ξ, η), which gives rise to

$$\begin{bmatrix} \dot{\eta} \\ \dot{\xi} \end{bmatrix} = \begin{bmatrix} \dfrac{\partial f(x_0, y_0)}{\partial x} & \dfrac{\partial f(x_0, y_0)}{\partial y} \\ \dfrac{\partial g(x_0, y_0)}{\partial x} & \dfrac{\partial g(x_0, y_0)}{\partial y} \end{bmatrix} \begin{bmatrix} \eta \\ \xi \end{bmatrix} \qquad (16)$$

This linear system is called the **linearized system** of system (15). The trajectories of the linearized system approximate the trajectories of nonlinear system (15) in a small region about the point (ξ, η) in much the same way as a tangent line approximates a curve near the point of tangency. For Problems 18–20, find the linearized approximation to the given nonlinear system about the given equilibrium point (x_0, y_0). Classify the equilibrium point of the linearized system.

18. $\dot{x} = x - 2xy$ $\qquad (x_0, y_0) = (0, 0)$
$\dot{y} = -4y + 2xy$

19. $\dot{x} = y$ $\qquad (x_0, y_0) = (0, 0)$
$\dot{y} = -\sin x$

20. $\dot{x} = y$ $\qquad (x_0, y_0) = (0, 0)$
$\dot{y} = \epsilon(x^2 - 1)y - x$ (Van der Pol equation)

21. Linearized Duffing's Equation Given Duffing's equation

$$\ddot{x} + x - 0.25x^2 = 0$$

(a) Rewrite the equation as a system of two first-order equations.

(b) Show that the two equilibrium points of the first-order system occur at $(0, 0)$ and $(4, 0)$.

(c) Find the linearized system about each equilibrium point.

(d) Classify each equilibrium point and verify that the trajectories near each equilibrium point appear as in Figure 8.30.

22. Classification of Trajectories Verify the behavior of the trajectories in Figure 8.29.

23. Journal Entry Summarize the nature of the trajectories of the 2×2 linear system.

Figure 8.30
Linearized Duffing's equation

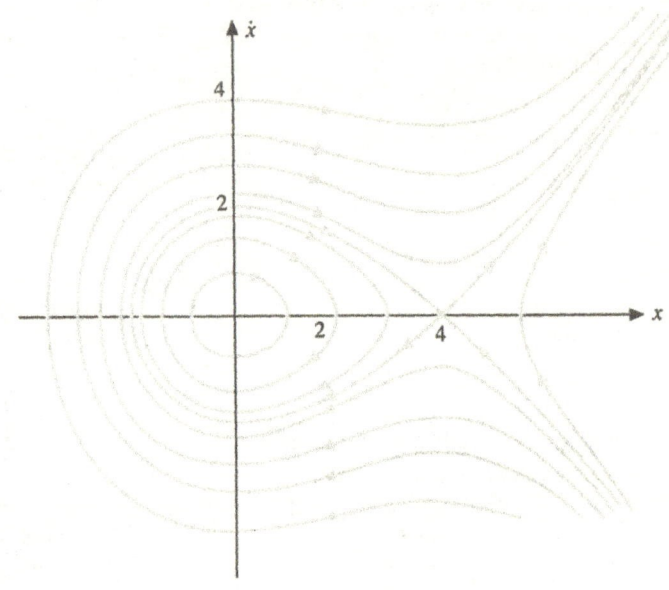

8.3   STABILITY: ALMOST LINEAR SYSTEMS

PURPOSE
To determine the type and stability of an equilibrium point in a class of nonlinear but almost linear systems by *linearizing* the system about an equilibrium point. We will see that the stability of an equilibrium point of a nonlinear system is generally the same as the stability of the equilibrium point of the linearized system.

TRANSFORMING
EQUILIBRIUM POINTS
TO THE ORIGIN

Often, nonlinear systems of differential equations can be approximated in a region about an equilibrium point by a certain *linear* system of equations. In those cases it is possible to analyze the stability of the point by analyzing the stability of the point of the corresponding linearized system. To illustrate this idea, consider the autonomous system

$$\frac{dx}{dt} = P(x, y) \qquad \frac{dy}{dt} = Q(x, y) \tag{1}$$

where $P(x_e, y_e) = Q(x_e, y_e) = 0$ defines an equilibrium point (x_e, y_e). We assume that the equilibrium point is an **isolated equilibrium point;** that is, there is a neighborhood about it that contains no other equilibrium point. We assume too that the equilibrium point under study is located at $(0, 0)$. We assume this without loss of generality, since we can always make the substitution $u = x - x_e$, $v = y - y_e$. Doing this and observing that $dx/dt = du/dt$, $dy/dt = dv/dt$, we see that the original system (1) is equivalent to the new system

$$\frac{du}{dt} = P(u + x_e, v + y_e) = P_1(u, v)$$
$$\frac{dv}{dt} = Q(u + x_e, v + y_e) = Q_1(u, v) \tag{2}$$

in which the equilibrium point is now located at $(0, 0)$ in the (u, v)-plane.

Example 1   **Transforming Equilibrium Points to the Origin**   Find the equilibrium points of the system

$$\frac{dx}{dt} = y$$
$$\frac{dy}{dt} = 2x - x^2 \tag{3}$$

and transform each nonzero equilibrium point to the origin.

Solution   The system has two equilibrium points, which are easily found to be $(0, 0)$ and $(2, 0)$. Transforming the nonzero equilibrium point $(2, 0)$ to the origin, we make the substitution $u = x - 2$, $v = y$ $(x = u + 2, y = v)$, getting the new system

$$\frac{du}{dt} = v$$
$$\frac{dv}{dt} = 2(u + 2) - (u + 2)^2 = -(u^2 + 2u) \tag{4}$$

Clearly, this new system has a critical point at $(0, 0)$. The trajectories of the two systems (3) and (4) are the same except that one is translated from the other. See Figure 8.31.

∎

Figure 8.31

Transformations of the form
$u = x - a, v = y - b$
translate trajectories.

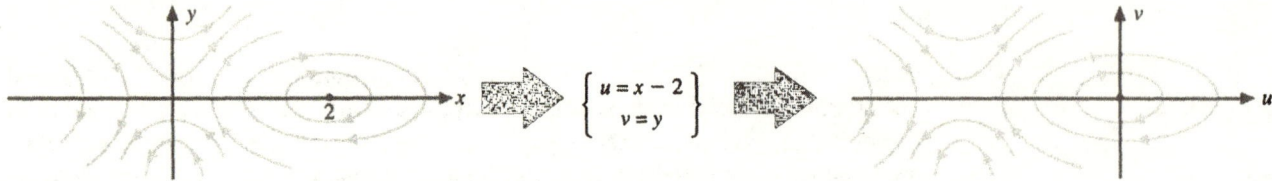

DETERMINING STABILITY OF ALMOST LINEAR SYSTEMS

Since we can always transform an equilibrium point to the origin, we assume that the origin $(0, 0)$ is an isolated equilibrium point of system (3). Using Taylor's series and noting that $P(0, 0) = Q(0, 0) = 0$, we can rewrite system (3) in the useful form

$$\frac{dx}{dt} = P(x, y) = ax + by + f(x, y)$$

$$\frac{dy}{dt} = Q(x, y) = cx + dy + g(x, y)$$

(5)

where $a = P_x(0, 0)$, $b = P_y(0, 0)$, $c = Q_x(0, 0)$, and $d = Q_y(0, 0)$. We also assume that the original functions $P(x, y)$ and $Q(x, y)$ give rise to functions $f(x, y)$ and $g(x, y)$ that are "small" near the origin in the sense that

$$\lim_{(x,y)\to(0,0)} \left(\frac{f(x, y)}{\sqrt{x^2 + y^2}}\right) = \lim_{(x,y)\to(0,0)} \left(\frac{g(x,y)}{\sqrt{x^2 + y^2}}\right) = 0$$

(6)

Roughly, conditions (6) say that $f(x, y)$ and $g(x, y)$ are small in comparison to the quantity $r = \sqrt{x^2 + y^2}$ near the origin. A nonlinear system (5) that satisfies Eq. (6) is called an **almost linear system**, and the linear part associated with the nonlinear system, that is,

$$\frac{dx}{dt} = P_x(0, 0)x + P_y(0, 0)y$$

$$\frac{dy}{dt} = Q_x(0, 0)x + Q_y(0, 0)y$$

(7)

is called the **associated linear system** (or **linear approximation**) expanded about the equilibrium point $(0, 0)$. Also, in order that the associated linearized system (7) has an *isolated* equilibrium point at $(0, 0)$, it must hold that

$$|P_x(0, 0)Q_y(0, 0) - P_y(0, 0)Q_x(0, 0)| \neq 0$$

It is our hope that we can determine the *type* and *stability* of a given equilibrium point (x_e, y_e) of system (3) by analyzing the type and stability of the equilibrium point $(0, 0)$ of system (7). The answer is (partially) given in the following theorem by Poincaré.

THEOREM 8.2: Stability of an Almost Linear System

Let λ_1 and λ_2 be the eigenvalues of the associated linear system (7) of the almost linear system

$$\frac{dx}{dt} = P(x, y) = P_x(0, 0)x + P_y(0, 0)y + f(x, y)$$
$$\frac{dy}{dt} = Q(x, y) = Q_x(0, 0)x + Q_y(0, 0)y + g(x, y)$$
(8)

Then (as long as λ_1 and λ_2 are *not* real and equal nor pure imaginary) the trajectories near the equilibrium point (0, 0) of the almost system (6) are of the same *type* and *stability* as the trajectories near the associated linear system.

Note: In the two (relatively unimportant) cases in which $\lambda_1 = \lambda_2$ are real equal eigenvalues or $\lambda_1, \lambda_2 = \pm i\beta$ are pure imaginary, we have

- If $\lambda_1 = \lambda_2$ are real and equal, then the equilibrium point (0, 0) of system (8) is either a node or a spiral point; and if $\lambda_1 = \lambda_2 < 0$, the equilibrium point is asymptotically stable; otherwise, for $\lambda_1 = \lambda_2 > 0$, it is unstable.

- If λ_1 and λ_2 are pure imaginary, then the equilibrium point (0, 0) is either a center point or a spiral point, which may be either asymptotically stable, stable, or unstable.

The importance of Theorem 8.2 lies in the fact that as long as $\lambda_1 \neq \lambda_2$ and λ_1 and λ_2 are not pure imaginary, it is possible to determine the nature of the trajectories near an equilibrium point and its stability by studying its associated linear system.

We summarize the results of Theorem 8.1 and Theorem 8.2 in Table 8.2.

Table 8.2
Type and Stability of Almost Linear Systems

λ_1, λ_2	Associated linear system		Almost linear system	
	Type*	Stability	Type*	Stability
$\lambda_1 > \lambda_2 > 0$	IN	Unstable	IN	Unstable
$\lambda_1 < \lambda_2 < 0$	IN	Asymptotically stable	IN	Asymptotically stable
$\lambda_2 < 0 < \lambda_1$	SP	Unstable	SP	Unstable
$\lambda_1 = \lambda_2 > 0$	PN or IN	Unstable	PN, IN, or SpP	Unstable
$\lambda_1 = \lambda_2 < 0$	PN or IN	Asymptotically stable	PN, IN, or SpP	Asymptotically stable
$\lambda_1, \lambda_2 = \alpha \pm i\beta$				
$\quad \alpha > 0$	SpP	Unstable	SpP	Unstable
$\quad \alpha < 0$	SpP	Asymptotically stable	SpP	Asymptotically stable
$\lambda_1, \lambda_2 = \pm i\beta$	CP	Stable	CP or SpP	Indeterminate

* The key for the symbols are: IN (improper node), PN (proper node), SP (saddle point), SpP (spiral point), and CP (center point).

Example 2 **Linearization about an Equilibrium Point** Verify that $(0, 0)$ is an isolated equilibrium point of the system

$$\frac{dx}{dt} = x + y - 2xy$$

$$\frac{dy}{dt} = -2x + y + 3y^3 \tag{9}$$

and determine its type and stability.

Solution The origin $(0, 0)$ is clearly an equilibrium point, since $x = 0$, $y = 0$ satisfies the equation $dx/dt = dy/dt = 0$. It is also an isolated equilibrium point, since

$$\begin{vmatrix} a & b \\ c & d \end{vmatrix} = \begin{vmatrix} 1 & 1 \\ -2 & 1 \end{vmatrix} = 3 \neq 0$$

Using polar coordinates, we can verify that system (9) is almost linear near $(0, 0)$ by observing that

$$\lim_{(x,y)\to(0,0)} \frac{|f(x, y)|}{\sqrt{x^2 + y^2}} = \lim_{r\to 0} \frac{|2r^2 \cos\theta \sin\theta|}{r} \leq \lim_{r\to 0} 2r = 0$$

$$\lim_{(x,y)\to(0,0)} \frac{|g(x, y)|}{\sqrt{x^2 + y^2}} = \lim_{r\to 0} \frac{|3r^3 \sin^3\theta|}{r} \leq \lim_{r\to 0} 3r^2 = 0$$

The eigenvalues of the associated linear system

$$\frac{dx}{dt} = x + y$$

$$\frac{dy}{dt} = -2x + y \tag{10}$$

are $\lambda_1 = -1$, $\lambda_2 = 3$. Hence by Theorem 8.1 from Section 8.2, we conclude that the equilibrium point $(0, 0)$ of the associated linear system (10) is an unstable saddle point. Hence by Theorem 8.2 (Poincaré's theorem) we conclude that $(0, 0)$ is an unstable saddle point of the nonlinear system (9). ∎

Example 3 **Phase Plane Analysis** Discuss the type and stability of the equilibrium points of the system corresponding to the equation

$$\ddot{x} + x - 0.25x^2 = 0 \tag{11}$$

Use this information to help sketch the trajectories in the phase plane.

Solution   Letting $y = \dot{x}$, we obtain the equivalent system

$$\dot{x} = y$$
$$\dot{y} = -x + 0.25x^2 \tag{12}$$

Setting $P(x, y) = y = 0$ and $Q(x, y) = -x + 0.25x^2 = 0$, we find the equilibrium points $(0, 0)$, and $(4, 0)$.

We first analyze the equilibrium point $(0, 0)$. In this case we do not have to make a preliminary translation to place the equilibrium point at the origin. It is also easily verified that system (12) is almost linear and that the equilibrium points of the associated linearized system

$$\begin{bmatrix} \dot{x} \\ \dot{y} \end{bmatrix} = \begin{bmatrix} 0 & 1 \\ -1 & 0 \end{bmatrix} \begin{bmatrix} x \\ y \end{bmatrix} \tag{13}$$

are isolated. Finding the eigenvalues of (13), we get $\lambda_1, \lambda_2 = \pm i$. Since the eigenvalues are pure imaginary, we first use Theorem 8.1 and conclude that $(0, 0)$ is a *stable center point* of the associated linear system (13). We then use Theorem 8.2 to conclude that $(0, 0)$ is either a *center point* or a *spiral point* of the nonlinear system (3). Unfortunately, Theorem 8.2 tells us nothing about the stability of $(0, 0)$ of the nonlinear system (3).

To analyze the trajectories near $(4, 0)$, we first translate $(4, 0)$ to the origin by letting $u = x - 4$, $v = y$ $(x = u + 4, y = v)$. When we do this, Eqs. (12) become

$$\frac{du}{dt} = v$$
$$\frac{dv}{dt} = -(u + 4) + 0.25(u + 4)^2 = u + \left(\frac{1}{4}u^2 + 4\right) \tag{14}$$

This is an almost linear system with associated linearized system

$$\begin{bmatrix} \dot{u} \\ \dot{v} \end{bmatrix} = \begin{bmatrix} 0 & 1 \\ 1 & 0 \end{bmatrix} \begin{bmatrix} u \\ v \end{bmatrix} \tag{15}$$

The isolated eigenvalues are easily found to be $\lambda_1 = -1$, $\lambda_2 = +1$. Since the eigenvalues are real and of opposite sign, we have that $(0, 0)$ is an *unstable saddle point* of the associated linear system (15). By Theorem 8.2 we conclude that $(4, 0)$ is an *unstable saddle point* of the nonlinear system (3).

Figure 8.32 shows the trajectories for Eqs. (12) and the trajectories of the linearized systems around the equilibrium points $(0, 0)$ and $(4, 0)$. ■

Figure 8.32
Trajectories near the
equilibrium points (0, 0) and
(4, 0): (a) trajectories of
almost linear system,
(b) trajectories of system
linearized around (0, 0),
(c) trajectories of system
linearized around (4, 0).

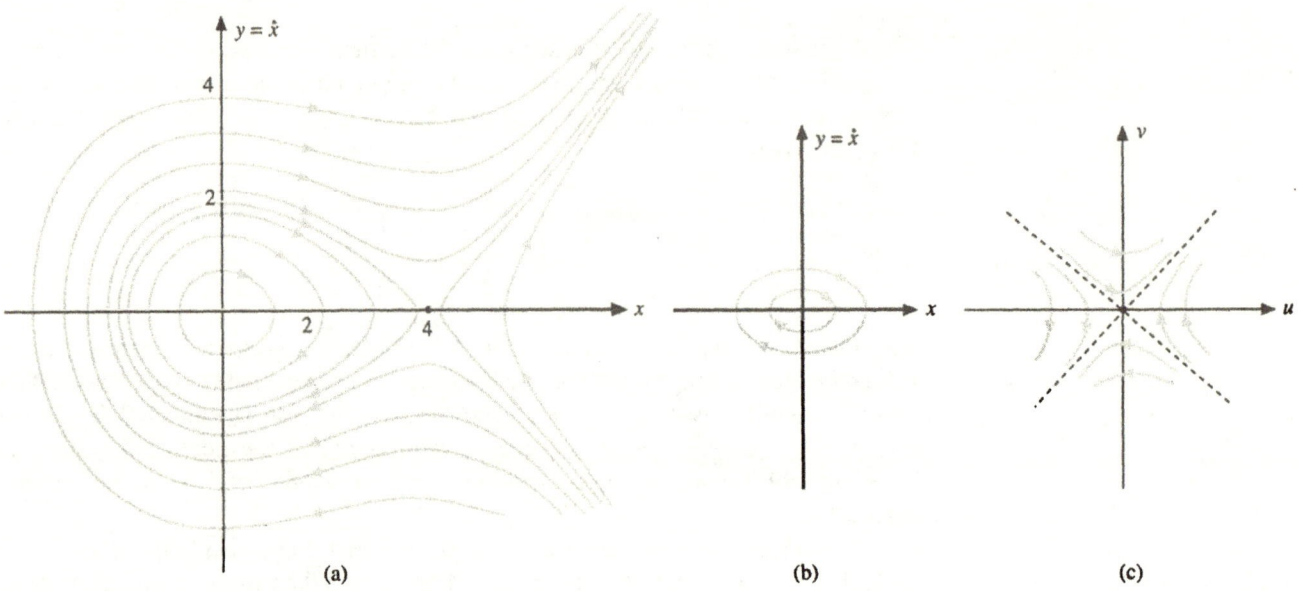

(a)   (b)   (c)

PROBLEMS: Section 8.3

*For Problems 1–4, verify that (0, 0) is an equilibrium point,
show that the system is almost linear, and discuss the type and
stability of the equilibrium point (0, 0).*

1. $\dot{x} = -2x + 3y + xy$
$\dot{y} = -x + y - 2xy^2$

2. $\dot{x} = -y - x^3$
$\dot{y} = x - y^3$

3. $\dot{x} = x + y + 2xy$
$\dot{y} = -2x + y + y^3$

4. $\dot{x} = y$
$\dot{y} = -\sin x - y$

*For Problems 5–7, determine the type and stability of each real
equilibrium point (x_e, y_e). Begin by first making the substitution*

*$u = x - x_e, v = y - y_e$ to transform the equilibrium point(s)
to the origin (0, 0).*

5. $\dot{x} = x + 2y + x^2 + y^2$
$\dot{y} = 2x - 2y - 3xy$

6. $\dot{x} = x - 3y + 2xy$
$\dot{y} = 4x - 6y - xy$

7. $\dot{x} = 1 - xy$
$\dot{y} = x - y^3$

8. Important Stability Result Consider the equation

$$\ddot{x} + g(x)\dot{x} + f(x) = 0 \qquad (16)$$

where $f(0) = 0$ and $x f(x) > 0$ for $x \neq 0$. If $g(x) \geq 0$ in
some interval $|x| < h$, then show that the equilibrium point
at $x = 0, \dot{x} = 0$ is stable.

9. **The Pendulum Equation** Show that the pendulum equation

$$\ddot{x} + \sin x = 0 \qquad (17)$$

has center points at $(2k\pi, 0)$ and saddle points at the points $(2(k + 1)\pi, 0)$, $k = 0, \pm 1, \pm 2, \ldots$. Sketch the trajectories of this equation in the phase plane, showing the equilibrium points $(-2\pi, 0)$, $(-\pi, 0)$, $(0, 0)$, $(\pi, 0)$, and $(2\pi, 0)$. To help in drawing the trajectories, you can use a computer to sketch the direction field of $dy/dx = \dot{y}/\dot{x}$.

10. **An Equation We've Seen Before** Find the equilibrium points of the equation

$$\ddot{x} + x - x^2 - 2x^3 = 0$$

and determine their type and stability.

11. **Prey-Predator Equation** An interesting application of stability theory lies in the interaction of two biological species. A typical example is the prey-predator situation, in which one species (predator) feeds on the other (prey). A standard example is the rabbit-fox system. The system of equations that describes the interaction of rabbits and foxes is the system

$$\frac{dr}{dt} = ar - brf$$
$$\frac{df}{dt} = -cf + drf \qquad (18)$$

where a, b, c, and d are positive constants and $r = r(t)$ and $f = f(t)$ denote the number of rabbits and foxes, respectively. There are two equilibrium points of this system at $(0, 0)$ and $(c/d, a/b)$, which represent the constant solutions $x(t) \equiv 0$, $y(t) \equiv 0$ and $x(t) \equiv c/d$, $y(t) \equiv a/b$.

(a) Substitute $u = x - c/d$, $v = y - a/b$ to transform the equilibrium point $(c/d, a/b)$ to the origin $(0, 0)$.

(b) Find the linearized system of the almost linear system found in part (a).

(c) Determine the type and stability of the equilibrium point $(0, 0)$ of the system found in part (b), thereby, it is hoped, determining the type and stability of the equilibrium point $(c/d, a/b)$.

(d) Interpret the trajectories of this system as shown in Figure 8.33.

Computer Problems

12. **Stable Limit Cycle** Using a computer approximate the solutions of the Van der Pol equation

$$\dot{x} = y$$
$$\dot{y} = -\epsilon(x^2 - 1)y - x \qquad (19)$$

for large ϵ (say $\epsilon = 10$) and verify that the trajectories look like the ones in Figure 8.34(a). The *closed curve* represents periodic solution of the system and is called a **limit cycle.** Since the nearby trajectories approach the limit cycle, it is called a **stable limit cycle.** Van der Pol's equation describes a *relaxation oscillator*, a two-phase phenomenon in which during the first phase energy is stored, and then during the second phase energy is discharged rapidly when a critical threshold is reached. Figure 8.34(b) shows a typical relaxation oscillator.

Figure 8.33
Linearized trajectories about an equilibrium population for the rabbit-fox equations

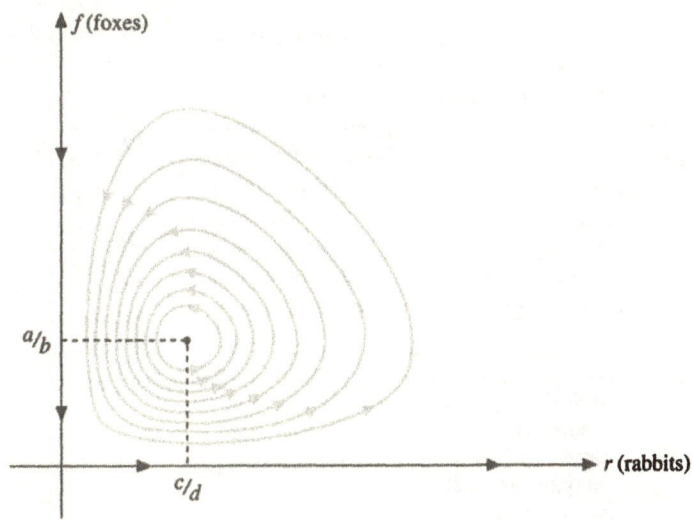

Figure 8.34
Van der Pol's equation: (a) All trajectories approach the stable Van der Pol limit cycle. (b) A typical relaxation oscillator. When the water fills the higher container, the two containers reverse positions with the water flowing out of the filled container and into the empty container. This process is then repeated periodically. Relaxation oscillators often occur in electrical circuits.

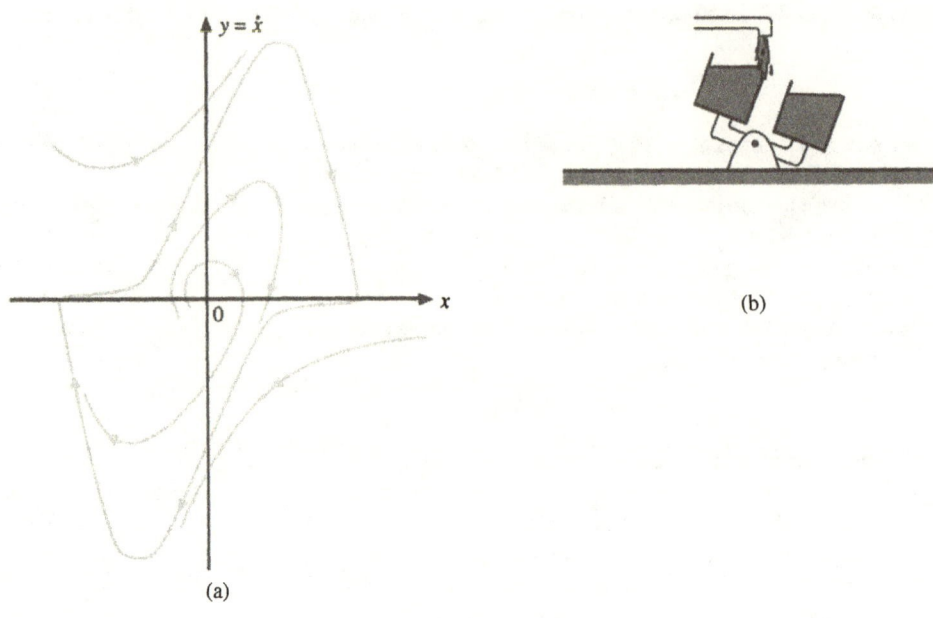

(a)

(b)

For Problems 13–16, rewrite the second-order equations as two first-order equations by letting $\dot{x} = y$, and then use a computer package to help you sketch the trajectories of the equation by drawing the direction field of the equation $dy/dx = \dot{y}/\dot{x}$.

13. $\ddot{x} + \sin x = 0$
14. $\ddot{x} + x - 0.1(x^2 + 2x^3) = 0$
15. $\ddot{x} - (1 - x^2)\dot{x} + x = 0$
16. $\ddot{x} + x - 0.25x^2 = 0$

The Liapunov Function

*We have seen in this section how it is often possible to determine the stability of an equilibrium point of an almost linear system by analyzing the stability of the corresponding linear system. However, if the system is not almost linear, the method breaks down. Another approach is known as **Liapunov's* second method** and **direct method.** (It is sometimes called the direct method, since it does not require knowledge of the solution of the system of differential equations.) Liapunov's basic theorem gives conditions that will ensure that an isolated equilibrium point is stable. Specifically, **Liapunov's theorem** states: If*

---

* Aleksandr M. Liapunov (1857–1918) was a Russian mathematician whose so-called *second method* was the conclusion of his doctoral dissertation, originally published in 1892. Liapunov argued intuitively that if a system of differential equations describes a physical system and if an equilibrium point is asymptotically stable, then it corresponds to a point of minimum potential energy, and so the potential energy must decrease as it approaches the point.

$$\frac{dx}{dt} = P(x, y) \qquad \frac{dy}{dt} = Q(x, y) \qquad (20)$$

has an isolated equilibrium point at (0, 0), and if there exists a continuous positive definite* function L(x, y) whose partial derivatives L_x and L_y are continuous and whose derivative

$$\frac{dL}{dt} = \frac{\partial L}{\partial x}P + \frac{\partial L}{\partial y}Q$$

is negative definite on some domain containing (0, 0), then (0, 0) is an asymptotically stable equilibrium point.

For Problems 17–18, use Liapunov's direct method (i.e., use Liapunov's theorem) to show that the origin is an asymptotically stable equilibrium point of the given system by verifying that the given function L(x, y) is a legitimate Liapunov function.

17. $\dfrac{dx}{dt} = y - 2x^3 \quad \dfrac{dy}{dt} = -2x - 3y^5$
   $L(x, y) = 2x^2 + y^2$

18. $\dfrac{dx}{dt} = xy^2 - \dfrac{1}{2}x^3 \quad \dfrac{dy}{dt} = -\dfrac{1}{2}y^3 + \dfrac{1}{5}yx^2$
   $L(x, y) = x^2 + 2y^2$

19. Journal Entry Summarize what you know about equilibrium points, almost linear systems, stability of almost linear systems, and so on.

---

* A function $L(x, y)$ is called *positive definite* on a domain D containing the origin if $L(0, 0) = 0$ and $L(x, y) > 0$ at all other points in D. A function $L(x, y)$ is called *negative definite* on D if $L(0, 0) = 0$ and $L(x, y) < 0$ at all other points in D.

8.4   CHAOS, POINCARÉ SECTIONS, AND STRANGE ATTRACTORS

PURPOSE
To introduce at an elementary level some recent ideas that are revolutionizing nonlinear dynamics and, consequently, nonlinear differential equations. We introduce the concepts of chaotic solutions of differential equations and the related ideas of the *Liapunov exponent, the Poincaré section,* and *strange attractors.*

INTRODUCTION TO CHAOS

Linear differential equations describe a world of *Platonic purity,* a world in which physical phenomena behave with clockwork regularity. However, scientists know that outside the classroom the world is seldom linear. Turbulence, irregularity, and random-like behavior are much closer to the norm. In the past, scientists always held to the time-honored belief that such erratic phenomena were nothing more than ''noise'' and had nothing to do with the underlying or ''real'' phenomena being investigated. Hence the philosophy was to study either linear problems or, at worst, problems that could be linearized.

In recent years, however, scientists are realizing that the turbulent and unpredictable behavior of many phenomena are not due to extraneous noise at all, but are the result of deeper relationships that were not previously considered. In particular, erratic and randomlike behavior, or **chaotic motion,*** has emerged from completely *deterministic* systems that have had nothing to do with extraneous noise. Examples of chaotic motion have been observed in a wide variety of scientific disciplines including ecology, economics, physics, chemistry, engineering, and fluid dynamics, to name a few.[†] Specific examples include the yearly variation of insect populations, the convection of a heated fluid, and stirred chemical reactor reactions. New examples of chaotic motion are continually being discovered.

Parallel to the observations of chaotic behavior in the real world has been the mathematical discovery that many solutions of (nonlinear) differential and difference equations also exhibit chaotic behavior.

To put the subject of ''chaos'' in historical perspective, in 1963 an MIT meteorologist, Edward Lorenz, was modeling thermally induced air convection and used a computer to approximate the solution of the following system of three nonlinear equations:[‡]

$$
\begin{aligned}
\dot{x} &= -10x + 10y & x(0) &= x_0 \\
\dot{x} &= 28x - y + xz & y(0) &= y_0 \\
\dot{z} &= -\frac{8}{3}z + xy & z(0) &= z_0
\end{aligned}
\tag{1}
$$

---

* Although there is no universally accepted definition of chaotic motion the term refers to motion that is irregular and unpredictable. It also has the characteristic that it is so sensitive to initial conditions that for all practical purposes its future cannot be determined.

[†] A compilation of the important papers leading to the theory of chaos has been given in the book *Chaos,* by Hao Bai-Lin (World Scientific Publishing Co. Ltd, 1984).

[‡] The dependent variables *x, y,* and *z* represent relevant wind velocities and air temperatures.

To check his results, he reran his computer program a second time, starting from the same initial conditions. The second time, however, he used only three decimal places for the initial conditions x_0, y_0, and z_0 instead of the six that he had used on the first run. He keyed the data into the computer and left the room for a cup of coffee. When he came back, he was in for a shock. The new result displayed on the screen wasn't even close to the first forecast. Checking to see that he had keyed in the correct values, he eventually realized that the difference was caused by the small difference in initial conditions. The differential equations were so sensitive to the initial conditions that after a short period of time the second solution was totally different from the first. See Figure 8.35. Lorenz later told *Discover* magazine, "I knew then that if the real atmosphere behaved like this [mathematical model], long-range weather prediction was impossible."

We do not mean to give the impression that nothing was known about chaotic systems before Lorenz's discovery in 1963. In fact, the eminent mathematician Poincaré was well aware that the solutions to some differential equations in celestial mechanics were extremely sensitive to initial conditions and speculated that the motion of orbiting planets could not be predicted.

Figure 8.35
The sensitivity of the solutions of Lorenz's equation to initial conditions

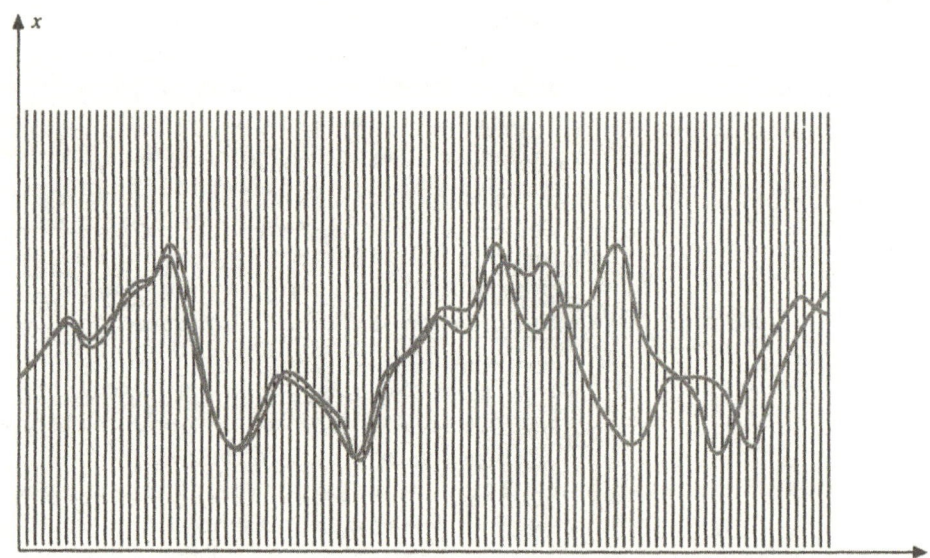

FORCED DISSIPATIVE CHAOS

Most interest in chaotic systems lies in (nonlinear) systems known as *forced dissipative systems,* which are systems with friction that are driven by an external force. Of course, if a dissipative system has no input, its motion will slow down and stop. However, if such a system is driven by an external force, interesting things can happen. A standard example that comes to mind when one thinks about this type of system is turbulence. The eddies in a stream and the irregular airflow behind an airplane wing are examples of chaotic motion of a forced dissipative system.

The following examples illustrate chaotic motion of forced dissipative systems.

Example 1 **Ball on a Vibrating Warped Surface**   A typical example of chaotic motion of a forced dissipative system is the back-and-forth motion of a ball on a warped surface with two depressions that vibrates with a periodic motion. The ball will continue to vibrate from one depression to the other, and no matter how accurate the initial conditions of the ball, the system is so sensitive that it is impossible to predict its future. See Figure 8.36. ∎

Figure 8.36
No matter how accurately you know the initial conditions of the ball, its sensitivity to these conditions makes its future motion impossible to predict.

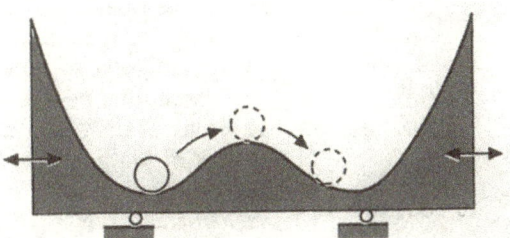

Example 2 **The Chaotic Triple Pendulum**   A simple example of chaotic motion of a forced dissipative system is a system of three linked pendulums that can be purchased in novelty shops. It consists of one large pendulum and two attached smaller ones. The larger one is driven by electrical pulses generated by an electron-magnetic assembly situated in the base of the toy, and the smaller ones are driven by repelling magnets. See Figure 8.37. If the reader has ever experimented with one of these toys, he or she will agree as to the sensitivity of the long-term behavior to the initial conditions.

Figure 8.37
The triple pendulum toy is so sensitive to initial conditions that its future is unpredictable.

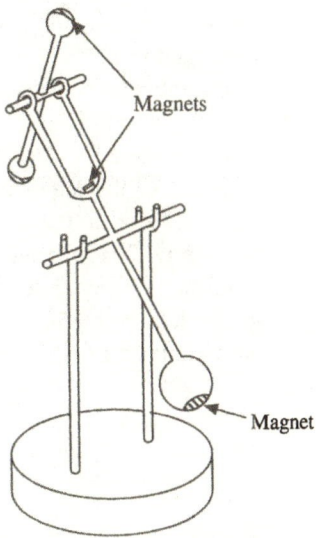

Magnets

Magnet

When the larger pendulum moves past a sensor located at the base of the toy, a magnetic field is created, which in turn causes a battery to generate a current, which in turn generates a torque to the pendulum. When the system is started with a slight disturbance, it starts oscillating, and before long its motion is completely randomlike or chaotic.* ∎

HISTORICAL NOTE

James Yorke (1937–   ) represents one of the new breed of mathematician/scientist whose research covers a wide range of disciplines. It was his study of the works of the mathematician Stephen Smale and meteorologist Edward Lorenz that led Yorke to some fundamental discoveries in chaos theory. It was in a 1970 paper that he first coined the word "chaos." Since 1966 he has been on the faculty of the University of Maryland where he now holds the position of Professor of Mathematics and Director of the Institute for Physical Science and Technology. His principal scientific contributions have been in the field of chaotic dynamics, where he has done pioneering work bridging the gap between rigorous mathematics and the sciences.

IDENTIFYING CHAOTIC BEHAVIOR: THE LIAPUNOV EXPONENT

First off, let us say that *linear* systems do *not* exhibit chaotic behavior, so we are solely in the domain of nonlinear phenomena. Physically, we are talking about systems that have nonlinear components, such as nonlinear springs, nonlinear damping, nonlinear resistance, or nonlinear transistors. We would like to know when a physical system that contains nonlinear components exhibits chaotic behavior and for what values of the parameter(s), if any, the system exhibits chaotic behavior. These questions are important to physical scientists because chaos implies loss of predictability. These are also nontrivial questions, since chaotic behavior can be confused with other types of behavior that *appears* chaotic, such as periodic behavior with *long* periods and *quasiperiodic behavior.*[†] Although there are several ways for engineers and mathematicians to detect legitimate chaos, such as Fourier analysis and pseudo–phase plane methods, we will focus on the Liapunov exponent.

As Lorenz learned from his computer experiments, chaos implies a sensitive dependence on initial conditions in which two trajectories that are close together move away from each other. The Liapunov exponent measures the degree of this divergence. To understand this concept, suppose the initial position x and velocity $y = \dot{x}$ of a

---

* Professor Alan Wolf and colleagues at Cooper Union College in New York have analyzed the motion of this pendulum and verified that its motion is chaotic. There are ways to determine whether the motion is chaotic and not simply periodic with a long period. One way is to carry out a *Fourier analysis* to verify chaotic motion. Interested readers should consult the book *Chaotic Vibrations,* by Francis C. Moon (John Wiley & Sons, New York, 1987).

[†] The combined oscillation $y = R_1 \cos \omega_1 t + R_2 \cos \omega_2 t$, where the ratio ω_1/ω_2 of the frequencies is an irrational number, is called *quasiperiodic.* This function is *not* periodic (although it looks periodic), and its behavior is often mistaken for chaotic motion.

mechanical system can be measured with accuracy Δx and Δy, respectively. In other words, we know that the starting point lies in the phase plane somewhere in a given box of area $\Delta x \, \Delta y$. See Figure 8.38. Of course, this is what always occurs in the real world, where there is a *region of uncertainty* due to the uncertainties in measurements. However, if the motion of the system is chaotic, then after time t the uncertainty of the system grows exponentially to $N(t)$ boxes, where

$$N(t) = e^{\lambda t} \qquad (t \geq 0) \tag{2}$$

The constant $\lambda > 0$ is called the **Liapunov exponent,** * which measures the *growth rate* of uncertainty of the system. In the phase plane it measures the rate at which two trajectories with different initial conditions separate. This concept fits naturally in the study of turbulent fluids, since points that are initially close together quickly wind up in totally different regions of the fluid. In fact it is the rapid, unpredictable separation of neighboring points that characterizes turbulence. We should point out, however, that there is no simple formula for finding the Liapunov exponent, and for the most part it is found by using a numerical algorithm and a computer.

Figure 8.38
The Liapunov exponent measures the exponential separation in time of two adjacent trajectories in phase space with different initial conditions. It is one of the ways in which chaotic motion is quantified.

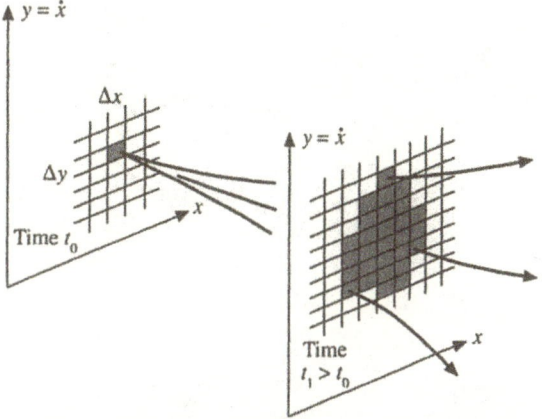

HISTORICAL NOTE

The Soviet mathematician, Aleksandr Mikhailovich Liapunov (1857–1918) was one of the pioneers of geometric or qualitative methods of differential equations. He was a student of Chebyshev at the University of St. Petersburg and taught at Kharkov University from 1885 to 1901. His so-called *second method*, which under certain conditions determines if an equilibrium point is stable, was part of his most influential work, *General Problem of Stability of Motion*, published in 1892 as part of his doctoral dissertation.

* For a more complete discussion on the Liapunov exponent, the reader should consult the text *Chaotic Vibrations*, by Francis C. Moon (John Wiley & Sons, New York, 1987).

POINCARÉ SECTIONS AND STRANGE ATTRACTORS

Integration of chaotic systems causes special problems because of their sensitive dependence on initial conditions, and since error is always present in any integration algorithm, it would seem that integration of any chaotic system would be meaningless. Often, however, it is necessary to know not the transient trajectory of the system for given initial condition, but the limiting behavior of the trajectory. For example, observe how the two orbits in Figure 8.39 of **Duffing's equation***

$$\ddot{x} + 0.05\dot{x} + x^3 = 7.5 \cos t \tag{3}$$

diverge, starting from almost identical initial conditions. The trajectory of Duffing's equation is so randomlike that it looks almost like the path of a fly buzzing around in your bedroom.

Figure 8.39
Divergence of orbits of Duffing's equation that start at nearby starting points *A* and *B*

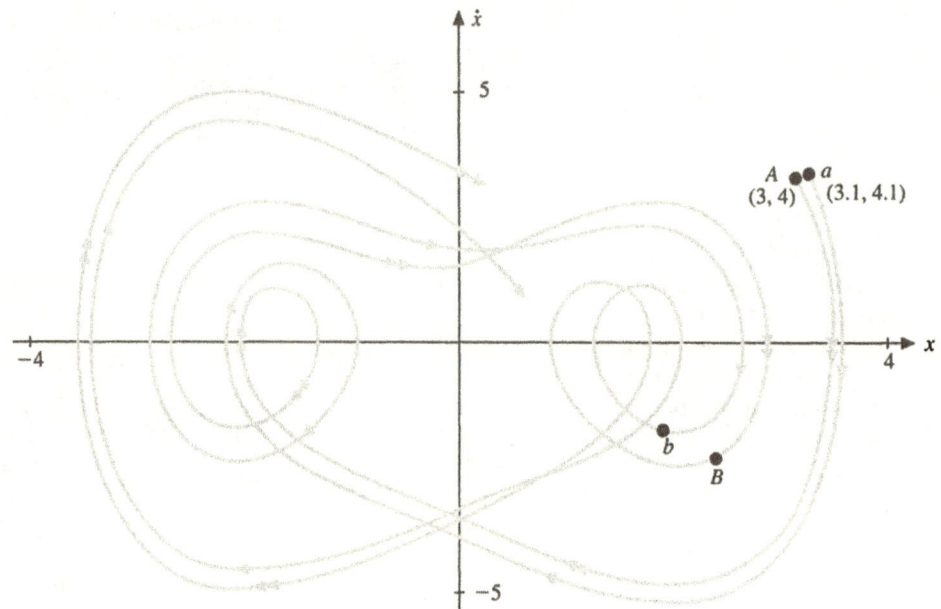

However, there is a way to analyze the behavior of forced dissipative equations such as Duffing's equation by simply forgetting all about the *continuous* motion of the trajectory and zeroing in on the solution at certain points in time. In other words, we look at the solution under a strobe light, thus freezing the solution at certain discrete points of time. If we observe the strobed solution at times[†] $t = 0, 2\pi, 4\pi, ...$, which we label as *A, B, C* (see Figure 8.40) and if we erase the continuous orbit and display only the *strobed* points, this set of points *A, B, C, ...*, which provides a discrete representation of the continuous solution, is called a **Poincaré section** of the equation.

---

* The general Duffing's equations is $\ddot{x} + \alpha\dot{x} + \beta x + \gamma x^3 = f(t)$ and is well known among mechanical engineers. It describes the forced vibrations of a mass-spring system with a nonlinear restoring force. This equation is often used to describe the vibrations of a spring when the deflections are large. Figures 8.39 and 8.40 are reprinted with permission from *Chaotic Vibrations,* by Francis C. Moon (John Wiley & Sons, New York, 1987).

† Although there are no definite rules for determining when to flash the strobe light, we have chosen the times $t = 0, \pi, 2\pi, . . .$ since they match the period of the forcing term.

Figure 8.40
The trajectory of Duffing's equation traces out a *continuous* curve in the phase plane. However, if a strobe light flashes periodically and the locations of the trajectory are recorded, a set of *isolated* points *A, B, C, ...*, called a Poincaré section, will begin to emerge and gradually fill up the plane in a distinctive pattern.

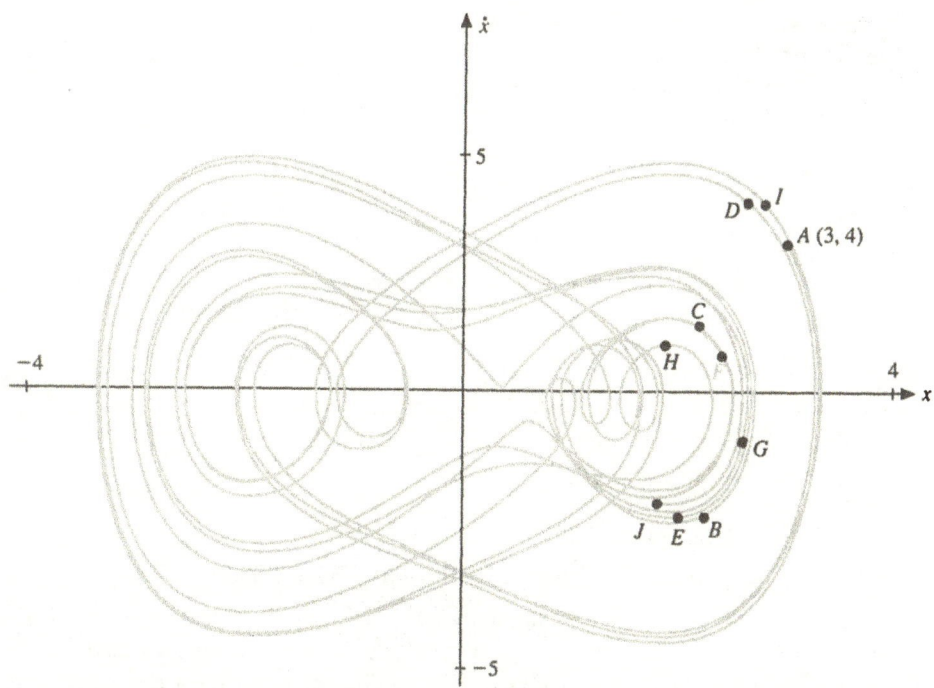

Gradually, as more and more strobed points are added, a fascinating *pattern* emerges before your eyes as the strobed points flash periodically. In fact, even more amazing is the fact that no matter *what the initial conditions*, the points *A, B, C, ...* in the Poincaré section quickly approach the same unique pattern. This settling down to a *limiting set of points* defines what is called the **strange** (or **chaotic**) **attractor.*** (From a practical point of view, if we omit the first few points or initial transient in the Poincaré section, the remaining points are essentially the same as the strange attractor.) See Figure 8.41.

The strange attractor was given its name because in a vague limiting sense it "attracts" the strobed points and also because of its generally strange appearance. What is interesting here is that before the discovery of strange attractors, it was the general consensus among mathematicians that the only types of (bounded) motion exhibited by differential equations were trajectories that were either

- equilibrium points or converged to equilibrium points,
- periodic or converged to periodic trajectories, or
- almost periodic or converged to almost periodic trajectories.

Strange attractors are completely different. Here is a completely different kind of set in the phase plane where the solution "lives."[†] This contrasts with the traditional

* In other words, for a chaotic system with given initial conditions, although you cannot determine the future trajectory of the system, you *can* determine the eventual states where the system will reside. In other words you can tell *where* you will eventually be in the future, you just can't tell *when* you will be there.

[†] The limiting solution is what one generally sees in nature and not the transient solutions. Thus we say that solutions "live" in the limiting solution.

Figure 8.41
The strange attractor of Duffing's equation emerges from the Poincaré section.

types of "lived-in" sets, such as *points* (equilibrium points), *closed curves* (periodic solutions), or, more rarely, *almost closed curves* (almost periodic solutions).

Since strange attractors were discovered, the solutions of many (nonlinear, of course) differential equations have been discovered to "move" toward strange attractors. Upon closer examination of these newly found strange attractors, it was discovered that they are in fact **fractals.*** To study fractals in any depth would take us too far afield from an introduction in chaotic motion of nonlinear differential equations. Roughly, fractals are geometric objects that have the ultimate broken or jagged structure. They are sometimes defined as geometric objects that have **fractional dimension.** They have also been defined as geometric objects or curves that exhibit increasing detail ("jaggedness") with increasing magnification. The reader may have seen the "snowflake" curve or the "Peano monster" curve. These curves are shown in Mandelbrot's book *The Fractal Nature of Geometry*.

HISTORICAL NOTE

David Ruelle (1935–   ) was born in the northern Belgium town of Ghent, Belgium, and attained a Ph.D. in physics at the Free University of Brussels. Ruelle, along with Dutch mathematician Floris Takens, introduced the concept of the *strange attractor* into dynamical systems in a 1971 paper on the nature of turbulence. In addition to Ruelle's many research papers and technical books on chaotic dynamical systems, he has written the general audience book, *Chance and Chaos* (Princeton University Press, N.J. 1991), which would make excellent reading for the reader of this book.

* Fractals were introduced into mathematics by Benoît Mandelbrot. For a further study of fractals, see *The Fractal Geometry of Nature*, by Benoît Mandelbrot (W. H. Freeman, New York, 1983) or *Fractals Everywhere*, by Michael Barnsley (Academic Press, New York, 1988).

EPILOGUE

By periodically sampling the solution of a differential equation at different times, it is possible to ultimately know a great deal about how the system behaves. (Periodically sampling of the solution might be compared to recording an animal population on the first day of spring every year in contrast to keeping a continual record.) The idea is that after many points are tabulated, an overall structure of the system begins to unfold. One does not know the *order* in which the points were added to the strange attractor or at what value of *time* the points were added, but since all continuous trajectories ''converge'' to the attractor, one knows a great deal about the trajectories of the system.

Although the trajectories of differential equations are continuous curves, the strange attractor is not smooth. The strange attractor is a ''limiting set'' and not itself a trajectory. However, many strange attractors appear to have ''jagged'' lines in them, and one can imagine that the trajectories of the system will pass along in the general region of these ''lines.''

The subject of chaotic nonlinear dynamics is a vigorous, expanding area of research that holds promise as a means of understanding many nonlinear phenomena.

PROBLEMS: Section 8.4

1. **Pseudo-Phase Plane** Plotting the points (x_0, x_1), (x_1, x_2), (x_2, x_3), ... in the plane, where $\{x_n\}$ is the solution of a first-order difference equation $x_{n+1} = f(x_n)$, defines a **discrete trajectory** in the **pseudo-phase plane**. One can show that a closed trajectory in the $(x, \dot{x})$ plane is also periodic in the pseudo-phase plane (x_n, x_{n+1}). Also, chaotic motion in the phase plane appears to be chaotic in the pseudo-phase plane. Describe the general nature of discrete trajectories in the pseudo-phase plane for the following cases.
   (a) When the solution $\{x_n\}$ converges to a point.
   (b) When the limiting solution is periodic with period 2.
   (c) When the limiting solution is periodic with period 3.
   (d) When the solution is chaotic.

2. **Poincaré Map for Periodic Functions** Find the Poincaré map of

$$x(t) = \sin 2t + \sin t$$

by plotting the first 1000 or so terms of the sequence

$$\big((x(t), \dot{x}(t)): t = 0, \pi, 2\pi, ...\big)$$

Note that the points of the sequence consist of $(0, 3)$ and $(0, 1)$. This tells us that $x(t)$ is periodic and does not display chaotic behavior.

3. **Poincaré Map for Incommensurate Frequencies** Find the Poincaré map of

$$x(t) = \sin 2\pi t + \sin t$$

by plotting the first 1000 or so terms of the sequence

$$\big((x(t), \dot{x}(t)):t = 0, 1, 2, ...\big)$$

Note that the resulting points in the sequence lie on a *continuous closed* curve in the $x\dot{x}$-plane. This tells us that $x(t)$ is quasiperiodic with two incommensurate frequencies present and does not display chaotic behavior.

Stagecoach Wheels and the Poincaré Section

The Poincaré section can often determine the natural frequency ω_0 of a system. For example, consider the simple dynamical system consisting of a point moving clockwise on the unit circle as shown in Figure 8.42. We can visualize this as the hand of a clock (not necessarily keeping good time). * *However, suppose*

---

* The reader might recall watching old Western movies in which the wheels on the stagecoach always appear to be rotating backward. The same principle holds there as it does in these problems.

the clock is not observed continuously, but is viewed under a strobe light flashing with a given frequency ω_s. When we have $\omega_0 = \omega_s$, the strobe light flashes at the same frequency as the hand of the clock, and the strobed motion (Poincaré section) will appear to be motionless. On the other hand, when we have $\omega_0 = 3\omega_s/4$, the strobe light will flash three-quarters of the way through each rotation of the clock hand, thus making the strobed motion of the hand move counterclockwise in jumps of 3 hours. Problems 4–7 ask about important relationships between the natural frequency of the clock hand, ω_0, and the strobe frequency, ω_s.

Figure 8.42

What is the Poincaré section or the strobed motion of the hour hand when the strobe light flashes with different frequencies?

4. How Will the Clock Appear? How will the hand of the clock appear when the relationship between the hand's frequency, ω_0, and the strobed frequency, ω_s, is as follows?

(a) $\omega_0 = \dfrac{3}{4}\omega_s$

(b) $\omega_0 = \dfrac{2}{4}\omega_s = \dfrac{1}{2}\omega_s$

(c) $\omega_0 = \dfrac{1}{4}\omega_s$

5. Hmmmmm Now that you have solved Problem 4, determine the strobed motion of the hand of the clock that is strobed at a frequency ω_s satisfying $\omega_0 = (p/q)\omega_s$, where $p/q < 1$ is a fraction reduced to lowest terms. *Hint:* Try solving this problem using specific fractions.

6. The Answer Revealed We will tell you the answer to Problem 5. If the natural frequency of the clock and the strobed frequency are related by $\omega_0 = (p/q)\omega_s$, where the ratio $p/q \le 1$ is reduced to lowest terms, then the Poincaré section will consist of q equally spaced points on the circle

(that appear to be moving counterclockwise), and the *order* in which the points appear will be displayed in such a way that $[q - (p + 1)]$ positions will always be skipped. For parts (a)–(c), determine the strobed motion of the hour hand for the indicated relationship.

(a) $\omega_0 = (2/5)\omega_s$.

(b) $\omega_0 = (11/12)\omega_s$

(c) $\omega_0 = (99/100)\omega_s$

7. Power of the Poincaré Section By observing the strobed points or Poincaré section of a system, it is possible to determine the natural frequency of the system. In the case of the hand of the clock, if the strobed motion is periodic (it moves counterclockwise when $p/q < 1$) with period q, and if the motion is such that s positions are always being skipped, then the natural frequency can be found from the relationship $\omega_0 = [(q - s - 1)/q]\omega_s$. For parts (a)–(c), find the natural frequency of a rotating hand of a clock in terms of the strobed frequency.

(a) If the (counterclockwise) strobed motion of the hand is periodic with period 4, moving in 6-hour jumps (say, starting at 12 o'clock).

(b) If the (counterclockwise) strobed motion of the hand is periodic with period 4, moving in 9-hour jumps (say, starting at 12 o'clock).

(c) If the (counterclockwise) strobed motion of the hand is periodic with period 12, moving in 5-hour jumps (say, starting at 12 o'clock).

Close Attractors of a Strange Kind

8. Strange Attractor of Duffing's Equation Write a computer program or use an existing program to find the strange attractor of Duffing's equation

$$\ddot{x} + 0.05\dot{x} + x^3 = 7.5 \cos t$$

Hint: Plot the solution in the phase plane for values of time $t = 0, 2\pi, 4\pi, \ldots$ but use a smaller step size to numerically integrate the equation. To get a good look at this attractor, you should look at at least 1000 points. The initial conditions are unimportant, since you should automatically throw out the first 100 computed points.

9. Strange Attractor of Van der Pol's Equation Write a computer program or use an existing program to find the strange attractor of Van der Pol's equation

$$\ddot{x} + 0.10(x^2 - 1)\dot{x} + x = 12 \cos t$$

Hint: Plot the solution in the phase plane for values of time $t = 0, 2\pi, 4\pi, \ldots$, but use a smaller step size to numerically integrate the equation. To get a good look at this attractor, as shown in Figure 8.43, you should look at at least 1000 points. The initial conditions are unimportant, since you will be throwing out the first 100 computed points.

Figure 8.43
The strange attractor of the
Van der Pol equation

10. **Negative Resistor Oscillator** One of the first examples of chaos in an electrical network was the behavior of a negative resistor oscillator discovered by Ueda and Akamatsu (1981), modeled by the equation

$$\ddot{x} + (x^2 - 1)\dot{x} + x^3 = \cos t \qquad (4)$$

Hint: Write a computer program or use an existing program to find the strange attractor for this equation. Plot the solution in the phase plane for $t = 0, 2\pi, 4\pi, \ldots$, but use a smaller step size to numerically integrate the equation. To get a good look at this attractor, as shown in Figure 8.44, you should look at at least 1000 points. The initial conditions are unimportant, since you should throw out the first 100 computed points.

11. **The Lorenz Differential Equations** The Lorenz differential equations for describing convection currents in the atmosphere,

$$\dot{x} = \sigma(y - x)$$
$$\dot{y} = \rho x - y - xz$$
$$\dot{z} = -\beta z + xy$$

where σ, ρ, and β are constants, are probably the most studied autonomous equations of recent times. Although they were not considered an adequate model for studying the convection considered, they did provide an excellent example showing the complicated behavior that is possible in three-dimensional autonomous systems. Run the following program LORENZ, which projects the long-run solution of Lorenz's equation on the xy-plane.

```
10   REM PROGRAM TO PLOT THE XY-PROJECTION
     OF LORENZ'S EQUATIONS
20   INPUT SIGMA, RHO, BETA, N, H, X, Y, Z
30   SCREEN 2
40   WINDOW (-2,2) - (2,2)
50   FOR I = 1 TO N
```

Figure 8.44
Strange attractor of electrical
system

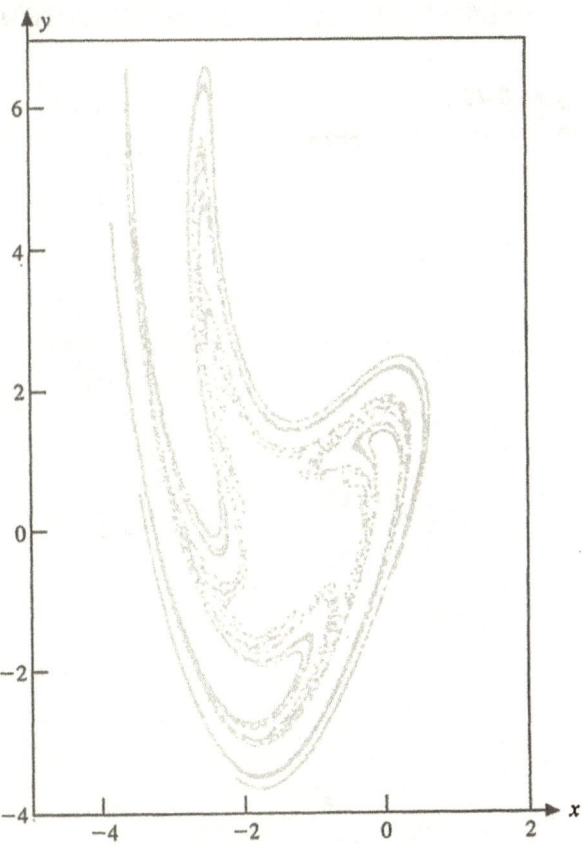

```
60     X = X + H * (SIGMA * (Y − X))
70     Y = Y + H * (RHO * X − Y − X * Z))
80     Z = Z + H * (−BETA * Z + X * Y))
90     PRINT X, Y
100    NEXT I
110    END
```

12. The Hénon Strange Attractor If you are given a system of two difference equations, you can obtain its Poincaré section by simple iteration. The *Hénon equations,*

$$x_{n+1} = 1 - ax_n^2 + y_n \qquad (5)$$
$$y_{n+1} = bx_n$$

are a natural extension of the logistic mapping to two dimensions and have chaotic behavior for certain values of a and b. Write a computer program or use an existing program to find the Hénon strange attractor for the parameters $a = 1.4$ and $b = 0.3$ The figure should appear like the one in Figure 8.45.

13. Fractal Basin Boundaries Fixed points and periodic solutions (closed curves) are called attractors when nearby points approach them. The collection of points in the phase plane that tend toward a given attractor is called the **basin of attraction** of the attractor, and the boundary between two or more basins of attraction is called a **basin boundary.** See Figure 8.46. If the basin boundary is a *fractal* (very jagged), it is called a **fractal basin boundary.** The existence of fractal basin boundaries has important implications for the study of dynamical systems because it means that small uncertainties in initial conditions lead to very different outcomes. Write a computer program or use an existing program to find the three distinct basins of attraction of the roots of the complex equation $z^3 - 1 = 0$ using Newton's method. The equation has three roots at $+1$, $-0.5 + 0.866i$, and $-0.5 - 0.866i$, called the **three roots of unity.** You can solve this equation by replacing the complex equation by two real equations $f(x, y) = 0$, $g(x, y) = 0$ and then using the iteration formula

$$\begin{bmatrix} x_{n+1} \\ y_{n+1} \end{bmatrix} = \begin{bmatrix} f_x(x_n, y_n) & f_y(x_n, y_n) \\ g_x(x_n, y_n) & g_y(x_n, y_n) \end{bmatrix}^{-1} \begin{bmatrix} f(x_n, y_n) \\ g(x_n, y_n) \end{bmatrix} \qquad (6)$$

These are the three basins of attraction (and a fourth if you consider initial points whose iterates diverge).

14. Journal Entry Summarize the concepts of the Poincaré section and strange attractors.

Figure 8.45
The Hénon strange attractor

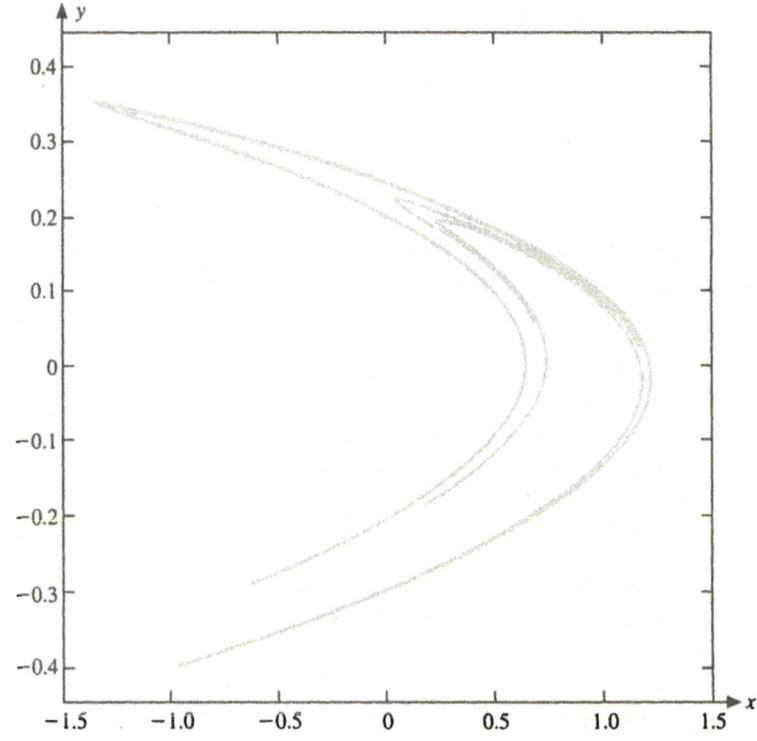

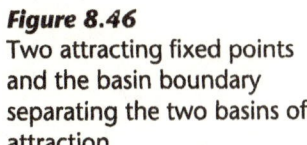

Figure 8.46
Two attracting fixed points
and the basin boundary
separating the two basins of
attraction

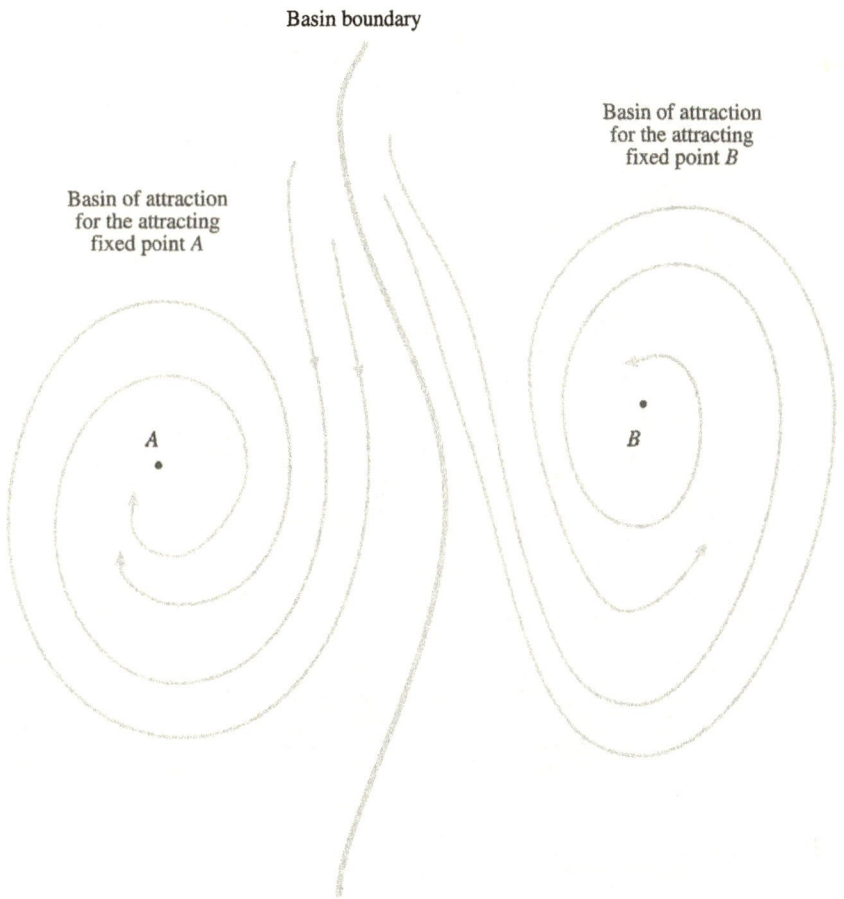

Basin boundary

Basin of attraction
for the attracting
fixed point A

Basin of attraction
for the attracting
fixed point B

A

B

ADDITIONAL READING ON CHAOS AND STRANGE ATTRACTORS

1. *Does God Play Dice?* by Ian Stewart (Basil Blackwell, London, 1989). An excellent expository on chaotic behavior by one of the best mathematical expositors.

2. *Chaotic Vibrations* by Francis C. Moon (John Wiley & Sons, New York, 1987). An excellent introductory textbook on chaos. Excellent for students in both engineering and mathematics.

3. *Differential Equations: A Dynamical Systems Approach* by J. H. Hubbard and B. H. West (Springer-Verlag, New York, 1991). One of the first textbooks intended for advanced undergraduates in which the qualitative theory of differential equations is stressed more than the quantitative theory. Written by two pioneers in the theory of dynamical systems.

9

Partial Differential Equations

9.1 FOURIER SERIES

9.2 FOURIER SINE AND COSINE SERIES

9.3 INTRODUCTION TO PARTIAL DIFFERENTIAL EQUATIONS

9.4 THE VIBRATING STRING: SEPARATION OF VARIABLES

9.5 SUPERPOSITION INTERPRETATION OF THE VIBRATING STRING

9.6 THE HEAT EQUATION AND SEPARATION OF VARIABLES

9.7 LAPLACE'S EQUATION INSIDE A CIRCLE

9.1 FOURIER SERIES

PURPOSE

To provide an introduction to the subject of Fourier series, which is instrumental in the solution of partial differential equations. We first state Dirichlet's theorem, which gives sufficient conditions for determining when a function can be represented by a Fourier series, and then give Euler's formulas, which allow us to find the Fourier coefficients of a Fourier series.

Periodic phenomena are widespread in physics and engineering. Forces, electric currents, voltages, wave motion, sound waves, and such are often periodic in nature. Although many phenomena exhibit simple harmonic motion, which can be represented as a single sine or cosine function, other phenomena, such as heartbeats, ocean tides, and musical sounds, exhibit more complex periodic behavior. See Figure 9.1.

Figure 9.1
Continuous periodic waveforms generated by musical instruments. (From D. C. Miller, *Sound Waves and Their Uses.* Macmillan: New York, 1938. Reprinted by permission.)

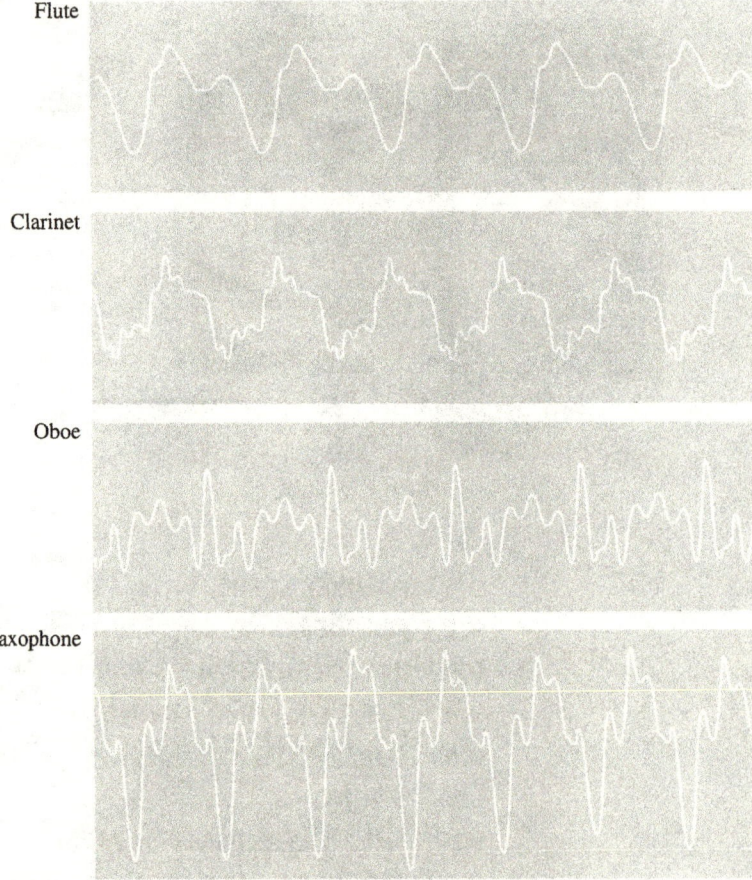

Flute

Clarinet

Oboe

Saxophone

One of the most important discoveries in mathematics was made by the French mathematician Joseph Fourier (1768–1830), who in 1807 hypothesized that *any* periodic function f of period 2π could be expressed by a series of the form

$$f(x) = \frac{a_0}{2} + a_1 \cos x + a_2 \cos 2x + \cdots + a_n \cos nx + \cdots$$
$$+ b_1 \sin x + b_2 \sin 2x + \cdots + b_n \sin nx + \cdots \tag{1}$$

Although Fourier's hypothesis is not completely correct without a few more restrictions on f, his ideas led to one of the most important areas of mathematics, known as **Fourier analysis.**

Before we begin our study of the Fourier series, it will be convenient to introduce a few important concepts.

PERIODIC FUNCTIONS

Most students of calculus know that a function $f(x)$ is **periodic** with period L if the condition $f(x + L) = f(x)$ holds for all x in the domain of f; the smallest value of L satisfying this condition is called the **fundamental period.** See Figure 9.2.

For example, the functions $\sin x$ and $\cos x$ are both periodic with periods 2π, 4π, 6π, ..., the period 2π being the fundamental period. On the other hand, the function $\tan x$ is periodic with periods π, 2π, 3π, ... with fundamental period π. Finally, any constant function is a periodic function with an arbitrary period.

Figure 9.2
Periodic function

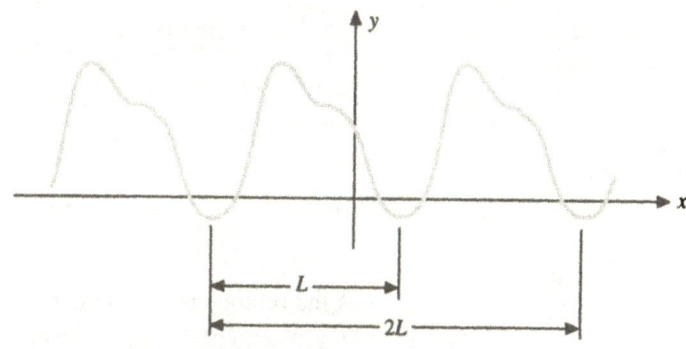

ORTHOGONALITY OF THE SINE AND COSINE FUNCTIONS

The collection of linearly independent functions

$$\left\{ 1, \cos\left(\frac{\pi x}{L}\right), \cos\left(\frac{2\pi x}{L}\right), \cos\left(\frac{3\pi x}{L}\right), ..., \sin\left(\frac{\pi x}{L}\right), \sin\left(\frac{2\pi x}{L}\right), \sin\left(\frac{3\pi x}{L}\right), ... \right\}$$

form what is called a mutually orthogonal set of functions, analogous to a mutually perpendicular set of vectors in vector analysis. In the case of functions, we say that two functions f and g are **orthogonal** on an interval $[a, b]$ if

$$\int_a^b f(x) g(x)\, dx = 0 \tag{2}$$

A set of functions is called **mutually orthogonal** if each distinct pair of functions is orthogonal. The above collection of sine and cosine functions (along with a constant function) forms a mutually orthogonal set of functions on the interval $[-L, L]$, as can be seen from the identities

$$\int_{-L}^{L} \cos\left(\frac{m\pi x}{L}\right) \cos\left(\frac{n\pi x}{L}\right) dx = \begin{cases} 0 & (m \neq n) \\ L & (m = n \neq 0) \\ 2L & (m = n = 0) \end{cases} \tag{3a}$$

$$\int_{-L}^{L} \cos\left(\frac{m\pi x}{L}\right) \sin\left(\frac{n\pi x}{L}\right) dx = 0 \qquad (\text{all } m, n) \qquad (3b)$$

$$\int_{-L}^{L} \sin\left(\frac{m\pi x}{L}\right) \sin\left(\frac{n\pi x}{L}\right) dx = \begin{cases} 0 & (m \neq n) \\ L & (m = n) \end{cases} \qquad (3c)$$

The **orthogonality conditions** (3) can be proven by straightforward integration. For example, when $m \neq n$, Eq. (3c) can be verified as follows:

$$\int_{-L}^{L} \sin\left(\frac{m\pi x}{L}\right) \sin\left(\frac{n\pi x}{L}\right) dx = 2\int_{0}^{L} \sin\left(\frac{m\pi x}{L}\right) \sin\left(\frac{n\pi x}{L}\right) dx$$

$$= \int_{0}^{L} \left\{ \cos\left(\frac{(m-n)\pi x}{L}\right) - \cos\left(\frac{(m+n)\pi x}{L}\right) \right\} dx$$

$$= \frac{L}{\pi}\left\{ \frac{\sin(m-n)\dfrac{\pi x}{L}}{m-n} - \frac{\sin(m+n)\dfrac{\pi x}{L}}{m+n} \right\}\Bigg|_{0}^{L} = 0 \qquad (4)$$

When $m = n$, we have

$$\int_{-L}^{L} \sin\left(\frac{m\pi x}{L}\right) \sin\left(\frac{n\pi x}{L}\right) dx = 2\int_{0}^{L} \left\{ \sin\left(\frac{m\pi x}{L}\right) \right\}^2 dx$$

$$= \int_{0}^{L} \left\{ 1 - \cos\left(\frac{2m\pi x}{L}\right) \right\} dx$$

$$= \left\{ x - \frac{\sin(2m\pi x/L)}{(2m\pi/L)} \right\}\Bigg|_{0}^{L} = L \qquad (5)$$

This verifies the orthogonality condition (3c). The other conditions (3a) and (3b) can be established by similar computations.

HISTORICAL NOTE

In 1807 the French mathematician Joseph Fourier shocked the mathematical world with his announcement that an arbitrary periodic function can be represented by an infinite series of sines and cosines. The announcement caused skepticism, inasmuch as Fourier did not state any conditions under which his claims were valid. In fact, the eminent mathematician Joseph Louis Lagrange thought that some functions could *not* be represented by such a series. It was not until 1829 that the German mathematician P. G. Lejieune Dirichlet (1805–1859) gave sufficient conditions under which such a function can be represented by such a series of sines and cosines.

**THE FOURIER SERIES
AND EULER'S
EQUATIONS**

Let us consider the possibility of representing a function f by a series of the form

$$f(x) = \frac{a_0}{2} + \sum_{n=1}^{\infty} \left\{ a_n \cos\left(\frac{n\pi x}{L}\right) + b_n \sin\left(\frac{n\pi x}{L}\right) \right\} \tag{6}$$

where the coefficients* $a_0, a_1, a_2, ..., b_1, b_2, ...$ are to be determined. Inasmuch as the individual terms in the infinite series (6) are periodic with periods $2L, 2L/2, 2L/3, ...,$ the Fourier series is periodic with fundamental period $2L$. This means that only functions of period $2L$ can be represented by a series of the form (6). So, given a periodic function f with period $2L$, the problem remains how to find the coefficients a_n and b_n in the series (6).

The coefficients a_n and b_n can be found from the orthogonality conditions (3a), (3b), and (3c). Starting with Eq. (6), we multiply Eq. (6) by $\cos(m\pi x/L)$, where m is a *fixed* positive integer, and integrate† each resulting term with respect to x from $-L$ to L. This gives

$$\int_{-L}^{L} f(x) \cos\left(\frac{m\pi x}{L}\right) dx = \frac{a_0}{2} \int_{-L}^{L} \cos\left(\frac{m\pi x}{L}\right) dx$$

$$+ \sum_{n=1}^{\infty} a_n \int_{-L}^{L} \cos\left(\frac{n\pi x}{L}\right) \cos\left(\frac{m\pi x}{L}\right) dx$$

$$+ \sum_{n=1}^{\infty} b_n \int_{-L}^{L} \sin\left(\frac{n\pi x}{L}\right) \cos\left(\frac{m\pi x}{L}\right) dx \tag{7}$$

Remembering that m is kept fixed while n ranges over the positive integers, it follows from the orthogonality conditions (3) that the *only* nonzero term on the right of Eq. (7) is the term with $n = m$. Hence we have

$$\int_{-L}^{L} f(x) \cos\left(\frac{m\pi x}{L}\right) dx = La_m$$

or

$$a_m = \frac{1}{L} \int_{-L}^{L} f(x) \cos\left(\frac{m\pi x}{L}\right) dx \qquad (m = 1, 2, ...) \tag{8}$$

To determine a_0, simply integrate Eq. (6) with respect to x from $-L$ to L to get

$$\int_{-L}^{L} f(x) \, dx = \frac{a_0}{2} \int_{-L}^{L} dx + \sum_{n=1}^{\infty} a_n \int_{-L}^{L} \cos\left(\frac{n\pi x}{L}\right) dx + \sum_{n=1}^{\infty} b_n \int_{-L}^{L} \sin\left(\frac{n\pi x}{L}\right) dx = La_0$$

or

$$a_0 = \frac{1}{L} \int_{-L}^{L} f(x) \, dx \tag{9}$$

---

* The coefficient 1/2 of a_0 is a conventional device that allows for symmetry in the final formulas. Its purpose will be apparent shortly.

† Although term-by-term integration of a convergent infinite series with variable terms is not always a valid mathematical operation, it can be verified that the Fourier series of a piecewise continuous function f defined on $-L \le x \le L$ may always be integrated term by term, giving rise to the function defined by $F(x) = \int_c^x f(t) \, dt$.

It is also possible to determine the coefficients b_m in the Fourier series by multiplying Eq. (6) by $\sin(m\pi x/L)$, integrating each term with respect to x from $-L$ to L, and making use of the orthogonality conditions. This gives

$$b_m = \frac{1}{L}\int_{-L}^{L} f(x)\sin\left(\frac{m\pi x}{L}\right) dx \qquad (m = 1, 2, ...) \qquad (10)$$

Formulas (8), (9), and (10) are known as **Euler's formulas,** and the series (6) whose coefficients are determined by Euler's formulas is called the **Fourier series** of f.

However, we must not delude ourselves into thinking that *any* function f has a Fourier series that acts as a representation of the function. What we have proven is that *if* a function f can be represented as a Fourier series of the form of Eq. (6), *then* the coefficients of this Fourier series are given by Euler's formulas. As we mentioned, the exact conditions for the convergence of a Fourier series to f were answered by a famous theorem due to the German mathematician P. J. Dirichlet. We state this theorem here.

THEOREM 9.1: Dirichlet's Representation Theorem for Fourier Series

If f is a bounded function of period $2L$, and if in any single period the function f has at most a finite number of maxima and minima and a finite number of discontinuities, then the Fourier series of f converges to $f(x)$ at all points x where f is continuous and converges to the average of the right- and left-hand limits of $f(x)$ at those points where $f(x)$ is discontinuous. Hence at the points of continuity of f the Fourier series converges to $f(x)$, and we can write

$$f(x) = \frac{a_0}{2} + \sum_{n=1}^{\infty}\left\{a_n\cos\left(\frac{n\pi x}{L}\right) + b_n\sin\left(\frac{n\pi x}{L}\right)\right\}$$

where the coefficients a_n and b_n are given by Euler's formulas*

$$a_0 = \frac{1}{L}\int_{-L}^{L} f(x)\, dx$$

$$a_n = \frac{1}{L}\int_{-L}^{L} f(x)\cos\left(\frac{n\pi x}{L}\right) dx$$

$$b_n = \frac{1}{L}\int_{-L}^{L} f(x)\sin\left(\frac{n\pi x}{L}\right) dx$$

$$(n = 1, 2, ...)$$

Example 1 **Fourier Series Containing Only Sine Terms** Find the Fourier series for the square wave

---

* Note that the coefficient a_0 can be found by letting $n = 0$ in the formulas for a_n. Note too that since the integrands in the integrals for a_n and b_n are all periodic with period $2L$, the integrals will be unchanged by integrating over *any* interval of length $2L$. The interval $[-L, L]$ is chosen for convenience.

$$f(x) = \begin{cases} -1 & (-\pi \le x < 0) \\ 1 & (0 \le x < \pi) \end{cases}$$

$$f(x + 2\pi) = f(x)$$

See Figure 9.3.

Figure 9.3
Square wave

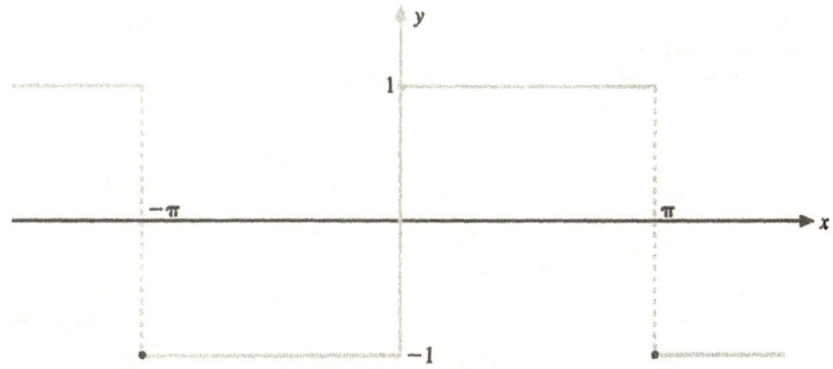

Solution   Since the period of f is $2L = 2\pi$, we have $L = \pi$. Using the formulas for the Euler formulas, we have

$$a_0 = \frac{1}{\pi} \int_{-\pi}^{\pi} f(x)\, dx = \frac{1}{\pi}\left(-\pi\right) + \frac{1}{\pi}\left(\pi\right) = 0 \tag{11}$$

$$\begin{aligned} a_n &= \frac{1}{\pi} \int_{-\pi}^{\pi} f(x) \cos\left(\frac{n\pi x}{\pi}\right) dx \\ &= \frac{1}{\pi} \int_{-\pi}^{0} (-\cos nx)\, dx + \frac{1}{\pi} \int_{0}^{\pi} \cos nx\, dx \\ &= \frac{1}{\pi}\left(-\frac{1}{n}\sin nx \Big|_{-\pi}^{0}\right) + \frac{1}{\pi}\left(\frac{1}{n}\sin nx \Big|_{0}^{\pi}\right) \\ &= 0 \qquad (n = 1, 2, \dots) \end{aligned} \tag{12}$$

$$\begin{aligned} b_n &= \frac{1}{\pi} \int_{-\pi}^{\pi} f(x) \sin\frac{n\pi x}{\pi}\, dx \\ &= \frac{1}{\pi} \int_{-\pi}^{0} (-\sin nx)\, dx + \frac{1}{\pi} \int_{0}^{\pi} \sin nx\, dx \\ &= \frac{1}{\pi}\left(\frac{1}{n}\cos nx \Big|_{-\pi}^{0}\right) + \frac{1}{\pi}\left(-\frac{1}{n}\cos nx \Big|_{0}^{\pi}\right) \\ &= \begin{cases} \dfrac{4}{n\pi} & (n \text{ odd}) \\[2mm] 0 & (n \text{ even}) \end{cases} \end{aligned} \tag{13}$$

Hence the Fourier series consists only of the sine terms

$$f(x) = \frac{4}{\pi}\left\{ \sin x + \frac{1}{3}\sin 3x + \frac{1}{5}\sin 5x + \cdots + \frac{1}{2n-1}\sin[(2n-1)x] + \cdots \right\} \quad (14)$$

Figure 9.4 shows the square wave and the first three partial sums of the Fourier series. ∎

Figure 9.4
First three partial sums of the
Fourier series of the square
wave

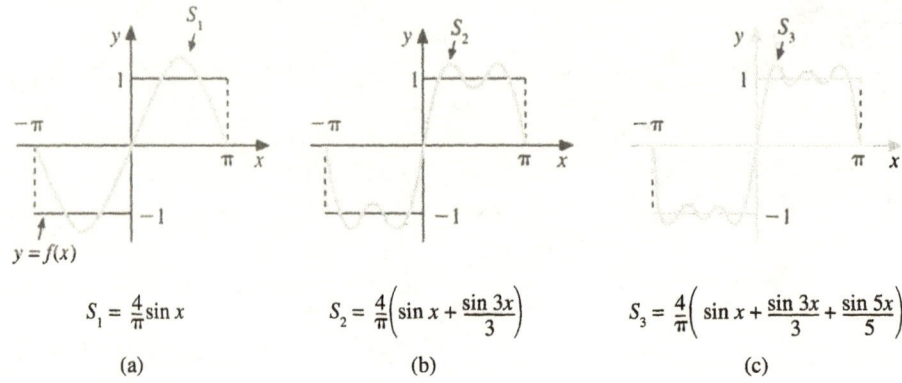

$$S_1 = \frac{4}{\pi}\sin x \qquad S_2 = \frac{4}{\pi}\left(\sin x + \frac{\sin 3x}{3}\right) \qquad S_3 = \frac{4}{\pi}\left(\sin x + \frac{\sin 3x}{3} + \frac{\sin 5x}{5}\right)$$

(a)   (b)   (c)

Example 2   Fourier Expansion Containing Both Sine and Cosine Terms   Find the Fourier series expansion of the triangular wave function f defined by

$$f(x) = \begin{cases} 0 & (-1 \le x < 0) \\ x & (0 \le x < 1) \end{cases}$$

$$f(x + 2) = f(x)$$

See Figure 9.5.

Figure 9.5
Triangular wave

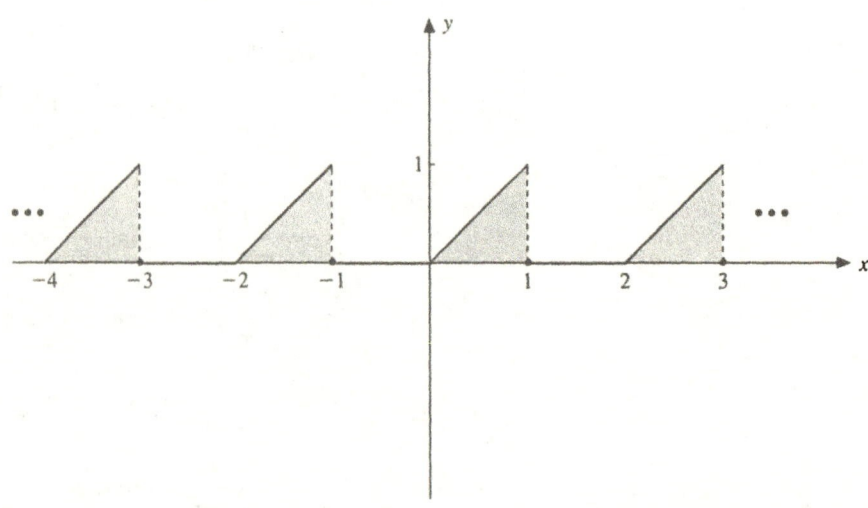

Solution   Since f is periodic with period $2L = 2$, we have $L = 1$. Hence the Fourier series of f has the form

$$f(x) = \frac{a_0}{2} + \sum_{n=1}^{\infty} \{a_n \cos n\pi x + b_n \sin n\pi x\} \tag{15}$$

where a_n and b_n are determined from the formulas*

$$a_0 = \int_{-1}^{1} f(x)\,dx \qquad = \int_{-1}^{0} 0\,dx + \int_{0}^{1} x\,dx = \frac{1}{2} \tag{16}$$

$$a_n = \int_{-1}^{1} f(t) \cos n\pi x\,dx = \int_{0}^{1} x \cos n\pi x\,dx$$

$$= \left\{ \frac{1}{(n\pi)^2} \cos n\pi x + \frac{x}{n\pi} \sin n\pi x \right\}\Big|_{0}^{1}$$

$$= \begin{cases} -2/(n\pi)^2 & (n \text{ odd}) \\ 0 & (n \text{ even}) \end{cases} \tag{17}$$

$$b_n = \int_{-1}^{1} f(x) \sin n\pi x\,dx = \int_{0}^{1} x \sin n\pi x\,dx$$

$$= \left\{ \frac{1}{(n\pi)^2} \sin n\pi x - \frac{x}{n\pi} \cos n\pi x \right\}\Big|_{0}^{1}$$

$$= (-1)^{n+1} \frac{1}{n\pi} \qquad (n = 1, 2, \ldots) \tag{18}$$

Hence the Fourier series representation of the triangular wave is

$$f(t) = \frac{1}{4} - \frac{2}{\pi^2} \sum_{n=1}^{\infty} \frac{\cos(2n-1)\pi x}{(2n-1)^2} + \frac{1}{\pi} \sum_{n=1}^{\infty} \frac{(-1)^{n+1}}{n} \sin n\pi x$$

$$= \frac{1}{4} - \frac{2}{\pi^2} \left\{ \cos \pi x + \frac{1}{9} \cos 3\pi x + \frac{1}{25} \cos 5\pi x + \cdots \right\}$$

$$+ \frac{1}{\pi} \left\{ \sin \pi x - \frac{1}{2} \sin 2\pi x + \frac{1}{3} \sin 3\pi x + \cdots \right\} \tag{19}$$

A partial sum of this representation has been drawn by using a computer and is shown in Figure 9.6. ∎

Figure 9.6
A triangular wave and a
partial sum of its Fourier
series

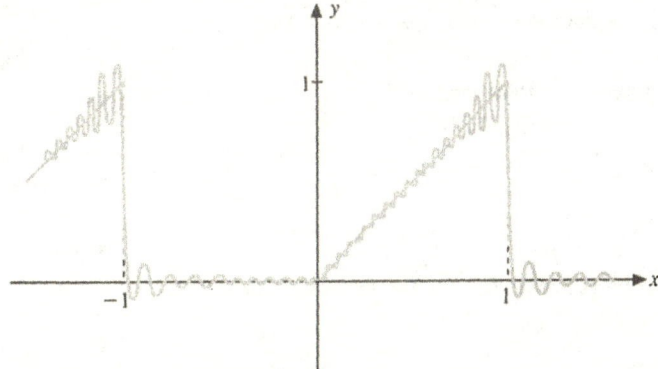

---

* The integral that defines the a_n and b_n can be evaluated by integration by parts. Keep in mind that $f(x) \geq 0$ on $-1 \leq x < 0$.

The list of integrals in Table 9.1 should be useful in finding Euler coefficients for a Fourier series.

Table 9.1
Useful Integrals in Finding Fourier Series

1. $\displaystyle\int_0^L \sin\left(\frac{n\pi x}{L}\right) dx = \begin{cases} 2L/n\pi & (n \text{ odd}) \\ 0 & (n \text{ even}) \end{cases}$

2. $\displaystyle\int_0^L \cos\left(\frac{n\pi x}{L}\right) dx = \begin{cases} L & (n = 0) \\ 0 & (n = 1, 2, \ldots) \end{cases}$

3. $\displaystyle\int_0^L x \sin\left(\frac{n\pi x}{L}\right) dx = \frac{(-1)^{n+1}L^2}{n\pi} \qquad (n = 1, 2, \ldots)$

4. $\displaystyle\int_0^L x \cos\left(\frac{n\pi x}{L}\right) dx = \begin{cases} -2(L/n\pi)^2 & (n \text{ odd}) \\ 0 & (n \text{ even}) \end{cases}$

5. $\displaystyle\int_0^L x^2 \sin\left(\frac{n\pi x}{L}\right) dx = \frac{(-1)^n 2L^3}{(n\pi)^2} \qquad (n = 1, 2, \ldots)$

6. $\displaystyle\int_0^L x^2 \cos\left(\frac{n\pi x}{L}\right) dx = \begin{cases} L^3/3 & (n = 0) \\ (-1)^n 2L/n^2 & (n = 1, 2, \ldots) \end{cases}$

PROBLEMS: Section 9.1

For Problems 1–10, determine whether the given function is periodic. If so, find the fundamental period.

1. 1

2. $\sin \pi x$

3. $\sin 3\pi x$

4. $\cos \pi x + 1$

5. $\cos \pi x + x$

6. $\cosh x$

7. $\sin^2 x$

8. $|\sin x|$

9. $\sin x + \cos 2x + \sin 3x$

10. $\sin 2x + \sin 3x + \sin 4x$

For Problems 11–13, verify that the given set of functions is mutually orthogonal over the given interval. Sketch the graphs of the functions.

Functions	Interval
11. $\{1, \sin x\}$	$[0, 2\pi]$
12. $\{\sin x, \cos x\}$	$[-\pi, \pi]$
13. $\left\{1, x, \frac{1}{2}(3x^2 - 1)\right\}$	$[-1, 1]$ Legendre polynomials

For Problems 14–22, find the Fourier series* of the given function.

14. $f(x) = 1$
  $(0 \le x < 1)$ $f(x + 1) = f(x)$

15. $fx = \sin x + \cos x$
  $(-\pi \le x < \pi)$ $f(x + 2\pi) = f(x)$

16. $fx = 1 + \sin x + \dfrac{1}{2}\sin 2x$
  $(0 \le x < 2\pi)$ $f(x + 2\pi) = f(x)$

17. $f(x) = 1 + \cos x + \dfrac{1}{3}\sin 2x$
  $(0 \le x < 2\pi)$ $f(x + 2\pi) = f(x)$

18. $f(x) = \begin{cases} 0 & (-1 \le x < 0) \\ 1 & (0 \le x < 1) \end{cases}$ $f(x + 2) = f(x)$
  See Figure 9.7.

19. $f(x) = \begin{cases} -x & (-1 \le x < 0) \\ x & (0 \le x < 1) \end{cases}$ $f(x + 2) = f(x)$
  See Figure 9.8.

20. $f(x) = x$ $(-1 \le x < 1)$ $f(x + 2) = f(x)$
  See Figure 9.9.

21. $f(x) = x^2$ $(-1 \le x < 1)$ $f(x + 2) = f(x)$
  See Figure 9.10.

22. $f(x) = \begin{cases} 1 & (-1 \le x < 0) \\ x & (0 \le x < 1) \end{cases}$ $f(x + 2) = f(x)$
  See Figure 9.11.

* Table 9.1 lists useful integrals for the evaluation of Euler coefficients.

Figure 9.7

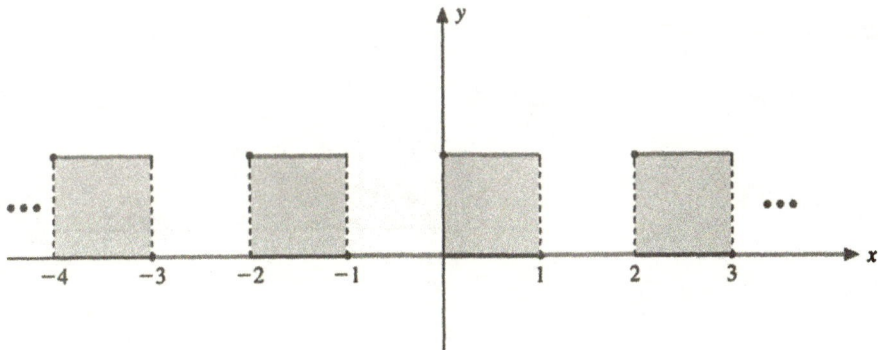

Figure 9.8

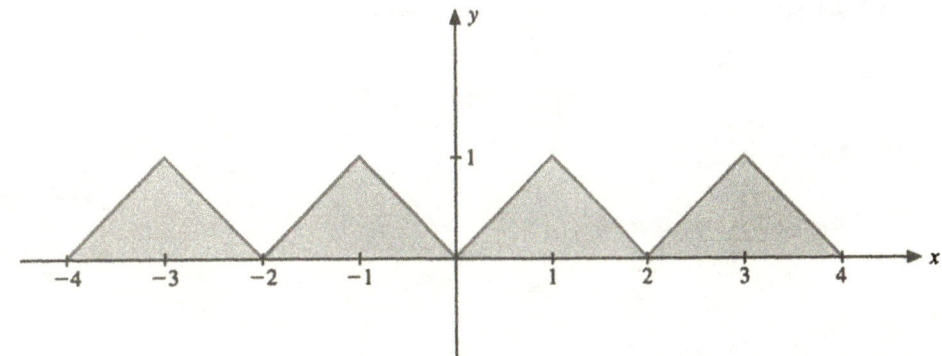

Figure 9.9

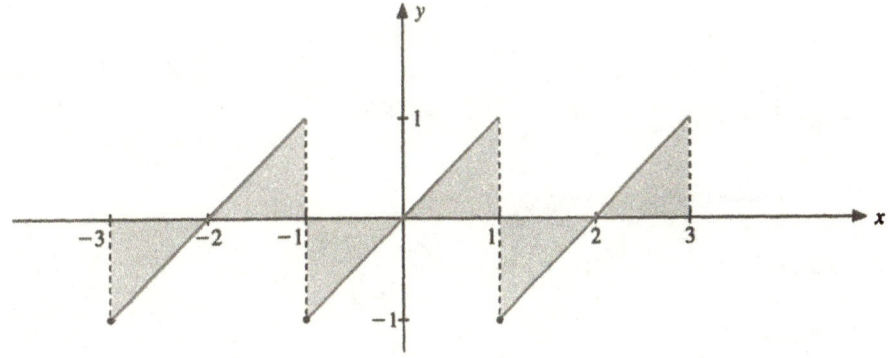

Figure 9.10

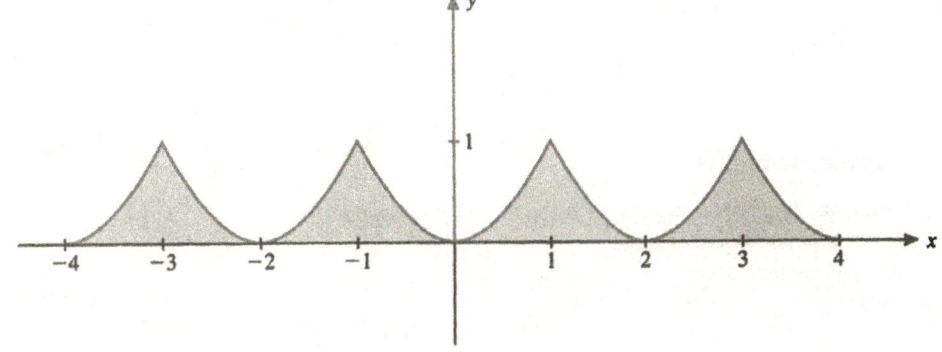

Figure 9.11

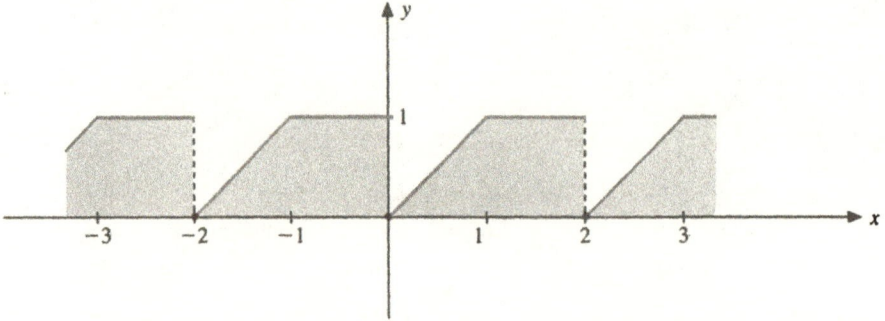

23. Complex Form of the Fourier Series

(a) Use Euler's formula $e^{ix} = \cos x + i \sin x$ to show that

$$\cos nx = \frac{e^{inx} + e^{-inx}}{2}$$

$$\sin nx = \frac{e^{inx} - e^{-inx}}{2i}$$ (20)

(b) Show that the Fourier series

$$f(x) = \frac{a_0}{2} + \sum_{n=1}^{\infty} \{a_n \cos nx + b_n \sin nx\}$$

can be rewritten as

$$f(x) = c_0 + \sum_{n=1}^{\infty} \{c_n e^{inx} + c_{-n} e^{-inx}\}$$

where

$$c_0 = \frac{a_0}{2} \qquad c_n = \frac{a_n - ib_n}{2} \qquad c_{-n} = \frac{a_n + ib_n}{2}$$

(c) Use the results from part (b) to write

$$f(x) = \frac{a_0}{2} + \sum_{n=1}^{\infty} a_n \cos\left(\frac{n\pi x}{L}\right) + b_n \sin\left(\frac{n\pi x}{L}\right)$$

in complex form

$$f(x) = \sum_{n=-\infty}^{\infty} c_n e^{inx} \qquad (-\pi < x < \pi)$$ (21)

where

$$c_n = \frac{1}{2\pi} \int_{-\pi}^{\pi} f(x) e^{-inx} \, dx$$

Computer Problems

For Problems 24–27, the graph of a function and its Fourier series are given. Use a computer to draw the graph of the function and the first, second, third, and fourth partial sums of the Fourier series.

24. $f(t) = \dfrac{4}{\pi}\left\{\sin \pi t + \dfrac{1}{3}\sin 3\pi t + \dfrac{1}{5}\sin 5\pi t + \cdots\right\}$

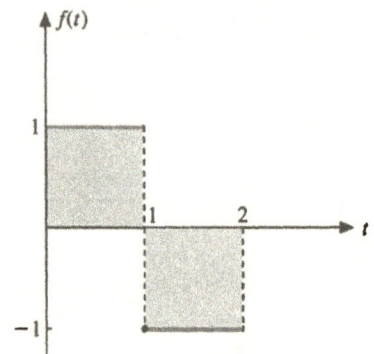

25. $f(t) = \dfrac{2}{\pi}\left\{\sin \pi t - \dfrac{1}{2}\sin 2\pi t + \dfrac{1}{3}\sin 3\pi t - \cdots\right\}$

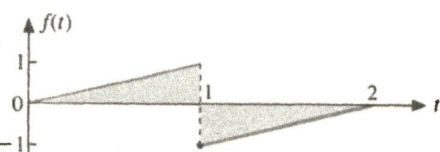

26. $f(t) = \dfrac{1}{2} - \dfrac{4}{\pi^2}\left\{\cos \pi t + \dfrac{1}{9}\cos 3\pi t + \dfrac{1}{5^2}\cos 5\pi t + \cdots\right\}$

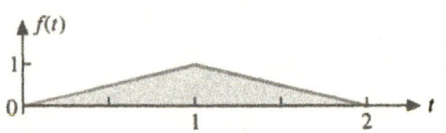

27. $f(t) = \dfrac{1}{2} - \dfrac{1}{\pi}\left\{\sin \pi t + \dfrac{1}{2}\sin 2\pi t + \dfrac{1}{3}\sin 3\pi t + \cdots\right\}$

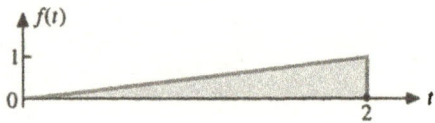

28. **Some Interesting Graphs** Use a computer to graph the functions $\sin n\pi x$ and $\cos n\pi x$ on the interval $[0, 2\pi]$ for n ranging from 1 to 10.

29. **Gibbs' Phenomenon** This phenomenon, named after physicist Josiah Willard Gibbs (1839–1903), refers to the fact that near points of discontinuity of a function f the partial sums S_n of the Fourier series do not converge as rapidly as they do near the points of continuity. Instead, the graphs of S_n tend to "overshoot" the jump discontinuities as if they can't turn the corner. Figure 9.12 illustrates this phenomenon for the function

$$f(x) = \begin{cases} 1 & (0 \le x < 1) \\ -1 & (1 \le x < 2) \end{cases} \qquad f(x + 2) = f(x)$$

Given the Fourier series of f:

$$f(x) = \frac{4}{\pi}\left\{ \sin \pi x + \frac{1}{3} \sin 3\pi x + \cdots \right. $$
$$\left. + \frac{1}{2n - 1} \sin (2n - 1)\pi x + \cdots \right\} \qquad (22)$$

use a computer to illustrate Gibbs' phenomenon by graphing the first five, seven, and nine terms of the Fourier series of f.

30. **Computer Problem** Use a computer to sketch the first few terms of the Fourier series of the square wave given by Eq. (14).

31. **Computer Problem** Use a computer to sketch the graphs of the first few terms of the Fourier series of the triangular wave given by Eq. (19). By the first n terms we mean all terms involving a_n and b_n for $n = 0, 1, 2, ..., n$.

32. **Journal Entry** Summarize your understanding of the Fourier series.

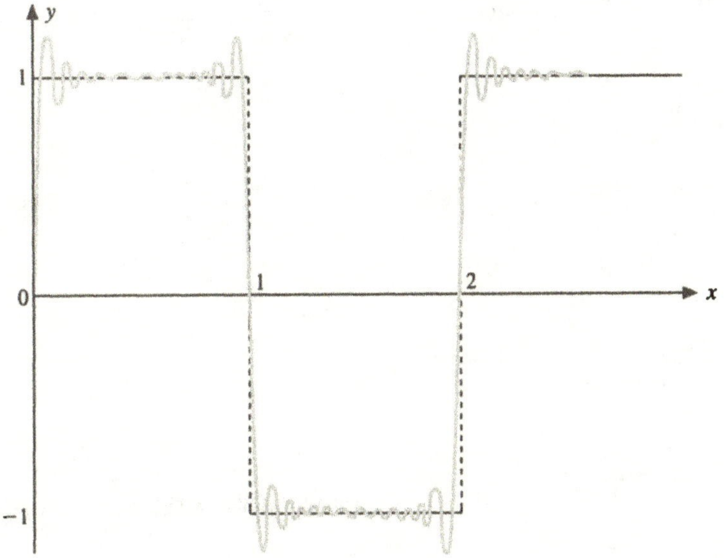

Figure 9.12
The square wave and Gibbs' phenomenon. Note how the graph of the partial sum overshoots the square wave at the points of discontinuity.

9.2   FOURIER SINE AND COSINE SERIES

PURPOSE
To show how to expand a function f defined on a finite interval $[0, L]$ either as a Fourier series containing only sine terms, called a Fourier sine series, or as a Fourier series containing only cosine terms, called a Fourier cosine series. These types of expansions will be crucial when we solve partial differential equations.

The reader might find it surprising, but it is possible to represent a function defined on an interval $(0, L)$ as *either* a Fourier series containing only sine terms or a Fourier series containing only cosine terms. To understand how this is possible, consider the following functions:

$$f(x) = x \quad (-1 \le x < 1) \quad f(x + 2) = f(x)$$
$$g(x) = |x| \quad (-1 \le x < 1) \quad g(x + 2) = g(x) \tag{1}$$

See Figure 9.13.

Figure 9.13
The functions f and g have the same values on the interval $[0, 1)$, but the Fourier series of f contains only sine terms, whereas the Fourier series of g contains only cosine terms.

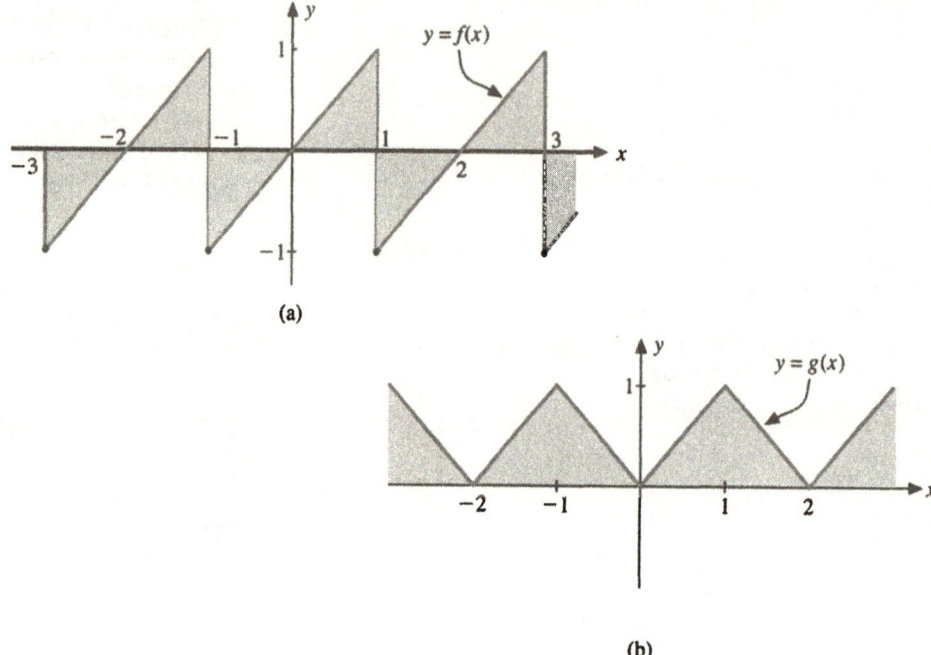

(a)

(b)

The Fourier series expansions of these functions are

$$f(x) = \frac{2}{\pi}\left\{ \sin \pi x - \frac{1}{2}\sin 2\pi x + \frac{1}{3}\sin 3\pi x - \cdots \right\} \tag{2a}$$

$$g(x) = \frac{1}{2} - \frac{4}{\pi^2}\left\{ \cos \pi x + \frac{1}{3^2}\cos 3\pi x + \frac{1}{5^2}\cos 5\pi x + \cdots \right\} \tag{2b}$$

But since $f(x) = g(x) = x$ for $0 \le x < 1$, it follows that on the interval $[0, 1)$ the function $h(x) = x$ can be represented by either Eq. (2a) or Eq. (2b). Of course, *outside* the interval $[0, 1)$ the two series (2) converge to different values, as is indicated in Figure 9.14.

This example leads us to wonder how we can represent any function defined on an interval of the type $(0, L)$ as a Fourier series containing only sine or cosine terms. Before we show how this is done, it is convenient to introduce two important classes of functions, the even and odd functions.

Figure 9.14
Although the different Fourier series representations of the function $f(x) = x$ converge to the same value on the interval [0, 1), they converge to different values on [−1, 0).

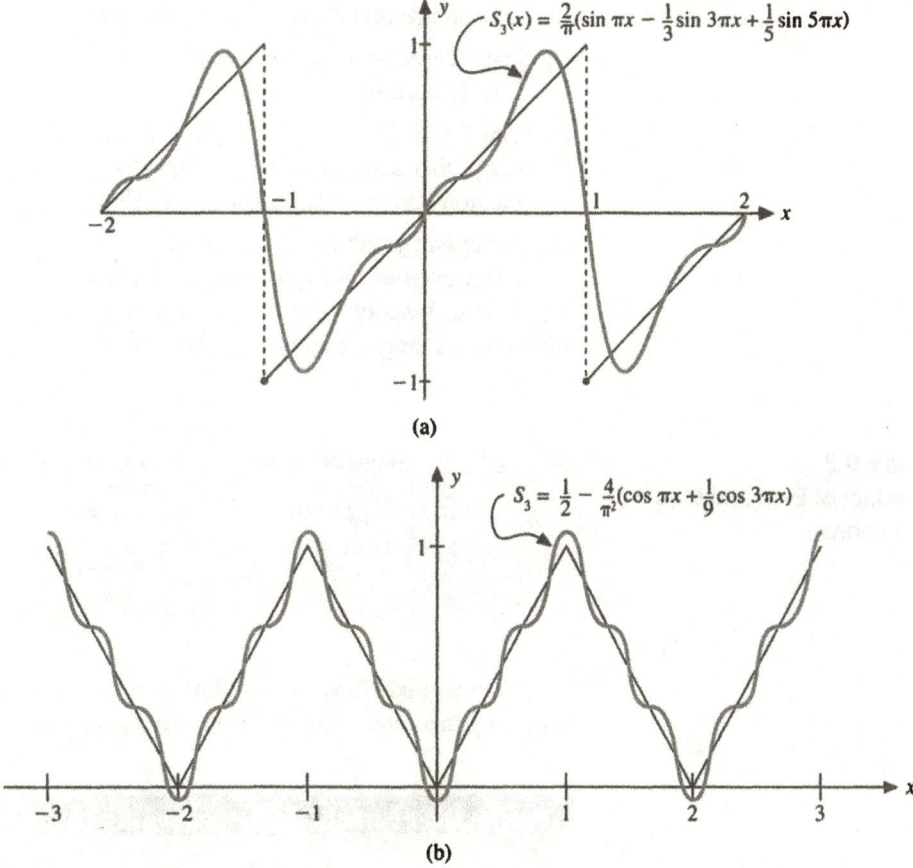

EVEN AND ODD FUNCTIONS

A function f is called an **even function** if $f(x) = f(-x)$ for all x in the domain of f, whereas f is called an **odd function** if $f(x) = -f(-x)$ for all x in its domain. The functions 1, x^2, x^4, ... are examples of even functions, and the functions x, x^3, x^5, ... are examples of odd functions. Also, $\sin x$ is an odd function, whereas $\cos x$ is an even function.

Example 1 **Even and Odd Functions** Determine whether each of the following functions is even, odd, or neither.

$$\text{(a) } f(x) = 1 + \frac{1}{x^2} \qquad \text{(b) } f(x) = x^2 \sin x \qquad \text{(c) } f(x) = |x| + x$$

Solution (a) Since

$$f(-x) = 1 + \frac{1}{(-x)^2} = 1 + \frac{1}{x^2} = f(x)$$

we conclude that $f(x)$ is an even function.

(b) Since $f(-x) = (-x)^2 \sin(-x) = -x^2 \sin x = -f(x)$, we conclude that $f(x)$ is an odd function.

(c) Here $f(-x) = |-x| + (-x) = |x| - x$. The reader can easily verify that the only x that satisfies $f(x) = f(-x)$ is $x = 0$; also, the only x that satisfies the equation $f(x) = -f(-x)$ is $x = 0$. Hence $f(x) = |x| + x$ is neither even nor odd.

The reader can graph these functions. ■

When even or odd functions are multiplied together, their product is even if and only if both functions are even or both functions are odd; otherwise, their product is odd. This statement is summarized in Table 9.2.

Table 9.2
Product of Even and Odd
Functions

Function type	g even	g odd
f even	fg even	fg odd
f odd	fg odd	fg even

Even and odd functions are important in the study of Fourier series because of the following theorem, which we state without proof.

THEOREM 9.2: Integral Properties of Even and Odd Functions

If f is an even integrable function defined on $[-L, L]$, then

$$\int_{-L}^{L} f(x)\, dx = 2 \int_{0}^{L} f(x)\, dx \tag{3a}$$

If f is an odd integrable function on $[-L, L]$, then

$$\int_{-L}^{L} f(x)\, dx = 0 \tag{3b}$$

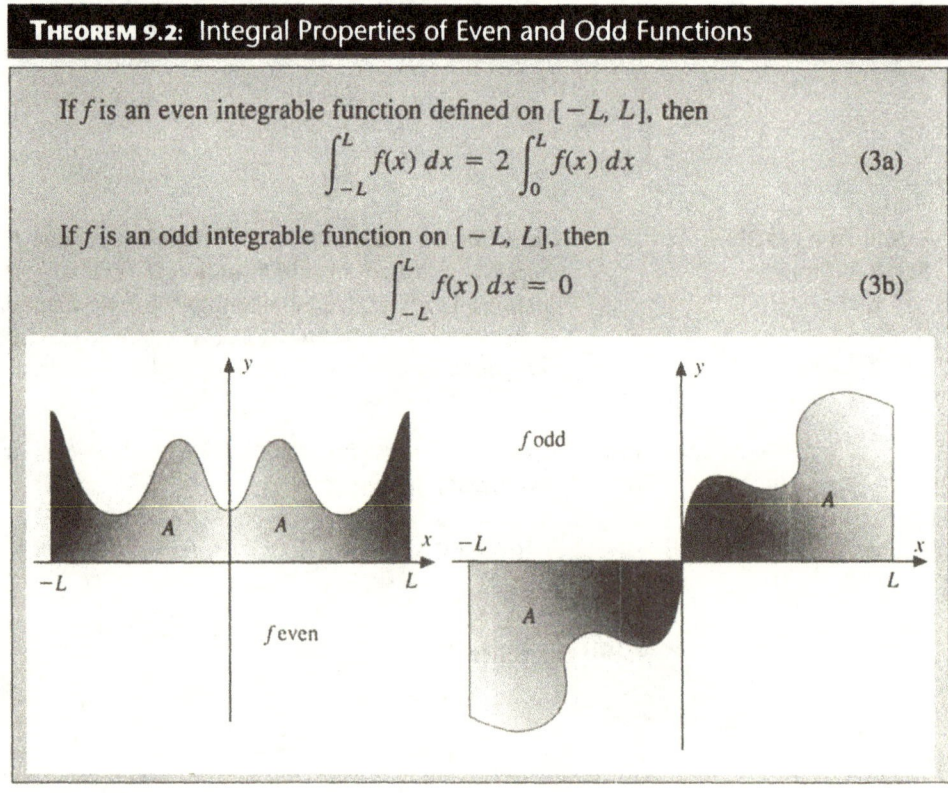

The proof of Theorem 9.2 is a simple consequence of the properties of the integral.

Example 2   **important Example**   Show the following properties of even and odd functions.

(a) The Fourier series of an odd function f defined on $(-L, L)$ contains only sine terms.

(b) The Fourier series of an even function contains only cosine terms (and possibly a constant term).

Solution   The coefficients of the Fourier series

$$f(x) = \frac{a_0}{2} + \sum_{n=1}^{\infty} a_n \cos\left(\frac{n\pi x}{L}\right) + b_n \sin\left(\frac{n\pi x}{L}\right) \tag{4}$$

are determined by

$$a_n = \frac{1}{L} \int_{-L}^{L} f(x) \cos\left(\frac{n\pi x}{L}\right) dx \qquad (n = 0, 1, ...) \tag{5a}$$

$$b_n = \frac{1}{L} \int_{-L}^{L} f(x) \sin\left(\frac{n\pi x}{L}\right) dx \qquad (n = 1, 2, ...) \tag{5b}$$

We observe that $\cos n\pi x/L$ are even functions for $n = 0, 1, ...$ and $\sin n\pi x/L$ are odd functions for $n = 1, 2, ...$.

(a) If $f(x)$ is odd on $[-L, L]$, we have that $f(x) \cos n\pi x/L$ is also odd, and thus the coefficients $a_n = 0$. Hence the Fourier series of f is a sine series.

(b) If $f(x)$ is even on $[-L, L]$, we have that $f(x) \sin n\pi x/L$ is odd, and thus the coefficient $b_n = 0$. Hence the Fourier series of f is a cosine series. ∎

This brings us to the most important topic of this section.

FOURIER SINE AND COSINE SERIES

In partial differential equations it is often necessary to represent a function that is defined on a finite interval as a Fourier series containing only sine terms or cosine terms. Suppose, for example, that we wish to represent a function f defined on the open interval $(0, L)$ as a Fourier series containing only sine terms. We can do this by recalling that the Fourier series of an odd function contains only sine terms. Hence we momentarily introduce a new function $F_{\text{odd}}(x)$ of period $2L$ defined by

$$F_{\text{odd}}(x) = \begin{cases} f(x) & (0 < x < L) \\ -f(-x) & (-L < x < 0) \end{cases} \tag{6}$$

called the **odd periodic extension** of $f(x)$. See Figure 9.15(a). The reasons for our interest in $F_{\text{odd}}(x)$ lie first in the fact that it is an odd function, and hence its Fourier series contains only sine terms, and second because $F_{\text{odd}}(x) = f(x)$ on $(0, L)$. In other words, we have found a Fourier sine series that converges to $f(x)$ for $0 < x < L$. We accomplished this by simply finding the Fourier series of the *odd periodic extension* of f.

Figure 9.15
Odd and even periodic
extensions of $f(x)$

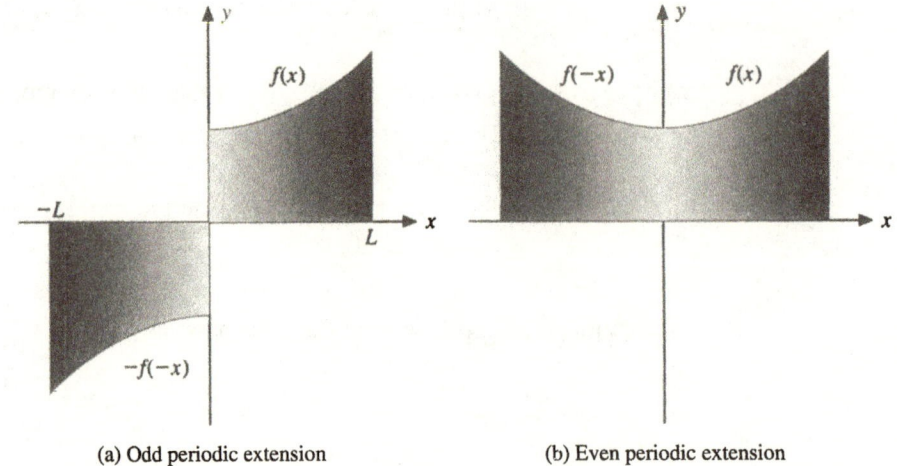

(a) Odd periodic extension   (b) Even periodic extension

In the same manner we can find a Fourier series that converges to $f(x)$ on $(0, L)$ that contains only cosine terms by finding the Fourier series of the **even periodic extension** of f, defined by

$$F_{\text{even}}(x) = \begin{cases} f(x) & (0 < x < L) \\ f(-x) & (-L < x < 0) \end{cases} \tag{7}$$

See Figure 9.15(b). These ideas are summarized in the following definition.

DEFINITION: Fourier Cosine and Sine Series

Let f be a piecewise continuous function defined on $(0, L)$. The **Fourier cosine series** of f on $(0, L)$ is

$$f(x) = \frac{a_0}{2} + \sum_{n=1}^{\infty} a_n \cos\left(\frac{n\pi x}{L}\right) \tag{8}$$

where

$$a_n = \frac{2}{L}\int_0^L f(x)\cos\left(\frac{n\pi x}{L}\right) dx \qquad (n = 0, 1, 2, \ldots) \tag{9}$$

The **Fourier sine series** of f on $(0, L)$ is defined by

$$f(x) = \sum_{n=1}^{\infty} b_n \sin\left(\frac{n\pi x}{L}\right) \tag{10}$$

where

$$b_n = \frac{2}{L}\int_0^L f(x)\sin\left(\frac{n\pi x}{L}\right) dx \qquad (n = 1, 2, \ldots) \tag{11}$$

Example 3   Fourier Expansion   Given the function

$$f(x) = x \qquad (0 < x < 1)$$

(a) Find the Fourier cosine series for f.

(b) Find the Fourier sine series for f.

Solution   (a) We compute

$$a_0 = \frac{2}{L} \int_0^L f(x)\, dx = 2 \int_0^1 x\, dx = 1$$

$$a_n = \frac{2}{L} \int_0^L f(x) \cos\left(\frac{n\pi x}{L}\right) dx$$

$$= 2 \int_0^1 x \cos(n\pi x)\, dx$$

$$= \begin{cases} -4/(n\pi)^2 & (n \text{ odd}) \\ 0 & (n \text{ even}) \end{cases} \tag{12}$$

Hence the Fourier cosine series of f is

$$f(x) = \frac{1}{2} - \frac{4}{\pi^2}\left\{ \cos(\pi x) + \frac{1}{3^2} \cos(3\pi x) + \frac{1}{5^2} \cos(5\pi x) + \cdots \right\} \tag{13}$$

(b) To find the Fourier sine series, we compute

$$b_n = \frac{2}{L} \int_0^L f(x) \sin\left(\frac{n\pi x}{L}\right) dx$$

$$= 2 \int_0^1 x \sin(n\pi x)\, dx$$

$$= \frac{2}{n\pi}(-1)^{n+1} \qquad (n = 1, 2, \ldots) \tag{14}$$

Hence the Fourier sine series for f is

$$f(x) = \frac{2}{\pi}\left\{ \sin(\pi x) - \frac{1}{2} \sin(2\pi x) + \frac{1}{3} \sin(3\pi x) - \cdots \right\} \tag{15}$$

■

The reader can get a chance to graph the partial sums of the Fourier sine and cosine series in the problem set.

PROBLEMS: Section 9.2

For Problems 1–10, determine whether the given function is even, odd, or neither.

1. $f(x) = |x|$

2. $f(x) = \sin^2 x$

3. $f(x) = x\,|x|$

4. $f(x) = \sin x \sin 2x$

5. $f(x) = x^2 + \sin x$

6. $f(x) = \ln|\cos x|$

7. $f(x) = x + |x|$

8. $f(x) = 3x + x^2 \sin x$

9. $f(x) = x \sin n\pi x \quad (n = 1, 2, \ldots)$

10. $f(x) = x^2 \sin n\pi x \quad (n = 1, 2, ...)$

11. A Difficult Integral? Evaluate the integral

$$\int_{-1}^{1} \{x^3 \cos 3\pi x + \sin \pi x\}\, dx$$

Hint: Don't spend more than five seconds.

12. Even and Odd Functions Prove the following statements.
(a) If f and g are even functions, then so is their product fg.
(b) If f and g are odd functions, then their product is even.
(c) If f is an even and g is an odd function, then their product fg is odd.

13. Evaluation of Integrals Use the properties of even and odd functions to evaluate the following integrals.

(a) $\int_{-1}^{1} x\, dx$

(b) $\int_{-1}^{1} x^6\, dx$

(c) $\int_{-\pi}^{\pi} x \sin x\, dx$

(d) $\int_{-\pi}^{\pi} x \cos x\, dx$

(e) $\int_{-\pi}^{\pi} \sin\left(\frac{n\pi x}{L}\right) \cos\left(\frac{m\pi x}{L}\right) dx$

(f) $\int_{-\pi}^{\pi} x^2 \sin x\, dx$

For Problems 14–17, find the even and odd periodic extensions of the given function and sketch their graphs.

14. $f(x) = 1 \qquad (0 < x < 1)$
15. $f(x) = x^2 \qquad (0 < x < 1)$
16. $f(x) = 1 - x \quad (0 < x < 1)$
17. $f(x) = \sin x \quad (0 < x < \pi)$

For Problems 18–21, find the Fourier sine series of the given function with the specified period. If a computer with graphing capabilities is available, sketch the graph of the function and the first few terms of its Fourier sine series.

18. $f(x) = \sin x \qquad (0 < x < \pi) \quad$ (period 2π)
19. $f(x) = 1 \qquad (0 < x < \pi) \quad$ (period 2π)
20. $f(x) = x \qquad (0 < x < 1) \quad$ (period 2)
21. $f(x) = 1 - x \quad (0 < x < 1) \quad$ (period 2)

For Problems 22–25, find the Fourier cosine series of the given function with the specified period. If a computer with graphing capabilities is available, sketch the graph of the given function and a few terms of its Fourier cosine series.

22. $f(x) = 1 \qquad (0 \le x \le \pi) \qquad$ (period 2π)
23. $f(x) = 1 - x \quad (0 \le x \le 1) \qquad$ (period 2)
24. $f(x) = \begin{cases} 1 & (0 < x < 1) \\ 0 & (1 < x < 2) \end{cases} \qquad$ (period 4)
25. $f(x) = \sin x \quad (0 < x < \pi) \qquad$ (period 2π)

Half-Range Expansions

*Consider a function f defined on $(0, L)$ and extend it from L to $2L$ by reflecting it across the line $x = L$. This new **reflected***

function F_{ref} is defined on $(0, 2L)$ by

$$F_{ref}(x) = \begin{cases} f(x) & (0 < x < L) \\ f(2L - x) & (L < x < 2L) \end{cases} \qquad (16)$$

See Figure 9.16. If one then computes the odd extension of F_{ref}, getting

$$G_{oddref}(x) = \begin{cases} F_{ref}(x) & (0 < x < 2L) \\ -F_{ref}(-x) & (-2L < x < 0) \end{cases}$$

*the Fourier series of $G_{oddref}(x)$, which is called the **half-range sine expansion** of $f(x)$, converges to $f(x)$ on $(0, L)$ and contains only odd sine terms whose coefficients b_n are determined by*

$$b_n = \begin{cases} 0 & (n \text{ even}) \\ \dfrac{2}{L}\displaystyle\int_0^L f(x) \sin\left(\dfrac{n\pi x}{2L}\right) dx & (n \text{ odd}) \end{cases} \qquad (17)$$

Figure 9.16
Reflection of a graph through $x = L$

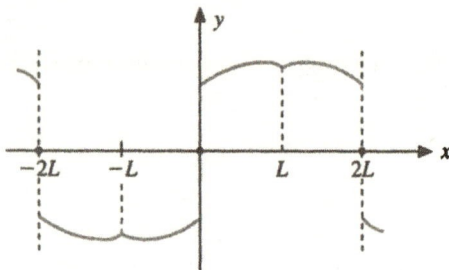

For Problems 26–27, find the half-range expansion sine of the given function.

26. $f(x) = x \quad (0 < x < 2\pi)$

27. $f(x) = \begin{cases} 0 & (0 \le t < 1) \\ 1 & (1 \le t \le 2) \end{cases}$

28. Periodic Forcing Terms Find the formal solution* of the initial-value problem

$$\ddot{x} + \omega^2 x = \sum_{n=1}^{\infty} b_n \sin nt \qquad x(0) = 0 \quad \ddot{x}(0) = 0 \quad (18)$$

where $\omega > 0$ is not equal to a positive integer.

---

* A solution of a differential equation is called a *formal solution* if it has the *form* of a solution but is not necessarily mathematically valid, since infinite series may not converge.

29. **Computer Problem** Use a computer to sketch the graphs of various partial sums of the Fourier cosine series of the function in Example 3.

30. **Computer Problem** Use a computer to sketch the graphs of various partial sums of the Fourier sine series of the function in Example 3.

31. **Journal Entry** Summarize the concepts of the Fourier sine and cosine series.

9.3   INTRODUCTION TO PARTIAL DIFFERENTIAL EQUATIONS

PURPOSE

To show what partial differential equations are, how they are solved, and how they can be used to predict phenomena in engineering and science.

WHAT IS A PARTIAL DIFFERENTIAL EQUATION?

Quite simply, a **partial differential equation** is an equation that contains partial derivatives of an unknown function. This is in contrast to an ordinary differential equation, in which the unknown function depends on a single variable and all the derivatives are ordinary derivatives. In a partial differential equation the unknown function u, or **dependent variable,** depends on two or more **independent variables.** Usually, in describing natural phenomena the dependent variable u will depend on one or more space variables x, y, z, and time t. Sometimes the dependent variable will depend on only space variables. For simplicity we use subscript notation to denote partial derivatives. For example, we write

$$u_x = \frac{\partial u}{\partial x} \qquad u_t = \frac{\partial u}{\partial t} \qquad u_{xx} = \frac{\partial^2 u}{\partial x^2} \qquad u_{xy} = \frac{\partial^2 u}{\partial y \partial x} \qquad \cdots \qquad (1)$$

Typical partial differential equations are

$$u_t = u_{xx} \qquad \text{(cne-dimensional heat equation)} \qquad (2a)$$

$$u_{tt} = u_{xx} \qquad \text{(one-dimensional wave equation)} \qquad (2b)$$

$$u_{rr} + \frac{1}{r}u_r + \frac{1}{r^2}u_{\theta\theta} = 0 \qquad \text{(Laplace's equation in polar coordinates)} \qquad (2c)$$

$$u_{tt} = \alpha^2(u_{xx} + u_{yy}) \qquad \text{(two-dimensional wave equation)} \qquad (2d)$$

In Eqs. (2) we have not said explicitly what the independent variables are, but unless stated otherwise, it is understood that the independent variables are those variables, other than the dependent variable, that appear in the equation. For example, in the one-dimensional heat equation (2a) the independent variables are understood to be x and t, whereas in the Laplace's equation (2c) the independent variables are r and θ. We also have not specified the domain over which the equations hold. Later, we will be careful to define the domain over which a partial differential equation is defined.

KINDS OF PARTIAL DIFFERENTIAL EQUATIONS

Partial differential equations come in many shapes and forms. Classification is important because the general theory and method of solution often apply only to a certain type of equation. Six basic classifications are the following.

1. *Order of a PDE.* The **order** of a PDE is the order of the highest partial derivative that appears in the equation. For example,

$$\begin{array}{lll}
\text{(a)} & u_t = u_x & \text{(first-order)} \\
\text{(b)} & u_t = \alpha^2 u_{xx} & \text{(second-order)} \\
\text{(c)} & u_t = u u_{xxx} + \sin x & \text{(third-order)}
\end{array}$$

2. *Number of Variables.* The **number of variables** is the number of independent variables. For example,

$$\begin{array}{lll}
\text{(a)} & u_t = \alpha^2 u_{xx} & \text{(two independent variables)} \\
\text{(b)} & u_{xx} + u_{yy} + u_{zz} = 0 & \text{(three independent variables)} \\
\text{(c)} & u_t = \alpha^2 (u_{xx} + u_{yy} + u_{zz}) & \text{(four independent variables)}
\end{array}$$

3. *Linear or Nonlinear.* Partial differential equations are classified as being either linear or nonlinear. Roughly, a partial differential equation is linear if the dependent variable u and its derivatives appear in a *linear fashion* (they are not multiplied, raised to powers, etc.). More precisely, the general **second-order linear equation** in two variables (which we choose to be x and y) is a partial differential equation that can be written in the form

$$Au_{xx} + Bu_{xy} + Cu_{yy} + Du_x + Eu_y + Fu = G \tag{3}$$

where A, B, C, D, E, F, and G are given functions of x and y. Of course, when we say that they are functions, we do not rule out the possibility that they may be constant functions, in which case we simply refer to them as constants. A second-order partial differential equation in two variables that is not of the form of Eq. (3) is a **nonlinear equation.** Examples of linear and nonlinear equations are

$$\begin{array}{lll}
\text{(a)} & u_t = \alpha^2 u_{xx} & \text{(linear)} \\
\text{(b)} & u_{xx} + u_{yy} + u = \sin x & \text{(linear)} \\
\text{(c)} & u u_x + y u_y + u = 1 & \text{(nonlinear)} \\
\text{(d)} & u_{xx} + u_{yy} + u^2 = 0 & \text{(nonlinear)}
\end{array}$$

Note that in Eq. (a) the variable t plays the role of y in the general equation (3). In this case we would identify the constants $A = \alpha^2$, $B = 0$, $C = 0$, $D = 0$, $E = -1$, $F = 0$, and $G = 0$. (We could just as well identify the constants by changing their signs; i.e., $A = -\alpha^2$, $E = 1$, and the other constants 0.)

4. *Homogeneous Linear Equation.* The general linear equation (3) is **homogeneous** if $G = 0$ for all x and y in the domain of the equation; otherwise, it is called

nonhomogeneous. Of course, if G is a constant, the equation is homogeneous if and only if $G = 0$. For example,

(a) $u_t = \alpha^2 u_{xx}$ (homogeneous)

(b) $u_t = \alpha^2 u_{xx} + 1$ (nonhomogeneous)

(c) $u_{xx} + x u_{yy} = 0$ (homogeneous)

5. *Constant or Variable Coefficients.* If the coefficients A, B, C, D, E, and F in Eq. (3) are *all* constants, then the equation is said to have **constant coefficients;** otherwise, it has **variable coefficients.** For example,

(a) $u_{xx} + u_{yy} = \sin x$ (constant coefficients)

(b) $u_t = \alpha^2 \left(u_{rr} + \dfrac{1}{r} u_r + \dfrac{1}{r^2} u_{\theta\theta} \right)$ (variable coefficients)

(c) $x u_x + y u_y = 1$ (variable coefficients)

6. *Three Basic Types of Linear Equations.* An important classification of linear partial differential equations is the one that classifies equations as being either **parabolic, hyperbolic,** or **elliptic.** The second-order linear equation (3) is classified according to the following rules:

- **Parabolic:** Equation (3) is parabolic if $B^2 - 4AC = 0$. Parabolic equations often describe heat flow and diffusion phenomena, such as heat flow through the earth's surface.

- **Hyperbolic:** Equation (3) is hyperbolic if $B^2 - 4AC > 0$. Hyperbolic equations often describe wave motion and vibrating phenomena, such as violin strings and drumheads.

- **Elliptic** Equation (3) is elliptic if $B^2 - 4AC < 0$. Elliptic equations are typically used to describe steady state phenomena and thus do not depend on time. Elliptic equations are important in the study of electricity and magnetism.

Examples are

(a) $u_t - u_{xx} = 0$ $B^2 - 4AC = 0^2 - 4(-1)(0) = 0$ (parabolic equation)

(b) $u_{xx} + u_{yy} = 0$ $B^2 - 4AC = 0^2 - 4(1)(1) = -4$ (elliptic equation)

(c) $u_{tt} - u_{xx} = 0$ $B^2 - 4AC = 0^2 - 4(-1)(1) = 4$ (hyperbolic equation)

WHY PARTIAL DIFFERENTIAL EQUATIONS ARE USEFUL

We associate electromagnetism with **Maxwell's equations,** fluid dynamics with the **Navier-Stokes equations,** and quantum mechanics with **Schrödinger's equation.** The common factor in these equations is that they are all systems of partial differential equations. In fact, many of the fundamental equations of physics are partial differential equations. The reason partial differential equations play such a crucial role in the description of physical phenomena is that partial derivatives are commonly used to describe changes in physical quantities (force, friction, flux, voltage, current, velocity, acceleration, and so on). Hence a partial differential equation can be interpreted as an equation that relates physical quantities.

As an example, suppose a rectangular slab of metal is being heated on one side. The resulting temperature distribution in the slab is determined by a *partial differential equation* that describes how heat flows from hot regions to cold regions in the slab, by the *initial temperature* of the slab, and by the *boundary conditions* on the other faces (e.g., these faces might be kept insulated or held at a fixed temperature). Other physical phenomena are described by other partial differential equations, initial conditions, and boundary conditions. The goal is to find the function (i.e., the solution) that satisfies the partial differential equations *and* the initial and boundary conditions.

HOW TO SOLVE A PARTIAL DIFFERENTIAL EQUATION

Recall from calculus that integration of a partial derivative results in an arbitrary function rather than an arbitrary constant. For example, if $u = u(x, y)$ is a function of x and y, then the solution of the partial differential equation $\partial u / \partial x = 0$ is $u = \phi(y)$, where ϕ is an arbitrary differentiable function of y.

Example 1 Solution of a PDE Find a solution $u = u(x, y)$ of the second-order equation

$$u_{yy} = 0 \tag{4}$$

Solution Integrating twice with respect to y gives

$$u_y = f(x) \tag{5}$$

$$u(x, y) = yf(x) + g(x) \tag{6}$$

where f and g are arbitrary differentiable functions of x. The reader can try different functions f and g in Eq. (6) to see that Eq. (4) reduces to an identity. ∎

Example 2 Solution of a PDE Find a solution $u = u(x, y)$ of the partial differential equation

$$u_{xy} = x + y \tag{7}$$

Solution We first integrate with respect to y (treating x as a constant). This gives

$$u_x = xy + \frac{y^2}{2} + F(x) \tag{8}$$

where $F(x)$ is an arbitrary differentiable function of x. Next we integrate with respect to x (treating y as a constant), getting

$$u(x, y) = \frac{x^2 y}{2} + \frac{xy^2}{2} + \int^x F(s)\, ds + g(y) \tag{9}$$

where $\int^x F(s)\, ds$ is any antiderivative of $F(x)$ with respect to x and $g(y)$ is an arbitrary differentiable function of y. Now, by letting $f(x) = \int^x F(s)\, ds$, we can write the general solution in the form

$$u(x, y) = \frac{x^2 y}{2} + \frac{xy^2}{2} + f(x) + g(y) \tag{10}$$

where $f(x)$ and $g(y)$ are arbitrary differentiable functions of x and y, respectively. In other words, the solution contains two *arbitrary functions* in contrast to solutions of ordinary differential equations that contain arbitrary *constants*. The reader should verify that Eq. (10) satisfies Eq. (7).   ∎

THE DECLINE OF THE GENERAL SOLUTION

Note that the general solution (10) of the partial differential equation $u_{xy} = x + y$ contains two arbitrary *functions*. This contrasts with the general solution of a second-order *ordinary* differential equation that contains two arbitrary *constants*. We shall see in the following sections that the general solution of a partial differential equation is not as important to partial differential equation theory as was the general solution of an ordinary differential equation in ordinary differential theory. If the reader will recall, the strategy used in finding the solution of an ordinary differential equation with initial condition(s) was to first find the general solution of the differential equation and then use the initial condition(s) to determine the arbitrary constant(s). Unfortunately, the process of substituting the general solution of a partial differential equation into some side conditions (initial and boundary conditions) is complicated by the fact that arbitrary *functions* are more unwieldy than arbitrary constants.

We will also see that very few partial differential equations can be solved as easily as the ones in the previous examples. Normally, one wishes to find the solution of a partial differential equation that also satisfies given initial and boundary conditions. The strategy is to find "enough" solutions of the partial differential equation that can be "pieced together" to produce a solution of the partial differential equation that also satisfies the initial and boundary conditions.

We list a few methods that are commonly used for solving partial differential equations with side conditions.

1. *Separation of Variables.* This technique reduces a partial differential equation (PDE) in n independent variables to n ordinary differential equations (ODE's). It is this method that will be studied later in this chapter.

2. *Integral Transforms.* This technique reduces a PDE in n independent variables to a PDE in $n - 1$ independent variables. Hence a PDE in two variables would be transformed into an ODE. In this case the ODE is then solved, and the inverse transform then finds the solution to the PDE. In solving ODE's we have seen that the most common transform is the Laplace transform. However, when solving PDE's, one also uses the **Fourier transform,** the **Mellin transform,** the **Hankel transform,** and a whole host of other transforms.

3. *Numerical Methods.* Numerical methods (often) change a PDE to a system of *difference equations* that can be solved by using a computer. For nonlinear equations this is often the only technique that will provide (approximate) solutions.

We could list a dozen other methods, but this would take us too far afield of our goal for this introductory chapter.

PROBLEMS: Section 9.3

For Problems 1–10, classify the given equation according to order, number of variables, and linearity. If the equation is linear, indicate whether it is homogeneous or nonhomogeneous and whether it has constant or variable coefficients.

1. $u_x + u_y = 0$
2. $u_t = u_{xx} + u_x + u$
3. $u_{tt} = u_{xx} + uu_x + 1$
4. $u_t = u_{rr} + u_r$
5. $u_{xx} + u_{yy} + \sin u = 0$
6. $x^2 u_x + u_y = \sin x$
7. $u_t = u_{xx} + u_{yy} + u_{zz} + 1$
8. $u_x + u_y + u_z = 0$
9. $u_{rr} + \dfrac{1}{r}u_r + \dfrac{1}{r^2}u_{\theta\theta} + u_{zz} = 0$
10. $u_t = u_{rr} + \dfrac{2}{r}u_r \dfrac{1}{r^2}u_{\theta\theta} + \dfrac{\cos\theta}{r^2}u_\theta + \dfrac{1}{r^2\sin\theta}u_{\phi\phi}$

For Problems 11–15, verify that the given function is a solution of the given partial differential equation. The functions f and g are arbitrary differential functions and α is a constant.

11. $u_t = \alpha^2 u_{xx}$ $u(x, t) = e^{-\alpha^2 t}\sin x$
12. $u_{tt} = \alpha^2 u_{xx}$ $u(x, t) = \sin \alpha t \cos x$
13. $u_{xx} + u_{yy} = 0$ $u(x, y) = \ln\left\{\dfrac{1}{x^2 + y^2}\right\}$
14. $u_{xx} + u_{yy} + u_{zz} = 0$ $u(x, y, z) = \dfrac{1}{\sqrt{x^2 + y^2 + z^2}}$
15. $u_{xx} = 0$ $u(x, y) = xf(y) + g(y)$

For Problems 16–21, assume that u is a function of the two variables x and y. Integrate the given equation to obtain the general solution (all solutions).

16. $u_x = 0$
17. $u_x = 1$
18. $u_x = 2x$
19. $u_x = y$
20. $u_{xy} = 0$
21. $u_{xy} = x^2 + y^2$
22. **A Tip of the Iceberg** How many solutions of the "heat equation" $u_t = u_{xx}$ can you find? Try solutions of the form $u(x, t) = e^{ax+bt}$ and find a and b.
23. **As Always, Superposition of Solutions** If $u_1(x, y)$ and $u_2(x, y)$ satisfy the linear homogeneous equation

$$Au_{xx} + Bu_{xy} + Cu_{yy} + Du_x + Eu_y + Fu = 0$$

then show that $c_1 u_1(x, y) + c_2 u_2(x, y)$ also satisfies the equation, where c_1 and c_2 are arbitrary constants.

24. **Classification of PDE's** Determine whether the following partial differential equations are hyperbolic, parabolic, or elliptic.

(a) $u_{xx} + 6u_{xy} + 12u_{yy} = 0$ (b) $u_{xx} + u_{yy} + 2u = 0$
(c) $u_{xx} + 9u_{xy} + 4u_{yy} = 0$ (d) $u_t = u_{xx} + u = 0$
(e) $u_{xx} + 8u_{xy} + 16u_{yy} = 0$ (f) $u_{tt} = u_{xx} + u_x + u$

25. **Canonical Wave Equation** To solve the **canonical wave equation** $u_{xy} = 0$, integrate first with respect to y and second with respect to x* to obtain the solution

$$u(x, y) = f(x) + g(y)$$

where f and g are arbitrary differentiable functions. Verify that this function is a solution, and list a few particular solutions.

26. **Solving a Partial Differential Equation** Solve the partial differential equation

$$u_{xx} - y^2 u = e^x$$

by the following steps.
(a) Solve the corresponding homogeneous equation given by $u_{xx} - y^2 u = 0$ by treating y as a constant and use the method learned in Chapter 3 to get

$$u_h(x, y) = f(y)e^{xy} + g(y)e^{-xy}$$

(b) Use the method of undetermined coefficients to find a particular solution of the form

$$u_p(x, y) = A(y)e^x$$

to get

$$u_p(x, y) = \dfrac{e^x}{1 - y^2}$$

(c) Verify that the function

$$u(x, y) = f(y)e^{xy} + g(y)e^{-xy} + \dfrac{e^x}{1 - y^2}$$

satisfies the nonhomogeneous partial differential equation.

27. **Journal Entry** Discuss the role of partial differential equations in your major area of interest and research.

---

* One can interchange the order of integration and achieve the same solution.

9.4 THE VIBRATING STRING: SEPARATION OF VARIABLES

PURPOSE
We solve the vibrating string equation using the method of separation of variables. We show that the solution, which describes the overall vibration of the string, can be interpreted as the sum of simple harmonic vibrations.

One of the most useful partial differential equations in applied mathematics is the wave equation. Various forms of this equation can be found in the study of water waves, acoustic waves, electromagnetic waves, seismic waves, and so on. Although a thorough study of these topics would take us too far afield from a course in differential equations, we introduce the reader to the simplest application of the wave equation, an analysis of the vibrations of a vibrating string.

THE VIBRATING STRING   By a vibrating string, we have in mind something like a piano string, a violin string, or a guitar string. It is assumed that the tension of the string is sufficiently great that we can neglect the troublesome effects of gravity and air friction. The string also moves in a plane, and the vibrations of the string are transverse. That is, the motion of the string is perpendicular to the line of equilibrium of the string. We let the x-axis be the line of equilibrium of the string (see Figure 9.17), and for simplicity we take the length of the string to be 1.

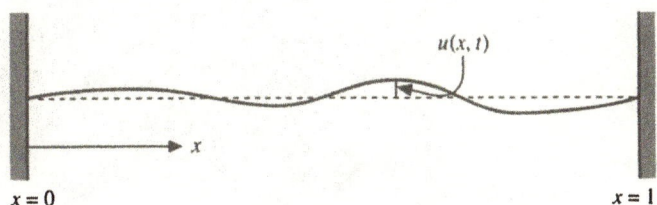

Figure 9.17
Vibrating string, such as piano, guitar, or violin string

The vertical displacement of the string from the line of equilibrium is denoted by $u(x, t)$, which will depend on both position x and time t. Under these conditions and assuming that the height of the vibrations is relatively small, the function $u(x, t)$ will satisfy the **one-dimensional wave equation**

$$u_{tt} = \alpha^2 u_{xx} \tag{1}$$

for $0 < x < 1$ and $t > 0$. The constant α^2 is a physical parameter and is given by $\alpha^2 = T/\rho$, where T is the tension of the string and ρ is the linear density of the string.*

To predict the future motion $u(x, t)$, we must introduce the concept of **boundary conditions.** For the vibrating string it is necessary to specify conditions at the ends of the string. In this problem we will assume that the ends of the string are *fixed,* thus giving the two boundary conditions (BC):

$$\text{BC:} \quad \begin{cases} u(0, t) = 0 \\ u(1, t) = 0 \end{cases} \quad (t \geq 0) \tag{2}$$

In addition, since the partial derivative u_{tt} that appears in the wave equation is of second order, we must specify *two* **initial conditions** (IC), which are the **initial position** and **initial velocity** of the string. These are written as

$$\text{IC:} \quad \begin{cases} u(x, 0) = f(x) \\ u_t(x, 0) = g(x) \end{cases} \quad (0 \leq x \leq 1) \tag{3}$$

where f and g represent the given initial position and velocity, respectively. The problem is to find the solution of the partial differential equation that also satisfies the boundary and initial conditions. Such a problem is called an **initial-boundary-value problem.** We summarize the above ideas.

Initial-Boundary-Value Problem for the Vibrating String

The initial-boundary-value problem that describes the vibrating string with fixed endpoints and arbitrary initial conditions is the problem of finding the function $u(x, t)$ that satisfies

$$\text{PDE:} \quad u_{tt} = \alpha^2 u_{xx} \quad (0 < x < 1) \quad (t > 0)$$

$$\text{BC:} \quad \begin{cases} u(0, t) = 0 \\ u(1, t) = 0 \end{cases} \quad (t \geq 0)$$

$$\text{IC:} \quad \begin{cases} u(x, 0) = f(x) \\ u_t(x, 0) = g(x) \end{cases} \quad (0 \leq x \leq 1)$$

where f and g are given functions.

Although an in-depth derivation of the wave equation would take us too far afield from a first course in differential equations, the following discussion demonstrates the plausibility of the wave equation.[†]

---

* In the English system of measurement the tension T would have units of pounds and linear density ρ would have units of slugs/inch.

[†] The interested reader can find a derivation of the wave equation in *Partial Differential Equations for Scientists and Engineers,* by S. J. Farlow (Dover Publications, New York, 1993).

Plausibility of the Wave Equation $u_{tt} = \alpha^2 u_{xx}$

The partial differential equation $u_{tt} = \alpha^2 u_{xx}$ is called the one-dimensional wave equation, and it expresses mathematically the physical fact that (at any point x and any time t) the *vertical acceleration* u_{tt} of the string is proportional to the *concavity* u_{xx} of the shape of the string. This statement is plausible if we recall that the second derivative* u_{xx} can be approximated by

$$u_{xx}(x, t) \doteq \frac{u(x + 2h, t) - 2u(x, t) + u(x - 2h, t)}{h^2}$$

$$= \frac{2}{h^2}\left(\frac{u(x + 2h, t) + u(x - 2h, t)}{2} - u(x, t)\right)$$

In other words, $u_{xx}(x, t)$ is proportional to the *difference* between the height of the string $u(x, t)$ and the *average* of u at the neighboring points $x + 2h$ and $x - 2h$. Hence if the height of the string is *less* than the average height at two neighboring points, then $u_{xx} > 0$, and hence $u_{tt} = \alpha^2 u_{xx}$ implies that the acceleration u_{tt} of the string is *upward*. Since acceleration is proportional to force, the force on the string is also upward. By the same token, if the height of the string is *greater* than the average height at two neighboring points, then $u_{xx} < 0$, and hence $u_{tt} = \alpha^2 u_{xx}$ implies that the acceleration u_{tt} of the string is downward.

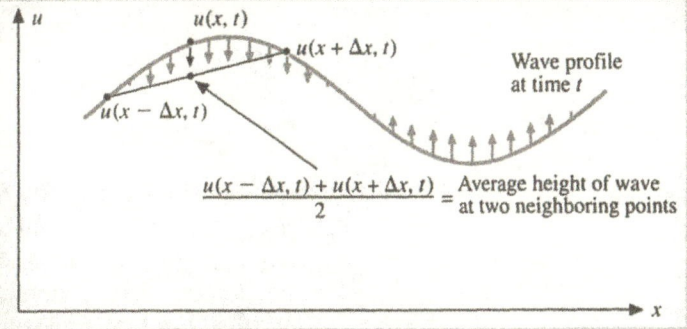

SEPARATION OF VARIABLES

We now find the solution $u(x, t)$ of the initial-boundary-value problem

$$\text{PDE:} \quad u_{tt} = \alpha^2 u_{xx} \qquad (0 < x < 1) \qquad (t > 0)$$

$$\text{BC:} \quad \begin{cases} u(0, t) = 0 \\ u(1, t) = 0 \end{cases} \qquad (t \geq 0)$$

$$\text{IC}^\dagger: \quad \begin{cases} u(x, 0) = f(x) \\ u_t(x, 0) = 0 \end{cases} \qquad (0 \leq x \leq 1) \qquad (4)$$

* Later, we will see that the second derivative u_{xx} is the one-dimensional version of what is known as the *Laplacian,* one of the most important concepts in mathematical physics.

† We have set the initial velocity $u_t(x, 0)$ of the string to zero to simplify the analysis. This will not interfere with the basic understanding of the problem.

The method of solution is known as **separation of variables,** or **Fourier's method.** We first present an overview of this powerful technique.

HISTORICAL NOTE

Although the method of separation of variables is often referred to as *Fourier's method,* it was first derived and studied by Jean d'Alembert (1717?–1783) in 1746. It was Leonhard Euler who first applied trigonometric analysis to the continuous string problem.

Overview of Separation of Variables (for the Vibrating String)

The method of separation of variables looks for simple solutions or *vibrations* of the wave equation of the form

$$u(x, t) = X(x)T(t)$$

where $X(x)$ is a function that depends only on x and $T(t)$ is a function that depends only on t. We say "simple" because any vibration of this form retains its fundamental *shape** for all $t \geq 0$.

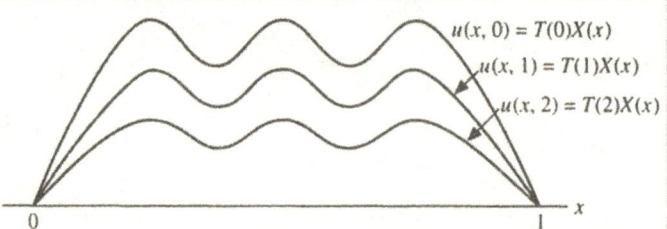

Such vibrations exhibit the property that all points on the string vibrate with the same frequency. The general idea is to find an *infinite number* of these simple vibrations (solutions) that also satisfy the given boundary conditions $u(0, t) = u(1, t) = 0$. These vibrations $u_n(x, t) = X_n(x)T_n(t)$ that satisfy the BC's are called **fundamental solutions** by mathematicians and **standing waves** by physicists. They are the *building blocks* of the overall solution. The overall solution is then found by taking linear combinations of the fundamental solutions

$$u(x, t) = \sum_{n=1}^{\infty} a_n X_n(x) T_n(t) \tag{5}$$

The coefficients a_n are chosen so that $u(x, t)$, which already satisfies the partial differential equation and the boundary conditions, *also* satisfies the *initial conditions.* With the coefficients a_n properly determined, $u(x, t)$ is the solution of the initial-boundary-value problem.

---

* For any two different values of time t_1 and t_2 the functions $X(x)T(t_1)$ and $X(x)T(t_2)$ are constant multiples of each other.

We now solve the vibrating problem by the method of separation of variables.

SEPARATION OF VARIABLES SOLUTION OF THE VIBRATING STRING PROBLEM

Step 1 (Finding Product Solutions). We begin by seeking solutions of the wave equation of the form $u(x, t) = X(x)T(t)$. Substituting this expression into the wave equation $u_{tt} = \alpha^2 u_{xx}$ gives

$$X(x)T''(t) = \alpha^2 X''(x)T(t) \tag{6}$$

If we now divide Eq. (6) by $\alpha^2 X(x)T(t)$, we obtain the equivalent expression

$$\frac{T''(t)}{\alpha^2 T(t)} = \frac{X''(x)}{X(x)} \tag{7}$$

in which the variables x and t are said to be *separated*. That is, the left-hand side of Eq. (7) depends only on t, and the right-hand side depends only on x. However, in order for Eq. (7) to be valid for $0 < x < 1$, $t > 0$, it is necessary that both sides of Eq. (7) be equal to a *constant*. Otherwise, by keeping one independent variable (say x) fixed and varying the other (say t), one side of the equation will change and the other side will remain unchanged, violating the fact the two sides are always equal. Hence we have

$$\frac{T''}{\alpha^2 T} = \frac{X''}{X} = -k \tag{8}$$

where k is any constant, called the **separation constant.** From Eq. (8) we get the two equations

$$X'' + kX = 0 \tag{9a}$$
$$T'' + k\alpha^2 T = 0 \tag{9b}$$

If we can solve these two ordinary differential equations, we can multiply their solutions together, obtaining a standing wave solution $u(x, t) = X(x)T(t)$. However, the form of the solutions of Eqs. (9) change for $k < 0$, $k = 0$, and $k > 0$. Solving Eqs. (9) separately for these three cases, we obtain

$$k < 0 \quad \begin{cases} X(x) = Ae^{\sqrt{-k}x} + Be^{-\sqrt{-k}x} \\ T(t) = Ce^{\sqrt{-k}\alpha t} + De^{-\sqrt{-k}\alpha t} \end{cases} \tag{10a}$$

$$k = 0 \quad \begin{cases} X(x) = Ax + B \\ T(t) = Ct + D \end{cases} \tag{10b}$$

$$k > 0 \quad \begin{cases} X(x) = A\cos(\sqrt{k}x) + B\sin(\sqrt{k}x) \\ T(t) = C\cos(\sqrt{k}\alpha t) + D\sin(\sqrt{k}\alpha t) \end{cases} \tag{10c}$$

By forming the respective products $X(x)T(t)$ we obtain standing wave solutions of the wave equation. The reader can verify that each of these respective products satisfy the wave equation. The results from Step 1 are summarized in Figure 9.18.

Figure 9.18
Standing wave solutions for
different values of *k*

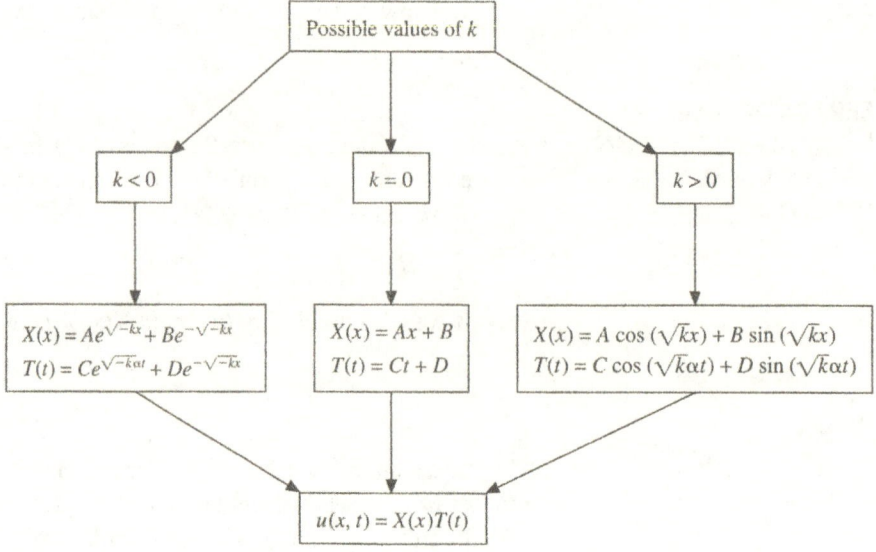

Step 2 (Finding Standing Waves). Thus far we have found solutions of the wave equation with all kinds of wavelengths. It is now up to the boundary conditions to pick out those wavelengths that conform to the reality of the situation. We do this by picking out those standing waves $X(x)T(t)$ that satisfy the boundary conditions. Substituting $X(x)T(t)$ into the boundary conditions gives

$$\text{BC:} \quad \begin{cases} u(0, t) = X(0)T(t) = 0 \\ u(1, t) = X(1)T(t) = 0 \end{cases} \quad (t \geq 0) \tag{11}$$

However, if the boundary conditions (11) are to be satisfied for *all* $t \geq 0$, we must have

$$X(0) = 0$$
$$X(1) = 0 \tag{12}$$

We now substitute the three solutions $X(x)$ in Eqs. (10) into the boundary conditions $X(0) = X(1) = 0$. For $X(x)$ in Eqs. (10a) and (10b) we get $X(x) \equiv 0$, which in turn leads to the trivial solution $u(x, t) \equiv 0$. However, substituting $X(x)$ from Eq. (10c) into the boundary conditions gives

$$X(0) = A \cos (\sqrt{k} \cdot 0) + B \sin (\sqrt{k} \cdot 0) = 0 \Rightarrow A = 0$$
$$X(1) = A \cos \sqrt{k} + B \sin \sqrt{k} = 0 \qquad \Rightarrow k = (n\pi)^2 \tag{13}$$

Hence the separation constant k must be chosen as one of the specific values $k_n = (n\pi)^2$ for $n = 1, 2, \ldots$ in order for $X(x)T(t)$ to satisfy the boundary conditions.

These values of k_n are called **eigenvalues** of the problem, and the corresponding solutions $X_n(x) = \sin(n\pi x)$ are called the **eigenfunctions** of the problem.

We now multiply $X(x)T(t)$ from Eq. (10c), substituting $k_n = (n\pi)^2$ for k and changing the notation of the constants to get the **fundamental solutions**

$$u_n(x, t) = X_n(x)T_n(t)$$
$$= \sin(n\pi x)\{a_n \cos(n\pi\alpha t) + b_n \sin(n\pi\alpha t)\} \tag{14}$$

where a_n and b_n are arbitrary constants. Figure 9.19 displays the graphs of a few standing waves and how they change over time.

We have found an infinite number of functions, each of which satisfies the wave equation *and* the boundary conditions. We now carry out the final step by finding the *single* function that also satisfies the initial conditions. Here is where the Fourier series makes its appearance.

Figure 9.19
Fundamental solutions of the vibrating string. Note that each position along a given standing wave vibrates with the same frequency.

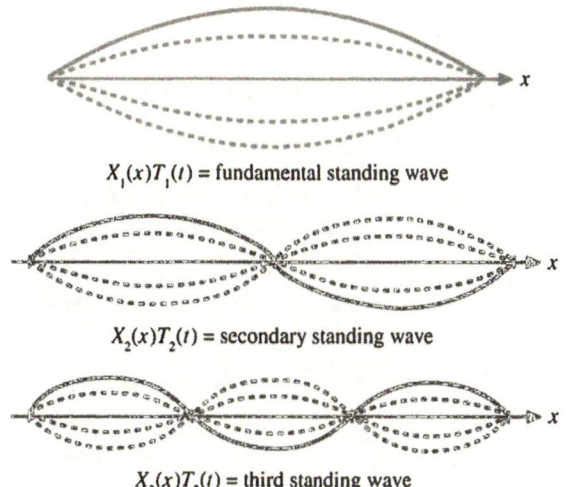

$X_1(x)T_1(t)$ = fundamental standing wave

$X_2(x)T_2(t)$ = secondary standing wave

$X_3(x)T_3(t)$ = third standing wave

Step 3 (Summing the Standing Waves). One may get the impression that the fundamental solutions $u_n(x, t)$ are the only functions that satisfy the wave equation and boundary conditions. This is not true. Any *sum*

$$u(x, t) = \sum_{n=1}^{\infty} \sin(n\pi x)\{a_n \cos(n\pi\alpha t) + b_n \sin(n\pi\alpha t)\} \tag{15}$$

also satisfies these equations.* This fact leads to the final step, which consists in finding the coefficients a_n and b_n that makes Eq. (15) satisfy the initial conditions

---

* Although it can easily be verified that any *finite* sum of functions that satisfies the wave equation and boundary conditions also satisfies these conditions, it takes a little more mathematical rigor to prove that an *infinite* sum of solutions satisfies these conditions.

$u(x, 0) = f(x)$ and $u_t(x, 0) = 0$. Substituting Eq. (15) into the initial conditions, we obtain

$$u(x, 0) = \sum_{n=1}^{\infty} b_n \sin(n\pi x) = f(x) \tag{16a}$$

$$u_t(x, 0) = \sum_{n=1}^{\infty} (n\pi\alpha)a_n \sin(n\pi x) = 0 \tag{16b}$$

First, from Eq. (16b) we determine that all the coefficients a_n must be zero in order for the equation to hold for all $0 \le x \le 1$. Second, note that Eq. (16a) is the Fourier sine series of $f(x)$. If we rewrite this equation in the familiar form as

$$f(x) = \sum_{n=1}^{\infty} b_n \sin(n\pi x) \qquad (0 \le x \le 1) \tag{17a}$$

the coefficients b_n are

$$b_n = 2 \int_0^1 f(x) \sin(n\pi x)\, dx \tag{17b}$$

We are done. We now summarize our results.

Solution of the Vibrating String Problem

The solution $u(x, t)$ of the initial-boundary-value problem

$$\text{PDE:} \quad u_{tt} = \alpha^2 u_{xx} \qquad (0 < x < 1) \qquad (t > 0)$$

$$\text{BC:} \quad \begin{cases} u(0, t) = 0 \\ u(1, t) = 0 \end{cases} \qquad (t \ge 0)$$

$$\text{IC:} \quad \begin{cases} u(x, 0) = f(x) \\ u_t(x, 0) = 0 \end{cases} \qquad (0 \le x \le 1)$$

is

$$u(x, t) = \sum_{n=1}^{\infty} b_n \sin(n\pi x) \cos(n\pi\alpha t) \tag{18}$$

where the coefficients b_n are the coefficients in the Fourier sine expansion

$$u(x, 0) = f(x) = \sum_{n=1}^{\infty} b_n \sin(n\pi x)$$

That is,

$$b_n = 2 \int_0^1 f(x) \sin(n\pi x)\, dx$$

Note: The solution $u(x, t)$ can be found by first expanding the initial condition $u(x, 0) = f(x)$ as a Fourier sine series, getting

$$f(x) = \sum_{n=1}^{\infty} b_n \sin (n\pi x)$$

Then, to find the solution $u(x, t)$, simply "insert" the factor $\cos (n\pi\alpha t)$ in each term, getting

$$u(x, t) = \sum_{n=1}^{\infty} b_n \sin (n\pi x) \cos (n\pi\alpha t) \tag{19}$$

Example 1   **Simple Vibrating String Problem**   Solve the initial-boundary-value problem

PDE: $u_{tt} = u_{xx}$ $(0 < x < 1)$ $(t > 0)$

BC: $\begin{cases} u(0, t) = 0 \\ u(1, t) = 0 \end{cases}$ $(t \geq 0)$

IC: $\begin{cases} u(x, 0) = \sin (\pi x) + \dfrac{1}{3} \sin (3\pi x) \\ u_t(x, 0) = 0 \end{cases}$ $(0 \leq x \leq 1)$

Solution   The only task is to compute the coefficients b_n that appear in the solution. Using the orthogonality properties of the functions $\{\sin (\pi x), \sin (2\pi x), \sin (3\pi x), ...\}$ on the interval $(0, 1)$, we get

$$b_1 = 2 \int_0^1 \left\{ \sin (\pi x) + \frac{1}{3} \sin (3\pi x) \right\} \sin (\pi x)\, dx = 1$$

$$b_2 = 2 \int_0^1 \left\{ \sin (\pi x) + \frac{1}{3} \sin (3\pi x) \right\} \sin (2\pi x)\, dx = 0$$

$$b_3 = 2 \int_0^1 \left\{ \sin (\pi x) + \frac{1}{3} \sin (3\pi x) \right\} \sin (3\pi x)\, dx = \frac{1}{3}$$

$$b_n = 2 \int_0^1 \left\{ \sin (\pi x) + \frac{1}{3} \sin (3\pi x) \right\} \sin (n\pi x)\, dx = 0 \qquad (n \geq 4)$$

Hence the solution is

$$u(x, t) = \sin (\pi x) \cos (\pi t) + \frac{1}{3} \sin (3\pi x) \cos (3\pi t) \tag{20}$$

Note: Since the initial condition $u(x, 0)$ is already written as a sine series, we can write down the Fourier sine coefficients by simply reading off the coefficients of $u(x, 0) = \sin (\pi x) + 1/3 \sin (3\pi x)$. That is, $b_1 = 1$, $b_2 = 0$, $b_3 = 1/3$, $b_n = 0$ for $n > 3$.

The initial shape of the wave and its future motion are illustrated in Figure 9.20. ■

Figure 9.20
Motion of a vibrating string
for $0 \leq t \leq 1$. Motion of the
string is periodic with period
2.

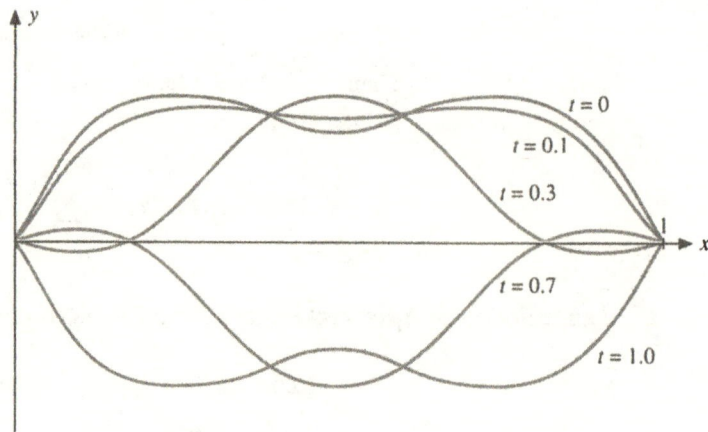

PROBLEMS: Section 9.4

For Problems 1–5, assume that the given partial differential equation has a solution of the form $u(x, t) = X(x)T(t)$. Show that the given equation "separates" into two ordinary differential equations and find the differential equation that $X(x)$ and $T(t)$ must satisfy.

1. $u_t - u_x = 0$
2. $u_t = u_{xx}$
3. $u_t = u_{xx} + 2u_x + u$
4. $u_{tt} - u_{xx} = 0$
5. $u_{tt} = u_{xx} + u_x + u$

For Problems 6–11, show that the given product solution satisfies the particular wave equation, where A, B, C, and D are arbitrary constants.

Equation	Solution
6. $u_{tt} = \alpha^2 u_{xx}$	$u(x, t) = \sin 2x \cos 2\alpha t$
7. $u_{tt} = 4u_{xx}$	$u(x, t) = e^{3x} e^{-6t}$
8. $u_{tt} = 9u_{xx}$	$u(x, t) = (x + 1)(t - 1)$
9. $u_{tt} = \alpha^2 u_{xx}$	$u(x, t) = (Ax + B)(Ct + D)$
10. $u_{tt} = \alpha^2 u_{xx}$	$u(x, t) = e^{3\alpha t}(Ae^{3x} + Be^{-3x})$
11. $u_{tt} = \alpha^2 u_{xx}$	$u(x, t) = \cos x \left(A \sin \alpha t + B \cos \alpha t\right)$

For Problems 12–13, verify that the given solution satisfies the wave equation and the boundary conditions.

12. PDE: $u_{tt} = \alpha^2 u_{xx}$

BC: $\begin{cases} u(0, t) = 0 \\ u(1, t) = 0 \end{cases}$

Solution: $u(x, t) = \sin (\pi x) \left(3 \sin \alpha \pi t + 2 \cos \alpha \pi t\right)$

13. PDE: $u_{tt} = \alpha^2 u_{xx}$

BC: $\begin{cases} u(0, t) = 0 \\ u(1, t) = 0 \end{cases}$

Solution:
$u(x, t) = \sin (\pi x) \left(a_n \sin n\pi\alpha t + b_n \cos n\pi\alpha t\right)$
(a_n and b_n are arbitrary constants, $n = 1, 2, ...$)

14. Simulation of the Wave Equation In Figure 9.21 the shape of a vibrating string has been drawn at a given instant of time. At each point on the string, draw small vertical arrows, pointing upward or downward, indicating the direction and magnitude of the acceleration (or force) at the given point. *Hint:* Consider the local concavity of this string.

Figure 9.21
Why does the vibrating string
behave the way it does?

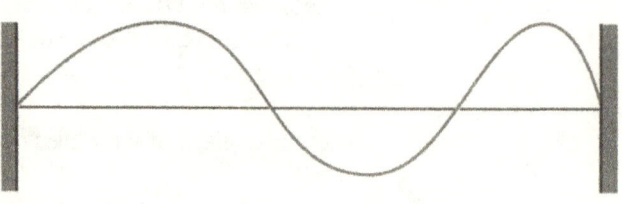

For Problems 15–17, show that the given function $u(x, t)$ satisfies the wave equation, $u_{tt} = \alpha^2 u_{xx}$, and the given initial conditions.

Initial Conditions	**Solution**

15. $u(x, 0) = \sin(\pi x)$ $u(x, t) = \sin(\pi x) \cos(\pi \alpha t)$
$u_t(x, 0) = 0$

16. $u(x, 0) = 2 \sin(3\pi x)$ $u(x, t) = 2 \sin(3\pi x) \cos(3\pi \alpha t)$
$u_t(x, 0) = 0$

17. $u(x, 0) = \sin(\pi x)$ $u(x, t) = \sin(\pi x) \cos(\pi \alpha t)$
$\qquad + \dfrac{1}{2} \sin(2\pi x)$ $\qquad + \dfrac{1}{2} \sin(2\pi x) \cos(2\pi \alpha t)$
$\qquad u_t(x, 0) = 0$

18. Not All Equations Can Be Separated Show that the method of separation of variables fails for the equation

$$u_{xy} + u_{xx} + u = 0$$

19. Magnetic Field from a Solenoid If a copper wire is wound with a wire carrying an electrical current, the resulting magnetic field H is described by the partial differential equation.

$$H_t = \gamma \left(H_{rr} + \frac{1}{r} H_r \right) \tag{21}$$

where r measures the distance from the center of the copper

rod, γ is a constant measuring the conductivity of the rod, and t is time.

(a) Assuming that $H(r, t) = R(r) T(t)$, find the differential equations that R and T must satisfy. (Call $-k$ the separation constant instead of k.)

(b) Find $T(t)$.

Computer Problems

For Problems 20–23, find the solution of the vibrating string problem ($\alpha = 1$) with the given initial conditions. Sketch the solution $u(x, t)$ over $0 \le x \le 1$ for $t = 0, 0.2, 0.4, ..., 1.8, 2$.

20. $u(x, 0) = \sin(\pi x) + \dfrac{1}{2} \sin(2\pi x) + \dfrac{1}{3} \sin(3\pi x)$
$\qquad u_t(x, 0) = 0$

21. $u(x, 0) = \sin(\pi x) + \dfrac{1}{3} \sin(3\pi x) + \dfrac{1}{9} \sin(5\pi x)$
$\qquad u_t(x, 0) = 0$

22. $u(x, 0) = \sin(\pi x) + \dfrac{1}{3} \sin(3\pi x) + \dfrac{1}{5} \sin(5\pi x) + \cdots$
$\qquad u_t(x, 0) = 0$

23. $u(x, 0) = 1$
$\qquad u_t(x, 0) = 0$

24. Computer Problem Use a computer to sketch the graph of the initial string position in Example 1. Then sketch the graph of the solution [Eq. (20)] for various times in the interval $0 \le t \le 2\pi$.

25. Journal Entry Summarize the results of this section.

9.5   SUPERPOSITION INTERPRETATION OF THE VIBRATING STRING

PURPOSE
To interpret the separation of variables solution of the vibrating string problem as a superposition of simple harmonic motions. We also solve a few specific vibrating string problems.

THE VIBRATING STRING AS A SUM OF SIMPLE VIBRATIONS

If there is one principle that lies behind most methods for solving linear partial differential equations, it is the **principle of superposition.** Roughly, this important principle states that the solution of a "linear problem" can be written as a sum of solutions of "simpler" linear problems. For example, a linear partial differential equation, written symbolically as $Lu = f$, can theoretically be solved by first rewriting the nonhomogeneous term f as a sum of simpler terms $f = \Sigma f_i$ and then solving each of the equations $Lu_i = f_i$ for u_i. Then, by summing these individual solutions $u = \Sigma u_i$, one obtains the solution u of the original equation $Lu = f$. Figure 9.22 illustrates this idea.

Figure 9.22
Most methods for finding solutions of linear differential equations are based on some form of the principle of superposition. Roughly stated, this principle says that the solution of a linear equation can be written as the sum of simpler solutions.

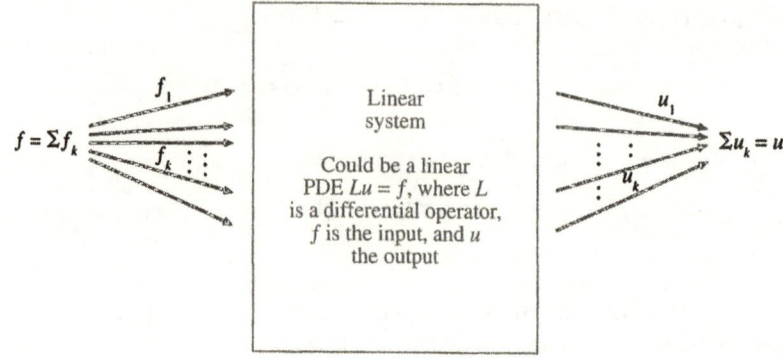

To understand how the motion of the vibrating string (defined in Section 9.4) can be interpreted as a sum of simple harmonic motions, first remember that a string, initially at rest $[u_t(x, 0) = 0]$, having an initial shape $u(x, 0) = b_n \sin(n\pi x)$ will have subsequent motion described by

$$u_n(x, t) = b_n \sin(n\pi x) \cos(n\pi\alpha t)$$

See Figure 9.23.

In other words, we have the "basic input-output" relationship

$$b_n \sin(n\pi x) \quad \rightarrow \quad b_n \sin(n\pi x) \cos(n\pi\alpha t) \tag{1}$$

$$\underbrace{\qquad\qquad}_{\uparrow} \qquad \underbrace{\qquad\qquad\qquad}_{\uparrow}$$

Initial shape   Resulting vibration

Figure 9.23
The fundamental vibrations of the vibrating string

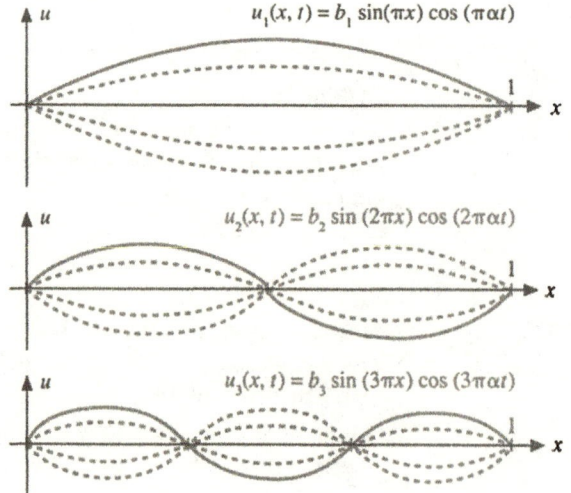

In general, the initial conditions of

$$u(x, 0) = b_1 \sin (\pi x) + b_2 \sin (2\pi x) + b_3 \sin (3\pi x) + \cdots$$

$$u_t(x, 0) = 0 \tag{2}$$

give rise to the general vibration described by

$$u(x, t) = b_1 \sin (\pi x) \cos (\pi \alpha t) + b_2 \sin (2\pi x) \cos (2\pi \alpha t) + \cdots \tag{3}$$

But this says that the general vibration is an infinite sum of simple vibrations given by $b_n \sin (n\pi x) \cos (n\pi \alpha t)$, each of which is the result of the corresponding component, $b_n \sin (n\pi x)$, of the initial condition $u(x, 0)$. We summarize these ideas.

Superposition Principle for the Vibrating String Problem
The solution of the vibrating string problem

$$\text{PDE:} \quad u_{tt} = \alpha^2 u_{xx} \qquad (0 < x < 1) \qquad (t > 0)$$

$$\text{BC:} \quad \begin{cases} u(0, t) = 0 \\ u(1, t) = 0 \end{cases} \qquad (t \geq 0) \tag{4}$$

$$\text{IC:} \quad \begin{cases} u(x, 0) = f(x) \\ u_t(x, 0) = 0 \end{cases} \qquad (0 \leq x \leq 1)$$

can be found (or interpreted) as follows.

1. Decompose the initial shape $u(x, 0) = f(x)$ into its basic components, getting

$$f(x) = b_1 \sin (\pi x) + b_2 \sin (2\pi x) + \cdots$$

2. Find the simple harmonic motion caused by each component, or

$$b_n \sin (n\pi x) \rightarrow b_n \sin (n\pi x) \cos (n\pi \alpha t)$$

3. Sum the simple harmonic motions, obtaining the overall vibration, getting

$$u(x, t) = b_1 \sin (\pi x) \cos (\pi \alpha t) + b_2 \sin (2\pi x) \cos (2\pi \alpha t) + \cdots$$

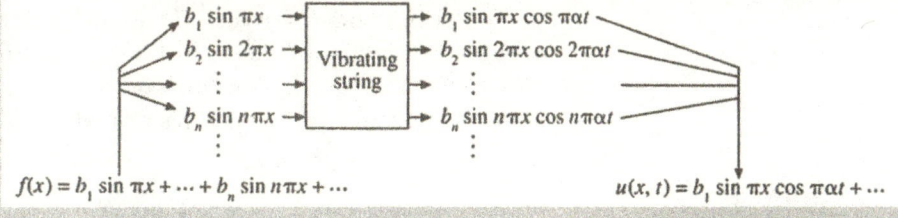

Example 1   Superposition of Solutions   Find the motion of a string that is initially at rest and whose initial position is given by

$$u(x, 0) = \sin (\pi x) + \frac{1}{3} \sin (3\pi x) + \frac{1}{5} \sin (5\pi x) \qquad (0 \leq x \leq 1) \tag{5}$$

Solution   Since $u(x, 0)$ is already expanded as a Fourier sine series, the Fourier coefficients are given by $b_1 = 1, b_3 = 1/3, b_5 = 1/5$, and all other b_n's are zero. Hence the solution is

$$u(x, t) = \sin \pi x \cos \pi \alpha t + \frac{1}{3} \sin 3\pi x \cos 3\pi \alpha t + \frac{1}{5} \sin 5\pi x \cos 5\pi \alpha t \quad (6)$$

∎

Note that the first term or **first harmonic** vibrates at the slowest frequency, the second term or **second harmonic** vibrates at three times the fundamental frequency, and the next term vibrates at five times the fundamental frequency.

Example 2   General Superposition   Find the motion of the vibrating string defined by*

$$\text{PDE:}\quad u_{tt} = u_{xx} \qquad (0 < x < 1) \qquad (t > 0)$$

$$\text{BC:}\quad \begin{cases} u(0, t) = 0 \\ u(1, t) = 0 \end{cases} \qquad (t > 0) \qquad\qquad (7)$$

$$\text{IC:}\quad \begin{cases} u(x, 0) = 1 \\ u_t(x, 0) = 0 \end{cases} \qquad (0 \le x \le 1)$$

Solution   Finding the Fourier series expansion of $u(x, 0) = 1$, we find

$$1 = \frac{4}{\pi}\left(\sin \pi x + \frac{1}{3} \sin 3\pi x + \frac{1}{5} \sin 5\pi x + \cdots \right) \qquad (8)$$

which means $b_{2n-1} = 4/(2n - 1)\pi$ and $b_{2n} = 0$ for $n = 1, 2, \dots$. Hence the solution is

$$u(x, t) = \frac{4}{\pi}\left(\sin \pi x \cos \pi t + \frac{1}{3} \sin 3\pi x \cos 3\pi t + \frac{1}{5} \sin 5\pi x \cos 5\pi t + \cdots \right) \qquad (9)$$

∎

Note that not every frequency appears in the overall vibration. (If it did, a plucked string would yield not a musical sound, but *noise*, or a jumble of all frequencies.[†]) The first term of the solution, called the **fundamental harmonic,** undergoes one oscillation every 2 units of time. The second term of the solution, or **second harmonic,** vibrates with a frequency three times faster than the fundamental harmonic; the **third harmonic** vibrates with a frequency five times the frequency of the first harmonic, and so on. The higher harmonics have frequencies that are *integer multiples* of the fundamental harmonic.

---

* This initial condition is not zero at the boundaries $x = 0$ and $x = 1$, which makes it physically "incompatible" with the *zero* boundary conditions $u(0, t) = 0$ and $u(1, t) = 0$. However, we settle with this initial condition for the sake of simplicity. The resolution of this problem involves concepts beyond the scope of this text.

[†] A good discussion on the mathematics of music can be found in *The World of Mathematics*, Volume 4: *Mathematics of Music*, by Sir James Jean (Simon & Schuster, New York, 1980).

Example 3 **The Plucked String** Find the resulting vibration of a string that is set in motion by pulling its midpoint aside the distance of $\epsilon/2$ and then releasing it. See Figure 9.24.

Solution The initial-boundary-value problem that defines this problem is

$$\text{PDE:}\quad u_{tt} = \alpha^2 u_{xx} \qquad (0 < x < 1) \qquad (t > 0)$$

$$\text{BC:}\quad \begin{cases} u(0, t) = 0 \\ u(1, t) = 0 \end{cases} \qquad (t \geq 0)$$

$$\text{IC:}\quad \begin{cases} u(x, 0) = f(x) \\ u_t(x, 0) = 0 \end{cases} \qquad (0 \leq x \leq 1) \tag{10}$$

where the plucked string is defined by

$$f(x) = \begin{cases} \epsilon x & (0 \leq x \leq \tfrac{1}{2}) \\ \epsilon(1 - x) & (\tfrac{1}{2} \leq x \leq 1) \end{cases}$$

The nth Fourier coefficient of $f(x)$ is

$$\begin{aligned} b_n &= 2 \int_0^1 f(x) \sin(n\pi x)\, dx \\ &= 2\epsilon \int_0^{1/2} x \sin(n\pi x)\, dx + 2\epsilon \int_{1/2}^1 (1 - x) \sin(n\pi x)\, dx \\ &= \frac{4\epsilon}{n^2\pi^2} \sin\left(\frac{n\pi}{2}\right) \\ &= \begin{cases} (-1)^{n+1} 4\epsilon/n^2\pi^2 & (n \text{ odd}) \\ 0 & (n \text{ even}) \end{cases} \end{aligned}$$

Hence the solution is

$$u(x, t) = \frac{4\epsilon}{\pi^2}\left\{ \cos(\pi t) \sin(\pi x) - \frac{1}{3^2} \cos(3\pi t) \sin(3x) + \cdots \right\} \qquad\blacksquare$$

Note that the period of the fundamental harmonic is 2, the period of the second harmonic is 2/3, and so on. The reader should think about the graph (see Problem 25) of this function for different values of time for one complete oscillation $0 \leq t \leq 2$.

Figure 9.24
Initial position of a plucked string

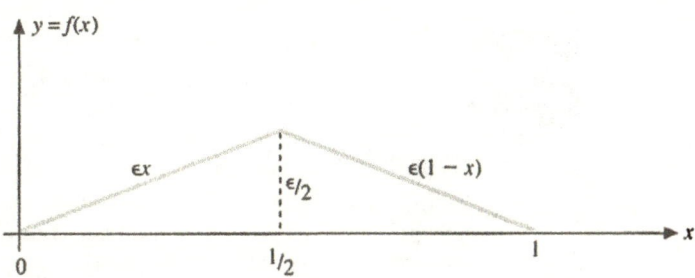

PROBLEMS: Section 9.5

For Problems 1–5, find the solution of the vibrating string problem with the given initial conditions. Assume the condition $u_t(x, 0) = 0$. Note that in these problems, one can simply pick off the Fourier sine coefficients of $u(x, 0)$ from the terms of the initial conditions. If you have access to a computer with a graphing package, plot the solution $u(x, t)$ over the closed interval $0 \le x \le 1$ for different values of t.

1. $u(x, 0) = \sin(\pi x)$
2. $u(x, 0) = \sin(\pi x) + \sin(2\pi x)$

3. $u(x, 0) = \sin(2\pi x) + \dfrac{1}{5}\sin(3\pi x)$

4. $u(x, 0) = \sin(\pi x) + \dfrac{1}{3^2}\sin(3\pi x) - \dfrac{1}{5^2}\sin(5\pi x) + \cdots$

5. $u(x, 0) = \sin(\pi x) - \dfrac{1}{2^2}\sin(2\pi x) + \dfrac{1}{3^2}\sin(3\pi x) - \cdots$

General Initial Position of the String

For Problems 6–8, find the solution of the vibrating string problem with $\alpha = 1$ and for given initial position $u(x, 0)$, where the Fourier sine series for each function $u(x, 0)$ is given. Assume that $u_t(x, 0) = 0$. If you have access to a computer with a graphing package, sketch the graphs of the solution $u(x, t)$ over the interval $0 \le x \le 1$ for different values of t.

6. $u(x, 0) = \begin{cases} 1 & (0 \le x < 0.5) \\ -1 & (0.5 \le x \le 1) \end{cases}$

$= \dfrac{4}{\pi}\left\{ \sin(2\pi x) + \dfrac{1}{3}\sin(6\pi x) + \dfrac{1}{5}\sin(10\pi x) + \cdots \right\}$

$(0 \le x \le 1)$

See Figure 9.25.

Figure 9.25
Square wave

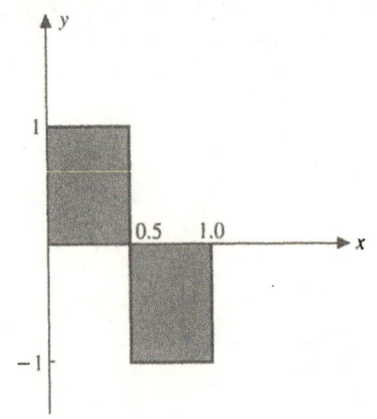

7. $u(x, 0) = x$

$= \dfrac{2}{\pi}\left\{ \sin(\pi x) - \dfrac{1}{2}\sin(2\pi x) + \dfrac{1}{3}\sin(3\pi x) + \cdots \right\}$

$(0 \le x \le 1)$

See Figure 9.26.

Figure 9.26
Linear wave

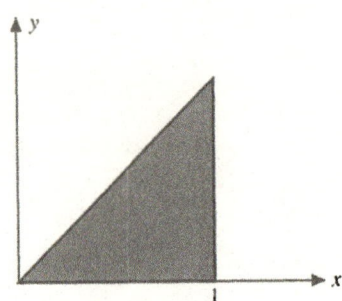

8. $u(x, 0) = \begin{cases} x & (0 \le x < 0.5) \\ 1 - x & (0.5 \le x \le 1) \end{cases}$

$= \dfrac{8}{\pi^2}\left\{ \sin(\pi x) - \dfrac{1}{9}\sin(3\pi x) + \dfrac{1}{25}\sin(5\pi x) + \cdots \right\}$

See Figure 9.27.

Figure 9.27
Triangular wave

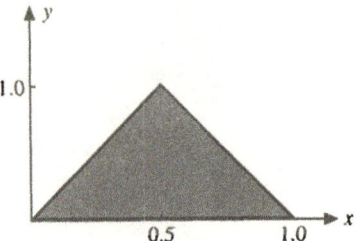

Initial Velocity Specified

The solution of the vibrating string problem with initial conditions $u(x, 0) = 0$ and $u_t(x, 0) = g(x)$ is

$$u(x, t) = \sum_{n=1}^{\infty} \left(\dfrac{b_n}{n\pi\alpha} \right) \sin(n\pi x) \sin(n\pi\alpha t) \qquad (11)$$

where b_n are the coefficients of the Fourier sine series of $g(x)$. This can be shown by substitution of the general solution (15) in Section 9.4 into the initial conditions and solving for the coefficients a_n and b_n. For Problems 9–12, use solution (11) to find the solution of the vibrating string with given initial conditions and $\alpha = 1$.

9. $u(x, 0) = 0 \quad u_t(x, 0) = \sin\pi x$
10. $u(x, 0) = 0 \quad u_t(x, 0) = \sin 3\pi x$

11. $u(x, 0) = 0 \quad u_t(x, 0) = \sin(\pi x) + \frac{1}{3}\sin(3\pi x)$

12. $u(x, 0) = 0 \quad u_t(x, 0) = 1$

13. More Superposition Show that the solution of the vibrating string problem with initial conditions $u(x, 0) = f(x)$ and $u_t(x, 0) = g(x)$ is the sum of the solutions of the vibrating string problem with the following initial conditions:

(A) $\begin{cases} u(x, 0) = f(x) \\ u_t(x, 0) = 0 \end{cases}$ (B) $\begin{cases} u(x, 0) = 0 \\ u_t(x, 0) = g(x) \end{cases}$

Solving the General Vibrating String Problem

For Problems 14–18, use the solution (11) for given string velocity and the superposition result from Problem 13 to find the solution of the vibrating string problem with given initial conditions and $\alpha = 1$.

14. $u(x, 0) = \sin(\pi x)$
$u_t(x, 0) = \sin(\pi x)$

15. $u(x, 0) = \sin(2\pi x)$
$u_t(x, 0) = \sin(2\pi x)$

16. $u(x, 0) = \sin(\pi x) + \frac{1}{3}\sin(3\pi x)$

$u_t(x, 0) = \frac{1}{2}\sin(4\pi x)$

17. $u(x, 0) = \sin(\pi x) + \frac{1}{3}\sin(3\pi x) + \cdots$

$u_t(x, 0) = 2\sin(3\pi x)$

18. $u(x, 0) = 1$
$u_t(x, 0) = 1$

Eigenvalue Problems

*An important class of problems known as **eigenvalue problems** arise in solving partial differential equations using the method of separation of variables. A typical eigenvalue problem would be to find all real values of λ, called **eigenvalues,** such that the boundary-value problem**

$$y'' + \lambda y = 0 \quad (0 < x < \pi)$$
$$y(0) = 0 \quad\quad\quad (12)$$
$$y(\pi) = 0$$

*has a nonzero solution y. The nonzero solution(s) y, if there are any, are called **eigenfunctions**. For Problems 19–22, find the eigenvalues and eigenvectors of the given boundary-value problem.*

19. $y'' + \lambda y = 0 \quad\quad y(0) = y(\pi) = 0$
20. $y'' + \lambda y = 0 \quad\quad y'(0) = y(\pi) = 0$
21. $y'' + \lambda y = 0 \quad\quad y(-\pi) = y(\pi), y'(-\pi) = y'(\pi)$
22. $y'' + 2y' + \lambda y = 0 \quad y(0) = y(\pi) = 0$

23. Computer Problem Use a computer to sketch the graphs $u(x, t)$ of the solution of the vibrating string problem described in Example 1 for different values of t in the interval $[0, 2]$.

24. Computer Problem Use a computer to sketch the graphs of the solution of the vibrating string problem from Example 2 for different values of t in the interval $[0, 2]$.

25. Computer Problem Use a computer to sketch the graph of the plucked string described in Example 3 for different values of t in the interval $[0, 2]$.

26. Journal Entry Summarize the principle of superposition.

9.6 THE HEAT EQUATION AND SEPARATION OF VARIABLES

PURPOSE
To introduce the one-dimensional heat equation, $u_t = \alpha^2 u_{xx}$ and solve one of the fundamental problems of heat flow by the method of separation of variables. Again, we will see that the solution can be interpreted as a superposition of simple solutions.

Although we introduce the new topic of one-dimensional heat flow in this section, the reader will get a sense of *déja vu*. The steps used here in solving the heat equation are basically the same steps that we used when we solved the vibrating string problem. For that reason the discussion is streamlined, and some of the details are left for the reader.

* We have seen this eigenvalue problem as a result of applying separation of variables to the vibrating string problem.

HEAT FLOW IN A ROD

The modern theory of heat conduction dates to the early 1800's and the work of Joseph Fourier.* For our part we will solve the fundamental problem of heat flow: heat flow in an insulated rod whose ends are kept fixed at the constant temperature of 0°C. We imagine a thin wire or rod in which heat flows only back and forth through the rod and not across the lateral boundary. See Figure 9.28.

Figure 9.28
A thin insulated rod or wire in which heat flows only back and forth through the rod and not across the lateral boundary

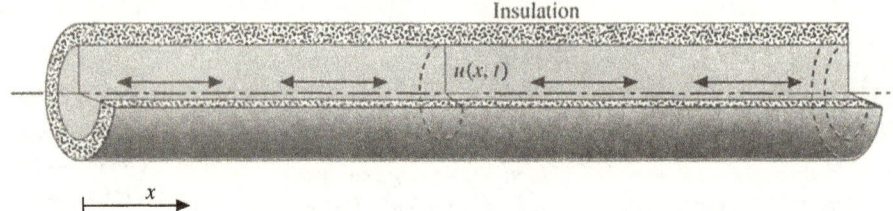

We denote the temperature along the rod by $u(x, t)$, where x measures the distance from one end of the rod and t represents time. To keep extraneous mathematical notation to a minimum, we take the length of the rod to be 1. The problem is to find the future temperature of the rod, given the initial temperature of the rod and the fact that the ends of the rod are always kept at a fixed temperature of 0°C. We summarize the above ideas by the following initial-boundary-value problem:[†]

$$\text{PDE:} \quad u_t = \alpha^2 u_{xx} \qquad (0 < x < 1) \qquad (t > 0)$$

$$\text{BC:} \quad \begin{cases} u(0, t) = 0 \\ u(1, t) = 0 \end{cases} \qquad (t \geq 0) \tag{1}$$

$$\text{IC:} \quad u(x, 0) = f(x) \qquad (0 \leq x \leq 1)$$

The partial differential equation, $u_t = \alpha^2 u_{xx}$, called the (one-dimensional) heat equation is a *parabolic* partial differential equation. The parameter α^2 is called the **thermal diffusivity,** and it depends on the material from which the rod is made. Some values of α^2 are listed in Table 9.3.

Table 9.3
Typical Thermal Diffusivity for Common Materials

Material	α^2 (ft²/sec)
Silver	0.00184
Copper	0.00115
Aluminum	0.00093
Granite	0.0000118
Water	0.0000016

Heat), which gave the foundation for today's theory of heat conduction. It was in connection with these studies that the subject of Fourier series was born.

[†] For a derivation of the heat equation, see the text *Partial Differential Equations for Scientists and Engineers,* by S. J. Farlow (Dover Publications, New York, 1993).

In comparing the heat conduction problem with the vibrating string problem from Section 9.4, it is instructive to note that the heat equation contains only a *first derivative* u_t with respect to time, which contrasts to the wave equation that contains the second derivative u_{tt}. This means that the heat equation requires only *one* initial temperature condition $u(x, 0) = f(x)$ to determine the future temperature $u(x, t)$.

Note too that the two boundary conditions $u(0, t) = u(1, t) = 0$ in the heat flow problem (1) are the same as the boundary conditions in the string problem, although in the context of heat flow we are keeping the ends of the rod fixed at the constant temperature of 0°C, whereas in the string problem we were keeping the ends of the string in a fixed position. Before solving the one-dimensional heat flow problem (1), we give an intuitive interpretation of the heat equation.

Intuitive Interpretation of the Heat Equation

The heat equation $u_t = \alpha^2 u_{xx}$ expresses the physical fact that the *rate of* change in temperature u_t at (x, t) is proportional to the second derivative of u with respect t x. (If $u_t > 0$ at a point, the temperature is increasing; whereas if $u_t < 0$, the temperature is decreasing.) We have seen that u_{xx} can be written as the difference between $u(x, t)$ and the average value of u at $x - 2h$ and $x + 2h$.

Hence the heat equation implies that when the temperature $u(x, t)$ at a point is *less* than the average temperature at its neighboring points, the temperature at the point is increasing. On the other hand, if the temperature at a point is *greater* than the average temperature at its neighboring points, the temperature is falling. What's more, the rate of change in temperature is proportional to this difference, the proportionality constant being α^2.

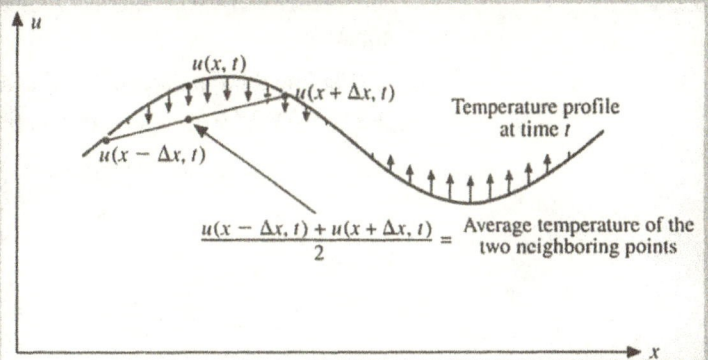

SOLVING THE HEAT EQUATION BY SEPARATION OF VARIABLES

The following steps describe the method of separation of variables applied to the heat flow problem (1).

Step 1   (Finding Product Solutions). We seek solutions of the special separable form $u(x, t) = X(x)T(t)$ by substituting this expression into the heat equation, getting

$$X(x)T'(t) = \alpha^2 X''(x)T(t) \tag{2}$$

Dividing each side of this equation by $\alpha^2 X(x)T(t)$, we obtain the separated form

$$\frac{T'(t)}{\alpha^2 T(t)} = \frac{X''(x)}{X(x)} \tag{3}$$

Inasmuch as x and t are independent of each other, each side of Eq. (3) must be a constant, say k (or minus k). Hence we write

$$\frac{T'(t)}{\alpha^2 T(t)} = \frac{X''(x)}{X(x)} = -k \tag{4}$$

In effect, we have replaced the heat equation by the two ordinary differential equations

$$X'' + kX = 0 \tag{5a}$$
$$T' + k\alpha^2 T = 0 \tag{5b}$$

These two equations can be solved for any value of k, and their product will be a solution of the heat equation. As in the analysis of the wave equation, we have three cases:

$$k < 0 \, \triangleright \, \begin{cases} X(x) = Ae^{\sqrt{-k}\,x} + Be^{-\sqrt{-k}\,x} \\ T(t) = Ce^{-k\alpha^2 t} \end{cases} \tag{6a}$$

$$k = 0 \, \triangleright \, \begin{cases} X(x) = Ax + B \\ T(t) = C \end{cases} \tag{6b}$$

$$k > 0 \, \triangleright \, \begin{cases} X(x) = A\cos\sqrt{k}x + B\sin\sqrt{k}x \\ T(t) = Ce^{-k\alpha^2 t} \end{cases} \tag{6c}$$

The reader can verify that each of the above products $X(x)\,T(t)$ satisfies the heat equation.

Step 2   (Finding the Fundamental Solutions) In order that the above product solutions $u(x, t) = X(x)\,T(t)$ satisfy the boundary conditions, we must have

$$u(0, t) = X(0)T(t) = 0 \tag{7a}$$
$$u(1, t) = X(1)T(t) = 0 \tag{7b}$$

or

$$X(0) = 0 \tag{8a}$$
$$X(1) = 0 \tag{8b}$$

Substituting the solutions $X(x)$ from Eqs. (6) into these boundary conditions gives

$$(k < 0) \quad \begin{cases} X(0) = A + B = 0 \\ X(1) = Ae^{\sqrt{-k}} + Be^{-\sqrt{-k}} = 0 \end{cases} \quad \triangleright X(x) \equiv 0 \quad \triangleright u(x, t) \equiv 0 \tag{9a}$$

$$(k = 0) \quad \begin{cases} X(0) = A(0) + B = 0 \\ X(1) = A + B = 0 \end{cases} \quad \triangleright X(x) \equiv 0 \quad \triangleright u(x, t) \equiv 0 \tag{9b}$$

$$(k > 0) \quad \begin{cases} X(0) = A\cos(0) + B\sin(0) = 0 \\ X(1) = A\cos\sqrt{k} + B\sin\sqrt{k} = 0 \end{cases} \quad \triangleright \lambda = \pm\, n\pi \tag{9c}$$

Hence the only nonzero solutions we find are the fundamental solutions

$$u_n(x, t) = b_n e^{-(n\pi\alpha)^2 t} \sin (n\pi x) \qquad (n = 1, 2, ...) \tag{10}$$

where the coefficients b_n are arbitrary constants. These fundamental solutions, each of which satisfies the heat equation and the boundary conditions, are the building blocks for the final solution. See Figure 9.29.

Figure 9.29

The fundamental solutions

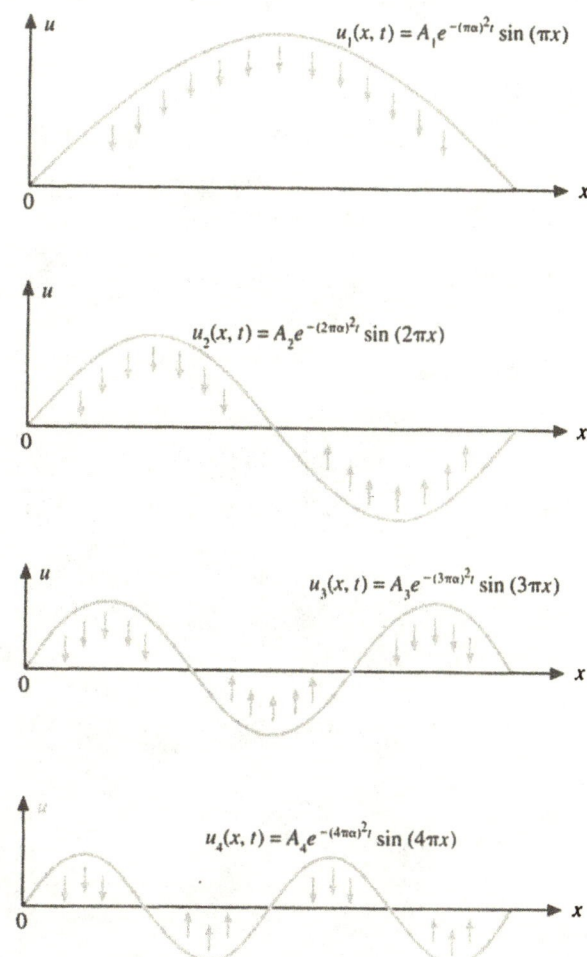

Step 3 (Summing the Fundamental Solutions). It now can be shown that any *sum** of fundamental solutions

$$u(x, t) = \sum_{n=1}^{\infty} b_n e^{-(n\pi\alpha)^2 t} \sin (n\pi x) \tag{11}$$

---

* Although it can easily be verified that any *finite sum* of fundamental solutions satisfies the heat equation, it takes a little more rigor to prove that an *infinite sum* of fundamental solutions satisfies the heat equation. The proof requires some knowledge of uniform convergence, and the proof is left to more advanced texts in partial differential equations.

also satisfies the heat equation and the boundary conditions for *any* constants b_n. Hence the final step is to determine the coefficients b_n so that Eq. (11) satisfies the initial condition $u(x, 0) = f(x)$. Substituting Eq. (11) into the initial condition gives

$$u(x, 0) = \sum_{n=1}^{\infty} b_n \sin (n\pi x) = f(x) \tag{12}$$

But this equation says that the coefficients b_n are nothing more than the coefficients in the Fourier sine series of $f(x)$. Hence

$$b_n = 2 \int_0^1 f(x) \sin (n\pi x) \, dx \qquad (n = 1, 2, ...) \tag{13}$$

In other words, the solution is Eq. (11), where the coefficients are given by (13).

> *Note: In plain English,* if the reader will look carefully at the solution (11) and then the Fourier sine expansion (12) of the initial condition, one can deduce the "rule of thumb" methodology for finding the solution: Find the Fourier sine series for the initial temperature, then insert the factor $e^{-(n\pi\alpha)^2 t}$ in the nth term of this expansion.
>
> We now summarize the previous results.

Solution of the One-Dimensional Heat Flow Problem

The separation of variables solution of the one-dimensional heat flow problem

$$\text{PDE:} \quad u_t = \alpha^2 u_{xx} \qquad (0 < x < 1) \qquad (t > 0)$$

$$\text{BC:} \quad \begin{cases} u(0, t) = 0 \\ u(1, t) = 0 \end{cases} \qquad (t \geq 0) \tag{14}$$

$$\text{IC:} \quad u(x, 0) = f(x) \qquad (0 \leq x \leq 1)$$

is the infinite series

$$u(x, t) = \sum_{n=1}^{\infty} b_n e^{-(n\pi\alpha)^2 t} \sin (n\pi x) \tag{15}$$

where

$$b_n = 2 \int_0^1 f(x) \sin (n\pi x) \, dx \tag{16}$$

Note: In plain English, from Eq. (15) the procedure for finding $u(x, t)$ is to expand the initial temperature $f(x)$ as a Fourier sine series

$$f(x) = b_1 \sin (\pi x) + b_2 \sin (2\pi x) + b_3 \sin (3\pi x) + \cdots$$

where $b_n = 2 \int_0^1 f(x) \sin (n\pi x) \, dx$, and then insert the factor $e^{-(n\pi\alpha)^2 t}$ in the nth term.

Example 1 **Solution of the Heat Equation** Find the solution of the one-dimensional heat flow problem:

$$
\begin{aligned}
&\text{PDE:} && u_t = u_{xx} && (0 < x < 1) \quad (t > 0)\\[4pt]
&\text{BC:} && \begin{cases} u(0, t) = 0 \\ u(1, t) = 0 \end{cases} && (t \ge 0) && (17)\\[4pt]
&\text{IC:} && u(x, 0) = \sin(\pi x) + \frac{1}{3}\sin(3\pi x) && (0 \le x \le 1)
\end{aligned}
$$

Solution   Since the initial temperature already has the form of a Fourier sine series, we simply insert the factor $e^{-(n\pi)^2 t}$ in each term, getting

$$
u(x, t) = e^{-\pi^2 t}\sin(\pi x) + \frac{1}{3}e^{-(3\pi)^2 t}\sin(3\pi x) \tag{18}
$$

Note that in Eq. (18) the second term decays to zero much faster than the first term, and after a long period of time the temperature "profile" over $0 \le x \le 1$ will look more and more like the graph of $\sin(\pi x)$. See Figure 9.30. ■

Figure 9.30
One-dimensional temperature flow

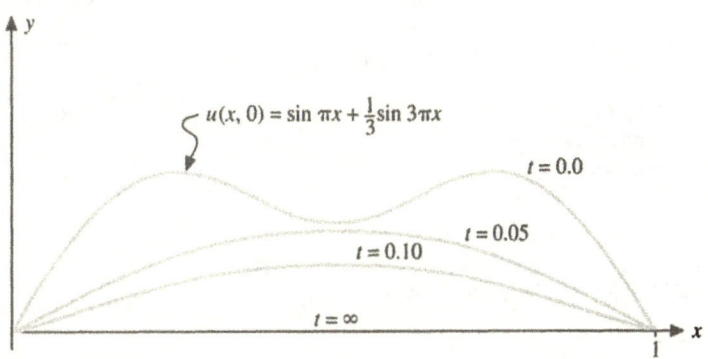

Example 2   Heat Conduction Problem   Solve the heat conduction problem

$$
\begin{aligned}
&\text{PDE:} && u_t = \alpha^2 u_{xx} && (0 < x < 1) \quad (t > 0)\\[4pt]
&\text{BC:} && \begin{cases} u(0, t) = 0 \\ u(1, t) = 0 \end{cases} && (t \ge 0) && (19)\\[4pt]
&\text{IC:} && u(x, 0) = 1 && (0 \le x \le 1)
\end{aligned}
$$

Solution   We expand the initial temperature $u(x, 0) = 1$ as a Fourier sine series, getting

$$
1 = \frac{4}{\pi}\left(\sin \pi x + \frac{1}{3}\sin(3\pi x) + \frac{1}{5}\sin(5\pi x) + \cdots\right) \tag{20}
$$

where the coefficients are $b_{2n-1} = 4/(2n - 1)\pi$, $b_{2n} = 0$ for $n = 1, 2, \ldots$. Hence the solution is

$$
u(x, t) = \frac{4}{\pi}\left(e^{-(\pi)^2 t}\sin(\pi x) + \frac{1}{3}e^{-(3\pi)^2 t}\sin(3\pi x) + \frac{1}{5}e^{-(5\pi)^2 t}\sin(5\pi x) + \cdots\right) \tag{21}
$$

This solution is shown for several values of time* in Figure 9.31.

■

Figure 9.31
One-dimensional temperature flow with the ends fixed at zero

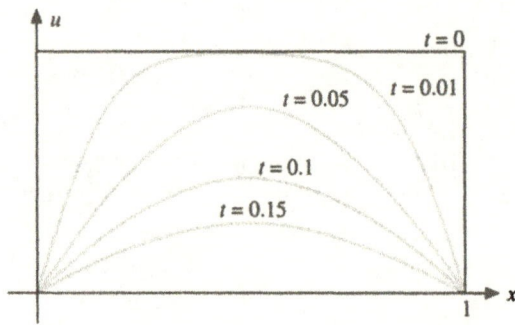

PROBLEMS: Section 9.6

For Problems 1–7, find the solution of the one-dimensional heat flow problem discussed in this section with given initial condition and $\alpha = 1$. If you have access to a computer with a graphing package, sketch the graph of the solution for different values of t.

1. $u(x, 0) = \sin (\pi x)$
2. $u(x, 0) = 3 \sin (2\pi x)$
3. $u(x, 0) = 2 \sin (3\pi x)$
4. $u(x, 0) = \sin (\pi x) + \dfrac{1}{4} \sin (2\pi x)$
5. $u(x, 0) = \sin (\pi x) - \dfrac{1}{2} \sin (2\pi x) + \dfrac{1}{3} \sin (3\pi x)$
6. $u(x, 0) = 1$
7. $u(x, 0) = x$

8. **Transformation to Homogeneous Boundary Conditions**
Suppose the heat flow problem (14) has nonhomogeneous boundary conditions $u(0, t) = T_1$ and $u(1, t) = T_2$, reflecting the fact that the ends of the rod are kept at temperatures T_1 and T_2, instead of at 0°C. Show that by letting

$$u(x, t) = (T_2 - T_1)x + T_1 + w(x, t) \qquad (22)$$

the function $w(x, t)$ satisfies the initial-boundary-value problem with homogeneous boundary conditions:

PDE: $w_t = \alpha^2 w_{xx} \qquad (0 < x < 1) \qquad (t > 0)$

BC: $\begin{cases} w(0, t) = 0 \\ w(1, t) = 0 \end{cases} \qquad (t \geq 0)$

IC: $w(x, 0) = f(x) - (T_2 - T_1)x - T_1$
$\qquad (0 \leq x \leq 1)$

9. **Problem with Nonhomogeneous BC** Use the result from Problem 8 to find the solution of

PDE: $u_t = u_{xx} \qquad (0 < x < 1) \qquad (t > 0)$

BC: $\begin{cases} u(0, t) = 1 \\ u(1, t) = 2 \end{cases} \qquad (t \geq 0)$

IC: $u(x, 0) = x + 1 + \sin (\pi x) \qquad (0 \leq x \leq 1)$

Hint: This problem is simpler than it looks.

Rod with Insulated Ends

If the ends of the rod in the one-dimensional heat flow problem (14) are insulated instead of fixed at zero, the boundary conditions are $u_x(0, t) = 0$ and $u_x(1, t) = 0$. For Problems 10–14, first use the method of separation of variables to show that the solution to (14) with insulated boundaries is

$$u(x, t) = \frac{a_0}{2} + \sum_{n=1}^{\infty} a_n e^{-(n\pi\alpha)^2 t} \cos (n\pi x) \qquad (23)$$

where the constants a_n are the coefficients in the Fourier cosine series of $f(x)$:

$$f(x) = \frac{a_0}{2} + \sum_{n=1}^{\infty} a_n \cos (n\pi x)$$

that is,

$$a_n = 2 \int_0^1 f(x) \cos (n\pi x)\, dx \qquad (n = 0, 1, ...)$$

---

* The value of α is directly related to the time scale. If t represents time in seconds or minutes, the value of α will normally be much smaller (such as $\alpha = 0.001$). We simply pick $\alpha = 1$ for demonstration to simplify matters; normally, if $\alpha = 1$, we would be talking about time measured in centuries or something.

Then use this solution to find the solution of the given initial-boundary-value problem. Also, find the steady state temperature (after a long time) for each problem.

10. $u(x, 0) = 1$

11. $u(x, 0) = \cos(\pi x)$

12. $u(x, 0) = 1 + \cos(\pi x) + \frac{1}{2} \cos(3\pi x)$

13. $u(x, 0) = \frac{1}{2} - \frac{4}{\pi^2} \left\{ \cos(\pi x) + \frac{1}{3^2} \cos(3\pi x) \right.$
$$\left. + \frac{1}{5^2} \cos(5\pi x) + \cdots \right\}$$

14. $u(x, 0) = x$

15. Heat Flow on $0 \le x \le L$ In this section we solved the one-dimensional heat flow problem on the *unit interval* $0 < x < 1$. It is possible to find the solution to the same problem on a general interval $0 < x < L$ by simply replacing $x \to x/L$ and $t \to t/L^2$ in the solution (15) in the text. To understand why this strategy works, consider the following two substitutions.

(a) Introduce the new "space" variable $\xi = x/L$ and show that this substitution transforms the original heat equation $u_t = \alpha^2 u_{xx}$ to

$$u_t = \frac{\alpha^2}{L^2} u_{\xi\xi} \qquad (0 < \xi < 1) \qquad (t > 0) \quad (24)$$

(b) Introduce the new "time" variable $\tau = (1/L^2)t$ and show that this substitution transforms Eq. (24) to

$$u_\tau = \alpha^2 u_{\xi\xi} \qquad (0 < \xi < 1) \qquad (\tau > 0) \quad (25)$$

But this equation is simply the heat equation written in terms of the new variables ξ and τ. Hence we solve this equation (as we have done in the text) and then write the solution in terms of x and t.

(c) Use the substitutions $x \to x/L$ and $t \to t/L^2$ to find the solution of the one-dimensional heat flow problem on the interval $[0, 2]$ with initial condition

$$u(x, 0) = \sin \pi x + \frac{1}{3} \sin 3\pi x$$

16. Simulation of the Heat Flow Problem An insulated metal rod whose initial temperature is graphed in Figure 9.32 has its endpoints fixed at 0°C. Use the fact that each term in the infinite series solution Eq. (15) of the heat flow problem damps to zero faster than the preceding one to sketch rough graphs of the solution at different values of time. What is the steady state solution, and what will the solution "look like" just before reaching steady state?

Typical One-Dimensional Heat Flow Problems

17. Finding the Initial-Boundary-Value Problem State the initial-boundary-value problem that describes the temperature of an insulated copper rod of length 10 inches if the rod has an initial temperature of 50°C and its ends are held fixed at 20°C and 70°C.

18. Radiation Boundary Conditions A more general type of boundary condition than has been discussed in this section occurs when heat flow through one (or both) of the ends of the rod is proportional to the temperature. Figure 9.33 illustrates this type of problem in which we fix the top of a rod at $u(x, 0) = 0$ and immerse the bottom of the rod in a solution of water fixed at the same temperature of zero. (Zero refers to some reference temperature.) By using Newton's law of cooling the boundary conditions can be shown to be

$$\text{BC:} \quad \begin{cases} u(0, t) = 0 \\ u_x(1, t) - hu(1, t) = 0 \end{cases} \qquad (t \ge 0)$$

where $h > 0$ is a constant that depends on the rate of heat flow across the boundaries. Find the steady state temperature of the insulated rod shown in Figure 9.33 if $h = 1$.

Figure 9.32
Simulation of heat flow

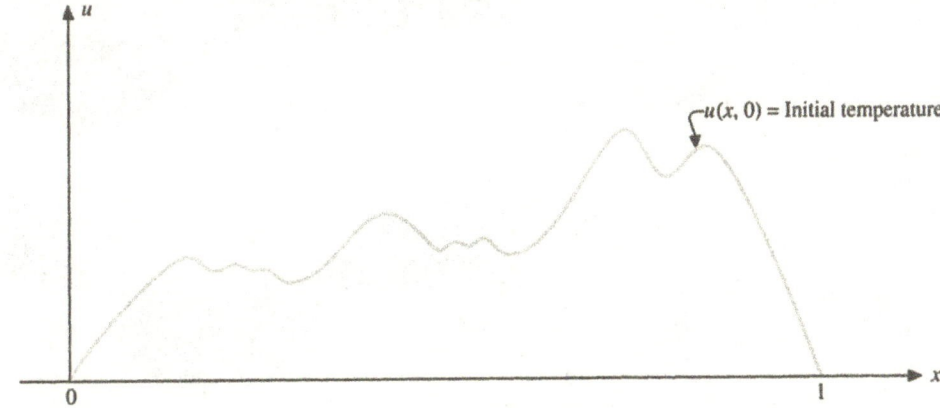

$u(x, 0) =$ Initial temperature

Figure 9.33
Here the temperature is fixed at 0 at the top ($x = 0$) and the rod immersed in water of temperature 0 at the bottom. The boundary condition at the bottom doesn't say that the temperature *is* 0 at the bottom, but heat will flow across that boundary depending on whether the temperature of the rod at the boundary is higher or lower than 0. This is the type of boundary condition that occurs at the surface of a window.

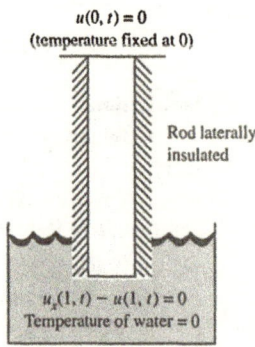

$u(0, t) = 0$
(temperature fixed at 0)

Rod laterally insulated

$u_x(1, t) - u(1, t) = 0$
Temperature of water = 0

Figure 9.34
Graphs of the first four fundamental solutions for different values of t. These are the shapes of sound waves formed when air is blown through a hollow tube when one end ($x = 0$) of the rod is closed and the other end ($x = 1$) is open.

19. **Mixed Boundary Conditions** Consider an insulated metal rod in which one boundary condition is $u(0, t) = 0$ (temperature fixed at zero) and the other boundary condition is $u_x(1, t) = 0$ (insulated boundary). Show that the fundamental solutions (solution of the heat equation and the boundary conditions) are

$$u_n(x, t) = e^{-(2n-1)^2 \pi^2 \alpha^2 t/4} \sin\left(\frac{(2n-1)\pi x}{2}\right) \quad (26)$$

for $n = 1, 2, \ldots$. The general shapes of a few of these fundamental solutions are shown in Figure 9.34 for different values of t.

20. **Computer Problem** Use a computer to sketch the graphs of some of the solutions of Problems 1–7 over certain time intervals.

21. **Journal Entry** Summarize why the solution of the temperature in a rod can be interpreted as a superposition.

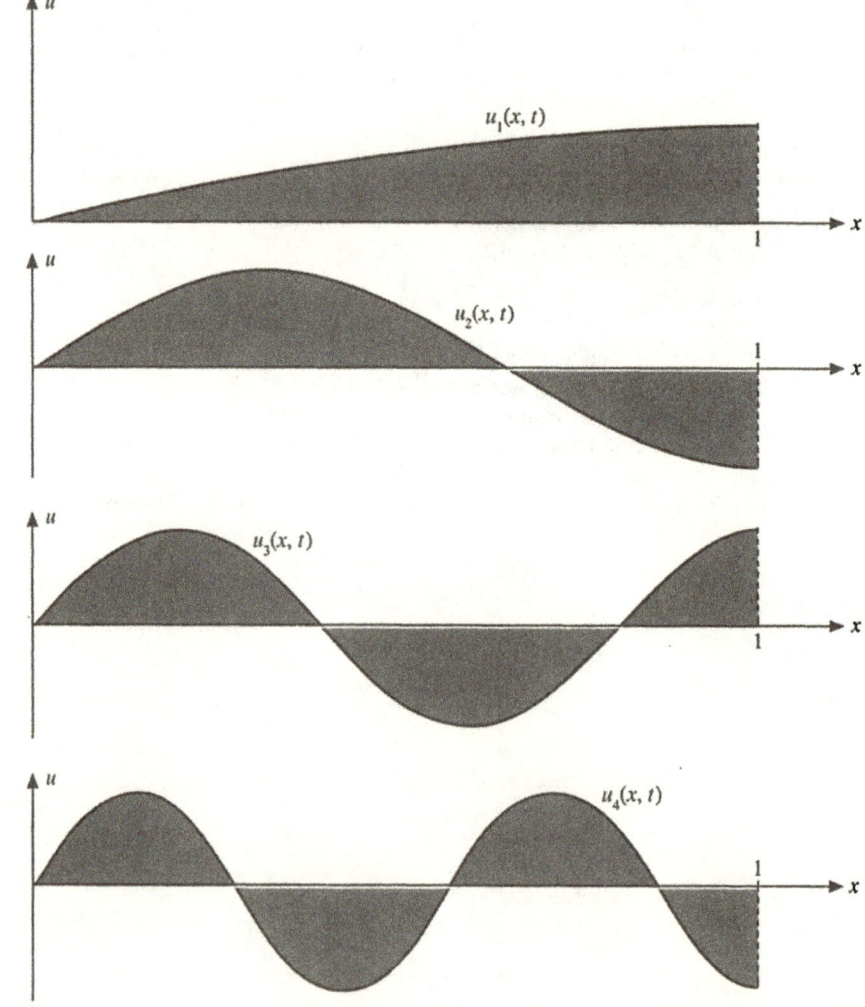

9.7   LAPLACE'S EQUATION INSIDE A CIRCLE

PURPOSE

To introduce the Laplacian in two and three dimensions and give an intuitive interpretation of this important concept. We then solve Laplace's equation in the interior of a circle when the solution is given on the boundary of the circle. We again use separation of variables but with some new twists.

Thus far in our introduction to partial differential equations we have introduced the most famous hyperbolic equation (wave equation) and the most famous parabolic equation (heat equation). We now introduce the most famous *elliptic* equation, Laplace's equation. In our study of the wave and heat equations we analyzed phenomena that changed with time. In the study of elliptic equations, we study phenomena that do not depend on time, but only on space—that is, *steady state* phenomena.

THE LAPLACIAN

We have seen that the second derivative u_{xx} plays an important role in the heat and wave equations. The reason for its importance lies in the fact that u_{xx} is proportional to the difference between the value of the function u at a point and the average of the function at two neighboring points. The second partial derivative u_{xx} is sometimes called the *one-dimensional Laplacian,* and its generalization to higher dimensions leads to one of the most important concepts in mathematical physics. In two and three dimensions the **Laplacian** $\nabla^2 u$ is defined as*

$$\nabla^2 u = \frac{\partial^2 u}{\partial x^2} + \frac{\partial^2 u}{\partial y^2} \qquad \text{(two-dimensional Laplacian)} \qquad (1a)$$

$$\nabla^2 u = \frac{\partial^2 u}{\partial x^2} + \frac{\partial^2 u}{\partial y^2} + \frac{\partial^2 u}{\partial z^2} \qquad \text{(three-dimensional Laplacian)} \qquad (1b)$$

The question is, why should the sum of the second-order partial derivatives be so important in the mathematical description of the universe? The answer is the same as in one dimension: It compares the value of a function at a point with the average of the function at neighboring points. *Roughly, the Laplacian, $\nabla^2 u$, measures the difference between the local and average values of u in an infinitesimal neighborhood of a point.* This interpretation hints at why the Laplacian occurs in the wave and heat equations. The change or motion of these types of physical systems often depends on just such a quantity.

---

* More accurately, what we give here is the Laplacian in *Cartesian coordinates.* The Laplacian can also be written in other coordinate systems.

Intuitive Interpretation of the Laplacian $\nabla^2 u$

The Laplacian $\nabla^2 u(x_0, y_0)$ of a function $u(x, y)$ at a point (x_0, y_0) measures the *discrepancy* between $u(x_0, y_0)$ and the average of $u(x, y)$ over a "small" circle* around (x_0, y_0). Basically, we are interested whether the Laplacian is positive, zero, or negative. In particular, we have

$(\nabla^2 u > 0)$:   If $\nabla^2 u > 0$ at a point (x_0, y_0), then $u(x_0, y_0)$ is *less* than the average of u on a small circle around (x_0, y_0).

$(\nabla^2 u = 0)$:   If $\nabla^2 u = 0$ at a point (x_0, y_0), then $u(x_0, y_0)$ is *equal* to the average of u on a small circle around (x_0, y_0).

$(\nabla^2 u < 0)$:   If $\nabla^2 u < 0$ at a point (x_0, y_0) then $u(x_0, y_0)$ is *greater* than the average of u on a small circle around (x_0, y_0).

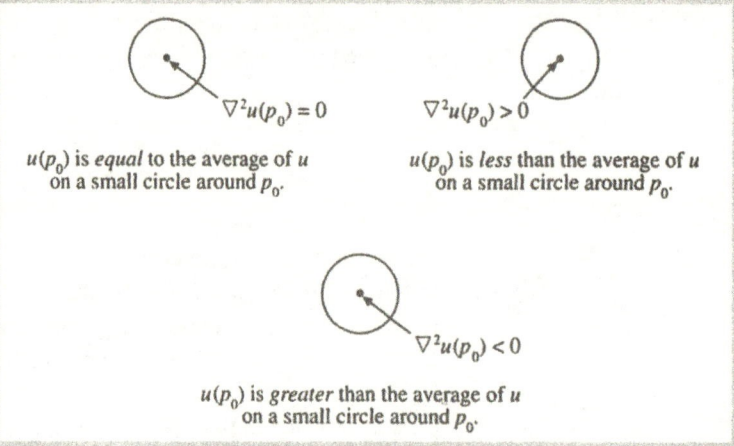

$\nabla^2 u(p_0) = 0$

$u(p_0)$ is *equal* to the average of u on a small circle around p_0.

$\nabla^2 u(p_0) > 0$

$u(p_0)$ is *less* than the average of u on a small circle around p_0.

$\nabla^2 u(p_0) < 0$

$u(p_0)$ is *greater* than the average of u on a small circle around p_0.

In three dimensions the Laplacian $\nabla^2 u$ of a function $u(x, y, z)$ of three variables has a similar interpretation with circles being replaced by spheres.

Note: You might think of having some kind of meter, like a geiger counter, but in this case it would be a "Laplacian meter." The meter would measure the Laplacian of a given function $u(x, y, z)$. You could go around measuring the *Laplacian* of the function (say, the temperature, electric field, and so on) at different points in space. If you were measuring the Laplacian in a temperature field, then a negative Laplacian at a point would mean that the temperature at the given point would be greater than the average temperature in a small sphere around the point. Hence the temperature at that point would decrease. See Figure 9.35.

* More precisely, the Laplacian measures the *limiting* difference between the average of $u(x, y)$ on a circle surrounding (x_0, y_0) and $u(x_0, y_0)$ as the radius of the circle approaches zero.

Figure 9.35
The sign of the Laplacian of a function at a point indicates whether the function is *less than, equal to,* or *greater than* the average of the function at neighboring points.

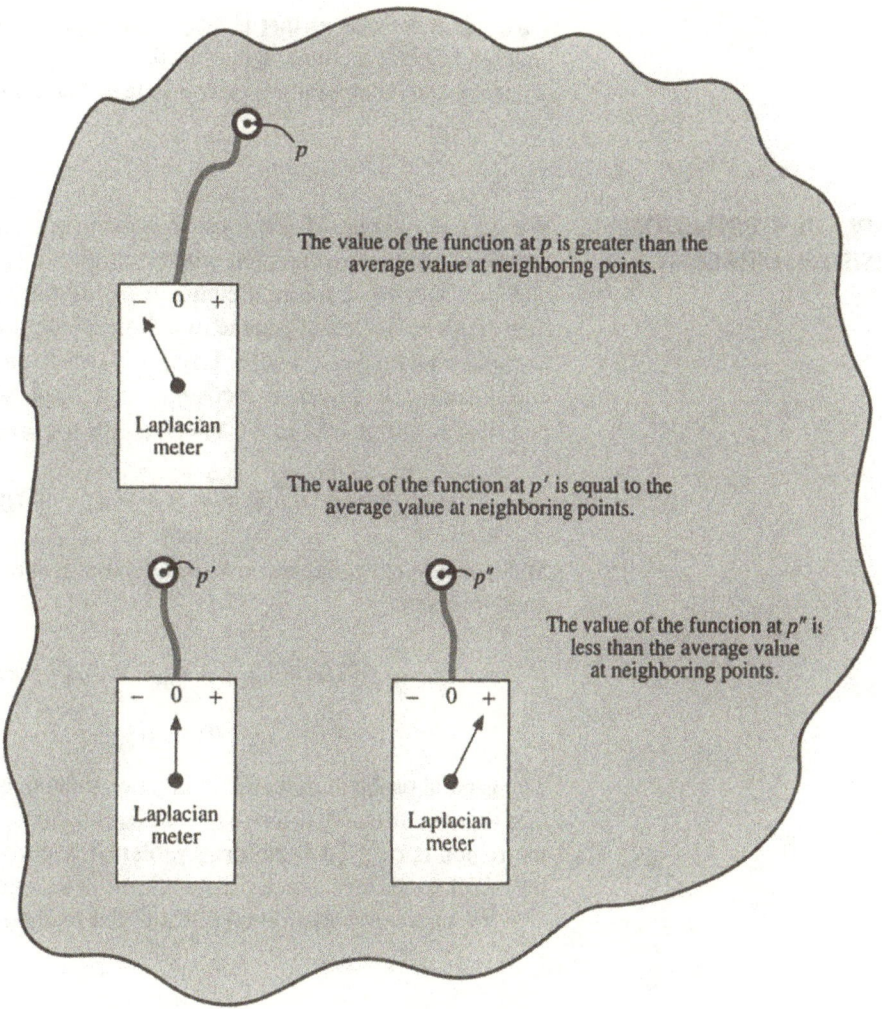

The value of the function at p is greater than the average value at neighboring points.

Laplacian meter

The value of the function at p' is equal to the average value at neighboring points.

The value of the function at p'' is less than the average value at neighboring points.

Laplacian meter

Laplacian meter

LAPLACE'S EQUATION

One of the most important partial differential equations in mathematical physics is the two-dimensional **Laplace's equation** in Cartesian coordinates:

$$\nabla^2 u \equiv u_{xx} + u_{yy} = 0 \tag{2}$$

defined over some region of the xy-plane. The solution $u(x, y)$ of Laplace's equation (2) defines a function $u = u(x, y)$ in which the value of u at any point is always equal to the *average* of u on a small circle around the point. A solution of Laplace's equation defines a surface in which its height at every point is equal to the average height over a ''small'' region around the point. You have seen surfaces like this many times. A stretched rubber membrane describes a function $u = u(x, y)$ that (approximately) satisfies Laplace's equation. Also, a soap film stretched inside a twisted wire loop forms a surface $u = u(x, y)$ that is another approximate solution of Laplace's equation. In general, Laplace's equation defines the steady state solution of many phenomena, such as steady state temperature, steady state electric fields, and steady state concentrations.

LAPLACE'S EQUATION INSIDE A CIRCLE

We now solve one of the most important boundary-value problems in physics: the problem of finding a function whose values are specified on a circle (say of radius 1) and that satisfies Laplace's equation *inside* the circle. Before solving this problem, however, we first transform the two-dimensional *Cartesian* Laplacian $\nabla^2 u \equiv u_{xx} + u_{yy}$ to polar coordinates r and θ. Using the usual transformation from Cartesian to polar coordinates, given by $x = r \cos \theta$, $y = r \sin \theta$, we can show (see Problem 20 at the end of this section) that the Laplacian in Cartesian coordinates is

$$\nabla^2 u = u_{rr} + \frac{1}{r}u_r + \frac{1}{r^2}u_{\theta\theta} \qquad \text{(Laplacian in polar form)} \qquad (3)$$

Stated mathematically, we now seek the function $u = u(r, \theta)$ that satisfies the boundary-value problem

$$\text{PDE:} \quad u_{rr} + \frac{1}{r}u_r + \frac{1}{r^2}u_{\theta\theta} = 0 \qquad (0 < r < 1) \qquad (4a)$$

$$\text{BC:} \quad u(1, \theta) = f(\theta) \qquad (0 \le \theta < 2\pi) \qquad (4b)$$

The general problem of finding a function that takes on specified values on the boundary of a region and satisfies an elliptic differential equation (like Laplace's equation) inside the region is called a **Dirichlet problem.*** See Figure 9.36. Note that the problem has no initial conditions, since time does not enter into the discussion.

We now solve this problem using the method of separation of variables.

Figure 9.36
An important problem in mathematical physics is to find the function that takes on given values on a circle and satisfies Laplace's equation interior to the circle.

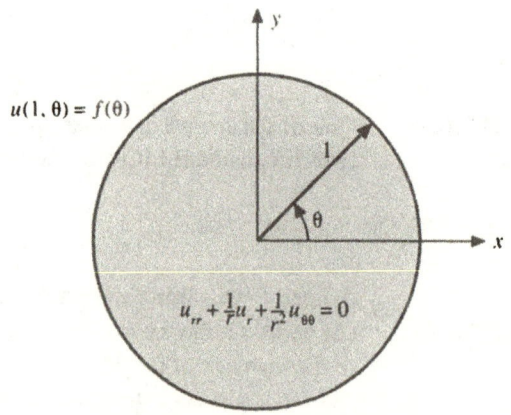

* The Dirichlet (pronounced "deer-a-klay") problem is named in honor of the German mathematician P. G. L. Dirichlet, who first studied the problem.

SEPARATION OF VARIABLES OF LAPLACE'S EQUATION INSIDE A CIRCLE

We now satisfy Laplace's equation inside a circle with given boundary conditions using the method of separation of variables.

Step 1 (Separating Variables). We begin by trying a solution of the form

$$u(r, \theta) = R(r)\,\Theta(\theta) \tag{5}$$

where $R(r)$ is a function of r alone and $\Theta(\theta)$ is a function of θ alone. Substituting $u(r, \theta) = R(r)\Theta(\theta)$ into Laplace's equation (4a), we obtain

$$R''(r)\Theta(\theta) + \frac{1}{r}R'(r)\Theta(\theta) + \frac{1}{r^2}R(r)\Theta''(\theta) = 0 \tag{6}$$

If we now divide each side of this equation by $R(r)\Theta(\theta)$ and then multiply by r^2, we find

$$r^2\frac{R''}{R} + r\frac{R'}{R} = -\frac{\Theta''}{\Theta} = k \tag{7}$$

Note that we have streamlined our notation by calling $R = R(r)$ and $\Theta = \Theta(\theta)$. Now, since the left-hand side of Eq. (7) depends only on r and the right-hand side depends only on θ, it is necessary that both sides are equal to some constant. Otherwise, by keeping one independent variable constant and varying the other, one side would remain unchanged and the other would vary. Calling the separation constant k as we have written in (7), we next get

$$r^2 R'' + rR' - kR = 0 \tag{8a}$$

$$\Theta'' + k\Theta = 0 \tag{8b}$$

Step 2 (Finding the Fundamental Solutions). The solutions of Eqs. (8) depend on whether $k < 0$, $k = 0$, or $k > 0$. We must look for the solutions of Eqs. (8) for each case. When $k < 0$, we see that $\Theta(\theta)$ is

$$\Theta(\theta) = c_1 e^{\sqrt{k}\theta} + c_2 e^{-\sqrt{k}\theta} \tag{9}$$

But $\Theta(\theta)$ must be periodic with period 2π in order for $u(r, \theta)$ to be periodic in θ. But this happens only when $c_1 = c_2 = 0$, which implies that $u(r, \theta) \equiv 0$. Hence there are no (nonzero) solutions when k is negative.

When $k = 0$, the differential equation (8a) is known as *Euler's equation* and it can be shown (see Problem 10 at the end of this section) that the general solution is

$$R(r) = k_1 + k_2 \ln r \tag{10}$$

In order that $u(r, \theta)$ remain bounded inside the circle, the logarithmic term cannot be accepted, and so we are left with the constant solution $R(r) = k_1$. Solving for $\Theta(\theta)$ in Eq. (8b) with $k = 0$ gives

$$\Theta(\theta) = c_1 + c_2\theta \tag{11}$$

However, in order for $\Theta(\theta)$ to be periodic, we must have $c_2 = 0$ and hence have $\Theta(\theta) = c_1$. In conclusion, the only (nonzero) solution that we find corresponding to $k = 0$ is

$$u(r, \theta) = \text{constant} \tag{12}$$

Finally, when $k > 0$, the differential equation (8a) is also called Euler's equation, and it can be shown (see Problem 10 at the end of this section) to have the general solution

$$R(r) = k_1 r^{\sqrt{k}} + k_2 r^{-\sqrt{k}} \tag{13a}$$

In order that $R(r)$ remain bounded as $r \to 0$, we must discard the term $r^{-\sqrt{k}}$ and hence are left with $R(r) = k_1 r^{\sqrt{k}}$. Now, solving for $\Theta(\theta)$ gives

$$\Theta(\theta) = c_1 \cos \sqrt{k}\theta + c_2 \sin \sqrt{k}\theta \tag{13b}$$

for which, in order that $\Theta(\theta)$ be periodic with period 2π, it is necessary that the separation constant take on the values $k = n^2$, where $n = 1, 2, \ldots$. Hence we have (finally) found the solutions

$$u_n(r, \theta) = R(r)\Theta(\theta) = r^n \cos (n\theta) \tag{14a}$$
$$u_n(r, \theta) = R(r)\Theta(\theta) = r^n \sin (n\theta) \qquad (n = 1, 2, \ldots) \tag{14b}$$

These solutions, together with the constant solution $u_0(r, \theta) = 1$ (that we found when $k = 0$), form the fundamental set of solutions. They constitute the set of all solutions of Laplace's equation of the form $R(r)\Theta(\theta)$. Of course, they are not the *only* solutions of Laplace's equation, since more solutions can be found by forming the sums

$$u(r, \theta) = \frac{a_0}{2} + \sum_{n=1}^{\infty} r^n \{a_n \cos (n\theta) + b_n \sin (n\theta)\} \tag{15}$$

where a_0, a_n, and b_n are arbitrary constants. Since any sum of the form of Eq. (15) is a solution of Laplace's equation, the final strategy is to pick the coefficients a_0, a_n, and b_n in such a way that Eq. (15) also satisfies the given boundary condition $u(1, \theta) = f(\theta)$.

Last Step (Boundary Conditions). Substituting Eq. (15) into the boundary conditions $u(1, \theta) = f(\theta)$ gives

$$f(\theta) = \frac{a_0}{2} + \sum_{n=1}^{\infty} \left\{ a_n \cos (n\theta) + b_n \sin (n\theta) \right\} \tag{16}$$

But this series is the Fourier series of $f(\theta)$ whose coefficients a_n and b_n are given by Euler's formulas

$$a_n = \frac{1}{\pi} \int_0^{2\pi} f(\theta) \cos (n\theta) \, d\theta \qquad (n = 0, 1, 2, ...) \qquad (17a)$$

$$b_n = \frac{1}{\pi} \int_0^{2\pi} f(\theta) \sin (n\theta) \, d\theta \qquad (n = 1, 2, ...) \qquad (17b)$$

We are done. The solution is given by the series (15), where the coefficients a_n and b_n are given by Eqs. (17). The above steps are summarized in Figure 9.37.

Figure 9.37
Steps in solving Laplace's equation inside a circle

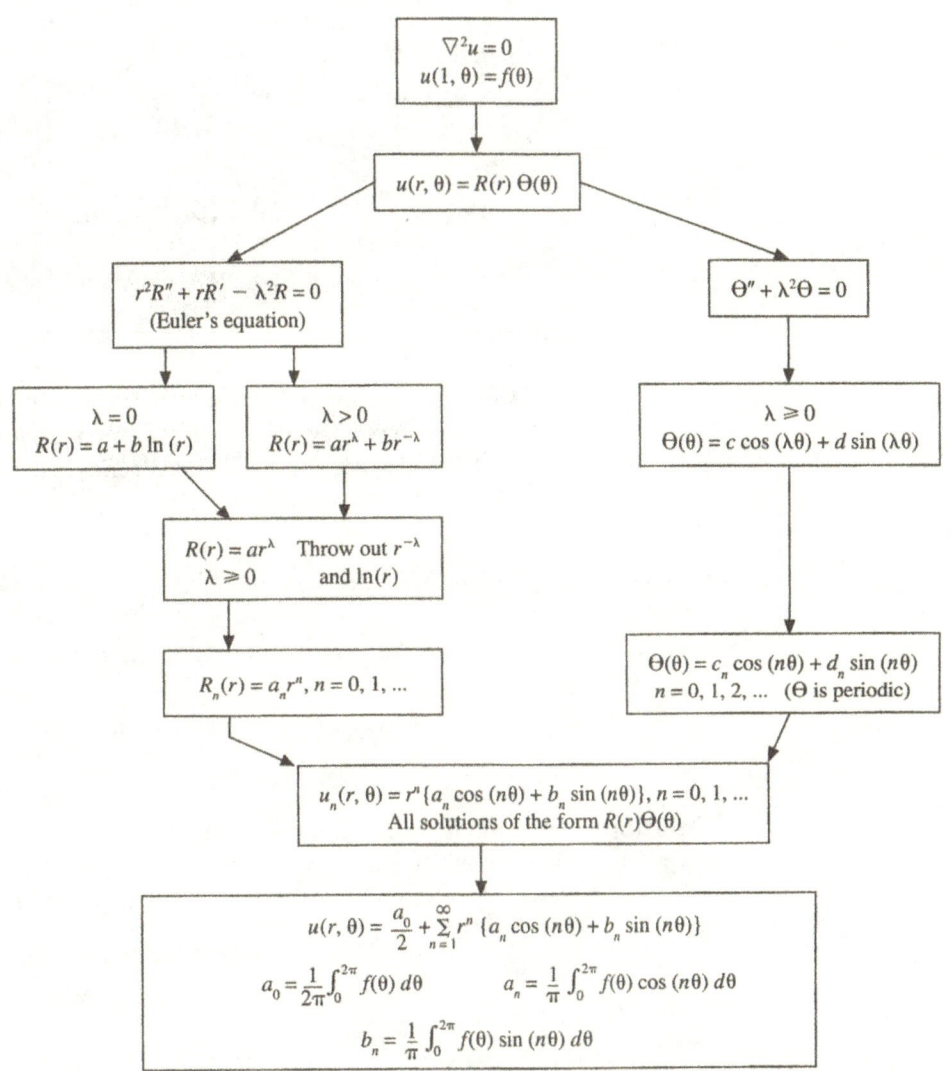

We summarize the results from the previous page.

Solution of Laplace's Equation Inside a Circle

The solution of Laplace's equation inside a circle:

$$\text{PDE:} \quad u_{rr} + \frac{1}{r}u_r + \frac{1}{r^2}u_{\theta\theta} = 0 \qquad (0 < r < 1)$$

$$\text{BC:} \quad u(1, \theta) = f(\theta) \qquad\qquad (0 \le \theta < 2\pi)$$

is

$$u(r, \theta) = \frac{a_0}{2} + \sum_{n=1}^{\infty} r^n\{a_n \cos(n\theta) + b_n \sin(n\theta)\} \qquad (18)$$

where

$$a_0 = \frac{1}{\pi}\int_0^{2\pi} f(\theta)\, d\theta$$

$$a_n = \frac{1}{\pi}\int_0^{2\pi} f(\theta) \cos(n\theta)\, d\theta \qquad (n = 1, 2, ...) \qquad (19)$$

$$b_n = \frac{1}{\pi}\int_0^{2\pi} f(\theta) \sin(n\theta)\, d\theta$$

Note: In plain English, analysis of the solution (18) and the Fourier series expansion of the boundary condition $f(\theta)$ shows that the solution can be interpreted by simply expanding $f(\theta)$ as a Fourier series and then "inserting" the factor r^n in the nth term of the expansion.

Example 1 **Laplace's Equation Inside a Circle** Solve the boundary-value problem

$$\text{PDE:} \quad u_{rr} + \frac{1}{r}u_r + \frac{1}{r^2}u_{\theta\theta} = 0 \qquad\qquad (0 < r < 1)$$

$$\text{BC:} \quad u(1, \theta) = 1 + \sin\theta + \frac{1}{2}\cos\theta \qquad (0 \le \theta < 2\pi)$$

Solution Since the boundary condition $f(\theta)$ is already expanded as its Fourier series (16), we simply identify the coefficients $a_0 = 2$, $a_1 = 1/2$, $b_1 = 1$, and all other a_n's and b_n's zero. Hence to find the solution, we simply "insert" the factor r^n in the nth term of $f(\theta)$, getting

$$u(r, \theta) = 1 + r\sin\theta + \frac{1}{2}r\cos\theta$$

The reader can easily visualize this surface interior to the circle $r = 1$ or can use a computer for assistance. ∎

Example 2   **Laplace's Equation Inside a Circle**   Solve the boundary-value problem

$$\text{PDE:}\quad u_{rr} + \frac{1}{r}u_r + \frac{1}{r^2}u_{\theta\theta} = 0 \qquad (0 < r < 1)$$

$$\text{BC:}\quad u(1, \theta) = \sin(2\pi\theta) \qquad (0 \le \theta < 2\pi)$$

Solution   Again, the boundary function $f(\theta) = \sin(2\pi\theta)$ is already expanded as its Fourier series with $b_2 = 1$ and all other coefficients zero. Hence the solution is

$$u(r, \theta) = r^2 \sin(2\theta)$$

The reader can easily visualize this surface interior to the circle $r = 1$ or can use a computer for assistance.   ∎

Example 3   **A Little Harder Problem**   Solve the boundary-value problem

$$\text{PDE:}\quad u_{rr} + \frac{1}{r}u_r + \frac{1}{r^2}u_{\theta\theta} = 0 \qquad (0 < r < 1)$$

$$\text{BC:}\quad u(1, \theta) = \begin{cases} 1 & (0 \le \theta < \pi) \\ -1 & (\pi \le \theta < 2\pi) \end{cases} \qquad (0 \le \theta < 2\pi)$$

Solution   Here the problem is nontrivial, and we must expand the boundary condition. The Fourier series expansion of $u(1, \theta) = f(\theta)$ can be shown to be the Fourier sine series

$$f(\theta) = \frac{4}{\pi}\left\{\sin\theta + \frac{1}{3}\sin 3\theta + \frac{1}{5}\sin 5\theta + \cdots\right\}$$

Again, inserting r^n in the nth term gives the solution

$$u(r, \theta) = \frac{4}{\pi}\left\{r\sin\theta + \frac{1}{3}r^3\sin 3\theta + \frac{1}{5}r^5\sin 5\theta + \cdots\right\}$$

The reader can use a computer to sketch the surface of the first three or four terms of this function.   ∎

PROBLEMS: Section 9.7

In Problems 1–5, the function f is written in both Cartesian and polar coordinates. Verify that the corresponding Laplacians are the same.

Cartesian Form	**Polar Form**
1. $f(x, y) = x$	$f(r, \theta) = r\cos\theta$
2. $f(x, y) = x + y$	$f(r, \theta) = r\cos\theta + r\sin\theta$
3. $f(x, y) = x^2 + y^2$	$f(r, \theta) = r^2$
4. $f(x, y) = xy$	$f(r, \theta) = r^2\cos\theta\sin\theta$
5. $f(x, y) = x^2 - y^2$	$f(r, \theta) = r^2\cos^2\theta - r^2\sin^2\theta$

6. Solution's of Laplace Equation   Verify that the following functions satisfy Laplace's equation

$$u_{rr} + \frac{1}{r}u_r + \frac{1}{r^2}u_{\theta\theta} = 0$$

(a) $f(r, \theta) = 1$

(b) $f(r, \theta) = r^2\cos 2\theta$

(c) $f(r, \theta) = \dfrac{1}{r}\sin\theta$

(d) $f(r, \theta) = \ln r^2$

(e) $f(r, \theta) = r^3\sin 3\theta$

7. **Laplace's Equation in Cartesian Coordinates** Show that the functions (a)–(d) satisfy Laplace's equation

$$u_{xx} + u_{yy} = 0$$

Then write these functions in terms of polar coordinates r and θ and show that the polar form of the function satisfies Laplace's equation in polar coordinates.

(a) $f(x, y) = 1$
(b) $f(x, y) = \ln(x^2 + y^2)$
(c) $f(x, y) = xy$
(d) $f(x, y) = x^2 - y^2$

8. **The Laplacian Meter** You have just bought a "temperature" Laplacian meter and have measured the Laplacian inside the two-dimensional region D shown in Figure 9.38. On the basis of the values of the Laplacian at the indicated points, what can you say about the temperature at the given points compared to the average temperature in a small region surrounding these points?

9. **Poisson's Equation** If a constant heat source of magnitude $h > 0$ is uniformly applied to the interior of a circle of radius 1, and if the boundary of the circle is kept at a constant temperature of 0°C, then the steady state temperature $u(r, \theta)$ inside the circle will satisfy the boundary-value problem

$$\begin{aligned}\text{PDE:} \quad &\nabla^2 u = -h \quad &&(0 < r < 1)\\ \text{BC:} \quad &u(1, \theta) = 0 \quad &&(0 \le \theta < 2\pi)\end{aligned}$$

where $\nabla^2 u = -h$ is called **Poisson's equation.** What can you say about the temperature inside the circle based on what you know about the Laplacian $\nabla^2 u$?

10. **Euler's Equation** Euler's equation is given by

$$r^2 R'' + r R' - \lambda^2 R = 0$$

(a) When $\lambda \ne 0$, show that the general solution is

$$R(r) = c_1 r^\lambda + c_2 r^{-\lambda}$$

by trying a solution of the form $R(r) = r^k$, where k is an undetermined constant.

(b) When $\lambda = 0$, show that the general solution is

$$R(r) = c_1 + c_2 \ln r$$

where c_1 and c_2 are arbitrary constants.

For Problems 11–19, find the solution of Laplace's equation inside the circle $r < 1$ for the given boundary conditions $u(1, \theta) = f(\theta)$. Check your solutions.

11. $f(\theta) = 1$
12. $f(\theta) = 2$
13. $f(\theta) = \sin\theta$
14. $f(\theta) = \cos\theta$
15. $f(\theta) = \sin 2\theta$

16. $f(\theta) = \sin\theta + \dfrac{1}{2}\sin 2\theta$

Figure 9.38
The Laplacian of a function provides important information about the function.

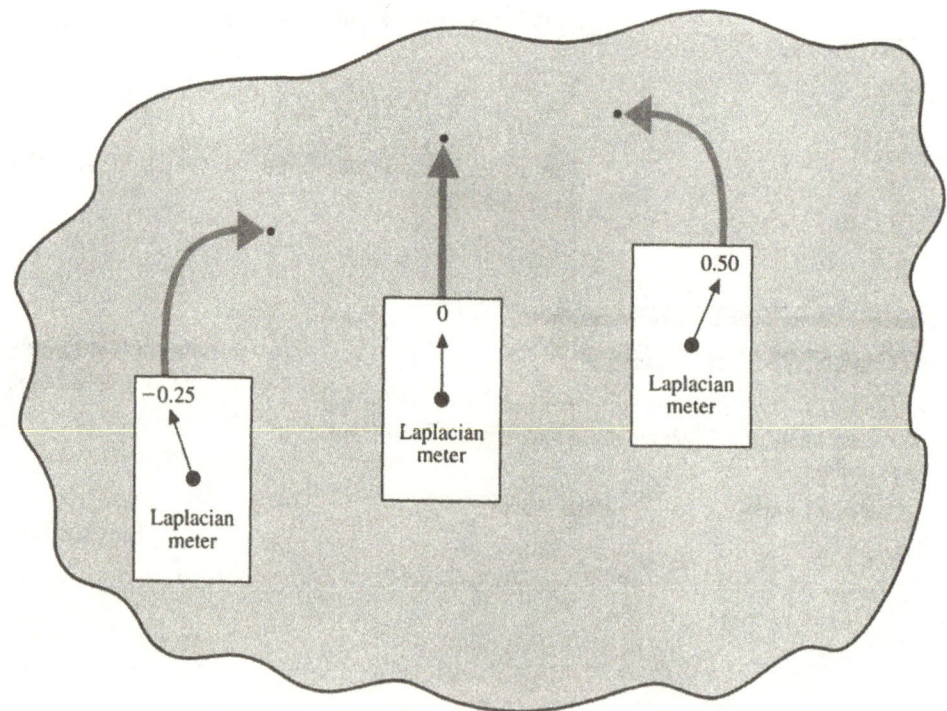

17. $f(\theta) = \cos\theta + \dfrac{1}{2}\cos 2\theta + \dfrac{1}{3}\sin 4\theta$

18. $f(\theta) = \sin(4\theta)$

19. $f(\theta) = \begin{cases} \theta & (0 \le \theta < \pi) \\ \pi - \theta & (\pi \le \theta < 2\pi) \end{cases}$

Hint: The Fourier series of $f(\theta)$ is

$$f(\theta) = \frac{\pi}{2} - \frac{4}{\pi}\left(\cos\theta + \frac{1}{3^2}\cos 3\theta + \frac{1}{5^2}\cos 5\theta + \cdots\right)$$

20. Laplace's Equation in Polar Coordinates Show that the transformation $x = r\cos\theta$, $y = r\sin\theta$ transforms the equation $u_{xx} + u_{yy} = 0$ to

$$u_{rr} + \frac{1}{r}u_r + \frac{1}{r^2}u_{\theta\theta} = 0$$

Other Boundary-Value Problems Involving Laplace's Equation

We have solved Laplace's equation inside a circle when the solution is given on the boundary. There are many other boundary-value problems involving Laplace's equation. For Problems 21–23, verify that the given solution satisfies the given boundary-value problem.

21. Laplace's Equation Outside a Circle

PDE: $u_{rr} + \dfrac{1}{r}u_r + \dfrac{1}{r^2}u_{\theta\theta} = 0 \qquad (r > 1)$

BC: $u(1, \theta) = \sin\theta + \sin 2\theta \qquad (0 \le \theta < 2\pi)$

Solution: $u(r, \theta) = \dfrac{1}{r}\sin\theta + \dfrac{1}{r^2}\sin 2\theta$

22. Laplace's Equation Inside a Square

PDE: $u_{xx} + u_{yy} = 0 \quad (0 < x < 1) \quad (0 < y < 1)$

BC: $\begin{cases} u(x, 0) = y(x, 1) = 0 & \text{(bottom and top)} \\ u(0, y) = 0 & \text{(left side)} \\ u(1, y) = \sin(\pi y) & \text{(right side)} \end{cases}$

Solution: $u(x, y) = \left\{\dfrac{\sinh \pi x}{\sinh \pi}\right\}\sin(\pi y)$

23. Laplace's Equation in an Annulus

PDE: $u_{rr} + \dfrac{1}{r}u_r + \dfrac{1}{r^2}u_{\theta\theta} = 0 \qquad (1 < r < 2)$

BC: $\begin{cases} u(1, \theta) = 0 \\ u(2, \theta) = \sin\theta \end{cases} \qquad (0 \le \theta < 2\pi)$

Solution: $u(r, \theta) = \dfrac{2}{3}\left(r - \dfrac{1}{r}\right)\sin\theta$

Neumann Boundary-Value Problem

*There are two basic kinds of boundary-value problems associated with Laplace's equation. We have already studied the Dirichlet problem, in which the solution is specified on the boundary. Another type of boundary-value problem is the **Neumann problem**, in which one specifies the outward normal derivative $\partial u/\partial n = g(x, y)$ on the boundary. For Problems 24–25, verify the solutions of the given Neumann problem. Note that on a circle, the outward normal derivative is simply $\partial u/\partial r$.*

24. $u_{rr} + \dfrac{1}{r}u_r + \dfrac{1}{r^2}u_{\theta\theta} = \quad (0 < r < 1)$

$\dfrac{\partial u}{\partial r}(1, \theta) = \sin\theta \qquad (0 \le \theta \le 2\pi)$

Solution: $u(r, \theta) = r\sin\theta$

25. $u_{rr} + \dfrac{1}{r}u_r + \dfrac{1}{r^2}u_{\theta\theta} = 0 \quad (0 < r < 1)$

$\dfrac{\partial u}{\partial r}(1, \theta) = \sin(2\theta) \qquad (0 \le \theta < 2\pi)$

Solution: $u(r, \theta) = \dfrac{1}{2}r^2\sin 2\theta$

Laplace's Equation Inside a Square

26. For parts (a)–(e) we will lead the reader step by step through the solution of the important boundary-value problem using the method of separation of variables.

PDE: $u_{xx} + u_{yy} = 0 \qquad (0 < x < 1) \qquad (0 < y < 1)$

BC: $\begin{cases} u(x, 0) = u(x, 1) = 0 & (0 < x < 1) \\ u(0, y) = 0 & (0 < y < 1) \\ u(1, y) = f(y) & (0 < y < 1) \end{cases}$

(a) Substitute $u(x, y) = X(x)Y(y)$ into Laplace equations, getting the two ordinary differential equations

$$X'' - kX = 0$$
$$Y'' + kY = 0$$

(b) Show the differential equations in part (a) must satisfy the three boundary conditions $X(0) = 0$, $Y(0) = 0$, $Y(1) = 0$ in order that the given boundary conditions $u(x, 0) = u(x, 1) = 0$, $u(0, y) = 0$ hold.

(c) Show that the only solutions of the equations in part (a) that satisfy the boundary conditions in part (b) are

$$\begin{aligned} X_n(x) &= \sinh(n\pi x) \\ Y_n(y) &= \sin(n\pi y) \end{aligned} \qquad (n = 1, 2, \ldots)$$

Hence the product solutions (fundamental solutions) of

Laplace's equation are

$$u_n(x, y) = \sinh(n\pi x) \sin(n\pi y) \qquad (n = 1, 2, ...)$$

(d) Show that the sum of the fundamental solutions

$$u(x, y) = \sum_{n=1}^{\infty} b_n u_n(x, y)$$

$$= \sum_{n=1}^{\infty} b_n \sinh(n\pi x) \sin(n\pi y)$$

is the solution of the boundary-value problem, provided that the coefficients b_n are chosen in such a way that $u(1, y) = f(y)$. Show that this is true when the values $a_n = b_n \sinh n\pi$ are the coefficients of the Fourier sine series of $f(y)$. That is,

$$b_n = \frac{2}{\sinh(n\pi)} \int_0^1 f(y) \sin(n\pi y)\, dy$$

(e) Use the solution found in part (d) to solve the boundary-value problem when $u(1, y) = \sin(\pi y)$. *Hint:* Very easy.

27. **Computer Problem** Use a computer to draw the three-dimensional graphs of some of the functions in Problems 1–5.

28. **Computer Problems** Use a computer to draw the three-dimensional graphs of some of the solutions of Problems 11–19.

29. **Journal Entry** Summarize what you know about partial differential equations.

APPENDIX

COMPLEX NUMBERS AND COMPLEX-VALUED FUNCTIONS

ALGEBRAIC
PRELIMINARIES

By a **complex number** we mean a number of the form

$$z = a + bi \tag{1}$$

where a and b are real numbers and i is the **imaginary unit** defined by

$$i = \sqrt{-1} \tag{2}$$

The real number a is called the **real component** (or **real part**) of z, and the real number b is called the **imaginary component** (or **imaginary part**) of z. We often denote the real and imaginary parts of z by

$$a = \text{Re}\,(z) \qquad b = \text{Im}\,(z)$$

The complex number $a + bi$ can be represented as a point in the **complex plane** with abscissa a and ordinate b. See Figure A.1. If the imaginary part of a complex number is zero, then the complex number is simply a real number, such as 3, 5.2, π, -8.53, 0, or -1. If the real part of a complex number is zero and the complex part is nonzero,

Figure A.1
The complex plane

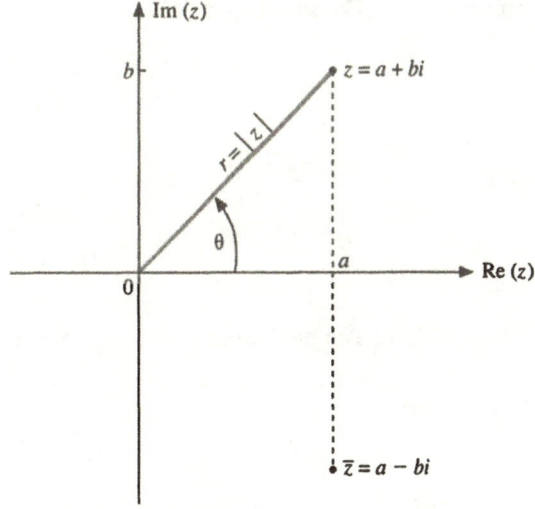

then we say that the complex number is **purely imaginary,** such as $3i$ or $-4i$. If both the real and imaginary parts of a complex number are zero, such as for $0 + 0i$, then we have the real number 0, and we write $0 = 0 + 0i$.

Two complex numbers are said to be equal when their real and imaginary parts are equal. For example, if

$$(x + y - 2) + (x - y)i = 1 - i$$

then

$$x + y - 2 = 1$$
$$x - y = -1$$

which is true if and only if $x = 1$, $y = 2$.

ADDITION, SUBTRACTION, MULTIPLICATION, AND DIVISION OF COMPLEX NUMBERS

The rules of arithmetic for complex numbers follow naturally from the usual rules for real numbers, with the provisions that all powers of i are to be reduced to lowest form according to the following rules:

$$i^2 = -1$$
$$i^3 = i^2 i = -i$$
$$i^4 = i^2 i^2 = 1$$
$$i^5 = i^4 i = i$$
$$\vdots$$

Addition and Subtraction

The sum and difference of two complex numbers $a + bi$ and $c + di$ are defined as

$$(a + bi) + (c + di) = (a + c) + (b + d)i \tag{3a}$$
$$(a + bi) - (c + di) = (a - c) + (b - d)i \tag{3b}$$

For example,

$$(3 + 2i) + (1 - i) = 4 + i$$
$$(1 - i) - (2 + 4i) = -1 - 5i$$

Multiplication

The product of two complex numbers $a + ib$ and $c + id$ is defined as

$$
\begin{aligned}
(a + bi)(c + di) &= a(c + di) + bi(c + di) \\
&= ac + adi + bci + bdi^2 \\
&= (ac - bd) + (bc + ad)i
\end{aligned}
\tag{4}
$$

For example,

$$
\begin{aligned}
(4 + i)(2 - 3i) &= 4(2 - 3i) + i(2 - 3i) \\
&= 8 - 12i + 2i - 3i^2 \\
&= 11 - 10i
\end{aligned}
$$

Division

The quotient of two complex numbers is obtained using the process of **rationalizing the denominator** and is defined by

$$\frac{a + bi}{c + di} = \frac{a + bi}{c + di} \cdot \frac{c - di}{c - di}$$

$$= \frac{a(c - di) + bi(c - di)}{c(c - di) + di(c - di)}$$

$$= \frac{ac - adi + bci + bdi^2}{c^2 - cdi + cdi - d^2 i}$$

$$= \left(\frac{ac - bd}{c^2 + d^2}\right) + \left(\frac{bc - ad}{c^2 + d^2}\right)i \qquad (5)$$

For example,

$$\frac{1 + 3i}{3 + 2i} = \frac{1 + 3i}{3 + 2i} \cdot \frac{3 - 2i}{3 - 2i}$$

$$= \frac{1(3 - 2i) + 3i(3 - 2i)}{3(3 - 2i) + 2i(3 - 2i)}$$

$$= \frac{3 - 2i + 9i - 6i^2}{9 - 6i + 6i - 4i^2}$$

$$= \frac{9 + 7i}{13}$$

$$= \frac{9}{13} + \frac{7}{13}i$$

ABSOLUTE VALUE AND AMPLITUIDE OF A COMPLEX NUMBER

The **absolute value** (or **modulus**) of a complex number $z = a + bi$ is defined by

$$r = |z| = \sqrt{a^2 + b^2} \qquad (6)$$

which from the point of view of the complex plane denotes the polar distance from the origin 0 to z. For example,

$$|3 + 2i| = \sqrt{3^3 + 2^2} = \sqrt{13}$$

The **polar angle** (or **amplitude**) of a complex number, $z = a + bi$, denoted θ, is defined by

$$\theta = \tan^{-1}\left(\frac{b}{a}\right) \qquad (7)$$

For example, $z = 1 + i$ has amplitude

$$\theta = \tan^{-1}(1) = \frac{\pi}{4}$$

Using the trigonometric formulas, $a = r \cos \theta$, $b = r \sin \theta$, we can write complex numbers in **polar** or **trigonometric form** as

$$z = a + bi = r \cos \theta + ir \sin \theta = r(\cos \theta + i \sin \theta) \tag{8}$$

For example, the complex number $1 + i$ can be expressed in polar form as

$$1 + i = r(\cos \theta + i \sin \theta)$$
$$= \sqrt{2}[\cos (\pi/4) + i \sin (\pi/4)]$$

We can also find **integer powers** of a complex number with the aid of **de Moivre's formula***

$$z^n = r^n(\cos n\theta + i \sin n\theta) \qquad (n = 1, 2, ...)$$

We also have rules relating the absolute value of a complex number. A few of the important ones are

$$|z_1 z_2| = |z_2||z_2|$$
$$\left|\frac{z_1}{z_2}\right| = \frac{|z_1|}{|z_2|}$$
$$|z_1 + z_2| \le |z_1| + |z_2| \qquad \text{(triangle inequality)}$$

See Figure A.2.

Figure A.2
The triangle inequality

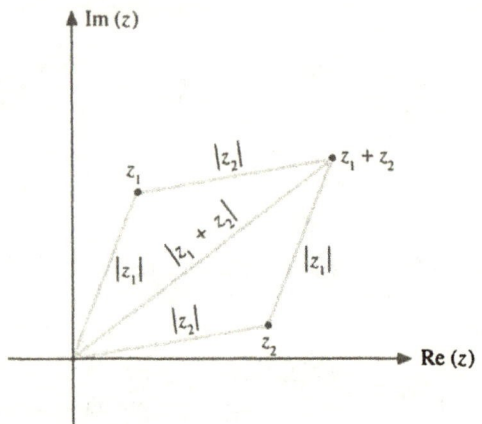

COMPLEX CONJUGATES   If two complex numbers differ only in the sign of their imaginary parts, the two complex numbers are called **complex conjugates** (or **conjugate** to each other). For example, the two complex numbers $3 + 2i$ and $3 - 2i$ are complex conjugates. The conjugate of a complex number, z, is usually denoted by $\bar{z}$. It is possible to write the magnitude of a complex number in terms of itself and its complex conjugate. For example, given the complex number $z = a + bi$, then

$$z\bar{z} = (a + bi)(a - bi) = a^2 + b^2 = |z|^2$$

* Named for the French mathematician Abraham de Moivre (1667–1754).

Thus we have

$$|z| = \sqrt{a^2 + b^2} = \sqrt{z\bar{z}} \qquad (9)$$

We also have

$$z + \bar{z} = (a + bi) + (a - bi) = 2a = 2\,\text{Re}\,(z)$$

or

$$\text{Re}\,(z) = \frac{z + \bar{z}}{2} \qquad (10)$$

Also,

$$z - \bar{z} = (a + bi) - (a - bi) = 2ib = 2i\,\text{Im}\,(z)$$

or

$$\text{Im}\,(z) = \frac{z - \bar{z}}{2i} \qquad (11)$$

COMPLEX-VALUED FUNCTIONS AND THEIR DERIVATIVES

In the study of differential equations we are introduced to **complex-valued functions of a real variable.** Typical complex-valued functions of a single real variable x are

$$F(x) = \cos 2x + i \sin 2x$$
$$G(x) = x^2 + ixe^x$$
$$H(x) = e^{2x} \cos x + ie^{2x} \sin x$$

In general, a complex-valued function of a real variable is an expression of the form

$$F(x) = f(x) + ig(x)$$

where f and g are real-valued functions of x, defined on some interval of interest. We define the **derivative** of the complex-valued function $F(x)$ by

$$F'(x) = f'(x) + ig'(x) \qquad (12)$$

provided that both f and g are differentiable over the domain of interest. Higher derivatives are defined in the same way, that is, $F''(x) = f''(x) + ig''(x)$, and so on. For example, we have

$$F(x) = \cos 2x + i \sin 2x$$
$$F'(x) = -2 \sin 2x + 2i \cos 2x$$
$$F''(x) = -4 \cos 2x - 4i \sin 2x$$
$$F'''(x) = 6 \sin 2x - 6i \cos 2x$$
$$\vdots$$

One of the most important complex-valued functions of a real variable that arises in the study of differential equations is the **complex exponential function** e^{ix}, defined by

$$e^{ix} = \cos x + i \sin x \qquad (13)$$

which is known as **Euler's formula.** More generally, we have the related function

$$e^{(a+bi)x} = e^{ax}(\cos bx + i \sin bx) \tag{14}$$

which is also a complex-valued function of the real variable x. When $x = 1$, we have the complex exponential

$$e^{a+bi} = e^{a}(\cos b + i \sin b) \tag{15}$$

This formula shows how to raise the constant e to a complex number. For example,

$$e^{i\pi/2} = \cos(\pi/2) + i \sin(\pi/2) = i$$
$$e^{3+2\pi i} = e^{3}(\cos 2\pi + i \sin 2\pi) = e^{3}$$
$$e^{2\pi i} = \cos 2\pi + i \sin 2\pi = 1$$
$$e^{i\pi/4} = \cos(\pi/4) + i \sin(\pi/4) = \frac{1}{\sqrt{2}} + i\frac{1}{\sqrt{2}}$$
$$e^{2} = e^{2}[\cos(0) + i \sin(0)] = e^{2}$$

As one might suspect, we have the usual rule of exponents for the complex exponential

$$e^{z_1}e^{z_2} = e^{z_1+z_2}$$

Hence we have

$$e^{2+3\pi i} = e^{2}e^{3\pi i}$$

DERIVATIVE OF A COMPLEX EXPONENTIAL

From the definition of the derivative of a complex-valued function of a real variable, we can write

$$\frac{d}{dx} e^{(a+bi)x} = \frac{d}{dx}\{e^{ax}(\cos bx + i \sin bx)\}$$

$$= \frac{d}{dx} e^{ax} \cos bx + i \frac{d}{dx} e^{ax} \sin bx$$

$$= e^{ax}(a \cos bx - b \sin bx) + ie^{ax}(a \sin bx + b \cos bx)$$

$$= (a + ib)e^{(a+ib)x}$$

Thus we have proven the important derivative from the calculus of complex-valued functions

$$\frac{d}{dx} e^{cx} = c\frac{d}{dx} e^{cx} \tag{16}$$

where c is a complex number. Examples of this rule are

$$\frac{d}{dx} e^{(1+i)x} = (1 + i)e^{(1+i)x}$$

$$\frac{d}{dx} e^{(2-3i)x} = (2 - 3i)e^{(2-3i)x}$$

$$\frac{d}{dx} e^{3ix} = 3ie^{3ix}$$

EXERCISES

1. Plot the complex numbers $3 + 2i$, $4i$, 2, $1 - i$, and 0 in the complex plane.

2. Evaluate each of the given expressions.
 (a) $(2 + 3i) + (4 - i)$
 (b) $(2 + 3i)(1 - i)$
 (c) $\dfrac{2 + i}{3 + i}$
 (d) $|5 - i|$

3. Verify that each of the two complex numbers given by $z = (1 \pm i\sqrt{3})/\sqrt{2}$ satisfies the equation $z^2 - z + 1 = 0$.

4. Show that all four numbers for all combinations of signs $(\pm 1 \pm i)/\sqrt{2}$ satisfy the equation $z^4 = -1$.

5. If $z = a + bi$, compute Re $(z^2 + 2z)$ and Im $(z^2 + 2z)$.

6. Compute $|z|$ of $z = 4 + 2i$ by using the general formula $|z| = \sqrt{z\bar{z}}$.

7. Find $F'(x)$ and $F''(x)$ for each of the following complex-valued functions of the real variable x.
 (a) $F(x) = e^{(1-i)x}$
 (b) $F(x) = e^{3ix}$
 (c) $F(x) = e^{2x}(\cos 3x + i \sin 3x)$

8. Write the following complex exponentials in trigonometric form.
 (a) $e^{1+\pi i}$
 (b) $e^{2+\pi i/2}$
 (c) $e^{i\pi}$
 (d) $e^{-i\pi}$

ANSWERS TO PROBLEMS*

CHAPTER 1

Section 1.1
1. First order, nonlinear   **2.** First order, linear, variable coefficients, nonhomogeneous
3. Second order, linear, variable coefficients, homogeneous   **4.** Second order, nonlinear
5. Third order, linear, constant coefficients, homogeneous   **6.** Third order, linear, constant coefficients, nonhomogeneous
7. Second order, linear, variable coefficients, nonhomogeneous   **8.** Second order, nonlinear
9. Second order, linear, variable coefficients, homogeneous   **10.** Second order, nonlinear
11. First order, one independent variable, ordinary, linear, constant coefficients, homogeneous
12. Second order, one independent variable, ordinary, linear, constant coefficients, homogeneous
13. Fourth order, one independent variable, ordinary, linear, constant coefficients, nonhomogeneous
14. Second order, two independent variables, partial, linear, variable coefficients, homogeneous
15. Third order, two independent variables, partial, nonlinear
16. Second order, one independent variable, ordinary, nonlinear
17. Second order, one independent variable, ordinary, linear, variable coefficients, homogeneous
18. (a) Nonhomogeneous (b) Nonhomogeneous (c) Nonhomogeneous (d) Nonhomogeneous

Section 1.2
For Problems 17–25 there are more solutions than the ones listed here, but these are the ones that a beginning student is likely to find. Later we will find all solutions.

17. $y(x) = e^x$ **18.** $y(x) = -\dfrac{1}{x}$ **19.** $y(x) = 1$ **20.** $y(x) = \dfrac{1}{2}e^x$ **21.** $y(x) = \dfrac{a}{2}e^x$ **22.** $y(x) = \dfrac{1}{1+a}e^x$

23. $y(x) = \dfrac{1}{x}$ **24.** $y(x) = \sin x$ **25.** $y(x) = e^x$ **26.** (a) $3x + c$ (c any constant) (b) $\dfrac{x^3}{3}$ (c any constant)

(c) $-\cos x + 1$ (d) $\dfrac{x^2}{2} + 1$ (e) $-\sin x + 2x$ **34.** (a) The sum of the terms on the left is positive for all x.

(b) The left-hand side is positive and the right-hand side is negative for all x. (c) The left-hand side is nonnegative and the right-hand side is negative for all x. (d) The value of the sine function can never be equal to 2.
36. Yes   **37.** Yes   **38.** Yes   **39.** Yes   **40.** Yes   **41.** Yes   **42.** No   **43.** No

44. (c) $f_y(x, y) = y^{-1/2}$ is not continuous for $y = 0$.   **45.** (a) $\partial f/\partial h = -\dfrac{1}{2}kh^{-1/2}$ is not continuous for $h = 0$.

(b) $h(t) = \begin{cases} 0.25(c - kt)^2 & (t < c/k) \\ 0 & (t \geq c/k) \end{cases}$ (c an arbitrary constant)

(c) The bucket is empty for $t \geq c/k$, and it is possible to travel up any of the quadratic curves when $t < c/k$.

48. $y(x) \doteq e^{-0.57x}$

CHAPTER 2

Section 2.1
1. $y = ce^{-2x}$ $(-\infty < x < \infty)$ **2.** $y = ce^{-2x} + e^x$ $(-\infty < x < \infty)$

3. $y = ce^x + \dfrac{1}{2}e^{3x}$ $(-\infty < x < \infty)$ **4.** $y = ce^{-x} + \dfrac{1}{2}\sin x - \dfrac{1}{2}\cos x$ $(-\infty < x < \infty)$

* Answers are not given where proofs are required.

5. $y = ce^{-x} + e^{-x} \tan^{-1}(e^x)$ $(-\infty < x < \infty)$ **6.** $y = ce^{-x^2} + 12$ $(-\infty < x < \infty)$

7. $y = ce^{-x^3} + \dfrac{1}{3}$ $(-\infty < x < \infty)$ **8.** $y = \dfrac{c}{x} + \dfrac{\ln x}{x}$ $(0 < x < \infty)$ **9.** $y = \dfrac{c}{x} + x$ $(0 < x < \infty)$

10. $y = c \cos x + \sin x$ $\left(-\dfrac{\pi}{2} < x < \dfrac{\pi}{2}\right)$ **11.** $y = cx^2 + x^2 \sin x$ $(0 < x < \infty)$ **12.** $y = \dfrac{c}{x^3} - \dfrac{\cos x}{x^3}$ $(0 < x < \infty)$

13. $y = \dfrac{c}{e^x + 1}$ $(-\infty < x < \infty)$ **14.** $y = \dfrac{c}{\sqrt{x^2 + 9}}$ $(-\infty < x < \infty)$ **15.** $y = \dfrac{c}{x}e^{-2x} + \dfrac{1}{2}xe^{-2x}$ $(0 < x < \infty)$

16. $y = 2e^x - 1$ **17.** $y = \dfrac{1}{2}(x^2 - 1) + e^{1-x^2}$ **18.** $y = x^4 + 3x^3$ **19.** $y = \dfrac{1}{2} + \dfrac{1}{2}e^{-x^2}$ **20.** $y = \dfrac{2}{e^x + 1}$

22. $y = -x - 1$ **23.** $xy^2 - e^y = c$ **24.** (a) $y = \exp\left\{ce^{bx} + \dfrac{a}{b}\right\}$ (b) $y = \exp\left\{ce^x + 1\right\}$

25. (b) $y = 0,\ y = \left\{ce^{-2x} - 1\right\}^{-1/2}$ **26.** (b) $y = 1 + (x + c)^{-1}$ **28.** (a) $y = 1 - e^{-x}$ (b) $y = \dfrac{1}{2} + \left(\dfrac{e^2}{2} - e\right)e^{-2x}$

29. (a) $y = 1 - e^{-x}$ (b) $y = (e - 1)e^{-x}$ **31.** $y = be^{x^2} + \dfrac{\sqrt{\pi}}{2}e^{x^2}\,\mathrm{erf}(x)$

Section 2.2

1. Separable   **2.** Separable   **3.** Not separable   **4.** Not separable   **5.** Separable   **6.** Not separable

7. Separable   **8.** Separable   **9.** Not separable   **10.** Separable   **11.** $y = \dfrac{1}{2}x^2 - \dfrac{1}{3}x^3 + c$ **12.** $y = cx^2$

13. $y = ce^{x^3/3}$ **14.** $2x^3y^2 + cy^2 + 3 = 0$ **15.** $e^x + e^{-y} = c$ **16.** $y = -2\sqrt{x} + c$ **17.** $\sqrt{x} - \sqrt{y} = c$

18. $y - \ln|y + 1| = x^2 + c$ **19.** $(x^2 - 1)e^{2y} - cx^2 = 0$ **20.** $c(1 - y)^{1/x^2} = 1 + y$ **21.** $r = c(1 - \cos\theta)$

22. $\dfrac{y^3}{3} - \sqrt{1 + x^2} = c$ **23.** $\dfrac{y}{1 - y} = ce^x$ **24.** $\sqrt{1 - x^2} - 2\sqrt{1 - y} = c$ **25.** $x \ln x + y \ln y = c$

26. $y = 1 + x$ **27.** $y = \sqrt{1 - x^2}$ **28.** $y = 2\left(\dfrac{1 + 2e^{4x}}{1 - 2e^{4x}}\right)$ **29.** $x^2 + \ln x - \ln y = 0$

30. $x^2 - 4\ln(1 + y^2) = 1$ **31.** $y = \dfrac{2}{\sqrt{\pi}}\displaystyle\int_0^x e^{-t^2}\,dt$ **32.** $1 = \left\{\dfrac{1}{2} - \displaystyle\int_1^x e^{-t^2}\,dt\right\}^{-1}$

33. $y = \tan\left(\displaystyle\int_0^x \sqrt{1 + \sin t}\,dt + \dfrac{\pi}{4}\right)$ **34.** $y = \left\{1 + 3\displaystyle\int_0^x e^{t^2}\,dt\right\}^{1/3}$ **35.** $y = \left\{1 + 3\displaystyle\int_0^x \dfrac{\sin t}{t}\,dt\right\}^{1/3}$

36. (a) $\dfrac{dy}{dx} - p(x)y = 0$ (b) $\dfrac{dy}{dx} - xy = 1$ (c) $\dfrac{dy}{dx} = xy^2$ (d) $\dfrac{dy}{dx} = x^2 + y^2$ **38.** (b) $y = cx - x^2$

39. (b) $e^{-x-y+1} + x = c$ (c) $\tan\left(\dfrac{1}{2}(\pi - x - y)\right) = -x + c$ **40.** (b) $y = \dfrac{1}{c - x}$

42. (a) $x = c$ (b) $x^2 - y^2 = c$ (c) $y^4 = cx$ (d) $x^2 + 2y^2 = c$ (e) $3x^2 + 2y^2 = c$ **44.** (b) $y = cx + \dfrac{1}{2}c^2$

45. $y = \sin(x + c)$ **46.** (a) $r = \dfrac{1}{2} - \dfrac{t}{24}$ $(0 \le t \le 12)$ (t measures time in months) (b) 12 months   **47.** 10:23 A.M.

Section 2.3

3. (a) $t_h = \dfrac{\ln 2}{k}$ **4.** $t_d = \dfrac{\ln 2}{k}$ **5.** $S(t) = \left(S_0 + \dfrac{d}{r}\right)e^{rt} - \dfrac{d}{r}$ **6.** $y(t) = y_0 e^{-kt}$ ⇨ $y(1/k) = y_0 e^{-1} \doteq \dfrac{y_0}{3}$

7. $Q(t) = 100 \exp\left(\dfrac{1}{50}\ln(3/4)t\right) \doteq 100e^{-0.00575t}$, $t_h = -\dfrac{50\ln 2}{\ln(3/4)} \doteq 120$ years

8. $5\dfrac{\ln 10}{\ln 2} \doteq 16.6$ hours   **9.** (a) $Q(t) = e^{\ln(0.8)t}$ (b) $t_h \doteq -\dfrac{\ln 2}{\ln (0.8)} = 3.1$ weeks (c) 0.107 gram   **10.** 4830 years

11. 54%   **12.** Four half-lives gives 1/16th, or 6.25%   **14.** 1115 years   **15.** $A(t) = 0.000004\,(1 - e^{-4.6209812t})$

16. (a) $0.2e^{\ln(0.90)t} \doteq 0.2e^{-0.105t}$ (b) 6.6 hours   **17.** Yes   **18.** $5e^{0.3\ln 2} \doteq 6.15$ grams

19. $-25\ln (300{,}000)/\ln(0.85) \doteq 1940$ ft   **20.** No, the alcohol percent was 0.098%.   **21.** $10\dfrac{\ln 3}{\ln 2} \doteq 15.8$ hr

22. $Q_0 = 5e^{-2\ln(8/5)} \doteq 1.95$ million   **23.** Pull the product after 34 days   **24.** $5e^{4\ln 2} = 80$ million

25. $12\dfrac{\ln 5}{\ln 2} \doteq 27.9$ hr   **26.** $100e^{\ln (1.02)t} \doteq 100e^{0.0198t}$ **27.** (a) $100e^{kt}$, $k = \dfrac{\ln 26}{490} \doteq 0.00665$ (b) 2088 A.D. (no)

28. (a) $N(t) = \dfrac{kN_0}{N_0 + (k - N_0)e^{-rt}}$, $r = 0.03$, $k = 200$, $N_0 = 4$ (b) $N(200) = 178.3$ million

29. (a) $M(t) = \begin{cases} e^{kt} & (0 \le t \le 10) \\ (25 - 3/k)e^{kt} + 6/k & (10 < t \le 20) \end{cases}$

where $k = \dfrac{\ln 2}{10} \doteq 0.0693$, M measures mice in thousands, and t measures time in years, $t = 0$ denoting 1980.

(b) $M(20) \doteq 13.4$ thousand mice in the year 2000   **30.** $S(t) = S_0e^{rt} ⊅ S(10) = e^{(0.10)(10)} = e$ **31.** $t_d = \dfrac{\ln 2}{r} \doteq \dfrac{70}{100r}$

32. $7382.39   **33.** $3e^{(0.08)(320)} \doteq 393{,}600{,}000{,}000$ bottles   **34.** $e^{(0.08)(366)} \doteq \$124{,}839{,}000{,}000{,}000$

35. $\dfrac{1}{50}\ln (18) \doteq 0.0578\ (5.78\%)$ **36.** (a) $S(t) = \dfrac{1000}{0.08}(e^{0.08t} - 1)$ (b) $3399.55 (c) $r = 0.09\ (9\%)$ (need a computer)

37. $S(t) = \dfrac{d}{r}(e^{rt} - 1)$, $d = \$15$, $r = 0.08/52 \doteq 0.00153846$ (b) After 4 years, $S(208) = \$3677$.   **38.** $\dfrac{\ln 5}{0.08} \doteq 20.1$ years

39. If $d = rS_0$, the amount of money in the bank remains constant (you are essentially withdrawing the interest); if $d > rS_0$, the amount of money in the bank will increase.   **40.** $432,332

43. (a) $F_0e^{-0.25\theta}$ (b) $4\ln (0.025) \doteq 14.75$ rad (2.3 times around the tree)

Section 2.4

1. (a) $Q(t) = -550e^{-t/100} + 600$ (b) $c(t) = -\dfrac{11}{6}e^{-t/100} + 2$ (c) 600 lb (d) 2 lbs/gal   **2.** (a) $Q(t) = 40e^{-t/25} + 10$

(b) $c(t) = 0.4e^{-t/25} + 0.10$ (c) 10 kg (d) 0.1 kg/liter   **3.** (a) $\dfrac{dQ}{dt} + \left(\dfrac{2}{2t + 50}\right)Q = 0.4$ $Q(0) = 0$

(b) $Q(t) = \dfrac{0.4t^2 + 20t}{2t + 50}$ $(0 \le t \le 25)$ (c) $c(t) = \dfrac{0.4t^2 + 20t}{(2t + 50)^2}$ $(0 \le t \le 25)$ (d) $\dfrac{dQ}{dt} + \dfrac{Q}{25} = 0.4$ $Q(25) = 7.5$

4. (a) $\dfrac{dQ}{dt} + \left(\dfrac{5}{100 - 2t}\right)Q = 300$ $Q(0) = 0$ $(0 \le t < 50)$ (b) $Q(t) = \dfrac{3}{1000}(100 - 2t)^{5/2}$ $(0 \le t < 50)$

5. $t = \dfrac{20}{3}\ln (7/3) \doteq 5.6$ min   **6.** 2 lb/gal   **7.** $\dfrac{\ln 2}{0.03} \doteq 23$ min

8. (a) $\dfrac{dV}{dt} + \dfrac{40}{100}V = 0.004$ $V(0) = 0.05$ (percent concentration) (b) $V(t) = 0.04e^{-0.4t} + 0.01$ (percent concentration)

(c) $2.5\ln 4 \doteq 3.5$ years   **9.** (a) $\dfrac{dQ_A}{dt} + \dfrac{Q_A}{50} = 0$ $Q_A(0) = 50$ lb

(b) $Q_A(t) = 50e^{-t/50}$ (c) $\dfrac{dQ_B}{dt} + \dfrac{Q_B}{50} = e^{-t/50}$ $Q_B(0) = 0$ (d) $Q_B(t) = te^{-t/50}$

11. $\dfrac{dQ}{dt} = k\left(a - \left(\dfrac{m}{m+n}\right)Q\right)\left(b - \left(\dfrac{n}{m+n}\right)Q\right)$

12. Limiting concentration: a/b, time to reach one-half limiting concentration: $\dfrac{\ln 2}{b}$

Section 2.5

1. $T(t) = T_0 e^{-kt} + M(1 - e^{-kt})$ **2.** $y' = -ky$ **4.** (a) 82.9°F (b) 1:09 P.M. **5.** (a) 36.7°F (b) 3.06 hours **6.** 30°F
7. $T(t) = 35e^{-0.0154t} + 70(1 - e^{-0.0154t})$, $T(20) \doteq 44.3°F$ **8.** 2.6 hours before 8 P.M. (5:24 P.M.)
9. $T(t) = 200e^{-3.8t} + 70(1 - e^{-3.8t})$; hence $T(t)$ reaches 90°F when $t = 0.49$. Hence the temperature reaches 90°F when time is $t = 0.49$ hr (29 min, 24 sec). Hence Professor Snarf has 36 seconds to get to class after drinking his coffee. **10.** John
11. 1:37 P.M. **12.** (a) $T' = -k(T - 50e^{-0.05t})$ $T(0) = 70$ (b) $T(t) = 58.3e^{-0.5t} + 11.7e^{-0.35t}$
(c) $T(t) = 98.6 \lozenge t = -2.83$ (using a computer) **13.** (a) $T' + T = 65 + 25\sin(2\pi t)$, $T(0) = 65$

(b) $T_{ss}(s) = 65 + A\cos 2\pi t + B\sin 2\pi t$, $A = \dfrac{-50\pi}{4\pi^2 + 1}$, $B = \dfrac{25}{4\pi^2 + 1}$ (c) $65 + \sqrt{A^2 + B^2} \doteq 67.3°F$

(d) $65 - \sqrt{A^2 + B^2} \doteq 62.7°F$
14. (a) $T' = -0.02(T - 70) + 2$ $T(0) = 100$ (b) $T(t) = -70e^{-0.02t} + 170$ (c) $T(8) = 110°F$
15. (a) $T' + 0.10T = 6$ $T(0) = 75$ (b) $T(t) = 15e^{-0.10t} + 60$

Section 2.6

1. $xy = c$ **2.** $x^2 y = c$ **3.** $xy^2 = c$ **4.** $x^2 y^2 = c$ **5.** $x\sin y = c$ **6.** $e^x \sin y = c$ **7.** $e^{xy} = c$
8. $\dfrac{x}{y} = c$ **9.** $\ln(x^2 + y^2) = c$ **10.** $\tan^{-1}\left(\dfrac{y}{x}\right) = c$ **11.** 125 mph

12. 12.5 ln 10 $\doteq$ 28.8 sec (He would have fallen 6280 ft.) **13.** (a) 220 ft/sec (b) 1247 ft (c) 20 ft/sec
15 The heavier parachutist will reach a limiting velocity twice that of the lighter parachutist.
17. (a) $v(t) = 17.2(1 - e^{0.0026t})$ ft/sec (b) 17.2 ft/sec (11.7 mph)

18. (a) 0.10 hr (6 min) (b) $\begin{array}{l} x(t) = 10t\cos\left(\sqrt{8}\ln(t/6)\right) \\ y(t) = 10t\sin\left(\sqrt{8}\ln(t/6)\right) \end{array}$ $(t \geq 0.10$ hr) (c) $r = 60e^{\theta/\sqrt{8}}$ **20.** (e) 16.8 sec

21. The surface obtained by revolving the curve $y = cx^4$ about the y-axis **23.** (a) 4.9 miles/sec (b) 85 min **24.** $y = \cosh x$
26. (e) $v_r < v_b$ $(k < 1)$ **27.** (b) $L = 100$ inches (c) $r = e^{-\theta}$
28. Using a computer, one can show that for all $t > 0$ the value of h in part (c) is never larger than 11; hence the tank never fills.
29. (d) $(x - c)^2 + y^2 = c^2 - 1$ $(c > 1)$ (circle)

Section 2.7

17. The manner in which the line segments approach the horizontal line $h = 0$ indicate that one can move unambiguously forward in time along a solution $h = h(t)$, eventually arriving at $h = 0$, but one cannot move backward in time starting on the line $h = 0$.
36. Constant solutions are $y = 0$, $y = 1$, $y = 2$. Other solutions approach $y = 0$, $y = 2$. Solutions move away from $y = 1$.

Section 2.8

19. $\phi_1(x) = 1 - x + \dfrac{1}{2}x^2$ $\phi_2(x) = 1 - x + x^2 - \dfrac{1}{6}x^3$ $\phi_3(x) = 1 - x + x^2 - \dfrac{1}{3}x^3 + \dfrac{1}{24}x^4$

20. $\phi_1(x) = \dfrac{1}{2!}x^2$ $\phi_2(x) = \dfrac{1}{2!}x^2 + \dfrac{1}{3!}x^3$ $\phi_3(x) = \dfrac{1}{2!}x^2 + \dfrac{1}{3!}x^3 + \dfrac{1}{4!}x^4$

21. $\phi_1(x) = x$ $\phi_2(x) = x + \dfrac{1}{3}x^3$ $\phi_3(x) = x + \dfrac{1}{2}x^2 + \dfrac{2}{15}x^5 + \dfrac{1}{63}x^7$

CHAPTER 3

Section 3.1
1. Linear, nonhomogeneous, variable coefficients   **2.** Linear, nonhomogeneous, variable coefficients   **3.** Nonlinear
4. Nonlinear   **5.** Nonlinear   **6.** Nonlinear   **7.** Linear, nonhomogeneous, constant coefficients   **8.** Nonlinear
9. Nonlinear   **10.** Linear, nonhomogeneous, variable coefficients   **14.** $y = \sin x$ **15.** $y = 1 - x$ **16.** $y = \cosh x$
17. $y = 2xe^{2x}$ **18.** $y = e^x \cos x$ **20.** $y'' = 0$ **21.** $y'' + 4y = 0$ **22.** $y'' - y = 0$
23. $y'' - y' - 2y = -2x - 1$ **24.** $y'' + 9y = 8 \sin x$ **25.** $y = \pm(1/3)(2x + c_1)^{3/2} + c_2$
26. $y = \frac{1}{4}x^2 + c_1 \ln x + c_2$ **27.** $y = \ln|\sec(x + c_1)| + c_2$ **28.** $y = c_1 \ln x + c_2 + x$ **29.** $y = c_1 e^{-x} + c_2 - xe^{-x}$
30. $y = \frac{1}{c_1} + \frac{c_1}{4}(x + c_2)^2$ **31.** $y = \frac{1}{k} \cosh kx$ **32.** $y = \pm(1/3)(2x + c_1)^{3/2} + c_2$ **33.** $y = c_1 \sin x + c_2 \cos x$
34. $y = c_1 e^x + c_2 e^{-x}$ **35.** $y = (1 - 6x)^{1/3}$ **36.** $y = c_2 - \ln|c_1 - x|$ or $y \equiv 0$
37. $y = (x + c_1) \ln(x + c_1) - x + c_2$ **39.** $(-\infty, \infty)$ for any x_0 **40.** $(-\infty, \infty)$ for any x_0 **41.** $(-\infty, \infty)$ for any x_0
42. $(-\infty, 0)$ if $x_0 < 0$, $(0, \infty)$ if $x_0 > 0$ **43.** $(-\infty, 0)$ if $x_0 < 0$, $(0, 1)$ if $0 < x_0 < 1$, $(1, \infty)$ if $x_0 > 1$
44. $(-\infty, \infty)$ for any x_0 **45.** $(-\infty, \infty)$ for any x_0

Section 3.2
1. Linearly dependent, $W(x) = 0$ **2.** Linearly dependent, $W(x) = 0$ **3.** Linearly dependent, $W(x) = 0$
4. Linearly independent, $W(x) = (m - n)e^{(m+n)x}$ **5.** Linearly independent, $W(x) = e^{2x}$
6. Linearly dependent, $W(x) = 0$ **7.** Linearly independent, $W(x) = e^{3x} \cos^2 x$ **8.** Linearly independent, $W(x) = x^4$
9. Linearly independent, $W(x) = 0$ **10.** Linearly dependent, $W(x) = 0$ **11.** (c) $y = c_1 + c_2 x$ (d) $y = x$
12. (c) $y = c_1 e^{2x} + c_2 e^{3x}$ (d) $y = 0$ **13.** (c) $y = c_1 e^{-2x} + c_2 e^{-3x}$ (d) $y = e^{-2x} - e^{-3x}$
14. (c) $y = c_1 e^{-2x} + c_2 e^{2x}$ (d) $y = \frac{1}{4}(e^{-2x} + 3e^{2x})$ **15.** (c) $y = c_1 e^{\sqrt{5}x} + c_2 e^{-\sqrt{5}x}$ (d) $y = \frac{1}{2}(e^{\sqrt{5}x} + e^{-\sqrt{5}x})$
16. (c) $y = c_1 \sin 2x + c_2 \cos 2x$ (d) $y = \frac{1}{2}(\sin 2x + 2 \cos 2x)$ **17.** (c) $y = c_1 \sinh x + c_2 \cosh x$ (d) $y = \sinh x$
18. (c) $y = c_1 e^x + c_2 x e^x$ (d) $y = e^x$ **19.** (c) $y = c_1 x + c_2 x^{-1}$ (d) $y = 0$
20. (c) $y = c_1 x^2 + c_2 x^{-1}$ (d) $y = \frac{1}{3}(2x^2 + x^{-1})$ **22.** (b) Yes, any linearly independent set $\{c_1 e^x + c_2 e^{-x}, c_3 e^x + c_4 e^{-x}\}$
28. $y = c_1 x + c_2 x^{-1}$ **29.** $y = c_1 x + c_2 x^{-2}$ **30.** $(x^2 - 2x)y = \frac{1}{4}e^{2x} + c_1 x + c_2$ **31.** Linearly independent
32. Linearly dependent   **33.** Linearly dependent   **34.** Linearly independent   **35.** Linearly independent
36. Linearly dependent   **37.** Linearly dependent   **38.** The sign and concavity of y are always of opposite sign.

Section 3.3
1. $y = c_1 e^x + c_2 e^{-x}$ **2.** $y = c_1 \sin x + c_2 \cos x$ **3.** $y = c_1 e^{2x} + c_2 x e^{2x}$ **4** $y = c_1 + c_2 e^{-x}$ **5.** $y = c_1 + c_2 \ln x$ **6.** $y = c_1 e^{2x} + c_2(2x^2 + 2x + 1)$ **7.** $y = c_1 x^3 + c_2 x^{-2}$ **8.** $y = c_1 x + c_2 x \ln x$ **9.** $y = c_1 x + c_2(x^2 - 1)$
10. $y = c_1 \frac{1}{\sqrt{x}} \sin x + c_2 \frac{1}{\sqrt{x}} \cos x$ **11.** $y_2(x) = xe^{bx}, y(x) = c_1 e^{bx} + c_2 x e^{bx}$ **12.** $y = \frac{x^2}{2}$ **13.** $y = \frac{x}{2}e^x$
14. $y = \frac{1}{2}e^x$ **15.** $y = -x \cos x - \sin x \ln|\sin x|$ **16.** $y = \frac{x}{2}\ln^2 x$ **17.** $y = \frac{x}{2}\ln x$
18. $y = (1 - 2x^2) \int (1 - 2x^2)^{-2} e^{-x^2} dx$ **19.** $y = x \int \frac{1}{x^2 \sqrt{1 - x^2}} dx$ **20.** $y = (x - 1) \int \frac{xe^{-x}}{(x - 1)^2} dx$
21. $y = \frac{\cos x}{\sqrt{x}}$ **22.** $y'' - 3y' + 2y = 0, y_2(x) = e^x, y_3(x) = e^{2x}$ **23.** $y'' - y = 0, y_2(x) = e^x, y_3(x) = e^{-x}$
25. (c) $y_2(x) = e^{-x}$ **26.** (b) e^x, e^{-x} (c) $y = c_1 x + c_2 x e^x + c_3 x e^{-x}$

Section 3.4

1. $y = c_1 + c_2 x$ **2.** $y = c_1 + c_2 e^x$ **3.** $y = c_1 e^{3x} + c_2 e^{-3x}$ **4.** $y = c_1 e^{x/2} + c_2 e^{-x/2}$ **5.** $y = c_1 e^x + c_2 e^{2x}$

6. $y = c_1 e^{-x} + c_2 e^{2x}$ **7.** $y = c_1 e^{-x} + c_2 x e^{-x}$ **8.** $y = c_1 e^{x/2} + c_2 x e^{x/2}$ **9.** $y = c_1 e^{x/2} + c_2 e^x$

10. $y = c_1 e^{3x} + c_2 x e^{3x}$ **11.** $y = c_1 e^{4x} + c_2 x e^{4x}$ **12.** $y = 0$ **13.** $y = \dfrac{2}{3} e^x + \dfrac{1}{3} e^{-2x}$

14. $y = x e^{-x}$ **15.** $y = \dfrac{1}{2} e^{3x} + \dfrac{1}{2} e^{-3x}$ **16.** $y = e^{3x} - 3x e^{3x}$ **17.** $y = e^{2x} - x e^{2x}$ **18.** $y = \dfrac{1}{2} e^{3x/2} - \dfrac{1}{2} e^{-x/2}$

19. $y = (-e^x + x e^x)/e$ **22.** (a) $y = c_1 + c_2 \ln x$ (b) $y = c_1 x^3 + c_2 x^{-4}$ (c) $y = c_1 x^{1/2} + c_2 x^{-3/2}$

23. (b) $y = c_1 x^2 + c_2 x^{-3}$

24. The curve whose rate of increase is equal to its height lies above the curve whose curvature is equal to its height.

Section 3.5

1. $y = c_1 \cos 3x + c_2 \sin 3x$ **2.** $y = e^{-x/2}\left(c_1 \cos \left(\dfrac{\sqrt{3}}{2} x \right) + c_2 \sin \left(\dfrac{\sqrt{3}}{2} x \right) \right)$ **3.** $y = e^{2x} (c_1 \cos x + c_2 \sin x)$

4. $y = e^{-x}\left(c_1 \cos (\sqrt{7} x) + c_2 \sin (\sqrt{7} x) \right)$ **5.** $y = e^{-x}\left(c_1 \cos (\sqrt{3} x) + c_2 \sin (\sqrt{3} x) \right)$

6. $y = e^{2x}\left(c_1 \cos (\sqrt{3} x) + c_2 \sin (\sqrt{3} x) \right)$ **7.** $y = e^{5x}(c_1 \cos x + c_2 \sin x)$

8. $y = e^{-2x/3}\left(c_1 \cos \left(\dfrac{\sqrt{23}}{3} x \right) + c_2 \sin \left(\dfrac{\sqrt{23}}{3} x \right) \right)$ **9.** $y = e^{x/2}\left(c_1 \cos \left(\dfrac{\sqrt{3}}{2} x \right) + c_2 \sin \left(\dfrac{\sqrt{3}}{2} x \right) \right)$

10. $y = e^{-x/2}\left(c_1 \cos \left(\dfrac{\sqrt{7}}{2} x \right) + c_2 \sin \left(\dfrac{\sqrt{7}}{2} x \right) \right)$ **11.** $y = -\dfrac{1}{2} \sin 2x$

12. $y = e^{2x}\left(\cos (3x) - \dfrac{2}{3} \sin (3x) \right)$ **13.** $y = e^{-x}(\cos x + \sin x)$ **14.** $y = e^{x/2}\left(\cos \left(\dfrac{\sqrt{3}}{2} x \right) - \dfrac{\sqrt{3}}{3} \sin \left(\dfrac{\sqrt{3}}{2} x \right) \right)$

15. $y = e^{2x}\left(-\cos \sqrt{3}\, x + \dfrac{2\sqrt{3}}{3} \sin \sqrt{3}\, x \right)$ **17.** $y = c \sin x$ **18.** (a) (i), (vi) (b) (iv), (v), (viii), (iii) (c) (vi) (d) (viii)

19. (b) $v' = x^2 - \dfrac{1}{x} v - v^2$, $v(x) = x$ (c) $y = e^{x^2/2}$ (d) $y = e^{-x^2/2}$

20. (a) $e^{i\pi/2}$ (b) $e^{3\pi i/2}$ (c) $3 e^{i\pi}$ (d) $\sqrt{2} e^{i\pi/4}$ (e) $\sqrt{2} e^{7\pi i/4}$ (f) $\sqrt{2} e^{3\pi i/4}$ **21.** $y = x$

Section 3.6

1. $y = c_1 + c_2 x + \frac{1}{2}x^2$ **2.** $y = c_1 + c_2 x + \frac{1}{2}x^2$ **3.** $y = c_1 + c_2 e^{-x} + x$ **4.** $y = c_1 \cos x + c_2 \sin x + 1$

5. $y = c_1 \cos x + c_2 \sin x + x$ **6.** $y = c_1 e^{-x} + c_2 e^{-2x} + \frac{1}{10}(\sin x - 3 \cos x)$

7. $y = e^{-x/2}\left(c_1 \cos\left(\frac{\sqrt{3}}{2}x\right) + c_2 \sin\left(\frac{\sqrt{3}}{2}x\right)\right) + x^2 - 2x$ **8.** $y = c_1 + c_2 e^{-x} + \frac{1}{3}x^3$ **9.** $y = c_1 e^{-2x} + c_2 e^{4x} - xe^x$

10. $y = c_1 e^{-x} + c_2 x e^{-x} + \frac{1}{12}x^4 e^{-x}$ **12.** (a) $y = -\cos 2x$ (b) $y = -2x^2$ (c) $y = -\frac{1}{3}\cos 2x + 3x^2$

13. (b) $y = -1 + \frac{c_1}{x^2}$ **14.** $y = c + x$ **15.** $y = c_1 + c_2 x + \frac{1}{2}x^2$ **16.** $y = ce^{-x} + 1$

17. $(x, y) = c(-2, 1) + (1, 0)$ **18.** $(x, y) = (0, 0) + (1, 0) = (1, 0)$ **19.** $(x, y, z) = c(-2, 1, 1) + (1, 0, 0)$

Section 3.7

1. $y = x + c$ **2.** $y = ce^{-x} + 1$ **3.** $y = ce^{-x} + x - 1$ **4.** $y = c_1 + c_2 x + \frac{1}{2}x^2$ **5.** $y = c_1 + c_2 e^{-4x} + \frac{x}{4}$

6. $y = c_1 \cos 2x + c_2 \sin 2x + \frac{1}{4}$ **7.** $y = c_1 + c_2 e^{-4x} + \frac{1}{8}x^2 - \frac{1}{16}$ **8.** $y = c_1 e^x + c_2 e^{-2x} + 3x$

9. $y = c_1 \cos x + c_2 \sin x + 3e^x + 3$ **10.** $y = c_1 e^{2x} + c_2 e^{-x} - 3e^x$ **11.** $y = c_1 + c_2 e^{-x} - \frac{6}{5}\sin 2x - \frac{3}{5}\cos 2x$

12. $y = e^{-2x}(c_1 \cos x + c_2 \sin x) + \frac{1}{5}e^x$ **13.** $y = c_1 + c_2 e^{-3x} + \frac{1}{2}(\sin x - \cos x)$

14. $y = c_1 e^{-2x} + c_2 x e^{-2x} + xe^{-x} - 2e^{-x}$ **15.** $y = c_1 e^x + c_2 e^{-x} - \frac{1}{2}(x \sin x + \cos x)$

16. $y = c_1 e^x + c_2 e^{2x} + \frac{1}{2}e^x(\cos x - \sin x)$ **17.** $y = c_1 e^{2x} + c_2 x e^{2x} + \frac{1}{6}x^3 e^{2x}$

18. $y = c_1 \cos x + c_2 \sin x - 2 \cos 2x + 6$ **19.** $y = c_1 e^x + c_2 e^{3x} + 2 \cos x - 4 \sin x$

20. $y = c_1 e^{-x} + e^x(c_2 - 2x + 2x^2)$ **21.** $y = c_1 e^{2x} + c_2 e^{3x} + \frac{1}{24}(7 \cosh x + 5 \sinh x)$ **22.** $y = e^x - 1$

23. $y = \cos x + 2x$ **24.** $y = e^x - \frac{1}{2}e^{-2x} - x - \frac{1}{2}$ **25.** $y = e^{2x} + xe^x$ **26.** $y = \frac{1}{2}x^2 e^{2x}$

27. $y = \frac{1}{3}\sin x - \frac{1}{12}\sin 4x$ **28.** $y = -\cos 2x + 3 \sin 2x$ **29.** $y_p = e^x(Ax^3 + Bx^2 + Cx)$

30. $y_p = e^x(Ax^2 + Bx + C)\cos x + e^x(Dx^2 + Ex + F)\sin x$

31. $y_p = e^x(Ax^3 + Bx^2 + Cx)\cos 2x + e^x(Dx^3 + Ex^2 + Fx)\sin 2x$ **32.** $y_p = (Ax^3 + Bx^2 + Cx)e^{3x} + D \cos x + E \sin x$

33. $y_p = A \cos x + B \sin x$ **34.** $y_p = Axe^x + Be^{-x} + Cx + D$

35. $y_p = e^{-x}(Ax^4 + Bx^3 + Cx^2)\cos 2x + e^{-x}(Dx^4 + Ex^3 + Fx^2)\sin 2x$

36. $y_p = Ax^2 + Bx + C + D \cos x + E \sin x + (Fx + G)e^{-x}$ **37.** $y_p = (Ax^4 + Bx^3 + Cx^2)e^x + D \cos x + E \sin x$

38. $y_p = (Ax + B)\cos x + (Cx + D)\sin x$ **39.** $y = c_1 e^x + c_2 x e^x + \cos x$

40. $y = c_1 \cos 5x + c_2 \sin 5x + \frac{1}{4}\sin x$ **41.** $y = c_1 \cos 5x + c_2 \sin 5x - 2x \cos 5x$

45. (a) $y_1 = y_0 - \frac{1}{2}gt^2$ (target) $y_2 = v_0 t - \frac{1}{2}gt^2$ (dart) (c) $y = y_0 - \frac{1}{2}g\left(\frac{x_0}{v}\right)^2, d = \sqrt{x_0^2 + y_0^2}$

Section 3.8

1. $y = c_1 + c_2 x + \dfrac{1}{2}x^2$ **2.** $y = c_1 \cos x + c_2 \sin x + 1$ **3.** $y = c_1 e^x + c_2 e^{-x} - x - 1$

4. $y = c_1 e^x + c_2 e^{-x} + \dfrac{5}{2}xe^x$ **5.** $y = c_1 + c_2 e^{-x} + 2x^2 - 4x$ **6.** $y = c_1 + c_2 e^{-x} - xe^{-x}$

7. $y = c_1 \cos x + c_2 \sin x - \dfrac{x}{2}\cos x$ **8.** $y = e^x\left(c_1 \cos x + c_2 \sin x - \dfrac{x}{2}\cos x\right)$

9. $y = c_1 \cos x + c_2 \sin x - x \cos x + \sin x \ln |\sec x|$ **10.** $y = c_1 \cos x + c_2 \sin x + x \cos x + \sin x \ln |\sin x|$

11. $y = c_1 e^{2x} + c_2 e^x + (e^x + e^{2x})\ln(1 + e^{-x})$ **12.** $y = c_1 e^{-x} + c_2 xe^{-x} + \dfrac{1}{2}x^2 e^{-x}\ln x - \dfrac{3}{4}x^2 e^{-x}$

13. $y_p = -x \sin x$ **14.** $y_p = \dfrac{1}{4}x^2 \ln x + \dfrac{1}{12}x^4$ **15.** $y_p = e^{-x}\left(\dfrac{1}{2} - x\right)$ **16.** $y_p = x^{1/2}$

17. $y_p = \displaystyle\int \sinh(x - s)e^{-s^2}\,ds$ **18.** $y = c + x$ **19.** $y = ce^{-2x} + \dfrac{1}{2}$ **20.** $y = ce^{-4x} + \dfrac{1}{5}e^x$ **21.** $y = \dfrac{c}{x^2} + x^2$

22. $y = \dfrac{c}{x} + x^3 + x$ **23.** $y = \dfrac{c}{x} - \dfrac{\cos x}{x}$ **25.** $y = c_1 e^x + c_2 e^{-x} + c_3 e^{2x} - \dfrac{1}{2}xe^x$

Section 3.9

1. $u = \cos t, \left(R = 1, \delta = 0, f = \dfrac{\omega_0}{2\pi} = \dfrac{1}{2\pi}, T = 2\pi\right)$ **2.** $u = \sin t, \left(R = 1, \delta = \dfrac{\pi}{2}, f = \dfrac{\omega_0}{2\pi} = \dfrac{1}{2\pi}, T = 2\pi\right)$

3. $u = \sqrt{2}\cos\left(t - \dfrac{\pi}{4}\right), \left(R = \sqrt{2}, \delta = \dfrac{\pi}{4}, f = \dfrac{\omega_0}{2\pi} = \dfrac{1}{2\pi}, T = 2\pi\right)$

4. $u = \sqrt{2}\cos\left(3t - \dfrac{\pi}{4}\right), R = \sqrt{2}, \delta = \dfrac{\pi}{4}, f = \dfrac{\omega_0}{2\pi} = \dfrac{3}{2\pi}, T = \dfrac{2\pi}{3}$

5. $u = \sqrt{2}\cos\left(2t - \dfrac{7\pi}{4}\right), \left(R = \sqrt{2}, \delta = \dfrac{7\pi}{4}, f = \dfrac{\omega_0}{2\pi} = \dfrac{1}{\pi}, T = \pi\right)$ **6.** (a) $\ddot{u} + 64u = 0 \quad u(0) = \dfrac{1}{4} \quad \dot{u}(0) = 1$

(b) $u = \dfrac{\sqrt{5}}{8}\left(\dfrac{2\sqrt{5}}{5}\cos 8t + \dfrac{\sqrt{5}}{5}\sin 8t\right) \doteq 0.28\cos(8t - 0.46)$

(c) $R = 0.28$ ft, $\delta = 0.46, f = \dfrac{\omega_0}{2\pi} = \dfrac{4}{\pi}$ oscillations/sec, $T = \dfrac{\pi}{4}$ sec   **7.** (a) $\ddot{u} + 64u = 0 \ u(0) = \dfrac{1}{3} \ \dot{u}(0) = -4$

(b) $\dot{u} + 64u = 0 \ u(0) = -\dfrac{1}{6} \ \dot{u}(0) = 1$

8. (a) $u = \dfrac{1}{6}\cos 16t$ (b) $R = \dfrac{1}{6}$ ft, $T = \dfrac{\pi}{8}$ sec, $f = \dfrac{8}{\pi}$ oscillations/sec (c) $\dfrac{\pi}{32}$ sec, $v = -\dfrac{8}{3}$ ft/sec

9. $u = -\dfrac{1}{3}\cos 8t + \dfrac{1}{4}\sin 8t \doteq \dfrac{5}{12}\cos(8t - 2.5)$ **10.** (a) $\sqrt{2}\cos\left(t - \dfrac{\pi}{4}\right)$ (b) $\sqrt{2}\cos\left(t - \dfrac{7\pi}{4}\right)$ (c) $\sqrt{2}\cos\left(t - \dfrac{3\pi}{4}\right)$

(d) $\sqrt{2}\cos\left(t - \dfrac{5\pi}{4}\right)$ **11.** $R = \sqrt{c_1^2 + c_2^2}, \sin\delta = -\dfrac{c_1}{R}, \cos\delta = \dfrac{c_2}{R}\left(\tan\delta = -\dfrac{c_1}{c_2}\right)$

15. Periods are the same.   **16.** $T = 2\pi\sqrt{\dfrac{L}{g}}$ **18.** $f = \dfrac{1}{2\pi}\sqrt{\dfrac{kR^2}{mR^2 + I}}$ cycles/unit time   **20.** 653 lb

21. (a) $x(t) = \sqrt{R^2 - d^2}\cos\sqrt{\dfrac{g}{R}}t$ (b) 2552 seconds (42 min)

Section 3.10

1. Damped amplitude $= 2e^{-t}$, damped natural frequency $= \dfrac{1}{\pi}$, damped period $= \pi$, time constant $= 1$

2. Damped amplitude $= 3e^{-2t}$, damped natural frequency $= \dfrac{\sqrt{3}}{2\pi}$, damped period $= \dfrac{2\pi}{\sqrt{3}}$, time constant $= \dfrac{1}{2}$

3. Damped amplitude $= 5e^{-0.25t}$, damped natural frequency $= \dfrac{1}{2\pi}$, damped period $= 2\pi$, time constant $= 4$

4. Damped amplitude $= e^{-2t}$, damped natural frequency $= \dfrac{2}{\pi}$, damped period $= \dfrac{\pi}{2}$, time constant $= \dfrac{1}{2}$ **5.** Underdamped

6. Critically damped   **7.** Critically damped   **8.** Critically damped   **9.** Critically damped   **10.** Overdamped
11. Underdamped

12. (a) $u = e^{-2t}\left(\dfrac{1}{2}\cos 2\sqrt{3}t + \dfrac{\sqrt{3}}{6}\sin 2\sqrt{3}t\right) = \dfrac{\sqrt{3}}{3}e^{-2t}\cos\left(2\sqrt{3}t - \dfrac{\pi}{6}\right)$

(b) Damped amplitude $= \dfrac{\sqrt{3}}{3}e^{-2t}$ (ft), damped natural frequency $= \dfrac{\sqrt{3}}{\pi}$ (oscillations/sec), damped period $= \dfrac{\pi}{\sqrt{3}}$ (sec), damped

time constant $= 0.5$ (sec)   **13.** (a) $u = \dfrac{\sqrt{2}}{4}e^{-8\sqrt{2}t}\cos\left(8\sqrt{2}t - \dfrac{\pi}{4}\right)$

(b) Damped amplitude $= \dfrac{\sqrt{2}}{4}e^{-8\sqrt{2}t}$, damped natural frequency $= \dfrac{4\sqrt{2}}{\pi}$, damped period $= \dfrac{\pi}{4\sqrt{2}}$

14. After $\dfrac{1}{16}$ sec the mass reaches a maximum displacement of 0.303 ft.

15. Define $b = d/m$ and $c = k/m$ and solve the two given equations for b and c. The differential equation is $\ddot{u} + b\dot{u} + cu = 0$.

18. (c) $d = 2m\,\Delta/T_d$ (m, Δ, and T_d can be observed directly.)   **19.** $\doteq 32$ mph   **20.** $q = \dfrac{\sqrt{10}}{3}q_0e^{-20t}\cos(60t - 0.32)$

21. $y'' = 0$ (The shortest path between $(0, 0)$ and $(1, 1)$ is $y = x$.)
22. $y'' - y = 0$ ($y = \cosh x$ gives the smallest surface area of revolution.)

23. $m\ddot{x} + kx = 0$ $x(t) = c_1\cos \omega t + c_2\sin \omega t$ $(w^2 = k/m)$ **27.** (a) $\dfrac{dy}{d\tau} - ay = 0$ $y(0) = 1$ $y(\tau) = e^{a\tau}$

(b) $\dfrac{d^2y}{d\tau^2} + y = 0$ $y(0) = 1$ $y'(0) = 0$ $y(\tau) = \cos\tau$ (c) $\dfrac{d^2y}{d\tau^2} - 3\dfrac{dy}{d\tau} + 2y = 0$ $y(0) = 0$ $y'(0) = -1$ $y(\tau) = e^{\tau} - e^{2\tau}$

Section 3.11

1. $-2\sin t\sin 2t$ **2.** $2\sin t\cos 2t$ **3.** $u_{ss} = \dfrac{1}{5}\cos(t - \delta)$, $\delta = \tan^{-1}(1.33) \doteq 0.93$ rad

4. $u_{ss} = \dfrac{2\sqrt{5}}{5}\cos(t - \delta)$, $\delta = \tan^{-1}(2) \doteq 1.1$ rad

5. $u_{ss} = \dfrac{4}{\sqrt{73}}\cos(t - \delta)$, $\delta = \tan^{-1}(-3/8) = \pi - \tan^{-1}(3/8) \doteq 2.78$ rad

6. (a) $\omega = 2\sqrt{3}$ rad/sec (b) $u = \dfrac{2}{3}\sin(2\sqrt{2}t) - \dfrac{4\sqrt{3}}{3}t\cos(2\sqrt{3}t)$

8. (a) $h(t) = \dfrac{49}{80\pi^2} \cos\left(\dfrac{2\pi t}{7}\right) + 3$ (buoy above the water line) (b) No

9. $u = \dfrac{3}{2} \sin 6t$ **10.** $u = -\dfrac{1}{4}e^{-2t}(3 \sin 4t + 4 \cos 4t) + \dfrac{1}{2}(\sin 2t + 2 \cos 2t)$

11. Kick with the same frequency as the swing.

Section 3.12

1. $W(1, x, x^2) = 2$ **2.** $W(1, x, x^2, x^3, ..., x^n) = n!$ **3.** $W(e^x, xe^x, e^{-x}) = 4e^x$

4. $W(1, \cos x, \sin x) = 1$ **5.** $y = c_1 + c_2 x + c_3 x^2$ **6.** $y = c_1 + c_2 x + c_3 x^2 + c_4 x^3$

7. $y = c_1 + c_2 \cos x + c_3 \sin x$ **8.** $y = c_1 + c_2 x + c_3 \cos x + c_4 \sin x$ **9.** $y = c_1 + c_2 x + c_3 e^x$

10. $y = c_1 e^{-x} + e^{x/2}\left(c_2 \cos\left(\dfrac{\sqrt{3}}{2}x\right) + c_3 \sin\left(\dfrac{\sqrt{3}}{2}x\right)\right)$ **11.** $y = c_1 + c_2 e^x + c_3 e^{2x}$

12. $y = c_1 e^{4x} + c_2 e^{2x} + c_3 e^{-x}$ **13.** $y = c_1 e^{4x} + c_2 e^{-x} + c_3 x e^{-x}$ **14.** $y = c_1 e^{3x} + c_2 e^{-2x} \cos x + c_3 e^{-2x} \sin x$

15. $y = c_1 e^{-x} + c_2 e^x + c_3 x e^x$ **16.** $y = c_1 e^x + c_2 x e^x + c_3 x^2 e^x$ **17.** $y = c_1 e^x + c_2 e^{2x} + c_3 e^{3x}$

18. $y = e^{\sqrt{2}x/2}\left(c_1 \cos\left(\dfrac{\sqrt{2}}{2}x\right) + c_2 \sin\left(\dfrac{\sqrt{2}}{2}x\right)\right) + e^{-\sqrt{2}x/2}\left(c_3 \cos\left(\dfrac{\sqrt{2}}{2}x\right) + c_4 \sin\left(\dfrac{\sqrt{2}}{2}x\right)\right)$

19. $y = c_1 e^{4x} + c_2 e^{-x} + c_3 x e^{-x} + c_4 x^2 e^{-x}$ **20.** $y = c_1 \sin x + c_2 \cos x + c_3 e^{2x} + c_4 e^{-2x}$

21. $y = c_1 e^x + c_2 x e^x + c_3 e^{-x} + c_4 x e^{-x}$ **22.** $y_p = 2x + 1$ **23.** $y_p = x^2$ **24.** $y_p = \sin x$ **25.** $y_p = \cos x$

26. $y_p = e^x$ **27.** $y_p = \sin x$ **28.** $y_p = -\dfrac{1}{2}x^2$ **29.** $y_p = \dfrac{1}{2}x^2 + 2x$ **30.** $y_p = -\dfrac{1}{24}x^{-1}$

32. $y_p = x(A_0 x^3 + A_1 x^2 + A_2 x + A_3) + Bx^2 e^x$ **33.** $y_p = x(A_0 x^2 + A_1 x + A_2) + (B_0 x + B_1)\cos x + (C_0 x + C_1)\sin x$

34. $y_p = \dfrac{1}{2}\int\left(\sinh(x - t) - \sin(x - t)\right)f(t)\, dt$ **35.** $y = \dfrac{w_0 x^2}{24\,EI}\left(x^2 - 4x + 6\right)$, Sag $= \dfrac{w_0}{8\,EI}$

36. $y = \dfrac{w_0 x}{24\,EI}\left(x^3 - 2x^2 + 1\right)$, Sag $= \dfrac{5w_0}{384\,EI}$

37. $y^{(n)} = y \diamond y(x) = \sum\limits_{k=1}^{n} c_k e^{r_{nk}x}$ where $r_{nk} = e^{(2\pi i/n)k}$ and c_k is an arbitrary constant. For $n = 1, 2, 3$, and 4 this reduces to

$n = 1 \diamond y(x) = c_1 e^x$
$n = 2 \diamond y(x) = c_1 e^{-x} + c_2 e^x$

$n = 3 \diamond y(x) = c_1 e^{-x/2} \cos\left(\dfrac{\sqrt{3}}{2}x\right) + c_2 e^{-x/2} \sin\left(\dfrac{\sqrt{3}}{2}x\right) + c_3 e^x$

$n = 4 \diamond y(x) = c_1 \cos x + c_2 \sin x + c_3 e^{-x} + c_4 e^x$
where the c_k's are arbitrary constants.

CHAPTER 4

Section 4.1

1. $R = \infty, (-\infty, \infty)$ **2.** $R = \infty, (-\infty, \infty)$ **3.** $R = 1, [-1, 1)$ **4.** $R = 1, (-1, 1)$ **5.** $R = 1, (-1, 1)$

6. $R = 1, (-1, 1)$ **7.** $R = 1, [-1, 1]$ **8.** $R = \sqrt{5}, (-\sqrt{5}, \sqrt{5})$ **9.** $2\sum\limits_{n=0}^{\infty} \dfrac{x^{2n}}{(2n)!}$ **10.** $\sum\limits_{n=0}^{\infty}\left(1 + \dfrac{1}{n}\right)x^n$

11. (a) $f'(x) = 1 + 2x + 3x^2 + 4x^3 + \cdots + nx^{n-1} + \cdots = \dfrac{1}{(1-x)^2}$

(b) $f''(x) = 2 + 6x + 12x^2 + \cdots + n(n-1)x^{n-2} + \cdots = \dfrac{2}{(1-x)^3}$

(c) $\displaystyle\int f(x)\,dx = x + \dfrac{x^2}{2} + \dfrac{x^3}{3} + \cdots + \dfrac{x^n}{n} + \cdots = -\ln(1-x)$ **18.** $1 + 2x + x^2$

19. $4 + 4(x-1) + (x-1)^2$ **20.** $(x+1)^2$ **21.** $1 + (x-1)$ **22.** $1 + 2(x-1) + (x-1)^2$

23. $1 - x + x^2 - x^3 + \cdots$ **24.** $\dfrac{1}{2} - \dfrac{1}{4}(x-1) + \dfrac{1}{8}(x-1)^2 - \dfrac{1}{16}(x-1)^3 + \cdots$

25. $1 - (x-1) + (x-1)^2 - (x-1)^3 + \cdots$ **26.** $x + x^2 + x^3 + \cdots$

27. $e + e(x-1) + \dfrac{e}{2!}(x-1)^2 + \dfrac{e}{3!}(x-1)^3 + \cdots$ **28.** $\dfrac{2}{\sqrt{\pi}} \displaystyle\sum_{n=0}^{\infty} \dfrac{(-1)^n x^{2n+1}}{(2n+1)n!}$

29. $\displaystyle\sum_{n=0}^{\infty} \dfrac{(-1)^n x^{2n+1}}{(2n+1)!\,(2n+1)}$

Section 4.2

1. All ordinary points **2.** $x = 0$ singular points, all other points are ordinary points
3. $x = \pm 1$ singular points, all other points are ordinary points **4.** $x = \pm i$ singular points, all other points are ordinary points
5. $x = 0$ is a singular point, all other points are ordinary points

6. $\displaystyle\sum_{n=0}^{\infty} \left((n+1)a_{n+1} - (n+1)a_n \right) x^n = 0 \quad a_{n+1} = a_n \quad (n = 0, 1, 2, \ldots)$

7. $\displaystyle\sum_{n=0}^{\infty} \left(2(n+2)(n+3)a_{n+2} + (n+1)a_n \right) x^n = 0 \quad a_{n+2} = -\dfrac{n+1}{2(n+2)(n+3)}a_n \quad (n = 0, 1, 2, \ldots)$

8. $a_0 + \dfrac{1}{2}x^2$ **9.** $a_0 + x + x^2 + \dfrac{1}{3}x^3$ **10.** $a_0 + (a_0+1)x + \dfrac{1}{2!}(a_0+1)x^2 + \cdots + \dfrac{1}{n!}(a_0+1)x^n + \cdots = ce^x - 1$

11. $a_0 \displaystyle\sum_{n=1}^{\infty} \dfrac{2}{n!}x^{2n}$ **12.** $a_0\left(1 - 2x + \dfrac{3}{2}x^2 - \dfrac{1}{3}x^3 + \cdots\right)$

13. $a_0\left(1 + \dfrac{x^2}{2!} + \dfrac{x^4}{4!} + \cdots\right) + a_1\left(x + \dfrac{x^3}{3!} + \dfrac{x^5}{5!} + \cdots\right) = a_0 \cosh x + a_1 \sinh x$

14. $a_0\left(1 - \dfrac{1}{6}x^3 + \dfrac{1}{180}x^6 - \cdots\right) + a_1\left(x - \dfrac{1}{12}x^4 + \dfrac{1}{504}x^7 - \cdots\right)$ **15.** $a_0\left(1 + \dfrac{1}{12}x^4 + \cdots\right) + a_1\left(x + \dfrac{1}{20}x^5 + \cdots\right)$

16. $a_0(1 - x^2) + a_1\left(x - \dfrac{1}{6}x^3 - \dfrac{1}{120}x^5 - \dfrac{1}{1680}x^7 - \cdots\right)$

17. $a_0\left(1 - \dfrac{1}{2}x^2 - \dfrac{1}{6}x^3 + \dfrac{1}{12}x^4 + \dfrac{3}{40}x^5 - \cdots\right) + a_1\left(x + \dfrac{1}{2}x^2 - \dfrac{1}{8}x^4 - \dfrac{1}{40}x^5 + \cdots\right)$

18. $a_0\left(1 - 3x^2\right) + a_1\left(x - \dfrac{1}{3}x^3\right)$ **19.** $1 - 2x^2 + \dfrac{1}{3}x^4$

20. $a_0 + x + 3x^2$

Section 4.3

1. ∞ **2.** 1 **3.** 2 **4.** 1 **5.** $\sqrt{3/2}$

6. 1 **8.** $1, x, 2x^2 - 1, 4x^3 - 3x, 8x^4 - 8x^2 + 1$ **9.** $1, -x+1, \dfrac{1}{2}(x^2 - 4x + 2), -\dfrac{1}{6}(x^3 - 9x^2 + 18x - 6)$

10. $y_1(x) = 1 - \dfrac{1}{3!}x^3 + \dfrac{1}{5!}x^5 + \dfrac{1}{2\cdot3\cdot5\cdot6}x^6 + \cdots, \quad y_2(x) = x - \dfrac{1}{3\cdot4}x^4 + \dfrac{1}{2\cdot3\cdot5\cdot6}x^6 + \cdots$

11. $y_1(x) = 1 - \frac{1}{2}x^2 + \frac{1}{6}x^3 - \frac{1}{40}x^5 + \cdots$, $y_2(x) = x - \frac{1}{6}x^3 + \frac{1}{12}x^4 - \frac{1}{60}x^5 + \cdots$

12. $y = c\left(1 + \frac{x^2}{2} + \frac{x^4}{8}\right) + \frac{1}{2}x^2 + \frac{1}{16}x^4$

Section 4.4

1. No singular points   **2.** R.S.P. at $x = 0$ **3.** I.S.P. at $x = 0$ **4.** R.S.P. at $x = 0$ **5.** I.S.P. at $x = 0$
6. I.S.P. at $x = 0$ **7.** R.S.P. at $x = 0$ **8.** R.S.P. at $x = -1$, I.S.P. at $x = 0, 1$ **9.** R.S.P. at $x = 0$

10. R.S.P. at $x = 0, \pm i$ **11.** $y_1(x) = x^{1/3}\sum_{n=0}^{\infty} a_n x^n$ $y_2(x) = x^{-1/3}\sum_{n=0}^{\infty} b_n x^n$

12. $y_1(x) = \sum_{n=0}^{\infty} a_n x^n$ $y_2(x) = \sqrt{x}\sum_{n=0}^{\infty} b_n x^n$ **13.** $y_1(x) = \sum_{n=0}^{\infty} a_n x^n$ $y_2(x) = \frac{1}{x^2}\sum_{n=0}^{\infty} b_n x^n$

14. $y_1(x) = \sqrt{x}\sum_{n=0}^{\infty} a_n x^n$ $y_2(x) = \frac{1}{\sqrt{x}}\sum_{n=0}^{\infty} b_n x^n$ **15.** $y_1(x) = cx$

16. $y_1(x) = \frac{1}{x^2}\left(1 + x^2 + \frac{1}{2!}x^4 + \frac{1}{3!}x^6 + \cdots\right) = \frac{1}{x^2}e^{x^2}$ **17.** $y_1(x) = \sum_{n=0}^{\infty} \frac{(-1)^n}{(2n)!}x^n$ $y_2(x) = \sqrt{x}\sum_{n=0}^{\infty} \frac{(-1)^n}{(2n+1)!}x^n$

18. $y_1(x) = 1 + \sum_{n=1}^{\infty} \frac{1}{3^n n! [5\cdot 8\cdot 11\cdots(3n+2)]}x^{3n}$ $y_2(x) = x^{-2}\left(1 + \sum_{n=1}^{\infty} \frac{1}{3^n n! [1\cdot 4\cdot 7\cdots(3n-2)]}x^{3n}\right)$

19. $y_1(x) = \sum_{n=0}^{\infty} \frac{1}{(2n+1)(2n-1)}x^n$, $y_2(x) = 1 + x^{-1/2}\sum_{n=1}^{\infty} \frac{1}{(2n)(2n-1)}x^n$

20. $y_1(x) = x\left(1 - \frac{1}{2^3}x^2 + \frac{1}{2^6\cdot 3}x^4 - \cdots\right)$, $y_2(x) = x^{-1}\left(1 + x^2 - \frac{1}{2^3}x^4 + \frac{1}{2^6\cdot 3}x^6 - \cdots\right)$

21. $y_1(x) = x^3$ $y_2(x) = x^{-4}$ **22.** (c) $y = c_1 x^2 + c_2 x^{-3}$

23. $y = 1 + x + \frac{1}{2}x^2 + \frac{1}{3}x^3 + \frac{1}{24}x^4 + \cdots$ $(-\infty < x < \infty)$

24. There is a Frobenius solution expanded about $x_0 = 0$ that converges for all x.
25. There is a Frobenius solution expanded about $x_0 = 0$ that converges for all x.
26. There is a Frobenius solution expanded about $x_0 = 0$ that converges at least in $-1 < x < 1$.
27. There is a Frobenius solution expanded about $x_0 = 0$ that converges at least in $-2 < x < 2$.
28. There is a Frobenius solution expanded about $x_0 = 0$ that converges at least in $-1 < x < 1$.

Section 4.5

1. $y = c_1 J_{1/2}(x) + c_2 J_{-1/2}(x)$ **2.** $y = c_1 J_1(x) + c_2 Y_1(x)$ **3.** $y = c_1 J_3(x) + c_2 Y_3(x)$ **4.** $y = c_1 J_{2/3}(x) + c_2 J_{-2/3}(x)$
5. $y = c_1 J_{\pi}(x) + c_2 J_{-\pi}(x)$ **6.** $y = c_1 J_5(x) + c_2 Y_5(x)$ **8.** $y = c_1 x^{1/2} J_{1/2}(x) + c_2 x^{1/2} J_{-1/2}(x)$
12. $y = c_1 J_1(2x) + c_2 Y_1(2x)$ **14.** $y = c_1 x^{-1} + c_2 x^{-2}$ **15.** $y = c_1 x^2 + c_2 x^2 \ln x$
16. $y = c_1 x^{-1} \cos(2\ln x) + c_2 x^{-1}\sin(2\ln x)$

CHAPTER 5

Section 5.1

1. $\frac{5}{s}$ **2.** $\frac{1}{s^2}$ **3.** $\frac{1}{s-2}$ **4.** $\frac{1}{s+1}$ **5.** $\frac{2}{s^2+4}$ **6.** $\frac{s}{s^2+9}$ **7.** $\frac{10}{s^2-4}$ **8.** $\frac{s-1}{(s-1)^2+4}$ **9.** $\frac{e^{-s}}{s}$

10. $\frac{1}{s}(1-e^{-s})$ **11.** $\frac{1+e^{-s\pi}}{s^2+1}$ **12.** $\frac{1}{s^2}(1-e^{-s})$ **13.** $\frac{1}{(s-a)^2}$ **14.** $\frac{n!}{(s-a)^{n+1}}$ **15.** $\frac{2as}{(s^2+a^2)^2}$

16. $\frac{s^2+a^2}{(s-a)^2(s+a)^2}$ **17.** $\frac{a}{s} + \frac{b}{s^2} + \frac{2c}{s^3}$ **18.** $\frac{2s+1}{s(s+1)}$ **19.** $\frac{2s}{s^2-4}$ **20.** $\frac{3}{s} + \frac{1}{s^2} + \frac{2}{(s+1)^2+4}$

21. $\frac{1}{(s+2)^2} + \frac{6}{(s+1)^3}$ **22.** $\frac{6}{(s+3)^4} + \frac{4(s+1)}{(s+1)^2+9}$ **23.** $\frac{s+3a}{s^2-a^2}$ **24.** $\frac{1}{(s+3)^2} + \frac{2}{s^2+1}$ **25.** Continuous

26. Discontinuous at $t = 2$ **27.** Discontinuous at $t = 1$ **28.** Discontinuous at $t = 1$ **29.** Discontinuous at $t = 1, 2$
30. Discontinuous at $t = 1, 2$ **31.** Discontinuous at $t = 2$ **32.** Exponential order   **33.** Exponential order
34. Exponential order   **35.** Exponential order   **36.** Not exponential order   **37.** Exponential order

38. Exponential order   **39.** Exponential order but the derivative is not   **41.** (a) $\dfrac{a}{s^2 - a^2}$

(b) $\dfrac{s}{s^2 - a^2}$ (c) $\dfrac{b}{[s - (a + b)][s - (a - b)]}$ (d) $\dfrac{s - a}{[s - (a + b)][s - (a - b)]}$ **42.** Answers in Table 5.1   **44.** No

47. $\mathcal{L}\{f'\} = s\mathcal{L}\{f\} - f(0)$ **48.** (a) Not continuous at $t = 0$ (b) Not continuous at $t = 0$ **49.** (b) $\dfrac{1}{s} + \dfrac{2}{s^2} - \dfrac{2}{s^3} + \dfrac{12}{s^4}$

(c) Approximate a function $f(t)$ by a polynomial $P(t)$ and use the given formula to find the Laplace transform of $P(t)$.

50. (c) $\Gamma\left(\dfrac{3}{2}\right) = \dfrac{1}{2}\sqrt{\pi},\ \Gamma\left(\dfrac{5}{2}\right) = \dfrac{3}{2}\cdot\dfrac{1}{2}\sqrt{\pi}$ **51.** $\mathcal{L}\{J_0\} = \dfrac{1}{\sqrt{s^2 + 1}}$ (closed form)

Section 5.2

1. $\dfrac{2a}{s^3} + \dfrac{b}{s^2} + \dfrac{c}{s}$ **2.** $\dfrac{2}{s^3} + \dfrac{1}{s - 2} - \dfrac{2}{s}$ **3.** $\dfrac{2}{s^3} - \dfrac{18}{s^2} + \dfrac{81}{s}$ **4.** $\dfrac{e^{-1}}{s - 2}$ **5.** $\dfrac{1}{s} + \dfrac{2}{s - 1} + \dfrac{1}{s - 2}$ **6.** $\dfrac{6s}{(s^2 + 1)^2}$

7. $\dfrac{2(3s^2 - 4)}{(s^2 + 4)^3}$ **8.** $\dfrac{3}{(s + 2)^2 + 9}$ **9.** $\dfrac{5(s - 5)}{(s - 5)^2 + 4}$ **10.** $\dfrac{2}{(s + 3)^3}$ **11.** $\dfrac{s(s - 2)}{(s^2 - 2s + 2)^2}$ **12.** $\dfrac{2(3s^2 - 4s + 2)}{(s^2 - 2s + 2)^3}$

13. $\dfrac{8s}{s^4 + 64}$ **14.** $\dfrac{1}{2s} - \dfrac{s}{2(s^2 + 4)}$ **15.** $\dfrac{3}{s^2 - 9}$ **16.** $\dfrac{3s}{4(s^2 + 1)} + \dfrac{s}{4(s^2 + 9)}$ **17.** $\dfrac{1}{2s^2} + \dfrac{s^2 - 4}{2(s^2 + 4)^2}$

18. $\dfrac{m + n}{2[s^2 + (m + n)^2]} + \dfrac{n - m}{[s^2 + (n - m)^2]}$ **19.** $\dfrac{1}{s^2 + 1}$ **20.** $\dfrac{3}{s(s^2 + 9)}$ **21.** $\dfrac{b}{(s + 1)^2 + b^2}$ **22.** $\dfrac{2}{(s + 2)^3}$

23. $\dfrac{1}{(s - 2)^2}$ **24.** $\dfrac{2}{(s + 1)^3}$ **25.** $\dfrac{2bs}{(s^2 + b^2)^2}$ **26.** $\dfrac{2(3s^2 - 4)}{(s^2 + 4)^3}$ **27.** $\dfrac{1}{(s + 2)^2}$ **28.** $\dfrac{2}{(s - 1)^3}$ **29.** $\dfrac{b}{(s - a)^2 + b^2}$

32. (b) $3\left(\dfrac{1}{(s - 2)^4} + \dfrac{1}{(s + 2)^4}\right)$ (c) $\dfrac{1}{2}\left(\dfrac{1}{(s - 1)^2 + 1} + \dfrac{1}{(s + 1)^2 + 1}\right)$ **33.** (b) $\dfrac{1}{(s - 3)^3} - \dfrac{1}{(s + 3)^3}$

(c) $\dfrac{1}{2}\left(\dfrac{s - 1}{(s - 1)^2 + 1} + \dfrac{s + 1}{(s + 1)^2 + 1}\right)$ **35.** $\dfrac{2}{s^2 + 4}$ (c) $\dfrac{1}{s - 5}$ **36.** (a) $\dfrac{1}{s(s^2 + 1)}$ (b) $\dfrac{1}{s(s - 2)}$ **37.** $\cot^{-1} s$

38. $\dfrac{1}{(s - 1)(s + 1)^2}$ **39.** $\dfrac{2(s + 2)}{(s^2 + 4s + 20)^2}$ **40.** $\dfrac{2(s + 1)}{s(s^2 + 2s + 2)^2}$ **41.** $\dfrac{2(3s^2 + 6s + 2)}{s(s^2 + 2s + 2)^3}$

42. $\dfrac{1}{s^2(s + 1)^2}$ **43.** $\dfrac{1}{s(s - 1)}$

Section 5.3

1. $\dfrac{1}{2}t^2$ **2.** $2 + 3e^t + \dfrac{7}{2}t^2$ **3.** $\dfrac{5}{\sqrt{3}}\sin\sqrt{3}t$ **4.** $3e^{3t} + 4e^{-3t}$ **5.** $\dfrac{1}{3} - \dfrac{1}{3}e^{-3t}$ **6.** $e^{-t}\cos 3t$ **7.** te^{-2t}

8. $e^{3t}\left(3\cos 4t + \dfrac{7}{2}\sin 4t\right)$ **9.** $\dfrac{1}{3}e^{-2t} + \dfrac{2}{3}e^t$ **10.** $e^{2t} - e^{-3t}$ **11.** $3e^t - e^{-t}$ **12.** $\dfrac{7}{\sqrt{3}}e^{-2t}\sin 3t$

13. $2e^{-2t}\cos 3t + 4e^{-2t}\sin 3t$ **14.** t^3e^{-2t} **15.** $8e^{2t} - e^{-t}\cos 2t + 3e^{-t}\sin 2t$ **16.** $t - \dfrac{1}{2}\sin 2t$

17. $\sin t - \dfrac{1}{2}\sin 2t$ **18.** $2e^{-t} - 3e^{-2t} + e^{3t}$ **19.** $-\dfrac{2}{7} + \dfrac{1}{6}e^{-t} + \dfrac{47}{42}e^{-7t}$ **20.** $2e^t + 3te^t - e^{-3t}$ **21.** $\dfrac{\sin t}{t}$

22. $\dfrac{e^{bt} - e^{at}}{t}$ **23.** t **24.** $e^t - 1$ **25.** $1 - \cos t$ **26.** $(t - 1)e^t + 1$ **30.** $5e^{-t} - 18e^{-2t} + 15e^{-3t}$

31. (b) $\dfrac{3}{s - 2} - \dfrac{2}{s + 1}$

Section 5.4

1. $t + 1$ **2.** e^t **3.** $(t + 1)e^t$ **4.** $(t + 1)e^{-t}$ **5.** $e^t - t$ **6.** $2e^t - e^{2t}$ **7.** $3e^{-2t} + 2t - 2$

8. $2e^{-t} + \sin 3t - 2\cos 3t$ **9.** $\cos 3t - \frac{1}{3}\sin 3t + \frac{1}{6}t\sin 3t$ **10.** $y(t) \equiv 1$ **11.** $e^t(2\cos 2t + \sin 2t)$

12. $10te^{-5t}$ **13.** $3 - 4e^{-t} + e^{-2t}$ **14.** $-\frac{1}{2}\sin t + 2\cos t - \frac{1}{2}t\cos t$ **15.** $\frac{3}{4}e^t - \frac{3}{2}te^t + \frac{3}{2}t^2 e^t - \frac{3}{4}e^{-t}$

16. $e^{-2t} + \cos t + 1$ **17.** $\cos t$ **23.** $Y' + s^2 Y = 1$ **24.** $c_1 \sin t + c_2 \cos t + 1$

25. The solution of $y' = ky$ grows faster ($e^t > \cosh kt$).

Section 5.5

1. $au(t) + (b - a)u(t - 1) + (c - b)u(t - 2)$ **2.** $u(t) + (e^t - 1)u(t - 1) + (2 - e^t)u(t - 2)$

3. $u(t) - (4t - t^2 - 1)u(t - 1) + (1 - 4t + t^2)u(t - 2)$ **4.** $\sin \pi t\, u(t - 1) - \sin \pi t\, u(t - 2)$ **5.** $\frac{1}{s} - \frac{e^{-s}}{s}$

6. $\frac{1}{s} - \frac{2e^{-s}}{s} + \frac{e^{-2s}}{s}$ **7.** $\frac{e^{-s}}{s^2}$ **8.** $\frac{2e^{-2s}}{s^3}$ **9.** $\frac{e^{-\pi s}}{s^2 + 1}$ **10.** $\frac{e^{3-3s}}{s - 1}$ **11.** $\frac{1}{s}$

12. $\left(\frac{4}{s} + \frac{4}{s^2} + \frac{2}{s^3}\right)e^{-2s}$ **13.** $\frac{-se^{-\pi s}}{s^2 + 1}$ **14.** $\frac{1 + e^{-\pi s}}{s^2 + 1}$ **15.** $u(t - 1)$ **16.** $u(t - 1)(t - 1)$ **17.** $u(t - 2)e^{3(t-2)}$

18. $\frac{1}{2}u(t - 3)\sin 2(t - 3)$ **19.** $u(t - 4)e^{-4(t-4)}$ **20.** $u(t - 4)(t - 4)e^{-4(t-4)}$ **21.** $u(t - 1)(1 - e^{-(t-1)})$

22. $\cos t - u(t - \pi)\sin t$ **23.** $u(t - 1) - 2u(t - 2) + 2u(t - 3) - u(t - 4)$

24. $u(t - 1)(t - 1) - 2u(t - 2)(t - 2) + 2u(t - 3)(t - 3) - u(t - 4)(t - 4)$ **25.** $\frac{e^{-as}}{s} - \frac{e^{-bs}}{s}$ **26.** $\frac{e^{-s}}{s} - \frac{e^{-2s}}{s}$

27. $\frac{e^{-s}}{s^2} - \frac{e^{-3s}}{s^2}$ **28.** $\frac{e^{-\pi s}}{s^2 + 1} - \frac{e^{-2\pi s}}{s^2 + 1}$ **30.** $\frac{1}{s(1 + e^{-s})}$ **31.** $\frac{-e^s}{s(1 + e^{-s})}$ **32.** $\frac{1 - (1 - s)e^{-s}}{s^2(1 - e^{-s})}$

33. $\frac{1 + 2s\,e^{-s} - 2\,e^{-s} + e^{-2s}}{s^2(1 - e^{-2s})}$ **34.** $\frac{1}{(s^2 + 1)(1 - e^{-\pi s})}$

Section 5.6

1. $t - u(t - 1)(t - 1)$ **2.** $t - 2u(t - 1)(t - 1) + u(t - 2)(t - 2)$ **3.** $u(t - 1)(1 - e^{-(t-1)})$

4. $(t - \sin t) - u(t - 1)((t - 1) - \sin(t - 1))$ **5.** $\frac{1}{2}t^2 - \frac{1}{2}u(t - 1)(t - 1)^2$

6. $\sin t + u(t - 3)1 - \cos(t - 3)$ **7.** $\sin t + u(t - \pi)(1 - \cos(t - \pi)) - u(t - 2\pi)(1 - \cos(t - 2\pi))$

8. $\frac{1}{4}t - \frac{1}{8}\sin 2t - u\left(t - \frac{\pi}{2}\right) - u\left(t - \frac{\pi}{2}\right)\left(\frac{1}{4}\left(t - \frac{\pi}{2}\right) - \frac{1}{8}\sin 2\left(t - \frac{\pi}{2}\right)\right)$

9. $\cos 2t + \frac{1}{6}u(t - 2\pi)(2\sin(t - 2\pi) - \sin 2(t - 2\pi))$

10. $y(t) = \frac{1}{16}(\sin 2t - 2t\sin 2t) - \frac{1}{16}u(t - 2\pi)(\sin 2t - 2(t - 2\pi)\sin 2t)$

11. $\frac{1}{3}u(t - 2)\left(1 - e^{-3(t-2)/2}\cos\frac{\sqrt{3}}{2}(t - 2) - 2\sqrt{3}\,e^{-3(t-2)/2}\sin\frac{\sqrt{3}}{2}(t - 2)\right) + \frac{2}{\sqrt{3}}e^{-3t/2}\sin\frac{\sqrt{3}}{2}t$

12. $\frac{5}{2}(1 - e^{-t}(\cos 2t + \sin 2t) - u(t - \pi)[1 - e^{-(t-\pi)}(\cos 2t + \sin 2t)])$

13. $t + (e^{-t} - 1) + u(t - 1)(e^{-(t-1)} - 1) + u(t - 2)(e^{-(t-2)} - 1) + \cdots$

14. $1 - \cos t - 2u(t - \pi)(1 - \cos (t - \pi)) + 2u(t - 2\pi)(1 - \cos (t - 2\pi)) - \cdots$

15. $1 - u(t - \pi)(1 - \cos (t - \pi)) + u(t - 2\pi)(1 - \cos (t - 2\pi)) - \cdots$

16. $\frac{1}{2}t^2 - u(t - 1)(t - 1)^2 + \frac{1}{2}u(t - 2)(t - 2)^2$

Section 5.7

5. $y(t) \equiv 1$ **6.** $1 - u(t - 1)$ **7.** $u(t - 2\pi) \sin t$ **8.** $\sin t + u(t - \pi) \sin t + u(t - 2\pi) \sin t$

9. $\cos t + u(t - 2\pi) \sin t$ **10.** $e^{-2t} \sin t$ **11.** $e^{-t}(\cos t + 2 \sin t) - u(t - \pi)e^{-(t-\pi)} \sin t$

12. $-u(t - \pi)e^{-(t-\pi)} \sin t$ **13.** $2e^{-2t} \sin t - u(t - \pi)e^{-2(t-\pi)} \sin t + u(t - 2\pi)e^{-2(t-2\pi)} \sin t$

14. $\frac{A}{m}(u(t - a)(t - a) + u(t - 2a)(t - 2a) + u(t - 3a)(t - 3a) + \cdots)$

16. The solution of the "marching soldier" differential equation is $\sin t + u(t - 2\pi) \sin t + u(t - 4\pi) \sin t + \cdots$, which oscillates with larger and larger amplitude.

17. They both have the solution $y(t) = \sin t$.

18. The solution is $y(t) = Au(t - t_0) \sin (t - t_0)$, whose derivative is $y'(t) = Au(t - t_0) \cos (t - t_0)$.

19. e^{-t} **20.** $1 - e^{-t}$ **21.** $1 - e^{-t} - u(t - 1)(1 - e^{-(t-1)})$ **22.** $e^{-t} + u(t - 1)e^{-(t-1)}$

23. $e^{-t} + u(t - 1)e^{-(t-1)} + u(t - 2)e^{-(t-2)}$ **24.** $e^{-t} + u(t - 1)e^{-(t-1)} + u(t - 2)e^{-(t-2)} + \cdots$

Section 5.8

4. t **5.** $\frac{1}{2}t^2$ **6.** $\frac{1}{2}t^2$ **7.** $\frac{1}{6}t^3$ **8.** $\frac{t^{2k-1}}{(2k - 1)!}$ **9.** $\sinh t$ **10.** $\frac{1}{a} \sinh at$

11. No, $a * t = b$ is the same as $\frac{a}{2}t^2 = b$.   **12.** $y = 1 * f(t)$ **13.** $y = 1 + 1 * f(t)$ **14.** $y = e^{-t} * f(t)$

15. $y = e^{-t} + e^{-t} * f(t)$ **16.** $y = \cos t + \sin t * f(t)$ **17.** $y = \frac{1}{2}[\sin t + \sin t * f(t)]$ **18.** $y = (e^{-t} - e^{-2t}) * f(t)$

19. $y = e^{-2t}(\cos t + 4 \sin t) + e^{-2t} \sin t * f(t)$

20. $1 * 1 = t$ **21.** $1 * t = \frac{1}{2}t^2$ **22.** $1 * e^{-t} = 1 - e^{-t}$ **23.** $4t * e^{2t} = e^{2t} + 2t - 1$ **24.** $t * \sin t = t - \sin t$

25. $\sin t * \sin t = \frac{1}{2}(\sin t - t \cos t)$ **26.** (a) $\frac{2}{\sqrt{\pi}}t^{1/2}$ (b) $\frac{4}{3\sqrt{\pi}}t^{3/2}$ (c) $\frac{2}{\sqrt{\pi}}\left(\frac{8a}{15}t^{5/2} + \frac{2b}{3}t^{3/2} + ct^{1/2}\right)$

28. (a) $\frac{1}{\sqrt{\pi}}t^{-1/2}$ (b) $\frac{2}{\sqrt{\pi}}t^{1/2}$ (c) $\frac{1}{\sqrt{\pi}}\left(\frac{8}{3}at^{3/2} + 2bt^{1/2} + ct^{-1/2}\right)$ **29.** $y(t) = e^{-t}$ **30.** $y(t) = \sin t$

31. $y(t) = t^3 + \frac{1}{20}t^5$ **32.** $y(t) = e^{2t}$ **33.** $y(t) \equiv 1$ (surprise)   **36.** $y(t) = \sin t$ **37.** $y(t) = \sin t - \frac{1}{2}t \sin t$

38. $y(t) = \frac{1}{2}t^2$ **39.** $y(t) = t$ **40.** TF $= \frac{1}{s}$ IRF $= 1$ $y(t) = \int_0^t f(\tau) d\tau$

41. TF $= \frac{1}{s + a}$ IRF $= e^{-at}$ $y(t) = \int_0^t e^{-a(t-\tau)} f(\tau) d\tau$ **42.** TF $= \frac{1}{s^2 + 1}$ IRF $= \sin t$ $y(t) = \int_0^t \sin (t - \tau)f(\tau) d\tau$

43. TF $= \frac{1}{(s + 2)^2 + 1}$ IRF $= e^{-2t} \sin t$ $y(t) \int_0^t e^{-2(t-\tau)} \sin (t - \tau)f(\tau) d\tau$

44. (b) $y(t) = \int_0^t \sinh (\tau)f(t - \tau) d\tau$

CHAPTER 6

Section 6.1

1. $x_1' = x_2$ $(x_1 = y)$ $x_2' = 1$ $(x_2 = y')$ **2.** $x_1' = x_2$ $(x_1 = y)$ $x_2' = x_1 + 1$ $(x_2 = y')$
3. $x_1' = x_2$ $x_1(0) = 1$ $(x_1 = y)$ $x_2' = -x_1 + t$ $x_2(0) = 0$ $(x_2 = y')$
4. $x_1' = x_2$ $x_1(0) = 1$ $(x_1 = y)$ $x_2' = -2x_1 - 3x_2 + 1$ $x_2(0) = 0$ $(x_2 = y')$
5. $x_1' = x_2$ $(x_1 = y)$ $x_2' = x_3$ $(x_2 = y')$ $x_3' = -rx_1 - qx_2 - px_3 + f(x)$ $(x_3 = y'')$
6. $x_1' = x_2$ $(x_1 = y)$ $x_2' = x_3$ $(x_2 = y')$ $x_3' = x_4$ $(x_3 = y'')$ $x_4' = x_1$ $(x_4 = y''')$ **7.** $x(t) = c_1 e^t$ $y(t) = c_2 e^t$
8. $x(t) = c_1 e^t + c_2 e^{-t}$ $y(t) = c_1 e^t - c_2 e^{-t}$ **9.** $x(t) = c_1 e^t + c_2 e^{2t}$ $y(t) = c_1 e^t + 2c_2 e^{2t}$

10. $x(t) = c_1 e^t + c_2 e^{-t}$ $y(t) = c_1 e^t + \dfrac{1}{2}c_2 e^{-t}$ **11.** $x(t) = c_1 e^t + c_2 e^{2t} - \dfrac{3}{2}t - \dfrac{5}{4}$ $y(t) = c_1 e^t + 2c_2 e^{2t} - t - \dfrac{3}{2}$

12. $x(t) = c_1 e^{-t} + c_2 t e^{-t} + 1$ $y(t) = c_1 e^{-t} + c_2 \left(te^{-t} - \dfrac{1}{6}e^{-t}\right) + 1$

13. $x(t) = c_1 e^{2t} - c_2 e^{-2t} - 1$ $y(t) = c_1 e^{2t} - c_2 e^{-2t} + 1$
14. $x(t) = c_1 e^{2t} + c_2 t e^{2t} - e^t + 1$ $y(t) = (c_1 - c_2)e^{2t} + c_2 t e^{2t} - 2e^t - 1$
15. $x(t) = c_1 e^t + c_2 e^{-t} + c_3 \cos t + c_4 \sin t$ $y(t) = c_1 e^t - c_2 e^{-t} - c_3 \sin t + c_4 \cos t$
$z(t) = c_1 e^t + c_2 e^{-t} - c_3 \cos t - c_4 \sin t$ $w(t) = c_1 e^t - c_2 e^{-t} + c_3 \sin t - c_4 \cos t$
16. $x(t) = -e^{-t} + 1$ $y(t) = -e^{-t} + 1$ **17.** $x(t) = e^{-2t} - 1$ $y(t) = -e^{-2t} + 1$
18. $x(t) = e^{2t} - 2te^{2t} - e^t + 1$ $y(t) = 3e^{2t} - 2te^{2t} - 2e^t - 1$
19. $x_1(t) = c_1 \cos(-t + c_2)$ $x_2(t) = c_1 \sin(-t + c_2)$

21. $x(t) = c_1 t + \dfrac{c_2}{t}$, $y(t) = c_1 t + \dfrac{c_2}{2t}$

22. $x(t) = c_1 e^{2t} + c_2 e^{-t}$, $y(t) = \dfrac{1}{2}c_1 e^{2t} + 2c_2 e^{-t}$ **23.** $x(t) = c_1 e^{2t} \cos t + c_2 e^{2t} \sin t$ $y(t) = -c_1 e^{2t} \sin t + c_2 e^{2t} \cos t$

24. $r(t) = c_1 \sin(\sqrt{ab}\, t) + c_2 \cos(\sqrt{ab}\, t)$, $j(t) = c_1 \sqrt{\dfrac{b}{a}} \cos(\sqrt{ab}\, t) - c_2 \sqrt{\dfrac{b}{a}} \sin(\sqrt{ab}\, t)$ (these reduce if $a = b = 1$)

25. (a) Romeo is turned off by Juliet's advances $(a_{12} < 0)$ but is still spurred on by his own feelings $(a_{11} > 0)$. He is the "particular lover." (b) Romeo is excited by Juliet's love $(a_{12} > 0)$ but is afraid of his own feelings $(a_{11} < 0)$. He is the "insecure lover." (c) Romeo is turned off by Juliet's advances $(a_{12} < 0)$ and is afraid of love $(a_{11} < 0)$. He is the "engineer."

Section 6.2

1. $2A = \begin{bmatrix} -2 & 0 & 6 \\ 4 & 2 & 4 \\ -2 & 0 & 2 \end{bmatrix}$ **2.** $A + 2B = \begin{bmatrix} 1 & 6 & 3 \\ 2 & 3 & 2 \\ -1 & 0 & 3 \end{bmatrix}$ **3.** $2C - D$ is not defined.   **4.** $AB = \begin{bmatrix} -1 & -3 & 3 \\ 2 & 7 & 2 \\ -1 & -3 & 1 \end{bmatrix}$

5. $BA = \begin{bmatrix} 5 & 3 & 9 \\ 2 & 1 & 2 \\ -1 & 0 & 1 \end{bmatrix}$ **6.** $CD = \begin{bmatrix} 3 & -1 & 0 \\ 8 & -1 & 2 \\ 9 & 2 & 6 \end{bmatrix}$ **7.** $DC = \begin{bmatrix} 1 & -1 \\ 6 & 7 \end{bmatrix}$ **8.** $(DC)^T = \begin{bmatrix} 1 & 6 \\ -1 & 7 \end{bmatrix}$

9. $C^T D$ is not defined.   **10.** $D^T C$ is not defined.   **11.** $A^2 = \begin{bmatrix} -2 & 0 & 0 \\ -2 & 1 & 10 \\ 0 & 0 & -2 \end{bmatrix}$ **12.** $|A| = 2$ **13.** $|C|$ is not defined.

14. $|D|$ is not defined.   **15.** T   **16.** F   **17.** T   **18.** T   **19.** T   **20.** T   **21.** F   **22.** T   **23.** T

24. $A + 2B = \begin{bmatrix} 3 + i & 0 \\ 2 + 4i & 4 - i \end{bmatrix}$ **25.** $AB = \begin{bmatrix} -3 + i & -1 + i \\ 8 + 4i & 5 - 3i \end{bmatrix}$ **26.** $BA = \begin{bmatrix} 1 - i & -3 \\ 4i & 1 - i \end{bmatrix}$

27. $A' = \begin{bmatrix} 0 & \cos t \\ 1 & 0 \end{bmatrix}$ **28.** $B' = \begin{bmatrix} 1 & 0 \\ 0 & 2t \end{bmatrix}$ **29.** $(AB)' = \begin{bmatrix} 1 & 2t \sin t + t^2 \cos t \\ 2t & 1 \end{bmatrix}$ **30.** $(A + B)' = \begin{bmatrix} 1 & \cos t \\ 1 & 2t \end{bmatrix}$ **33.** -9

35. $A^{-1} = \begin{bmatrix} -5 & 2 \\ 3 & -1 \end{bmatrix}$ **36.** $e^A = \begin{bmatrix} 1 & 1 & \frac{1}{2} & \frac{1}{6} \\ 0 & 1 & 1 & \frac{1}{2} \\ 0 & 0 & 1 & 1 \\ 0 & 0 & 0 & 1 \end{bmatrix}$ **38.** First derivative: $\begin{bmatrix} 0 & 0 & 0 \\ 2 & 0 & 0 \\ 0 & 1 & 0 \end{bmatrix}$

39. Second derivative: $\begin{bmatrix} 0 & 0 & 0 \\ 0 & 0 & 0 \\ 2 & 0 & 0 \end{bmatrix}$ **40.** Third derivative: $\begin{bmatrix} 0 & 0 & 0 \\ 0 & 0 & 0 \\ 0 & 0 & 0 \end{bmatrix}$ **41.** First derivative: $\begin{bmatrix} 0 & -1 \\ 1 & 0 \end{bmatrix}$

42. $\begin{bmatrix} 0 & 0 & 0 \\ 2 & 0 & 0 \\ 0 & 1 & 0 \end{bmatrix}\begin{bmatrix} a \\ b \\ c \end{bmatrix} = \begin{bmatrix} 0 \\ 1 \\ 2 \end{bmatrix}$ $y = \frac{1}{2}t^2 + 2t + c$ **43.** $\begin{bmatrix} 0 & 0 & 0 \\ 0 & 1 & 0 \\ 0 & 1 & 1 \end{bmatrix}\begin{bmatrix} a \\ b \\ c \end{bmatrix} = \begin{bmatrix} 0 \\ 1 \\ 0 \end{bmatrix}$ $y = ae^{-t} + t - 1$

44. $\begin{bmatrix} 0 & 0 & 0 & 0 \\ 0 & 0 & 0 & 0 \\ 6 & 0 & 0 & 0 \\ 0 & 2 & 0 & 0 \end{bmatrix}\begin{bmatrix} a \\ b \\ c \\ d \end{bmatrix} = \begin{bmatrix} 0 \\ 0 \\ 1 \\ 0 \end{bmatrix}$ $y = \frac{1}{6}t^3 + ct + d$

45. $\begin{bmatrix} 0 & 0 & 0 \\ 0 & 0 & 0 \\ 0 & 0 & 1 \end{bmatrix}\begin{bmatrix} a \\ b \\ c \end{bmatrix} = \begin{bmatrix} 0 \\ 0 \\ 1 \end{bmatrix}$ $y = a\sin t + b\cos t + 1$

46. $\begin{bmatrix} 0 & 0 & 0 & 0 \\ 0 & 0 & 0 & 0 \\ 0 & 0 & 1 & 0 \\ 0 & 0 & 0 & 1 \end{bmatrix}\begin{bmatrix} a \\ b \\ c \\ d \end{bmatrix} = \begin{bmatrix} 0 \\ 0 \\ 1 \\ 0 \end{bmatrix}$ $y = a\sin t + b\cos t + t$

47. $\begin{bmatrix} 0 & 0 & 0 & 0 & 0 & 0 \\ 0 & 0 & 0 & 0 & 0 & 0 \\ 0 & -2 & 0 & 0 & 0 & 0 \\ 2 & 0 & 0 & 0 & 0 & 0 \\ 0 & 0 & 0 & 0 & 1 & 0 \\ 0 & 0 & 0 & 0 & 0 & 1 \end{bmatrix}\begin{bmatrix} a \\ b \\ c \\ d \\ e \\ f \end{bmatrix} = \begin{bmatrix} 0 \\ 0 \\ 1 \\ 0 \\ 0 \\ 0 \end{bmatrix}$ $y = -\frac{1}{2}t\cos t + c\sin t + d\cos t$ **48.** $y(t) = c_1\cos t + c_2\sin t - \frac{t}{2}\cos t$

49. $y = a + \frac{1}{3}(1 + 6a)t + \frac{1}{2}t^2$ (*a* arbitrary)

Section 6.3

1. Linearly independent   **2.** Linearly independent   **3.** Linearly independent   **4.** $\begin{bmatrix} x_1' \\ x_2' \end{bmatrix} = \begin{bmatrix} 0 & 1 \\ -1 & 0 \end{bmatrix}\begin{bmatrix} x_1 \\ x_2 \end{bmatrix} + \begin{bmatrix} 0 \\ 0 \end{bmatrix}$

5. $\begin{bmatrix} x_1' \\ x_2' \end{bmatrix} = \begin{bmatrix} 2 & 3 \\ 1 & -1 \end{bmatrix}\begin{bmatrix} x_1 \\ x_2 \end{bmatrix} + \begin{bmatrix} 0 \\ 0 \end{bmatrix}$ **6.** $\begin{bmatrix} x_1' \\ x_2' \end{bmatrix} = \begin{bmatrix} 2 & 1 \\ 4 & 1 \end{bmatrix}\begin{bmatrix} x_1 \\ x_2 \end{bmatrix} + \begin{bmatrix} 1 \\ \sin t \end{bmatrix}$ **7.** $\begin{bmatrix} x_1' \\ x_2' \end{bmatrix} = \begin{bmatrix} 2 & 1 \\ 1 & -1 \end{bmatrix}\begin{bmatrix} x_1 \\ x_2 \end{bmatrix} + \begin{bmatrix} e^t \\ 1 \end{bmatrix}$

8. $\begin{bmatrix} x_1' \\ x_2' \end{bmatrix} = \begin{bmatrix} 0 & 1 \\ -1 & 0 \end{bmatrix}\begin{bmatrix} x_1 \\ x_2 \end{bmatrix}$ $x_1 = y, x_2 = y'$ **9.** $\begin{bmatrix} x_1' \\ x_2' \end{bmatrix} = \begin{bmatrix} 0 & 1 \\ -1 & -2 \end{bmatrix}\begin{bmatrix} x_1 \\ x_2 \end{bmatrix} + \begin{bmatrix} 0 \\ 1 \end{bmatrix}$ $x_1 = y, x_2 = y'$

10. $\begin{bmatrix} x_1' \\ x_2' \\ x_3' \end{bmatrix} = \begin{bmatrix} 0 & 1 & 0 \\ 0 & 0 & 1 \\ -2 & 1 & -1 \end{bmatrix} \begin{bmatrix} x_1 \\ x_2 \\ x_3 \end{bmatrix} + \begin{bmatrix} 0 \\ 0 \\ e^t \end{bmatrix}$ $x_1 = y, x_2 = y', x_3 = y''$

11. $\begin{bmatrix} x_1' \\ x_2' \\ x_3' \\ x_4' \end{bmatrix} = \begin{bmatrix} 0 & 1 & 0 & 0 \\ 0 & 0 & 1 & 0 \\ 0 & 0 & 0 & 1 \\ -1 & 0 & 0 & 0 \end{bmatrix} \begin{bmatrix} x_1 \\ x_2 \\ x_3 \\ x_4 \end{bmatrix} = \begin{bmatrix} 0 \\ 0 \\ 0 \\ 1 \end{bmatrix}$ $x_1 = y, x_2 = y', x_3 = y'', x_4 = y'''$ **12.** $x_1' = x_1 + 2x_2$
$x_2' = 4x_1 - x_2$

13. $x_1' = x_1$ **14.** $x_1' = 4x_1 + 3x_2 + e^{-t}$ **15.** $x_1' = x_2$
$\quad\ \ x_2' = -x_2 + 1$ $\qquad\ x_2' = -x_1 - x_2$ $\qquad\ x_2' = x_3$
$\qquad\qquad\qquad\qquad\qquad\qquad\qquad\qquad\qquad\qquad\ x_3' = -2x_1 + x_2 + 3x_3 + \sin t$

16. $W = -e^{3t}$ (linearly independent)   **17.** $W = -5e^{2t}$ (linearly independent)
18. $W = -e^{2t}$ (linearly independent)   **19.** $W = 2e^{8t}$ (linearly independent)

Section 6.4

1. $\lambda_1 = 1, \lambda_2 = 2, \xi^{(1)} = \begin{bmatrix} 0 \\ 1 \end{bmatrix}, \xi^{(2)} = \begin{bmatrix} 1 \\ 0 \end{bmatrix}$ **2.** $\lambda_1 = -1, \lambda_2 = 4, \xi^{(1)} = \begin{bmatrix} -1 \\ 2 \end{bmatrix}, \xi^{(2)} = \begin{bmatrix} 2 \\ 1 \end{bmatrix}$

3. $\lambda_1 = 1, \lambda_2 = 1, \xi^{(1)} = \begin{bmatrix} 0 \\ 1 \end{bmatrix}, \xi^{(2)} = \begin{bmatrix} 1 \\ 0 \end{bmatrix}$ **4.** $\lambda_1 = 1, \lambda_2 = 2, \xi^{(1)} = \begin{bmatrix} -1 \\ 1 \end{bmatrix}, \xi^{(2)} = \begin{bmatrix} 0 \\ 1 \end{bmatrix}$

5. $\lambda_1 = -1, \lambda_2 = 1, \lambda_3 = 2, \xi^{(1)} = \begin{bmatrix} 0 \\ 1 \\ 0 \end{bmatrix}, \xi^{(2)} = \begin{bmatrix} 0 \\ -1 \\ 1 \end{bmatrix}, \xi^{(3)} = \begin{bmatrix} -1 \\ -1 \\ 1 \end{bmatrix}$

6. $\lambda_1 = 1, \lambda_2 = 2, \lambda_3 = 3, \xi^{(1)} = \begin{bmatrix} -1 \\ 1 \\ 2 \end{bmatrix}, \xi^{(2)} = \begin{bmatrix} -2 \\ 1 \\ 4 \end{bmatrix}, \xi^{(3)} = \begin{bmatrix} -1 \\ 1 \\ 4 \end{bmatrix}$

7. $\lambda_1 = -1, \lambda_2 = -3, \lambda_3 = 5, \xi^{(1)} = \begin{bmatrix} -2 \\ 1 \\ 1 \end{bmatrix}, \xi^{(2)} = \begin{bmatrix} 0 \\ -1 \\ 1 \end{bmatrix}, \xi^{(3)} = \begin{bmatrix} 1 \\ 1 \\ 1 \end{bmatrix}$

8. $\lambda_1 = 2, \lambda_2 = -1, \lambda_3 = -1, \xi^{(1)} = \begin{bmatrix} 1 \\ 1 \\ 1 \end{bmatrix}, \xi^{(2)} = \begin{bmatrix} 1 \\ 0 \\ -1 \end{bmatrix}, \xi^{(3)} = \begin{bmatrix} 0 \\ 1 \\ -1 \end{bmatrix}$

9. $x(t) = c_1 \begin{bmatrix} 1 \\ 2 \end{bmatrix} + c_2 e^{-5t} \begin{bmatrix} 2 \\ -1 \end{bmatrix}$ **10.** $x(t) = c_1 e^{2t} \begin{bmatrix} 1 \\ -1 \end{bmatrix} + c_2 e^{3t} \begin{bmatrix} 1 \\ -2 \end{bmatrix}$

11. $x(t) = c_1 e^{2t} \begin{bmatrix} 1 \\ 3 \end{bmatrix} + c_2 e^{4t} \begin{bmatrix} 1 \\ 1 \end{bmatrix}$ **12.** $x(t) = c_1 e^t \begin{bmatrix} 1 \\ 2 \end{bmatrix} + c_2 e^{2t} \begin{bmatrix} 0 \\ 1 \end{bmatrix}$

13. $x(t) = c_1 e^t \begin{bmatrix} 1 \\ 0 \\ -1 \end{bmatrix} + c_2 e^{2t} \begin{bmatrix} -2 \\ 1 \\ 0 \end{bmatrix} + c_3 e^{3t} \begin{bmatrix} 0 \\ 1 \\ -1 \end{bmatrix}$ **14.** $x(t) = c_1 e^{-t} \begin{bmatrix} -1 \\ 0 \\ 1 \end{bmatrix} + c_2 e^{3t} \begin{bmatrix} 1 \\ 4 \\ 3 \end{bmatrix} + c_3 e^{-2t} \begin{bmatrix} 1 \\ -1 \\ 3 \end{bmatrix}$

15. $x(t) = c_1 e^{3t} \begin{bmatrix} 1 \\ 0 \\ 1 \end{bmatrix} + c_2 e^{3t} \begin{bmatrix} -1 \\ 1 \\ 0 \end{bmatrix} + c_3 e^{-3t} \begin{bmatrix} -1 \\ -1 \\ 1 \end{bmatrix}$ **16.** $x_1(t) = 5e^{2t}$ **17.** $x_1(t) = \dfrac{5}{2}e^{2t} - \dfrac{1}{2}$
$\qquad\qquad\qquad\qquad\qquad\qquad\qquad\qquad\qquad\qquad\quad\ x_2(t) = 4e^{3t}$ $\qquad\quad x_2(t) = \dfrac{5}{2}e^{2t} + \dfrac{1}{2}$

18. $x_1(t) = 2e^{3t} - e^{-t}$
$x_2(t) = 2e^{2t} + e^{-t}$

19. $x_1(t) = 2e^{2t} - e^{3t}$
$x_2(t) = -2e^{2t} + 2e^{3t}$

20. $x_1(t) = \dfrac{3}{4} - \dfrac{3}{8}e^{2t} - \dfrac{3}{8}e^{-2t}$
$x_2(t) = \dfrac{3}{4} + \dfrac{3}{8}e^{2t} - \dfrac{9}{8}e^{-2t}$
$x_3(t) = \dfrac{1}{4} + \dfrac{3}{8}e^{2t} + \dfrac{3}{8}e^{-2t}$

21. $x_1(t) = 3e^{2t} - 1$
$x_2(t) = 3e^{2t} + 1$
$x_3(t) = 2e^{-t}$

22. (a) $\lambda^2 - 1 = 0$ (b) $y(t) = c_1 e^t + c_2 e^{-t}$

23. (a) $\lambda^2 + 3\lambda + 2 = 0$ (b) $y(t) = c_1 e^{-t} + c_2 e^{-2t}$

24. (a) Double root of $\lambda = 1$ (b) $\mathbf{x}^{(1)}(t) = e^t \begin{bmatrix} 2 \\ 1 \end{bmatrix}$ (c) $\mathbf{x}^{(2)}(t) = te^t \begin{bmatrix} 2 \\ 1 \end{bmatrix} + e^t \begin{bmatrix} 1 \\ 0 \end{bmatrix}$

25. $\mathbf{x}(t) = c_1 e^{5t} \begin{bmatrix} 1 \\ -2 \end{bmatrix} + c_2 \left(te^{5t} \begin{bmatrix} 1 \\ -2 \end{bmatrix} + e^{5t} \begin{bmatrix} 0 \\ 1 \end{bmatrix} \right)$

26. (b) $\mathbf{x}^{(1)}(t) = e^t \begin{bmatrix} 1 \\ -2 \\ 1 \end{bmatrix}$ (c) $\mathbf{x}^{(2)}(t) = te^t \begin{bmatrix} 1 \\ -2 \\ 1 \end{bmatrix} + e^t \begin{bmatrix} -1 \\ 1 \\ 0 \end{bmatrix}$ (d) $\mathbf{x}^{(3)}(t) = \dfrac{1}{2}t^2 e^t \begin{bmatrix} 1 \\ -2 \\ 1 \end{bmatrix} + te^t \begin{bmatrix} -1 \\ 1 \\ 0 \end{bmatrix} + e^t \begin{bmatrix} 1 \\ 0 \\ 0 \end{bmatrix}$

29. $\mathbf{x}(t) = c_1 t^{-1} \begin{bmatrix} 1 \\ 2 \end{bmatrix} + c_2 t^2 \begin{bmatrix} 2 \\ 1 \end{bmatrix}$

30. (a) $A = \begin{bmatrix} 0 & -1 \\ -1 & 0 \end{bmatrix}$ (c) $x_1(t) = \dfrac{1}{2}e^t + \dfrac{1}{2}e^{-t}$ (d) $y_1(t) = -\dfrac{1}{2}e^t + \dfrac{1}{2}e^{-t}$
$x_2(t) = \dfrac{1}{2}e^t - \dfrac{1}{2}e^{-t}$ $\quad y_2(t) = \dfrac{1}{2}e^t + \dfrac{1}{2}e^{-t}$

(e) The vectors $\mathbf{x} = \mathbf{x}(t)$ and $\mathbf{y} = \mathbf{y}(t)$ are perpendicular for all $t \geq 0$.

31. (a) $\dfrac{dI}{dt} = -k_1 I \quad k_1 = \dfrac{\ln 2}{6.7} \doteq 0.1034548$
$\dfrac{dX}{dt} = k_1 I - k_2 X \quad k_2 = \dfrac{\ln 2}{9.2} \doteq 0.0753420$
$\begin{bmatrix} I(t) \\ X(t) \end{bmatrix} = c_1 e^{-k_1 t} \begin{bmatrix} 1 \\ -1 \end{bmatrix} + c_2 e^{-k_2 t} \begin{bmatrix} 0 \\ 1 \end{bmatrix}$

Section 6.5

1. $\mathbf{x}(t) = c_1 \begin{bmatrix} \sin t \\ \cos t \end{bmatrix} + c_2 \begin{bmatrix} -\cos t \\ \sin t \end{bmatrix}$

2. $\mathbf{x}(t) = c_1 e^{-t} \begin{bmatrix} \sin 2t \\ 2\cos 2t \end{bmatrix} + c_2 e^{-t} \begin{bmatrix} -\cos 2t \\ 2\sin 2t \end{bmatrix}$

3. $\mathbf{x}(t) = c_1 e^t \begin{bmatrix} \cos 2t \\ -\sin 2t \end{bmatrix} + c_2 e^t \begin{bmatrix} \sin 2t \\ \cos 2t \end{bmatrix}$

4. $\mathbf{x}(t) = c_1 \begin{bmatrix} 0 \\ e^{2t} \\ 0 \end{bmatrix} + c_2 e^t \begin{bmatrix} -\sin t \\ 0 \\ \cos t \end{bmatrix} + c_3 e^t \begin{bmatrix} \cos t \\ 0 \\ \sin t \end{bmatrix}$

5. $\mathbf{x}(t) = c_1 e^t \begin{bmatrix} 2 \\ -3 \\ 2 \end{bmatrix} + c_2 e^t \begin{bmatrix} 0 \\ \cos 2t \\ \sin 2t \end{bmatrix} + c_3 e^t \begin{bmatrix} 0 \\ \sin 2t \\ -\cos 2t \end{bmatrix}$

6. $\mathbf{x}(t) = 2e^{2t} \begin{bmatrix} \cos t \\ -\sin t \end{bmatrix} - 3e^{2t} \begin{bmatrix} \sin t \\ \cos t \end{bmatrix}$

7. $\mathbf{x}(t) = 3e^t \begin{bmatrix} -1 \\ 0 \\ 1 \end{bmatrix} + 2 \begin{bmatrix} \sin t - \cos t \\ \sin t - \cos t \\ 2\cos t \end{bmatrix} + 4 \begin{bmatrix} \cos t + \sin t \\ \cos t + \sin t \\ -2\sin t \end{bmatrix}$

8. (a) $\lambda^2 + 1 = 0$ (b) $y(x) = c_1 \cos x + c_2 \sin x$

9. (a) $\lambda^2 + 2\lambda + 5 = 0$ (b) $y(x) = c_1 e^{-x} \cos 2x + c_2 e^{-x} \sin 2x$

10. (a) $\lambda^3 + 1 = 0$ (b) $y(x) = c_1 e^{-x} + e^{x/2}\left(c_2 \cos\left(\dfrac{\sqrt{3}}{2}x\right) + c_3 \sin\left(\dfrac{\sqrt{3}}{2}x\right)\right)$

11. (a) $\lambda^4 - 1 = 0$ (b) $y(x) = c_1 e^x + c_2 e^{-x} + c_3 \cos x + c_4 \sin x$

13. $e^{At} = \begin{bmatrix} \cosh t & \sinh t \\ \sinh t & \cosh t \end{bmatrix}$ **14.** $e^{At} = \begin{bmatrix} \cos t & \sin t \\ -\sin t & \cos t \end{bmatrix}$ **18.** $X(t) = \begin{bmatrix} e^t & e^{-t} \\ e^t & 3e^{-t} \end{bmatrix}$

19. $X(t)\,X^{-1}(0)x(0) = \dfrac{1}{2}\begin{bmatrix} e^t & e^{-t} \\ e^t & 3e^{-t} \end{bmatrix}\begin{bmatrix} 3 & -1 \\ -1 & 1 \end{bmatrix}\begin{bmatrix} 1 \\ 0 \end{bmatrix} = \dfrac{1}{2}\begin{bmatrix} 3e^t - e^{-t} \\ 3e^t - 3e^{-t} \end{bmatrix}$

Section 6.6

1. $e^{At} = \begin{bmatrix} e^{-t} & 0 \\ 0 & e^{2t} \end{bmatrix}$ **2.** $e^{At} = \begin{bmatrix} e^{2t} & 0 \\ 0 & e^{3t} \end{bmatrix}$

$x(t) = \begin{bmatrix} e^{-t} & 0 \\ 0 & e^{2t} \end{bmatrix}\begin{bmatrix} c_1 \\ c_2 \end{bmatrix} + \begin{bmatrix} 1 \\ 0 \end{bmatrix}$ $x(t) = \begin{bmatrix} e^{2t} & 0 \\ 0 & e^{3t} \end{bmatrix}\begin{bmatrix} c_1 \\ c_2 \end{bmatrix} + \begin{bmatrix} 0 \\ -2 \end{bmatrix}$

3. $e^{At} = \begin{bmatrix} \cosh t & \sinh t \\ \sinh t & \cosh t \end{bmatrix}$ **4.** $e^{At} = \begin{bmatrix} \cos t & \sin t \\ -\sin t & \cos t \end{bmatrix}$

$x(t) = \begin{bmatrix} \cosh t & \sinh t \\ \sinh t & \cosh t \end{bmatrix}\begin{bmatrix} c_1 \\ c_2 \end{bmatrix} + \begin{bmatrix} 1 \\ 1 \end{bmatrix}$ $x(t) = \begin{bmatrix} \cos t + 2 \sin t \\ 2 \cos t - \sin t - 1 \end{bmatrix}$

5. $x(t) = c_1 e^{-t}\begin{bmatrix} 2 \\ -1 \end{bmatrix} + c_2 e^{3t}\begin{bmatrix} 2 \\ 1 \end{bmatrix} - \begin{bmatrix} -10 \\ 1 \end{bmatrix}$ **6.** $x(t) = c_1\begin{bmatrix} 1 \\ 1 \end{bmatrix} + c_2 e^t\begin{bmatrix} 3 \\ 2 \end{bmatrix} - t\begin{bmatrix} 11 \\ 11 \end{bmatrix} - \begin{bmatrix} 15 \\ 10 \end{bmatrix}$

7. $x(t) = c_1 e^{3t}\begin{bmatrix} 1 \\ 1 \end{bmatrix} + c_2 e^{5t}\begin{bmatrix} 1 \\ 3 \end{bmatrix} + \dfrac{1}{2}\begin{bmatrix} e^{5t} \\ e^{5t} \end{bmatrix}$ **8.** $x(t) = c_1 e^t\begin{bmatrix} 1 \\ -1 \end{bmatrix} + c_2 e^{3t}\begin{bmatrix} 1 \\ -3 \end{bmatrix} + \begin{bmatrix} 1 \\ t \end{bmatrix}$

9. (a) $e^{At} = \begin{bmatrix} \cosh t & \sinh t \\ \sinh t & \cosh t \end{bmatrix}$ (c) $x(t) = e^{At}\,c$ **12.** (a) $x(t) = c_1 e^{3t} + c_2 e^{-t}$ (b) $x_p(t) = 1$ (c) $x(t) = x_h(t) + x_p(t)$

                                                    $y(t) = 2c_1 e^{3t} - 2c_2 e^{-t}$ $y_p(t) = 2$ $y(t) = y_h(t) + y_p(t)$

13. (a) $x_h(t) = c_1 e^t + c_2 e^{3t}$ (b) $x_p(t) = 1$ (c) $x(t) = x_h(t) + x_p(t)$

       $y_h(t) = -c_1 e^t - 3c_2 e^{3t}$ $y_p(t) = t$ $y(t) = y_h(t) + y_p(t)$

14. (a) $x_h(t) = c_1 e^{-2t} + 2c_2 e^{5t}$ (b) $x_p(t) = -2te^{-2t}$ (c) $x(t) = x_h(t) + x_p(t)$

       $y_h(t) = -3c_1 e^{-2t} + c_2 e^{5t}$ $y_p(t) = 6te^{-2t} - e^{-2t}$ $y(t) = y_h(t) + y_p(t)$

Section 6.7

1. $x(t) = \sin t$ **2.** $x(t) = e^{3t} - 2e^{2t}$ **3.** $x(t) = 6e^t - 7e^{2t} + 2e^{4t}$ **4.** $x(t) = t \sin t$

   $y(t) = \cos t$ $y(t) = -2e^{3t} + 2e^{2t}$ $y(t) = 6e^t - 14e^{2t} + 8e^{4t}$ $y(t) = \sin t + t \cos t$

5. $x(t) = -e^t + e^{2t}$ **6.** $x(t) = -\dfrac{3}{2}e^t + \dfrac{1}{2}e^{3t} + 1$ **7.** $x(t) = c_1 e^{3t}\begin{bmatrix} 1 \\ 2 \end{bmatrix} + c_2 e^{-t}\begin{bmatrix} 1 \\ -2 \end{bmatrix}$

   $y(t) = -e^t + 2e^{2t} - e^{3t}$

                              $y(t) = \dfrac{3}{2}e^t - \dfrac{3}{2}e^{3t} + t$

8. $x(t) = c_1 e^{-t}\begin{bmatrix} 2 \cos 2t \\ \sin 2t \end{bmatrix} + c_2 e^{-t}\begin{bmatrix} -2 \sin 2t \\ \cos 2t \end{bmatrix}$ **9.** $x(t) = \dfrac{11}{8}e^{4t} - \dfrac{1}{2}t + \dfrac{5}{8}$

                                        $y(t) = -\dfrac{11}{4}e^{4t} + t^2 - t + \dfrac{7}{4}$

10. $x_1(t) = e^t,\ x_2(t) = e^t - t^2,\ x_3(t) = e^t - 2t,\ x_4(t) = e^t - 2$

13. (a) $x_h(t) = c_1 e^{3t} + c_2 e^{-t}$ (b) $x_p(t) = 1$ (c) $x(t) = x_h(t) + x_p(t)$
     $y_h(t) = 2c_1 e^{3t} - 2c_2 e^{-t}$ $y_p(t) = 2$ $y(t) = y_h(t) + y_p(t)$

14. (a) $x_h(t) = c_1 e^{t} + c_2 e^{3t}$ (b) $x_p(t) = 1$ (c) $x(t) = x_h(t) + x_p(t)$
     $y_h(t) = -c_1 e^{t} - 3c_2 e^{3t}$ $y_p(t) = t$ $y(t) = y_h(t) + y_p(t)$

15. (a) $x_h(t) = c_1 e^{2t} + 2c_2 e^{5t}$ (b) $x_p(t) = -2te^{-2t}$ (c) $x(t) = x_h(t) + x_p(t)$
     $y_h(t) = -3c_1 e^{2t} + c_2 e^{5t}$ $y_p(t) = 6te^{-2t} - e^{-2t}$ $y(t) = y_h(t) + y_p(t)$

16. $m_1(t) = e^{-t}$
    $m_2(t) = te^{-t}$

Section 6.8

1. (a) $\dot{x} = -0.75x + 0.50y + 20$
     $\dot{y} = 0.50x - 0.70y$
   (b) $x_{ss} = 51°F$, $y_{ss} = 36°F$

2. (a) $\dot{x} = -0.50x + 0.25y + 13$
     $\dot{y} = 0.25x - 0.75y + 50$
   (b) $x_{ss} = 71°F$, $y_{ss} = 90°F$

3. $\dot{x} = -0.70x + 0.50y + 20$
    $\dot{y} = 0.50x - 0.60y + 5$
   $x_{ss} = 85°F$, $y_{ss} = 79°F$

4. $65°F$

5. $x_1' = 3 - 0.05x_1 + 0.02x_2$ $x_1(0) = 0$
    $x_2' = 0.05x_1 - 0.05x_2$ $x_2(0) = 0$

6. (a) $\begin{bmatrix} x_1' \\ x_2' \\ x_3' \end{bmatrix} = \begin{bmatrix} 0 & 0.02 & 0 \\ 0.06 & 0 & 0.05 \\ 0.06 & 0.01 & 0 \end{bmatrix} \begin{bmatrix} x_1 \\ x_2 \\ x_3 \end{bmatrix}$

7. $\begin{bmatrix} x_1' \\ x_2' \\ x_3' \end{bmatrix} = \begin{bmatrix} 0 & 0.05 & 0.15 \\ 0.30 & 0 & 0 \\ 0 & 0.20 & 0 \end{bmatrix} \begin{bmatrix} x_1 \\ x_2 \\ x_3 \end{bmatrix}$, $\begin{bmatrix} x_1(0) \\ x_2(0) \\ x_3(0) \end{bmatrix} = \begin{bmatrix} 500 \\ 100 \\ 250 \end{bmatrix}$

9. $\dot{x}_2 = k_{12}x_1 - (k_{21} + k_{23})x_2 - sx_2 + k_{32}x_3$
    $\dot{x}_3 = k_{23}x_2 - k_{32}x_3$

10. (a) $m_1\ddot{x}_1 = -k_1x_1 + k_2(x_2 - x_1)$ (b) $x_1(t) = 2\cos t - \cos 2t$
      $m_2\ddot{x}_2 = -k_2(x_2 - x_1)$ $x_2(t) = 4\cos t + \cos 2t$

12. $x(t) = \dfrac{Em}{H^2 e}\left(1 - \cos\left(\dfrac{He}{m}t\right)\right)$ $y(t) = \dfrac{Em}{H^2 e}\left(\dfrac{He}{m}t - \sin\left(\dfrac{He}{m}t\right)\right)$

Section 6.9

1. $x(0.1) = 1.2$ **2.** $x(0.1) = 1.0053$ **3.** $x(0.1) = 1.0161$ **4.** $x(0.1) = 0.9950$ **5.** $x(0.1) = 0.9950$
   $y(0.1) = 1.005$ $y(0.1) = 1.1054$ $y(0.1) = 0.6823$ $y(0.1) = -0.0990$ $y(0.1) = -0.0990$

6. $x(0.1) = 0.9957$ **7.** $x(0.1) = 1.0001$ **8.** $x(0.1) = 0.9998$
   $y(0.1) = -0.0840$ $y(0.1) = 0.0049$ $y(0.1) = -0.0050$

9. $x_1' = x_2$ $x_1(0) = 1$ $(x_1 = y, x_2 = y', x_3 = t)$
   $x_2' = -x_1^2 + x_3^2$ $x_2(0) = 0$
   $x_3' = 1$ $x_3(0) = 0$

10. $x_1' = x_2$ $x_1(0) = 0$ $(x_1 = y, x_2 = y', x_3 = t)$
    $x_2' = \sin x_3 - x_1 - x_2x_3$ $x_2(0) = 0$
    $x_3' = 1$ $x_3(0) = 0$

11. $x_1' = x_2$ $x_1(0) = 2$ $(x_1 = y, x_2 = y', x_3 = t)$
    $x_2' = x_3 - x_1 - \left(1 - x_1^2\right)x_2$ $x_2(0) = -1$
    $x_3' = 1$ $x_3(0) = 0$

12. $x_1' = x_2$ $x_1(0) = 1$ $x_1 = y$
    $x_2' = x_3$ $x_2(0) = 0$ $x_2 = y'$
    $x_3' = \cos x_4 - x_1x_4 - 2x_3$ $x_3(0) = 2$ $x_3 = y''$
    $x_4' = 1$ $x_4(0) = 0$ $x_4 = t$

CHAPTER 7

Section 7.1

1. $\left\{3, \dfrac{3}{2}, \dfrac{3}{4}, \dfrac{3}{8}, \dfrac{3}{16}, \ldots\right\}$ **2.** $\{2, 1, 5, 7, 17, \ldots\}$ **3.** $\{2, 0, 2, 0, 2, \ldots\}$ **4.** $\{1, 1, -1, -1, 1, \ldots\}$

5. $\left\{1, -\dfrac{1}{2}, \dfrac{1}{4}, -\dfrac{1}{8}, \dfrac{1}{16}, \ldots\right\}$ **6.** $\{1, 1, 1, 1, 1, \ldots\}$ **7.** $\{1, 1, 1, 1, 1, \ldots\}$ **8.** $\{1, 2, 4, 8, 16, \ldots\}$

6. $\{1, -1, 1, -1, 1, \ldots\}$ **7.** $\{1, 2, 4, 8, 16, \ldots\}$ **8.** $\{1, 3, 9, 27, 81, \ldots\}$
Linearly independent   Linearly independent   Linearly independent

9. $\{0, 2, 4, 6, 8, \ldots\}$ **10.** $\{1, -1, 1, -1, 1, \ldots\}$ **11.** $\{1, 6, 11, 16, 21, \ldots\}$
$\{1, 2, 3, 4, 5, \ldots\}$ $\{1, 1, 1, 1, 1, \ldots\}$ $\{4, 6, 8, 10, 12, \ldots\}$
Linearly independent   Linearly independent   Linearly independent

12. $\{1, 6, 11, 16, 21, \ldots\}$ **13.** $\{1, -1, 1, -1, 1, \ldots\}$ **14.** $\{1, -2, 4, -8, 16, \ldots\}$
$\{b, a + b, 2a + b, 3a + b, 4a + b, \ldots\}$ $\{1, 1, 1, 1, 1, \ldots\}$ $\{1, -1, 1, -1, 1, \ldots\}$
Linearly independent if $a/b \neq 5$ Linearly independent   Linearly independent

15. $\{1, 2, 4, 8, 16, \ldots\}$ **16.** $y_1 - y_0 = 0$ $y_0 = 1$ **17.** $y_1 - y_0 = 1$ $y_0 = 2$
$\{0, 2, 8, 24, 64, \ldots\}$ $y_2 - y_1 = 0$ $y_1 = 1$ $y_2 - y_1 = 1$ $y_1 = 3$
Linearly independent $y_3 - y_2 = 0$ $y_2 = 1$ $y_3 - y_2 = 1$ $y_2 = 4$
    $y_4 - y_3 = 0$ $y_3 = 1$ $y_4 - y_3 = 1$ $y_3 = 5$
    $y_5 - y_4 = 0$ $y_4 = 1$ $y_5 - y_4 = 1$ $y_4 = 6$

18. $y_1 - y_0 = 0.25y_0$ $y_0 = 1$ **19.** $y_1 = \dfrac{1}{2}\left(1 + \dfrac{2}{y_0}\right)$ $y_0 = 1$ **20.** $y_1 = 1 + \dfrac{1}{y_0}$ $y_0 = 1$
$y_2 - y_1 = 0.25y_1$ $y_1 = 1.25$
$y_3 - y_2 = 0.25y_2$ $y_2 = 1.5625$ $y_2 = \dfrac{1}{2}\left(1 + \dfrac{2}{y_1}\right)$ $y_1 = 1.5$ $y_2 = 1 + \dfrac{1}{y_1}$ $y_1 = 2$
$y_4 - y_3 = 0.25y_3$ $y_3 = 1.953123$
$y_5 - y_4 = 0.25y_4$ $y_4 = 2.4414063$ $y_3 = \dfrac{1}{2}\left(1 + \dfrac{2}{y_2}\right)$ $y_2 = 1.1667\ldots$ $y_3 = 1 + \dfrac{1}{y_2}$ $y_2 = 1.5$

$y_4 = \dfrac{1}{2}\left(1 + \dfrac{2}{y_3}\right)$ $y_3 = 1.357\ldots$ $y_4 = 1 + \dfrac{1}{y_3}$ $y_3 = 1.6667\ldots$

$y_5 = \dfrac{1}{2}\left(1 + \dfrac{2}{y_4}\right)$ $y_4 = 1.237\ldots$ $y_5 = 1 + \dfrac{1}{y_4}$ $y_4 = 1.6$

21. $y_1 - y_0 = 1$ $y_0 = 0$ **22.** $y_2 - y_1 + y_0 = 1$ $y_0 = 1$ **23.** $y_2 + y_0 = 0$ $y_0 = 1$
$y_2 - y_1 = 2$ $y_1 = 1$ $y_3 - y_2 + y_1 = 1$ $y_1 = 0$ $y_3 + y_1 = 0$ $y_1 = 0$
$y_3 - y_2 = 3$ $y_2 = 3$ $y_4 - y_3 + y_2 = 1$ $y_2 = 0$ $y_4 + y_2 = 0$ $y_2 = -1$
$y_4 - y_3 = 4$ $y_3 = 6$ $y_5 - y_4 + y_3 = 1$ $y_3 = 1$ $y_5 + y_3 = 0$ $y_3 = 0$
$y_5 - y_4 = 5$ $y_4 = 10$ $y_6 - y_5 + y_4 = 1$ $y_4 = 2$ $y_6 + y_4 = 0$ $y_4 = 1$

24. $y_2 - y_0 = 0$ $y_0 = 0$ **25.** $y_2 = y_1 + y_0$ $y_0 = 1$ **26.** $c_1 = 1$, fundamental set
$y_3 - y_1 = 0$ $y_1 = 1$ $y_3 = y_2 + y_1$ $y_1 = 1$
$y_4 - y_2 = 0$ $y_2 = 0$ $y_4 = y_3 + y_2$ $y_2 = 2$
$y_5 - y_3 = 0$ $y_3 = 1$ $y_5 = y_4 + y_3$ $y_3 = 3$
$y_6 - y_4 = 0$ $y_4 = 0$ $y_6 = y_5 + y_4$ $y_4 = 5$

27. $c_1 = 0$, not a fundamental set   **28.** $c_1 = -\dfrac{7}{2}$, fundamental set   **29.** $c_1 = -3$, fundamental set

30. $c_1 = 1$, fundamental set

Section 7.2

1. $y_n = c(0.5)^n$ **2.** $y_n = c_1 + c_2(-1)^n$ **3.** $y_n = c_1 2^n + c_2(-3)^n$ **4.** $y_n = c_1 + c_2(-2)^n$
5. $y_n = c_1 + c_2 n$ **6.** $y_n = c_1(-2)^n + c_2 3^n$ **7.** $y_n = c_1(-5)^n + c_2 n(-5)^n$

8. $y_n = c_1 2^n \cos\left(\dfrac{n\pi}{3}\right) + c_2 2^n \sin\left(\dfrac{n\pi}{3}\right)$ **9.** $y_n = c_1 2^n \cos\left(\dfrac{n\pi}{2}\right) + c_2 2^n \sin\left(\dfrac{n\pi}{2}\right)$ **10.** $y_n = c_1 \cos\left(\dfrac{2n\pi}{3}\right) + c_2 \sin\left(\dfrac{2n\pi}{3}\right)$

11. $y_n = 2^n$ **12.** $y_n = (0.5)^n$ **13.** $y_n = \sin\left(\dfrac{n\pi}{2}\right)$ **14.** $y_n = 2^{n-1} \sin\left(\dfrac{n\pi}{2}\right)$ **15.** $y_n = \alpha^n \cos\left(\dfrac{n\pi}{2}\right)$

16. $y_n = 2 \cdot 3^n + 3n \cdot 3^n$ **17.** $y_n = 2^{n/2}\left(\cos\left(\dfrac{n\pi}{4}\right) - \sin\left(\dfrac{n\pi}{4}\right)\right)$ **18.** $y_n = \left(1 + \dfrac{n}{2}\right)2^n$ **19.** $y_n = 1 + 2n$

20. $y_n = \dfrac{2}{\sqrt{3}} \sin\left(\dfrac{2n\pi}{3}\right)$ **21.** $y_n = \dfrac{2^n}{\sqrt{3}} \sin\left(\dfrac{n\pi}{3}\right)$ **22.** $y_n = c_1(-1)^n + c_2 2^n + c_3(-2)^n$

23. $y_n = c_1 + c_2 2^{n/2} \cos\left(\dfrac{n\pi}{4}\right) + c_3 2^{n/2} \sin\left(\dfrac{n\pi}{4}\right)$ **24.** $y_n = c_1 + c_2 n + c_3 n^2 + c_4 n^3$

25. $y_n = c_1 + c_2(-1)^n + c_3 \cos\left(\dfrac{n\pi}{2}\right) + c_4 \sin\left(\dfrac{n\pi}{2}\right)$ **26.** $y_n = c_1 \cos\left(\dfrac{n\pi}{2}\right) + c_2 \sin\left(\dfrac{n\pi}{2}\right) + c_3 n \cos\left(\dfrac{n\pi}{2}\right) + c_4 n \sin\left(\dfrac{n\pi}{2}\right)$

27. $y_n = c_1 2^n + c_2 n 2^n + c_3(-2)^n + c_4 n(-2)^n$ **28.** $p_n = \dfrac{N - n}{N}$

34. Forward Euler: $y_n = y_0(1 - h)^n$ Backward Euler: $y_n = y_0\left(\dfrac{1}{1 + h}\right)^n$

35. $F_n = \dfrac{1}{\sqrt{5}}\left\{\left(\dfrac{1 + \sqrt{5}}{2}\right)^{n+1} - \left(\dfrac{1 - \sqrt{5}}{2}\right)^{n+1}\right\}$, golden mean $\doteq 1.618...$

Section 7.3

1. $y_n = c_1(-1)^n + 1$ **2.** $y_n = c 2^n - n^2 - 2n - 3$ **3.** $y_n = c_1 \cos\left(\dfrac{n\pi}{2}\right) + c_2 \sin\left(\dfrac{n\pi}{2}\right) + \dfrac{1}{2}n^2 - n$

4. $y_n = c_1(-1)^n + c_2 n(-1)^n + \dfrac{1}{4}$ **5.** $y_n = c_1 + c_2 n + n^3 - 3n^2$ **6.** $y_n = c_1 3^n + c_2(-2)^n - n^2 - \dfrac{1}{3}n - \dfrac{5}{9}$

7. $y_n = c_1 2^n + c_2 n 2^n + n^2 2^n$ **8.** $y_n = 2^n\left(c_1 \cos\left(\dfrac{n\pi}{2}\right) + c_2 \sin\left(\dfrac{n\pi}{2}\right)\right) + \dfrac{\cos(n - 2) + 4\cos n}{17 + 8\cos 2}$ **9.** $y_n = 2^n - 1$

10. $y_n = \dfrac{3}{2} \sin\left(\dfrac{n\pi}{2}\right) + \dfrac{1}{2}n^2 - n$ **11.** $y_n = -\dfrac{1}{4}(-1)^n + \dfrac{1}{2}n(-1)^n + \dfrac{1}{4}$ **12.** $y_n = \dfrac{3}{5}(3)^n - \dfrac{2}{45}(-2)^n - n^2 - \dfrac{1}{3}n - \dfrac{5}{9}$

16. (a) $y_n = b_{n-1}b_{n-2} \cdots b_1 b_0 y_0$ **20.** $M_n = 2^n - 1$ **21.** (a) $T_{n+1} = T_n + n + 1 \quad T_1 = 1$

  (b) $y_n = n!$ (b) $T_n = \dfrac{n(n+1)}{2}$

22. (a) $S_{n+1} = S_n + 2n + 1 \quad S_1 = 1$ **23.** (a) $P_{n+1} = P_n + 3n + 1 \quad P_1 = 1$

  (b) $S_n = n^2$ (b) $P_n = \dfrac{n(3n-1)}{2}$

24. $F(z) = \dfrac{1}{z(z - 1)}$ **25.** $F(z) = \dfrac{z + 1}{z^3}$ **26.** $F(z) = \dfrac{z}{z + 1}$ **27.** $F(z) = \dfrac{az}{z - e^{-k}}$ **28.** $F(z) = \dfrac{z + 2}{z^2}$

32. $\{0, 1, 0, 0, ...\}$ **33.** $\{0, 0, 1, 0, 0, ...\}$ **34.** $\{1, 1, 1, ...\}$ **35.** $\left\{0, -\dfrac{1}{2}, -\dfrac{1}{4}, -\dfrac{1}{8}, ...\right\}$

36. $\{1, 1, 1, ...\}$ **37.** $\left\{0, 0, \dfrac{1}{2^2}, 0, \dfrac{1}{2^4}, 0 \cdots\right\}$ **38.** $y_n = n^2 + 1$

39. $y_n = \dfrac{1}{4}\left(2^n + 3(-2)^n\right)$ **40.** $y_n = (-1)^n(1 - n)$ **41.** $y_n = \dfrac{1}{2} \sin\left(\dfrac{n\pi}{2}\right) + \dfrac{1}{2}n^2 - n$

Section 7.4

1. $s_n = s_{n-1} + n \quad s_0 = 0$ **2.** $s_n = s_{n-1} + n^2 \quad s_0 = 0$ **3.** $s_n = s_{n-1} + 2^n \quad s_0 = 1$ **4.** $s_n = s_{n-1} + 3^n \quad s_0 = 1$

5. $S_n = \$1000 (1.0002)^n$ after n days   **6.** (a) $S_{n+1} = 1.08S_n - 30{,}000 \quad S_0 = 200{,}000$
After 1 day: $S_1 = \$1000.20$ (b) $S_n = -175{,}000(1.08)^n + 375{,}000$ (S_n becomes zero in 9.9 years.)
After 10 days: $S_{10} = \$1002.00$
After 365 days: $S_{365} = \$1075.72$

7. (a) $S_n = 12.5d[(1.08)^n - 1]$ **8.** John: $\$573{,}770.16$ **9.** (a) $S_{n+1} = 1.10S_n + 1000(1 + 0.05)^n$
  (b) $\$1742.86$ Ann: $\$683{,}368.42$ (b) $S_n = 21{,}000(1.10)^n - 20{,}000 (1.05)^n$
                                                    $S_4 = \$6{,}435.98$
                                  (c) $S_{50} = \$2{,}235{,}859.92$

10. (a) $S_{n+1} = 1.002S_n + 25 \quad S_0 = 0$
  (b) $S_n = 12{,}500[(1.002)^n - 1]$
  (c) $S_{208} = \$6440.70$

11. (a) $P_n = S(1 + i)^n - d\left(\dfrac{(1 + i)^n - 1}{i}\right)$

  (b) $d = S\left(\dfrac{i}{1 - (1 + i)^{-N}}\right)$

  (c) $\$1028.61$

12. (a) $S_{n+1} = 1.02S_n - 1 \quad S_0 = 100$ **13.** (a) $-50{,}000(1.10)^n + 150{,}000$
  (b) $S_n = 50[(1.02)^n + 1]$ (b) 10,000 per year
  (c) No fish after 25 years

14. (a) $-200(1.25)^n + 1200$ **15.** (a) $400[1 - (0.75)^n]$
  (b) Will be 0 after 8 years   (b) Approaches 400 grams

16. (b) $-300(0.75)^n + 400$ **17.** (a) $y_{n+1} = 0.63y_n + d \quad y_0 = d$ **18.** (c) $\doteq 206$ mg
  (c) 400 mg   (b) $\doteq 2.7$ d   (d) $\doteq 7$ or 8 days
  (d) 200 mg   (c) $\doteq 130$ mg

19. (a) $9000 + 1000(3)^n$ **20.** $y_n = \dfrac{1}{r - 1}\left(y_0 r - y_1 + (y_1 - y_0)r^n\right)$
  (b) 4 years

21. $y_n = \dfrac{n^2 + n + 2}{2}$ **22.** (b) $T_n = 2^n - 1$, $T_{15} = 32{,}767$ ways (you win a new car)

24. (a) $A_1 = \$10{,}000, A_{n+1} = A_n + 1000 \quad (n = 1, 2, ...)$ $B_1 = 10{,}000, B_{n+1} = B_n + 2(300)$
     $A_n = 10{,}000 + 1000(n - 1) \quad (n = 1, 2, ...)$ $B_n = 10{,}000 + 600 (n - 1)$
     $A_{20} = \$29{,}000$ $B_{20} = 21{,}400$

Section 7.5

1. 0, 1 fixed points, 0 attracting, 1 repelling   **2.** $\dfrac{1}{2} \pm \dfrac{\sqrt{5}}{2}$, both repelling

3. $0, \pm 1$ fixed points, 0 attracting, ± 1 repelling   **4.** $0.73 \cdots$ fixed point, attracting

Problems 5–19 consist of computer projects, so no answers are given.

20. both (a) and (b) are folding and stretching maps and both are chaotic   **21.** (a) 4-cycle
                                                                (b) 3-cycle
                                                                (c) $\lim\limits_{x \to \infty} = 2$

Section 7.6
This section consists of computer projects, so no answers are given.

CHAPTER 8

Section 8.1

1. $\dot{x} = y$
$\dot{y} = -\dfrac{g}{L}\sin x$ (autonomous)

2. $\dot{x} = y$
$\dot{y} = -\dfrac{g}{L}\sin x + \sin t$ (nonautonomous)

3. $\dot{x} = y$
$\dot{y} = -\omega^2 x - \beta x^3$ (autonomous)

4. $\dot{x} = y$
$\dot{y} = -\omega^2 x - \beta x^3 + F_0 \cos \gamma t$ (nonautonomous)

5. $\dot{x} = y$
$\dot{y} = -(\alpha + \beta \cos t)x$ (nonautonomous)

6. $\dot{x} = y$
$\dot{y} = -x - \epsilon(x^2 - 1)y + F_0 \cos \gamma t$ (nonautonomous)

7. $(n\pi, 0)$: $n = 0, \pm 1, \pm 2, \ldots$ **8.** $(0, 0)$ **9.** $(n\pi, 0)$: $n = 0, \pm 1, \pm 2, \ldots$ **10.** $(0, 0)$

12. (a) $\dot{x} = y$
$\dot{y} = -\dfrac{g}{m}\sin x$
(b) $(n\pi, 0)$, $n = 0, \pm 1, \pm 2, \ldots$
(c) $y^2 - \dfrac{2g}{m}\cos x = C$

13. (a) $\dot{x} = y$
$\dot{y} = -x - x^3$
(b) $(0, 0)$
(c) $2x^2 + 2y^2 + x^4 = C$

16. (a) $(0, 0), \left(\dfrac{2}{3}, \dfrac{1}{3}\right), \left(0, \dfrac{1}{2}\right), (1, 0)$

(c) The point $\left(\dfrac{2}{3}, \dfrac{1}{3}\right)$ represents a possible "coexistence" point.

17. $\dfrac{1}{2}y^2 - \cos x = C$ **18.** $\dfrac{1}{2}y^2 + \dfrac{1}{3}x^3 = C$ **19.** $\dfrac{1}{2}y^2 + \dfrac{1}{2}x^2 + \dfrac{1}{4}x^4 = C$

21. (a) Trajectories are horizontal lines where parallelograms remain parallelograms of constant height and base. This isn't a complete proof, but if one considers approximating areas by parallelograms, it becomes feasible.
(b) Areas are simply rotated, so areas remain constant.

23. $y = Ae^{-x} - x + 1$
$y = B + \dfrac{1}{2}x^2$

Section 8.2

1. $\lambda_1 = 1, \lambda_2 = 2$, improper node, unstable   **2.** $\lambda_1 = -1, \lambda_2 = 1$, saddle point, unstable

3. $\lambda_1 = \lambda_2 = -1$, proper node, asymptotically stable   **4.** $\lambda_1, \lambda_2 = -2 \pm i$, spiral point, asymptotically stable

5. $\lambda_1 = -5, \lambda_2 = 7$, saddle point, unstable

6. $\lambda_1, \lambda_2 = \pm 4i$, center, stable   **7.** $\lambda_1 = -1, \lambda_2 = 2$, saddle point, unstable

8. $\lambda_1, \lambda_2 = 1 \pm 2i$, spiral point, unstable   **9.** $(1, -1)$, proper node, unstable

10. $(1, 1)$, spiral point, asymptotically stable   **11.** $(-2, 2)$, center, stable   **12.** $(0, 2)$, saddle point, unstable

13. $\left(\dfrac{dh - bk}{ad - bc}, \dfrac{ak - ch}{ad - bc}\right)$ same type and stability of the system $\dot{x} = ax + by, \dot{y} = cx + dy$

14. $d^2 - 4mk < 0 \Diamond$ spiral point (asymptotically stable)
$d^2 - 4mk = 0 \Diamond$ proper node (asymptotically stable)
$d^2 - 4mk > 0 \Diamond$ improper node (asymptotically stable)

15. (b) All points on the line $-x + y = 0$
(c) $x(t) = c_1$
$y(t) = c_2 e^t + c_1$
(d) Trajectories are vertical lines $x = c$

17. $x(t) = c_1 \begin{bmatrix} a \\ b \end{bmatrix} + c_2 t \begin{bmatrix} c \\ d \end{bmatrix}$

18. Linearized equations: $\dot{x} = y, \dot{y} = -4y$
$(0, 0)$ is a saddle point.

19. Linearized equations: $\dot{x} = y, \dot{y} = -x$
$(0, 0)$ is a center point.

20. Linearized equations: $\dot{x} = y$, $\dot{y} = -x - \epsilon y$
$\epsilon \leq -2 \lozenge (0, 0)$ is an improper node, unstable
$-2 < \epsilon < 0 \lozenge (0, 0)$ is a stable spiral point,
$\epsilon = 0 \lozenge (0, 0)$ is an unstable point,
$0 < \epsilon < 2 \lozenge (0, 0)$ is a stable spiral point,
$2 \leq \epsilon \lozenge (0, 0)$ is a stable improper node.

21. (a) $\dot{x} = y$
$\dot{y} = 0.25 x^2 - x$
(c) Linearization about $(0, 0)$: $\dot{x} = y$, $\dot{y} = -x$
Linearization about $(4, 0)$: $\dot{x} = y$, $\dot{y} = x$
(d) $(0, 0)$ is a center point,
$(4, 0)$ is a saddle point.

Section 8.3

1. $(0, 0)$ is a stable spiral point.

2. Nothing can be said about the stability of the nonlinear system at $(0, 0)$ when the corresponding linear system has a center point at $(0, 0)$.

3. $(0, 0)$ is an unstable spiral point.   **4.** $(0, 0)$ is an asymptotically stable spiral point.   **5.** $(0, 0)$ is an unstable saddle point.

6. $(0, 0)$ is an asymptotically stable node.

7. $(0, 0)$ is a proper node, an improper node, or a spiral point and is asymptotically stable.

10. $(0, 0)$ is a center point of the linearized system, so Theorem 8.2 says nothing about the nature of $(0, 0)$ of the nonlinear system.

$(-1, 0)$ is a saddle point of the linearized system, so the nonlinear system has a saddle point there.

$\left(\dfrac{1}{2}, 0\right)$ is the same as $(0, 0)$ — is a center point of the linearized system so nothing is known about the nonlinear system at $(1/2, 0)$.

11. (a) (b) (c) $\left(\dfrac{c}{d}, \dfrac{a}{b}\right)$ is a center point or a spiral point, which may be either asymptotically stable, stable, or unstable.

Section 8.4

1. (a) The point converges to some point on the 45° line.
(b) The point oscillates between two points symmetric around the 45° line.
(c) The solution follows a closed path of length 3.
(d) The solution appears chaotic.

4. (a) Counterclockwise motion making 3-hour jumps
(b) Counterclockwise motion making 6-hour jumps
(c) Counterclockwise motion making 9-hour jumps

5. Motion is counterclockwise with period q jumping over $[q - (p + 1)]$ positions, i.e., the Poincaré section consists of q equally spaced points on the unit circle.

6. (a) Motion has period 5, jumping over $5 - 3 = 2$ positions on each strobe flash.
(b) Motion has period 12, jumping over $12 - 12 = 0$ positions on each strobe flash (i.e., it makes hourly counterclockwise jumps before repeating).
(c) Motion has period 100, jumping over $100 - 100 = 0$ positions on each strobe flash (i.e., it moves counterclockwise on the circle, each jump changing by $360/100 = 3.6°$).

7. (a) $\omega_0 = (2/4)\omega_s$
(b) $\omega_0 = (1/4)\omega_s$
(c) $\omega_0 = (7/12)\omega_s$

CHAPTER 9

Section 9.1

1. Periodic, but no fundamental period   **2.** $T = 2$ **3.** $T = \dfrac{2}{3}$ **4.** $T = 2$ **5.** Not periodic   **6.** Not periodic

7. $T = \pi$ **8.** $T = \pi$ **9.** $T = 2\pi$ **10.** $T = \pi$ **14.** $f(x) = 1$ **15.** $f(x) = \sin x + \cos x$

16. $f(x) = 1 + \sin x + \dfrac{1}{2}\sin 2x$ **17.** $f(x) = 1 + \cos x + \dfrac{1}{3}\sin 2x$ **18.** $f(x) = \dfrac{2}{\pi}\sum_{n=1,3,5,\ldots}^{\infty}\dfrac{1}{n}\sin (n\pi x) + \dfrac{1}{2}$

19. $f(x) = \dfrac{1}{2} - \dfrac{4}{\pi^2}\sum_{n=1,3,5,\ldots}^{\infty}\dfrac{1}{n^2}\cos (n\pi x)$ **20.** $f(x) = \dfrac{2}{\pi}\sum_{n=1}^{\infty}\dfrac{(-1)^{n+1}}{n}\sin (n\pi x)$

21. $f(x) = \dfrac{1}{3} + 4 \displaystyle\sum_{n=1}^{\infty} \dfrac{(-1)^n}{n^2\pi^2} \cos n\pi x$

22. $f(x) = \dfrac{3}{4} + \displaystyle\sum_{n=1}^{\infty} \dfrac{(-1)^{n+1}}{n\pi} \sin (n\pi x) - \displaystyle\sum_{n=1,3,\cdots} \dfrac{2}{n\pi}\left(\sin n\pi x + \dfrac{1}{n\pi}\cos n\pi x\right)$

Section 9.2

1. Even   **2.** Even   **3.** Odd   **4.** Even   **5.** Neither   **6.** Even   **7.** Neither   **8.** Odd   **9.** Even   **10.** Odd

11. 0   **13.** (a) 0 (b) $\dfrac{2}{7}$ (c) 2π (d) 0 (e) 0 (f) 0

14. Even periodic extension: $F_e(x) = \begin{cases} 1 & (0 < x < 1) \\ 1 & (-1 < x < 0) \end{cases}$

      Odd periodic extension: $F_o(x) = \begin{cases} 1 & (0 < x < 1) \\ -1 & (-1 < x < 0) \end{cases}$

15. Even periodic extension: $F_e(x) = \begin{cases} x^2 & (0 < x < 1) \\ x^2 & (-1 < x < 0) \end{cases}$

      Odd periodic extension: $F_o(x) = \begin{cases} x^2 & (0 < x < 1) \\ -x^2 & (-1 < x < 0) \end{cases}$

16. Even periodic extension: $F_e(x) = \begin{cases} 1 - x & (0 < x < 1) \\ 1 + x & (-1 < x < 0) \end{cases}$

      Odd periodic extension: $F_o(x) = \begin{cases} 1 - x & (0 < x < 1) \\ -1 - x & (-1 < x < 0) \end{cases}$

17. Even periodic extension: $F_e(x) = \begin{cases} \sin x & (0 < x < \pi) \\ -\sin x & (-\pi < x < 0) \end{cases}$

      Odd periodic extension: $F_o(x) = \begin{cases} \sin x & (0 < x < \pi) \\ \sin x & (-\pi < x < 0) \end{cases}$

18. $f(x) = \sin x$ **19.** $f(x) = \dfrac{4}{\pi}\displaystyle\sum_{n=1}^{\infty}\dfrac{1}{2n-1}\sin(2n-1)x$ **20.** $f(x) = \dfrac{2}{\pi}\displaystyle\sum_{n=1}^{\infty}\dfrac{(-1)^{n+1}}{n}\sin n\pi x$

21. $f(x) = \dfrac{2}{\pi}\displaystyle\sum_{n=1}^{\infty}\dfrac{1}{n}\sin(n\pi x)$ **22.** $f(x) = 1$ **23.** $f(x) = 1 + \dfrac{4}{\pi^2}\displaystyle\sum_{n=1}^{\infty}\dfrac{1}{(2n-1)^2}\cos[(2n-1)\pi x]$

24. $f(x) = 1 + \dfrac{2}{\pi}\displaystyle\sum_{n=1}^{\infty}\dfrac{(-1)^{n-1}}{2n-1}\cos\left(\dfrac{(2n-1)\pi x}{2}\right)$ **25.** $f(x) = \dfrac{2}{\pi} + \dfrac{2}{\pi}\displaystyle\sum_{n=1}^{\infty}\left(\dfrac{1}{2n+1} + \dfrac{1}{2n-1}\right)\cos 2nx$

26. $f(x) = 8\displaystyle\sum_{n=2,4,\cdots}\dfrac{(-1)^{n/2+1}}{n}\sin\left(\dfrac{nx}{4}\right) + \dfrac{32}{\pi}\displaystyle\sum_{n=1,3,\cdots}\dfrac{(-1)^{(n+1)/2\,+\,1}}{n}\sin\left(\dfrac{nx}{4}\right)$

27. $f(x) = \displaystyle\sum_{n=1}^{\infty} b_n \sin\left(\dfrac{n\pi x}{4}\right)$ $b_n = \dfrac{(-1)^{(n+1)/2\,+\,1}}{n\pi}$ (n odd), $b_n = \dfrac{(-1)^{n/2+1}}{n\pi}$ (n even)

28. $x(t) = \displaystyle\sum_{n=1}^{\infty} b_n\left(\dfrac{\omega\sin nt - n\sin\omega t}{\omega(\omega^2 - n^2)}\right)$ ($\omega \neq 1, 2, 3, \ldots$)

Section 9.3

1. First order, two variables, linear, homogeneous, constant coefficients
2. Second order, two variables, linear, homogeneous, constant coefficients   **3.** Second order, two variables, nonlinear
4. Second order, two variables, linear, homogeneous, constant coefficients   **5.** Second order, two variables, nonlinear
6. First order, two variables, linear, nonhomogeneous, variable coefficients
7. Second order, four variables, linear, nonhomogeneous, constant coefficients

8. First order, three variables, linear, homogeneous, constant coefficients

9. Second order, three variables, linear, homogeneous, variable coefficients

10. Second order, four variables, linear, homogeneous, variable coefficients

16. $u(x, y) = f(y)$, where f is an arbitrary, differentiable function of y

17. $u(x, y) = x + g(y)$, where g is an arbitrary, differentiable function of y

18. $u(x, y) = x^2 + g(y)$, where g is an arbitrary, differentiable function of y

19. $u(x, y) = xy + f(y)$, where f is an arbitrary, differentiable function of y

20. $u(x, y) = f(x) + g(y)$, where f and g are arbitrary, differentiable functions of x and y, respectively

21. $u(x, y) = \dfrac{1}{3} x^3 y + \dfrac{1}{3} xy^3 + f(x) + g(y)$, where f and g are arbitrary, differentiable functions of x and y, respectively

22. The strategy gives solutions $u(x, t) = e^{ax + a^2 t}$, where a is any constant. However, there are more solutions.

24. (a) Elliptic (b) Elliptic (c) Hyperbolic (d) Parabolic (e) Parabolic (f) Hyperbolic

Section 9.4

1. $T'(t) - kT(t) = 0$ **2.** $T'(t) - kT(t) = 0$ **3.** $T' - kT = 0$
$X'(x) - kX(x) = 0$ $X''(x) - kX(x) = 0$ $X'' + 2X' + (1 - k)X = 0$

4. $T''(t) = kT(t)$ **5.** $T''(t) - kT(t) = 0$ **19.** (a) $T' + kT = 0$

$\quad X''(x) = kX(x)$ $X''(x) + X'(x) + (1 - k)X = 0$ $R'' + \dfrac{1}{r} R' + \dfrac{k}{\gamma} R = 0$

(b) $T(t) = Ce^{-kt}$

Section 9.5

1. $u(x, t) = \sin \pi x \cos \pi \alpha t$ **2.** $u(x, t) = \sin \pi x \cos \pi \alpha t + \sin 2\pi x \cos 2\pi \alpha t$

3. $u(x, t) = \sin 2\pi x \cos 2\pi \alpha t + \dfrac{1}{5} \sin 3\pi x \cos 3\pi \alpha t$

4. $u(x, t) = \sin \pi x \cos \pi \alpha t + \dfrac{1}{3^2} \sin 3\pi x \cos 3\pi \alpha t - \dfrac{1}{5^2} \sin 5\pi x \cos 5\pi \alpha t + \cdots$

5. $u(x, t) = \sin \pi x \cos \pi \alpha t - \dfrac{1}{2^2} \sin 2\pi x \cos 2\pi \alpha t + \dfrac{1}{3^2} \sin 3\pi x \cos 3\pi \alpha t + \cdots$

6. $u(x, t) = \dfrac{4}{\pi} \left(\sin 2\pi x \cos 2\pi t + \dfrac{1}{3} \sin 6\pi x \cos 6\pi t + \cdots \right)$

7. $u(x, t) = \dfrac{2}{\pi} \left(\sin \pi x \cos \pi t - \dfrac{1}{2} \sin 2\pi x \cos 2\pi t + \dfrac{1}{3} \sin 3\pi x \cos 3\pi t + \cdots \right)$

8. $u(x, t) = \dfrac{8}{\pi^2} \left(\sin \pi x \cos \pi t - \dfrac{1}{9} \sin 3\pi x \cos 3\pi t + \dfrac{1}{25} \sin 5\pi x \cos 5\pi t + \cdots \right)$

9. $u(x, t) = \dfrac{1}{\pi} \sin \pi x \sin \pi t$ **10.** $u(x, t) = \dfrac{1}{3\pi} \sin 3\pi x \sin 3\pi t$ **11.** $u(x, t) = \dfrac{1}{\pi} \sin \pi x \sin \pi t + \dfrac{1}{9\pi} \sin 3\pi x \sin 3\pi t$

12. $u(x, t) = \displaystyle\sum_{n=1}^{\infty} \left(\dfrac{2}{(2n - 1)\pi} \right)^2 \sin (2n - 1)\pi x \sin (2n - 1)\pi t$ **14.** $u(x, t) = \sin \pi x \cos \pi t + \dfrac{1}{\pi} \sin \pi x \sin \pi t$

15. $u(x, t) = \sin 2\pi x \cos 2\pi t + \dfrac{1}{2\pi} \sin 2\pi x \sin 2\pi t$

16. $u(x, t) = \sin \pi x \cos \pi t + \dfrac{1}{3} \sin 3\pi x \cos 3\pi t + \dfrac{1}{8\pi} \sin 4\pi x \sin 4\pi t$

17. $u(x, t) = \sin \pi x \cos \pi t + \dfrac{1}{3} \sin 3\pi x \cos 3\pi t + \cdots$

18. $u(x, t) = \displaystyle\sum_{n=1,3,\cdots}^{\infty} \left(\dfrac{4}{n\pi} \right) \sin n\pi x \cos n\pi t + \sum_{n=1,3,\cdots}^{\infty} \left(\dfrac{4}{n^2 \pi^2} \right) \sin n\pi x \sin n\pi t$

19. $\lambda_n = n^2, n = 1, 2, ...$

  $y_n(x) = A \sin nx, n = 1, 2, ...$

20. $\lambda_n = (n - 1/2)^2, n = 1, 2, ...$

  $y_n(x) = A \cos [(n - 1/2)x], n = 1, 2, ...$

21. $\lambda_n = n^2$

  $y_n = A \cos nx + B \sin nx, n = 0, 1, 2, ...$

22. $\lambda_n = n^2 + 1$

  $y_n = Ae^{-x} \sin nx, n = 2, 3, ...$

Section 9.6

1. $u(x, t) = e^{-\pi^2 t} \sin \pi x$

2. $u(x, t) = 3e^{-4\pi^2 t} \sin 2\pi x$

3. $u(x, t) = 2e^{-9\pi^2 t} \sin 3\pi x$

4. $u(x, t) = e^{-\pi^2 t} \sin \pi x + \dfrac{1}{4}e^{-4\pi^2 t} \sin 2\pi x$

5. $u(x, t) = e^{-\pi^2 t} \sin \pi x - \dfrac{1}{2}e^{-4\pi^2 t} \sin 2\pi x + \dfrac{1}{3}e^{-9\pi^2 t} \sin 3\pi x$

6. $u(x, t) = \dfrac{4}{\pi} \sum\limits_{n=1}^{\infty} \dfrac{1}{(2n-1)} e^{-[(2n-1)\pi]^2 t} \sin [(2n - 1) \pi x]$

7. $u(x, t) = \dfrac{2}{\pi} \sum\limits_{n=1}^{\infty} \dfrac{(-1)^{n+1}}{n} e^{-(n\pi)^2 t} \sin n\pi x$

9. $u(x, t) = x + 1 + e^{-\pi^2 t} \sin \pi x$

10. $u(x, t) = 1$

11. $u(x, t) = e^{-\pi^2 t} \cos \pi x$

12. $u(x, t) = 1 + e^{-\pi^2 t} \cos \pi x + \dfrac{1}{2}e^{-9\pi^2 t} \cos 3\pi x$

13. $u(x, t) = \dfrac{1}{2} - \dfrac{4}{\pi^2}\left(e^{-\pi^2 t} \cos \pi x + \dfrac{1}{3^2}e^{-9\pi^2 t} \cos 3\pi x + \dfrac{1}{5^2}e^{-25\pi^2 t} \cos 5\pi x + \cdots\right)$

14. $u(x, t) = \dfrac{1}{2} - \dfrac{4}{\pi^2} \sum\limits_{n=1,3,...}^{\infty} \dfrac{1}{n^2}e^{-(n\pi)^2 t} \cos n\pi x$

15. $u(x, t) = e^{-(\pi/2)^2 t} \sin \left(\dfrac{x}{2}\right) + \dfrac{1}{3}e^{-(3\pi/2)^2 t} \sin \left(\dfrac{3\pi x}{2}\right)$

16. As $t \to \infty$, the solution looks more and more like a multiple of $\sin \pi x$.

17. PDE: $u_t = 0.00115u_{xx}$ $(0 < x < 5/6)$

  BC: $\begin{cases} u(0, t) = 20 \\ u(5/6, t) = 70 \end{cases}$ $(0 < t < \infty)$

  IC: $u(x, 0) = 50$ $(0 \le x \le 5/6)$

18. $u(x, t) = \sum\limits_{n=1}^{\infty} a_n e^{-(\lambda_n \alpha)^2 t} \sin (\lambda_n x)$, where a_n are the Fourier sine coefficients of $u(x, 0) = f(x)$ and λ_n are the solutions of $\tan \lambda = -\lambda$.

Section 9.7

7. (a) $f(r, \theta) = 1$

  (b) $f(r, \theta) = 2 \ln r$

  (c) $f(r, \theta) = r^2 \sin \theta \cos \theta$

  (d) $f(r, \theta) = r^2 (\cos^2 \theta - \sin^2 \theta)$

8. When the Laplacian is negative at a point, the temperature is greater than the average temperature of its neighbors (hence falling); when it is zero, it is the same as the average of its neighbors (hence unchanging); and when it is positive, it is less than the average of its neighbors (hence rising).

9. The temperature will be greater than zero inside the circle.

11. $u(r, \theta) = 1$

12. $u(r, \theta) = 2$

13. $u(r, \theta) = r \sin \theta$

14. $u(r, \theta) = r \cos \theta$

15. $u(r, \theta) = r^2 \sin 2\theta$

16. $u(r, \theta) = r \sin \theta + \dfrac{1}{2}r^2 \sin 2\theta$

17. $u(r, \theta) = r \cos \theta + \dfrac{1}{2}r^2 \cos 2\theta + \dfrac{1}{3}r^4 \sin 4\theta$

18. $u(r, \theta) = r^4 \sin 4\theta$

19. $u(r, \theta) = \dfrac{\pi}{2} - \dfrac{4}{\pi}\left(r \cos \theta + \dfrac{1}{3^2}r^3 \cos 3\theta + \dfrac{1}{5^2}r^5 \cos 5\theta + \cdots\right)$

26. (e) $u(x, y) = \dfrac{1}{\sinh \pi} \sinh \pi x \sin \pi y$

Appendix (Complex Numbers and Complex-Valued Functions)

2. (a) $6 + 2i$

(b) $5 + i$

(c) $\dfrac{7}{10} + i\dfrac{1}{10}$

(d) $\sqrt{26}$

5. $\operatorname{Re}(z^2 + 2z) = a^2 - b^2 + 2a$

$\operatorname{Im}(z^2 + 2z) = 2b(a + 1)$

6. $2\sqrt{5}$

7. (a) $F'(x) = (1 - i)\, e^{(1-i)x}$

$F''(x) = -2i\, e^{(1-i)\,x}$

(b) $F'(x) = 3i\, e^{3ix}$

$F''(x) = -9\, e^{3ix}$

(c) $F'(x) = (2 + 3i)\, e^{(2+3i)x}$

$F''(x) = (-5 + 12i)\, e^{2x}\,[\cos 3x + i \sin 3x]$

8. (a) $-e$

(b) ie^2

(c) -1

(d) -1

INDEX

Abel's identity, 123
Adjoint equation(s)
   Lagrange's adjoint equation, 117
   for a single equation, 111
   for a system, 353
Airy's equation, 211
Airy's function, 211
Airy, George, 211
Algebraic multiplicity, 343
Almost linear equation, 480
Almost periodic function, 237
Amplitude factor, 186
Analytic function, 204
Annihilator method, 197
Annuity, 53, 423
Asymptotic stability, 475
Attractor
   definition, 455
   limit cycle, 455
   strange, 495
Autonomous system, 454

Basis functions, 332
Beam problem
   cantilever beam, 195
   clamped beam, 195
   hinged beam, 195
   uniformly loaded beam, 195
   Young's modulus of elasticity, 195
Beats, 192
Bernoulli equation, 36
Bessel's equation
   definition of, 232
   general solution, 236
   parametric, 241
Bessel function
   of the first kind, 234
   Neuman function, 235
   of the second kind, 235
   Weber function, 235
Bifurcation diagram of logistic solution, 435

Boundary conditions of a partial differential equation, 530
Branched function, 277

Calculus of variations, 180
Casoratian, 401
Cauchy, Augustin-Louis, 327
Cauchy-Euler equation, 231, 242
Cauchy-Euler system, 353
Chaotic motion
   additional reading, 439, 501
   bifurcation diagram, 435
   chaos number, 435
   cobweb diagram, 430
   definition of, 489
   doubling sequence, 436
   Feigenbaum number, 437
   forced dissipative chaos, 490
   fractals, 443, 496
   geometric interpretation, 438
   Liapunov exponent, 492
   logistic equation, 429
   Lorenz equation, 489
   period-doubling phenomenon, 435
   strange attractor, 495
   universal constant, 437
Characteristic equation
   complex roots, 135
   of a linear system, 343
   real and unequal roots, 131, 343
   repeated roots, 132
   second-order differential equation,
    130
Characteristic root, 343
Characteristic vector, 343
Clairaut's equation, 44
Closed loop control, 465
Cobweb diagram, 430
Compartmental analysis, 370, 382
Complete set of solutions, 337
Complex number
   absolute value, 569
   argument of, 567

arithmetic operations, 568
complex conjugate, 570
complex-valued function, 571
definition of, 567
de Moivre formula, 570
derivative of complex valued function, 571
Euler's formula, 136
function of, 135
imaginary part, 567
modulus of, 569
polar angle, 569
polar form, 570
real part, 567
roots of, 136
solving differential equations using,
   156
Compound interest, 51
   annuity, 53
   continuous compounding, 53
   doubling time, 54
   future value, 53
Conservative system, 467
Continuous compounding interest, 53
Control theory
   closed loop control, 465
   control function, 461
   feedback control, 465
   open loop control, 465
   spaceship problem, 458, 461
   switching line, 463
Convolution integral, 303
   interpretation, 304
   Laplace transform of, 305
   properties, 305
Cooling equation, 68
Cooling, Newton's law of, 67
Coupled springs, 384
Critical point (*see* Equilibrium point)
Critically damped motion, 175

d'Alembert, Jean le Rond, 125
d'Alemert's reduction of order method,
   126

Damped vibration, 165
Decay, radioactive, 49
Deflection of a beam, 195
Delayed function, 278
de Moivre formula, 570
Dependence (*see* Linear dependence)
Determinants (*see* Matrices)
Difference equation
  applications of, 420
  definition, 396
  existence of solutions, 399
  initial-value problem, 398
  fundamental set of solutions, 401
  linear equation, 397
  nonhomogeneous linear equation, 403
  solution of, 404
  undetermined coefficients (method of), 411
  $\mathbb{Z}$-transform, 413
Differential equation
  constant coefficient, 12
  definition, 12
  homogeneous linear, 12
  independent variable, 10
  initial-value problem, 20, 113
  linear, 12
  linear fractional equation, 98
  matrix representation, 330
  missing x, 117
  missing y, 116
  nonlinear, 12
  order, 11
  ordinary, 12
  solution
    explicit, 13
    implicit, 13
  system (*see* Systems of differential equations)
Differential equation (special kinds)
  Airy's equation, 211
  Bernoulli's equation, 36
  Bessel's equation, 232
  Chebychev's equation, 129
  Clairaut's equation, 44
  decay equation, 47
  delay equation, 27
  growth equation, 47
  Hermite's equation, 221
  hypergeometric equation, 232

Laguerre's equation, 129
  logistic equation, 57
  Ricatti's equation, 37
  separable equation, 38
  Stephan's equation, 99
  Torricelli's equation, 84
Differential operator, 275, 317
Diffusion equation (*see* Heat equation)
Dirac delta function, 293
Dirac, Paul, 292
Direction element, 91
Direction field, 91
Dirichlet, P. G. Lejeune, 506
Dirichlet problem
  inside a circle, 558
  inside a rectangle, 565
Dirichlet's theorem, 508
Discontinuous forcing function, 284
Distribution (generalized function), 296
D operator, 317
Dynamical system
  definition, 440
  fractal, 443
  Julia set, 441
  Mandelbröt set, 446
  orbit, 440

Eigenfunction, 535
Eigenvalues of a boundary value problem, 535
Eigenvalue of a matrix, 343
  algebraic multiplicity, 343
  characteristic equation, 343
  complex, 354
  definition, 343
  geometric multiplicity, 343
  repeated, 350
Eigenvector of a matrix, 343
  definition, 343
Elimination, method of, 314
Equilibrium point
  center node, 474
  definition, 454
  improper node, 471, 473
  isolated, 481
  of logistic equation, 432
  proper node, 472
  saddle point, 472

  spiral point, 474
  stable, 455
Errors
  cumulative discretization, 97
  total error in a numerical method, 97
Escape time algorithm for Julia sets, 444
Euler, Leonhard, 137
Euler's equation in the calculus of variations, 180
Euler's formula for e^{ix}, 136, 572
Euler's method, 94
Even function, 517
Existence and uniqueness of solutions
  difference equation, 22
  first-order differential equation, 114
  higher-order differential equation, 189
  Picard's theorem, 22
  second-order differential equation, 189
  system of linear differential equations, 335
Explicit solution, 17
Exponential growth, 47
Exponential order, 249

Falling body problem, 77
Feigenbaum cascade, 436
Feigenbaum universal number, 437
First-order differential equation, 30
Fixed point of logistic equation
  attracting, 434
  repelling, 434
Forced vibrations, 165, 185
Fourier, Joseph, 506
Fourier series
  complex form, 514
  convergence of, 508
  cosine series, 520
  definition, 507
  Dirichlet's theorem, 508
  Euler equation, 507
  Fourier coefficients, 507
  Gibb's phenomenon, 515
  sine series, 520
Fractal, 443, 496
Fractional calculus, 311

Frequency
frequency response curve, 186
simple harmonic motion, 167
Frobenius, Ferdinand Georg, 227
Frobenius, method of, 226
Fundamental harmonic, 542
Fundamental matrix, 337
Fundamental set of solutions
higher-order differential equation, 190
partial differential equation, 532, 559
single differential equation, 122
system of differential equations, 253
Future value
annuity, 53
deposits, 51

Galloping Gertie, 2
Game of bob, 188
Gamma function, 233
General solution, 24
Generalized function, 296
Gibb's phenomenon, 515
Growth equation, 48

Half-life, 50
Half-range expansions, 522
Hamiltonian, 467
Hamilton's principle, 181
Harmonics, 542
Heat equation
heat flow problem, 550
interpretation of, 547
one-dimensional, 545
separation of variable solution, 547
Heating equation in ordinary differential equations, 68
Higher-order differential equation, 189
Homogeneous differential equation
(*see* Differential equations)
Hubbard's empty bucket, 27

Identity matrix, 324
Implicit function theorem, 26
Implicit solution, 17
Impulse function, 293
Impulse response function, 308
Independence (*see* Linear independence)

Indicial equation, 228
Initial condition(s), 20
Initial-value problem
definition, 20
first-order, 34
second-order, 113
systems, 315
Input-output system, 141
Integrating factor, 30
Integrating factor method, 30
Inverse Laplace transform
definition, 260
linearly, 262
properties of, 266
Inverse matrix, 328

Julia, Gaston, 444
Julia set
computer program, 445
definition of, 442
drawings of, 443
escape time algorithm, 444

Kernel of the Laplace transform, 244

Lagrange, Joseph Louis, 159
Laplace, Marquis Pierre Simon de, 246
Laplace's equation, 557
Laplace transform
branched function, 277
of a convolution, 305
definition of, 245
of a delayed function, 278
of a derivative, 255
of the Dirac delta function, 297
of a discontinuous function, 284
exponential order, 249
of an impulse function, 297
inverse of, 260
kernel of, 244
linearity of, 248
of a periodic function, 282
properties of, 258
solution of a differential equation, 268
solution of a system of differential equations, 367
step function, 276
tables of, 251

transfer function, 308
translation property, 256
Laplacian
definition, 555
interpretation of, 556
Law of mass action, 66
Liapunov, Aleksandr M., 493
Liapunov exponent, 403
Liapunov function, 488
Limit cycle, 455
Linear combination, 112
Linear dependence
of a function, 118
of a sequence, 400
of a vector of functions, 336
Linear independence
Casoratian, 401
of a function, 118
of a sequence, 400
of a vector of functions, 336
Wronskian, 119, 338
Linear ordinary differential equation
characteristic equation, 343
definition, 12
first-order, 12
general solution of, 15, 348
higher-order, 12
homogeneous, 12, 110
nonhomogeneous, 12
particular solution of, 24, 362
second-order, 110
standard form, 110
superposition principle for, 112
variable coefficients, 110
Linearization of a nonlinear system, 479
Logistic equation
cobweb diagram for, 430
definition, 429

Maclaurin's series, 204
Mandelbrot, Benoit, 448
Mandelbrot set
computer program for finding, 448
definition of, 446
Mass spring system, 165, 384
Matrices
addition of, 324
column matrix, 325
complex matrix, 333

Matrices *(Cont.)*
   definition of, 323
   derivative of, 328
   determinant of, 326
   difference of, 324
   eigenvalues of, 343
   eigenvectors of, 343
   equality of, 323
   exponential, 329, 358
   functions of, 328
   fundamental matrix, 337
   identity matrix, 324
   inverse matrix, 328
   representation of a derivative, 331
   transpose matrix, 324
   zero matrix, 324
Matrix exponential, 329, 358
Matrix representation of analytic operations, 321
Mechanics
   falling body, 27
   mechanical vibrations, 163
   Newton's second law, 163
   simple harmonic motion, 166
   underdamped vibrations, 165
   unforced vibrations, 165
Method of elimination, 314
Method of Frobenius, 226
Method of undetermined coefficients, 155
Method of variation of parameters, 158
Mixing problems
   equal input and output rate of flow, 61
   interconnected tanks, 376
   single tank, 61
   unequal input and output rate of flow, 63

Newton, Sir Isaac, 15
Newton's law of cooling, 68
Newton's second law of motion, 77, 164
Node, 239
Nonhomogeneous differential equation
   solution of a single equation
     Laplace transform, 268
     undetermined coefficients, 140
     variation of parameters, 158

   solution for a system
     Laplace transform, 367
     theory, 340, 359
     undetermined coefficients, 366
     variation of parameters, 363
Nonlinear differential equation
   autonomous, 455
   systems of, 451
Numerical methods, 92, 385
   Euler method, 94
   method of tangents, 94
   numerical versus analytical solution, 98
   Runge-Kutta method, 103
   three-term Taylor series method, 100

Odd function, 517
Open loop control, 465
Operator, D, 317
Operator method, 275
Optimal control theory, 89
Orbit, 440, 453
Order of a differential equation, 11
Ordinary point, 209
Orthogonal functions, 505
Orthogonal trajectories, 44
Overdamped motion, 175

Partial differential equation
   definition, 523
   elliptic, 525
   heat equation, 545
   hyperbolic type, 525
   Laplace's equation, 555
   parabolic type, 525
   wave equation, 529
Partial fraction decomposition, 363
Particular solution, 24
Periodic extension, 519
Periodic function, 505
Periodic motion, 504
Phase angle, 167
Phase plane, 453
Picard, Charles Emilé, 23
Picard's method, 108
Picard's theorem, 22
Poincaré, Henri Jules, 454
Poincaré section, 494
Power series
   analytic functions, 204

center of convergence, 200
convergence of, 200
defining a function, 202
definition of, 200
expansion about an ordinary point, 217
interval of convergence, 201
properties, 203
radius of convergence, 201
ratio test, 201
reversion of, 206
shift-of-index theorem for power series, 205
term by term differentiation, 203
term by term integration, 203
Power series solutions, 207
Principle of superposition for homogeneous equations, 112
Proper node, 472

Qualitative theory of differential equations, 452
Quantitative theory of differential equations, 452

Radioactive decay, 49
Radiocarbon dating, 49
Radius of convergence, 202
Recurrence relation, 210
Reduction of order, 125
Regular point, 455
Regular singular point, 225
Resonance, 183
Ricatti's differential equation, 37, 140
Richardson's extrapolation, 99
RLC equation, 180
Runge, Carl, 106
Runge-Kutta method
   computer program, 105
   single equation, 103
   systems, 388

Saddle point, 472
Scholastic journal (directions), 14
Schwartz, Laurent, 296
Separable equations (ordinary differential equations)
   definition of, 38
   transforming to, 39

Separation of variables (partial differential equations)
  for Laplace's equation in polar coordinates, 559
  for one-dimensional heat equations, 547
  for one-dimensional wave equations, 533
Series (*see* Power series)
Series solution
  Frobenius, method of, 226
  indicial equation, 228
  ordinary point, solution about, 207
  recurrence relation, 210
  regular singular point, solution about, 225
Simple harmonic motion, 166
Singular matrix, 328
Singular point, 209
  irregular, 225
  regular, 225
Singular solution, 24
Solution of a differential equation, 15
  explicit, 17
  general, 15
  implicit, 17
  matrix exponential solution, 358
  n-parameter family of solutions, 24
  particular, 24
  singular, 24
  spaceship problem, 458
  steady state, 182
  transient, 182
Stability of an equilibrium point
  asymptotic, 475
  definition of, 475
Standing waves, 239, 532
Steady state solution, 182
Steinmetz, Charles, 157
Step function, 276
Stephan's law of cooling, 99
Strange attractor, 495
Sturm's problem, 124
Submarine search problem, 78
Superposition principle for a linear differential equation

for a homogeneous equation, 112
for a nonhomogeneous equation, 142
for a partial differential equation, 539
for a system of equations, 335
Sylvester, James, 329
Systems of differential equations
  adjoint system, 353
  definition, 314
  eigenvalue (solution by), 341
  fundamental matrix, 337
  general solution, 337
  initial-value problem, 315
  linear system, 335
  method of elimination, 318
  nonhomogeneous systems, 339
  numerical solutions of, 385
  reducing to a system from an nth order equation, 316
  theory of linear systems, 144, 334
  undetermined coefficients, 366
  variation of parameters, 362

Table of Laplace transforms, 251
Taylor series method of solution, 100, 385
Three-term Taylor series method, 100
Torricelli's equation, 84
Trajectories
  finding, 455
  phase plane, 453
Transfer function, 308
Transform
  inverse Laplace, 260
  Laplace, 245
Transient solution, 182
Tratrix, 46

Undetermined coefficients (method of)
  for difference equations, 411
  for differential equations, 140
  for systems, 366
  table, 153

Uniqueness of solution for
  first-order equation, 21
  higher-order equation, 189
  linear system of equations, 335
  linear second-order equation, 189
  Picard's theorem, 22
  second-order equation, 114
Unit impulse function, 293
Unit step function, 276
Universal constant, 437

Variation of parameters
  for single equation, 158
  for systems of equations, 362
Vector (*see* Matrices)
Vibrations
  critically damped, 175
  damped, 165
  forced, 165, 185
  frequency response curve, 186
  overdamped, 175
  simple harmonic motion, 166
    amplitude, 167
    natural frequency, 167
    period, 167
    phase angle, 167
  for systems, 384
  undamped, 165
  underdamped, 173

Wave equation, 529
  canonical, 528
  initial-value problem (vibrating string), 529
  interpretation of, 531
Wronski, Josef Marie Hoene, 120
Wronskian
  Abel's theorem, 123
  for a single equation, 119
  for a system, 190, 338

$\mathbb{Z}$-transform, 413
  properties, 415
  tables, 414

A CATALOG OF SELECTED
DOVER BOOKS
IN SCIENCE AND MATHEMATICS

Mathematics

FUNCTIONAL ANALYSIS (Second Corrected Edition), George Bachman and Lawrence Narici. Excellent treatment of subject geared toward students with background in linear algebra, advanced calculus, physics and engineering. Text covers introduction to inner-product spaces, normed, metric spaces, and topological spaces; complete orthonormal sets, the Hahn-Banach Theorem and its consequences, and many other related subjects. 1966 ed. 544pp. 6⅛ x 9¼.   0-486-40251-7

ASYMPTOTIC EXPANSIONS OF INTEGRALS, Norman Bleistein & Richard A. Handelsman. Best introduction to important field with applications in a variety of scientific disciplines. New preface. Problems. Diagrams. Tables. Bibliography. Index. 448pp. 5⅜ x 8½.   0-486-65082-0

VECTOR AND TENSOR ANALYSIS WITH APPLICATIONS, A. I. Borisenko and I. E. Tarapov. Concise introduction. Worked-out problems, solutions, exercises. 257pp. 5⅜ x 8¼.   0-486-63833-2

AN INTRODUCTION TO ORDINARY DIFFERENTIAL EQUATIONS, Earl A. Coddington. A thorough and systematic first course in elementary differential equations for undergraduates in mathematics and science, with many exercises and problems (with answers). Index. 304pp. 5⅜ x 8½.
0-486-65942-9

FOURIER SERIES AND ORTHOGONAL FUNCTIONS, Harry F. Davis. An incisive text combining theory and practical example to introduce Fourier series, orthogonal functions and applications of the Fourier method to boundary-value problems. 570 exercises. Answers and notes. 416pp. 5⅜ x 8½.
0-486-65973-9

COMPUTABILITY AND UNSOLVABILITY, Martin Davis. Classic graduate-level introduction to theory of computability, usually referred to as theory of recurrent functions. New preface and appendix. 288pp. 5⅜ x 8½.   0-486-61471-9

ASYMPTOTIC METHODS IN ANALYSIS, N. G. de Bruijn. An inexpensive, comprehensive guide to asymptotic methods–the pioneering work that teaches by explaining worked examples in detail. Index. 224pp. 5⅜ x 8½   0-486-64221-6

APPLIED COMPLEX VARIABLES, John W. Dettman. Step-by-step coverage of fundamentals of analytic function theory–plus lucid exposition of five important applications: Potential Theory; Ordinary Differential Equations; Fourier Transforms; Laplace Transforms; Asymptotic Expansions. 66 figures. Exercises at chapter ends. 512pp. 5⅜ x 8½.   0-486-64670-X

INTRODUCTION TO LINEAR ALGEBRA AND DIFFERENTIAL EQUATIONS, John W. Dettman. Excellent text covers complex numbers, determinants, orthonormal bases, Laplace transforms, much more. Exercises with solutions. Undergraduate level. 416pp. 5⅜ x 8½.   0-486-65191-6

RIEMANN'S ZETA FUNCTION, H. M. Edwards. Superb, high-level study of landmark 1859 publication entitled "On the Number of Primes Less Than a Given Magnitude" traces developments in mathematical theory that it inspired. xiv+315pp. 5⅜ x 8½.   0-486-41740-9

CALCULUS OF VARIATIONS WITH APPLICATIONS, George M. Ewing. Applications-oriented introduction to variational theory develops insight and promotes understanding of specialized books, research papers. Suitable for advanced undergraduate/graduate students as primary, supplementary text. 352pp. 5⅜ x 8½. 0-486-64856-7

COMPLEX VARIABLES, Francis J. Flanigan. Unusual approach, delaying complex algebra till harmonic functions have been analyzed from real variable viewpoint. Includes problems with answers. 364pp. 5⅜ x 8½. 0-486-61388-7

AN INTRODUCTION TO THE CALCULUS OF VARIATIONS, Charles Fox. Graduate-level text covers variations of an integral, isoperimetrical problems, least action, special relativity, approximations, more. References. 279pp. 5⅜ x 8½. 0-486-65499-0

COUNTEREXAMPLES IN ANALYSIS, Bernard R. Gelbaum and John M. H. Olmsted. These counterexamples deal mostly with the part of analysis known as "real variables." The first half covers the real number system, and the second half encompasses higher dimensions. 1962 edition. xxiv+198pp. 5⅜ x 8½. 0-486-42875-3

CATASTROPHE THEORY FOR SCIENTISTS AND ENGINEERS, Robert Gilmore. Advanced-level treatment describes mathematics of theory grounded in the work of Poincaré, R. Thom, other mathematicians. Also important applications to problems in mathematics, physics, chemistry and engineering. 1981 edition. References. 28 tables. 397 black-and-white illustrations. xvii + 666pp. 6⅛ x 9¼. 0-486-67539-4

INTRODUCTION TO DIFFERENCE EQUATIONS, Samuel Goldberg. Exceptionally clear exposition of important discipline with applications to sociology, psychology, economics. Many illustrative examples; over 250 problems. 260pp. 5⅜ x 8½. 0-486-65084-7

NUMERICAL METHODS FOR SCIENTISTS AND ENGINEERS, Richard Hamming. Classic text stresses frequency approach in coverage of algorithms, polynomial approximation, Fourier approximation, exponential approximation, other topics. Revised and enlarged 2nd edition. 721pp. 5⅜ x 8½. 0-486-65241-6

INTRODUCTION TO NUMERICAL ANALYSIS (2nd Edition), F. B. Hildebrand. Classic, fundamental treatment covers computation, approximation, interpolation, numerical differentiation and integration, other topics. 150 new problems. 669pp. 5⅜ x 8½. 0-486-65363-3

THREE PEARLS OF NUMBER THEORY, A. Y. Khinchin. Three compelling puzzles require proof of a basic law governing the world of numbers. Challenges concern van der Waerden's theorem, the Landau-Schnirelmann hypothesis and Mann's theorem, and a solution to Waring's problem. Solutions included. 64pp. 5⅜ x 8½. 0-486-40026-3

THE PHILOSOPHY OF MATHEMATICS: AN INTRODUCTORY ESSAY, Stephan Körner. Surveys the views of Plato, Aristotle, Leibniz & Kant concerning propositions and theories of applied and pure mathematics. Introduction. Two appendices. Index. 198pp. 5⅜ x 8½. 0-486-25048-2

INTRODUCTORY REAL ANALYSIS, A.N. Kolmogorov, S. V. Fomin. Translated by Richard A. Silverman. Self-contained, evenly paced introduction to real and functional analysis. Some 350 problems. 403pp. 5⅜ x 8½.   0-486-61226-0

APPLIED ANALYSIS, Cornelius Lanczos. Classic work on analysis and design of finite processes for approximating solution of analytical problems. Algebraic equations, matrices, harmonic analysis, quadrature methods, much more. 559pp. 5⅜ x 8½.   0-486-65656-X

AN INTRODUCTION TO ALGEBRAIC STRUCTURES, Joseph Landin. Superb self-contained text covers "abstract algebra": sets and numbers, theory of groups, theory of rings, much more. Numerous well-chosen examples, exercises. 247pp. 5⅜ x 8½.   0-486-65940-2

QUALITATIVE THEORY OF DIFFERENTIAL EQUATIONS, V. V. Nemytskii and V.V. Stepanov. Classic graduate-level text by two prominent Soviet mathematicians covers classical differential equations as well as topological dynamics and ergodic theory. Bibliographies. 523pp. 5⅜ x 8½.
0-486-65954-2

THEORY OF MATRICES, Sam Perlis. Outstanding text covering rank, nonsingularity and inverses in connection with the development of canonical matrices under the relation of equivalence, and without the intervention of determinants. Includes exercises. 237pp. 5⅜ x 8½.   0-486-66810-X

INTRODUCTION TO ANALYSIS, Maxwell Rosenlicht. Unusually clear, accessible coverage of set theory, real number system, metric spaces, continuous functions, Riemann integration, multiple integrals, more. Wide range of problems. Undergraduate level. Bibliography. 254pp. 5⅜ x 8½.
0-486-65038-3

MODERN NONLINEAR EQUATIONS, Thomas L. Saaty. Emphasizes practical solution of problems; covers seven types of equations. ". . . a welcome contribution to the existing literature...."—*Math Reviews*. 490pp. 5⅜ x 8½.   0-486-64232-1

MATRICES AND LINEAR ALGEBRA, Hans Schneider and George Phillip Barker. Basic textbook covers theory of matrices and its applications to systems of linear equations and related topics such as determinants, eigenvalues and differential equations. Numerous exercises. 432pp. 5⅜ x 8½.
0-486-66014-1

LINEAR ALGEBRA, Georgi E. Shilov. Determinants, linear spaces, matrix algebras, similar topics. For advanced undergraduates, graduates. Silverman translation. 387pp. 5⅜ x 8½.   0-486-63518-X

ELEMENTS OF REAL ANALYSIS, David A. Sprecher. Classic text covers fundamental concepts, real number system, point sets, functions of a real variable, Fourier series, much more. Over 500 exercises. 352pp. 5⅜ x 8½.   0-486-65385-4

SET THEORY AND LOGIC, Robert R. Stoll. Lucid introduction to unified theory of mathematical concepts. Set theory and logic seen as tools for conceptual understanding of real number system. 496pp. 5⅜ x 8¼.   0-486-63829-4

Math–Decision Theory, Statistics, Probability

ELEMENTARY DECISION THEORY, Herman Chernoff and Lincoln E. Moses. Clear introduction to statistics and statistical theory covers data processing, probability and random variables, testing hypotheses, much more. Exercises. 364pp. 5⅜ x 8½.   0-486-65218-1

STATISTICS MANUAL, Edwin L. Crow et al. Comprehensive, practical collection of classical and modern methods prepared by U.S. Naval Ordnance Test Station. Stress on use. Basics of statistics assumed. 288pp. 5⅜ x 8½.   0-486-60599-X

SOME THEORY OF SAMPLING, William Edwards Deming. Analysis of the problems, theory and design of sampling techniques for social scientists, industrial managers and others who find statistics important at work. 61 tables. 90 figures. xvii +602pp. 5⅜ x 8½.   0-486-64684-X

LINEAR PROGRAMMING AND ECONOMIC ANALYSIS, Robert Dorfman, Paul A. Samuelson and Robert M. Solow. First comprehensive treatment of linear programming in standard economic analysis. Game theory, modern welfare economics, Leontief input-output, more. 525pp. 5⅜ x 8½.
         0-486-65491-5

PROBABILITY: AN INTRODUCTION, Samuel Goldberg. Excellent basic text covers set theory, probability theory for finite sample spaces, binomial theorem, much more. 360 problems. Bibliographies. 322pp. 5⅜ x 8½.   0-486-65252-1

GAMES AND DECISIONS: INTRODUCTION AND CRITICAL SURVEY, R. Duncan Luce and Howard Raiffa. Superb nontechnical introduction to game theory, primarily applied to social sciences. Utility theory, zero-sum games, n-person games, decision-making, much more. Bibliography. 509pp. 5⅜ x 8½.   0-486-65943-7

INTRODUCTION TO THE THEORY OF GAMES, J. C. C. McKinsey. This comprehensive overview of the mathematical theory of games illustrates applications to situations involving conflicts of interest, including economic, social, political, and military contexts. Appropriate for advanced undergraduate and graduate courses; advanced calculus a prerequisite. 1952 ed. x+372pp. 5⅜ x 8½.
         0-486-42811-7

FIFTY CHALLENGING PROBLEMS IN PROBABILITY WITH SOLUTIONS, Frederick Mosteller. Remarkable puzzlers, graded in difficulty, illustrate elementary and advanced aspects of probability. Detailed solutions. 88pp. 5⅜ x 8½.   0-486-65355-2

PROBABILITY THEORY: A CONCISE COURSE, Y. A. Rozanov. Highly readable, self-contained introduction covers combination of events, dependent events, Bernoulli trials, etc. 148pp. 5⅜ x 8¼.   0-486-63544-9

STATISTICAL METHOD FROM THE VIEWPOINT OF QUALITY CONTROL, Walter A. Shewhart. Important text explains regulation of variables, uses of statistical control to achieve quality control in industry, agriculture, other areas. 192pp. 5⅜ x 8½.   0-486-65232-7

TABLE OF INTEGRALS

1. $\displaystyle \int a\, du = au + C$

2. $\displaystyle \int (au + b)\, du = \frac{a}{2}u^2 + bu + C$

3. $\displaystyle \int u^n\, du = \frac{1}{n+1}\, u^{n+1} + C$

4. $\displaystyle \int \frac{1}{u}\, du = \ln|u| + C$

5. $\displaystyle \int \frac{f'(u)}{f(u)}\, du = \ln|f(u)| + C$

6. $\displaystyle \int (a + bu)^n\, du = \frac{1}{(n+1)b}\, (a + bu)^{n+1} + C \quad (n \neq -1)$

7. $\displaystyle \int \frac{du}{a + bu} = \frac{1}{b}\ln|a + bu| + C$

8. $\displaystyle \int \frac{du}{(a + bu)^2} = -\frac{1}{b(a + bu)} + C$

9. $\displaystyle \int \frac{u\, du}{(a + bu)^2} = \frac{1}{b^2}\left(\ln|a + bu| + \frac{a}{a + bu} + C\right)$

10. $\displaystyle \int \frac{du}{u(a + bu)} = -\frac{1}{a}\ln\left|\frac{a + bu}{u}\right| + C$

11. $\displaystyle \int u\sqrt{a + bu}\, du = -\frac{2(2a - 3bu)(a + bu)^{3/2}}{15b^2} + C$

12. $\displaystyle \int \frac{u\, du}{\sqrt{a + bu}} = -\frac{2(2a - bu)}{3b^2}\sqrt{a + bu} + C$

13. $\displaystyle \int \frac{du}{a^2 - u^2} = \frac{1}{2a}\ln\left|\frac{a + u}{a - u}\right| + C$

14. $\displaystyle \int \frac{du}{u^2 - a^2} = \frac{1}{2a}\ln\left|\frac{u - a}{u + a}\right| + C$

15. $\displaystyle \int \frac{du}{\sqrt{a^2 - u^2}} = \sin^{-1}\left(\frac{u}{a}\right) + C$

16. $\displaystyle \int \frac{du}{\sqrt{u^2 \pm a^2}} = \ln\left|u + \sqrt{u^2 \pm a^2}\right| + C$

17. $\displaystyle \int \sqrt{u^2 \pm a^2}\, du = \frac{1}{2}\left(u\sqrt{u^2 \pm a^2}\right) \pm a^2\ln\left(u + \sqrt{u^2 \pm a^2}\right)$